19th International Conference on Ion Sources (ICIS2021)

Journal of Physics: Conference Series Volume 2244

Online
20 – 24 September 2021

ISBN: 978-1-7138-5498-2
ISSN: 1742-6588

Printed from e-media with permission by:

Curran Associates, Inc.
57 Morehouse Lane
Red Hook, NY 12571

Some format issues inherent in the e-media version may also appear in this print version.

Institute of Physics
Dirac House, Temple Back
Bristol BS1 6BE UK

Phone: 44 1 17 929 7481
Fax: 44 1 17 920 0979

techtracking@iop.org

Additional copies of this publication are available from:

Curran Associates, Inc.
57 Morehouse Lane
Red Hook, NY 12571 USA
Phone: 845-758-0400
Fax: 845-758-2633
Email: curran@proceedings.com
Web: www.proceedings.com

TABLE OF CONTENTS

Journal of Physics: Conference Series **2244** (2022) 011001 doi:10.1088/1742-6596/2244/1/011001

Proceedings of the 19th International Conference on Ion Sources, Vancouver, BC, Canada, 2021

The 19th International Conference on Ion Sources was hosted by TRIUMF in Vancouver, BC, Canada, September 20 to September 24, 2021.

The biennial conference series covers the physics and technology of ion sources for scientific research and applications. With the conference being the main event within the ion sources community, it offers a forum for the exchange of ideas and the establishing of collaborations. New developments and advancements have been presented and discussed. As in previous conferences of this series, all aspects of ion sources have been covered in the categories

- Fundamental processes in ion sources and plasmas
- Production of high intensity ion beams
- Production of highly charged ion beams
- Negative ion sources
- Ion sources for fusion
- Radioactive ion sources, charge breeders and polarized beams
- Beam formation, extraction, transport, and diagnostics
- Applications of ion sources
- Key technologies for ion sources

Due to restrictions on traveling and meeting sizes caused by health regulations during the COVID-19 pandemic, the conference was held as a virtual conference. Talks have been presented via ZOOM and two poster sessions were held within the remote meeting platform "Gather town", which was also open for discussions among participants outside the plenary sessions.

The main session on electron beam ion sources was dedicated to the memory of Professor Dr. Evgeny Denisovich Donets who, to our great regret, passed away in June 2021. Professor Donets was the inventor of the electron beam ion source and one of the initiators of this conference series. He was a longtime member of the international advisory committee. His memory will be highly regarded within the ion source community.

A total of 238 participants have registered for the conference, 61 of whom were students. The international advisory committee selected 14 invited talks and the scientific program committee an additional 42 oral and 117 poster contributions from the submitted abstracts covering all above-mentioned topics. The following proceedings contain 109 papers which have been peer reviewed by conference delegates and finally accepted by the editorial team. 34 students have participated in the student poster competition. Members of both the international advisory committee and the scientific program committee acted as judges and three winners have been awarded:

1st prize: Joseph Adegun, University of Victoria and TRIUMF, Canada
Improvement of the efficiency and beam quality of the TRIUMF charge state booster (CSB)

2nd prize: Ippei Yamada, Doshisha University, Japan
Non-destructive beam spatial profile measurement for a high-intensity ion beam using a gas sheet monitor

3rd prize: Ryo Kato, Keio University, Japan
Numerical analysis of isotope effect in NIFS negative ion source

Sincere thanks also to the sponsors of this event: BWXT Medical Ltd, D-Pace, ISOBAREX Corp., Pantechnik, and Pfeiffer Vacuum. We want to acknowledge their important contribution to our field of research.

The 20th International Conference on Ion Sources will be hosted again by TRIUMF and will take place in Victoria, BC, Canada in September 2023.

For the ICIS2021 organizers Oliver Kester and Friedhelm Ames

Screenshot of participants during ZOOM session at ICIS2021

International advisory committee:

Oliver Kester (chair)	TRIUMF	Canada
Jose Alonso	LBNL	USA
Edward Beebe	BNL	USA
E. D. Donets	JINR	Russia
Dan Faircloth	RAL	UK
Ursel Fantz	IPP, Univ. Augsburg	Germany
Santo Gammino	INFN-LNS	Italy
Yong-Seok Hwang	Seoul Nat'l Univ.	Korea
Ralph Hollinger	GSI	Germany
Alexander Ivanov	BINP	Russia
Atsushi Kitagawa	NIRS-QST	Japan
Daniela Leitner	LBNL	USA
Renan Leroy	GANIL	France
Jacques Lettry	CERN	Switzerland
Marc Loiselet	CRC	Belgium
Takahide Nakagawa	RIKEN	Japan
Shixiang Peng	PKU	China
Beatrix Schunke	ITER	France
Panagiotis Svarnas	Univ. Patras	Greece
Katsuyoshi Tsumori	NIFS	Japan
Rick Vondrasek	ANL	USA
Robert Welton	ORNL	USA
Daniel Z Xie	LBNL	USA
Hongwei Zhao	IMP	China

Scientific program committee

Friedhelm Ames (Chair)	TRIUMF	Canada
Clayton Dickerson	ANL	USA
Mainak Bandyopadhyay	ITER	India
Luigi Celona	INFN-LNS	Italy
Yong-Sub Cho	KAERI	South Korea
Baoqun Cui	CIAE	China
Akiyoshi Hatayama	Keio Univ.	Japan
Chundong Hu	IPP	China
Ivan Izotov	IAP	Russia

Keerthi Jayamanna	TRIUMF	Canada
Hannu Koivisto	JYFL	Finland
Alain Lapierre	NSCL-FRIB	USA
Guillaume Machicoane	FRIB	USA
Masahiro Okamura	BNL	USA
Liangting Sun	IMP	China
Thomas Thuillier	LPSC	France
Fredrik Wenander	CERN	Switzerland
Christian Wimmer	IPPI	Germany
Mi-Sook- Won	KBSI	South Korea

Local organizing committee

Friedhelm Ames (chair)
Silke Bergelt-Bruckner
Jana Thomson
Brad Schultz
Christopher Charles
Carla Babcock
Francis Pau

Peer Review Statement

All papers published in this volume have been reviewed through processes administered by the Editors. Reviews were conducted by expert referees to the professional and scientific standards expected of a proceedings journal published by IOP Publishing.

- **Type of peer review:** Single Anonymous
- **Conference submission management system:** Morressier
- **Number of submissions received:** 115
- **Number of submissions sent for review:** 115
- **Number of submissions accepted:** 113
- **Acceptance Rate (Submissions Accepted / Submissions Received × 100):** 98.3
- **Average number of reviews per paper:** 1.14
- **Total number of reviewers involved:** 68
- **Contact person for queries:**
  Name: Friedhelm Ames
  Email: ames@triumf.ca
  Affiliation: TRIUMF

Cutoff limitation of left-hand polarization wave and candidates for further enhanced producing multicharged ions on ECRIS

Y Kato, W Kubo, S Harisaki, M Anan, K Tsuda, K Sato, I Owada and T Maenaka

Division of Electrical, Electronic and Infocommunications Engineering, Graduate School of Engineering, Osaka Univ., 2-1 Yamada-oka, Suita, Osaka 565-0871, Japan

kato@eei.eng.osaka-u.ac.jp

Abstract. Based on experimentally obtained plasma parameters in an electron cyclotron resonance (ECR) ion source (ECRIS) and theoretical considerations, it is turned out the essential factor that is currently presumed to define the increase in multicharged ion current in ECRIS is not simply the density limit of ordinary wave and right-hand cutoffs, but is also higher density one of left-hand cutoff. There are two response guidelines that can be considered to make it possible to overcome limitations, except for the conventional simply increasing the frequency and the magnetic field strength. One is advanced high-frequency resonance, *i.e.*, upper-hybrid resonance (UHR), which is conversion from electromagnetic to electrostatic wave essentially without cutoff. The others are due to the introduction of lower frequency waves than ECR's one, which has no density limit in a more essential sense. The latter is the introduction of lower-hybrid resonance (LHR) or ion cyclotron resonance (ICR). We will describe experimentally obtained plasma parameters, and will discuss these candidate applications.

1. Introduction

An Electron Cyclotron Resonance (ECR) Ion Source (ECRIS) has been used to today with dramatic developments in the relevant fields for many years since the founding of the group of Geller [1]. Based on the experimental scaling law of various high-yielding devices in various countries around the world, it can be said that the device constructing has reached the maturity period empirically. However, it is considered that the situation is still insufficient for understanding the phenomena in plasma in ECRIS, such as the reason of the scaling law itself and the factors that require higher frequencies of ECRIS. In particular, the recognition of wave propagation in magnetized plasma and the efficiency of ECR is insufficient, and there seems to be ample room for improvement in the introduction of electromagnetic waves. Of course, many efforts have been made to make full use of simulations to understand them [2-4]. However, since it is difficult to measure the plasma parameters of ECRIS, the phenomenon has not been fully understood, and it seems that the idea that leads to various drastic improvements has not been recalled. Historically, ECRIS has taken various empirical methods to increase the yield of multicharged ions [5]. It is considered that the understanding of the phenomenon is not sufficient for these as well.

In general, it is difficult to directly observe ECR plasmas in ECRIS's operating in normal accelerators. On the other hand, our device makes it possible to understand phenomena in existing plasma parameters by taking advantage of the ability to measure their distributions in the ECR plasma. As a result of experimental verification of the efficiency of ECR [6-8] and consideration of the accessibility condition of electromagnetic wave propagation in the magnetized plasma in ECRIS, the possibility of high-frequency upper hybrid resonance (UHR) heating was shown [9]. By the way, the UHR is the lowest-order mode resonance of the electron Bernstein wave (Bn wave) that converts the mode from electromagnetic waves to electrostatic waves [1]. Since it is difficult to distinguish the phenomenon of ECR and UHR with ordinary single frequency microwaves, we introduced microwaves for ECR and at the same time introduced high frequency microwaves far from that frequency in X-mode. As a result, we observed an increase in multicharged ion currents due to the UHR and simultaneously an increase on the high energy side of electron energy distribution function that supported it [10]. Furthermore, Dual ECR experiment was conducted and the yield of multicharged ions was successfully increased [11].

Simulation studies and plasma measurements cannot completely rule out the possibility of concomitant instability on ECRIS, but it is believed that the main cause is the cutoff density limit of left-hand polarization wave cutoff. On the other hand, in the high-density region, a resonance phenomenon using low-frequency electromagnetic waves (usually called Radio Frequency: RF) that does not have a similar cutoff density limit, that is, lower hybrid resonance (LHR), and also ion cyclotron resonance (ICR). The LHR heats the electrons or the ions, and the ICR the ions. In the case of ion heating, it is considered that increasing the multicharged ion current is obtained by increasing the ion cooling effect [5] by selective heating of low Z ions during low Z gas mixing, which has been conventionally performed. In addition, when the ICR is caused, it is considered that the potential well [12] formed by ECR can be relaxed, and the multicharged ions are considered to increase by the same effect as the well-known after grow effect [13]. In this paper, we will explain the observation fact that L-cutoff is the density limit, and propose these candidates that can overcome their limitations.

2. Experimental Apparatus

Figure 1 shows a top view of the entire ECRIS (Graduate School of Engineering, Osaka University) device used in this experiment. The magnetic field configuration is formed by a pair of mirror coils A and B and an auxiliary coil C for adjustment near the resonance point formed in the center of the mirror magnetic field, and superposed with an octupole magnetic field by a permanent magnet. The distribution of the magnetic field strength along the mirror axis with respect to a typical coil current is shown in the figure below in Fig.1. 2.45GHz microwaves for ECR discharge are coaxially converted from the waveguide to the inside of the vacuum vessel via a three-dimensional circuit independently from two magnetron sources with a maximum power of 1.3kW through the waveguide coaxial conversion windows. One is a semi-dipole antenna at the tip of the inner conductor where the mirror magnetic field is at the maximum position after being reduced to a taper shape while keeping the aspect ratio of the inner and outer diameters constant in vacuum and then inserted along the mirror magnetic field from the upstream end side of the mirror at z = -418mm [11]. On the other hand, the microwave power is supplied from the Ti rod antenna from the side of the vacuum vessel on the downstream side near the extraction electrode. Parameter measurements, the electron density n_e and the temperature T_e, are measured by using Langmuir probes LP1 and LP2 with respect to the distribution of x = 0 to 50 mm from the side of the vacuum vessel on the upstream side (z = -175 mm) and downstream side (z = 175 mm), respectively.

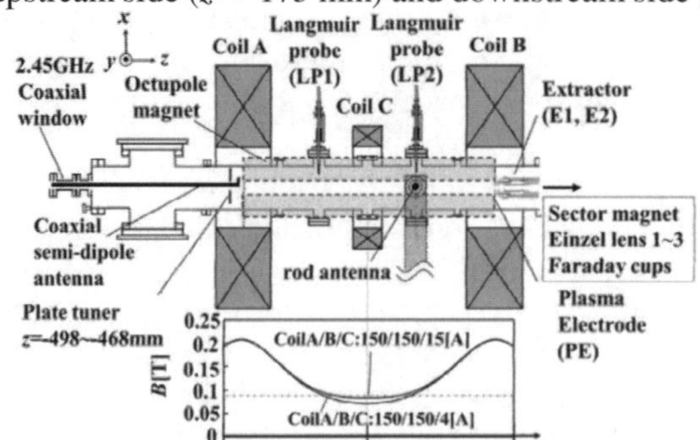

Figure 1. Top view of ECRIS (Osaka Univ.)

Journal of Physics: Conference Series **2244** (2022) 012001 doi:10.1088/1742-6596/2244/1/012001

The plasma parameters along the mirror axis (z-axis direction) were measured by preparing an L-shaped probe LP3 along the mirror axis direction when the upstream microwave introduction system was removed. The measurement range of LP3 is z = -451 to -51 mm. The tip of the LP3 probe can be measured on the z-axis at the centre of the chamber. The measurable microwave power varies depending on experimental conditions, but it is about 100-200 W or less, according to case by case.

3. Experimental results

Figure 2 shows the z-axis distribution of the n_e (Fig. 2(a)) and the T_e (Fig. 2(b)) obtained from LP3 under various experimental conditions. When the pressure is 2.5 or 5.0×10^{-4} Pa, the I_C is slightly changed with the coil current $I_{A,B}$ = 150A, and the incident microwave power is 20, 40, 80, 120, 200-210W. In Fig.2(a), when the microwave frequency is 2.45 GHz, the O-cutoff density and the L-cutoff density derived from the magnetic field strength corresponding to each position are shown by dotted lines. The n_e tends to increase as it approaches the mirror bottom near z = 0 mm, except for I_c = 0 A (green). In addition, the higher the pressure and the larger the microwave power, the higher the density tends to be. At 80 W or more, the n_e exceeding the O-cutoff density is measured in the region near the center, and it had been explained by the Whistler mode. We have frequently measured the n_e exceeding the O-cutoff density even at the positions of LP1 and LP2. At 200W, the L-cutoff density is approaching. Considering the insertion effect of LP3 on plasma, it is clear that the L-cutoff limit density has been reached. In the T_e, the parameter dependence is not as large as n_e, but the value tends to increase as the resonance point approaches. At low pressure and low power, I_c = 0A (green) becomes high temperature around z = -100mm, and at high pressure and high power, I_c = 12A (orange) becomes low temperature. The former is thought to be due to the close resonance point at I_c = 0 A, and the latter is thought to be due to the large effect of probe insertion as described later.

In inserting the probe into the plasma, it is necessary to monitor the degree of influence independently, and the reliability of the data is ensured based on the index. When LP3 was inserted, we monitored the ion saturation current value I_{is} that flows to the LP1 and LP2 applied voltage V_p = -45V at x = 50 mm, and the incident and reflected power of microwaves. As a result, it is suggested that when the microwave power is higher than about 100 W, the value corresponding to the L-cutoff limit density may be sufficiently reached near the center of the vacuum vessel near the resonance point. And fitting was performed on the incident microwave powers 40W, 50W, and 80W, and the microwave incident powers were scaled for the n_e distributions. It is highly possible that this cutoff is a constraint for the ion current extracted from ECRIS.

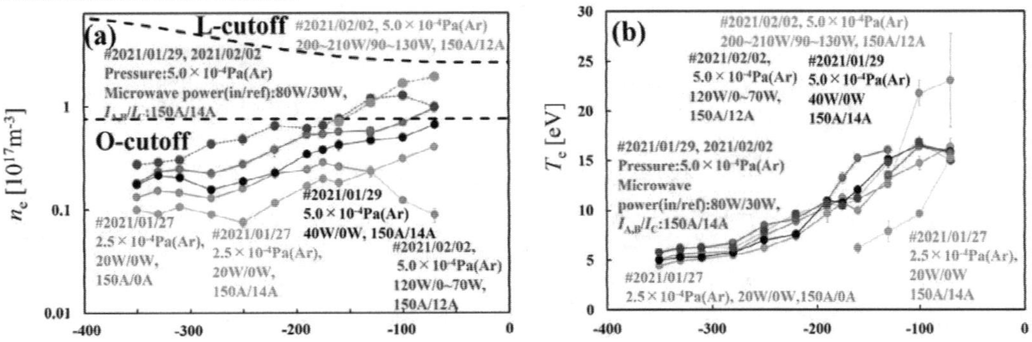

Figure 2. Measured axial (z) profiles of n_e(a) and T_e(b).

4. Discussions and perspective of future improvements

4.1. Accessibility conditions on real space in actual ECRIS and corresponding CMA diagram

We consider the correspondence between the accessibility condition of electromagnetic wave propagation in real space in actual ECRIS and that on the Clemmow-Mullaly-Allis (CMA) diagram [14] showing the state of electromagnetic wave propagation in magnetized plasma. Figures 3 (a) and (b) show the accessibility conditions from each antenna in the ECRIS real space when is microwave low power 40W and high power 200W, respectively. And the corresponding accessibility conditions on CMA diagram are shown in the right figures of Fig. 3 (a) and (b). In the real space and the CMA diagram, the orange line shows the wave propagation path from the coaxial semi-dipole antenna along to the z-axis from the upper stream of the ECRIS, and the blue line shows the wave propagation path from the rod antenna at the sidewall.

First, Fig. 3 (a) shows a typical case of low microwave power of 40 W. In the left figure, the ECR resonance point is reached from the strong magnetic field side without being blocked by the cutoff in the path of wave propagation from each antenna. In this situation, in the CMA diagram, all four basic modes propagate from the strong magnetic field side to the weak magnetic field side of the A region where they can propagate, all reach the ECR resonance region and the C region, and return to the A region again. Since the R-wave cannot exist in the C region, it is necessary to convert it to the L or O or X-wave. The R-wave becomes an evanescent wave, and cannot contribute to ECR at the opposite sides.

Next, let us consider a typical case of high microwave power of 200 W in Fig. 3 (b). The density becomes high, and a considerable region in the plasma becomes an overdense region beyond the O-cutoff, and the L-cutoff region covers the ECR resonance point. On the propagation paths of electromagnetic waves from each antenna, as can be seen from the CMA diagram, R-waves can propagate in each region of A, D, and F on the strong to the weak magnetic field side. It can be seen that the ECR resonance point can be reached without the propagation being hindered. After the ECR resonance point, the R-wave is a non-propagating region in the E region and propagation of a propagatable L or X wave is allowed [11]. After that, it was found that the route from any antenna reached the L-cutoff point, and that all modes reached the non-propagating G region. It was found that the main factor for the limitation of multicharged ion currents causes on microwave power dependence on ECRIS is not R- or O-cutoff, but the L-cutoff limit density for constraint of non-propagating on electromagnetic waves.

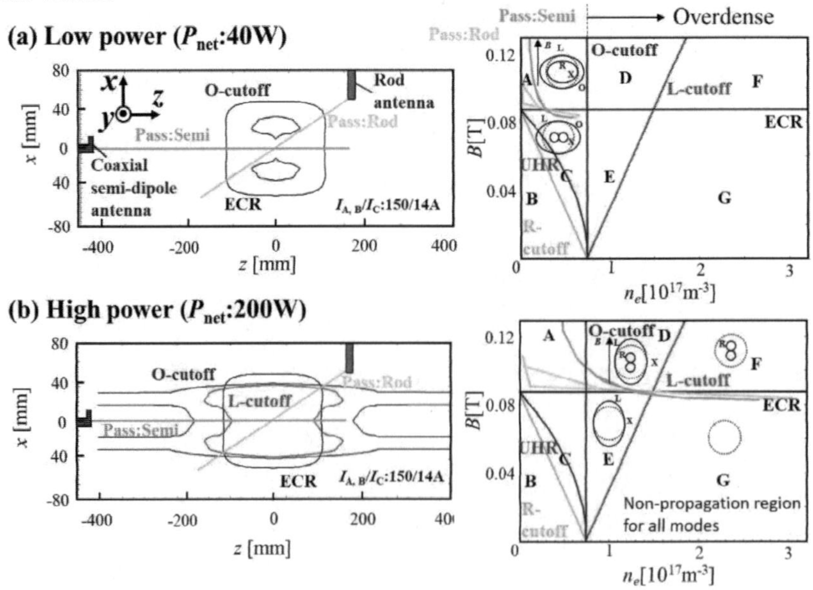

Figure 3. The accessibility condition in the real space of ECRIS (left figure) and in CMA diagram (right figures) at the microwave power 40 W (a) and 200 W (b), respectively.

4.2. Other candidates for furthermore improvements of multicharged ion beam currents

Kubo et al., Osaka Univ. group conducted an experiment based on the idea of Dual-ECR Heating, which simultaneously feeds microwaves from the strong magnetic field side on the upstream side and downstream side of the resonance point, which has not been performed in ECRIS in the past. As a result, it was shown that the multicharged ion current is usually saturated with each single microwave power supply, but the increased multicharged ion currents can be obtained by simultaneous power supply [11]. Kato et al., the same group conducted an UHR experiment with X-mode pioneeringly on 2.45GHz ECRIS superimposed by using 4-6GHz microwaves [10].

The possibility that the L-cutoff limit density is constraint for further increase of multicharged ion currents is shown from experimental data and consideration of electromagnetic wave propagation in the magnetized plasma. Therefore, we pay attention to the resonance phenomenon of low-frequency electromagnetic waves mediated by ions that are not restricted by L-cutoff, that is, lower hybrid resonance heating (LHR) and ion cyclotron resonance heating (ICR). The former contributes to the heating of electrons or ions, or both, and the latter contributes to the heating of ions. In the case of ion heating, the cooling of the ionic species of the light element gas is actively promoted to improve the efficiency in multicharged ion generation by mixing the light element gas, which has been empirically performed in ECRIS from long ago. Furthermore, it can be expected that the potential well formed by ECR can be relaxed by ICR, and an increase in the amount of multicharged ion currents with the same effect as the well-known afterglow effect. Currently, we are preparing for the introduction of these and have started the initial experiments.

References

[1] Geller R 1996 *Electron Cyclotron Resonance Ion Sources and ECR Plasmas* (first ed.; Institute of Physics, Bristol and Philadelphia, IOP Publishing Ltd., UK,)

[2] Castro G, Mascali D, Agnello R, Celona L, Leonardi O, Neri L, Nicolosi D, Torrisi G, and Gammino S, 2016 *Rev. Sci. Instrum.* **87**, pp.02A507-1-3

[3] Mascali D, Neri L, Celona L, Castro G, Torrisi G, Gammino S, Sorbello G, Giavola G, and Ciavola G 2014 *Rev. Sci. Instrum.* **85**, pp.02A511-1-5

[4] Torrisi G, Mascali D, Sorbello G, Castro G, Celona L, and Gammino S 2019 IEEE Trans. Anten. Prop. **67**, pp.2142-2148.

[5] Drentje A G 2003 *Rev. Sci. Instrum.* **742** pp.2631-2645.

[6] Kurisu Y, Kiriyama R, Takenaka T, Nozaki D, Sato F, Kato Y and Iida T 2012 *Rev. Sci. Instrum.* **83**, pp.02A310-1-3.

[7] Yano K, Kurisu Y, Nozaki D, Kimura D, Imai Y, Kumakura S, Sato F, Kato Y and Iida T 2014 *Rev. Sci. Instrum.* **85**, pp.02A937-1-3

[8] Kimura D, Kurisu Y, Yano K, Kumakura S, Imai Y, Sato F, Kato Y and Iida T 2014 *Rev. Sci. Instrum.* **85**, pp.02A938-1-3.

[9] Kato Y, Yano K, Nishiokada T, Nagaya T, Kimura D, Kumakura S, Imai Y, Hagino S, Otsuka T and Sato F 2016 *Rev. Sci. Instrum.* **87**, pp.02A710-1-4.

[10] Kato Y, Nishiokada T, Hamada K, Onishi K, Takeda T, Okumura K, Omori T, Kubo W, Ishihara M and Harisaki S 2020 *Rev. Sci. Instrum.* **91**, pp.013315-1-6.

[11] Kubo W, Harisaki S, Owada I, Sato K, Tsuda K and Kato Y 2021 *Rev. Sci. Instrum.* **92** pp.043514-1-9. https://doi.org/10.1063/5.0035631

[12] Kato Y and Ishii S 1998 *Rev. Sci. Instrum.* **69**, pp.1176-1178.

[13] Kato Y, Kobayashi S and Ishii S 2000 *Rev. Sci. Instrum.* **71**, pp.657-659.

[14] Lieberman A M and Lichtenberg J A 2005 *Principle of Plasma Discharges and Materials Processing*, 2nd Edit., A John Wiley & Son, Inc Publications, Chap.4, pp.110; Chen F F 1984 *Introduction to Plasma Physics and Controlled Fusion*, 2nd Edit., Volume 1: Plasma Physics, Plenum Press., Chap.14, pp.127-131; Geller R 1996 *Electron Cyclotron Resonance Ion Sources and ECR Plasmas, Institute of Physics*, pp.155.

Acknowledgments

The authors would like to thank the late Dr Geller R, Prof Yoshida Y (Toyo Univ.), Dr Kitagawa A (NIRS), and Dr Muramatsu M (NIRS), Emeritus Prof. Kawai Y (Kyushu Univ.) for provided informative discussion and continued encouragement. The authors would like to thank all the staff and graduates of Osaka Univ. for their great efforts in preparing and constructing this experimental device.

Numerical study of different electron density observed in Hydrogen and Deuterium ion source plasmas

T Shibata[1,+], R Kato[2], H Nakano[3], K Hayashi[2], J Sato[2], K Miyamoto[4], K Hoshino[2] and A Hatayama[2]

[1]Accelerator Laboratory, KEK, Tsukuba 300-3256, Japan
[2]Faculty of Science and Technology, Keio University, Yokohama 223-8522, Japan
[3]National Institute for Fusion Science, NINS, Toki 509-5292, Japan
[4]Graduate School of Education, Naruto University of Education, Naruto 772-8502, Japan
[+]Corresponding author: takanori.shibata@kek.jp

Abstract. Sequences of hydrogen (H) and deuterium (D) experiments have been done by NIFS research and development negative ion source (RNIS) for the deuterium NBI development. In the experiments, the co-extracted electron current with the negative ions and the electron density in the plasma generation region in the D experiment have been around three times higher than that in the H experiment. To explain the difference of the electron density in the RNIS driver region, a zero-dimensional numerical model is developed in the present study. The model only focuses on the isotope effect for vibrationally excited level of electronically grounded state molecules and its relevant cross-sections. The calculation results show that difference of the ionization channel numbers via molecular vibrationally excited states could be a reason to enhance ionization rate in D plasma.

1. Introduction

Isotope effect between hydrogen (H) and deuterium (D) plasmas in negative ion sources have been attracting interests especially for fusion plasma heating fields. Development of the deuterium NBI (Neutral Beam Injection) system is one of the next tasks to achieve large deuterium fusion plasma experiments. In order to investigate the difference of the negative hydrogen ion (H^-) and the negative deuterium ion (D^-) beam characteristics, sequences of NBI experiments have been done in NIFS (National Institute for Fusion Science) LHD (Large Helical Device) operations [1]. The ratio of co-extracted electron current to the negative ion current is up to 0.38 in the H^- beam case, while the ratio is up to 1.37 in the D^- beam operation. For further investigation of the difference in H and D cases, plasma diagnostics have been performed in NIFS-research and development negative ion source (RNIS). Also, the electron densities measured by Langmuir probes in the plasma generation region of D plasma is around three times higher than that of H plasma [1]. The results show that the isotope effects of H and D plasma appear in the electron density and result in the difference of the co-extracted electron current. To investigate the isotope effects in the plasma generation, the vibrationally excited states of hydrogen and deuterium molecules (H_2 and D_2) are focused. In the present study, a zero-dimensional (0D) model is developed for H_2/D_2 plasma. In the next section, the main reactions and the effects taken into account in the 0D model is explained. In section 3, the basic equations and the characteristic physics are shown. In section 4, the results of the density calculation for the H and D cases are explained. In the last section, we summarize the effect due to difference of the vibrationally excited states between H_2 and D_2 cases.

Table 1. Main reactions taken into account in the model. Description $N^{1,3}\Lambda_\sigma$ stands for the electronically excited states of molecules; B, B', B'', C, D, D'(singlet) and b, e (triplet). The abbreviations of each reaction names correspond to the subscripts of the reaction rate in Eqs. (2) – (8).

Reaction	Ref.
H_2 Dissociation (DISS,H2): $H_2(X^1\Sigma_g^+; v_i) + e^- \rightarrow H_2(N^{1,3}\Lambda_\sigma) + e^- \rightarrow 2H + e^-$	[2,3]
EV process (EV): $H_2(X^1\Sigma_g^+; v_i) + e^- \rightarrow H_2(B^1\Sigma_u^+, C^1\Pi_u) + e^- \rightarrow H_2(X^1\Sigma_g^+; v_f) + e^- + h\nu$	[2,4,5]
eV process (eV): $H_2(X^1\Sigma_g^+; v_i) + e^- \rightarrow H_2(X^1\Sigma_g^+; v_f = v_i \pm 1) + e^-$	[2,4]
H_2 Non-Dissociative Ionization (IOZ,H2): $H_2(X^1\Sigma_g^+; v_i) + e^- \rightarrow H_2^+(X^2\Sigma_g^+) + 2e^-$	[2,6]
H_2 Dissociative Ionization (DI,H2):$H_2(X^1\Sigma_g^+; v_i) + e^- \rightarrow H + H^+ + 2e^-$	[4]
H_2 Dissociative Electron Attachment (DA,H2): $H_2(X^1\Sigma_g^+; v_i) + e^- \rightarrow H_2^-(X^2\Sigma_u^+) \rightarrow H^- + H$	[4,7]
H2+ Dissociation (DISS, H2+): $H_2^+(X^2\Sigma_g^+) + e^- \rightarrow H + H^+ + e^-$	[2]
H_2^+ Dissociative Recombination (DR,H2+): $H_2^+(X^2\Sigma_g^+) + e^- \rightarrow H_2^{*Ryd}(N^{1,3}\Lambda_\sigma) \rightarrow 2H$	[2]
H_2^+ Dissociative Ionization (DI,H2+): $H_2^+(X^2\Sigma_g^+) + e^- \rightarrow 2H^+ + 2e^-$	[2]
H_3^+ Dissociation (DISS,H3+): $H_3^+ + e^- \rightarrow H_3^+(N^{1,3}\Lambda_\sigma) + e^- \rightarrow H^+ + H_2(X^1\Sigma_g^+; v_f) + e^-$	[2]
H_3^+ Dissociative Recombination (DR,H3+): $H_3^+ + e^- \rightarrow 3H$ or $H_2(X^1\Sigma_g^+; v_f) + H$	[2]
H_3^+ Production (H3+PROD): $H_2(X^1\Sigma_g^+; v_i) + H_2^+ \rightarrow H_3^+ + H$	[2,8]
H Ionization (IOZ,H): $H + e^- \rightarrow H^+ + 2e^-$	[9]
H Recombination (RC,H+): $H^+ + e^- \rightarrow H + h\nu$	[2]

2. Isotope Effect of the vibrationally excited levels in H_2 and D_2

In the present 0D model, densities of molecules, atoms, positive ions and electrons are solved. For the molecule, population density of vibrationally excited levels in the electronically grounded state; $H_2(X^1\Sigma_g^+; v = 0 - 14)$ or $D_2(X^1\Sigma_g^+; v = 0 - 20)$ is taken into account. The isotope effect between the H_2 and D_2 molecules is the difference in the cross-sections related to the vibrationally excited levels. The main reactions taken into account in the model are shown in table 1. The cross-sections for molecular ionization and dissociation processes depend on the eigenvalues for the vibrationally excited levels. From the mass scaling law [3,4], the fitting formulas for the D_2 cross-sections can be obtained from H_2 cross-sections by substituting pseudo-vibrational level

$$v' = \sqrt{\frac{\mu_{H2}}{\mu_{D2}}} \times v_i \tag{1}$$

corresponding to the D_2 vibrational level v_i. The constants μ_{H2} and μ_{D2} are the reduced mass for H_2 and D_2, respectively. Also, the Franck-Condon (FC) factors $q_{v-v'}^{M-N}$ and the A-coefficients of the spontaneous radiation process $A_{v-v'}^{M-N}$ for the transition between the electronically excited levels $M \rightarrow N$ and the vibrationally excited levels $v \rightarrow v'$ differs in the H_2 and D_2 molecules. The FC factors and the A-coefficients [5] are applied to the cross-section fitting formula of the EV process. The cross-section of the H_2 Dissociation process is calculated from the total excitation cross-section by applying the dissociation ratios. Differences of these ratios in the H_2 and the D_2 molecules are taken into the present model as the isotope effect. The peak cross-section of the H_2 Dissociative Electron Attachment process strongly differs in the low vibrationally excited level between the isotopes.

3. Zero-Dimensional model for H_2 and D_2

In the zero-dimensional model, time variations of the molecule, atom, positive ion and electron density are solved by the following rate equations;
Molecule (vibrationally excited state):

$$dn_{\mathrm{H2}}(v)/dt = S_{\mathrm{GAS}} + S_{\mathrm{eV}}(v' \to v) + S_{\mathrm{EV}}(v' \to v) - S_{\mathrm{eV}}(v \to v') - S_{\mathrm{EV}}(v \to v')$$
$$+S_{\mathrm{DR,H3+}} - S_{\mathrm{IOZ,H2}} - S_{\mathrm{DISS,H2}} - S_{\mathrm{DI,H2}} - S_{\mathrm{H3+PROD}} - n_{\mathrm{H2}}/\tau_{\mathrm{H2}}$$
$$+0.5 \times (1 - R_{\mathrm{N+}}) \times (n_{\mathrm{H+}}/\tau_{+}) + 0.5 \times (1 - R_{\mathrm{N0}}) \times (n_{\mathrm{H}}/\tau_{\mathrm{H}}). \quad (2)$$

Atom:

$$dn_{\mathrm{H}}/dt = S_{\mathrm{DISS,H2}} + S_{\mathrm{DI,H2+}} + S_{\mathrm{DISS,H2+}} + S_{\mathrm{RC,H+}} + S_{\mathrm{DR,H2+}} + S_{\mathrm{DR,H3+}} - S_{\mathrm{IOZ,H}} - n_{\mathrm{H}}/\tau_{\mathrm{H}}$$
$$+R_{\mathrm{N+}} \times (n_{\mathrm{H+}}/\tau_{+}) + R_{\mathrm{N0}} \times (n_{\mathrm{H}}/\tau_{\mathrm{H}}), \quad (3)$$

Positive Ion:

$$dn_{\mathrm{H+}}/dt = S_{\mathrm{IOZ,H}} + S_{\mathrm{DI,H2}} + S_{\mathrm{DISS,H2+}} + S_{\mathrm{DI,H2+}} - S_{\mathrm{RC,H+}} - n_{\mathrm{H+}}/\tau_{+}, \quad (4)$$
$$dn_{\mathrm{H2+}}/dt = S_{\mathrm{IOZ,H2}} - S_{\mathrm{DISS,H2+}} - S_{\mathrm{DI,H2+}} - S_{\mathrm{DR,H2+}} - S_{\mathrm{H3+PROD}} - n_{\mathrm{H2+}}/\tau_{+}, \quad (5)$$
$$dn_{\mathrm{H3+}}/dt = S_{\mathrm{H3+PROD}} - S_{\mathrm{DISS,H3+}} - S_{\mathrm{DR,H3+}} - n_{\mathrm{H3+}}/\tau_{+}, \quad (6)$$

Electron:

$$dn_{\mathrm{e}}/dt = S_{\mathrm{FIL}} + S_{\mathrm{IOZ,H2}} + S_{\mathrm{IOZ,H}} + S_{\mathrm{DI,H2}} + S_{\mathrm{DI,H2+}} - S_{\mathrm{DR,H2+}} - S_{\mathrm{DR,H3+}} - S_{\mathrm{RC,H+}} - n_{\mathrm{e}}/\tau_{\mathrm{e}}. \quad (7)$$

The variables $n_{\mathrm{H2}}(v)$, n_{H}, $n_{\mathrm{H+}}$, $n_{\mathrm{H2+}}$, $n_{\mathrm{H3+}}$ and n_{e} are density of molecules of vibrationally excited level, atoms, positive ions (H^+, H_2^+, H_3^+ or D^+, D_2^+, D_3^+) and electrons. Contributions from the reactions in table 1 to the density variation are described in the right side of the equation in the form of reaction rate S (m^{-3} s^{-1}). The subscripts of the reaction rate terms (such as $S_{\mathrm{IOZ,H}}$) correspond to the reaction names shown in table 1.For example, the rate of the H ionization is calculated as

$$S_{\mathrm{IOZ,H}} = n_{\mathrm{e}} n_{\mathrm{H}} \langle \sigma_{\mathrm{IOZ,H}} v \rangle = n_{\mathrm{e}} n_{\mathrm{H}} \times \int \sigma_{\mathrm{IOZ,H}}(E) v_{\mathrm{e}}(E) f_{\mathrm{e}}(E) dE \quad (8)$$

where $\sigma_{\mathrm{IOZ,H}}$ is the cross-section for the H ionization. The impact energy and the relative velocity is approximated by the electron energy E and the velocity $v_{\mathrm{e}}(E)$. The electron energy distribution function $f_{\mathrm{e}}(E)$ is assumed to be Maxwellian distribution with fixed electron temperature. All the reaction rates are calculated in the same manner as Eq. (8).

The gas injection rate is S_{GAS} is calculated from the mass flow rate in the experiment [1] with the relation 1 sccm $= 4.48 \times 10^{17}$ molecules s^{-1} and the volume of the ion source chamber V_{IS}. All molecules injected are assumed to be in the vibrationally grounded level $v = 0$. The electron injection rate S_{FIL} is calculated from the arc current I_{ARC}, which is nearly equal to the electron emission current from the filament as $S_{\mathrm{FIL}} = I_{\mathrm{ARC}}/qV_{\mathrm{IS}}$, where q is elementary charge.

The characteristic time of the particle transport loss τ is denoted in table 2. The transport loss times of the molecule and the atom are decided from chamber size (~ 1m) and thermal velocity ($v_{\mathrm{H}} \sim 10^3 \mathrm{m\ s}^{-1}$). The molecules are assumed to be reflected to the chamber walls but lost at the plasma grid (PG) surface while the atoms are lost at the chamber wall. Considering the ionization is the dominant process for electron and positive ion production in equations (4) and (7), we obtain the relation

$$\tau_{+}/\tau_{\mathrm{e}} = (S_{\mathrm{IOZ}} + S_{\mathrm{FIL}})/S_{\mathrm{IOZ}}. \quad (9)$$

From the electron transport analysis by KEIO-MARC code [10], the relation $S_{\mathrm{IOZ}} \sim S_{\mathrm{FIL}} = 3 \times 10^{22} \mathrm{m}^{-3}\mathrm{s}^{-1}$ and the average electron transport loss time $\tau_{\mathrm{e}} = 1.113 \times 10^{-5}$s are obtained [11]. According to equation (9), we simply set the transport loss times $\tau_{+} = \tau_{\mathrm{e}} = 1 \times 10^{-5}$s.

In the present model, it is assumed that the chamber wall surface is fully covered by H/D atoms due to the long duration of plasma discharge and gas injection. For these cases, atoms and positive ions incoming toward the chamber walls are reflected as atoms or molecules. Several reports show that the total reflection probability is nearly 100 % [12,13]. We assume that probability of the reflection as atoms for incoming atoms or H^+/D^+ ions are given by Eckstein and Verbeek [14], which are $R_{\mathrm{N0}} = 95.55$ % for incoming atoms and $R_{\mathrm{N+}} = 90.95$ % for incoming H^+/D^+ ions. The other part of the reflected particle is molecules via surface recombination with the probability given by $(1 - R_{\mathrm{N0}})$ and $(1 - R_{\mathrm{N+}})$. In the present study, the same values of R_{N0} and $R_{\mathrm{N+}}$ are applied for the H and the D calculations to make clear the effect of the difference between vibrationally excited levels of the H_2 and the D_2 molecules and its related reaction cross-sections. These models are self-made by Fortran program which

Table 2. Characteristic time of the particle transport loss.

Molecule	$\tau_{\mathrm{H2}} = 5 \times 10^{-3}$ s	**Positive Ion**	$\tau_{+} = 1 \times 10^{-5}$ s
Atom	$\tau_{\mathrm{H}} = 1 \times 10^{-3}$ s	**Electron**	$\tau_{\mathrm{e}} = 1 \times 10^{-5}$ s

Table 3. Calculation parameters.

Arc current	600 A	**Electron temperature**	$T_e = 5$ eV
Gas flow rate	850 sccm	**Positive ion temperature**	$T_+ = 1$ eV
Initial n_e, n_{H+}, n_{D+}	1×10^{18} m^{-3}	**Atom/Mol. temperature**	$T_0 = 1000$ K
Initial n_H, n_D	1×10^{18} m^{-3}	**Time resolution**	$\Delta t = 10^{-9}$ s
Initial n_{H2}, $n_{D2}(v = 0)$	1×10^{19} m^{-3}	**Time step number**	$N_{step} = 10^8$

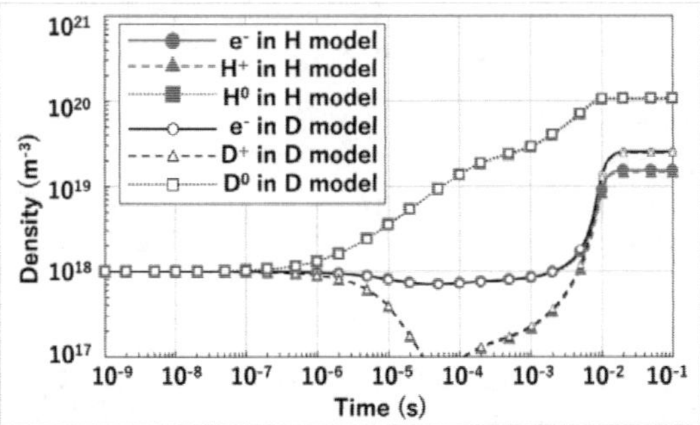

Figure 1. Time variation of the electron, the H$^+$/D$^+$ ion and the atomic density in the H and the D models.

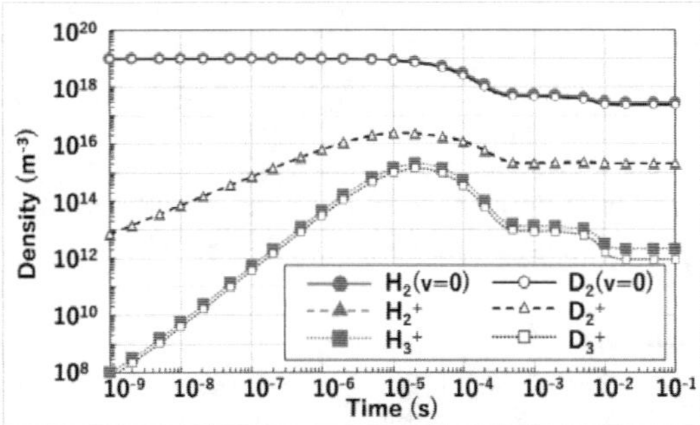

Figure 2. Time variation of the molecule (v = 0), H_2^+/D_2^+ and H_3^+/D_3^+ ion density in the H and the D models.

solves the simultaneous rate equations in explicit manner. The typical run-times are around 5 minutes on a laptop PC with a single CPU (Intel Core i7-10510U) and a 12 GB RAM. The 0D models are one of the powerful tools to obtain the physical picture of the plasma (related to atomic and molecular processes) with less calculation resources.

4. Numerical results

The calculation parameters corresponding to the NIFS-RNIS experiments are shown in table 3. The time variations of the electron, the H$^+$/D$^+$ ion and the atomic density are shown in figure 1. The density of the H and D atoms in the steady-state (at $t = 10^{-1}$ s) are almost same (up to 1.06 and 1.08 × 10^{20} m^{-3}, respectively. On the other hand, the dominant particle species of the charged particles are the electrons and the H$^+$/D$^+$ ions. As shown in figure 2, ratio of the molecular ion density is much lower than the proton density. The charge balance between positive and negative charge takes place both in the H and D cases. However, the steady-state electron densities are up to 1.54 × 10^{19} m^{-3} in the H model and up to 2.54 × 10^{19} m^{-3} in the D model. The only difference of the H/D models is the isotope effect of the vibrationally excited level of the H_2/D_2 molecules and the relevant cross-sections.

The higher plasma density in the D model is due to the higher ionization rate. Reaction rates of the ND ionization process of molecule which takes into account the contributions from all vibrationally excited levels are $S_{IOZ,H2} = 4.20 \times 10^{21}$m^{-3}s^{-1}in the H model and $S_{IOZ,D2} = 7.00 \times 10^{21}$m^{-3}s^{-1} in the D model. To exclude the effect of the density difference, the effective rate coefficients

$$R_{IOZ,H2} = S_{IOZ,H2}/n_e n_{H2}(v=0) = \sum_v n_{H2}(v)\langle\sigma_{IOZ,H2}(v)v_e\rangle/n_{H2}(v=0) \quad (10)$$

are compared between the two models. The values are $R_{IOZ,H2} = 9.88 \times 10^{-16}$m^{3}s^{-1} in the H model and $R_{IOZ,D2} = 1.17 \times 10^{-15}$m^{3}s^{-1}in the D model. The difference in the rate coefficient enhances the production rate of electrons. This leads to higher electron impact ionization rate of the atoms and the resultant plasma density in the D model than in the H model. This difference of the rate coefficients is due to the difference of the vibrationally excited level of $H_2(X^1\Sigma_g^+; v = 0 - 14)$ and $D_2(X^1\Sigma_g^+; v = 0 - 20)$ molecules. As shown in figure 3, the population density of vibrationally excited levels in H_2 and D_2 matches along the vibrational potential energy. Even in this match, the ionization rate differs in the two models. This is due to the number of ionization channels via the vibrationally excited levels.

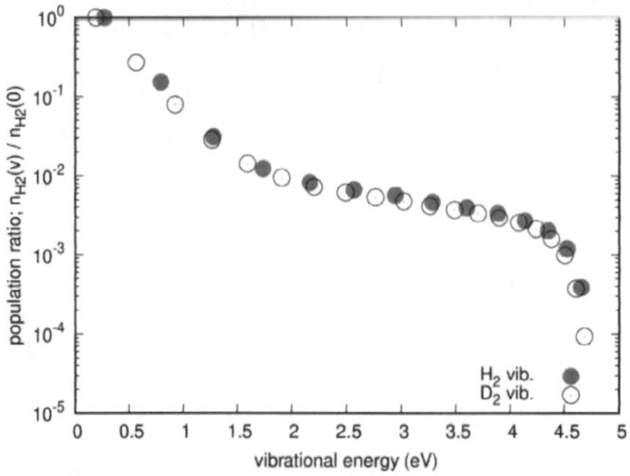

Figure 3. Population of the vibrationally excited levels of the H_2 and D_2 molecules in the steady-state. The horizontal axis is vibrational potential energy in eV units.

5. Summary

A zero-dimensional model based on the rate equations is developed in order to understand the electron density difference in the hydrogen (H) and deuterium (D) plasma experiments. In the present study, only differences between the H and D cases of the calculations are the vibrationally excited levels of electronically grounded state molecules and its relevant cross-sections (as mentioned in Sec.2). The numerical results show that the electron (plasma) density in the D case is higher than that in the H case by factor of around 1.65. Even the population density for vibrationally excited levels in H_2 and D_2 matches in the calculation, the rate coefficient of the non-dissociative ionization process is around 1.18 times higher in the D case than in the H case. From the calculation, it is shown that difference of the ionization channels from H_2 and D_2 molecules via the vibrationally excited levels results in the enhancement of the ionization rate coefficients from D_2. This ionization rate is one of the reasons for the electron density difference observed in the NIFS-RNIS experiments. However, the electron density ratio between H and D cases in the present calculation is 1.65 while the density ratio is around 3 in the experiments. On the other hand, there are other candidates for the plasma density difference in the plasma generation region, such as gas flow rate, electron temperature, reflection probability at wall or characteristic time of transport loss. Dependence of the electron density for these effects will be investigated individually by the zero-dimensional model in the near future.

Acknowledgement

This work was supported by KAKENHI 19H01883.

Reference

[1] H. Nakano, *et al.*, J. J. Appl. Phys. **59**, SHHC09 (2020).
[2] R. K. Janev, *et al.*, Berichite des Forschungszentrum Jülich; 4105 (2003).
[3] R. Celiberto, *et al.*, Phys. Rev. A **60**, 2091-2103 (1999).
[4] R. Celiberto, *et al.*, Atom. Data Nucl. Data Tables **77**, 161-213 (2001).
[5] U. Fantz and D. Wünderlich., Atom. Data Nucl. Data Tables **92**, 853-973 (2006).
[6] K. Sawada and T. Fujimoto, J. Appl. Phys. **78**, 2913 (1995).
[7] I. I. Fabrikant, Physica Scripta **T96**, 45-51 (2002).
[8] R. Ito, *et al.*, JAERI-M report 93-117 (1993).
[9] L. C. Johnson, Astrophys. J. **174**, 227-236 (1971).
[10] T. Shibata, *et al.*, J. Appl. Phys. **114**, 143301 (2013).
[11] R. Kato, *et al.*, these proceedings.
[12] I. Takagi, *et al.*, J. Nucl. Materials 290-293, 501-504 (2001).
[13] E. Vietzke, *et al.*, J. Nucl. Materials 266-269, 324-329 (1999).
[14] W. Eckstein and H. Verbeek, Nucl. Fusion 24, S9 (1984), Chap.2 (pp.12-28).

Redefining plasma chambers for ECR Ion Sources: the IRIS structure

D. Mascali[1], A. Galatà[2], S. Gallo[2], O. Leonardi[1], G. S. Mauro[1], E. Naselli[1], A. Pidatella[1], F. Russo[1], G. Sorbello[3] and G. Torrisi[1*]

[1] Istituto Nazionale di Fisica Nucleare - Laboratori Nazionali del Sud (INFN-LNS), Via S. Sofia 62, 95123 Catania, Italy.
[2] Istituto Nazionale di Fisica Nucleare - Laboratori Nazionali di Legnaro (INFN -LNL), Viale dell'Università, 2, 35020 Legnaro, Italy.
[3] Dipartimento di Ingegneria Elettrica, Elettronica e Informatica (DIEEI), Università degli Studi di Catania, Viale Andrea Doria 6, 95125, Catania

E-mail: giuseppe.torrisi@lns.infn.it

Abstract. One possible way to optimize microwave coupling and plasma confinement in Electron Cyclotron Resonance (ECR) Ion Sources is a revolutionary design strategy of plasma chambers, breaking the cylindrical symmetry. This contribution reports about the design and numerical validation of an innovative resonant cavity playing as plasma chamber of ECR ion sources. The new chamber, named IRIS (Innovative Resonators for Ion Sources), was argued starting from the 3D structure of the plasma and, therefore, fashioned to the twisting magnetic structure. The microwave launching scheme was radically changed as well, consisting of side-coupled slotted-waveguides with diffractive apertures smoothly matching the overall structure of the camera. This approach also enables a profound optimization of cooling systems and overall spaces in general (for gas feedings, oven systems, sputtering, etc.). Here we report on the conceptual study, electromagnetic design and PIC simulations of the electron heating in the novel resonant cavity, comparing results with those for standard (cylindrical) chamber, and also considering the impact of microwave feeding led by single aperture rectangular waveguides vs. waveguide-slotted antennas. Manufacture strategy, based on additive manufacturing techniques, will also be discussed.

1. Introduction and Motivation

Several efforts have been dedicated in the past decades to improve microwave coupling to ECR plasmas, alternative to the pure scaling of magnetic field and frequencies, most of them devoted to Frequency Tuning effect, Two Frequency Heating, and other techniques. In particular, the best performing machines such as SECRAL [?], VENUS [?] or SERSE [?], steadily improved their performances by means of:

- innovative magnetic systems [?];
- innovative launching-wave systems based on special a TE_{01}-to-HE_{11} mode converter, waveguides array or Vlasov launcher [?, ?, ?];
- unconventional variable cross section of the plasma chamber to optimize spaces, accesses, cooling, and maximize the radial confinement [?].

Journal of Physics: Conference Series **2244** (2022) 012003 doi:10.1088/1742-6596/2244/1/012003

In this paper, we present a study aimed at a substantial redefinition of the plasma chamber shape of an ECR ion source as a possible way to optimize microwave coupling to plasma, plasma structure, beam quality and, more in general, space and accesses to the ECR plasma chamber itself. The new shape of the plasma chamber, named IRIS (Innovative Resonators for Ion Sources), is inspired by the 3D structure of the plasma (that is in turn influenced by the magnetic field configuration). This new shape radically breaks the commonly used cylindrical symmetry: in classical cylindrical plasma chambers, in fact, the cylindrical symmetry implies - for most of the excited modes - a low electromagnetic field amplitude in the near axis region. From a theoretical point of view for circular waveguides and cylindrical cavities, the eigenmode solutions of Maxwell equations in terms of TE and TM modes are expressed in terms of Bessel functions (and first derivative): for higher order modes, $J_{n,\nu}$ functions (and their derivatives) are zero in the origin, except the J_0 [?]. In [?] this effect was investigated in connection the so-called RF plugging, i.e. the additional contribution to the plasma confinement inside the high density core (the plasmoid) enclosed by the ECR surface due to the RF heating. It was found that for low RF values the plasma density was lower as well, so that plasmoid can result depleted (in terms of density) in the near axis region. This can have a direct impact on the beam shapes, that can be hollow as it is for the plasmoid. In the new proposed approach, the cylindrical symmetry breaking should allow to concentrate the RF power more in the near axis region, making the coupling more robust and efficient and impacting on hollow ion beam structure observed so far in many sources. Also the microwave launching scheme was radically changed: the proposed new configuration consists of side-coupled slotted-waveguides with diffractive apertures smoothly matching the overall structure of the chamber walls (see Figure ??).

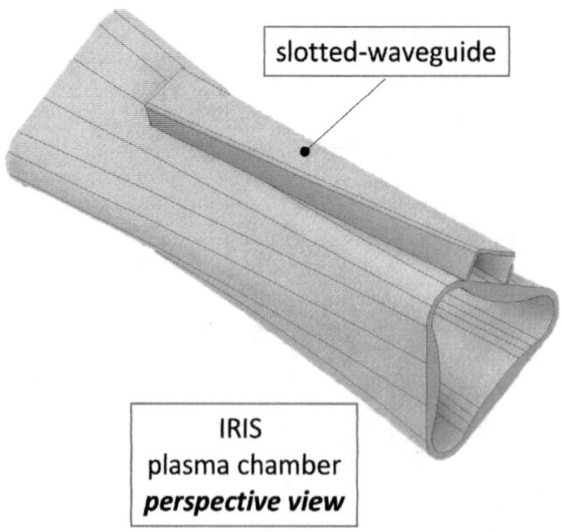

Figure 1: The proposed new configuration: IRIS plasma chamber with slotted-waveguide matching the structure of the chamber walls.

2. Design Process steps

The final version of the IRIS geometry results from different phases of the design which required a continuous "interplay" between the mechanical CAD modeling and the outputs of the electromagnetic simulations. The starting geometry has been obtained by the electrons trajectories as they move under the influence of the plasma-confining magnetic field. For all the following simulations:

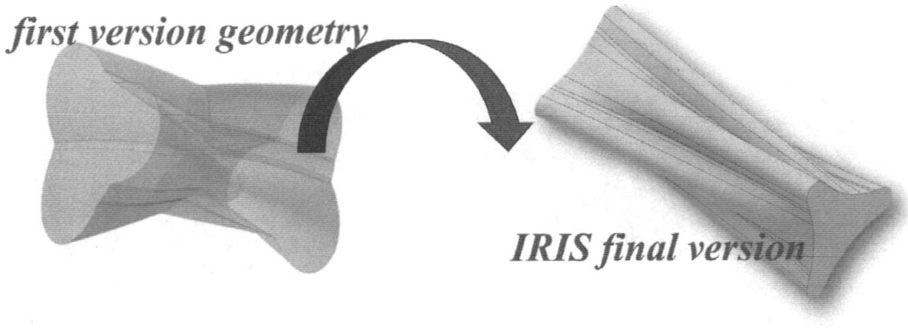

Figure 2: The first geometry is obtained from 3D electrons occupancy map: the typical "star-shaped" plasma structure impressed by the B-minim geometry has been iteratively optimized to obtain the final version.

- Electromagnetic (EM) Eigenmode;
- Driven full-wave electromagnetic simulation in empty and plasma-filled chamber;
- Self-consistent plasma PIC simulations.

we considered as reference scenario the ECR source called CAESAR [?] installed at INFN-LNS, to compare the IRIS EM simulations with a "real" cylindrical plasma chamber also concerning the typical size and operating frequency range.

Electromagnetic (EM) Eigenmode Simulations
The first goal was to get an EM mode with high electric field amplitude along the chamber axis. The cavity geometry was optimized by EM Eigenmode Simulations in empty and plasma-filled chambers.

Driven full-wave electromagnetic simulation in empty chamber
Driven full-wave electromagnetic simulations were performed in COMSOL and CST, by considering an axial rectangular waveguide as RF injection system. The reflection coefficient S_{11} has been used as a quantitative parameter in order to find the best solution in terms of microwave coupling and matching in the frequency range of [14; 14.5] GHz.

In the IRIS scenario, the number of modes having very good waveguide-to-cavity couplings ($S_{11} < -15$ dB) was higher than in the cylindrical chamber. Taking a given frequency, the related electric field profile has typically a maximum around the cavity center, as it can be seen in Figure **??** (b): this plot show the higher field achieved by IRIS if compared with the on-axis hollow structure of the electric field obtained with the CAESAR chamber (Figure **??** (a)).

The further step was to adopt a different launching, by the aforementioned slotted antenna. IRIS side-coupled slotted waveguide design details can be found in [?]. Simulation results show that the new microwave coupling scheme greatly improves the number of modes that can be coupled inside the plasma chamber. This could be advantageous when one applies the frequency tuning technique in order to improve the ion source performances. Moreover, by using the IRIS cavity, the side-coupled scheme employment results in a symmetrization of the desired mode(s) with respect to the chamber axis. E-field distributions typically present a maximum around the cavity axis. Its profile results now symmetric along the cavity axis with respect to the configuration shown relative to the axial injection system. Moreover, this new solution releases space on the end flanges of the plasma chamber for further ancillary equipment employment.

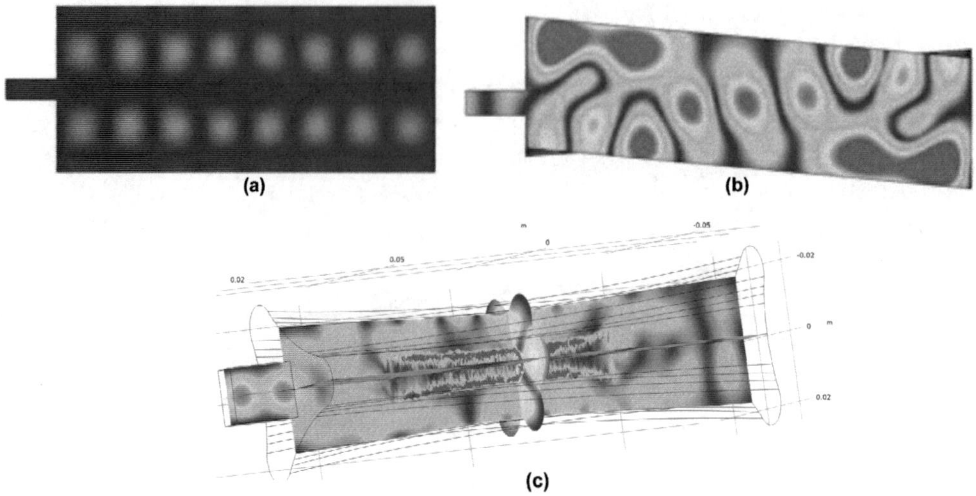

Figure 3: Driven full-wave electromagnetic simulation results: Electric Field intensity in empty CAESAR chamber (a), empty IRIS chamber (b), plasma filled IRIS chamber (c)

3. Self-consistent plasma PIC simulations

To validate the improvements obtained by the unconventional IRIS geometry and its launching system, 3D self-consistent simulations of the plasma generated by both CAESAR and IRIS have been carried out using a numerical tool developed by the INFN ion sources group . The self-consistent modelling is achieved by an iterative procedure that aims at solving the collisional kinetics equation; details can be found in [?,?]. The results of driven EM simulation in plasma filled IRIS chamber demonstrate that an enhancement of the electric field and impressive shortening of the wavelength - shown in Figure ?? (c) - is obtained in the IRIS scenario. Also the electromagnetic power deposition in the plasma becomes much more efficient in the reshaped configuration and involve also different regions with respect ECR showing a peculiar shape which is conformal to the cavity geometry. The new concept has been validated by self-consistent, stationary PIC simulations of the electron heating in both standard (cylindrical) chamber fed by single aperture rectangular waveguides vs IRIS geometry. IRIS geometry leads to an increase of the average energy, whose maximum concentrates now inside the plasmoid around the plasma chamber axis. The concentration of the plasma density inside the plasmoid reaches a factor of 6, higher in IRIS with respect to CAESAR. It is worth noting that the IRIS geometry implies a net increase of the maximum occupancy, in particular in radial direction where it is close to an order of magnitude.

4. Preliminary manufacturing and test

Due to the non-conventional shape of the novel "plasma-shaped" cavity and relative twisted slotted waveguide injection system, standard manufacturing techniques based on subtractive processes could be difficult to employ. To overcome these difficulties, for the future realization we plan to use Additive Manufacturing (AM) techniques, and in particular 3D metal printing

through high quality metal powder compound.
Two simple cylinders made through 3D metal printing have been produced by an external company. Vacuum tightness has been successfully tested.
Waveguides also have been manufactured by means of 3D metal printing, in order to verify the design parameters and in particular the effect of surface roughness on radio frequency transmission: the AM 3D printed slotted waveguides are matched below -10 dB in the frequency range but shows a shift in the resonance frequency and a reflection coefficient almost 3 dB greater than the standard process manufactured waveguide, due to tolerance and surface roughness.

5. Conclusions and Perspective

An innovative electromagnetic resonator to be employed as plasma chamber of ECRISs has been here presented. The structure presents some key advantages with respect to the classic cylindrical cavity: a) modes with a predominance of axial electrical component can be (relatively) easily excited and b) more space is available for the surrounding magnetic system (i.e. axial coils). In conjunction with the new plasma cavity, a novel microwave launching scheme for ECRIS plasma chambers, based on slotted waveguides placed along the chamber outer wall, has also been studied. The use of the side-coupled launch scheme greatly improves the number of modes that can be coupled into the plasma chamber: this could be an advantage in single frequency tuning operations. Other advantages with respect to the standard axial launch scheme are a) more symmetric field and power distribution from the multiple radiating waveguide slots and b) the possibility to have more space available on the injection flange for other ancillaries. Realization of the plasma chamber plus side-coupled injection system with additive manufacturing technique is ongoing and the prototype test is expected soon by using the CAESAR ion source setup at INFN-LNS. The overall new scheme allows to save/optimize space both axially and longitudinally, opening also the possibility to reshape and repositioning the hexapole coils in order to improve cooling and confinement.

Acknowledgments

The authors wish to thank the 3rd Nat. Comm. of INFN, under the PANDORA_ Gr3, for the financial support. Part of this work has been carried out within the Grant73/IRIS project, also supported by INFN (Italian patent pending n. 102020000001756).

References

[1] H. W. Zhao, L. T. Sun, J. W. Guo, W. Lu, D. Z. Xie, D. Hitz, X. Z. Zhang, and Y. Yang, "Intense highly charged ion beam production and operation with a superconducting electron cyclotron resonance ion source," *Phys. Rev. Accel. Beams*, vol. 20, p. 094801, Sep 2017.
[2] J. Y. Benitez, K. Y. Franzen, A. Hodgkinson, T. Loew, C. M. Lyneis, L. Phair, J. Saba, M. Strohmeier, and O. Tarvainen, "Recent progress on the superconducting ion source VENUS," *Review of Scientific Instruments*, vol. 83, no. 2, p. 02A311, 2012.
[3] D. Hitz, A. Girard, G. Melin, S. Gammino, G. Ciavola, and L. Celona, "Results and interpretation of high frequency experiments at 28 GHz in ECR ion sources, future prospects," *Review of Scientific Instruments*, vol. 73, no. 2, pp. 509–512, 2002.
[4] G. Torrisi, G. Sorbello, O. Leonardi, D. Mascali, L. Celona, and S. Gammino, "A new launching scheme for ecr plasma based on two-waveguides-array," *Microwave and Optical Technology Letters*, vol. 58, no. 11, pp. 2629–2634, 2016.
[5] J. W. Guo, L. Sun, W. Lu, W. H. Zhang, Y. C. Feng, Z. Shen, L. X. Li, J. B. Li, X. Z. Zhang, D. Hitz, and H. W. Zhao, "A new microwave coupling scheme for high intensity highly charged ion beam production by high power 24-28 GHz SECRAL ion source," *Review of Scientific Instruments*, vol. 91, no. 1, p. 013322, 2020.
[6] H. Koivisto, A. Ikonen, T. Kalvas, S. Kosonen, R. Kronholm, M. Marttinen, O. Tarvainen, and V. Toivanen, "A new 18 GHz room temperature electron cyclotron resonance ion source for highly charged ion beams," *Review of Scientific Instruments*, vol. 91, no. 2, p. 023303, 2020.

[7] G. Torrisi, G. Castro, L. Celona, A. Galatà, S. Gammino, D. Mascali, G. Mauro, M. Mazzaglia, E. Naselli, and G. Sorbello, "Plasma Heating and Innovative Microwave Launching in ECRIS: Models and Experiments," in *23rd International Workshop on ECR Ion Sources*, 2019, p. TUB2.

[8] D. Mascali, L. Neri, L. Celona, G. Castro, G. Torrisi, S. Gammino, G. Sorbello, and G. Ciavola, "A double-layer based model of ion confinement in electron cyclotron resonance ion source," *Review of Scientific Instruments*, vol. 85, no. 2, p. 02A511, 2014.

[9] S. Gammino and G. Ciavola, "The contribution of the infn-lns to the development of electron cyclotron resonance ion sources (invited)," *Review of Scientific Instruments*, vol. 71, no. 2, pp. 631–636, 2000.

[10] G. S. Mauro, G. Torrisi, O. Leonardi, A. Pidatella, G. Sorbello, and D. Mascali, "Design and analysis of slotted waveguide antenna radiating in a plasma-shaped cavity of an ECR ion source," *Telecom*, vol. 2, no. 1, pp. 42–51, 2021. [Online]. Available: https://www.mdpi.com/2673-4001/2/1/4

[11] A. Galatà, D. Mascali, C. S. Gallo, and G. Torrisi, "Self-consistent modeling of beam-plasma interaction in the charge breeding optimization process," *Review of Scientific Instruments*, vol. 91, no. 1, p. 013506, 2020.

[12] G. Torrisi, D. Mascali, G. Sorbello, G. Castro, L. Celona, and S. Gammino, "Numerical and experimental characterization of RF waves propagation in ion sources magnetoplasmas," *IEEE Transactions on Antennas and Propagation*, vol. 67, no. 4, pp. 2142–2149, 2019.

Improvement of Velocity Distribution Function Measurement of Hydrogen Atoms in ion Source Discharges

T Tokai [1], K Bito [1], Y Shimabukuro [1, 2] and M Wada [1]

[1]Graduate School of Science and Engineering, Doshisha University, Kyotanabe, Kyoto 610-0321, Japan
[2]Present address: Naka Fusion Institute, National institute for Quantum and Radiological Science and Technology, Naka, Ibaraki 311-0193, Japan

E-mail: ctwf0368@mail4.doshisha.ac.jp

Abstract. The efficiency of negative hydrogen (H^-) ion surface production in an ion source should be highly dependent upon the velocity distribution function of neutral atomic hydrogen (H^0) striking the plasm grid (PG). A system composed of forming an atom beam, rotating beam chopper, ionizer and pulse count unit has been constructed to measure H^0 velocity distribution functions. The background noise due to stagnation of hydrogen gas in the region between the beam formation skimmer and the ion source PG was improved by inserting a flow reflection structure. The short life time of the secondary electron emission detector (SEM) was prolonged by properly setting the hydrogen ion trajectories from the ionizer to the the first dynode of the SEM. The system is running stably to record the velocity distribution function but statistical noise synchronized to the trigger pulse of the chopper must be removed before the measurment of change in H^0 velocity distribution functions dependin upon the type of discharge.

1. Introduction

The principles mechanisms of negative hydrogen (H^-) ion production can be categorized into two processes: surface production and volume production [1]. In surface production process, H^- ions are considered formed from the surface of the electrode, called plasma grid (PG), with the extraction holes facing the plasma. The following reaction equations describe surface production due to hydrogen plasma particle impact.

$$H_n^+ + \text{surface} \rightarrow H^- \quad (1)$$

$$H^0 + \text{surface} \rightarrow H^- \quad (2)$$

An incident hydrogen ion or neutral should have the kinetic energy corresponding to the difference between the PG surface work function and the hydrogen electron affinity. For Cs covered Mo, the work function is 1.68 eV [2] and the corresponding energy difference should be 0.93 eV. For plasma ions accelerated across the sheath, this threshold energy can be easily acquired. However, hydrogen atoms are neutral and contribute to negative ion formation when their kinetic energy exceeds the threshold. Okuyama and Mori reported that collisions of hydrogen atoms effusing from the ion source plasma with the low work function surface caused surface production of H^- ions [3]. Meanwhile, Ivanov *et al* installed arc discharge atomic hydrogen sources to their H^- ion source to increase H^- surface production by H^0 impact [4]. Shimabukuro *et al* observed that the second mode jump can increase the power absorption by an electron cyclotron resonance (ECR) driven small plasma source to produce larger flux of fast H^0 atoms [5]. To clarify if these atom sources can create H^0 atoms fast enough to produce H^- ion efficiently by surface production process, a project to construct a system to measure the H^0 velocity distribution function was initiated [6]. The plan is to measure the difference in velocity distributions of hydrogen atoms from various plasma sources.

2. Experimental setup

Figure 1 shows a schematic diagram of the experimental setup for the velocity distribution function measurement. The experimental system is divided into the source part, pulse chamber part, flight chamber part, and secondary electron multiplier (SEM) chamber part. There parts are installed with individually separated pumping systems. The functions of components are explained in reference [6]. In this paper the principle of measurement is summarized highlighting the components being improved or will be improved.

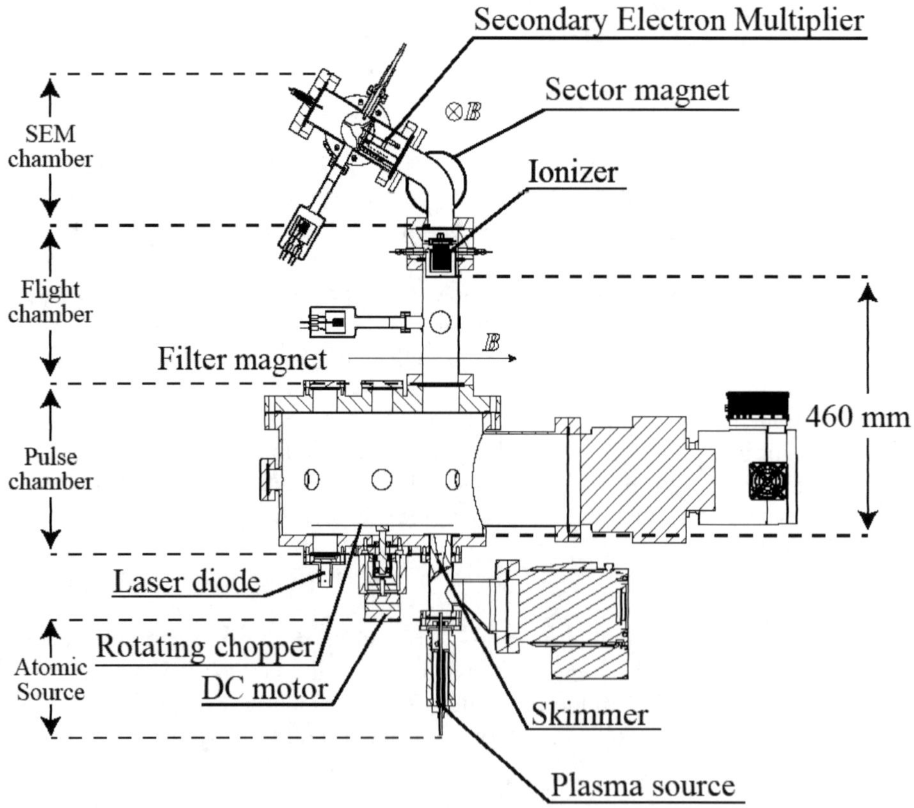

Figure 1. A schematic diagram of experimental setup of the velocity distribution function measurement.

2.1. *Hydrogen Atom Source*

The Localized Inductively Coupled Plasma (LICP) source [4] served as the test source for the system improvement. Flux of H^0 is produced by 2.45 GHz microwave power coupled to Nd-Fe magnets realizing ECR conditions for forming mode jumps from E-mode to H-mode, and H-mode to W-mode. Typical flow rate ranges from 1 sccm to 10 sccm. Steady state operation is possible up to 50 W, but cooling becomes often insufficient above 60 W microwave input power. For a mode change to W-mode, power input more than 30 W is necessary, but the power level can be reduced for maintaining the W-mode discharge down to 20 W.

2.2. *Pulse/Flight chambers*

The rotating chopper blade forms the pulse of H^0 beam delivered from the skimmer connecting the atom source to the pulse chamber. A collimated beam of laser light passes through the slit opened on the blade to generate a pulse signal synchronized to the opening of the shutter of the H^0 beam. The beam traverses the pulse chamber and fly additional distance to reach the orifice in front of the ionizer chamber.

2.3. *Ionizer/SEM detector*

Part of the H^0 beam is ionized by the electron impact ionization in the ionizer. The beam travels with the original speed as it departs the atom source, then partly ionized and the portion is electrostatically accelerated at the exit of the ionizer. The 3 kV extraction potential applied to the ionizer makes the flight time from the ionizer exit to the ion conversion plate negligible for the H^0 time-of-flight (ToF). After the mass separation by a magnetic momentum mass selector, H^+ and H_2^+ ions hit the surface of ion conversion plate, and the produced electron are multiplied by the dynode array in the SEM to produce signal pulse that are counted after a high-speed amplifier by a multi-channel scaler (Stanford Research Systems, SR430).

3. System improvement

3.1. *New skimmer design*

The original background count pulse before the system improvement was too large to construct the velocity distribution function from the measured data. Thus, a substantial change in the skimmer design was proposed with the gas flow reflector coupled to the H^0 aperture. The old skimmer shown in figure 3 (a) was attached to a short vacuum tube section and gas flow can easily become stagnant. The new design reflects molecules toward the pumping inlet of a turbo molecular pump enabling reduction of partial pressure due to stagnant flow.

(a) An old skimmer. (b) A new skimmer.

Figure 2. Skimmer for velocity distribution function measurement system.

The effectiveness of the new skimmer design can be recognized by confirming the small pressure change against the flow rate increase for the atom source. Figure 3 plots the pressure measured by vacuum ionization gauges at flight chamber and the SEM chamber. As shown in the figure, the leakage from the source chamber to the flight chamber is diminished from 30% to 9% by the new skimmer design. The pressure increase in the SEM region was negligible, while the old skimmer increased the pressure of SEM chamber at the rate 1.7×10^{-7} Pa/sccm.

3.2. *Optimization of SEM dynode voltage*

Figure 4 shows a simplified schematic of the SEM detection structure. We investigated the ion beam trajectory by computer simulation and found the bias voltage to the first dynode of the SEM affects the electron trajectories from the ion conversion plate (or conversion dynode in figure 4). Some electrons originating from the ion conversion plate do not pass through the entrance aperture to strike the first dynode when the bias to the dynodes was improper.

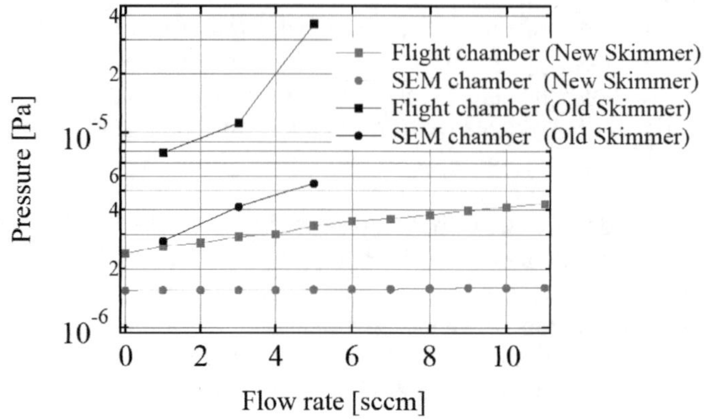

Figure 3. Gas pressure characteristics of each skimmer against flow rate.

The pulse height of the signal must change by the electron trajectories in dynodes. Thus, we plotted maximum pulse height as the function of bias applied to the first dynode by fixing the bias to the ion conversion plate to 3 kV. Figure 5 shows a graph of the result and found that the negative bias to the first dynode should be less than 2650 to realize a measurement stable against the slow drift of the incoming ion trajectory.

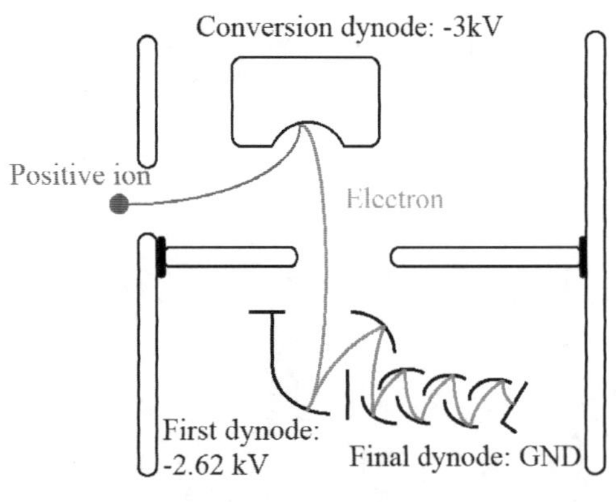

Figure 4. A schematic diagram of a simplified secondary electron multiplier.

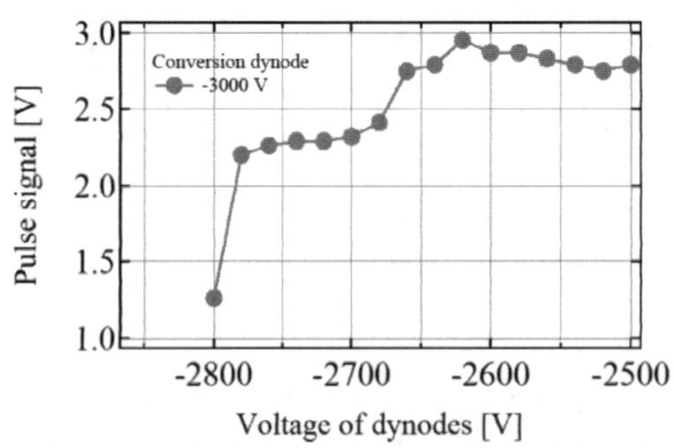

Figure 5. Voltage characteristics of dynode.

We conducted a test experiment after the above system improvement. The experiment was carried out with 3 sccm hydrogen gas flowing from the LICP source. After the hydrogen gas introduction, the gas pressures were 3.8×10^{-5} Pa in the pulse chamber part, 2.9×10^{-6} Pa in the flight chamber part, and 1.1×10^{-6} Pa in the SEM chamber part. Figure 6 (a) shows a typical raw data as atomic hydrogen ions (H^+) were detected as 16384 data points. Figure 6 (b) shows the smoothed data (averaged over binomial smoothing filter over about 10000 data points) obtained from the raw data in Figure 6 (a). Figure 7 shows the smoothed data at 50 W under the same conditions. The mean velocity of the hydrogen atoms $\langle v\rangle$ is calculated by the following equation.

$$\langle v\rangle = \sqrt{\frac{k_B T}{m}} \tag{3}$$

Where k_B is the Boltzmann constant. By obtaining the temperature T from fitting the measured data to a Maxwellian distribution (585 K) the mean velocity was obtained to be 2.2 $\times 10^3$ m/s.

A periodic noise was observed between 0 to 3 ms. Noise showing the similar waveform was confirmed not only during the plasma ignition but also during the plasma extinction, which suggests that the source of the noise is hydrogen leakage synchronized to the rotating chopper.

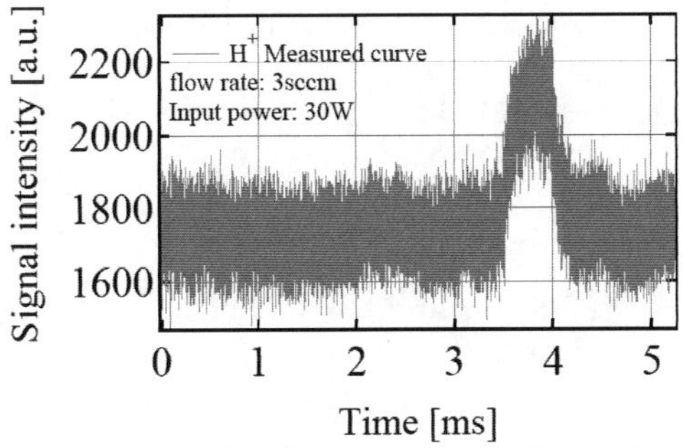

(a) An example of measured raw data at input microwave power in 30W.

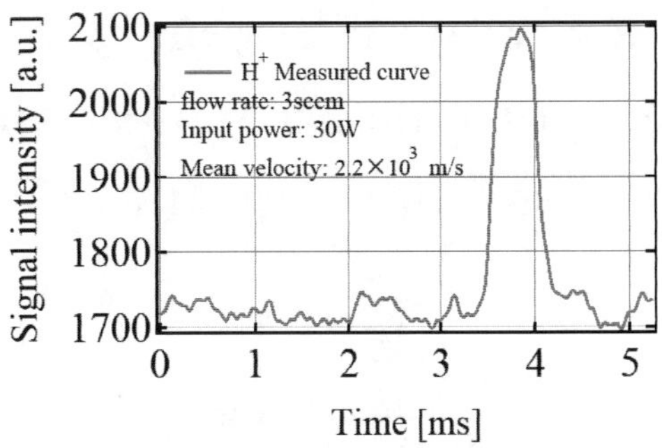

(b) An example of a smoothed result of raw data at input microwave power in 30 W.

Figure 6. An example of the ToF spectrum result for the LICP source.

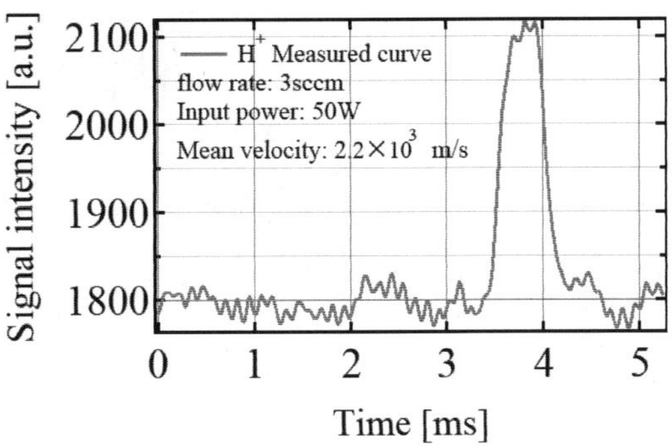

Figure 7. An example of a smoothed result of raw data at input microwave power in 50 W.

4. Conclusion

Two major improvements were made to the system for measuring the velocity distribution function of atomic hydrogen. One is to reduce the leakage flow rate from the ionizer section by improving the gas flow pattern around the skimmer. The other is the adjustment of the bias voltage of the ion conversion plate of the secondary electron multiplier that affected the electron trajectory reaching the first dynode. The results of the test experiments after the modification showed the presence of periodic noise synchronized to the chopper rotation. Velocity distribution functions from different types of hydrogen atom sources will be measured after diminishing the synchronized noise and further improving vacuum system in the ionizer section.

References

[1] Bacal M and Wada M 2015 *Appl. Phys. Rev.* **2**, 021305
[2] Aamodt R L, Brown L J and, Nichols B D 1962 *J. Appl. Phys.* **33**, 2080
[3] Okuyama T and, Mori Y 1992 *Rev. Sci. Instrum.* **63**, 2711
[4] Ivanov A A *et al* 2014 *Rev. Sci. Instrum.* **85** 02B129
[5] Shimabukuro Y, Takahashi H, Iwamoto S, Tanaka K, and Wada M 2020 *Plasma. Source Sci. Technol.* **29**, 015005
[6] Tokai T, Shimabukuro Y, Takahashi H, Bito K, and Wada M 2021 *AIP Conf. Proc.* **2372** 02003

Energy distribution and bremsstrahlung spectra of energetic electrons escaping from the ECR plasma with high energy input

E. M. Kiseleva[1,2], I. V. Izotov[1,2], V. A. Skalyga[1,2], S.S. Vybin[1,2], R.L. Lapin[1]

[1] Institute of Applied Physics, Russian Academy of Sciences (IAP RAS), Nizhny Novgorod, Russia

[2] Lobachevsky State University of Nizhny Novgorod - National Research University (UNN), Nizhny Novgorod, Russia

E-mail: kiseleva@ipfran.ru

Abstract. Electron energy distribution function (EEDF) is an important characteristic that defines the processes of plasma confinement and ionization. Understanding such processes is crucial for the proper tuning of ion sources. In the ECR plasma, the EEDF is significantly non-Maxwellian, and the exact shape remains still unknown. As the initial step towards the EEDF resolving, the energy distribution of the electrons lost from the magnetic mirror trap was directly measured in a wide range of neutral gas pressures and gyrotron powers along with bremsstrahlung spectra. A series of experiments were performed on the newly constructed Gasdynamic Ion Source for Multipurpose Operation (GISMO) facility allowing record-breaking specific energy input into the plasma. Obtained distributions showed unconventional behavior as the function of source parameters. The EEDF shape and the bremsstrahlung spectra showed a threshold qualitative change with the gyrotron power presumably as a result of the development of kinetic instabilities.

1. Introduction

Ion sources based on the electron cyclotron resonance discharge plasma are widely used in the field of accelerator physics due to their ability to generate highly charged ions. However, the requirements for such facilities are constantly increasing along with the performance of modern accelerators. The development process slows down as the fundamental limitations arise. Thus, understanding the physics of the processes occurring in the plasma of an ECR discharge is vital for the applications and improvements.

One of the important factors that determine processes occurring in the plasma is the electron energy distribution function (EEDF). In the ECR plasma, the EEDF is a sophisticated function of various parameters. In many models that evaluate the processes of confinement and ionization, the Maxwellian shape is used, which significantly lowers the accuracy of simulations. The real EEDF shape is still unknown.

In 2012, the idea was proposed at the Institute of Applied Physics of the Russian Academy of Sciences [1] to directly measure the distribution of the electrons lost from the magnetic mirror trap, which in turn can be used for the EEDF reconstruction. However, the exact procedure remains in the process of development. In the current work, with the use of this method the lost electron energy distributions (LEEDs) were measured in a wide range of neutral gas pressures and gyrotron powers.

The ECR plasma confined in a magnetic trap may contain a hot unstable electron component that significantly affects confinement processes. Due to this factor, a large amount of bremsstrahlung appears in such a plasma, which arises when energetic electrons are precipitated onto the wall. Thus, the bremsstrahlung spectra were obtained simultaneously with the LEEDs in the experiments described in this work. Determining the direct correlation between bremsstrahlung and the LEED can possibly give the answer for the reconstruction of the real EEDF.

2. Experimental setup

The experiments that are described in this work were conducted on the Gasdynamic Ion Source for Multipurpose Operation facility (GISMO). It is a gasdynamic ECR source with a high specific energy input recently created at the IAP RAS [2], [3]. The scheme and the photo of the facility are shown in Fig.1.

In GISMO, the plasma is confined in a simple mirror-like trap made of permanent magnets. The parameters of the trap are $B_{max} = 1.5$ T, $B_{min} = 0.25$ T. The mirror ratio, respectively, is 6. The specific feature of the magnetic configuration is the presence of a cusp, an area with $B = 0$, downstream the mirror trap.

The plasma is ignited by a high-power modern gyrotron generating radiation at a frequency of 28 GHz with power up to 10 kW. Such unique parameters, including the small size of the plasma chamber, allow to achieve record-breaking values of the specific energy input into the plasma of a continuous ECR discharge at a level of up to $50 - 100$ W/ cm^3, whereas for conventional min-B ECR ion sources this value does not exceed $1 - 5$ W / cm^3.

While we altered the gas pressures from 0.1 to 3 mTorr during our experiments, the facility has the ability to operate in a wide range of gas pressures in the discharge from 10^{-2} Torr to 10^{-6} Torr, which makes it possible to conduct research both in classical (collisionless) and quasi-gasdynamic (collisional) confinement modes.

The idea of measuring the energy distribution of electrons leaving the plasma was proposed and tested in the Institute of Applied Physics in 2012 [1]. The method is analogous to the common ion mass spectrometry with an inverted polarity of an electromagnet that deflects charged particles, as thoroughly described in [4]. However, to apply such method, it is necessary to make some important changes in the system. In our case, a secondary electron multiplier and a Stanford SR570 current preamplifier were used to be able to register electron currents as small as 1 pA. A voltage of -3.5 kV was applied to the cathode of the electron multiplier with respect to the chamber potential, which prohibited the registration of electrons with energies below 3.5 keV. The geometry of the system made it possible to allow scanning with the resolution of about 1 keV. Moreover, the presence of a cusp in the magnetic system makes it possible to observe only electrons originated from the plasma close to the symmetry axis of the system. A significant fraction of the electrons is lost on the chamber walls, thus generating bremsstrahlung.

There are many factors that affect the electron signal in the detector. Taking this into account, we made the following corrections for the electron current (Fig. 2): the coefficients of secondary emission and backscattering of the electron multiplier cathode (marked blue and green, correspondingly), as well as the transport function (marked red) that determines the fraction of electrons with a particular energy lost on the way from the plasma electrode to the detector due to the geometric and magnetic peculiarities of the system.

Simultaneously with the LEEDs, the bremsstrahlung spectra were obtained by the XR-100T-CdTe Amptek X-Ray spectrometer. It was installed in a way to observe the end of the plasma chamber, where the energetic particles precipitated. As a result, we could only detect the

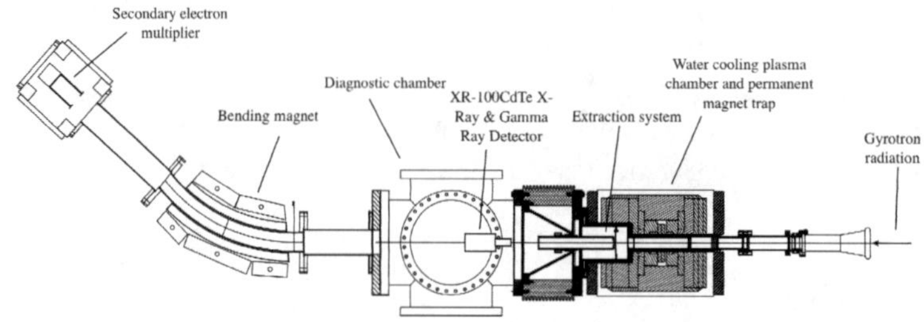

Figure 1. The experimental scheme

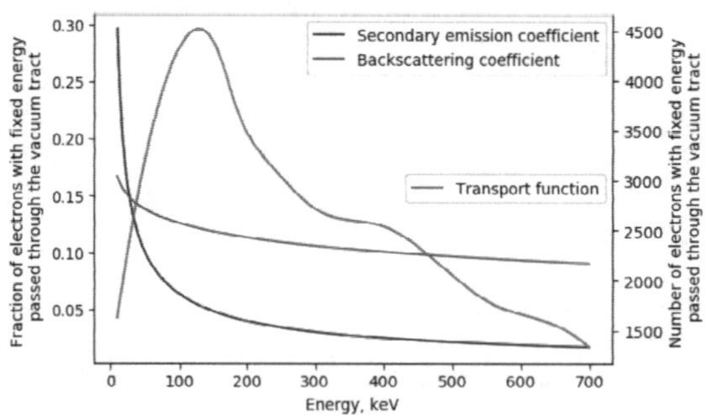

Figure 2. Factors that affected the electron signal

bremsstrahlung that could go through the metal, i.e. with energies exceeding 10-20 keV.

3. Experimental results

The distributions were measured in a wide range of hydrogen pressures and powers of the gyrotron radiation that sustains the ECR discharge. As it was mentioned before, the LEEDs make up a sophisticated function of external parameters.

Examples of obtained normalized distributions are presented as heatmaps in Fig.3 for different gyrotron powers, where the horizontal axis is the electron energy, the vertical axis is pressure, and the electron signal is shown in false colour.

A qualitative change in the LEED shape is observed with both power and pressure. At small pressure values ($0.1 - 0.2$ mTorr), the main fraction of particles is concentrated in around the values of $50 - 60$ keV and almost does not change with an increase in the gyrotron power. At $0.3 - 1.7$ mTorr, the fraction of electrons is formed in a wide range of $5 - 30$ keV and weakly depends on the power. LEEDs at high pressures are characterized by a qualitative change in

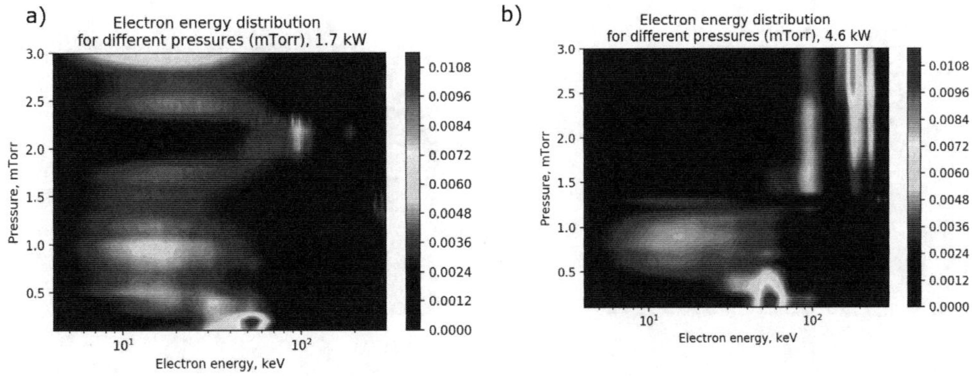

Figure 3. Lost electron energy distributions heatmaps measured for a) 1.7 b) 4.6 kW of gyrotron power

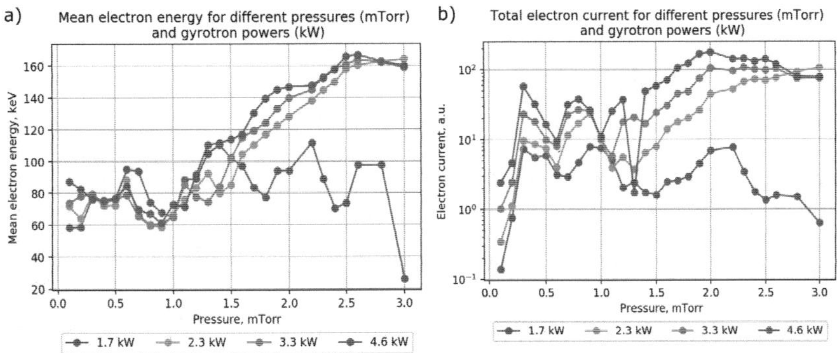

Figure 4. Integral LEED characteristics: a) Mean energy; b) Total electron current

shape and the appearance of high-energy (100−200 keV) fractions of energetic electrons when the power exceeds some threshold value. One can assume that the confinement conditions somehow change with a relatively small change in pressure. Such abrupt changes in LEED shape are presumably related to the onset of kinetic instabilities and will be further investigated.

It is important to add some integral characteristics of the obtained distributions, such as mean energy and the total electron current calculated mathematically with all the corrections that were described before (Fig.4). If the point 2.1 mTorr is taken out of consideration, it might be concluded that there is some major effect that changes the properties of the plasma between 1.7 and 2.3 kW as the behaviour of mean energy and electron current completely differs between those values. However, the dependence on pressure is such that it repeats the conclusions made out of distributions.

The obtained bremsstrahlung spectra are shown in Fig. 5, and the integral characteristics for spectra are shown in Fig. 6. It was impossible to get the proper ones for 1.7 kW of the gyrotron power due to low input rate. As the major part of the bremsstrahlung is lost on the way through the metal, the majority of information also gets missed. However, it is seen that the behaviour of the spectra and the integral characteristics is similar to the ones of the LEEDs. This fact allows to state that the main source of bremsstrahlung even in the gasdynamic ion source with high density might be hot electrons leaving the plasma and hitting the walls of the chamber.

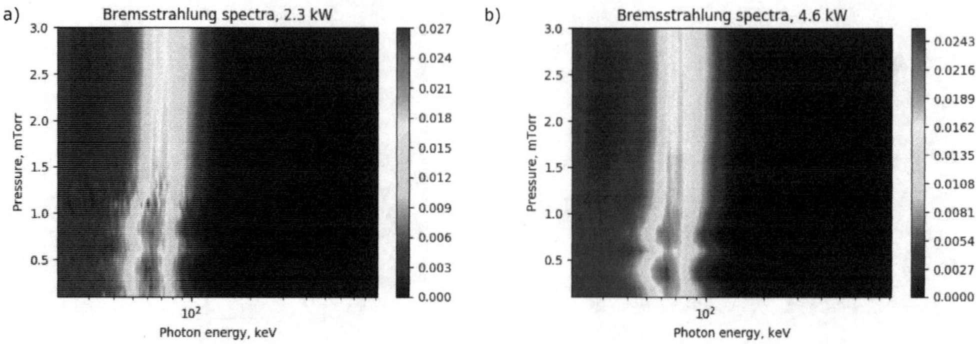

Figure 5. Bremsstrahlung spectra obtained for a) 2.3 b) 4.6 kW of gyrotron power

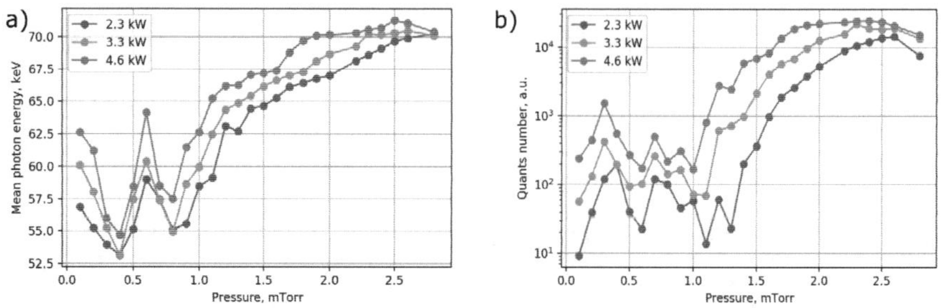

Figure 6. Integral bremsstrahlung characteristics: a) mean energy; b) input rate

4. Conclusion

In our experiments, the LEEDs for the plasma of a continuous ECR discharge with high energy input were obtained for the first time in a wide range of pressures and microwave powers along with the bremsstrahlung spectra. The observations showed that the LEED exhibits qualitative change when the power exceeds certain threshold. Bremsstrahlung is also strongly correlated with the above-mentioned process, which gives the potential for the future research in that direction.

5. Acknowledgements:

This research was supported by the grant of Russian Science Foundation (project number 19-12-00377).

[1] S.V. Golubev, I. V. Izotov, D.A. Mansfeld and V.E. Semenov, Experimental electron energy distribution function investigation at initial stage of electron cyclotron resonance discharge, Review of Scientific Instruments **83**, 2 (2012)

[2] V.A. Skalyga, S.V. Golubev, I. V. Izotov et al., New developments in the field of high current ECR ion sources at the IAP RAS, EPJ Web Conf. **187** (2018)

[3] V.A. Skalyga, A.F. Bokhanov, S.V. Golubev et al., Status of the gasdynamic ion source for multipurpose operation (GISMO) development at IAP RAS, Review of Scientific Instruments **90**, 12 (2019)

[4] I. Izotov, O. Tarvainen, V. Skalyga et al., Measurement of the energy distribution of electrons escaping minimum-B ECR plasmas, Plasma Sources Science and Technology **27** (2018)

Experimental study on dynamic effects of H^- and D^- negative ions in an ECR-plasma source

M Mitrou[1,2]**, P Svarnas**[1] **and S Béchu**[2]

[1]High Voltage Laboratory, Electrical and Computer Engineering Department, University of Patras, Rion–Patras GR–26504, Greece
[2]Université Grenoble Alpes, CNRS, Grenoble INP (Institute of Engineering), LPSC-IN2P3, 38000 Grenoble, France

svarnas@ece.upatras.gr

Abstract. The electrostatic probe-based photo-detachment technique with one or two laser beams can provide insight on the negative ion absolute densities and dynamics, respectively, in electronegative plasmas. In this work, this diagnostic is installed in the ECR-driven (2.45 GHz) negative ion source "Prometheus I" and details of the setup design along with main underlying physics are discussed. Potential origins of signal distortions and thus erroneous employment of this technique are demonstrated. Based on these observations, H^- and D^- negative ion densities and temperatures are measured in a reliable manner. The experiments are carried out for variable microwave power (2.45 GHz), at 1.33 Pa, unveiling two distinct kinetic equilibriums of the negative ions, in both gases, and an isotope effect.

1. Introduction

Hydrogen negative ions (H^-) are important in astrophysics, particle accelerators, and particularly nowadays in fusion science where they are being accelerated and then neutralized to form high energy atom beams for heating fusion plasmas [1]. However, in the latter case, and towards the international collaboration fusion project ITER, power level operation is being challenged with feeding deuterium to the ion sources. When the working gas of a negative ion source is changed from hydrogen to its isotope, deuterium, an "isotope effect" is observed; namely, several plasma characteristics such as the electron energy distribution, the atomic fraction, and the spectra of rovibrationally excited molecules change [2]. Inevitably, these changes affect the production and destruction processes of the D^- negative ions and hence their densities and energies. Thus, the application of diagnostic tools for probing and comparing directly properties related to H^- and D^- ions is mandatory.

Laser induced photo-detachment is a well-known and widely applied diagnostic technique for this purpose. In principle, a very modest laser pulse energy converts all the negative ions along the laser beam path into photo-detached electrons and atoms, which serve as "tags" of the negative ions from which they are originated [3]. The simplest technique for measuring the density of the tagged photo-detached electrons is the use of a Langmuir-type electrostatic probe, operated at the electron saturation part of its I-V curve. This concept has been broadly applied for many decades to measure the H^- and D^- negative ion densities.

On the other hand, quite surprisingly and at the same time challenging, a very limited number of works are devoted to the full exploitation of the photo-detachment technique, i.e., the determination of the negative ion energies-temperatures. This parameter is linked to the extracted negative ion beam quality and emittance, while it may be an appreciated input to numerical models of plasma kinetics in ion sources. The fundamental physical concepts underlying this method when an electrostatic probe is used for the electron detection, implies [4]: (i) the creation of a controlled perturbation in the negative ion density, followed by (ii) the measurement of the recovery of the negative ion density n_t^- towards its initial value n_0^-, and (iii) fitting the measured results to a physical model of the density recovery process, in which the ion thermal drift velocity plays a central role. In practice, a first laser shot detaches all negative ions in the irradiated volume and a reference, photo-detachment signal is acquired. A second laser shot is triggered after a delay time Δt and it detaches all negative ions that have flown into the laser irradiated volume during Δt. Thus, the probe generates a second photo-detachment signal, which is superposed with the reference one and lags that by Δt. By increasing the delay Δt, the amplitude of this second signal approaches the amplitude of the reference signal, progressively, providing thus information on the negative ion density recovery at different points in time [3]. Namely, by subtracting the reference signal from the mixed signal for each Δt, the second signal is isolated, and its peak is measured. The ratio between the isolated and the reference signal amplitudes equals to n_t^-/n_0^-, where n_0^- is conventionally estimated from the peak of the reference probe signal. Thus, n_t^- is determined.

To the best of our knowledge, the most indicative reports on the application of this process are the Refs. [4-9], concerning filament-driven sources. More recently, Kado et al. [10] proposed a novel and simple method to measure thermal and flow velocities of negative ions by making use of the photo-detachment technique in a steady-state linear divertor/edge plasma simulator MAP (material and plasma)-II. They stated that, flow measurements, especially for ions, is important in the divertor/edge region of fusion devices since the flow and its shear influence the plasma turbulence and the friction force against impurity contaminations.

Despite the afore mentioned works, there is a substantial lack of data on the energies of H^- and D^- negative ions formed in electron cyclotron resonance (ECR) driven sources. The present work is devoted to this topic. Experimental results are reported and commented, along with enlightening tips for avoiding experimental uncertainties. Overall, two distinct kinetic equilibriums of these ions are found, in both gases, and higher ion energies are estimated in the case of D^-, under the assumption of a ballistic model.

2. Experimental setup

The negative ion source "Prometheus I" and the electrostatic probe system are described elsewhere [11]. Note that the source is driven by five elementary ECR modules with permanent magnets (875 G) and all measurements are realized at 65 mm downstream of the ECR-zone middle planes (**Fig. 1**).

As regards the two-laser photo-detachment setup, **Fig. 1** depicts the actual concept. The beams of two independent Nd:YAG lasers (Quantel Brilliant EaZy) are coaxially aligned by means of appropriate optical components (see caption of **Fig. 1**). A multichannel pulse/delay generator (Berkley Nucleonics Corp; Model 525), coupled with a custom-made opto-isolating and amplifying unit, is employed for the flash lamp and Q-switch synchronization of the two lasers with a 4-ns resolution. A pyroelectric energy meter (Ophir Photonics; PE25BF-DIF-C), equipped with a diffuser for concentrated beams, is used for the beam energy determination. A fast photodetector (Ophir Photonics; FPS-10) allows for the monitoring of the temporal behavior of the laser pulses and its signal triggers the probe acquisition. To avoid any potential errors that arise from the use of conventional capacitive decoupling circuits, a wideband current transformer (Pearson Electronics 6585) provides the impulses generated by the photo-detached electrons. All signals are recorded on a wideband digital oscilloscope (LeCroyWaveSurfer 104Xs-A) and averaged in-situ to increase signal-to-noise ratio (>200 averaged samples). No further signal processing (filtering – smoothing) is applied during or after the acquisition.

The working gas is either H_2 (99.999%; AirLiquide) or D_2 (99.8%; Linde). The pressure is maintained constant at 1.33 Pa, based on previous optimization studies on the ratio between the electron and the negative ion densities. The total microwave (2.45 GHz) power is varied up to 0.9 kW. Two sets of measurements are conducted, providing mean values and standard deviations.

Source wall conditioning requires special attention since notable variations in the plasma properties are observed when the two gases are interchanged. This effect is attributed to the gas retention. To ensure wall recovery, i.e., eliminate this issue and achieve stable and reproducible plasma properties, various

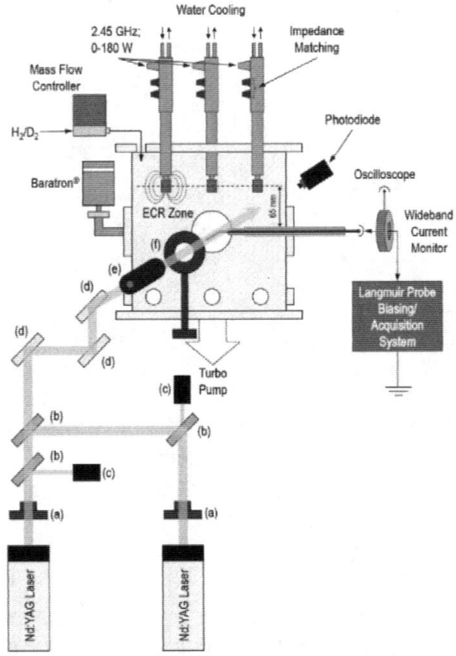

Figure 1. "Prometheus I" source and diagnostics. (a) Half wave plates on precision rotation mounts; (b) beam splitters; (c) laser beam dumps; (d) mirrors; (e) laser beam expander (8×), having Ø2.5 mm front and Ø5 mm rear diaphragms; (f) removable pyroelectric laser energy sensor.

trials led to the following protocol: each gas switch is followed by a bake-out of the source at 150°C, under base pressure (5.33×10^{-5} Pa), for at least 30 h. Throughout the entire experimental series, the temperature of the source walls is not controlled in any manner, therefore it is solely influenced by the operating parameters and the sequence of the scheduled measurements. However, in between measurements, a 20-min interval is allowed for the thermal stabilization of the source.

3. Results and Discussion

It has been established that, for the proper design of the photo-detachment (either one- or two-laser version), four main parameters must be selected properly: (i) the laser wavelength with respect to the cross section of the photo-detachment reaction of the respective negative ions; (ii) the laser beam diameter with respect to the probe collection radius; (iii) the laser power density with respect to the saturation of the ratio between the density of the photo-detached electrons and the background negative ion density; and (iv) the probe bias with respect to the plasma potential. These points have been clarified previously [3]. Moreover, it is here underlined that the relative position (vertically; horizontally; rotationally) of the L-shaped probe tip with respect to the laser beam axis is a very sensitive parameter which may lead to highly distorted photo-detachment signals and, consequently, to erroneous data. **Figs. 2(a)-(c)** depict the problem by giving an indicative case. In a similar manner, even if the probe is "perfectly" aligned with the laser beam axis, different beam profiles (in terms of shape or energy distribution) result to modified signals. Thus, a beam expander and two diaphragms (**Fig. 1**) are employed to provide an almost homogeneous flux of 1064 nm photons, along a 5 mm in diameter cross section path. It was observed that the amplitude of the first distinct peak of the photo-detachment signal, which is used for the calculation of the steady state negative ion density, n_0^-, is quite immune to any reasonable misalignment or asymmetry. Nevertheless, these imperfections become crucial when ion dynamic effects are studied, since the probe tip must "see" two identical irradiative scenarios. **Fig. 2(d)** demonstrates the matched photo-detachment signals achieved from the two lasers, after following the above precautions. Both signals are checked periodically during the experiments.

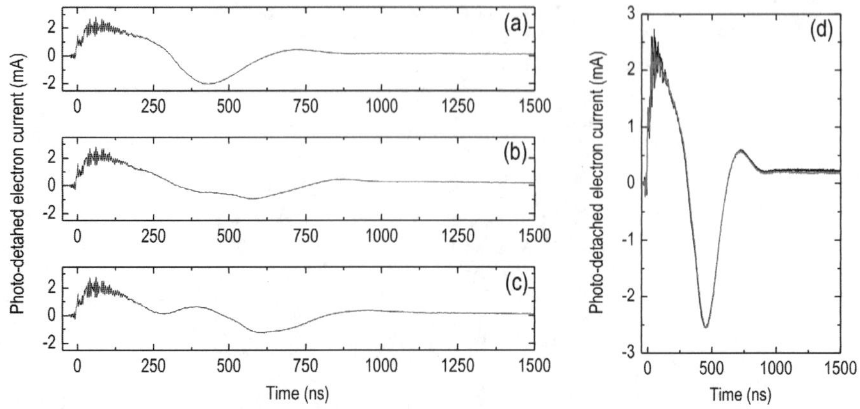

Figure 2. (a) Reference photo-detachment signal, when the probe tip and the laser beam are optimally aligned. Distorted signals due to vertical, lower displacement of the laser beam for about **(b)** 0.5 mm and **(c)** 1.1 mm, with respect to (a); horizontal and rotational alignments remain as in (a). **(d)** Matched signals induced by the two different laser beams, following the precautions discussed in the text; the red and the black curve are superposed. Operating conditions: H_2; 600 W.

When the two lasers are triggered properly, mixed photo-detachment signals are recorded. **Fig. 3** (middle frames) provides corresponding examples for two delay times Δt; 400 ns and 4 μs. By sweeping Δt between 200 ns and 5 μs, a bundle of such mixed signals is acquired. The subtraction of the reference signal (e.g., **Figs. 3(a1)** and **3(b1)**) from them, leads to the sought-out signals (e.g., **Figs. 3(a3)** and **3(b3)**). The amplitude of these signals is then divided by the amplitude of the reference signal and the recovery curves (n_t^- / n_0^- ratio vs. time) are constructed [3,4]. Two curve examples are given in **Fig. 4**.

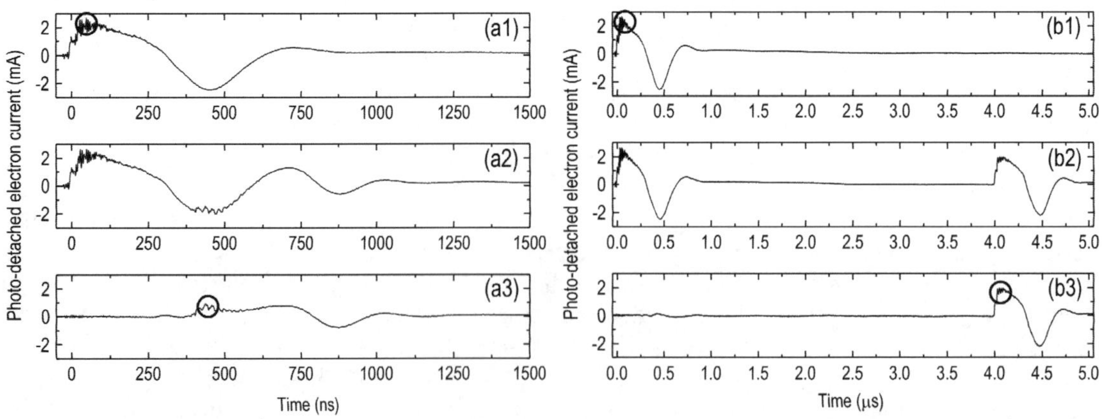

Figure 3. (a1) Single-laser photo-detachment signal. **(a2)** Double-laser photo-detachment signal, with Δt = 400 ns. **(a3)** Subtraction signal, i.e., (a2 – a1). **(b1)** to **(b3)** Similar, with Δt = 4 μs. The inset circles define the amplitudes used for the calculation of the n_t^- / n_0^- ratio. Operating conditions: H_2; 600 W.

It is here attempted to fit to the data a ballistic model [12], assuming two ionic populations of different temperatures [8]. Under the present conditions, the model traces acceptably the data up to about 2 μs (**Fig. 4**). Fitting for longer recovery times, increases the overall least squares error significantly. According to this model, "cold" ions have energy less than 0.05 eV in both H_2 and D_2 plasmas, whereas "hot" ions achieve higher energies in D_2. In any case, the "cold" negative ions dominate the "hot" ones, with a fraction of about 65–70% (**Fig. 4**). Besides, two electronic populations have been observed [11]. Indicative (900 W) plasma features, in H_2 | D_2, are: V_p= 7.1 | 8.4 V; n_e^{cold} = 2.3 | 3.8 (×10^{10} cm^{-3}); n_e^{hot} = 4.3 | 4.4 (×10^8 cm^{-3}); kT_e^{cold} = 0.7 | 1.0 eV; kT_e^{hot} = 14.6 | 14.5 eV; and n_0^- = 3.9 | 4.1 (×10^9 cm^{-3}).

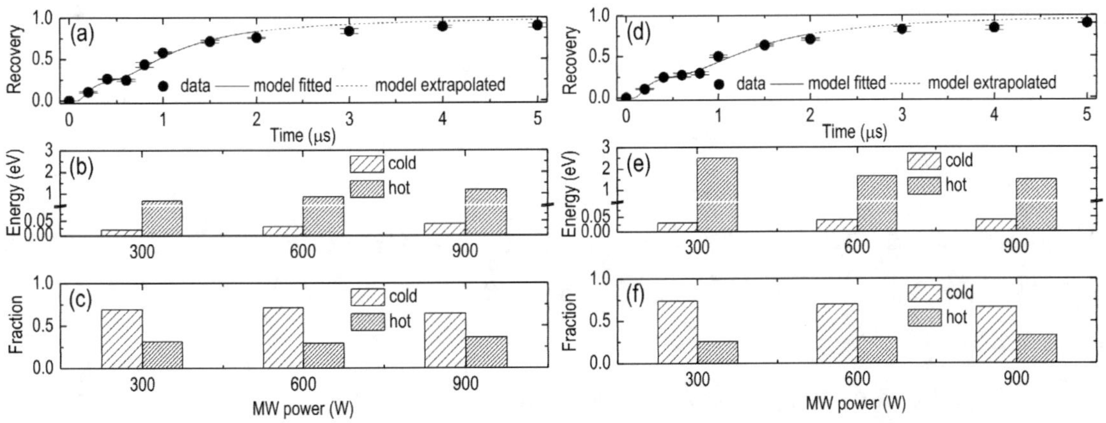

Figure 4. Ion density recovery (600 W): **(a)** H^- and **(d)** D^-. Ion energy versus MW power (be aware of the break on the vertical axis): **(b)** H^- and **(e)** D^-. Ion fraction versus MW power: **(c)** H^- and **(f)** D^-.

The production of two ionic populations and the elevated energy of the hot ions, should be traced back to their origin. In the ECR zones, inelastic collisions with hot electrons lead to vibrationally excited molecules which diffuse downstream, where the present measurements are realized. There, H^- or D^- ions are formed due to cold electron dissociative attachment (DA) on these molecules, which are expected to acquire only the energy of formation by DA, being thus cold. On the other hand, recombinative desorption processes on the metallic walls as a source of a wide spectrum of rovibrationally excited molecules, is broadly recognized. Ions formed near the walls, again by DA, are accelerated by the positive plasma potential difference, moving towards the bulk plasma, being thus hot(ter). This is in line with the constantly higher plasma potential measured in D_2 and the higher hot ion energy obtained, with respect to the H_2 case. Furthermore, we have measured increasing plasma potential versus the MW power, in both gases. Thus, higher power should lead to higher acceleration and finally increased energy for the hot ions. While this holds true in the H_2 case, the trend is different in the D_2 case. This implies that the accelerating ions are subject to various transport mechanisms in each gas–plasma, and it is another explicit difference with respect to the cold negative ions.

Conclusions

The two-laser photo-detachment technique was applied in both H_2 and D_2 ECR plasmas. Two H^- and D^- ion populations were detected, i.e., a cold and a hot one, presumably due to different origins. Experimental data were treated under the assumption of a ballistic model which must be further assessed.

Acknowledgments

This work was supported by the "Andreas Mentzelopoulos" Ph.D. Scholarships (M. Mitrou).

References

[1] Bacal M 2006 *Nucl. Fusion* **46** S250
[2] Bacal M and Wada M 2020 *Plasma Sources Sci. Technol.* **29** 033001
[3] Bacal M 2000 *Rev. Sci. Instrum.* **71** 3981
[4] Bacal M 1993 *Plasma Sources Sci. Technol.* **2** 190
[5] Devynck P, Auvray J, Bacal M, *et al.* 1989 *Rev. Sci. Instrum.* **60** 2873
[6] Bacal M, Berlemont P, Bruneteau A M, *et al.* 1991 *J. Appl. Phys.* **70** 1212
[7] Leroy R, Bacal M, Berlemont P, *et al.* 1992 *Rev. Sci. Instrum.* **63** 2686
[8] Ivanov A A Jr 2004 *Rev. Sci. Instrum.* **75** 1754
[9] Ivanov A A Jr, Bacal M, Rouillé C, *et al.* 2004 *Rev. Sci. Instrum.* **75** 1747
[10] Kado S, Kajita S, Shikama T, *et al.* 2006 *Contrib. Plasma Phys.* **46** 367
[11] Aleiferis S, Svarnas P, Béchu S, *et al.* 2018 *Plasma Sources Sci. Technol.* **27** 075015
[12] Stern R A, Devynck P, Bacal M, *et al.* 1990 *Phys. Rev. A* **41** 3307

H^- and D^- production efficiency in a multi-dipole ECR-plasma source as a function of gas pressure

M Mitrou[1,2]**, P Svarnas**[1] **and S Béchu**[2]

[1]High Voltage Laboratory, Electrical and Computer Engineering Department, University of Patras, Rion–Patras GR–26504, Greece
[2]Université Grenoble Alpes, CNRS, Grenoble INP (Institute of Engineering), LPSC-IN2P3, 38000 Grenoble, France

svarnas@ece.upatras.gr

Abstract. The electron cyclotron resonance (ECR) negative ion source "Prometheus I" is operated either with high purity H_2 (> 99.999%) or D_2 (> 99.8%) to probe H^- and D^- ions, respectively, and examine the isotope effect within a wide range of gas pressure. These ions are predominantly formed in the bulk plasma by dissociative attachment (DA) of low-energy (cold) electrons to highly ro-vibrationally excited molecules. The latter result mainly from the radiative decay and excitation (EV) process sustained by high-energy (hot) electrons heated in the ECR zones. Langmuir probe and laser photo-detachment measurements are realized within the pressure range 0.27 to 2.67 Pa under constant microwave power (0.9 kW). It is revealed that: (i) the plasma potential, cold electron temperature, and cold electron density tend to be higher in deuterium; (ii) no pronounced difference in the hot electron density and temperature is found between the two plasmas; and (iii) overall a similar H^- and D^- negative ion yield (up to 6×10^9 cm^{-3}; under the present conditions) is achieved. However, for equal plasma densities an isotope effect is exhibited showing higher H^- density over the entire pressure range. Finally, the n_{H^-} / n_e ratio is constantly higher than the n_{D^-} / n_e one and they both peak around 1.33 Pa.

1. Introduction

The heating and current drive of the ITER plasma will rely on the injection of a total of 33 MW of 1 MeV D^0. The formation of these powerful D^0 beams will be based on the neutralization of accelerated D^- beams since, at such high energies, the efficiency of this process remains acceptable (at around 56%) for negative ions but becomes unsuitably low for positive ions [1]. Considerable progress has been made at the prototype test facility ELISE, yet still only 66% of the ITER demand for extracted negative ion current density has been reached over a time span of 45 min [2]. Additionally, operation with H_2 is foreseen for the early, non-nuclear phases of ITER.

Thus, any fundamental study on a laboratory scale on a direct comparison between H_2 and D_2 plasma properties would be of added value for the general project of fusion application. Indeed, this task has triggered the interest of different research groups, regardless of the type of source employed. Nevertheless, most of the actual experiments refer to H_2 whereas the final fuel will be D_2 and not adequate number of works have been devoted to a direct comparison of the two plasma under identical conditions. This holds true especially in the case of electron cyclotron resonance (ECR) driven sources. Indicatively, such comparative tests have been realized in an ECR source with driven plasma rings [3], at the laboratory scale experiment HOMER (HOMogeneous Electron cyclotron Resonance plasma) [4, 5], and in the SCHEMEII+ source (Source of exCited HydrogEn MolEcules) [6].

Accordingly, the present work is devoted to the ECR volume source Prometheus I, extending its exploitation to D_2 since previously operation only with H_2 had been thoroughly carried out [7, 8]. Basic plasma properties are accessed via electrostatic probe and laser induced photo-detachment. The experiments are carried out throughout a wide range of working pressure, whereas the microwave power is maintained constant. The results are then compared and evaluated so as to identify similarities and differences between the ECR plasmas of the two isotopes. The isotope effect is mirrored on the electron energy distribution functions (EEDFs) of the H_2 and D_2 plasmas, and it is also noticeable on the negative ion to electron density ratio. Despite that, comparable H^- and D^- negative ion densities are achieved.

2. Experimental setup

A thorough description of the negative ion source Prometheus I is given elsewhere [7, 8]. Langmuir probe and laser photo-detachment are employed for plasma main parameters and negative ion densities measurements, respectively, as it has been previously analyzed [7, 8]. The source is driven by five elementary ECR modules with permanent magnets (875 G) and all measurements are realized at 65 mm downstream of the ECR zones middle planes. Four sets of measurements are conducted, providing mean values and standard deviations (see error bars in figures). Operation within the pressure range of 0.27 to 2.67 Pa (accuracy around ±0.25%) may be achieved by the insertion of either H_2 (99.999%; AirLiquide) or D_2 (99.8%; Linde) gas (digitally controlled mass flow). The microwave (2.45 GHz) power is fixed at 0.9 kW in total. Source wall conditioning when the two gases are interchanged requires special attention since notable variations in the plasma properties are observed. This effect can most likely be attributed to the adsorption of remainders of the formerly used gas into the interior surfaces of the source, also known as gas retention. In order to ensure wall recovery, i.e., eliminate this issue and achieve stable and reproducible plasma properties, various trials led to the following protocol: each gas switch is followed by a bake-out of the source at 150°C, under base pressure (5.33×10^{-5} Pa), for at least 30 h. Throughout the entire experimental series, the temperature of the source walls is not controlled in any manner, therefore it is solely influenced by the operational parameters and the measurement sequence. However, it is monitored using thermocouples for consistency reasons. In between measurements, a 20-min interval is allowed for the thermal stabilization of the source. Herein, the measurements are conducted in a descending order relative to pressure, i.e from 2.67 Pa to 0.27 Pa (flow rate: 3.9 to 23.4 sccm for H_2 and 4.8 to 29.2 sccm for D_2), during which the temperature of the lateral walls increases from about 89.5 to 118.7 °C and 104.1 to 125.3 °C for H_2 and D_2, respectively. EEDFs are constructed from the probe curves quasi-experimentally, i.e., by fitting a bi-Maxwellian function to the electron current component of the probe curve and its subsequent double derivation (see [8] for details).

3. Results

Fig. 1 clearly shows that two electron populations are produced due to ECR heating, a "cold" and a "hot" one, having about one order of magnitude difference in their temperatures. At the same time, **Fig. 2** gives the absolute density values of these populations. The low-density hot population is a fingerprint of the energetic electrons generated in the ECR zones, whereas the high-density cold population refers to the electrons escaping these zones. The former promotes ro-vibrational excitation of molecules (EV reaction), which diffuse away of the ECR zones. The latter promote H^- and D^- ion production through the dissociative attachment (DA) reaction between them and the diffused excited molecules, in the source volume.

Higher plasma potential (up to about 2 V difference; not shown here), and higher cold electron temperatures and densities are obtained in the case of D_2 over almost the entire pressure range, whereas the isotope effect practically ceases to exist in the case of hot electron temperatures and densities. These observations on the electrons are jointly mirrored on the EEDF patterns depicted in **Fig. 3**. The diverging features between the H_2 and D_2 EEDFs are more pronounced at the lower pressures.

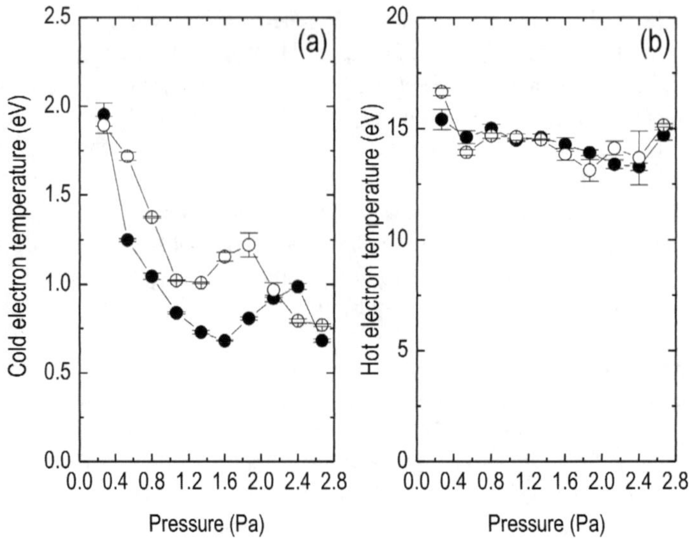

Figure 1: **(a)** Cold and **(b)** hot electron temperature vs. the operating gas pressure. ● H_2 ; ○ D_2 ; 0.9 kW.

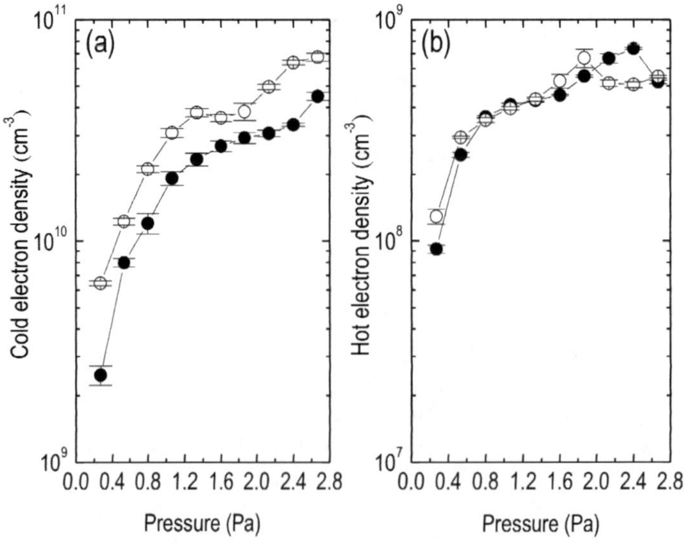

Figure 2: **(a)** Cold and **(b)** hot electron density vs. the operating gas pressure. ● H_2 ; ○ D_2 ; 0.9 kW.

On the other hand, **Fig. 4(a)** makes a direct comparison between the negative ion yields achieved in the two gases. At a first glance, the H^- and D^- densities have a qualitatively and quantitatively similar evolution vs. the operating pressure, i.e., a sharp increase is followed by a quasi-saturated phase. Despite that, an obvious isotope effect is exhibited by a different presentation of the data. In **Fig. 4(b)** the relative negative ion density is given, unveiling a strong isotope effect in favour of H_2 and a sharp optimum around 1.33 Pa in both gases. On the top of that, **Fig. 5(a)** compares the H^- and D^- densities under equal plasma densities (cold electrons), after pressure elimination. Higher values for the H^- density, are observed. The corresponding curves versus the hot electrons (**Fig. 5(b)**) are associated with closer experimental points in the two gases.

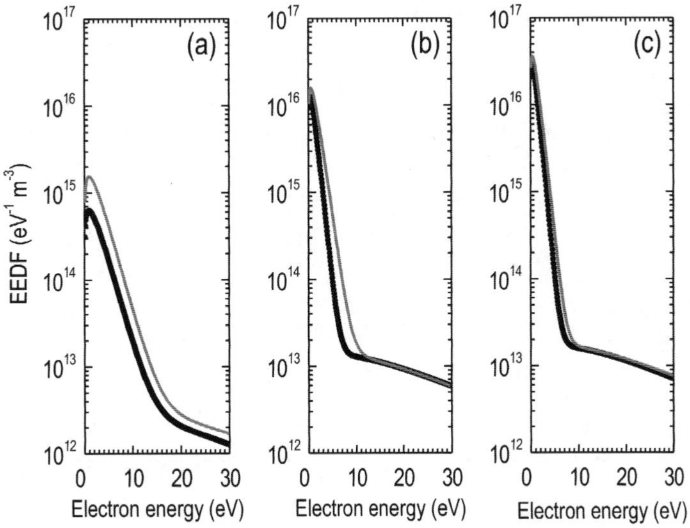

Figure 3: EEDFs under bi-Maxwellian assumption at **(a)** 0.27 Pa, **(b)** 1.33 Pa, and **(c)** 2.67 Pa. ●/black H_2 ; –/red D_2 ; 0.9 kW.

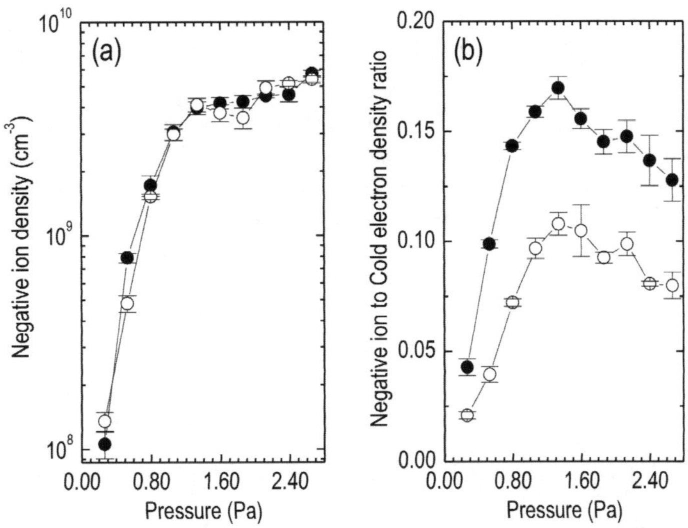

Figure 4: (a) H^- & D^- ion density and **(b)** H^- & D^- ion to cold electron density ratio vs. the operating gas pressure. ● H_2 ; ○ D_2 ; 0.9 kW.

4. Discussion

A rigorous interpretation of the isotope effects recorded here demands the consideration of the production and destruction paths of the H^- and D^- ions and their molecular precursors $H_2(v)$ and $D_2(v)$. Furthermore, the present data are perfectly linked to the cross sections of many production/destruction processes in H_2 and D_2 [9, 10], but this will be the object of another report. Briefly, the results support EV and DA as dominant production paths. Instead, main destruction paths include vibrational translational relaxation in collisions with atoms (V-t process) or molecules (V-T process) and electron detachment from H^- and D^- upon their collision with atoms. Atomic and molecular densities increase with higher pressure, affecting the above processes, and an optimum pressure for $H_2(v)$ and H^- production has also been observed in other ECR- or filament-driven sources [11].

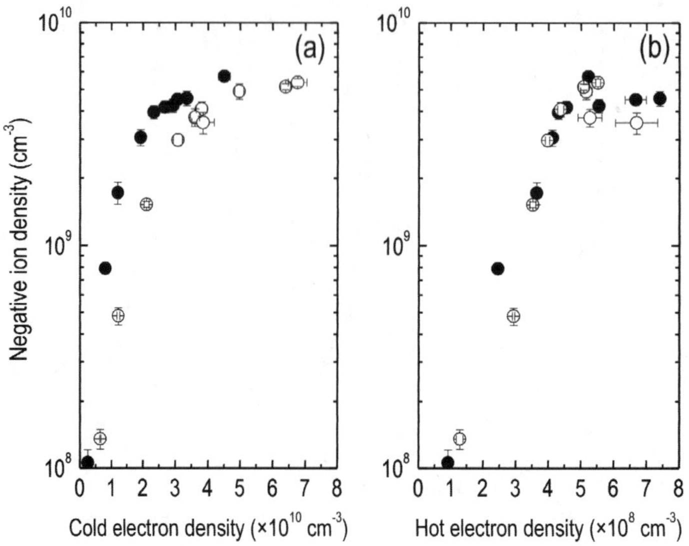

Figure 5: **H$^-$** & **D$^-$** ion density vs. **(a)** cold and **(b)** hot electron density. Data pairs are created with the pressure elimination. ● H_2 ; ○ D_2 ; 0.9 kW.

5. Conclusions

ECR-driven plasmas led to H$^-$ and D$^-$ ion production associated with densities within the same order of magnitude (10^9 to 10^{10} cm^{-3}) over a wide range of pressure (0.27 to 2.67 Pa). However, an isotope effect was demonstrated with respect to the plasma potential, cold electron temperature and density, all being generally higher in the D_2 case. Additionally, the isotope effect exists for the negative ions too, when they were compared under equal plasma densities, promoting in this case higher H$^-$ than D$^-$ ion densities. Finally, a distinct optimum of the negative ion to electron density ratio appeared at around 1.33 Pa, yielding a higher peak in the case of H_2 plasma.

Acknowledgments

This work was supported by the "Andreas Mentzelopoulos" Ph.D. Scholarships of the University of Patras (M. Mitrou).

References

[1] Hemsworth R S, Boilson D, Blatchford P, Dalla Palma M, Chitarin G, *et al.* 2017 *New J. Phys.* **19** 025005
[2] Wünderlich D, Riedl R, Mano I, Mimo A, Fantz U, *et al.* 2019 Rev. Sci. Instrum. **90** 113304
[3] Dougar-Jabon V D 2001 *Physica Scripta* **63** 322-325
[4] Friedl R, Kurutz U and Fantz U 2016 *AIP Conf. Proc.* **1869** 030022
[5] Kurutz U, Friedl R and Fantz U 2017 *Plasma Phys. Control. Fusion* **59** 075008
[6] Béchu S, Lemaire J L, Gavilan L, Aleiferis S, Shakhatov V, *et al.* 2020 *J. Quant. Spectrosc. Radiat. Transf.* **257** 107325
[7] Aleiferis S, Tarvainen O, Svarnas P, Bacal M and Béchu S 2016 *J. Phys. D: Appl. Phys.* **49** 095203
[8] Aleiferis S, Svarnas P, Béchu S, Tarvainen O and Bacal M 2018 *Plasma Sources Sci. Technol.* **27** 075015
[9] Celiberto R, Janev R K, Laricchiuta, Capitelli M, Wadehra J M and Atems D E 2001 *At. Data Nucl. Data Tables* **77(2)** 161-213
[10] Janev R K, Reiter D and Samm U 2003 Forschungszentrum Jülich JUEL 4105
[11] Svarnas P, Breton J, Bacal M and Mosbach T 2006 *Rev. Sci. Instrum.* **77** 03A532

Development and status of the FRIB 28 GHz SC ECRIS

Haitao Ren[1], Guillaume Machicoane[1], Nathan Bultman[1], Xing Rao[1], Phil Morrison[1], Mykola Omelayenko[1], Larry Tobos[1], Jesse Fogleman[1], Helene Felice[2], Ray Hafalia[2], Heng Pan[2], and Soren Prestemon[2]

[1]Facility for Rare Isotope Beams, Michigan State University, East Lansing, MI, 48824, USA
[2]Lawrence Berkeley National Laboratory, Berkeley, California 94720, USA

Work supported by the U.S. Department of Energy Office of Science under Cooperative Agreement DE-SC0000661.

ren@frib.msu.edu

Abstract. To meet a beam power requirement of 400 kW for heavy ions, a high intensity 28 GHz superconducting Electron Cyclotron Resonance (ECR) ion source is under development at the Facility for Rare Isotope Beams (FRIB) in collaboration with Lawrence Berkeley National Laboratory (LBNL). The magnet was built and tested at LBNL and integrated into the ion source cryostat on the FRIB high voltage platform. Magnet cooldown to 4.2K was completed successfully in December of 2020. The static heat load at 4.2K has been measured to be around 1.2 W, in good agreement with the design value. The heat load is managed through 2 Gifford-McMahon/Joule-Thomson (GM-JT) cryocoolers that have been in operation for several months. Magnet energization is scheduled in November of 2021. Warm components preparation and assembly test are ongoing in parallel. The ion source commissioning with 18 GHz Klystron shall start in December of 2021. Details of the ion source development, status, and commissioning plan will be presented in the paper.

1. FRIB Introduction

The Facility for Rare Isotope Beams (FRIB) is a high power heavy ion accelerator facility, founded by US Department of Energy Office of Science (DOE-SC) with contributions from Michigan State University (MSU). The FRIB project scope includes the linac accelerator, target system, pre-separator, and fragment separator, shown in Figure 1. It has become a DOE-SC User Facility since September 2020. The FRIB linac accelerator has the ability to accelerate ion species as heavy as Uranium to energies of $\geq$ 200 MeV/u and beam power up to 400 kW for rare isotope productions [1]. FRIB also has the capabilities for fast, stopped, and reaccelerated beams as integrated with the existing experimental areas, outlined in orange in Figure 1.

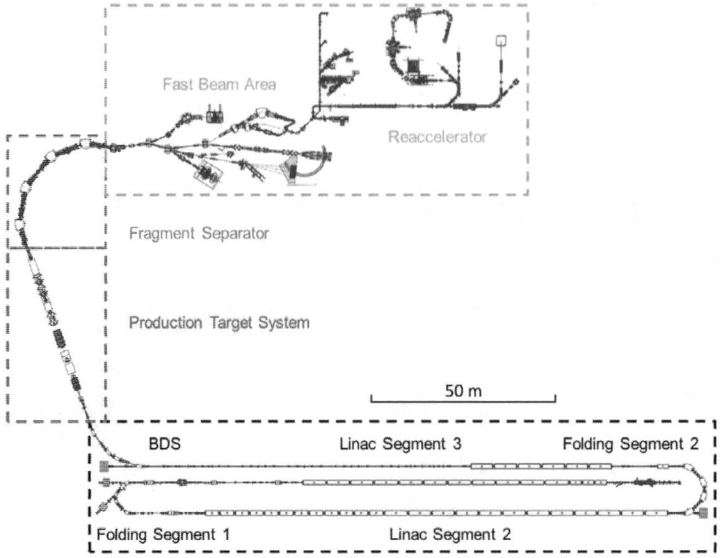

Figure 1. Layout of the Facility for Rare Isotope Beams (FRIB).

Figure 2 shows the timeline of the project. DOE-SC and MSU signed the Cooperative Agreement in June 2009. In August 2013, DOE-SC approved the project baseline and start of civil construction (CD2-3a) with a total project cost of $730M, funded by DOE, MSU and the State of Michigan. FRIB obtained CD-3b approval from DOE in August 2014, and the technical construction began in October 2014. The project completion (CD-4) is scheduled to be in June 2022, but the forecast is for an early completion by January 2022. The first user experiments are expected in summer 2022.

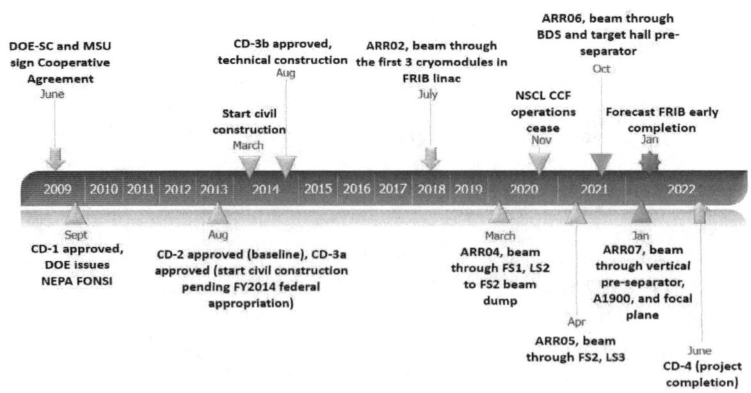

Figure 2. Timeline of the FRIB project.

The FRIB driver linac technical construction has been completed with the last equipment of liquid Lithium Stripper was installed in March 2021. The linac consists of 3 Linac Segments (LS1-3), 2 Folding Segments (FS1-2) and a Beam Delivery System (BDS), including 46 cryomodules, 4 superconducting dipoles, 242 warm magnets, 7 warm bunchers, 2 charge strippers, and 1 radio frequency quadrupole (RFQ) in the tunnel, shown in Figure 1.

At FRIB, phased beam commissioning was done in parallel with the technical installations in the tunnel. We planned seven beam commissioning stages with an Accelerator Readiness Review (ARR)

Journal of Physics: Conference Series **2244** (2022) 012008 doi:10.1088/1742-6596/2244/1/012008

proceeding each stage [2]. ARR 01-05 were for the beam commissioning of the linacs, and ARR 06-07 are planned for beam commissioning through the target and the secondary beamline.
Figure 3 shows the linac commissioning configuration and status with ARRs. The first ARR was the commissioning of the Front End system, including the Artemis ECR ion source, Low Energy Beam Transport (LEBT), RFQ, and Medium Energy Beam Transport (MEBT). ARR-01 was achieved in July 2017 with $^{40}Ar^{9+}$, $^{86}Kr^{17+}$ beams accelerated to 0.5 MeV/u through RFQ [3]. The first Superconducting Radio Frequency (SRF) acceleration was done with beams through the LS1 first three β=0.041 cryomodules in July 2018 (ARR02). The beam energy reached 2 MeV/u. ARR03 in February 2019 had beams accelerated through the entire LS1 and the first dipole in FS1, with energies up to 20.3 MeV/u. The FRIB energy goal (200 MeV/u) was demonstrated in March 2020 (ARR04), when $^{36}Ar^{18+}$ was accelerated through LS1 and LS2, reaching 204.4 MeV/u [4]. In May 2021 (ARR05), $^{86}Kr^{34+}$ was successfully accelerated through all 324 superconducting resonators to 212 MeV/u and delivered to the BDS dump.

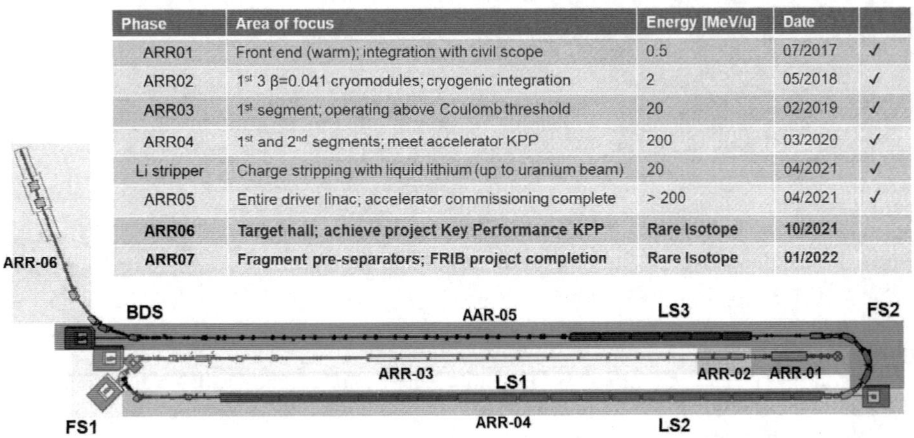

Phase	Area of focus	Energy [MeV/u]	Date	
ARR01	Front end (warm); integration with civil scope	0.5	07/2017	✓
ARR02	1st 3 β=0.041 cryomodules; cryogenic integration	2	05/2018	✓
ARR03	1st segment; operating above Coulomb threshold	20	02/2019	✓
ARR04	1st and 2nd segments; meet accelerator KPP	200	03/2020	✓
Li stripper	Charge stripping with liquid lithium (up to uranium beam)	20	04/2021	✓
ARR05	Entire driver linac; accelerator commissioning complete	> 200	04/2021	✓
ARR06	**Target hall; achieve project Key Performance KPP**	**Rare Isotope**	**10/2021**	
ARR07	**Fragment pre-separators; FRIB project completion**	**Rare Isotope**	**01/2022**	

Figure 3. FRIB linac phased beam commissioning configuration and status.

The Lithium Stripper is one of the key elements for FRIB's ultimate high power beam operation. The beam commissioning of the Lithium Stripper was completed in April 2021. It is the world's first liquid metal charge stripper for high power operations. As shown in Figure 4, the linac layout allows for both liquid lithium stripper and carbon foil stripper coexistence for enhanced availability. The lithium stripper has been commissioned with $^{124}Xe^{26+}$ at 17 MeV and $^{238}U^{33+}$ at 20 MeV. High power beam test of the lithium stripping was successfully achieved using $^{36}Ar^{10+}$ at 400 W with a duty cycle of 5.4%. A 100 mm diameter carbon foil was installed on the carbon stripper. The rotating foil has been tested at 20 rpm with beam.

Figure 4. Liquid lithium and rotating carbon foil charge strippers installed downstream of LS1.

2. Front End System

As shown in Figure 5, FRIB Front End consists of two ECR ion sources, LEBT, RFQ and MEBT. Artemis is a room temperature ECR ion source operating at 14 GHz. It has been in operation for FRIB beam commissioning and early operations. The other ion source is the 28 GHz superconducting ECR for future high-power beam operation and is expected to be commissioned in early 2022. The LEBT beamline transports 12 keV/u ion beams from the ion source to the RFQ in the tunnel. FRIB RFQ is a four-vane type room temperature cavity used to accelerate ion beams to 0.5 MeV/u. The MEBT, downstream of the RFQ, is for beam re-bunching and matching to the superconducting linac, with two room temperature bunchers.

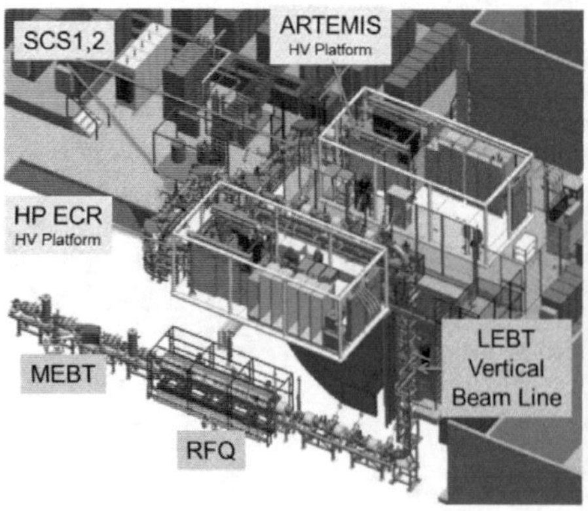

Figure 5. Layout of FRIB Front End system, and status of the two ion sources on high voltage platform.

For FRIB beam commissioning, the ion sources need to provide at least ^{36}Ar, ^{86}Kr, ^{129}Xe beams with intensity of 25 - 50 eμA and energy of 12 keV/u. With Artemis ECR ion source, the Front End has developed multi ion species to support the beam commissioning and early operation, including $^{40}Ar^{9+}$, $^{86}Kr^{17+}$, $^{20}Ne^{6+}$, $^{129}Xe^{27+}$, $^{238}U^{34+}$, $^{82}Se^{14+}$, $^{40}Ca^{9+}$, $^{124}Xe^{26+}$, $^{82}Pb^{30+}$. For FRIB operations, the ion sources are required to produce ion beams for a large base of stable elements ranging from Oxygen to Uranium. The beam intensity of single charge state should be 400 – 450 eμA for all elements from ion source to reach 400 kW on the target with M/Q > 7. According to the National Superconducting Cyclotron Laboratory (NSCL) operation experience at MSU, the beam intensity is required to be stable for several weeks.

3. FRIB Superconducting ECR Ion Source

3.1. Coldmass Design and Test

FRIB developed a 28 GHz superconducting ECR ion source, shown in Figure 6, in collaboration with LBNL. The magnet is based on the LBNL VENUS ECR [5] but with modified mechanical design to allow for swapping sextupole coils. LBNL's Bladder and Key Technology was used to preload the magnets in the structure [6]. The technology allows adjustments and reversibility in magnet preloading – also making it possible to replace magnet components in the structure.

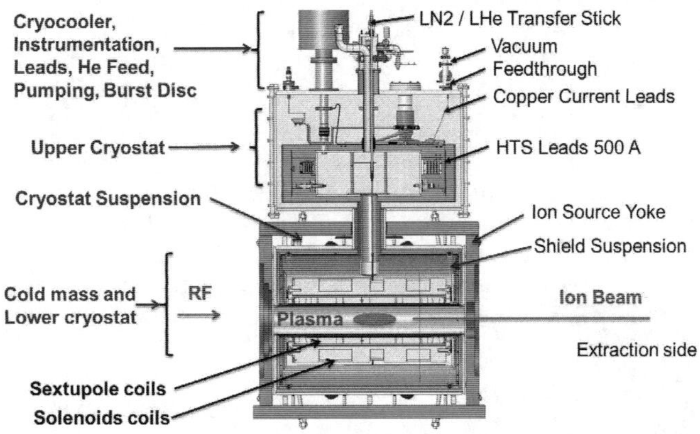

Figure 6. Cross section of the FRIB superconducting ECR ion source.

The superconducting coldmass was built and tested at Berkeley. Solenoids coils experienced a few training quenches and performed reliably after that. One of the sextupole coils was exhibited an excessive number of quenches during testing and was swapped with a new coil. With the new coil, the sextupole magnet reached the required field level in five training quenches. The training quenches were likely caused by the redistribution of stresses after the disassembly and reloading of the magnet support structure.

Figure 7 shows solenoid and sextupole field measured at Berkeley. Magnet field satisfied the requirements. Adjustment of B_{min} demonstrated without quenching, and field cycling from zero to nominal values was demonstrated, as well, without quenching.

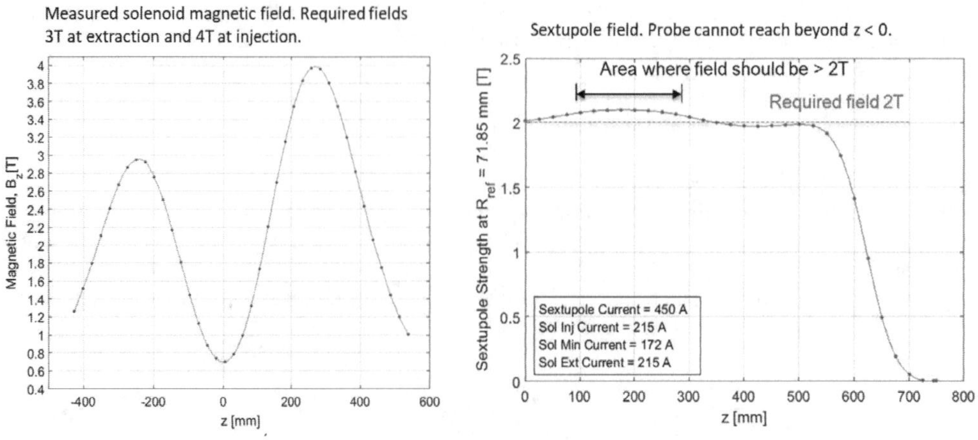

Figure 7. Solenoid and sextupole magnet test results at Berkeley.

3.2. Magent Cooldown at FRIB

Magnet coldmass was delivered to FRIB in January of 2018. The cryostat assembly with the coldmass was completed on the high voltage platform in November 2020. The magnet cooldown test was performed and reached to 4 K in December 2020. Coil resistance was measured during the cooldown process. All coils reached superconducting phase at 4 K, shown in Figure 8.

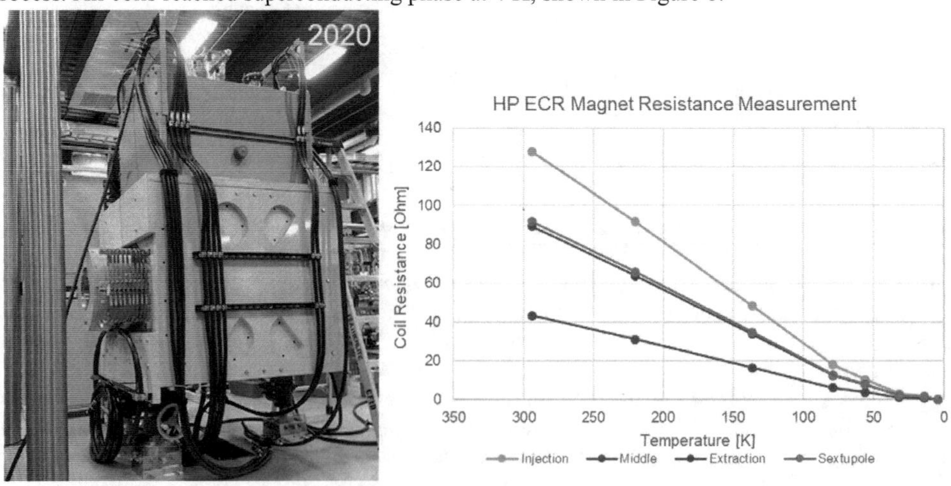

Figure 8. Magnet cooldown on platform, and coils resistance measurement during cooldown.

The measured static heat load at 4 K is about 1.2 W, in good agreement with the thermal analysis. The thermal analysis results are shown in Figure 9. The upper cryostat can accommodate two GM-JT cryocoolers (with total cooling capacity specified for 10 Watt at 4 K) to account for dynamic thermal loads that could be generated at 28 GHz operation at high RF power. Only one GM-JT cryocooler is presently installed on the cryostat for the cooldown and ion source commissioning at 18 GHz. This GM-JT cryocooler has been reliably operating for 9 months.

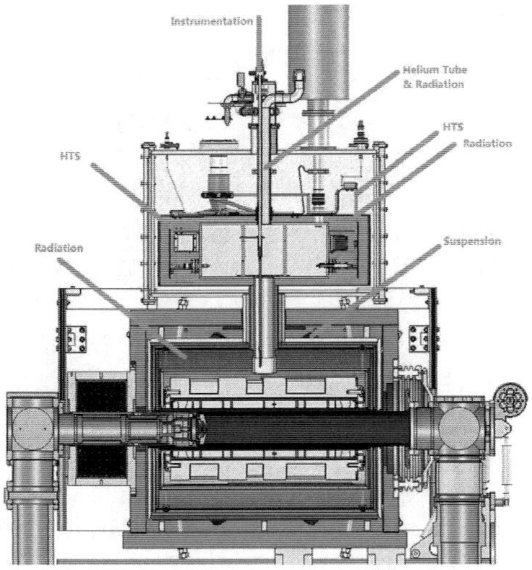

Thermal Analysis for Cryostat – 4.2 K Load

Heat Load Source	Heat Load mW
Radiation Heat Load	110
HTS	206
Helium Tube	41
Suspension System	152
Instrumentations	10
Relief Tube - Radiation	61
Relief Tube - Conduction	131
GM-JT Cryocooler	310 X 2
Total	**1333 mW**

Figure 9. Cryostat thermal analysis results.

3.3. Plasma Chamber Development
The plasma chamber housing the ECR plasma is floating at high voltage potential (around 40 kV) while the warm bore of cryostat is grounded. An insulator is needed between the chamber outer wall and cryostat warm bore. Polyetheretherketone (PEEK), Kapton, and Mylar were typical choices for insulator material. FRIB chose Kapton as a baseline material due to high cost of PEEK. However, Bremsstrahlung X-rays produced from the ECR plasma can damage the Kapton insulation. Tungsten tubes with 2 mm thickness were installed outside the plasma chamber to protect the insulator.
Thermal analysis has been performed to improve the water cooling design of the plasma chamber. According to the thermal analysis, spiral cooling channels can safely cool the plasma chamber for 9 kW RF power operation with nominal water flow of 0.95 liter per second. 3D printed channels were used to improve the water flow at the injection end of the chamber. Welding and pressure testing of the chamber have been completed. The chamber water flow testing is in progress. It will be installed on the magnet after that.

3.4. High Temperature Oven
To meet the demands of high intensity high charged uranium and other heavy metal ion beams for FRIB high power operation, we are developing a high temperature inductive oven, shown in Figure 10. For Uranium Oxide sublimation, the oven temperature is required > 2000 °C to produce enough vapor. The mechanical design has been completed based on the inductive oven used at NSCL for Nickel and Germanium beams development (1500 °C) [7]. Thermal calculations have been performed for the design and integration. The oven fabrication will start soon, and will be tested on the Artemis ion source first.

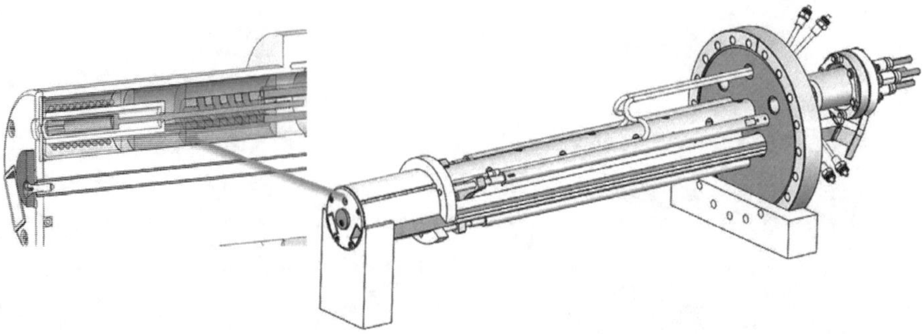

Figure 10. FRIB inductive oven design completed.

4. Conclusion

The FRIB superconducting ECR magnet was built and successfully tested at LBNL. The magnet with cryostat has been assembled on the ECR high voltage platform. The magnet cool-down test was completed, with coil resistance and cryostat heat load checked. The GM-JT cryocooler on the cryostat has been installed and functioning for 9 months in local working mode. Warm component assembly is in progress with the plasma chamber ready for water flow testing. We are approaching the point to energize the magnet when the control system is ready. The plasma test with 18 GHz Klystron is planned in December of 2021.

5. Acknowledgement

We would like to thank Janilee Benitez for her contributions to the magnet cooldown and energizing preparation.

6. References

[1] J. Wei et al., “FRIB Accelerator Status and Challenges,” in Proc. LINAC'12, Tel-Aviv, Israel, TU1A04, p. 417-421, Sep. 2012.
[2] J. Wei et al., “The FRIB SC-Linac - Installation and Phased Commissioning,” in Proc. SRF'19, Dresden, Germany, MOFAA3, p. 12-20, Jun.-Jul. 2019.
[3] E. Pozdeyev et al., “First Acceleration at FRIB,” in Proc. LINAC'18, Beijing, China, WE2A01, p. 615-619, Sep. 2018.
[4] P.N. Ostroumov et al., “First Simultaneous Acceleration of Multiple Charge States of Heavy Ion Beams in a Large-Scale Superconducting Linear Accelerator,” Phys. Rev. Lett., vol. 126, no. 11, p. 114801, 2021.
[5] C. Taylor et al., "Magnet system for an ECR ion source", in Proc. IEEE Trans on Applied Supercon-ductivity, vol. la, No. 1, Mar. 2000, pp. 224-227.
[6] H. Felice, et al., “Design of a Superconducting 28 GHz Ion Source Magnet for FRIB Using a Shell-Based Support Structure”, IEEE TRANSACTIONS ON APPLIED SUPERCONDUCTIVITY, VOL. 25, NO. 3, JUNE 2015
[7] P.A. Zavodszky et al., “Status report on the design and construction of the superconducting source for ions at the National Superconducting Cyclotron Laboratory/Michigan State University”, REVIEW OF SCIENTIFIC INSTRUMENTS 77, 03A334 (2006)

Measurement of ionization, charge exchange and ion confinement times in charge breeder ECR ion sources with short pulse 1+ injection of metal ions

M Luntinen[1], J Angot[2], O Tarvainen[3], V Toivanen[1], T Thuillier[2] and H Koivisto[1]

[1] University of Jyväskylä, Department of Physics, Survontie 9D, 40500 Jyväskylä, Finland
[2] Univ. Grenoble Alpes, CNRS, Grenoble INP, LPSC-IN2P3, 53 Avenue des Martyrs, 38000 Grenoble, France
[3] STFC ISIS Pulsed Spallation Neutron and Muon Facility, Rutherford Appleton Laboratory, Harwell, OX11 0QX, UK

E-mail: mailto:misapema@jyu.fi

Abstract. The Consecutive Transients (CT) method is used for estimating the characteristic times of ionization, charge exchange and confinement within the plasma of a Charge Breeder Electron Cyclotron Resonance Ion Source (CB-ECRIS). The method reveals differences in the characteristic times between different source configurations, with K^{9+} charge breeding efficiencies of 8.9 % and 20.4 %, and allows qualitative explanation of the improved breeding efficiency. The increase in K^{9+} efficiency is accompanied by a decrease in ionization time for low charge states, a decrease of charge exchange time for high charge states, and an overall decrease of the ion confinement time, which increases non-linearly with the charge state. The charge exchange time exhibits a minimum near charge state K^{8+}, indicating low neutral density near the plasma core. The CT-method yields a distribution of possible n_e and $\langle E_e \rangle$ corresponding to the spatial distribution of different charge state ions. The results hint at a non-uniform plasma electron density and energy distribution as well as a nested-layer distribution for the ion populations — hot and dense plasma with high charge state ions near the plasma core.

1. Introduction

The Consecutive Transients (CT) method has been used to obtain postdictions for the ionization, charge exchange and confinement times (τ^q_{inz}, τ^q_{cx} and τ^q) of charge state q ion populations in a CB-ECRIS plasma [1]. The method is based on measuring the extracted current transients prompted by short pulse injection of metal ions into the plasma, making fits to the transients of (minimum) five consecutive charge states, and an optimisation procedure to obtain the plasma electron density n_{e} and average energy $\langle E_{\text{e}} \rangle$ of the presumed Electron Energy Distribution (EED) as well as the times τ^q_{inz}, τ^q_{cx} and τ^q. The method is based on the balance equation [2, 3] describing the temporal evolution of the densities of each ion population. The method probes the plasma conditions (n_e, $\langle E_{\text{e}} \rangle$) of the support plasma, which determine τ^q_{inz}, τ^q_{cx} and τ^q of the injected ions. The results are spatially localised to the origin of the charge state q ion population. Measuring five consecutive charge state transients poses experimental limitations on the support/injected species combinations, since one must avoid q/m overlap in the Charge State Distribution (CSD). The method takes into account the uncertainty of the ionization cross section data and resulting

Table 1: The former (June 2020) and new (February 2021) charge breeder operating parameters.

	Configuration	
Parameter	Former	New
B_{inj} (T)	1.58	1.57
B_{min} (T)	0.45	0.44
B_{ext} (T)	0.83	0.84
μW power (W)	504	530
Support gas species	He	H_2
P_{inj} ($\times 10^{-8}$ mbar)	9.0	13.6
K^+ intensity (nA)	710	500
Injection pulse width (ms)	5	5

rate coefficients by the means of a Monte Carlo approach. No assumptions need to be made regarding the confinement scheme i.e. the functional dependence of τ^q.

Here we demonstrate that the CT-method can reveal the physical causes resulting in a change of the charge breeding efficiency. The comparison is made between the characteristic times obtained for potassium ions in the support plasma of a CB-ECRIS. The source configurations correspond to $^{39}K^{9+}$ efficiencies of 8.9 % and 20.4 %, respectively. Identifying the mechanisms underpinning the factor of > 2 improvement in the charge breeding efficiency of a stable isotope can guide the optimisation of Radioactive Ion Beam (RIB) production.

2. Experimental methods

The 1+→N+ test bench [4] at LPSC Grenoble is dedicated to the development and characterization of the Phoenix CB-ECRIS. The 1+ beam line is used for generating, characterizing and injecting the 1+ ion beam into the CB-ECRIS and the N+ beam line for analysing the extracted multicharged beams. The CB-ECRIS and the test bench are continuously upgraded to increase the global and charge state specific breeding efficiencies, Σ and η^q, and to reduce the charge breeding time and impurities in the N+ CSD. Short pulse injection of 1+ ions is used for the measurement of the charge breeding times τ_{CB}; the time when 90 % of all ions of charge state q have been extracted is designated as τ_{CB}[5]. In 2018-2019, the 1+→N+ test bench was upgraded to improve the vacuum and the alignment of the device [4]. The optimum efficiency of K^{10+} charge state measured in June 2020 was 10.6 % with τ_{CB}^{10+} of 132 ms. The first data using the CT-method to obtain the plasma parameters were taken in this configuration [1] with He support plasma.

The CB tuning and conditioning were then optimised to improve the CB efficiency to 20.4 % for K^{9+} with H_2 support gas (τ_{CB}^{9+} of 131 ms). The CT-method was applied in this configuration in February 2021 to estimate which parameters could have caused the efficiency improvement. Table 1 compares the two CB configurations. The main differences are the support gas species, the B_{min} value (+11.8%) and the microwave power (+5.2%). The charge breeding efficiencies η^q of the K charge states are shown in Table 2. A significant increase of K^{9+} efficiency was obtained (+11.5% absolute efficiency), accompanied by a small efficiency shift towards lower charge states. The extracted current transients of K^{3+}–K^{12+} were measured to obtain the characteristic times for K^{5+}–K^{10+} ion populations. The injected K^+ pulse was kept short (5 ms pulse width) and low in intensity (500 – 710 nA) to minimize the perturbance on the support plasma, while still resulting in high signal-to-noise ratio of the transient current.

Table 2: The charge breeding efficiencies η^q, charge breeding times τ^q_{CB} and the median values of the population confinement times τ^q. The global efficiency includes charge states K^{4+} – K^{12+}.

	η^q (%)		τ^q_{CB} (ms)		τ^q (ms)	
Ion	Former	New	Former	New	Former	New
K^{3+}	1.2	1.3	9	10		
K^{4+}	1.0	1.4	11	12		
K^{5+}	1.1	1.9	13	15	3	3
K^{6+}	1.2	3.0	16	49	3	4
K^{7+}	1.5	5.8	20	90	4	4
K^{8+}	2.7	11.1	45	119	15	12
K^{9+}	8.9	20.4	98	131	16	8
K^{10+}	10.6	11.8	132	138	24	12
K^{11+}	8.5	4.7	149			
K^{12+}	5.1	1.3	155			
Σ (%)	41.8	62.7				

3. Numerical methods

3.1. Principle of the method

The method for analyzing the measured beam current transients of the $q+$ ions has been extensively described in Ref. [1] and is only briefly recapitulated here: The balance equations [2, 3] governing the evolution in time of the ion population densities are defined by

$$\begin{aligned}\frac{\mathrm{d}n^q}{\mathrm{d}t} &= +n_e \langle\sigma v\rangle^{\mathrm{inz}}_{q-1\to q}\, n^{q-1} - n_e \langle\sigma v\rangle^{\mathrm{inz}}_{q\to q+1}\, n^q \\ &\quad +n_0 \langle\sigma v\rangle^{\mathrm{cx}}_{q+1\to q}\, n^{q+1} - n_0 \langle\sigma v\rangle^{\mathrm{cx}}_{q\to q-1}\, n^q - \frac{n^q}{\tau^q},\end{aligned} \tag{1}$$

where the ionization and charge exchange rate coefficients ($\langle\sigma v\rangle^{\mathrm{inz/cx}}$) together with the plasma electron and neutral densities (n_e, n_0) determine the characteristic times

$$\tau^q_{\mathrm{inz}} \equiv \left[n_e \langle\sigma v\rangle^{\mathrm{inz}}_{q\to q+1}\right]^{-1} \tag{2}$$

$$\tau^q_{\mathrm{cx}} \equiv \left[n_0 \langle\sigma v\rangle^{\mathrm{cx}}_{q\to q-1}\right]^{-1} \tag{3}$$

and the loss term $-n^q/\tau^q$ defines the ion confinement time τ^q. Here n^q refer to the injected species, while n_e and n_0 are properties of the support plasma, which are assumed to be constant in time. Using the following identity [6, 7, 1] for the charge state q beam current

$$I^q = \kappa F_B L S \frac{n^q q e}{\tau^q}, \tag{4}$$

where κ is the beamline transmission efficiency, F_B a factor dependent on the mirror ratio of the ion source, L the length of the plasma chamber and S the area of the extraction aperture, one can recast the balance equation in the form

$$\frac{\mathrm{d}}{\mathrm{d}t} I^q = a_q I^{q-1} - b_q I^q + c_q I^{q+1}, \tag{5}$$

where I^{q-1}, and I^{q+1} are the beam currents of charge states $q-1$ and $q+1$, respectively, and the coefficients a_q, b_q and c_q are defined as

$$a_q = n_e \langle\sigma v\rangle^{\text{inz}}_{q-1\to q} \frac{q}{q-1} \frac{\tau^{q-1}}{\tau^q} \tag{6}$$

$$b_q = \left(n_e \langle\sigma v\rangle^{\text{inz}}_{q\to q+1} + n_0 \langle\sigma v\rangle^{\text{cx}}_{q\to q-1} + 1/\tau^q\right) \quad \text{and} \tag{7}$$

$$c_q = n_0 \langle\sigma v\rangle^{\text{cx}}_{q+1\to q} \frac{q}{q+1} \frac{\tau^{q+1}}{\tau^q}. \tag{8}$$

The coefficients a_q, b_q and c_q can be determined by fitting Eq. 5 to the experimentally measured current I^q. The fitting is done by taking $I^{q-1}(t)$ and $I^{q+1}(t)$ as input parameters of Eq. 5, and solving the differential equation for the middle charge state current (numerical solution for the charge state q current denoted by $\mathfrak{I}^q(t)$). A least-squares method is then used to minimize the difference between $\mathfrak{I}^q$ and I^q to obtain a_q, b_q and c_q. Definitions 6, 7 and 8 then yield

$$\frac{q}{q+1} \frac{a_{q+1}}{n_e \langle\sigma v\rangle^{\text{inz}}_{q\to q+1}} = \frac{b_{q+1} - n_e \langle\sigma v\rangle^{\text{inz}}_{q+1\to q+2} - \frac{a_{q+1}c_q}{n_e\langle\sigma v\rangle^{\text{inz}}_{q\to q+1}}}{b_q - n_e \langle\sigma v\rangle^{\text{inz}}_{q\to q+1} - \frac{a_q c_{q-1}}{n_e\langle\sigma v\rangle^{\text{inz}}_{q-1\to q}}}, \tag{9}$$

which is an equation of two unknowns, n_e and $\langle E_e\rangle$, since the rate coefficients $\langle\sigma v\rangle^{\text{inz}}_{q'\to q''}$ may be calculated as a function of the average energy $\langle E_e\rangle$ of the EED. In lieu of experimental determination of the EED, we have thus far assumed the Maxwell-Boltzmann (MB) distribution. Equation 9 provides an infinitude of solutions for n_e and $\langle E_e\rangle$, which may be constrained on physical grounds such that

$$\begin{cases} \tau^q > 0, \\ \tau^{q+1} > 0, \\ 1/n_0 \langle\sigma v\rangle^{\text{cx}}_{q\to q-1} > 0, \\ n_{e,\text{low}} < n_e < n_{e,\text{co}}, \\ \langle E_e\rangle_{\text{low}} < \langle E_e\rangle < \langle E_e\rangle_{\text{high}}, \end{cases} \tag{10}$$

where τ^q, τ^{q+1} and $\left[n_0 \langle\sigma v\rangle^{\text{cx}}_{q\to q-1}\right]^{-1} \equiv \tau^q_{\text{cx}}$ can be calculated from the definitions of a_q, b_q and c_q (c.f. Ref. [1] Eqs. (16) and (19)). The upper bound of n_e is the cut-off frequency [8] $n_{e,\text{co}}$, while its lower bound $n_{e,\text{low}}$ is taken from the literature [9]. The lower limit of $\langle E_e\rangle$ is set to be on the order of the plasma potential, i.e., ~ 10 eV, while its upper limit can be constrained at 10 keV, since the fraction of electrons found with energies above a few keV is 20–50 % [7, 10]. We thus focus on the warm electron population, which is mainly responsible for the ionization.

3.2. Code enhancements

While the numerical method has remained unchanged in principle since the introduction of the method [1], the code has been updated improving the precision and speed of the algorithm: *(i)* The rate coefficients are now calculated as a function of the average energy $\langle E_e\rangle$ of the EED (not T_e), which facilitates the comparison between different EEDs in the future. *(ii)* In Ref. [1] we adopted an "umbrella" uncertainty bound of 60 % for the rate coefficient as reported by Voronov [11]. The code now uses the experimental uncertainties of the cross section data [12]. *(iii)* The handling of the constraints has been improved: The penalty function is first minimized within the n_e and $\langle E_e\rangle$ bounds. After the minimum has been found, the characteristic times are calculated and minima leading to negative (unphysical) characteristic times are discarded. All the following results have been computed using the improved version (v1.2) of the code.

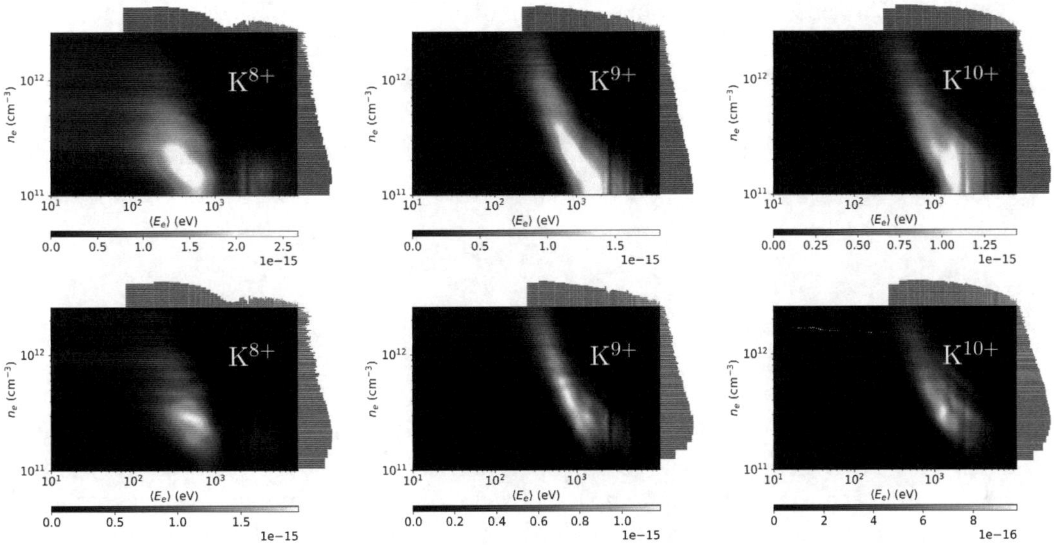

Figure 1: The $(n_e, \langle E_e \rangle)$ -solution sets of K8+–K10+ for the two charge breeder configurations Former (upper row) and New (lower row).

4. Results and analysis

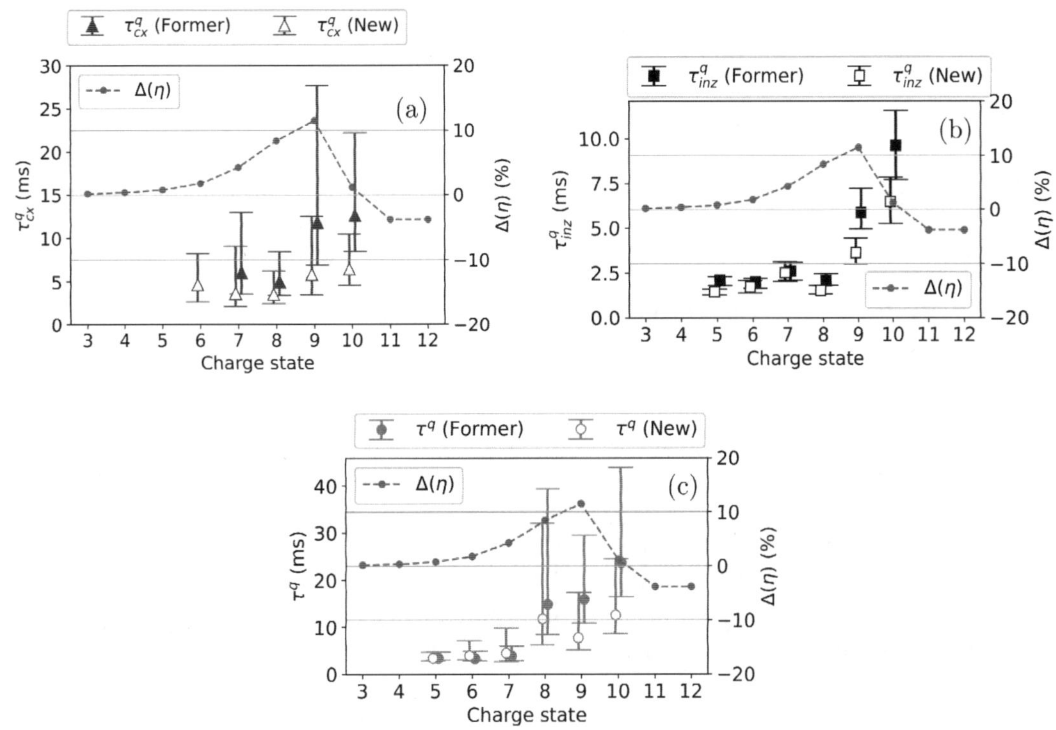

Figure 2: Comparisons of τ^q_{cx}, τ^q_{inz} and τ^q and the absolute difference $\Delta(\eta)$ of CB efficiencies.

Figure 1 shows heatmaps of the $(n_e, \langle E_e \rangle)$ i.e. viable solutions of Eq. 9 in both CB

Table 3: The most probable values of n_e and $\langle E_e \rangle$ in the two CB configurations.

	n_e ($\times 10^{11}$ cm^{-3})		$\langle E_e \rangle$ (eV)	
Ion	Former	New	Former	New
K^{5+}	1.0	1.0	96	129
K^{6+}	1.3	1.4	124	115
K^{7+}	1.2	1.0	156	168
K^{8+}	1.4	2.1	150	214
K^{9+}	1.0	2.6	288	357
K^{10+}	1.4	2.5	353	518

configurations for K^{8+}–K^{10+}. These examples represent the set of (n_e, $\langle E_e \rangle$)-pairs satisfying Eq. 9 in the vicinity of their respective ion population n^q. It must be recalled, that the I^q beams are extracted from the whole plasma, and the n_e and $\langle E_e \rangle$ values represent averages for a given ion population. The variation of the n_e and $\langle E_e \rangle$ as a function of q thus reflects the average spatial variation of each n^q. Hence, the characteristic times calculated at these (n_e, $\langle E_e \rangle$) also yield a distribution of results. Further constraining (if possible) of the n_e and $\langle E_e \rangle$ spans, and the uncertainty of the ionization cross sections, would decrease the uncertainty of the results. Histograms of the solution sets are projected onto the n_e and $\langle E_e \rangle$ axes (note the logarithmic scale).

Comparison of the histograms and the most probable values of n_e and $\langle E_e \rangle$ — tabulated in Table 3 — reveals that in the new configurations the local n_e and $\langle E_e \rangle$ are higher in particular for the HCIs. This is commensurate with the decrease of their τ^q_{inz}. Increase of n_e is also to be expected from the increased neutral gas input compared to the former configuration, as the increased n_0 allows for a higher n_e, particularly in the core of the plasma where the highly charged ions (HCIs) are believed to be confined in the potential dip [3], which itself is generated by the higher concentration of hot, well-confined electrons.

Figure 2 shows τ^q_{inz}, τ^q_{cx} and τ^q as a function of q for the two data sets. The characteristic times are overlaid with the absolute charge breeding efficiency change between the CB configurations. The uncertainty bounds enclose a one sigma (34.1 %) fraction of all results around the median value, represented by the data point. The distribution of values is typically sharply peaked around the lower values, with the most probable value lying below the median. The size of the uncertainty bounds is affected by both, the uncertainty of the experimentally measured ionization cross sections (e.g. at worst for K^{8+} the uncertainty is $\delta\sigma^{\mathrm{inz}}_{q\to q+1}/\sigma^{\mathrm{inz}}_{q\to q+1} = 200$ %) and the limits imposable upon the solution space ($n_e, \langle E_e \rangle$). Especially the K^{9+} efficiency is improved between the two configurations. Each of the characteristic times, in particular τ^q_{cx} and τ^q have decreased for the HCIs, although in most cases the uncertainty bounds overlap.

Molecular hydrogen has a larger diameter and smaller ionization potential than helium, which may explain the decrease in τ^q_{cx}, the charge exchange cross section being inversely proportional to the square of the ionization potential [13], i.e. $\sigma^{\mathrm{cx}}_{q\to q-1} \propto I^{-2}$. Hence, switching to a H_2 support plasma increases $\sigma^{\mathrm{cx}}_{q\to q-1}$. There is a minimum in τ^q_{cx} around K^{7+} and K^{8+} for both of the data sets. This minimum is attributable to the increase of $\sigma^{\mathrm{cx}}_{q\to q-1}$ with q and the nested-layer structure of the ECRIS plasma. The latter is consistent with simulations and experiments [14, 15, 16, 17, 18] suggesting that the HCIs originate in the plasma core near the plasma chamber axis, while lower charge states originate from the peripheral plasma forming an overlapping layered structure. The increase of τ^q_{cx} towards the HCIs implies a decrease of neutral density toward the plasma core, since the $\sigma^{\mathrm{cx}}_{q\to q-1}$ increasing with q would otherwise imply a further decrease in τ^q_{cx}. The decreased τ^q_{cx} of the HCIs is in line with the increased charge breeding efficiency of the lower charge states in the new configuration.

There is a small decrease of $\tau^q_{\rm inz}$ of the low charge states, while a prominent decrease, attributed to the increase of n_e implied by the solution sets (Fig. 1.), is seen for the HCIs in the new configuration. The effect of the faster ionization is to feed the high charge state populations. The electron shell closure at K^{9+} causes a discrete jump between $\tau^{8+}_{\rm inz}$ and $\tau^{9+}_{\rm inz}$.

The median of the confinement time τ^q solutions has decreased in particular for K^{9+} and K^{10+}. Especially the former data set shows a non-linear q-dependence for τ^q. A linear model for τ^q would yield negative confinement times for low charge states, while a power law fit is found to best represent the results [1]. In Ref. [7] a linearly increasing trend and much shorter values ($\sim$ 5 ms) of τ^q were found. Short τ^q has been used as an argument in favor of a predominantly collisional or ambipolar ion confinement[19]. However, recent optical measurements of the Doppler broadening of ion emission lines [20] have found $\sim$ 10 eV ion temperatures, unattainable with such short τ^q. There is mounting evidence from emittance [15] and current density measurements [17, 18], simulations [14, 3], and afterglow experiments [21, 22], that the HCIs are confined within a potential dip $\Delta\phi$ of the plasma potential distribution. The HCIs remain confined until they obtain enough energy to overcome the electrostatic barrier, i.e. in excess of $eq\Delta\phi$. The electrostatic confinement scheme is in accordance with the power law trend observed here.

It is important to note that it is the combination of $\tau^q_{\rm inz}$, $\tau^q_{\rm cx}$, and τ^q ($\forall q \in \{0, 1, 2, \ldots, q_{\rm max}\}$) that determines the steady-state CSD of the plasma. The change in the K^{9+} efficiency can thus be explained through the characteristic times: *(i)* The decrease in $\tau^q_{\rm cx}$ for the HCIs indicates faster rate of charge exchange and a consequent shift of the CSD towards lower charge states. In effect, the K^{9+} population is fed from the HCIs via charge exchange; *(ii)* The decrease in τ^q allows the K^{9+} ions — once produced — to escape more rapidly, which increases the extracted beam intensity; *(iii)* The shorter ionization time of the low charge states supplies to the K^{9+} population (but is counteracted by the simultaneous decrease of $\tau^q_{\rm cx}$). In short, the increase of the K^{9+} CB efficiency is explicable by the pile-up into $q = 9$ and its shorter confinement time.

5. Conclusion

It has been shown, that the CT-method can be used to discern differences between the characteristic times in different CB configurations. It therefore enables the analysis of fundamental processes in the ion source plasma. Here, the source tune has been changed by varying multiple operating parameters, and the characteristic times are affected by multiple physical effects. In future experiments, the evolution of the characteristic times will be studied by varying one parameter at a time.

It must be noted that an infinitude of $(n_e, \langle E_e \rangle)$ pairs satisfies Eq. 9. It would be a great benefit to the method, if it were possible to set stricter constraints on n_e and $\langle E_e \rangle$, and if the ionization cross section were known more precisely. The presumed EED also being a source of possible error, its experimental measurement in plasma is desired. In short term we plan to study the sensitivity of the results on the assumed EED.

The CT-method provides a breakdown of $\tau^q_{\rm CB}$ to its components: The characteristic times $\tau^q_{\rm inz}$, $\tau^q_{\rm cx}$ determine the rate at which ions of different charge states are generated, while τ^q gives the rate at which ions of a given population exit the plasma. It is important to note that $\tau^q_{\rm CB}$ is affected by the characteristic times of all charge states populations as a given particle may spend time as an ion at various different charge states before being extracted from the source. The $\tau^q_{\rm CB}$ is thus a measure of the cumulative confinement time [23] of an ion, including its charge state history, while τ^q measures the confinement time of ion population at charge state q, and is directly affected by changes in the ion confinement conditions.

Based on the discussion of the causes underlying the change in K^{9+} efficiency, it can be said that in order to maximize the charge breeding efficiency of a given charge state one must maximize the pile-up into that charge state population the charge states both above and below, i.e. one must minimize the ionization time from below, and minimize the charge exchange time from above. The best efficiencies can be obtained for charge states whose valence electron lies on a closed shell (such as K^{9+}), because the ionization from such a state is minimized and the population will not be depleted as much via ionization to higher states. At the same time one should minimize the confinement time of the desired charge state to efficiently extract the ions.

Data availability statement

The data that support the findings of this study are openly available at `DOI:10.6084/m9.figshare.14685432`, and the analysis code (release version 1.2.4) is publicly available at: `https://github.com/misapema-jyfl/popTauPy`.

Acknowledgments

We acknowledge grants of computer capacity from the Finnish Grid and Cloud Infrastructure (persistent identifier urn:nbn:fi:research-infras-2016072533).

References

[1] Angot J, Luntinen M, Kalvas T, Koivisto H, Kronholm R, Maunoury L, Tarvainen O, Thuillier T and Toivanen V 2020 *Plasma Sources Science and Technology* **30** 035018
[2] Shirkov G, Mühle C, Musiol G and Zschornack G 1991 *Nucl. Instrum. Methods Phys. Res. A: Accel. Spectrom. Detect. Assoc. Equip.* **302** 1 – 5 ISSN 0168-9002
[3] Shirkov G D 1993 *Plasma Sources Science and Technology* **2** 250–257
[4] Angot J, Galatà A, Maunoury L, Thuillier T, Baylac M, Migliore M and Sole P 2020 *ECRIS 2020 proceedings WEZZO02*
[5] Angot J, Tarvainen O, Thuillier T, Baylac M, Lamy T, Sole P and Jacob J 2018 *Phys. Rev. Accel. Beams* **21**(10) 104801
[6] West H J 1982 Calculation of ion charge state distribution in ecr ion source Tech. rep. Lawrence Livermore National Laboratory
[7] Douysset 1 G, Khodja H, Girard A and Briand J P 2000 *Phys. Rev. E* **61**(3) 3015–3022
[8] Geller R 1996 *Electron cyclotron resonance ion sources and ECR plasmas* (IOP) ISBN 9780750301077
[9] Tarvainen O, Koivisto H, Galatà A, Angot J, Lamy T, Thuillier T, Delahaye P, Maunoury L, Mascali D and Neri L 2016 *Phys. Rev. Accel. Beams* **19**(5) 053402
[10] Perret C 1998 *Caractérisation de la population électronique dans un plasma de source d'ions à résonance cyclotronique électronique* Ph.D. thesis Université Joseph Fourier, Grenoble
[11] Voronov G 1997 *At. Data Nucl. Data Tables* **65** 1 – 35 ISSN 0092-640X
[12] Lennon M A, Bell K L, Gilbody H B, Hughes J G, Kingston A E, Murray M J and Smith F J 1988 *Journal of Physical and Chemical Reference Data* **17** 1285–1363 ISSN 15297845
[13] Knudsen H, Haugen H K and Hvelplund P 1981 *Phys. Rev. A* **23**(2) 597–610
[14] Mironov V, Sergey B, Bondarchenko A, Efremov A and Loginov V 2015 *Physical Review Special Topics - Accelerators and Beams* **18**
[15] Wutte, D, Clark, D J, Laune, B, Leitner, M A, Lyneis C M 2001 *AIP Conference Proceedings* vol 445 pp 445–448
[16] Leitner M A, Wutte D and Lyneis C M 2001 *Proceedings of the 2001 Particle Accelerator Conference, Chicago* vol 5 ed Lucas P and Webber S Argonne National Laboratory and Fermi National Accelerator and Laboratory Oak Ridge National Laboratory (Chicago, Illinois U.S.A.: IEEE) pp 67–69
[17] Panitzsch L, Stalder M and Wimmer-Schweingruber R F 2011 *Rev. Sci. Instrum.* **82** 033302
[18] Panitzsch L, Peleikis T, Stalder M and Wimmer-Schweingruber R F 2011 *Rev. Sci. Instrum.* **82** 093302
[19] Melin G, Drentje A G, Girard A and Hitz D 1999 *Journal of Applied Physics* **86** 4772
[20] Kronholm R, Kalvas T, Koivisto H, Laulainen J, Marttinen M, Sakildien M and Tarvainen O 2019 *Plasma Sources Science and Technology* **28** 075006
[21] Sortais P 1992 *Rev. Sci. Instrum.* **63** 2801–2805
[22] Nakagawa T *et al.* 1998 *Review of Scientific Instruments* **69** 637–639
[23] Marttinen M *et al.* 2020 *Rev. Sci. Instrum.* **91** 013304

CERN hadron sources: status and innovation overview

F Wenander

Beams Department, CERN, CH-1217 Geneva, Switzerland

Email: fredrik.wenander@cern.ch

Abstract. The CERN accelerator complex is served by several different types of hadron sources, producing both high- and low-intensity negative H^- beams, highly charged positive stable ions, but also transforming low-charge radioactive ions to higher charges before post-acceleration. Apart from the operational sources, research and development is carried out at dedicated test stands for H^- production and pre-acceleration, ECRIS oven experiments and general EBIS development. In this paper, we will report on the performance and the latest development related to the sources. Efforts to enhance the theoretical understanding of the source physics will be briefly reviewed.

1. Introduction

The high-energy proton-proton collisions at the Large Hadron Collider (LHC) make up CERN's most high-profile experimental program. Nonetheless, CERN's accelerator complex as a whole provides the basis for a sprawling physics program involving different particle types with energies ranging from some keV to several TeV. The low energies are necessary for particle trapping, e.g. for the production of antihydrogen at the Antiproton Decelerator (AD), while in-ring and fixed-target collisions in most cases require higher energies. Although the primary particle type at CERN is protons, used either as the experimental collision-particle itself, or, as primary driver beam for production of exotic particles types, also different types of heavy ions belong to the standard particle assortment. In addition to stable ions, production of a large variety of radioactive ions and post-acceleration to a few MeV/u is part of CERN's experimental program.

The broad R&D program for ion sources at CERN aims first and foremost at assuring stable operational conditions in conjunction with highest performance of the sources, but also to extend their capabilities in order to pave the way for new physics opportunities. Within the last couple of years, the H^- source for LINAC4 has become operational but future requirements already impose continued studies to improve the source performance further. A low-intensity H^- source for the Extremely Low ENergy Antiproton (ELENA)) ring at the AD, initially foreseen to only be employed during the commissioning phase, will most likely become a permanent installation, so stability issues therefore need to be addressed. Likewise, an improvement of both the long-term and shot-to-shot stability of the Electron Cyclotron Resonance Ion Source (ECRIS) at LINAC3 is desirable. Finally, enhanced performance of the Electron Beam Ion Source (EBIS) charge breeder at REX/HIE-ISOLDE is requested in order to expand the physics reach. In the following sections, an overview of the performance and development programs of these four sources will be given.

However, first a brief homage to CERN's former proton workhorse: the duoplasmatron source providing protons for the 50 MeV LINAC2. Over the lifespan of 40 years, more than 700 million of proton pulses with a length up to 150 μs and with an instantaneous current of >200 mA were produced. Over the last two years of operation, amounting to >12000 h, the source was only inoperative for 6.5 h due to faults. At the end of 2018, with the start of long shutdown 2 (LS2), the exploitation of LINAC2 was discontinued, and the baton as principal linear accelerator at CERN was handed over to LINAC4.

2. LINAC4 H^- source

In the LINAC4 injector, H^- ions are now being accelerated to a final energy of 160 MeV before undergoing charge-exchange injection into the four rings of Proton Synchrotron Booster. From a 2 MHz caesiated RF source, the ions pass a 1.8 m low-energy beam transport section before being accelerated to 3 MeV in a 3 m-long 352.2 MHz Radio Frequency Quadrupole (RFQ). The repetition rate is 0.83 Hz, although operation at 2 Hz is possible.

The present source version, IS-03b, features a two-stage extraction system, where the co-extracted electrons are dumped onto the puller-dump electrode at an energy of ~10 keV, before the ions are further accelerated to 45 keV. The multi-cusp field has been removed as no significant difference in transverse emittance, required power and efficiency through the RFQ could be noted compared to having the multi-cusp present. The hole diameter of the plasma electrode has been enlarged from 6.5 mm to 7.5 mm, resulting in a larger transmission through the RFQ. Furthermore, the source is now exclusively operated in surface production mode, making use of caesiation to increase the H^- simultaneously, as the ratio of extracted e^- to H^- is reduced. The continuous low-flow caesiation is attained by keeping the caesium oven at 60 °C, while the transfer line stays at 80 °C to avoid cold spots.

2.1. Source performance and reliability

Being the main driver for the CERN accelerator complex, a very high reliability of >99% is requested, and a pulse-to-pulse stability of 2% (peak-to-peak) is specified after the LINAC. During 2021, the source availability has been 99.9%, with the gas system being the main reason for the short interruptions. The measured fluctuation of the source current amounts to 0.9% (1 σ), where half of the instability is due to RF power variations. The operational source, installed in May 2020, provides a beam current of 35 mA, of which 27 mA is transmitted through the RFQ, as well as through the remaining LINAC. Normally, the e^-/H^- ratio is below 4, while a peak RF power of 20-25 kW is required for generating this current. An autopilot acts on the RF power level, basing its adjustment on the current of the previous 10 extracted pulses. Similarly, the autopilot restarts the HV power converters automatically after trips and monitors the e^-/H^- ratio, which is a signature for the source caesium status. The occasional manual tuning focusses mainly on the RF pulse shape, the gas injection settings and the caesiation system temperatures. Since it was installed last year in the Linac4 tunnel, this source has been vented twice for a gas valve exchange - a preventative action performed every 4 months that takes 24 hours from beam stopped to beam back. In spite of the good reliability, two spare units that have been caesiated and tested, are stored under N_2 atmosphere.

2.2. Test stand activities

The layout of the LINAC4 test stand is identical to the on-line machine, although without a pre-chopper and with an emittance meter replacing the diagnostics tank and an acceptance mask at the position of the RFQ. The test stand is used for validation of spare units, source development and studies, as well as for target material irradiation studies, as endoscopic analysis of the LINAC4 RFQ showed signs of damage and degradation on the vanes.

The beam formation symmetry of the IS-03b H^- source operated in plasma caesiated surface mode is broken by the magnetic filter field inducing a left-right charge drift, and by radial asymmetry resulting from the combination of ions created in the plasma volume and those produced and directly extracted from the active caesiated surface. The goal of the beam formation studies is to provide detailed experimental evidence to validate the ONIX simulation, which serves as input for the IBSimu beam tracking, without the use of "free" plasma parameters. Optical emission spectroscopy serves to calibrate the ONIX plasma, while beam emittance spectroscopy, emittance and profile measurements contribute to distinguish asymmetries induced by beam optics from those of the meniscus ion-density distribution. For details see refs. [1],[2].

As a complement to the emittance meter, an installed set of four separated plates with square holes followed by a Faraday cup, acts as a compact "mask" that reproduces the transverse acceptance of the RFQ. Particle tracking has shown a good transmission agreement, although the mask overestimates the transmission slightly due to the ignorance of the longitudinal acceptance of the RFQ. Tests with the ISO-03 source exhibit a saturation in transmitted current at around 43 mA, while the RFQ only transmits 32 mA at most. The reason for the discrepancy is under investigation and may be due to a lack of space charge neutralization near the RFQ entrance.

2.3. Modified source extraction design

The present version of the H^- source is capable of providing a beam current of 27 mA, measured after the RFQ reliably. However, this value is below the goal of 45 mA to be provided for operational use from 2023 onwards. A large transverse beam emittance compared to the RFQ acceptance is believed to be the main obstacle. The source is using an electron dump design with a low impact energy since 45 keV electrons in the early tests with a DESY source type caused immediate damage to the dump. As the 2018 ion source review committee discouraged the use of pure volume mode operation in favour of caesiation, the anticipated amount of co-extracted electrons will be low, thereby again opening up for ion extraction options with higher electron beam dump energy, that should be beneficial for the emittance of the H^- beam.

For these reasons, a simplified extraction geometry has been studied where the intermediate puller-dump and Einzel lens are being suppressed [3]. In the geometry, now being 6 cm shorter, the 45 keV electrons are deflected onto a dedicated dump. The simulated peak power density is <100 W/mm^2 for <80 mA co-extracted electrons. A sketch of the design with the associated beam simulation is shown in Fig. 1. Initial experiments at the test bench resulted in a 52 mA extracted beam from the source, with 94% transmission through the RFQ mask. The measured normalized emittance is ~0.3 mm.mrad (1 σ) in both planes. Inspection of the source after a few weeks of operation did not reveal any trace or damage on the puller electrode nor on the electron dump made of INERMET(c).

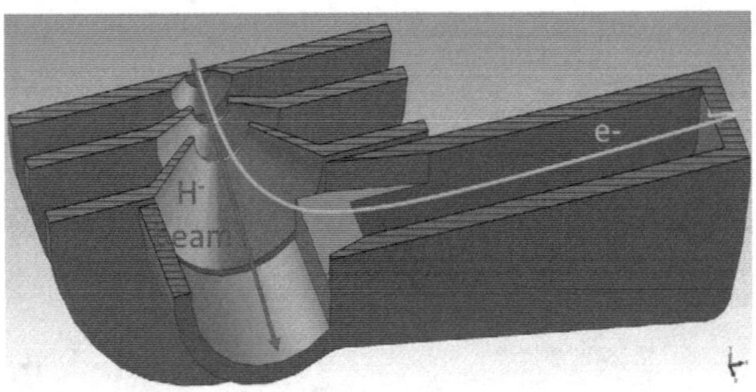

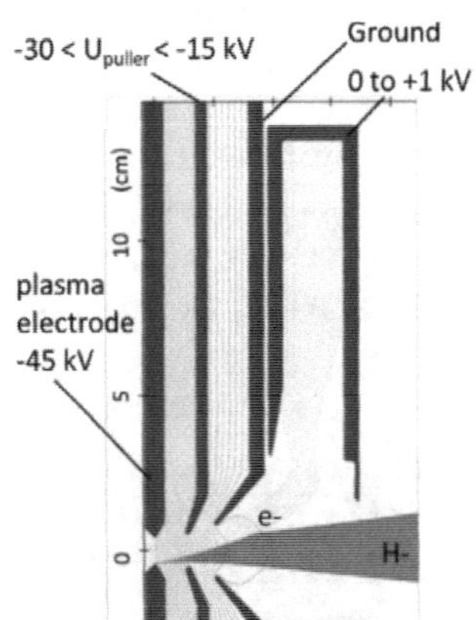

Figure 1. (Left) Schematic cross-section of the new source extraction design with electron dump close to ground potential. (Right) Corresponding IBISimu extraction simulation for a ~50 mA H^- current.

3. ELENA H^- source

ELENA is a compact ring that has been installed to complement the AD antimatter factory at CERN. With this recent adoption, the injected antiprotons can be decelerated from 5.3 MeV down to 100 keV and further cooled, before being expulsed to the experimental setups and thereby enhancing the trapping efficiency by two orders of magnitude compared to using degrader foils at AD extraction beam energy. For commissioning purposes of the ring and subsequent electrostatic transfer lines towards experiments, a local H^- source is connected to the ELENA ring. Using H^- particles with 100 keV, the antiproton beam at extraction energy can be mimicked. The source can provide higher intensity and more frequent injections than possible with the antiprotons from AD, which operates with a 110 s cycle time. In addition, it has facilitated progress with the commissioning phase also in the absence of antiprotons during LS2.

The H^- ion source was granted to the ELENA project as an in-kind contribution from the Jülich Forschungszentrum, Germany, and installed in 2015 at CERN. The multi-cusp volume-type source operates with a continuous arc discharge as well as H_2 gas injection. The pulsed beam structure is attained by rapid switching of the puller electrode potential. Details about the design of the source are found in ref. [4].

3.1. Source performance and recent improvements

Results from the initial time of operation include H^- pulses with a current of 50 to 100 μA and a length of a few μs. Since then, new important observations have been collected, facilitated by improved beam instrumentation. A recently installed Pearson 5753 wide-band beam current transformer (BCT) allows for single-shot measurements of the extracted beam, where the previous BCT required averaging over several pulses due to low sensitivity. With this, observations from pick-up elements in the ring of an intra-pulse instability that will be discussed below, could be confirmed. The introduction of an annular aluminum-doped scintillation screen (Chromox AF995R) in front of the BCT, in conjunction with a triggered digital camera and narrowband filter (694±5 nm), has permitted trajectory surveillance of the extracted particles. This tool has been essential for increasing the understanding of a slow transverse drift of the beam.

After long-standing reliability problems of the HV transformers (both of solid-state and oil-based types), -100 kV can now regularly be applied. Nonetheless, to reduce the risk of sparks inside the transformer, the HV is pulsed from -60 kV to -100 kV just in advance of the ion pulse.

3.2. Intra-pulse instability and transverse beam drift

The fast (about 1 MHz) intra-pulse instability, occurring for nominal operational parameters results in a poor intensity stability of the H^- beam injected into ELENA, as the acceptance window is around 650 ns. Numerous efforts, laid out in ref. [4], have been performed to identify the reason for the instability. Especially noteworthy is the observation of a stable proton pulse if the voltage configuration of the source is inversed. The intra-pulse instability is therefore understood to originate in the actual H^- production region, located after the magnetic filter region and therefore not easily visible on most of the external signals available, although the exact mechanism is presently unknown. The remedy is to either operate the source with a low arc voltage close to plasma extinction (~ 25 V), yielding a very stable although low extracted current, or by increasing the H_2 gas flow significantly (from 1.5 to 5 sccm).

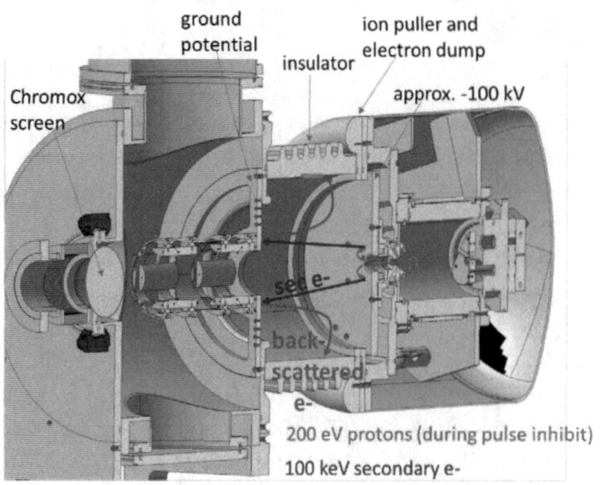

Figure 2. Low-energy protons escaping the ion puller and electron dump electrode are believed to create secondary electrons when reflected backwards onto the copper surface. The secondary electrons gain velocity and reach the ground electrode with an impact energy of nearly 100 keV. A large fraction of these electrons may backscatter at an energy of several ten keV, and will then be deflected onto the non-shielded main insulator and there build up a charge capable of steering the ion beam.

In addition, a slow drift of the beam position has been observed in the ring, and the cause of the orbit drift has been tracked back to the ion source assembly. On the previously mentioned scintillator screen, a vertical movement of the focused beam spot of up to 10 mm can be observed. The movement is sporadic (hours-days-weeks) and occurs over some minutes. Moreover, it can be provoked by applying higher negative voltage (-250 V or more) onto the electron dump during the H^- inhibiting phase. The beam position cannot be reestablished by switching off the plasma, nor by lowering the HV, only by venting the source or leaving it switched off for a night or longer. After careful exclusion of other causes, we believe the reason is a charging-up phenomenon of the HV insulator, not being shielded from the beam region (see Fig. 2 caption for a qualitative argument). Simulations support the findings and a modification of the insulator shielding is foreseen.

4. LINAC3 ECR ion source

The heavy ion beams, either used in LHC for high energy collision experiments or at the Super Proton Synchrotron North Area for fixed-target experiments, originates from the GTS-LHC ECR ion source. The produced highly charged ions are accelerated in LINAC3 to 4.2 MeV/u before being stripped in a carbon foil to even higher charge states and thereafter injected into the Low Energy Ion Ring (LEIR), where they are accumulated, cooled, and accelerated to provide intense beams for physics.

4.1. Source design and performance

The 14.5 GHz ECRIS is based on the original Grenoble Test Source, although with the addition of a center coil and a stronger Halbach-style 36-piece permanent magnet; operation commenced in 2005. The source operates in afterglow mode, with a 10 Hz RF heating cycle at 50% duty factor, and produces a typical pulse length of ~1 ms. At an energy of 2.5 keV/u, the desired charge state is selected in a spectrometer before injection into an RFQ cavity, which selects 200 μs of the pulse for acceleration. From two micro-ovens at the back of the source's plasma chamber, lead vapor evaporates into the plasma that is sustained by additional injection of oxygen gas. Delivered beams, their method of production and achieved intensities are summarized in Table 1.

Intensity and stability are key parameters for the beam. Therefore, throughout the years, several studies have been performed on the source, and it has undergone various modifications. For instance, operation at 18 GHz and at nearby frequencies of 14.5 GHz has been tested, as well as double-frequency injection using a TWTA (12-18 GHz) in addition to the 14.5 GHz. The extraction system has previously been modified with the addition of an Einzel lens and increase of the downstream beam pipe aperture. Operational experience shows that some of the source instabilities can be linked to the oven performance, hence modifications of the design have been studied over a longer period, both at the operational source, but primarily at a dedicated oven test stand. Below, the latest studies and alternations to the source are briefly introduced. For more details see refs. [5],[6].

Table 1. Beams produced in the ECRIS and their typical pulse intensity after charge selection in the spectrometer.

Beam	Intensity (μA)	Method	Year
$^{208}Pb^{29+}$	~170	Oven	All other years
$^{40}Ar^{11+}$	~100	Gas	2015
$^{129}Xe^{22+}$	~160	Gas	2017

4.2. Micro-oven investigations

The two oven crucibles are loaded with approximately 1.5 g enriched ^{208}Pb each, and resistively heated with a tantalum filament. The oven power is adjusted manually to attain an optimal and stable extracted beam current. In the past, oven refills and orifice cleaning were rescheduled every 2 weeks, as at that point, a high heating power (20 W) was required, or the oven orifice was blocked. Approximately only 1/3 of the lead had then been consumed. MolFlow+ simulations showed that 50% of the emitted lead vapor impinged on the relatively cold tantalum cover and in combination with the buffer gas oxygen, a lead oxide blockage at the tip of the oven disrupted the operation (see Fig. 3, left). With the addition of a so-called beak at the orifice of the ceramic crucible, interception of the lead vapor by the oven cover can be avoided, while the beak is still at a sufficiently high temperature to avoid condensation on itself. The design was supported with thermal simulations, and the first results have demonstrated an oven life-time of 59 days without any blockage (see Fig. 3, right). It still remains to consistently reproduce the long life-time.

Figure 3. (Left) Lead oxide blockage of the exit hole on the old oven design. (Right) No blockage after 18 days operation using the beak modification.

4.3. Other recent ECRIS studies

Aluminum in combination with oxygen are believed to create a layer on the plasma chamber with high secondary electron yield. However, the original plasma chamber made of aluminum had to be replaced in 2010 by a stainless steel chamber due to plasma stability and mechanical lifetime issues. Thereby, the rigidity increased at the cost of slightly reduced beam intensity and a factor of 3 higher required injected RF power. To reclaim the favorable plasma surface condition, while maintaining the mechanical strength, a stainless steel chamber has been coated with a 20 μm thick Al layer using cylindrical DC magnetron sputtering. After insertion at the GTS-LHC source, a very high lead intensity is rapidly attained, and the optimum RF power is only 400-600 W. The pulse-to-pulse stability is, however, slightly worse than for a stainless steel chamber, and after 6 weeks the RF power is back at its nominal value of 1000-1600 W. No aluminum was detected in the extracted beam. A post-mortem analysis of the removed and cut open chamber demonstrated a build-up of a lead oxide layer at chamber regions that are not constantly bombarded by loss electrons.

With the aim of decoupling the influence of the plasma meniscus shape on the beam Twiss parameters, and thereby allowing for increased flexibility in the tuning of the beam into the charge state spectrometer, a moveable extraction puller has recently been implemented. This allows for an external remote manipulation of the axial position of the puller, in unison with the intermediate electrode, of ±10 mm around the nominal setting. The initial tests have not yet demonstrated a higher extracted beam current and they will be continued.

On-going is also data extraction of the source settings in relation to extracted beam current, using clustering analysis. The goal is to identify reoccurring patterns, and link setting clusters with source performance, with the ultimate aim to automatically adjust the settings. Initial analysis suggests that more observables are necessary, and therefore a real-time optical spectroscopy to detect the lead vapor density is under implementation.

5. Charge breeding at REXEBIS

In difference to the above discussed sources, the EBIS at the radioactive nuclear beam facility REX/HIE-ISOLDE is not producing ions but performing charge breeding, that is transforming externally injected 1+ ions to n+, before passing on the ion bunch to a post-accelerator via an A/Q-spectrometer. The radioactive ions in question are spanning the full nuclear chart. In fact, REXEBIS has charge bred ions ranging from $^{6}He^{2+}$ to $^{228}Ra^{53+}$ during its 20 years of operation, covering 134 isotopes from 34 elements, and provided beam time at more than 200 occasions.

5.1. Non-adiabatic electron gun

REXEBIS, with its relative moderate electron beam parameters, belongs to the first generation radioactive charge breeders. Recent development has focused on the upgrade of the electron beam properties in the trapping region of the EBIS. A higher electron current density is desirable as the charge breeding time can be reduced, thereby permitting experiments access to short-lived ions and operation of the post-accelerator at a higher repetition rate, which in turn results in a lower instantaneous particle rate of the extracted bunches. With the introduction of an immersed electron gun making use of a non-adiabatic (NA) magnetic element, in combination with a cathode placed at a low field of 700 Gs, a significantly higher current density has been obtained. By placing the NA element on the descending phase of the electron beam oscillation, the cyclotron motion can be reduced, and thereby a ripple-free laminar beam can be produced, even for magneto-immersed guns positioned in a field with only a few hundred Gs and with cathode emissions exceeding 20 A/cm^2 [7].

5.2. Commissioning results from the non-adiabatic electron gun

During the commissioning campaign, a gun perveance of 0.87 μA/V$^{3/2}$ was measured, to be compared to a simulated design perveance of 0.73 μA/V$^{3/2}$. In spite of electron gun simulations indicating an enlargement in the radial oscillations with an increased deviation from the optimum electron current of 500 mA, an electron beam with a current ranging from a few mA to 420 mA could be transmitted with low losses (<150 μA). To be noted: the gun has constantly been operated in the thermionically limited regime due to inadequate emission, while the simulations were performed in the Child-Langmuir regime. The main part of the electron losses reaches the outer drift tube in front of the suppressor electrode, and it is presently believed to be caused by elastically reflected or back-scattered electrons from the collector region.

The charge breeding efficiency, i.e. the conversion efficiency of the injected 1+ ions to the desired charge state n+ is of paramount importance due to the scarcity of the radioactive ions, and was therefore carefully investigated. Measurements showed an efficiency matching the one of the original electron gun, in spite of operating at a lower electron beam current than foreseen. In fact, observations, for instance the efficiency-independence on the electron beam current, indicate that the emittance of the pulsed injected 1+ beam can be larger than the acceptance of the EBIS electron beam, while still capturing a large fraction of the injected ions. Tangible results include breeding efficiencies of approximately 30% for $^{23}Na^{9+}$, $^{39}K^{17+}$ and $^{205}Tl^{53+}$. These high values are attained thanks to closed atomic shell configurations (Na and K), and an ionization potential approaching the electron beam energy for the Tl case. Measurements of the global efficiency, i.e. the sum of all charge states leaving the EBIS, show values of 65-85%, independent on the confinement time. This points towards an excellent survival rate of the injected ions. Furthermore, the efficiency seems unrelated to the ion mass, suggesting that the magnetic field of the solenoid has no visible influence on the 1+ ion acceptance. See Fig. 4 for a selection of breeding efficiencies into a single charge state and Fig. 5 for global efficiency values.

Another important aspect is the vacuum level inside the trapping region, as with an increased electron beam current, higher outgassing rates may be expected from the gun and collector regions. Broad range A/Q-spectra have therefore been used to measure the rate of ions created from residual gas in the trapping region, and from those partial pressures of various gases have been calculated. A total pressure inside the trapping region of ~$3 \cdot 10^{-11}$ mbar has been found, with neon gas streaming from the Penning trap being the dominating constituent. At this pressure, <20% electron beam neutralization is reached after 300 ms breeding time.

The electron beam current density in the ion trapping was determined by recording the charge state evolution in REXEBIS for various elements and for varying beam currents, and using external ion injection as well as neutral gas-injection. The effective current density was thereafter deduced by fitting the measured charge state evolution with predictions of a simple theoretical model, including electron-impact ionization and radiative recombination, but excluding charge exchange and evaporative losses, as the two latter could be proven to have limited influence on the result. Apart from arriving at an effective current density of 265 A/cm^2 to 424 A/cm^2 for a 300 mA electron beam current, the investigation revealed strong dependence on the ion mass and breeding time, as well as difference between gas and externally injected ions.

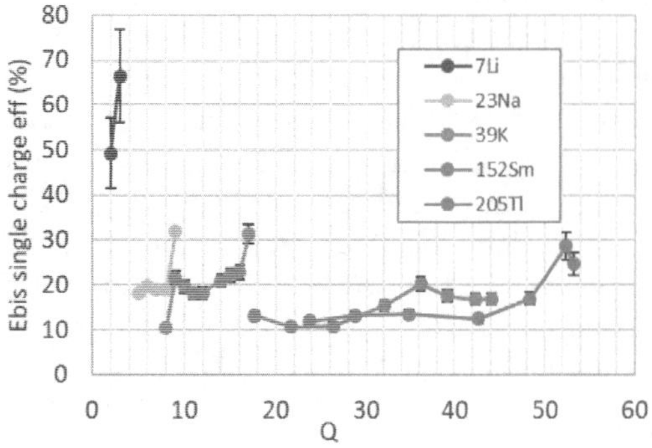

Figure 4. EBIS breeding efficiency into a single charge state with a 200 mA electron beam.

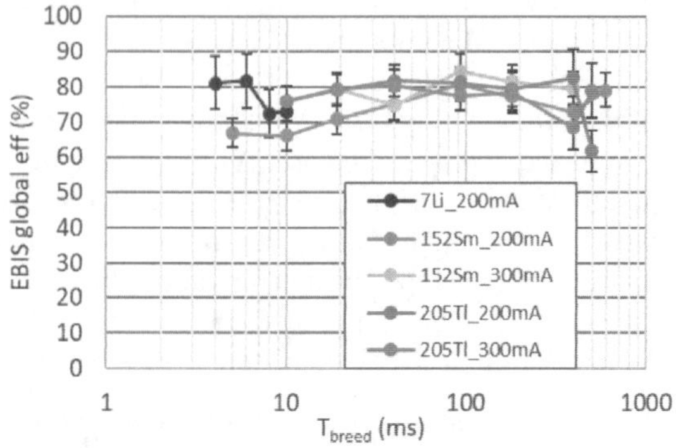

Figure 5. EBIS global efficiency for a selection of elements and different electron beam currents.

To shed light on the varying effective current densities, the axial energy spread of the trapped ions was probed and fitted to Maxwell-Boltzmann distributions. For all cases studied, the equivalent holding voltage $k_B T_{ion}/q$ did not exceed 40 V, and one can therefore assume strongly suppressed ion escape and evaporative cooling. The high and large variation in current density may therefore be explained by a non-uniform current density profile.

In order to sustain an operational electron current of more than 300 mA, we intend to test a novel scandium-doped dispenser-type cathode developed at Beijing University of Technology, from which a current density of 30 A/cm^2 at 950 °C, for more than 2000 h continuous operation, has been reported. For a detailed analysis of the commissioning results, see ref. [8].

6. Acknowledgments

The presented work is the result of common efforts within the Beams Department. The EBIS related work has been supported by the Research Foundation - Flanders, through the Big Science project Nr. G0C2813N. This work has also been sponsored by the Wolfgang Gentner Programme of the German Federal Ministry of Education and Research (grant no. 05E15CHA).

References

[1] Lettry J et al. 2021 Beam formation studies on the CERN IS03 H- Source, *Proc. ICIS 2021,* Vancouver, Canada

[2] Vnuchenko A, Lettry J, Mochalsky S, Wünderlich D, Fantz U, Revel A and Minea T 2021 Beam formation in the CERN negative ion source for the Linac4 accelerator, *Proc. ICIS 2021,* Vancouver, Canada

[3] Bellodi G, De Grandsaignes D'Hauterives H J, Noll D 2021 Linac4 source extraction re-design for higher current operation, *CERN-ACC-NOTE-2021-0009*

[4] Gamba D, Aguglia D, Lefort B, Machado C, Wenander F and Gebel R 2021 *AIP Conf. Proc.* 2373, 040005

[5] Bellodi G, Küchler D, Kövener T, Mahner E and Scrivens R 2021 Experiments to Improve the Performance of the GTS-LHC ECR Ion Source, *Proc. ICIS 2021*, Vancouver Canada

[6] Küchler D and Mihailescu M, 2021 Statistical analysis of operational data of the GTS-LHC ion source, *Proc. ICIS 2021*, Vancouver Canada

[7] Pikin A, Pahl H and Wenander F, 2020 *Phys. Rev. Acc. Beams* **23** 103502

[8] Pahl H, Bidault N, Khatri G, Pikin A and Wenander F, 2021 Non-adiabatic electron gun at REXEBIS: commissioning results and charge breeding investigations, submitted to *Phys. Rev Acc. Beams*

Journal of Physics: Conference Series 2244 (2022) 012011 doi:10.1088/1742-6596/2244/1/012011

Two-rod-antenna microwave injection system for production of circularly polarized microwaves in cylindrical ECRIS cavities

Alexandra Philipp, Maria Molodtsova, and Erik Ritter
DREEBIT GmbH, Southwallstrasse 5, 01900 Großröhrsdorf, Germany
E-mail: alexandra.philipp@dreebit.com

Abstract. In the design of Electron Cyclotron Resonance Ion Sources (ECRIS) three topics generally need to be addressed: (i) The feed of a microwave to generate a plasma, (ii) the creation of a magnetic field guiding the charged particles, and (iii) the design of an ion extraction system. In this report, we concentrate on improvements of aspect (i), a new microwave injection system introduced to the Dresden ECRIS 2.45M, a permanent-magnet 2.45 GHz ECRIS, applied, e.g., for the production of proton currents in medical particle therapy. The improvements include the replacement of its one-rod antenna system by two rod antennas on one cross-sectional circle of the cylindrical ECRIS cavity positioned under an angle of 90° towards each other. Feeding two microwave signals with a relative phase shift of 90°, a right-hand circularly polarized wave can be created inside the cavity to efficiently heat the electrons. Basic considerations, simulations, and results from first experiments are presented.

1. Introduction

Electron Cyclotron Resonance Ion Sources represent state of the art sources for low, intermediately, and also highly charged ions in many fields of basic and applied research. They supply ions for nuclear physics experiments [1, 2], materials science [3], and medical applications, e.g. for cancer treatment with proton therapy. [4].

Considering these developments, DREEBIT designed and built the Dresden ECRIS 2.45M, introduced in 2015 [2]. The prototype already delivered favorable results, yet, inquiries about providing higher proton currents with this permanent-magnet based, table-top sized ECRIS triggered an investigation to improve the energy transfer from the microwave generator to the electrons of the ECR plasma.

As known from literature [5], only the right-hand circularly polarized component of a microwave can be processed by an ECRIS through cyclotron absorption. In the Dresden ECRIS 2.45M, as in most other ECRIS setups, linearly polarized waves are injected, which consist of both left- and right-hand polarized parts, from which only the latter can directly be consumed.

For the now called Dresden ECRIS 2.45M-2A, a microwave injection system was designed comprising two rod antennas inside the ECRIS, set up in a way so that two 90° phase shifted microwave signals superimpose and form a circularly polarized wave propagating towards the plasma confinement region. Based on this concept, ideally, the full power can directly be absorbed [6].

2. Experimental Setup

The setup of the Dresden ECRIS 2.45M-2A, shown in Fig. 1, is based on the design of its predecessor [2]. The magnet system as well as the ion extraction lens ensemble remained unchanged. Therefore, the effect of the new microwave injection system can be evaluated comparing the ion source performance of both design stages.

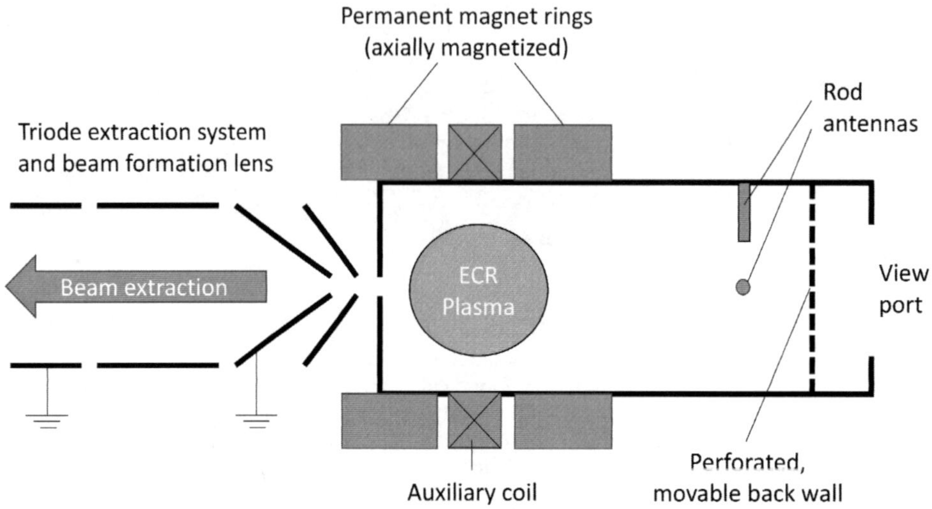

Figure 1. Schematic drawing of the Dresden ECRIS 2.45M-2A setup.

The two rod antennas emitting microwaves into the ECRIS cavity are positioned outside of the plasma confinement region. They are situated at a point where the antenna length plus distance to the back wall of the cylinder is approximately $\lambda/4$. In order to determine the optimum measures exactly, S-parameter calculations including sweeps over the antenna length and position were performed at Technical University of Dresden using the program ANSYS HFSS.

The antennas are connected to a solid state microwave generator via coaxial cables. The generator, which produces up to four 100 W output signals of the same frequency with full control of the phase shift variable from 0 to 360°, was custom made for DREEBIT GmbH by SAIREM SAS [7]. Additional time-resolved HF simulations carried out at SAIREM using the CST microwave studio software also validated the creation of circularly polarized waves inside this type of ion source, which helped to gain confidence to pursue the idea.

On the side of its extraction system, the ECRIS is mounted onto a test bench to analyze the provided beam current. The beamline includes a 90° bending magnet for ion species separation as well as einzel lenses and Faraday cups for beam transportation and current measurements.

3. H^+ production using one and two antennas

In order to be able to compare the ion source performance with the two-antenna microwave injection system to the previous version, the measurements were performed in several steps. First, a newly manufactured Dresden ECRIS 2.45M was commissioned until stable current output of 1.1 mA after three days of approximately eight hours operation time each. Then, the source was transformed to version 2.45M-2A by replacing only the back end of the vacuum vessel, while both the magnet and the extraction lens system stayed attached to the test bench.

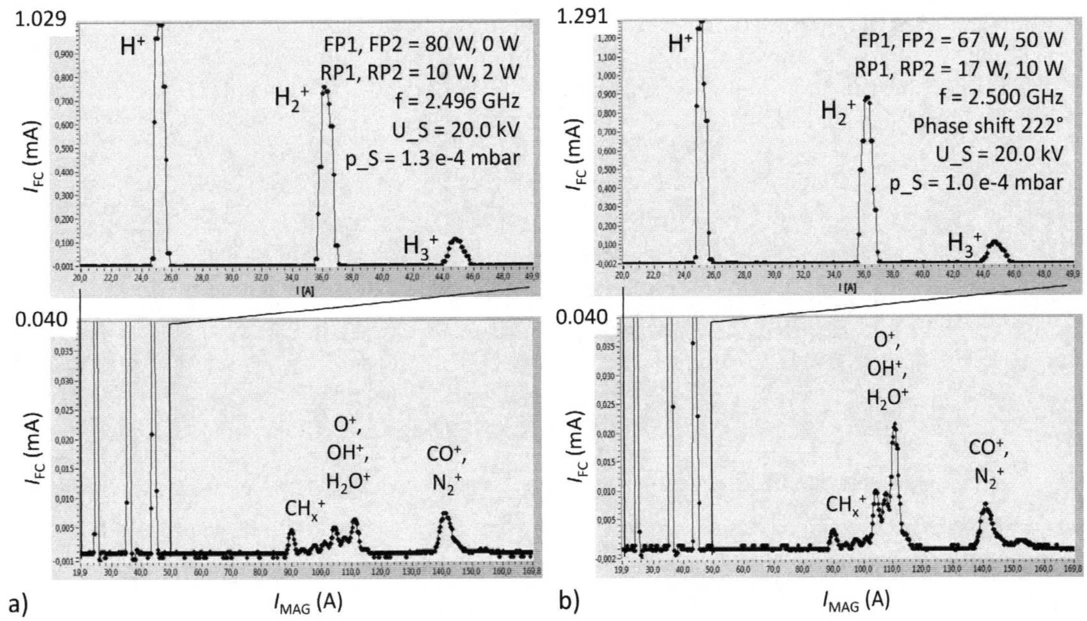

Figure 2. a) Spectrum of hydrogen ions (top) and residual gas components (bottom) extracted from the source operated in single antenna mode. b) Hydrogen (top) and residual gas spectrum (bottom) measured immediately after switching on the second antenna and several tuning iterations to find the optimum settings.

After the same initial commissioning period using only one of the now two installed antennas at similar optimized operational parameters, the current saturated resulting in spectra shown in Fig. 2 a). So far, no significant difference in achievable proton current and operational parameters was noticed. Variations of all operational parameters including pressure and microwave frequency were performed. Yet, forward microwave power values larger than 80 W did not improve the H^+ current, which reached a maximum of 1.1 mA.

Only when the second antenna was taken into operation, more power could be transferred to the plasma, resulting in spectra given in Fig. 2 b). Now, a maximum H^+ current of 1.3 mA could be reached, representing a 20 % improvement. An evaluation of the peaks of the residual gas components showed equal values for C^+ and N_2^+ but increased currents of O^+, OH^+, and H_2O^+. This indicates identical conditions in terms of vacuum contamination and leak rate but differences in the interaction of the plasma with surfaces containing water residues, which calls for further investigation.

4. Optimization of magnetic field distribution in relation to two-antenna microwave feed

Using the auxiliary coil installed between the permanent ring magnets, the ECRIS's B field distribution was altered in order to further enhance the interaction of the microwaves with the plasma. Tests in both coil current directions revealed improvements if the B field minimum on-axis in the center between the magnetic rings was lowered as presented in Fig. 4 a).

By this modification, the bubble-shaped ECR B field isosurface is stretched along the source axis and the pathways of electrons change. If the B field minimum along z stays above 0 mT, electrons following the field lines can pass through the center, resulting in mainly one on-axis magnetic bottle where the charged particles are stored and extracted from. If the B field direction changes along the source axis, the volume is split into three different confinement areas of interest, two approximately cone-shaped ones with a similar direction as the original on-axis bottle and one ring shaped confinement area around the B field center with an opposite direction of field lines.

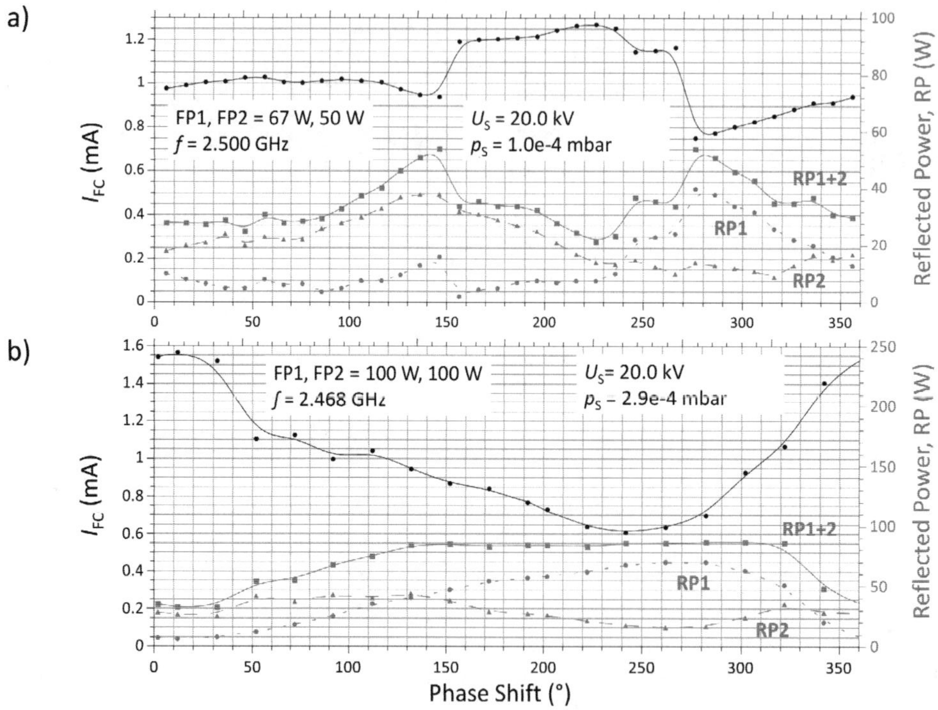

Figure 3. H^+ current and reflected power at antenna 1, 2 and 1+2 for operation with a) the original B field distribution and b) the modified B field using the auxiliary coil versus the phase shift of the two microwave signals applied to the rod antennas, measured at the output port of the microwave generator.

With the two-antenna microwave injection system, the direction of polarization can be varied to specifically heat certain areas within the respective B field distribution. This can be seen in Fig. 3 showing the effects of phase shift variation on the extracted H^+ current and the reflected microwave power. The phase shift values given in this graph were measured at the output ports of the microwave generator and may be different at the area of actual microwave absorption, e.g., due to manufacturing tolerances. Therefore, absolute numbers can be used as approximate guidance values only, while relative phase shift values can be used with good confidence.

In case a), the original B field distribution, a phase shift around 200° stands out with improved source performance coinciding with a minimum in total reflected power RP1+2. For comparison, in this experimental configuration, a phase shift of 270° would preferably heat the main on-axis magnetic bottle induced by the magnetization of the permanent magnet rings.

In case b), the opposite direction of polarization is clearly favored. A 90° phase shift would preferably heat the ring-shaped magnetic confinement area with an opposite sign of field lines, which expands more towards the center in radial direction if the B field minimum is lowered. After optimization of all the now controllable parameters, an H^+ current of 1.75 mA was reached resulting in a 70 % improvement compared to the previous ion source version, see Fig. 4 b).

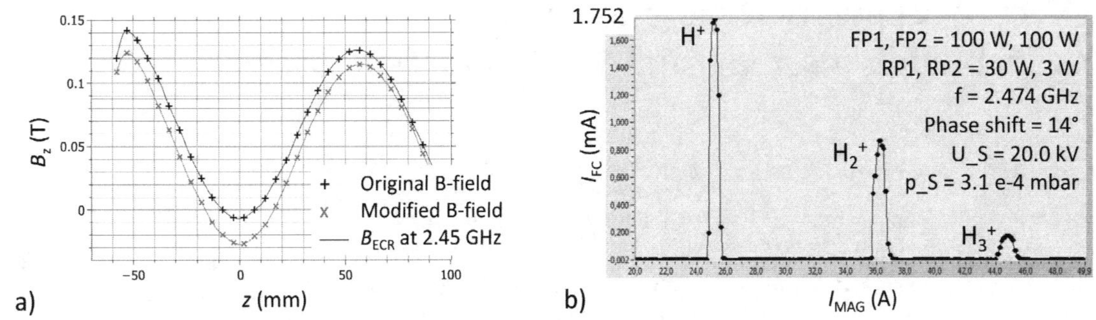

Figure 4. a) Hall probe scans of the longitudinal B field along the central source axis. b) Spectrum of hydrogen ions after optimization of the B field distribution and phase adjustment of the two-antenna microwave feed.

5. Summary and outlook

Having reached a 70 % improvement in H^+ current extracted from the Dresden ECRIS 2.45M-2A prototype compared to its predecessor, our investigations will continue with further optimization of the magnetic field distribution in relation to the now available control of the direction of the circular polarization of microwaves fed to the source. In the future, we plan to combine the information gained from ion extraction measurements with recording optical images of the radial plasma distribution inside the ECRIS cavity and using methods of optical spectroscopy for a better understanding of plasma confinement and particle interaction processes leading to further improvements of the source performance.

Acknowledgements

The authors thank Louis Latrasse at SAIREM SAS as well as Evgeniy Zakutin and Martin Laabs at Technical University of Dresden for consultations and contributed simulations.

References

[1] Pardo R 2019 *Rev. Sci. Instrum.* **90** 123312

[2] Kreller M, Baumgart C, Zschornack G, Czerski K, Kaczmarski M, Targosz-Sleczka N, Huke A, Guprecht G, Weißbach D 2016 *Proc. of International Workshop on ECR Ion Sources (ECRIS'16), Busan, Korea, August 28 - September 1, 2016* pp 59-63

[3] Jiang S, Khymyn R, Chung S, Le T Q, Herrera Diez L, Houshang A, Zahedinejad M, Ravelosona D, and Åkerman J 2020 *Appl. Phys. Lett.* **116** 072403

[4] Degiovanni A et al. 2018 *Proc. 9th International Particle Accelerator Conference (IPAC'18), Vancouver, BC, Canada, April 29 - May 4, 2018* pp 425-428

[5] Geller R 1996 *Electron Cyclotron Resonance Ion Sources and ECR Plasmas* (Bristol and Philadelphia: IoP Publishing) p 20

[6] Philipp A 2021 *ECR-Ionenquelle und Verfahren zum Betreiben einer ECR-Ionenquelle* German patent no. 102019111908 (filed May 8, 2019, published August 12, 2021), US patent no. 11094510, EP and CN pending

[7] For more information please contact welcome@sairem.com with reference to the Customized Prototype GMX solid state microwave generator developed for DREEBIT GmbH, 2018

Effect of repetitive irradiation on ion saturation current of laser ablation plasma using 1 kHz repetitive laser

K Igarashi[1], M Iwasa[1], K Takahashi[1], T Sasaki[1], T Kikuchi[1,2]

[1] Department of Electrical Engineering, Nagaoka University of Technology, Nagaoka, Niigata 940-2188, Japan
[2] Department of Nuclear System Safety Engineering, Nagaoka University of Technology, Nagaoka, Niigata 940-2188, Japan

kigarashi@stn.nagaokaut.ac.jp

Abstract. We demonstrated 100 shots of burst laser irradiation with an Al target using a 1 kHz repetitive pulse laser. By using the laser, we generated ablation plasmas with a repetition rate of 1 kHz. The time-evolution of ion saturation current density in the generated ablation plasma was measured using the orifice probe. The result shows that the waveforms on 100 spikes were observed. The ion saturation current density decreased rapidly from the first to the third shot and reached a minimum value at the fourth shot. After the fourth shot, it reached 1 μA/mm^2 at the 53rd shot and settled at 0.5 μA/mm^2. We extracted the time-evolution of ion saturation current densities of the first, 10th, 50th and 100th shots. The drift velocities were estimated by Time of Flight from the time-evolution of current. The result shows that the drift velocity of the ablation plasma increases with the increase in the number of shots.

1. Introduction

A laser ion source can supply heavy ions with high current density because the ion beam is extracted from the plasma expanding from the solid density target [1-9]. Therefore, high energy density physics, heavy-ion inertial fusion [10, 11], particle beam therapy using high current carbon beam [12], and accelerator-driven neutron generation systems [13] are studied as applications for the laser ion source. High power and high repetition rate operation is especially required for accelerator-driven systems and spot scanning in particle beam therapy to reduce the operation time [14].

For high power operation of the ion source, a high repetitive pulse laser is required. Pulse lasers in industrial applications are already capable of operating at a high repetition frequency [15]. A laser with a repetition rate of 100 kHz has also been demonstrated in the semiconductor lithography to generate extreme ultraviolet (EUV) light sources [16]. However, the method of target insertion in the EUV light sources is different. The target structure of conventional laser ion sources is used for a metallic plate. Therefore, for the high repetition frequency operation of the laser ion source, it is necessary to study the laser target and its peripheral technologies differently from the development of EUV light sources.

However, in the high repetitive laser ion source, problems are predicted to occur in the generation of the ablation plasma. Kashiwagi *et al.* have reported that the interaction between plasmas occurs when the pulse interval is shortened using two lasers. As a result, the total charge in the ablation plasma decreased [17]. An absorption due to plasma-laser interaction occurs in high-repetitive laser irradiation. The laser is irradiated the residual plasma generated by previous irradiation. Thus, the laser

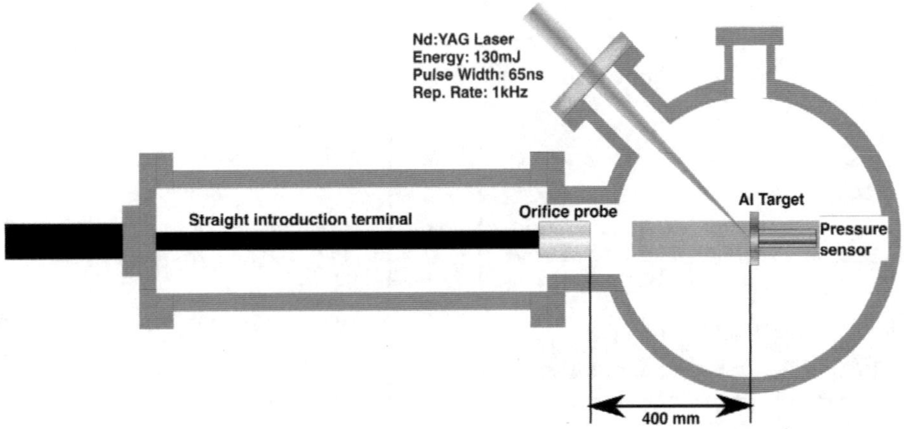

Figure 1. Experimental setup for ablation plasma generation and ion saturation current density measurement.

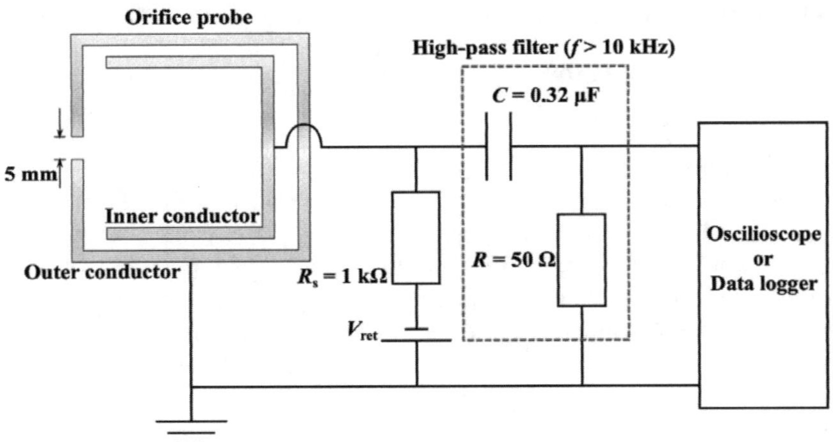

Figure 2. The structure of the orifice probe.

energy is absorbed with the plasma. The target surface condition also affects the ablation plasma in high-repetitive laser irradiation. Solid targets leave irradiation spots after the laser irradiation. The target surface condition can be kept a rolling-up tape target [18]. However, the target surface condition does not keep in high-repetitive laser irradiation. Consequently, the change of the target surface condition can affect the ablation plasma. Therefore, the effect of repeated laser irradiation on the ion source needs to be investigated experimentally.

To understand the repetitive laser ion source, we proposed to observe the ablation plasma generated by a high repetitive pulse laser. The ablation plasma was generated with the 1 kHz pulsed laser, and the effect of repetitive pulse irradiation was observed. The time-evolution of ion saturation current density in the plasma produced at high repetition rates was observed with an orifice probe [19]. Based on the observation results, we discuss the plasma dynamics of the repetitive ion source, especially the ion saturation current density and the drift velocity.

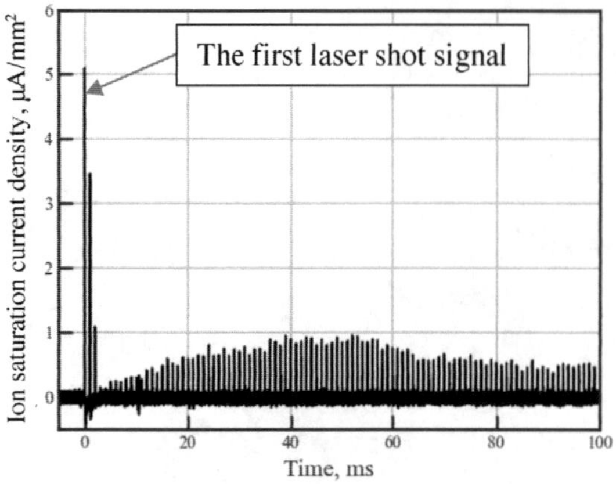

Figure 3. Time-evolution of ion saturation current density observed in the orifice probe. Time 0 seconds corresponds to the first shot of the laser. The repetition frequency was 1 kHz, and the total number of shots was 100.

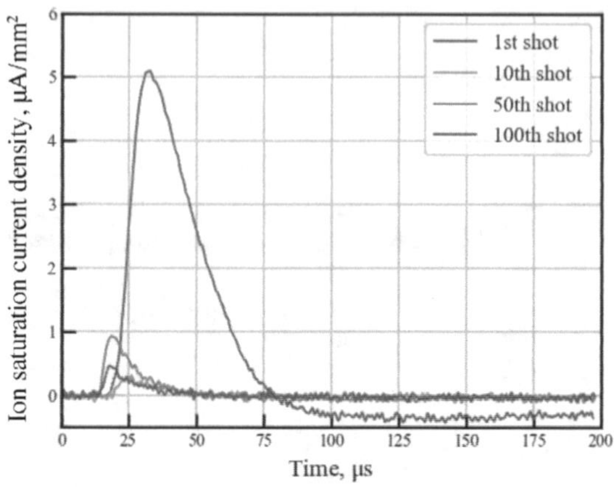

Figure 4. Ion saturation current density waveforms at first, 10th, 50th, and 100th shots, respectively. The laser irradiation was started from 0 s.

2. Experimental setup

In this study, we used a high repetitive pulse laser (Powerlase, AO16-60). The experimental setup to generate an ablation plasma is shown in Figure 1. The pulse laser consists of Nd:YAG laser, which parameters are 130 mJ of laser energy, 65 ns of pulse width, and 1064 nm of wavelength. The laser beam is focused with a lens having 700 mm of focal length. The laser intensity on the target was estimated to be 3.1×10^8 W/cm^2 evaluated from the laser spot size. A target used was an aluminum plate. The target size is 19 mm in height, 28 mm in width, and 1 mm in thickness.

The structure of the orifice probe is shown in Figure 2. The ablation plasmas were immediately generated when the laser was irradiated at Al target in a chamber. The chamber pressure was $1.4 \sim 2.6 \times 10^{-3}$ Pa during the experiment. The ions in the ablation plasma reached the electrode in the orifice probe, which was placed at 400 mm from the Al target introduced by the linear motion feedthrough. The orifice probe measured an ion saturation current density in the ablation plasma. A voltage of -50 V was applied to the inner conductor of the orifice probe, which is responsible for driving back electrons that enter the orifice probe. The voltage value is reasonable because an electron temperature of an ablation plasma is about 1 eV [20, 21]. The signal of ion saturation current density was received with the orifice probe and was transmitted to the atmosphere through a cable with BNC connectors. A drift velocity v_d of the plasma was estimated using the Time of Flight (ToF) method from the time-evolution of ion saturation current density. The drift velocity was calculated by the distance between the Al target and the orifice probe divided by the average flight time of ions.

The ion saturation current density in the ablation plasma was observed and recorded by a data logger (YOKOGAWA DL950, sampling rate: 200 MS/s) through a high-pass filter ($f >$ 10 kHz) and a termination resistor. The measurement system was suitable to measure typical timescales of laser duration and the ablation plasma.

3. Result and Discussion

In the experiment, we demonstrated 100 shots of burst laser irradiation. The repetition frequency of the laser was 1 kHz. The time-evolution of the ion saturation current density observed in the orifice probe is shown in Figure 3. It showed that the waveforms on 100 spikes were observed. The ion saturation current density signals corresponded to the laser shot signals. It means that the ion production can be repeated using a 1 kHz pulsed laser. The results also showed that the ion saturation current density decreased rapidly in the first to the third shot and reached a minimum value in the fourth shot. After the 54th shot, the peak ion saturation current density decreased and settled down to about 0.5 μA/mm^2. It means that the ion saturation current density extracted from the ablation plasma becomes unstable by repetitive laser irradiation to the same spot. In order to achieve stable operation of the repetitive ion source, it is necessary to introduce a mechanism to move the target in accordance with the repetition frequency.

Figure 4 shows waveforms of the ion saturation current density at each shot. The laser irradiation timing corresponds to 0 s. It indicated that the current density of the first shot had the highest peak value. On the other hand, that of the 10th, 50th, 100th shots have 8 ~ 20 % peak values of the first shot. Furthermore, the drift velocities were 9.81 km/s of the first shot, 13.3 km/s of the 10th shot, 17.2 km/s of the 50th shot, 15.4 km/s of the 100th shot. It showed that the drift velocity of the ablation plasma generated by the repetitive laser irradiation gradually increased. In the first shot, the ions were observed at 50 ~ 75 μs from the beginning of the laser irradiation. On the other hand, such ions signals were not monitored in the 10th, 50th, and 100th shots.

4. Conclusion

In this study, we generated the ablation plasma using the 1 kHz repetitive pulse laser for the laser ion source. The ion saturation current density was observed in all shots. The observed ion saturation current density decreased suddenly from the first to the third shot and reached a minimum value at the fourth shot. After the fourth shot, it increased up to 1 μA/mm^2 at the 53rd shot and settled at 0.5 μA/mm^2 of the ion density. The ion saturation current density waveform was confirmed by the ToF method. The result showed that the drift velocity of the ablation plasma gradually increased. In the first shot, the ions were observed at 50 ~ 75 μs from the beginning of the laser irradiation. On the other hand, such ions signals were not monitored in the 10th, 50th, and 100th shots. It is found that the ion beam can be extracted as the 1 kHz repetitive ion source.

References

[1] B. Yu. Sharkov, A. V. Shumshurov, V. P. Dubenkow, *et al.*, *Rev. Sci. Instrum.* **63**, 2841 (1992)
[2] L. Torrisi, S. Cavallaro, M. Cutroneo, *et al.*, *Rev. Sci. Instrum.* **83**, 02B310 (2012)
[3] M. Okamura, *Journal of Instrumentation*, BNL-112613-2016-JA(2016)
[4] H. Y. Zhao, J. J. Zhang, Q. Y. Jin *et al.*, *Rev. Sci. Instrum.* **87**, 02A917 (2016)
[5] R. Pardo, *Rev. Sci. Instrum.* **90**, 123312 (2019)
[6] K. Takahashi, M. Kuzumoto, Y. Matsumoto, T. Sasaki, T. Kikuchi *et al.*, *Rev. Sci. Instrum.* **91**, 033310 (2020)
[7] H. Kashiwagi, K. Yamada, and S. Kurashima, *Nucl. Instrum. Methods Phys. Res. B* **406**, pp. 256-259(2017)
[8] T. Karino, *Rev. Sci. Instrum.* **91**, 033316 (2020)
[9] S. Kondrashev, E. Beebe, T. Kanesue *et al.*, *Rev. Sci. Instrum.* **91**, 023320 (2020)
[10] M. Okamura, M. Sekine, K .Takahashi, K. Kondo, and T. Kanesue, *Nucl. Instrum. Methods Phys. Res. A* **733**, 97(2014)
[11] K. Horioka, *Matter Radiat. Extremes* **3**, 12(2018)
[12] T. Sako, A. Yamaguchi, K. Sato *et al.*, *Rev. Sci. Instrum.* **87**, 02C109 (2016).
[13] T. Kanesue, M. Okamura, and Y. Saito, in *Proceeding of IPAC2017*(MOPIC053, Copenhagen, Denmark), pp. 638-640
[14] S. Peggs, D. Barton, J. Beebe-Wang, J. Cardona *et al.*, in *Proceedings of EPAC2002*(Paris, France), pp. 2754-2756
[15] J. Schille, L. Schneider, P. Lickschat, U. Loeschner, R. Ebert, H. Exner, *et al.*, *J. Laser Appl.* **27**, S28007(2015)
[16] T. Higashiguchi, M. Hamada, and S. Kubodera, *Rev. Sci. Instrum.* **78**, 036106 (2007)

[17] H. Kashiwagi, and K. Yamada, *Rev. Sci. Instrum.* **91**, 033305(2020)
[18] T. Ono, Y. Uchida, H. Horisawa, and I. Funaki, Vacuum **83**, 213(2009)
[19] S. Shimamoto, T. Kasuya, and Y. Kimura, *Rev. Sci. Instrum.* **81**, 02A704(2010)
[20] J. M. Hendron, C. M. O. Mahony, T. Morrow, and W. G. Graham, *Journal of Applied Physics* **81**, 2131(1997)
[21] I. Weaver, G. W. Martin, W. G. Graham, T. Morrow, and C. L. S. Lewis, *Rev. Sci. Instrum.* **70**, 1801(1999)

New Commissioning Results of the MIST-1 Multicusp Ion Source

Daniel Winklehner, Janet Conrad, Joseph Smolsky, Loyd Waites, Philip Weigel

Massachusetts Institute of Technology, Cambridge, MA, USA

E-mail: winklehn@mit.edu

Abstract. For the sterile neutrino experiment IsoDAR (Isotope Decay-At-Rest), we have developed a compact particle accelerator system delivering a 10 mA, continuous wave (cw) proton beam at 60 MeV to a neutrino production target. The accelerator comprises a compact isochronous cyclotron, an RFQ embedded in the cyclotron yoke, and an ion source. To reduce space charge effects during injection and acceleration, we are accelerating H_2^+ instead of protons. To produce the needed cw H_2^+ beam current of 10 mA (nominal) at the required purity and quality, we have built a new filament driven, multicusp ion source (MIST-1). Here we report commissioning results for long-time running at reduced power, demonstrating the feasibility of the design. Highlights include an H_2^+ beam current density of $\approx$ 10 mA/cm^2, $\approx$ 80 % H_2^+ fraction, and extrapolated emittances of 0.05 π-mm-mrad (RMS, normalized) after extraction. We also present high fidelity simulations that are in good qualitative and quantitative agreement with emittance measurements in our test beam line.

1. Introduction

IsoDAR (Isotope Decay-At-Rest) [1–3] is a proposed experiment to search for so-called *sterile neutrinos.* These hypothetical particles could provide an explanation to anomalies seen in neutrino oscillation experiments. In order to be decisive within 5 years, we require a cw beam current of 10 mA of 60 MeV protons on target at $>$ 80 % duty factor. Furthermore, to reduce background in the detector, such an experiment has to be set up in an underground facility, placing stringent size constraints on the accelerator system. We have determined that compact cyclotrons are the best choice [4]. However, commercial cyclotrons to date only deliver about 1 mA of cw beam currents [5, 6]. The record holder for cw proton beams from cyclotrons is PSI Injector II with 2.7 mA [7].

Three innovative design choices make this increase in beam current from a compact cyclotron possible: 1. Accelerating H_2^+, 2. Injecting through a RadioFrequency Quadrupole (RFQ) embedded in the cyclotron yoke, 3. utilizing a beam dynamics effect that can occur in isochronous cyclotrons when accelerating high beam currents, the so-called *vortex effect* [8]. H_2^+ can be stripped into protons after extraction from the cyclotron, yielding twice the electrical beam current. Thus we need only accelerate 5 mA of H_2^+, alleviating space charge issues.

In these proceedings, we report on the Multicusp Ion Source at MIT (MIST-1), which we built to produce the H_2^+ ion beam for this experiment. The design is based on a multicusp ion source at LBNL [9, 10]. We chose this type to accommodate the following nominal ion source parameters necessary for RFQ direct injection: 1. H_2^+ fraction $>$ 80 %, 2. 10 mA of H_2^+, 3.

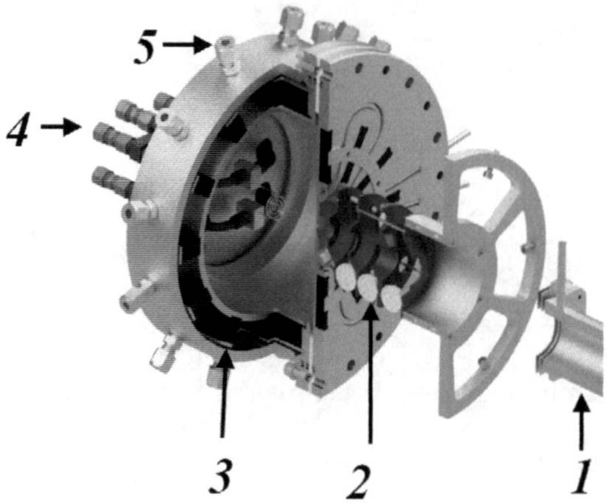

Figure 1. MIST-1 ion source with extraction system and faraday cup. 1. Faraday cup, 2. Extraction System, 3. Permanent magnets (Sm_2Co_{17}), 4. Water cooled Filament feedthroughs, 5. Water cooling fittings. From [11].

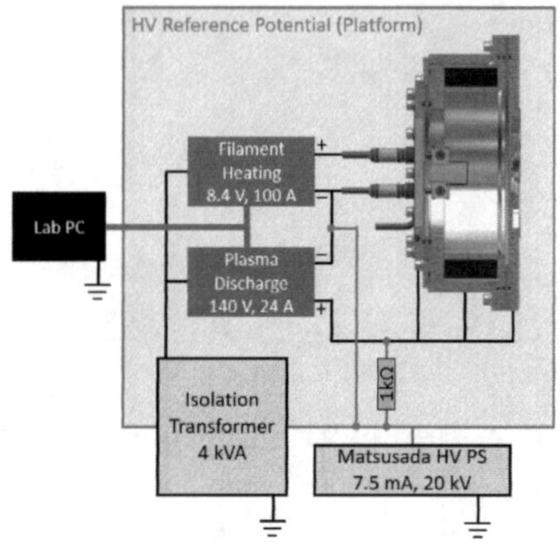

Figure 2. The wiring schematic of the ion source with the source high voltage reference potential in red. The control PC commmunicates with the power supplies through a USB-optical extension cable. PS maximum values are indicated. From [11].

RMS, normalized emittance < 0.1 π-mm-mrad, 4. Continuous DC operation. We also desire a low energy spread of a few eV.

2. The MIST-1 Ion Source

The mechanical and electrical design of the MIST-1 ion source have been described in detail elsewhere [11–14]. Here we summarize the main design parameters. The mechanical design is shown in FIG. 1 and the electrical wiring diagram in FIG. 2.

MIST-1 is a multicusp ion source, where a magnetic field created by 12 Sm_2Co_{17} magnets creates a radial confinement field supported by 12 front plate magnets and four stripes of back plate magnets. We keep the magnetic field in the extraction region below 5 Gauss to let ions escape. We feed gas into the source through a Mass Flow Controller (MFC) with a full range of 5 SCCM. A tungsten filament is heated to release electrons through thermionic emission. These electrons are accelerated away from the filament by a modest discharge voltage (40-150 V) appplied between the filament and the source chamber. Interaction of the electrons with neutral H_2 molecules leads to electron-impact ionization and the production of secondary electrons and ions, forming a discharge plasma. Typical discharge currents ($I_{dis.}$) in the ion source are currently between 4 and 10 A. However, the power supply can support up to 24 A and in future studies we will push to higher currents. The filament feedthroughs, the back plate, front plate and source body are all water-cooled with de-ionized water at a combined water flow of 2 l/min.

The plasma chamber length is short compared to proton sources to prevent collisions of the H_2^+ ion with H_2 gas molecules, which can produce H_3^+ via $H_2^+ + H_2 \rightarrow H_3^+ + H$. Furthermore, interactions with electrons can dissociate H_2^+ before it leaves the source. Bringing the filament close to the extraction aperture (≈ 25 mm distance for MIST-1), increases the H_2^+ fraction [9].

The ion source, MFC, and filament heating- and discharge power supplies are located on a high voltage (HV) platform that is raised to the source HV reference potential (max. 20 kV). The ion beam is shaped and guided by a pentode extraction system containing an Einzel lens element with maximum voltage of 20 kV and a puller electrode at -2 kV, providing protection

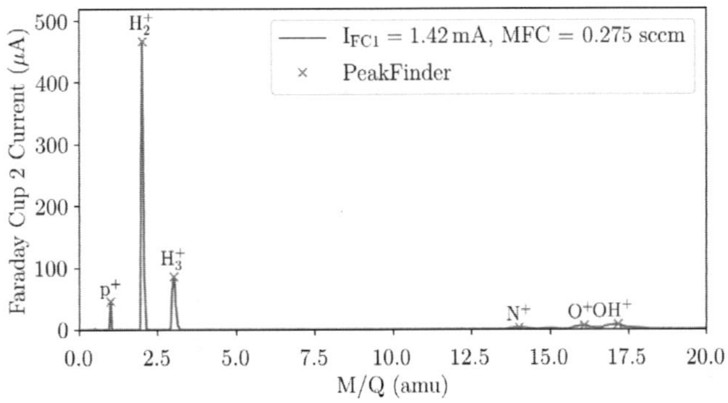

Figure 3. Mass spectrum indicating 76 % H_2^+ in the extracted beam. From [11].

from back-streaming electrons.

3. Measurements

We perform our measurements with a short diagnostic beam line that consists of a faraday cup directly behind the extraction system (FC1) for total beam current measurements, a dipole magnet combined with slits and a second faraday cup (FC2) for mass separation and a set of two Allison emittance scanners [15] after the dipole magnet to measure each species' horizontal and vertical emittance.

3.1. Summary of Recent Results

In [11] we reported our first commissioning results from a 4 mm diameter aperture. The highlights were:

- **Highest total current density:** 41 mA/cm^2 with 31 % H_2^+.
- **Highest H_2^+ fraction:** 76 % H_2^+, total current density: 11.4 mA/cm^2 (see FIG. 3),
- **Highest extracted H_2^+ current density with H_2^+ dominant:** 10.4 mA/cm^2 of H_2^+ at 62 % H_2^+ fraction.

During systematic variations of H_2 mass flow, discharge voltage and discharge current, we found no strong dependence of species fractions on the discharge voltage between 40 V and 150 V when holding the discharge current fixed at 4 A, and no strong dependence on the discharge current between 2 A and 8 A while holding the discharge voltage fixed at 140 V. In both cases the mass flow was 0.375 SCCM. Varying the mass flow showed the expected shift toward lower H_2^+ fraction and higher H_3^+ fraction with increased H_2 density in the source (see also FIG. 4).

3.2. New Measurements

The reason we kept the discharge current fixed in the previous measurements is that we are using a PID loop to stabilize the discharge by making small modifications to the filament heating current. In order to decouple the effects of increasing discharge current due to additional secondary electrons produced by having higher H_2 gas density in the source and the effect on species fraction, we performed a measurement where we deactivated the PID loop, increased the mass flow, let the discharge current rise to the source steady-state value and then re-activated the PID controller to stabilize at the new $I_{dis.}$. This measurement (Series 1), combined with data for a series where $I_{dis.}$ was kept fixed at 4 A (Series 2) and a third series, where $I_{dis.}$ was reduced to keep the total extracted current fixed at 1.25 mA (Series 3) is shown in FIG. 4. In Series 1,

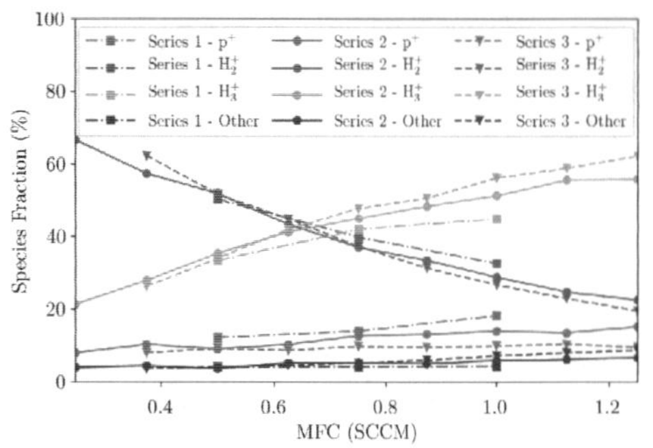

Figure 4. Three series varying gas flow, using different treatments of $I_{dis.}$. See text for a detailed explanation.

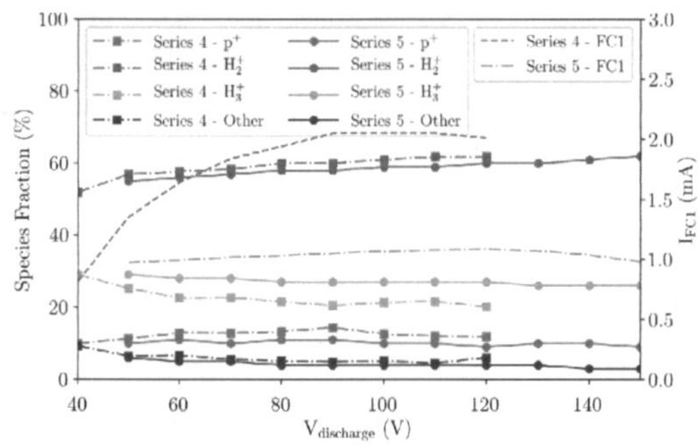

Figure 5. Two series varying discharge voltage, using different treatments of $I_{dis.}$. See text for a detailed explanation.

both total current and $I_{dis.}$ increased. In Series 2, the total current increased, but less quickly. In Series 3, we kept the total current constant by decreasing $I_{dis.}$. Within our measurement uncertainties, Series 2 and Series 3 show the same behaviour. Combined with Series 1, observe a trend of favoring H_2^+ production with higher $I_{dis.}$, indicating that another process beyond the neutral H_2 density plays a role in the species fractions, albeit a subtle one.

In a second measurement (FIG. 5 - Series 4), we varied the discharge voltage, and let $I_{dis.}$ find a steady-state using the same method as before. In Series 5, we used the PID loop to keep $I_{dis.}$ constant at 4 A. In Series 4, we observe an increase in total extracted current as $I_{dis.}$ increases, in Series 5, we observe only small variations in total current. It should be noted that after raising the discharge voltage to 60 V, the filament heating power supply reached its voltage limit of 8.4 V and stayed there for the remaining data points.

At this point, we cannot make strong statements about the subtle dependency of species fraction on discharge current and even subtler dependency on discharge voltage. The electron energy distribution (EED) in the plasma might offer an explanation, However, further investigations using plasma diagnostics like optical emission spectroscopy [16] or a Langmuir probe are necessary.

4. Simulations

To simulate the extraction from the ion source, we used code IBSimu [17], to simulate beam transport through the diagnostic beam line, we used Warp [18]. Both codes are well-established and well-benchmarked. In [11], we showed that we could accurately reproduce the beam dynamics of our beam line and match both species ratios as well as emittance scans after the dipole magnet with good qualitative and quantitative agreement (see for example FIG. 6). This let us extrapolate the emittance after the ion source from measurements to be ≈ 0.05 π-mm-mrad normalized, RMS.

5. Conclusion

We have presented commissioning results of the MIST-1 ion source, designed and built to deliver a high intensity H_2^+ beam (nominally 10 mA DC beams at 80 % H_2^+ fraction and < 0.1 π-mm-mrad normalized, RMS emittance). So far at the required purity, we could demonstrate a DC beam current of 1 mA of H_2^+ from the source at the required emittance from a 4 mm aperture. Extrapolating from these results to an 8 mm aperture, we are about a factor 2.5 away from the design goal. Further studies optimizing filament size, shape and position as well as magnet type

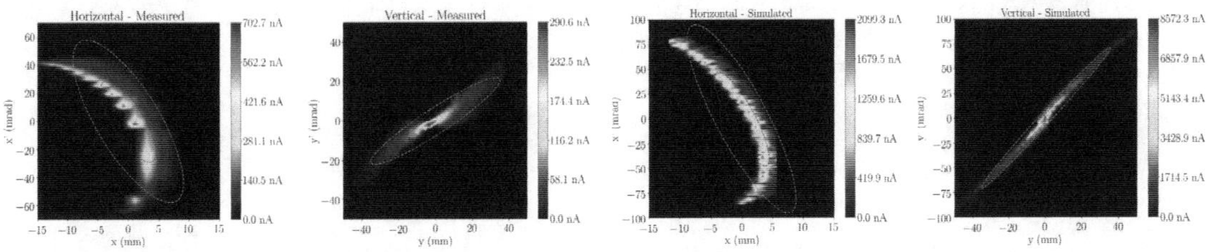

Figure 6. First emittance scans of a mass-separated H_2^+ beam. Measurement vs. simulation. From [11].

are ongoing. An upgrade is planned, where we replace front plate and back plate with materials that have higher heat transfer coefficients, letting us go to higher discharge power.

6. Acknowledgements

This work was supported by NSF grants PHY-1505858 and PHY-1626069, as well as the Heising-Simons Foundation. Winklehner was supported by funding from the Bose Foundation.

References

[1] Bungau A, Adelmann A, Alonso J R, Barletta W, Barlow R, Bartoszek L, Calabretta L, Calanna A, Campo D, Conrad J M, Djurcic Z, Kamyshkov Y, Shaevitz M H, Shimizu I, Smidt T, Spitz J, Wascko M, Winslow L A and Yang J J 2012 *Physical Review Letters* **109** 141802 publisher: American Physical Society URL `https://link.aps.org/doi/10.1103/PhysRevLett.109.141802`

[2] Winklehner D, Bahng J, Calabretta L, Calanna A, Chakrabarti A, Conrad J, D'Agostino G, Dechoudhury S, Naik V, Waites L and Weigel P 2018 *Nuclear Instruments and Methods in Physics Research Section A: Accelerators, Spectrometers, Detectors and Associated Equipment* **907** 231–243 ISSN 0168-9002 URL `http://www.sciencedirect.com/science/article/pii/S0168900218308714`

[3] Abs M *et al.* 2015 *arXiv:1511.05130 [hep-ex, physics:physics]* URL `http://arxiv.org/abs/1511.05130`

[4] Adelmann A *et al.* 2012 *arXiv* (*Preprint* 1210.4454) URL `http://arxiv.org/abs/1210.4454`

[5] 2013 Cyclone 70 (IBA – Ion Beam Applications S.A.) `http://www.iba-cyclotron-solutions.com/products-cyclo/cyclone-70`

[6] 2019 BEST 70p Cyclotron (Best Cyclotron Systems, Inc.) `http://www.bestcyclotron.com/product_70p.html`

[7] Grillenberger J, Humbel J M, Mezger A C, Seidel M and Tron W 2013 *Proceedings of Cyclotrons2013, Vancouver BC* p 3

[8] Baumgarten C 2011 *Phys. Rev. ST Accel. Beams* **14**(11) 114201 URL `https://link.aps.org/doi/10.1103/PhysRevSTAB.14.114201`

[9] Ehlers K and Leung K 1983 *Review of Scientific Instruments* **54** 677–680

[10] Ehlers K W and Leung K N 1979 *Review of Scientific Instruments* **50** 1353–1361 ISSN 0034-6748 URL `https://aip.scitation.org/doi/10.1063/1.1135725`

[11] Winklehner D, Conrad J M, Smolsky J and Waites L 2021 *"Under revision in RSI"* (*Preprint* 2008.12292) URL `http://arxiv.org/abs/2008.12292`

[12] Axani S, Winklehner D, Alonso J and Conrad J M 2016 *Review of Scientific Instruments* **87** 02B704

[13] Axani S, Winklehner D, Conrad J and IsoDAR/DAEdALUS Collaboration 2017 *Proceedings of 38th International Conference on High Energy Physics PoS(ICHEP2016)* (Sissa Medialab) p 484 URL `https://pos.sissa.it/282/484`

[14] Winklehner D, Axani S, Bedard P, Conrad J, Corona J, Hartwell F, Smolsky J, Tripathee A, Waites L, Weigel P, Wester T and Yampolskaya M 2017 *Proceedings of the 17th International Conference on Ion Sources*

[15] Allison P W, Sherman J D and Holtkamp D B 1983 *Nuclear Science, IEEE Transactions on* **30** 2204–2206 ISSN 0018-9499

[16] Komppula J and Tarvainen O 2015 *Plasma Sources Science and Technology* **24** 045008 ISSN 0963-0252 publisher: IOP Publishing URL `https://doi.org/10.1088/0963-0252/24/4/045008`

[17] Kalvas T, Tarvainen O, Ropponen T, Steczkiewicz O, Ärje J and Clark H 2010 *Review of Scientific Instruments* **81** 02B703 ISSN 0034-6748 publisher: American Institute of Physics URL https://aip.scitation.org/doi/full/10.1063/1.3258608
[18] 2021 Warp http://warp.lbl.gov/

Status and prospects of the 60 GHz SEISM ion source

Thomas Andre[1], Julien Angot[1], Maud Baylac[1], Pierre Olivier Dumont[1], Thierry Lamy[1], Patrick Sole[1], Thomas Thuillier[1], Francois Debray[2], Ivan Izotov[3], Vadim Skalyga[3]

[1] Université Grenoble-Alpes, CNRS-IN2P3, LPSC, 38000 Grenoble, France
[2] LNCMI, CNRS-UGA-UPS-INSA, 25, avenue des Martyrs, 38042 Grenoble, France
[3] Institute of Applied Physics, RAS, 46 Ulyanova St., 603950 Nizhny Novgorod, Russian Federation

E-mail: `thomas.andre@lpsc.in2p3.fr`

Abstract. SEISM is a unique ECR ion source operating at a frequency of 60 GHz. The prototype is based on a simple magnetic geometry, the cusp, allowing the use of polyhelix coils (developed with LNCMI, Grenoble) to generate the closed ECR surface at 2.14 T. The plasma is sustained by a high intensity HF pulse (up to 300 kW). Previous experiments at LNCMI have successfully demonstrated the establishment of the nominal magnetic field and the extraction of ion beams with a current density up to 1 A cm^{-2}. The presence of afterglow peaks was also observed, proving the existence of ion confinement in a CUSP ECR source. An experimental campaign is carried out in 2021 using a new transport line designed to improve the transmission of the beam to the new detectors. Short and long-term research plans are presented to transform this high current density into a high intensity ion beam that can be used for accelerators of the future.

1. Introduction

The interest of high frequency ECR ion source[1] is constantly growing from the last decades with several projects, such as the 45 GHz supraconducting ECR ion source from IMP[2] and Berkeley[3]. In this frame, in 2014, the 60 GHz ECR ion source, SEISM (**S**ixty gigah**E**rtz **I**on **S**ource using **M**egawatt magnets), produced its first ion beams with a current density up to $J \approx 1\,\mathrm{A\,cm^{-2}}$ [4] using a CUSP magnetic configuration[5] with a closed ECR surface of 2.14 T. These results were obtained with a 1 mm diameter plasma electrode to manage the beam in a low acceptance beam line. Such a high current density in a CUSP configuration is so far unique in the field of the ECR ion source and it opens new potential applications for next generation particle accelerators. The experiment suffered a set-back in 2014 by a short circuit created by a steel wire in the cooling system which destroyed two of the four coils. At the restart of the project in 2018, an analysis was launched to improve the implantation and the performances of the experiment installed in one of the Grenoble High Magnetic Field Laboratory (GHMFL) experimental hall.

2. Status of the SEISM experiment

2.1. Experiment upgrade

To be able to operate again the ion source, the damaged coils from the previous campaign were rebuilt. The results obtained show that the beam line originally composed of a single bending magnet had to be upgraded to improve the transmission of the beam line. From numerical simulations performed with the Tracewin code [6] and available equipments, a new beam line was installed. It is composed of a quadrupole triplet with 150 mm gap to ensure focusing of the ion beam and the original dipole with a 60 mm gap was replaced by a dipole with a larger gap of 90 mm. Figure 1 shows a picture of the overall experiment installed in the dedicated room inside the GHMFL. Transport simulations indicate that the theoretical transmission could reach 90 % for a 1 mm plasma electrode diameter.

Figure 1. Photo of the new implantation of the SEISM ion source at LNCMI laboratory. From left to right : 60 GHz Gyrotron and his microwave beam line, SEISM ion source, quadrupole triplet (grey), dipole (pink), beam diagnostics

Another upgrade concerns the range of pressure available in the source. The base pressure of the old plasma chamber was 1×10^{-5} mbar. A new plasma chamber was built with $AISi_7MgO_6$ aluminum alloy using additive method. The chamber is composed of two parts, which is required to place a polarized ring against the wall in the center of the plasma chamber, where the cusp field lines expand radially. The parts are tightened together with a plastic gasket in between to ensure the vacuum sealing. The 3D printing technique allowed to create a water-cooling channel of 1 mm size inside the 3 mm wall thickness. The vacuum level inside the chamber reaches now 1×10^{-7} mbar and is perfectly vacuum tight. This improvement will also allow to increase the voltage applied on the extraction of the source from 25 to 40 kV to improve the beam line transport.

2.2. Objectives of future experimental sessions

Experimental campaigns are scheduled in 2021 on the SEISM source. Thanks to the upgrade of the beam line, one should be able to reliably reproduce the data obtained in 2014. The objectives of theses future measurements are to characterize the plasma as a function of the source parameters such as, magnetic field, gas pressure, HF power and the radial biased ring voltage, located between the two CUSP coils.

The data obtained in [4] show the presence of afterglow (AFG) peaks [7] in the Faraday cup located after the dipole, that indicate the existence of ion confinement in an ECR CUSP magnetic field. However, one observed that the AFG peaks of the different charge states are not synchronous as in minimum-B magnetic field configurations.

Figure 2 shows the intensity of the different charge states versus time. The black line represents the HF pulse. In this plot, one can see that the AFG peaks occur at a specific

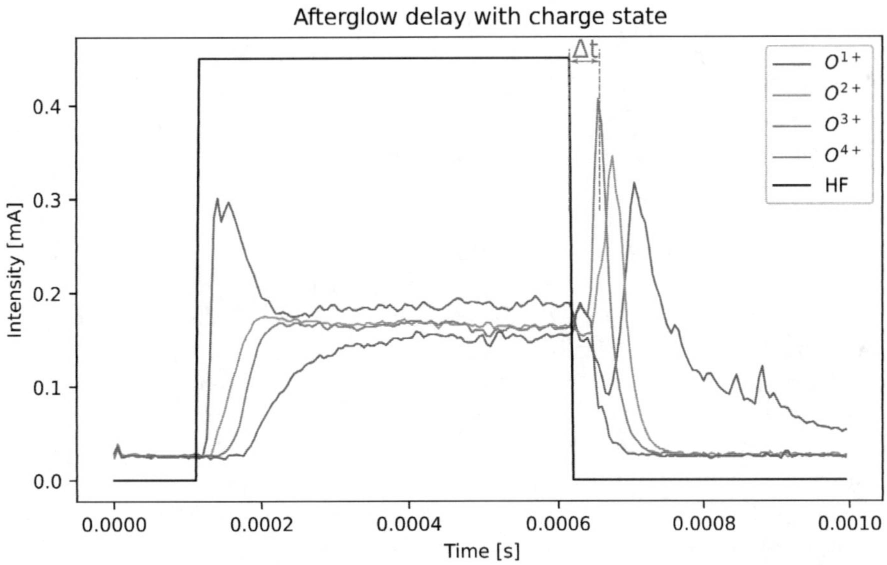

Figure 2. Oxygen charge state ion beam intensity evolution versus time. O^{1+} (blue), O^{2+} (orange), O^{3+} (green), O^{4+} (red), HF pulse (black)

time Δt with respect to the end of the HF pulse, which varies with the charge state of the ion beam and spans from $\approx 10\,\mu s$ to $\approx 80\,\mu s$.

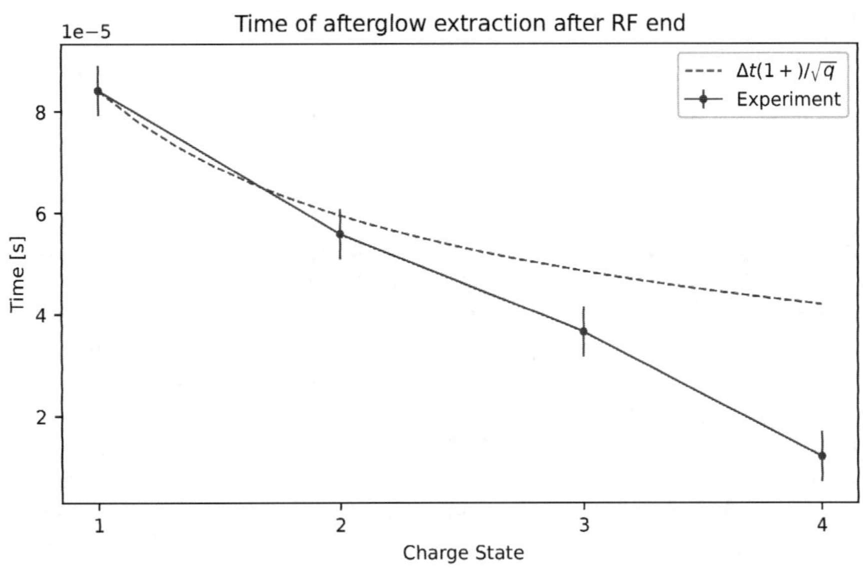

Figure 3. Time of the afterglow peak emission versus charge state of oxygen ion beam. Dotted solid lines (experiments), dashed lines ($t'(n+) = t(1+)/\sqrt{n}$).

Figure 3 shows the evolution of this time difference for the recorded charge states n =1,2,3,4, where solid lines represent the experimental data subtracted from the time of flight (TOF) of ions through the analysis beam line $t(n+) = \Delta t(n+) - TOF(n+)$. $t(n+)$ is the time at which each ion charge state population reaches its maximum at the plasma electrode after the RF pulse end. The dashed line plots the time of the $n+$ peak as $t'(n+) = t(1+)/\sqrt{n}$ which represents the theoretical time difference expected if all the ion species were present at the same time at low energy in the plasma chamber and accelerated by a common local electric field present at the plasma sheath, before the high voltage acceleration. The theoretical curve does not reproduce the data, suggesting that the origin of the AFG shift is not purely electrical. The data analysis excluded the possibility of a drop of the ion source high voltage and this hypothesis was discarded from the possible reasons of time delay between charge states collection in the faraday cup. A possibility would be that ions are confined in the cusp at different locations, and their time to reach the extraction hole would consequently be different. Another possibility would be that ions are electrostatically confined deep in the cusp and build up a temperature depending on their charge state, leading to a different drift time for different ion species to reach the extraction electrode. New measurement including a retarding field analyser will be performed in the upcoming campaign to get a better understanding of this phenomenon.

The plasma stability will be investigated using the technique described in [8], as well as the lost electron energy distribution [9] (LEED) which is unknown in an ECR CUSP configuration. Finally, the ion beam quality will be studied, with Faraday cups in different locations along the beam line for current measurement, giving information on the transport of the intense beam, and a pepperpot device [10] for emittance measurement.

Prior to the experiments, several tests were performed to guarantee the operation of the source. A new control-command was developed (Labview) and deployed. Water cooling of the coils was tested with pressure up to 17 bar and 41 L s^{-1} of flow. The coils were powered with a current of 18 kA so far and the stability was maintained for several hours, as well as the high voltage applied to the source that reaches 30 kV. Thanks to theses previous operations, first plasmas were ignited and ion beam extracted. Experimental campaigns will be performed late 2021 and 2022.

3. Long Term plans

Funding was recently approved by the French Agence Nationale de la Recherche for the SEISM project. The R&D will continue to enhance high intensity beams for the next generation accelerators.

The first priority is to upgrade the prototype with superconducting coils, in order to generate sufficient magnetic confinement with "standard" electrical current. The present prototype require a current supply 26 kA provided by GHMFL and must be approved by a program committee. A second priority is to upgrade the 60 GHz gyrotron high voltage power supply from 300 kW- 1 ms - 2 Hz pulse to 20 kW continuous working (CW) operation. Next, the main goal is to enlarge the plasma electrode extraction hole of SEISM to investigate the production of extreme intensity beams of medium charge state ions in pulsed and CW operation. Figure 4 shows a schematic of the PACIFICS beam line. The geometry for the superconductive coils of the ion source is also based on a CUSP topology, but with a symmetric configuration for the coils. The magnetic field produced should have an axial field B_z with a maximum value around 3 T and a radial component B_r of 4 T and produce a closed 2.14 T ECR surface. The constraints for the coils design and their cryostats is to allow a large diameter plasma chamber up to 150-200 mm diameter, and a large space between the cryostats ($\approx$ 80 mm) to be able to sets optical ports for plasma diagnostics. The extraction of the extremely intense beam from the source will be performed by a multi-electrode system develop in collaboration with the IRFU CEA Saclay. Finally the ion current intensity will be measured on a beam dump. Ports allowing to install beam diagnostics will also be present.

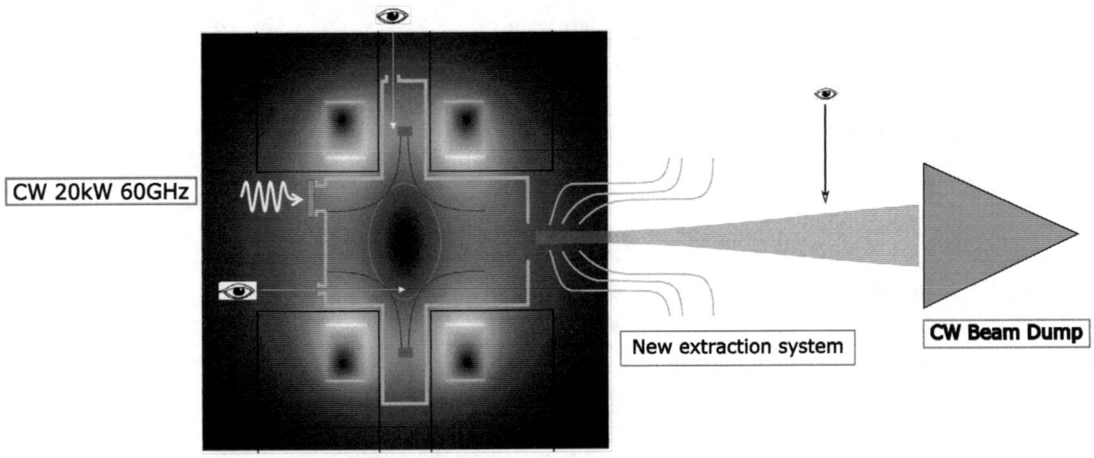

Figure 4. Scheme of the PACIFICS experiment. Map of the magnetic field, cryostats (black), ECR surface of 2.14 T (red), plasma envelope (dark blue), radial biased ring (green), plasma chamber and electrodes (grey), beam dump (orange).

Acknowledgments

The authors acknowledge the support of the LNCMI- CNRS, member of the European Magnetic Field Laboratory (EMFL).

References

[1] Geller R 2018 *Electron cyclotron resonance ion sources and ECR plasmas* (Routledge)
[2] Zhao H W, Sun L T, Guo J W, Zhang W H, Lu W, Wu W, Wu B M, Sabbi G, Juchno M, Hafalia A, Ravaioli E and Xie D Z 2018 *Review of Scientific Instruments* **89** 052301 URL `https://doi.org/10.1063/1.5017479`
[3] Xie D Z, Benitez J Y, Hodgkinson A, Loew T, Lyneis C M, Phair L, Pipersky P, Reynolds B and Todd D S 2016 *Review of Scientific Instruments* **87** 02A702 (*Preprint* `https://doi.org/10.1063/1.4931713`) URL `https://doi.org/10.1063/1.4931713`
[4] Lamy T, Jacob J, Angot J, Sole P, Thuillier T, Bakulin M, Eremeev A, Izotov I, Movshevich B, Skalyga V *et al.* 2015 *13th International Conference on Heavy Ion Accelerator Technology (HIAT 2015), Yokohama, Japan* pp 277–281
[5] Latrasse L, Marie-Jeanne M, Lamy T, Thuillier T, Giraud J, Fourel C, Trophime C, Debray F, Sala P and Dumas J 2010 *Review of Scientific Instruments* **81** 02A324
[6] Uriot D and Pichoff N 2014 *CEA Saclay, June*
[7] Sortais P 1992 *Review of Scientific Instruments* **63** 2801–2805 (*Preprint* `https://doi.org/10.1063/1.1142811`) URL `https://doi.org/10.1063/1.1142811`
[8] Tarvainen O, Kalvas T, Koivisto H, Kronholm R, Marttinen M, Sakildien M, Toivanen V, Izotov I, Skalyga V and Angot J 2019 *Review of Scientific Instruments* **90** 113321
[9] Izotov I, Tarvainen O, Skalyga V, Mansfeld D, Koivisto H, Kronholm R, Toivanen V and Mironov V 2020 *Review of Scientific Instruments* **91** 013502 (*Preprint* `https://doi.org/10.1063/1.5128322`) URL `https://doi.org/10.1063/1.5128322`
[10] Zhang M 1996 Emittance formula for slits and pepper-pot measurement Tech. rep. Fermi National Accelerator Lab., Batavia, IL (United States)

Journal of Physics: Conference Series **2244** (2022) 012015 doi:10.1088/1742-6596/2244/1/012015

A few ECRIS developments at LBNL

D Xie, J Benitez and D Todd

Nuclear Science Division, Lawrence Berkeley Laboratory, Berkeley, California 94720, USA

E-mail: zqxie@lbl.gov

Abstract. In addition to the ongoing investigation of bremsstrahlung radiation produced by energetic electrons in the superconducting Electron Cyclotron Resonance (ECR) ion source VENUS, the LBNL ion source group has pursued several other research topics since 2015 to explore potential advancements of ECR ion sources. Three of these activities are investigations into the benefits of non-cylindrical plasma chambers, the development of new high-temperature ovens that will efficiently produce the intense, metallic ion beams necessary for super heavy element production at the 88-Inch Cyclotron facility, and the finalization of the conceptual and engineering design of the superconducting NbTi magnet structure for the 4th generation, 45 GHz ECR ion source MARS-D. This article summarizes and briefly discusses the highlights of these ECRIS developments.

1. Introduction

The development of Electron Cyclotron Resonance ion sources (ECRIS) started in the early 1970s and its advancement since that time has followed the prediction of Geller's scaling law: $n_e \propto \omega^2 \propto B^2$ [1], where n_e, ω, and B are the electron density, heating frequency, and resonance field, respectively. VENUS is the first 3rd-generation superconducting ECRIS optimized for 28 GHz operation with magnetic field maxima of 4 T on axis and 2 T radial at the inner surfaces of the plasma chamber [2]. It has been operating since 2002 [3] and remains one of the top performing ECRISs with other leading sources being SECRAL-I and SECRAL-II at IMP/China, SCECR at RIKEN/Japan and SUSI at MSU/USA [4,5,6,7,8]. Optimizations, especially those conducted in the last decade, have substantially improved VENUS' performance and resulted in a better understanding of the ECR plasma [9,10].
Development of a 4th-generation 45 GHz ECRIS based on a novel magnet structure using NbTi conductor, MARS-D [11,12], has been continuously moving forward at LBNL. In this article, three of the developments of ECRIS at LBNL are presented and discussed.

2. Symmetry of ECRIS plasma chamber

Cylindrical plasma chambers that provide a resonance cavity for microwaves to energize ECR plasma have become the standard cavity for ECRISs. The third generation ECRISs operating at frequencies greater than 20 GHz have typically been built with cavity volumes of 3 - 8 L. The question arises as to whether a symmetric cavity is still necessary as the wavelengths decrease below 1.5 cm? The cylindrical plasma chamber of VENUS has a 14.2 cm diameter and a 50 cm length resulting in an 8 L cavity volume that can support at least thousands of wave modes. To investigate possible **ExB** damping at the plasma edge to improve source performance [13], tests were conducted in VENUS with an asymmetric aluminum chamber insert occupying the volume of the cylindrical chamber not occupied by the plasma that can be electrically biased to hundreds of volts relative to the chamber. Figure 1**a** shows the design concept and geometric cuts to avoid direct plasma contact, and Figure 1**b** shows installation of the aluminum insert into VENUS which decreased the plasma chamber volume from 8 L to 5 L.

Preliminary tests with this complex aluminum insert were very encouraging: no noticeable change of the microwave coupling was observed in comparison to operations without the insert, and improvements in the oxygen charge state distribution (CSD) with higher average charge that should benefit the production of higher ion charge states.

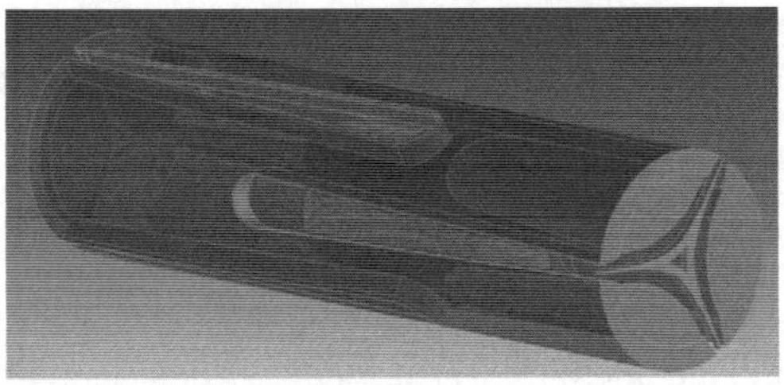

a

b

Figure 1. a. A 3D model of the chamber insert with cuts to avoid direct plasma contact in VENUS. A typical ECR volume (red ball) is positioned inside the insert model as well two end-triangle stars representing the plasma end shapes to assist the design. **b**. Installation of the aluminum insert into VENUS chamber.

Figure 2 shows an oxygen CSD optimized for O^{7+} production with the aluminum insert in VENUS but without plasma electric biasing. This is the first ECRIS oxygen CSD peaked at O^{7+} and produced an average charge $\langle q \rangle \approx 6.5$. For comparison, Figure 3 shows the best ratio of O^{7+} to O^{6+} ever achieved without the aluminum insert: 880 eμA of each charge state and approximately 600 eμA of O^{8+} with an average charge $\langle q \rangle \approx 5.8$. The improved oxygen CSD indicates the benefit of the aluminum insert on production of higher ion charge states. Microwave power was limited due to insufficient insert insulation during the preliminary tests. At 1 kW of 18 GHz wave power the best VENUS oxygen CSD was produced by simultaneously biasing both the aluminum insert and the source probe-disk, while biasing only the aluminum insert or the source probe-disk alone yielded poorer results. A full exploration of the aluminum insert will take place in the future, but the preliminary tests have demonstrated:

- A plasma cavity symmetry is not a necessity for third and future generations of ECRIS.
- An asymmetric plasma cavity could enhance ECRIS' production of higher charge state ion beams.

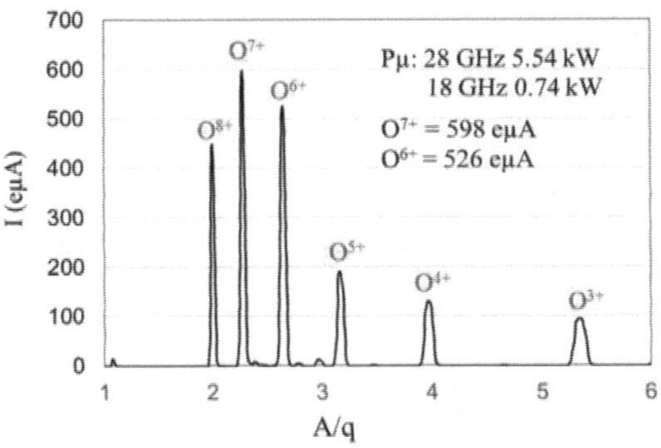

Figure 2. The best oxygen CSD peaked at O^{7+} produced with an aluminum chamber insert in VENUS at a total wave power of 6.3 kW of 28+18 GHz without plasma biasing. The ratio of electric current of O^{7+} to O^{6+} is ~ 1.14 and this CSD resulted in the highest oxygen average charge $\langle q \rangle \approx 6.5$ from an ECRIS.

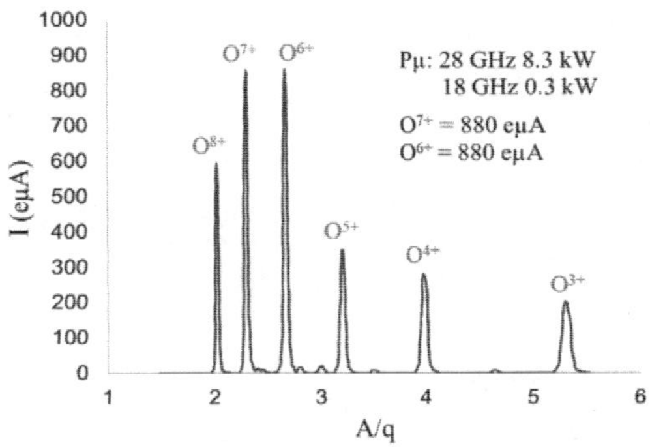

Figure 3. The best oxygen CSD VENUS produced without a chamber insert. The source was tuned to optimize O^{7+} at a total wave power of 8.6 kW of 28+18 GHz. Same current of 880 eμA of O^{7+} and O^{6+} and ~ 600 eμA of O^{8+} were produced resulting in an oxygen average charge $\langle q \rangle \approx 5.8$

3. New resistive-heating high temperature ovens for VENUS

The resistive-heating temperature ovens (HTOs) designed for VENUS [14], custom-made from Ta, W or Re with wall thicknesses of 0.005"-0.008", loading volume of ~ 0.3 cm^3, and costing 3-5 k$/each, cannot satisfactorily withstand the Lorentz forces in VENUS' high fields during high-temperature operation. These ovens last hours to a few days depending on temperature and field when producing ~ 100 eμA of refractory metal ion beams. Figure 4 shows a completely failed Ta oven after a few hours of operation. Future super heavy element research using the 88" cyclotron is expected to require at least 1 pμA of ^{50}Ti and ^{51}V which demands 130-200 eμA of $^{50}Ti^{12+}$ and $^{51}V^{12+}$ from VENUS and the expense of low-abundance metals like ^{50}Ti requires keeping consumption below 5 mg/hr for weeks-long operation. As the existing HTOs cannot meet these requirements, we have developed a new type of HTOs using commercial folded "boats" made of 0.010"-thick Ta and W with a loading volume of ~ 2 cm^3. These boats of cost $25/each are manufactured by RD Mathis Company (Part No: FB10-0.010 Ta or W). Figure 5 shows a folded tungsten boat with a large opening that should slightly reduce the temperature for the same vapor mass flow exit from a small aperture oven used in ECR ion sources [15]. Ti and V are directly loaded into the boats without any crucibles. Unfortunately, the boat-oven has to be set with a tilted face orientation at the location designed for the original ovens. The optimum face orientation of the boat-oven should be a vertical facing-up but that is not possible in VENUS without costly redesigning and refabricating the entire injection assembly

Figure 4. The left and right photos show a failed high temperature Ta oven after a few hours of operation. The Lorentz force resulted from VENUS magnetic fields completely destroyed this custom-made oven.

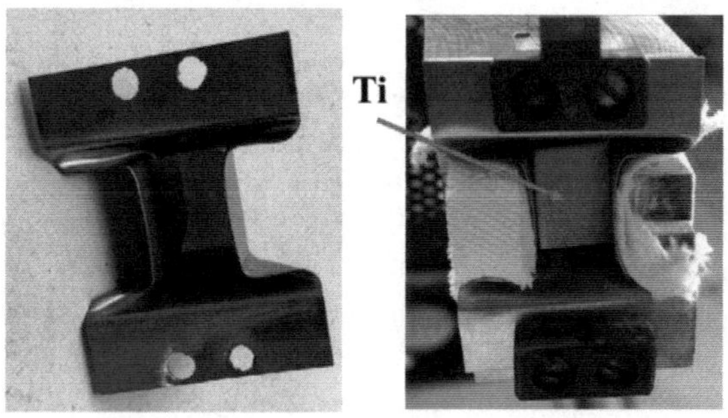

Figure 5. This shows a commercial 0.010" thick W boat (left). A Ti-loaded boat is installed in the housing (right). This boat is able to mitigate the Lorentz force with production of ~ 150 eμA of Ti or V for weeks with minimal deformation.

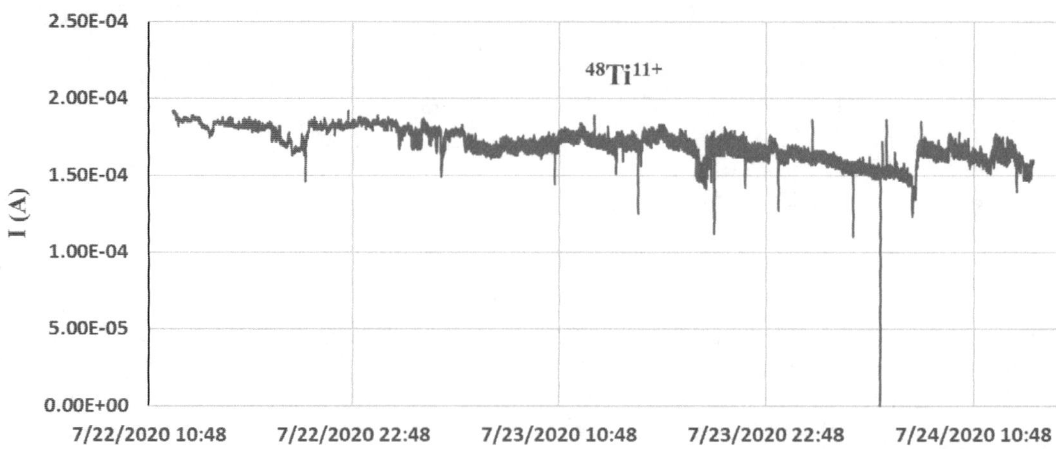

Figure 6. A chart record of two days for an average current of 170 eμA of $^{48}Ti^{11+}$ produced with VENUS using a commercial tungsten folded boat.

Figure 6 shows a chart record of two days where an average current of 170 eμA of $^{48}Ti^{11+}$ was maintained with VENUS using a commercial tungsten folded boat. To produce the needed beam currents the required oven heating powers were ~ 700-800 W and ~ 400-500 W for the tungsten and tantalum boats, respectively, when producing Ti and V beams. The ovens were tested on-and-off a few times with a total duration of a week or longer, and all showed minimal deformations. These tests suggested that these folded boats can operate in VENUS for weeks-long productions to deliver $^{48}Ti^{11+}$ and $^{51}V^{12+}$ currents in excess of 150 eμA. The consumptions of Ti and V in the tests were of 3-4 mg/hr that is slightly higher, especially compare to the ratio of consumption/beam-current, than what were reported by RIKEN ECR group [16]. The higher consumptions indicate that there is still room for improvements with VENUS and planned optimizations of the boat-oven location and aperture orientation will take place in the near future.

4. Progress on MARS-D ECR ion source

The next generation of ECR ion source will operate at higher magnetic fields and microwave frequencies. The most critical challenge moving forward is the construction of superconducting magnet systems capable of producing the needed high magnetic fields.

MARS-D, a fourth-generation, demonstrator ECRIS operating at 45 GHz, was proposed and its conceptual design was initiated in 2015. Since then the development of this new ion source has continued moving forward [12, 17]. The MARS-D ECR ion source will be based on a new and efficient superconducting magnet geometry: Mixed Axial and Radial field System (MARS). This new superconducting magnet geometry coverts the conventional sextupole of six racetrack coils into a single closed-loop-coil for efficient magnetic field generation and utilization. Instead of employing a higher current superconductor to reach the high fields necessary for 4th-generation operation, MARS-D's magnet will still employ the NbTi conductor that has well-proven reliability with decades of operation in ECR ion sources. A NbTi MARS magnet has been designed to produce a minimum-B configuration with field maxima of 5.6 T axially and 3.2 T radially for 45 GHz operations. Optimizations of the MARS-D conceptual design have been completed and efforts are now focused on finalizing the mechanical and electrical details of this new ion source.

Stress analysis of the NbTi MARS magnet was completed recently after finalizing the mechanical dimensions of the magnet cold-mass. The analysed stress and system stored energy of the NbTi MARS magnet are listed in table 1 with comparison to two other magnets: the 28 GHz FRIB VENUS source and IMP's 45 GHz FECR. The comparison shows that even though it is generating fields sufficient for 45 GHz operation, the maximum stress in the NbTi MARS magnet is just slightly over one half of that in the other two magnets and lower stress greatly benefits the magnet operation reliability. In addition to the substantially lower coil stress, the NbTi MARS magnet stores the least energy, ~ 1/3 of IMP FECR (45 GHz) and ~ 3/4 of VENUS (28 GHz). The magnet system stored energy E is proportional to $(LI)^2$, L is the length of the conductor and I the current flow through it. This lower stored energy indicates a smaller footprint of the MARS-D magnet.

Table 1. Maximum coil stress and system stored energy of three sc magnets.

	MARS-D[a] (45 GHz)	IMP FECR[b] (45 GHz)	FRIB VENUS[c] (28 GHz)	Unit
Sextupole coil	80	144	140	MPa
Solenoid coils	55	100	90	MPa
Sys. stored Energy	560	1600	720	kJ
Magnet geometry	MARS	Sext-In-Sol	Sext-In-Sol	
Superconductor	NbTi	Nb_3Sn	NbTi	

[a] M Juchno, Recent MARS-D NbTi magnet stress analysis (June 2021, unpublished).
[b] M Juchno, IMP 45 GHz ECRIS magnet design review (Dec 2016, unpublished).
[c] H Felice, Review on the FRIB VENUS magnet (Sept 2014, unpublished).

With the finalization of the magnet dimensions, more accurate magnet coil loading can be computed. Figure 7 shows and compares the designed superconductor loading of three NbTi magnets for the 28 GHz sources VENUS at FRIB and SECRAL at IMP and the 45 GHz MARS-D. This plot shows that the NbTi MARS magnet can support 45 GHz operation at about the same loading factor as the two 28 GHz sources, VENUS/LBNL and SECRAL/IMP, that have run reliably for over a decade.
The two aforementioned comparisons show the advantages of a MARS magnet geometry:

- More efficient magnet design
- Smaller cold mass (lower magnet cost)
- Continues the well proven NbTi magnet in ECRIS

These advantages have suggested that MARS is a good magnet scheme not only able to extend the NbTi magnet to 45 GHz operations but also a great magnet geometry for future generations of ECRIS.

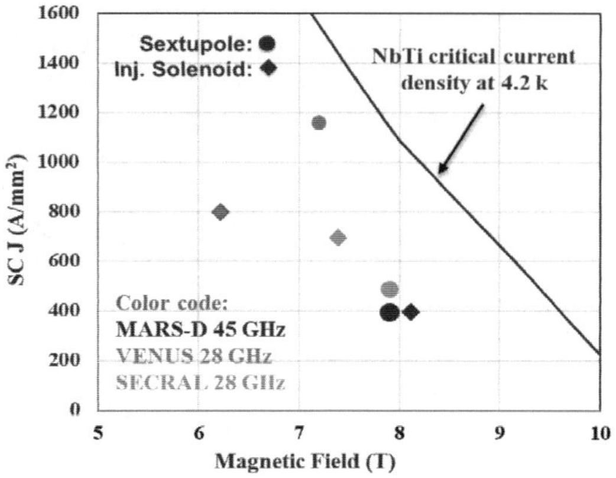

Figure 7. This figure shows the designed NbTi conductor loadings of the magnet major coils of MARS-D, VENUS and SECRAL ECR ion sources. It clearly shows that these design loadings among these three magnets are rather similar. However, the MARS-D magnet is able to generate much higher magnetic fields for operation at 45 GHz while the other two operating at 28 GHz. This loading comparison indicates that the MARS magnet geometry is a more efficient magnet scheme.

5. References

[1] Geller R *et al.* 1987 *Proc. 8th Int. Workshop on ECR Ion Sources* (E. Lansing, MI, USA) p 1
[2] Lyneis C and Xie Z 1995 *Proc. 12th Int. Workshop on ECR Ion Sources* (RIKEN, Japan) p 119
[3] Leitner D, Abbott S, Dwinell R, Leitner M, Taylor C and Lyneis C 2003 *Proc. Particle Accelerator Conference* (Portland, Oregon U.S.A) p 86
[4] Lyneis C, Leitner D, Leitner M, Taylor C and Abbott S 2010 Rev. Sci. Instrum. **81** 02A201
[5] Zhao H W *et al.* 2006 Rev. Sci. Instrum. **77** 03A333
[6] Zhao H W *et al.* 2012 Rev. Sci. Instrum. **83** 02A321
[7] Nakagawa T, Higurashi Y, Ohnishi J and Aihara T 2010 Rev. Sci. Instrum. **81** 02A320
[8] Machicoane G *et al.* 2014 *Proc. 21st Int. Workshop on ECR Ion Sources* (Nizhny, Russia)
[9] Xie D, Lu W, Benitez J, Lyneis C and Todd D 2016 *Proc. 22nd Int. Workshop on ECR Ion Sources* (Busan, S. Korea)
[10] Benitez J, Lyneis C, Phair L, Todd D and Xie D 2017 IEEE. Trans. Pla. Science **45** 7 p 1746
[11] Xie D 2012 Rev. Sci. Instrum. **83** 02A302
[12] Xie D *et al.* 2016 Rev. Sci. Instrum. **87** 02A702
[13] Craig D *et al.* 1997 Phy. Rev. Let. **79**, No. 10
[14] Loew T, Abbott S, Galloway M, Leitner D and Lyneis C 2007 *Proc. Particle Accelerator Conference* (Albuquerque, New Mexico, USA) p 1742
[15] Huang W, Xie D and Sun L 2019 Rev. Sci. Instrum. **90** 123301
[16] Nagatomo T *et al.* 2020 Rev. Sci. Instrum. **91** 023318
[17] Xie D, Benitez J, Kireeff Covo M, Hodgkinson A, Juchno M, Phair L, Todd D and Lu W 2020 *Proc. 24th Int. Workshop on ECR Ion Sources* (E. Lansing, Michigan, USA)

Acknowledgments

This work was supported in part by the U.S. Department of Energy, Office of Science, Office of Nuclear Physics under contract number DE-AC02-05CH11231.

Journal of Physics: Conference Series **2244** (2022) 012016 doi:10.1088/1742-6596/2244/1/012016

Experiment of two-frequency heating for production of multiply charged ion at compact ECR ion source

Masayuki Muramatsu[1], Shinji Nomura[2], Shuhei Harisaki[3], Yushi Kato[3], Yasushi Abe[4], Mihiro Watanabe[5], Ken Katagiri[1], and Atsushi Kitagawa[1]

1 National Institutes for Quantum and Radiological Science and Technology (QST), 4-9-1 Anagawa, Inage, Chiba 263-8555, Japan

2 Technology Research Center, Sumitomo Heavy Industries Ltd, Yokosuka, Kanagawa, 237-8555, Japan

3 Division of Electrical, Electronic and Infocommunications Engineering, Graduate School of Engineering, Osaka University, 2-1 Yamadaoka, Suita-shi, Osaka 565-0871, Japan

4 RIKEN Nishina Center, RIKEN, 2-1 Hirosawa, Wako-shi, Saitama 351-0198, Japan

5 Kochi University of Technology, 185 Miyanokuchi, Tosayamada, Kami City, Kochi, 782-8502, Japan

muramatsu.masayuki@qst.go.jp

Abstract. Several research and developments have been performed at a compact ECR ion source (Kei3) with a fixed magnetic field at HIMAC. The Kei3 is a testing machine for application of Kei series developed for high-energy carbon-ion radiotherapy. It is difficult for Kei3 to produce ions that are heavier than carbon ion because it uses the same magnetic field distribution as the Kei series. In order to improve the beam current of multiply charged ions such as neon and argon, the two-microwave-frequency heating method was tested in the Kei3. Microwaves are introduced into the plasma chamber with a WR-90 rectangular waveguide. A traveling wave tube (TWT) amplifier made by the NEC (LD79X75A1) is used as the microwave source. The frequency band is 9.75-10.25 GHz and the maximum output power is 750 W. The WR-75 waveguide is also introduced from the upper part of the upstream vacuum chamber and installed on top of the existing the WR-90 waveguide. A TWT amplifier manufactured by the Xicom (XTRD-300IJ) was used as a microwave source for the two-microwave-frequency heating. The frequency band is 10-18 GHz and the maximum output power is 300 W. A beam test with oxygen, neon, and argon was performed in order to confirm the effect of two microwave frequency heating. In this test, we focused on O^{6+}, Ne^{7+}, and Ar^{9+} ions. We measured the dependence of the XTRD-300IJ microwave frequency and the microwave power of the LD79X75A1 on a charge state distributions.

Journal of Physics: Conference Series **2244** (2022) 012016 doi:10.1088/1742-6596/2244/1/012016

1. Introduction

Carbon-ion radiotherapy using the Heavy Ion Medical Accelerator in Chiba (HIMAC) has been carried out at the National Institutes for Quantum and Radiological Science and Technology (QST) since 1994 [1]. The multi-ion radiotherapy with dose distribution and Liner Energy Transfer optimization is being studied at QST [2,3]. Helium, carbon, oxygen and neon ions are considered as ion species for multi-ion therapy. However, in the case of facility for high-energy heavy-ion radiotherapy, it is desirable to use with one electron cyclotron resonance (ECR) ion source with all permanent magnets from the viewpoint of cost, operation and maintenance. Currently we are developing a compact ECR ion source for multi-ion irradiation. A prototype compact ECR ion source, named Kei3, based on the Kei series has been developed for production of various ions at the QST [4]. In order to increase a beam current of multiply charged ion such as oxygen, neon and argon, we tested two-microwave-frequency heating method [5,6,7].

2. Improvement of beam current of multiply charged ions by two-frequency heating

In order to confirm the effectiveness of the two-microwave-frequency heating method for the production of multiply charged ions in a compact ECR ion source with a fixed magnetic field, a beam test was performed at the Kei3 with all permanent magnets. The results of this test will determine the specifications of the microwave amplifier. The Kei3 is a testing machine for application of the Kei series developed for treatment. It is difficult for the Kei3 to produce ions that are heavier than carbon ion because it uses the same magnetic field distribution as the Kei series. We have improved the beam intensity of multiply charged ions by the gas mixing method and optimization of microwave introduction using a plate tuner [8]. A beam test with oxygen, neon, and argon was performed in order to confirm the effect of two-frequency heating.

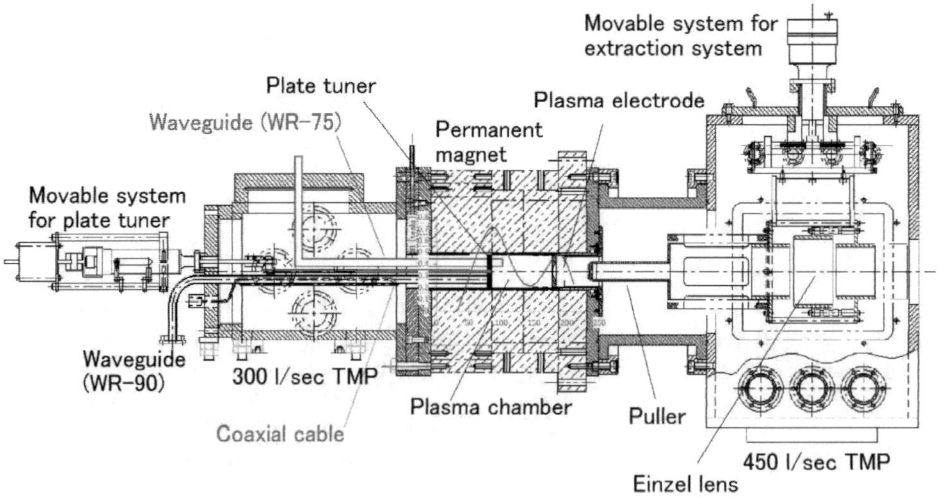

Figure 1: Schematic view of the Kei3 source with additional waveguide and coaxial cable. There are Becr (magnetic field of ECR) of 6.04-20.98 GHz in plasma chamber.

Figure 1 shows schematic drawing of the Kei3. Microwaves are introduced from the vacuum chamber on the upstream side of the permanent magnet through the WR-90 rectangular waveguide. A traveling wave tube (TWT) amplifier made by the NEC (LD79X75A1: TWT1) is used as the microwave source. The frequency band is 9.75-10.25 GHz and the maximum output power is 750 W. The WR-75 waveguide is introduced from the upper part of the upstream vacuum chamber and installed on top of the existing the WR-90 waveguide. A TWT amplifier manufactured by the Xicom (XTRD-300IJ: TWT2) was used as a microwave source for two-frequency heating. The frequency band is 10-18 GHz and the maximum output power is 300 W. A plate tuner is installed at the end of the WR-90 waveguide (upstream mirror magnetic field peak) and can be moved 30 mm upstream from that position by remote control using a drive device. Ionization gas is also introduced from the upstream vacuum chamber as well as the microwave. A voltage of 10 kV is applied for beam extraction to the permanent magnet, plasma chamber, upstream vacuum chamber, and TWT1. TWT2 is installed on the ground using a waveguide DC break. The extracted ions from the extraction electrode are measured by the Faraday cup through the Einzel-lens and the 90-degree analyzing electromagnet.

2.1. Two-frequency heating experiment for oxygen and neon at two rectangular waveguides

In this test, we focused on O^{6+} and Ne^{7+} ions. First, parameters of the Kei3 (gas flow rate, microwave power, extraction electrode position) were adjusted so that each ion was maximized only used TWT1. After that, the microwave of TWT2 was introduced there, and the frequency dependence of the ion beam current was measured. Next, the microwave power dependence of TWT1 was measured. Oxygen gas was used as a mixing gas for the production of neon ions.

Figure 2 shows the microwave frequency dependence of O^{6+} and Ne^{7+}. The beam current of O^{6+} and Ne^{7+} were confirmed when the frequency of TWT1 was fixed at 10 GHz and the frequency of TWT2 was changed from 12 to 18 GHz in 0.1 GHz steps. The maximum beam current was obtained close the TWT2 frequency of 15.8 GHz for O^{6+}, 15.3 GHz for Ne^{7+}. The beam currents of O^{6+} and Ne^{7+} at this frequency were 45 and 10.5 eμA. The beam current of O^{6+} was increased in the high frequency region (15.2-18.0 GHz). In the case of neon, it was lower than the maximum value of TWT1 alone except 15.2-15.4 GHz. Also, no beams were obtained at 13.6-14.9 GHz and 16.0 GHz. It seems that the plasma has disappeared because the current of the extraction voltage power supply was also not observed.

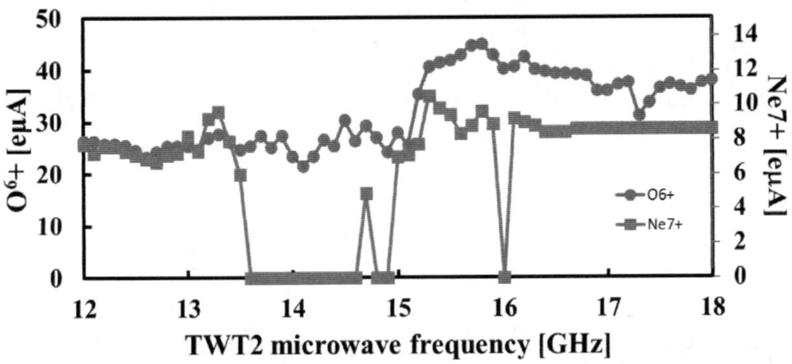

Figure 2. Dependence of beam current of O^{6+} and Ne^{7+} as a function of microwave frequency of TWT2. The frequency of TWT1 is 10 GHz and the power is 601W. Microwave power were 99.6 W for O^{6+} and 57 W for Ne^{7+}.

Figure 3 shows the microwave power dependence (TWT1) of O^{6+} (a) and Ne^{7+} (b). In the case of oxygen, the beam current became maximum at 289 W with TWT1 alone (35.2 eμA). The beam current tended to decrease even if the power of TWT1 was increased to over 289 W. The beam current was the maximum (43.8 eμA) at the output power of TWT1 was 676 W when the microwave frequency of 15.8 GHz at TWT2 was used. The plasma became unstable and the beam current decreased when the output of TWT1 was 700 W. In the case of neon, the beam current is a increased with two-microwave-frequency heating compared to the case of TWT1 alone. However, the effect was not so large as compared with oxygen.

From these results, it was found that even if the output power of TWT2 is about 100 W, a sufficient effect can be obtained for the production of multiply charged ions. It is considered that the vacuum conductance can be improved by changing the rectangular waveguide of TWT2 to a coaxial cable.

Journal of Physics: Conference Series **2244** (2022) 012016 doi:10.1088/1742-6596/2244/1/012016

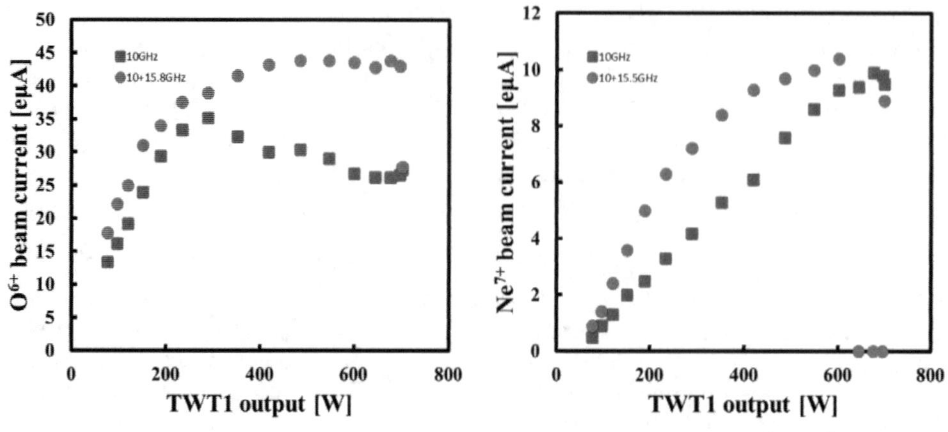

(a) (b)

Figure 3. Dependence of beam current of O^{6+} (a) and Ne^{7+} (b) as a function of microwave power of TWT1. The TWT1 has a frequency of 10 GHz and a power range of 76-700 W, and the TWT2 has a frequency of 15.8 GHz and a power of 99.6 W for O^{6+}, 15.5 GHz and 57.0 W for Ne^{7+}. The position of the plate tuner is the end of the waveguide (peak of the upstream mirror magnetic field).

2.2. Microwave injection by coaxial cable

An ECR ion source that forms a confinement magnetic field with an all-permanent magnet tends to have a smaller inner diameter of the magnet in order to make a strong magnetic field. Two rectangular waveguides installed in the compact ECRIS for two-microwave-frequency heating adversely affects the vacuum. From the results of the beam tests, it is possible to introduce microwaves using a coaxial cable without using a waveguide because the output of TWT2 is 100 W or less and a sufficient beam current is obtained. In this experiment, the end of the coaxial cable is a semi-dipole antenna. The length of the antenna was 5 mm (λ/4 at 15 GHz, see figure 4). This coaxial cable is connected to the TWT2.

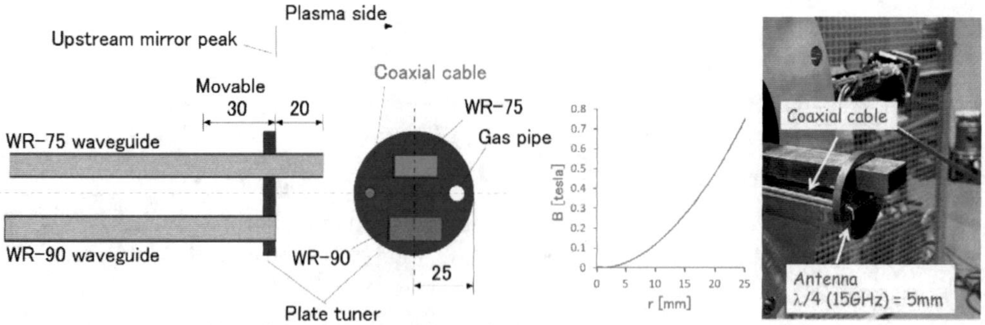

Figure 4. Experimental setting of waveguide and antenna. Inner diameter of plasma chamber is 50 mm. There is Becr of 18 GHz on r=23 mm.

Figure 5 shows dependence TWT2 frequency at comparison between two waveguide and coaxial cable (a) and charge state distribution (CSD) of the argon (b). In the frequency dependence, the beam current of Ar^{9+} is higher when microwaves are introduced using an antenna than the rectangular waveguide. The beam current was maximized at 15 GHz with the antenna, probably due to the fixed antenna length. Maximum beam current of Ar^{9+} was 16.2 eμA at TWT2 frequency of 14.9 GHz. In the CSD of argon, charge state was shifted to a higher charge when two-frequency heating. In addition, the CSD shifted to a higher charge when the microwave was introduced by the antenna rather than by the rectangular waveguide.

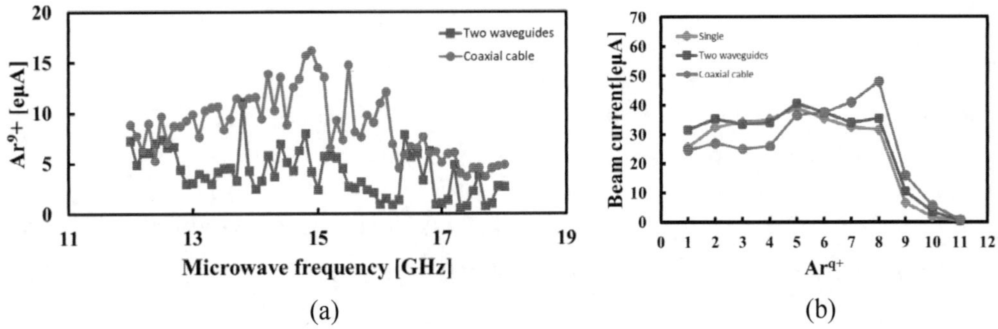

Figure 5. (a) Dependence of Ar^{9+} current and microwave frequency of TWT2. The frequency of TWT1 is 10 GHz and the power is 601W. The TWT2 has a frequency range of 12-18 GHz. (b) CSD of argon. Comparison between single heating (10.0 GHz) and two-frequency heating (13.8 GHz for waveguide, 14.8 GHz for coaxial cable).

3. Conclusion and next step

The two-frequency-heating was tested in the Kei3 ion source. Oxygen and argon were effective in production of multiply charged ion, but neon did not obtain significant change. The effect of the two-frequency-heating seemed to change depending on the difference in ion species and the difference in conditions such as the combined use of the gas mixing method. In the future, we plan to investigate in detail the dependence on the amount of ionization gas and the effect of two-frequency-heating with different ion species. The beam current was higher when microwaves are introduced using an antenna than the rectangular waveguide. We plan to optimize the length and position of the antenna. In addition, another antenna will be installed to investigate the microwaves introduced into the plasma chamber.

References

[1] Y. Hirao et al., Nuclear Physics A 538, 541-550 (1992).
[2] T. Inaniwa, N. Kanematsu, Physics in Medicine & Biology 61 542-550 (2016).
[3] T. Inaniwa, N. Kanematsu, K. Noda and T. Kamada, Physics in Medicine & Biology 62 5180 (2017).
[4] M. Muramatsu et al., Review of Scientific Instruments 87, 02C110 (2016).
[5] Z. Q. Xie and C.M. Lyneis, “Improvements on the LBL AECR source”, ECRIS1995, RIKEN, April 1995, INS-J-182 p. 24 (1995).
[6] S. Biri et al., Review of Scientific Instruments 85, 02A931 (2014).
[7] A. Kitagawa et al., Review of Scientific Instruments 87, 02C107 (2016).
[8] M. Muramatsu et al., proceeding of 2018 22nd International Conference on Ion Implantation Technology (IIT), 200-202, (2018).

Metal evaporation study using the GANIL high temperature oven for intense metal ion beams production.

O. Bajeat, C. Barué, M. Dubois, F. Lemagnen, L. Maunoury, M. Michel, B. Osmond, M. Royet.

Grand Accélérateur National d'Ions Lourds, bd Henri Becquerel, BP 55027, F-14076 Caen cedex 05, France.

bajeat@ganil.fr

Abstract. A new resistive oven dedicated to the production of metal ion beams using ECR ion sources has been developed at GANIL. It aims to produce beams such as $^{58}Ni^{q+}$, $^{50}Ti^{q+}$, $^{50}V^{q+}$ or $^{238}U^{q+}$ with ECR4 ion source on cyclotrons injectors and with Phoenix V3 ion source on Spiral 2 facility. Typical intensity about 1.2 10^{13} pps is requested for the Super Separator Spectrometer (S3). Evaporated atom fluxes and their angular distributions have been measured on a dedicated test bench thanks to a quartz microbalance. These investigations were carried out using three crucibles with various exit caps geometries for which the fluxes and angular distributions display distinct behaviors. Based on those results, crucible is designed to obtain a high flux of atoms overlapping the ECR plasma while minimizing the atoms losses on the ion source chamber.

1. Introduction

Metallic ion beams are produced at Ganil using three methods: thermal evaporation with resistive oven [1], ion sputtering [2] and metals ions from volatile organic compounds (MIVOC) method [3]. The evaporation ovens [4] currently used are made of a tungsten wire and operate up to 1600°C. In order to increase refractory metallic beams intensities included uranium beams on GANIL-Cyclotrons facility and to foresee the future requests of metallic ion beams for S3 on Spiral 2, a new resistive oven able to reach 2000°C has been developed [5]. This work is dedicated to the measurement of the flux and angular distribution of metallic atoms from several crucible geometries.

2. The high temperature oven

2.1. Specification

For S3, the oven will be mounted in the Phoenix V3 ECRIS [6] while on GANIL-Cyclotrons, it will be used with the ECR4 ion sources. The oven must operate at least 2 or 3 weeks without any breakup and the oven head must be removable in order to be replaced quickly in operation. It must be unresponsive to the high magnetic field at injection of the source which is about 1 Tesla or over in the two ECRISs. According to the targeted refractory element, the requested temperature sets between 1500°C and 2000°C corresponding to the temperature for which the vapor pressure is about 10^{-2} mbar; typical value required for producing high intensity of multicharged ions. Table 1 presents data on the targeted temperatures for species to be produced on the GANIL facilities.

Table 1 : Fusion temperatures and required temperatures to reach vapor pressure of 10^{-2} mbar.

Element	T fusion °C	T (10^{-2} mbar) °C
V	1910	1826
Ti	1668	1728 (liquid)
Ni	1455	1505 (liquid)
U	1132	2100 (liquid)
UO_2	2827	1990

2.2. The oven design

The oven shown on figure 1 is constituted of a removable head connected on two concentric copper tubes made of copper alloy (CuCr1Zr) used for the current supply. Copper tubes are cooled down by water circulation at their tips. The crucible is fixed on a sliding rod and is inserted in the centre of the resistor.

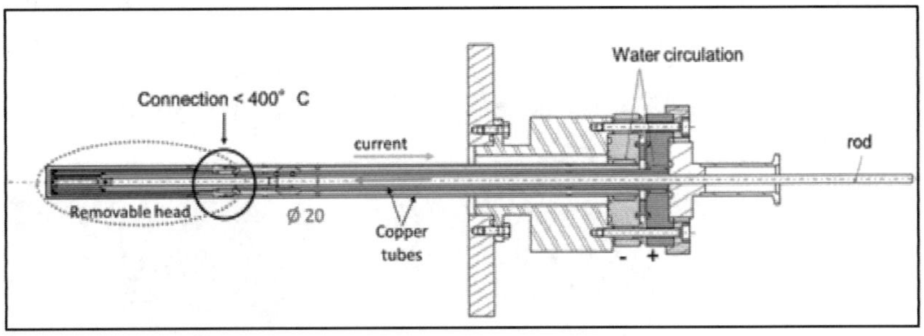

Figure 1: General view of the oven

Type of materials, thickness and lengths of all tubes have been optimized thanks to ANSYS thermal simulations (figure 2) in order to reach 2000°C within the crucible and less than 300°C at the connection between the removable head and the cooled copper current supplies.

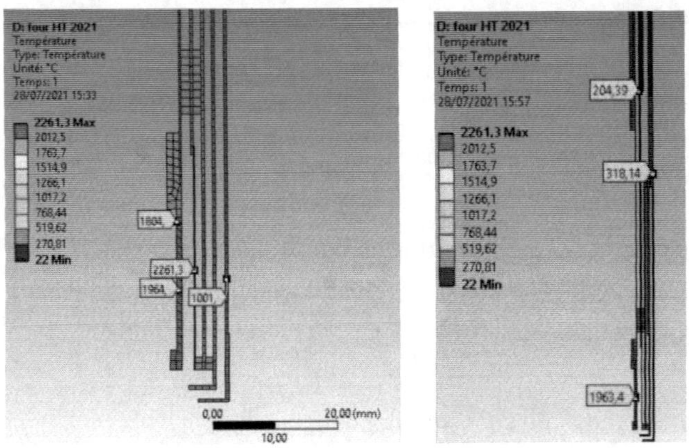

Figure 2: ANSYS thermal simulations with 350 W heating power. Left: the temperature in the head (2261°C on the resistor/ 1964°C in the crucible). Right: estimation of the temperatures at the intersection between the head and the copper tubes.

2.2.1. The sample-holder. Some identified elements affect and can damage the crucible through specific chemical processes. Two crucibles materials are considered: WL20 (machinable tungsten alloy with 2%

of La_2O_3) is used for all metals excepted uranium and Yttrium oxide (Y_2O_3) will be carried out for uranium. The crucible is fixed on a molybdenum tube itself embedded on a stainless steel rod (figure 3). The main function of the molybdenum tube is to reduce the thermal conduction between the crucible and the rod.

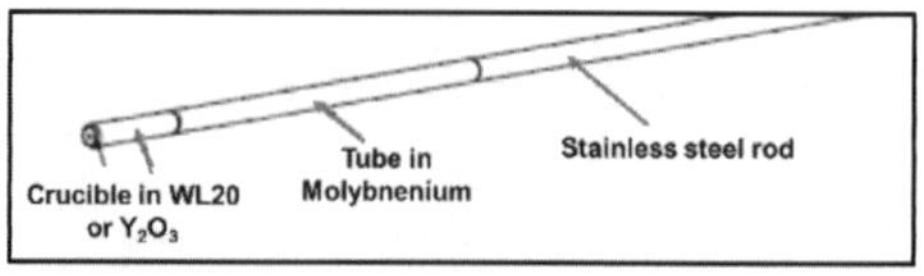

Figure 3: the sample holder

3. Evaporation rate measurements

A quartz microbalance (Q-pod) is used on a dedicated test bench with a quartz crystal detector on a cooled rotating support (figure 4). The crystal characteristic frequency is 6 MHz when applying a static electric field. During metallic atoms evaporation deposited on the quartz surface, its mass increases and thus its mechanical resonance frequency changes. The quartz frequency measurement is done by a dedicated Inficon controller which displays the instantaneous frequency and deduces the mass per cm^2 accumulated during a programmable integration time. This setup had been successfully utilized for calcium evaporation measurement with a low temperature oven [7]. The quartz detector can rotate around a vertical axis passing through the exit of the oven. Measurements are done for angles between 0 and 90°. The distance between the quartz detector and the oven extremity is 5 cm at all angles.

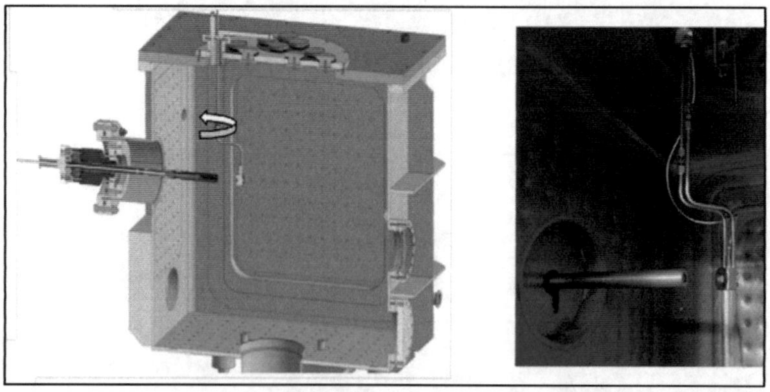

Figure 4: The oven and the quartz microbalance in the vacuum chamber

3.1. *Results using distinct caps*

Three different caps have been tested (figure 5): Φ hole diameter 1, L length 2 or 3 mm (cap labels Φ1L2 and Φ1L3) and hole diameter 2 length 2 (cap label Φ2L2).

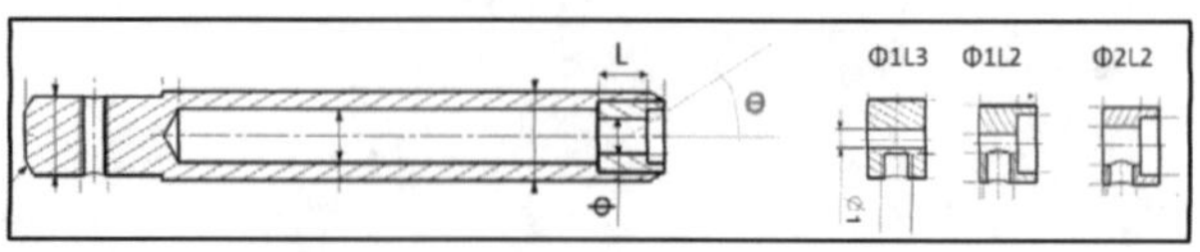

Figure 5: The crucible and the three caps investigated

3.2. Angular distribution

According to atomic effusion theory [8], angular distribution is wider if the ratio Φ over L is larger. Moreover, the total flux should be proportional to the section of the hole. The aim of these measurements is to confirm this trend in order to evaluate the flux of atoms overlapping the ECR plasma core and to select the best cap to obtain a high flux while minimizing the losses of atoms on the wall of the plasma chamber. The figure 6 shows measured angular distributions for iron at 1500°C with the three cap geometries and for vanadium at 2000°C using 2 caps geometries. Charts on the left display absolute measurements while on the right, the curves are normalized at 0° to compare the angular distributions. The measurements reveal that for both iron and vanadium flux there is not a large difference in the angular distributions as a function of the different caps.

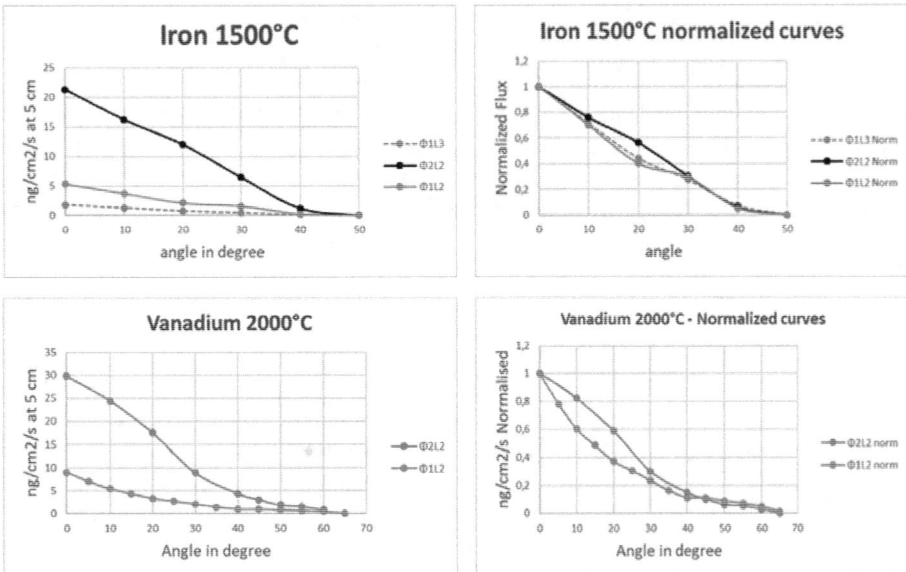

Figure 6: angular distribution (left) and normalized angular distribution (right) for Iron at 1500°C (top) and Vanadium at 2000°C (bottom)

The flux emitted in a α angle cone is defined as:

$$\Phi = 2\pi D^2 \int_0^{\alpha} sin\theta \, f(\theta) d\theta$$

With D: distance from the crucible exit to the quartz detector and f(θ): function representing the at/cm2/s measured on the quartz detector versus angle θ.

We consider the useful flux is the one emitted in a cone of 20° corresponding approximatively to the best superposition with ECR plasma. On the table 2 are reported the values of flux and fraction of vanadium emitted in a 20° cone thanks to the crucible heated at 2000°C. It appears that the difference concerning the two fractions is not critical. For the Vanadium, the cap ϕ2L2 seems to be the most suitable one to deliver high flux of metal atoms towards the ECR plasma.

Table 2: flux and fractions of vanadium atoms emitted in a 20° cone

Vanadium 2000°C	Total flux within a 20° cone (at.s^{-1})	Fraction of atoms emitted inside the 20° cone
Cap ϕ2L2	2.3 10^{15}	~ 37%
Cap ϕ1L2	5.2 10^{14}	~ 33%

3.3. Consumption rates

The flux measured by the quartz microbalance is not constant. Vanadium consumption rate deducted from the quartz measurements was 1.9 mg/h at the beginning of the test and 0.85 mg/h after 19 hours. The real consumption weighed after 22 hours of heating was 2.3 mg/h. That discrepancy can be partially explained by the loss of metal onto the front end screens at angles greater than 50° (figure 7).

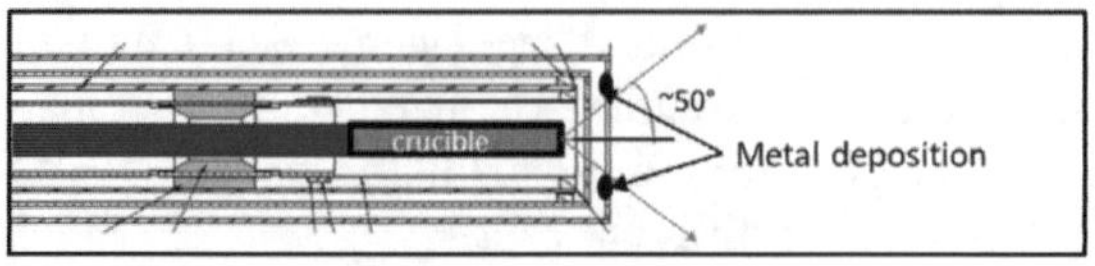

Figure 7: loss of metals on screens for large angles

4. Vanadium beams production with ECR4 on Ganil injector

The oven has been mounted inside the ECR4 ion source on the GANIL-Cyclotrons injector for vanadium beam production test. The source operated with helium as buffer gas. When heating the power of the oven at about 320 W (105 A) for a crucible temperature higher than 2000°C we reach 10 eμA on muticharged vanadium ions with a maximum intensity on the V^{9+} ions as shown on figure 8.

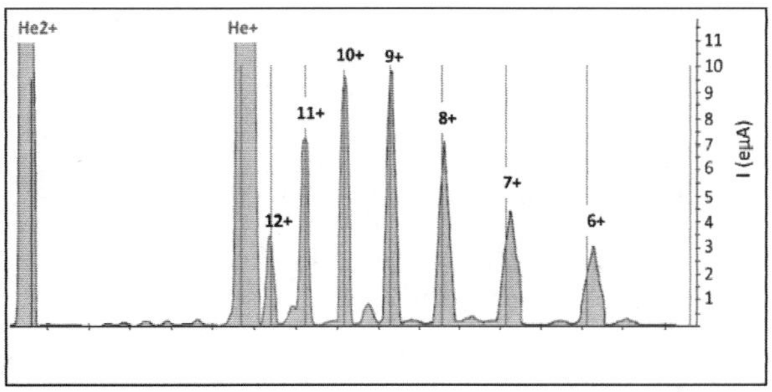

Figure 8: an example of Vanadium spectrum using the high temperature oven on ECR4 ion source.

5. Conclusion

Off-line flux measurement of vanadium atoms and on-line tests of the oven in ECR4 have validated the last version of the high temperature oven for vanadium beam production. Flux and angular distributions measurement for different caps geometries showed that the difference on losses at large angles between a cap with hole diameter 1 and 2 mm is not critical. For low vapor pressure metals the best choice will be to employ a cap with the larger hole to get the highest atom flux. This oven will be now mounted in the Phoenix V3 ion source on Spiral 2 facility.

References

[1] P. Sortais, *Nucl. Instr. and Meth. in Phys. Res. B 98 (1995) pp508-516*
[2] M. P. Bourgarel et al., *Rev. Sci. Instr., vol. 63 (1992), pp 2854*
[3] H. Koivisto, J. Ärje and M. Nurmia, *Nucl. Instr. and Meth. in Phys. Res. B 94 (1994)*
[4] P. Lehérissier et al., Rev. Sci. Instrum. 77, 03A318 (2006).
[5] O. Bajeat et al., A new resistive high temperature oven for metallic beams production. *24th ECRIS conference. 2020.*
[6] T. Thuillier et al. Status of the SPIRAL2 injector commissioning. *Rev. Sci. Instr. 87, 02A733 (2016)*
[7] A. Leduc et al. Angular distribution measurement of atoms evaporated from a resistive oven applied to ion beam production. *Journal of Instrumentation April 2021, vol.16, no.4, P04001.*
[8] N.F. Ramsey. Molecular beams. *Clarendon Press. Oxford.*

Progress of the High-Current EBIS Charge Breeder for the FRIB

Hyock-Jun Son[1], Alain Lapierre[1], Antonio C.C. Villari[1], Ana Henriques[1], Charisse Supangco[1], Cody Knowles[1], Daniel Crisp[1], Samuel Nash[1], and Edward N Beebe[2]

[1] Facility for Rare Isotope Beams, Michigan State University, 640 South Shaw Lane, East Lansing, MI 48824, USA
[2]Brookhaven National Laboratory, Upton, NY 11973, USA

E-mail: son@frib.msu.edu

Abstract. The ReA post-accelerator of the National Superconducting Cyclotron Laboratory employs an Electron-Beam Ion Trap (EBIT) as a charge breeder to reaccelerate rare-isotope beams to several MeV/u. The Facility for Rare-Isotope Beams (FRIB) is near completion and will provide RIB rates expected to exceed in some cases 10^{10} particles/s. The ReA EBIT operates with an electron current of 300 – 600 mA, corresponding to an electron current density of 170 – 340 A/cm^2 and a maximum trap capacity of 10^{10} elementary charges, which can be insufficient to handle high FRIB rates. A High-Current Electron-Beam Ion Source (HCEBIS) has been constructed based on the backbones of the TEST EBIS from the Brookhaven National Laboratory. By using a 4-A electron beam, a current density of 298 A/cm^2 and a maximum trap capacity of 2.4×10^{11} elementary charges can be achieved. This paper presents the status of the HCEBIS. It also includes results of electron- and ion-beam simulations and discusses electron-beam commissioning plans.

1. Introduction

The ReA post-accelerator [1] of the National Superconducting Cyclotron Laboratory (NSCL) at Michigan State University reaccelerates rare-isotope beams (RIB) after production to several MeV/u for nuclear-physics experiments. One of its essential components is an Electron-Beam Ion Trap (EBIT) employed as a charge breeder. The Facility for Rare-Isotope Beams (FRIB) is near completion and expected to provide the first RIB by the end of 2021. At full power, the FRIB will provide RIB rates expected to exceed in some cases 10^{10} particles per second (pps). Moreover, experiments at ReA have requested beams of stable isotopes of comparable high intensities. The ReA EBIT normally operates with an electron current of 300 – 600 mA for an estimated density of 170 – 340 A/cm^2 [2]. This current corresponds to a maximum capacity of 10^{10} elementary charges, which, depending on the charge states needed for reacceleration, can be insufficient to handle RIB of high rates.

Parts of the TEST Electron-Beam Ion Source (EBIS) [3] from the Brookhaven National Laboratory (BNL) were transferred to the NSCL in 2018 to build a High-Current Electron-Beam Ion Source (HCEBIS). The HCEBIS utilizes a LaB_6 dispenser cathode (diameter of 9.2 mm) capable of emitting a current of more than 10 A [4]. However, for its implementation at the FRIB, a target operating electron-beam current of 4 A was chosen. This corresponds to a charge capacity of 2.4×10^{11} elementary charges and an electron-beam current density simulated to be 298 A/cm^2 in a 5 T magnetic field in the trap center. The assembly of the HCEBIS has recently been completed from the electron gun to the collector with a diagnostic beamline. Implementation of the power supplies and control system is being finalized. The specifications of the HCEBIS are summarized in Table 1. This paper presents the status of the HCEBIS, including the result of simulations, a schematic of the power supplies and high-voltage platform configuration, and also discusses electron-beam commissioning plans.

Table 1. HCEBIS specification

Electron-beam current	≤ 4 A
Electron-beam energy	≤ 20 keV
Magnetic field (trap center)	≤ 5 T
Electron-beam current density	≤ 300 A/cm^2
Trap length	0.7 m

2. Electron- and ion-beam simulations

Electron-beam simulations have been performed using TRAK [5] to find the optimum parameters for efficient electron-beam transport from the electron gun to the collector as well as maximize the current density in the trap center. Figure 1 shows the magneto-immersed electron gun as well as the electric potential and magnetic field distributions along the electron-beam axis in the HCEBIS. A single axisymmetric mesh domain was used to prevent discontinuities. The emitted beam current converged to 4 A with a potential difference of 20.6 kV between the cathode and anode. The electron-beam energy was 14 keV in the trap center. The electron-beam radius varied with the magnetic field along the beam axis, and the electron beam was compressed by the maximum magnetic field of 5 T. An axial magnetic field of 0.1 T near the cathode surface (positioned at z=0 in Figure 1) is the minimum value to maximize magnetic compression of the electron beam without beam reflections due to magnetic mirroring. With these operating parameters, the estimated maximum current density is around 298 A/cm^2. This beam compression corresponds to a minimum beam radius of 0.63 mm, as shown in Figure 1. This optimum magnetic field is comparable to the result of 0.14 T published in Ref [6].

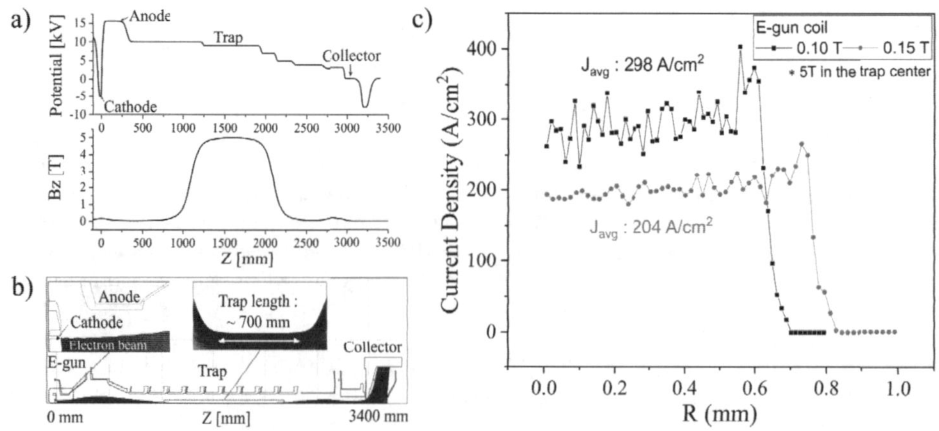

Figure 1. a) Axial electric potential (no electron-beam space charge) and magnetic field distributions along the electron-beam propagation axis, b) electron-beam trajectories from the cathode to the collector, and c) electron-beam current density in the trap center.

When the HCEBIS replaces the ReA EBIT, efficient beam transport through the post-accelerator will be important to reach the requested rates on target. As an input for beam transport simulations through the ReA LINAC, the phase-space of Ne^{8+} beams extracted from the HCEBIS was simulated at an interface before the charge-over-mass (Q/A) separator. The acceleration voltage was set to 31 kV to reach an ion beam energy of 248 keV for Ne^{8+}, needed to match the injection velocity of the ReA radiofrequency quadrupole (RFQ) accelerator. It is important to generate a realistic initial ion distribution in the trap for the extraction simulations. In Ref [7], Marrs notes that the equilibrium trapped-ion temperature for highly charged species is approximately 10 % of the trapping potential experienced by the ions. The ion temperature can then be approximated as $T_i \approx 0.1qeV_{well}$, where T_i is the ion temperature in eV energy unit, q is the ion charge, e is the elementary charge, and V_{well} is the axial potential for trapping the ions. In Ref [8], this factor was obtained from simulations to be between 0.3 – 0.5 for highly charged Ar (11+ and 18+), but was measured with highly charged Ti (20+) in Ref [9] to be in the range 0.065. We assume in the present work a factor of 0.1. In the trap, the radius of the ion cloud confined within the electron beam can be estimated as [7].

$$\sigma_i = \sigma_e \sqrt{\frac{T_i + qeU_i}{qeU_e}} \tag{1}$$

where σ_i, σ_e, U_i, and U_e represent the ion-cloud radius, electron beam radius, ion space-charge potential, and electron-beam space-charge potential, respectively. In Ref [7], the electron-beam space-charge (radial) potential was written as a series expansion of $(r^2/(2\sigma_e^2))^n$ terms, where r is the radial position. Eq. (1) is valid under the assumption that the second-order term (and higher terms) can be neglected, such that $r^2/(8\sigma_e^2) << 1$. For the parameters studied in the present work, the ion-cloud radius calculated with Eq. (1) is estimated to deviate from that using a more exact description of the space-charge potential by more than 10% for ion-cloud radii exceeding the electron-beam radius.

Figures 2 and 3 show the variation of the ion-cloud radius and emittance of Ne^{8+} beams for different axial potentials and compensation ratios (U_i/U_e) of the electron-beam space-charge potential by the space charge of the trapped ions. In the emittance simulations, this effect was taken into account by reducing the electron-beam current to lower its space-charge potential by the compensation ratio. As the electron space-charge potential is compensated with positive charges, so is the axial potential provided by the trap's potential barriers. Except for one test case discussed below, we assumed that for the different compensation ratios, the electrostatic potential applied to the barriers was adjusted to keep the axial potential experienced by the ions fixed. The electron-beam radius and space-charge potential in the trap center were 0.796 mm and 516 V, respectively, with a magnetic field of 0.15 T at the cathode surface. In Figure 2, the ion-cloud radius increases with the axial potential and compensation ratio. The ion-cloud radius becomes larger than the electron-beam radius before reaching the compensation ratio of 0.9. A ratio of 1 was excluded in the figures because of large ion losses in the extracted beams of up to ~35%. For ratios lower than 1, the losses are less than 6%. With the axial potential set to 1.6 kV, the ion-cloud radius becomes the same as the electron-beam radius at the compensation ratio of 0.69.

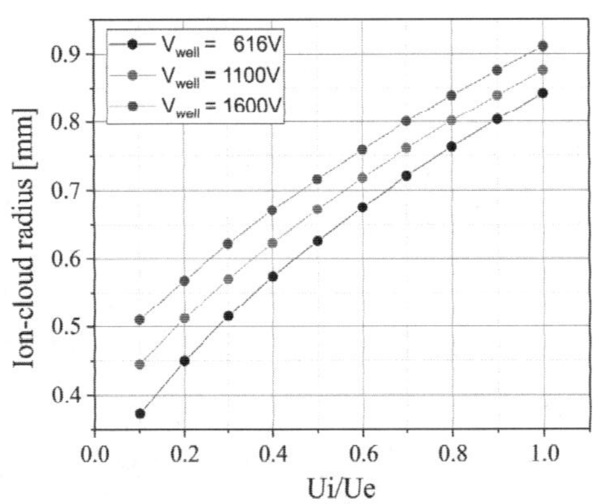

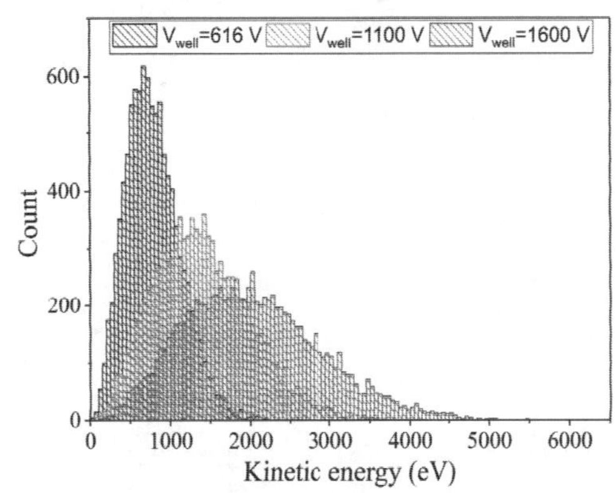

Figure 2. (left) Ion-cloud radius in the trap for different axial potentials and compensation ratios. (right) Maxwell-Boltzmann ion energy distribution used in the simulations.

A Maxwell-Boltzmann distribution with 3-degree-of-freedom [10] was used for generating the initial ion energy distribution presented in Figure 2. A most probable ion kinetic energy per charge was set to be 10% of each axial potential, as $1.5T_i/q \approx 0.1eV_{well}$. The range of the angular distribution was $-\pi/2 \leq x' \leq \pi/2$ to generate an isotropic beam (0-degree velocity vector pointing towards the electron collector). Then, those ions were launched from the inner edge of the upstream potential barrier in the presence of the electron-beam space charge potential.

The trajectory of extracted Ne^{8+} ions is shown in Figure 3 (a) along with the z position of 4303 mm, where information on the extracted ions was collected. Figure 3 (b) and (c) show the phase-space of an extracted beam and variation of the 4·RMS normalized emittance with the compensation ratio for different initial ion energy distributions. The 4·RMS normalized emittance was calculated using Eq. (2) [11]:

$$\varepsilon_{normalized} = \beta\gamma\varepsilon_{4\cdot RMS} = \beta\gamma\left(4\cdot\sqrt{\overline{x^2}\cdot\overline{x'^2} - \overline{x\cdot x'}^2}\right)\ mm\cdot mrad \qquad (2)$$

The emittance increases with the compensation ratio. For instance, when the ion-cloud radius is equal to the electron-beam radius, the normalized emittance is 0.312 mm·mrad for $V_{well} = 1.6$ kV, $U_i/U_e = 0.69$, and $T_i/q = 160$ eV. To study how the ion space-charge potential modifies the axial potential, Figure 3 (c) shows the beam emittance for a case in which the potential applied to the barriers is *not* adjusted for the compensation ratios. It is kept constant. In this particular case, the axial potential experienced by the ions varies with the different compensation ratios as the ion space charge increases between the trapping barriers. For a low ratio, the axial potential is high, which leads to a high ion temperature and large beam emittance. However, for higher ratios, the axial potential decreases, lowering the ion temperature (only cold ions remain trapped) and hence reducing the beam emittance. At $U_i/U_e = 0.5$, the axial potential is decreased by the ion space-charge potential down to 380 V. For even higher ratios, the axial potential no longer confines ions.

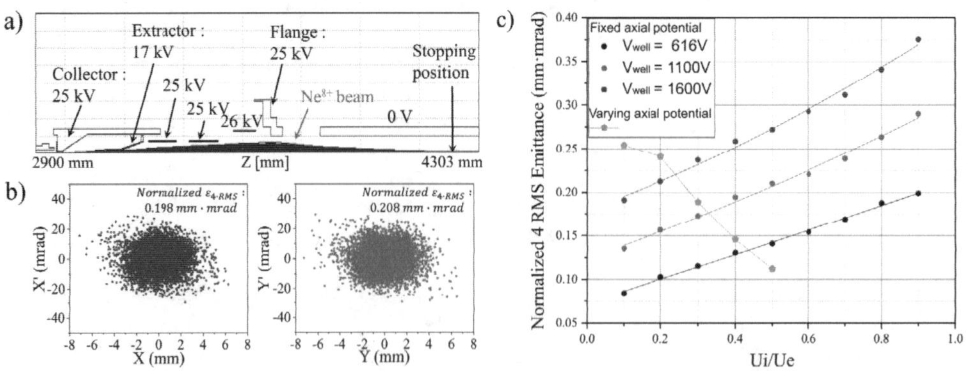

Figure 3. Ne^{8+} beam extraction simulations: a) extracted beam trajectories, b) phase space at the interface position for $V_{well} = 616$V, $U_i/U_e = 0.9$, and $T_i/q = 61.6$ eV, and c) 4 RMS normalized emittance for different barrier potentials and compensation ratios.

3. Power supplies and high-voltage schematic, and plans for electron-beam commissioning

Figure 4 shows a power supplies and high-voltage schematic. The cathode heater, collector bias, suppressor, and reflector supplies are located on a high-voltage "cathode" platform that can be floated up to 12 kV for electron acceleration. A fast high-voltage (HV) 20-kV amplifier (Matsusada AMP-20B20) for the anode is employed for pulse-mode operation. For an electron beam of 4 A to be emitted from the cathode, the required voltage difference between the cathode and anode is around 20.6 kV according to the previously measured perveance of 1.12×10^{-6} $A/V^{3/2}$; the simulated perveance was 1.45×10^{-6} [12] (in Ref [13], 1.49×10^{-6} was estimated). Therefore, the anode HV amplifier is located on ground to provide a potential difference of more than 20 kV. For the collector, a power supply (Glassman

LQ5P2.0) having a maximum output of 5 kV and 2 A is implemented. Therefore, the target electron beam current is 2 A for the first electron-beam commissioning. Another power supply will be used for 4 A in the future. The drift tube can be biased up to 10 kV, to provide a maximum electron-beam energy of up to 20 keV. There are three current transformers (CTs) installed, two AC-CT (Pearson Current Monitor 110) and one DC-CT (LEM LA 25-NP), to measure the emission, collection, and loss of electron beam current, as shown in Figure 4. All power supplies are currently installed in the ground and high-voltage platform racks.

The assembly of the HCEBIS has been completed from the electron gun to the corrector with the diagnostic beamline, as shown in Figure 4. The diagnostic beamline consists of a 2.7 m long vacuum pipe for Time-of-Flight (ToF) measurements. This beamline incorporates a Faraday cup (FC) and a microchannel plate (MCP). During electron-beam commissioning, residual gas or injected gas such as neon will be charge bred. The diagnostic beamline will be used to measure the total current of the charge-bred ion beam in the FC and relative charge state distributions with the MCP. In order to measure charge state distributions, one of the steering electrodes will be used as a chopper with a fast HV switch, which is capable of producing short ion pulses of several tens ns.

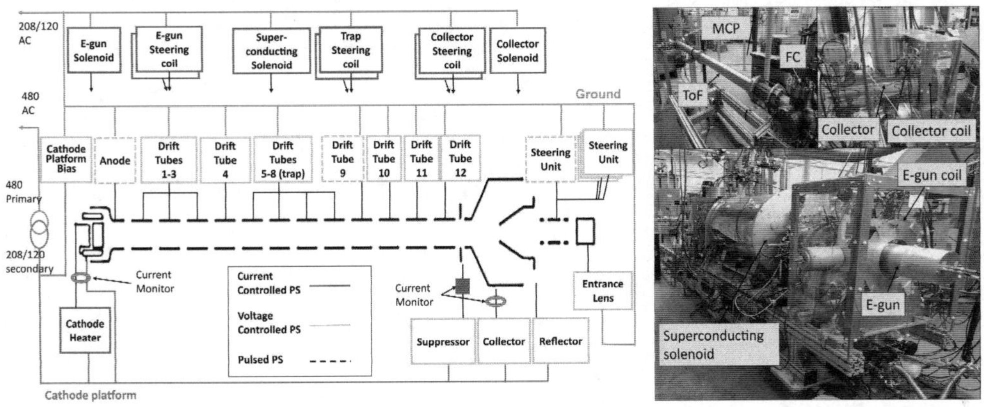

Figure 4. (left) Power supplies and platform schematic for the HCEBIS. (right) Photographs of the HCEBIS.

4. Conclusion

Electron- and ion-beam simulations have been performed to find the optimum parameters for efficient electron-beam transport, maximize the current density in the trap center, and estimate the emittance of extracted ion beams. The HCEBIS has been assembled from the electron gun to the diagnostic beamline. First electron-beam commissioning and charge breeding tests with residual gas are expected to be performed by the end of the calendar year.

Acknowledgements

Work supported by National Science Foundation, under contract No. PHY-1565546.

References

[1] Villari A C C *et al* 2016 *Proc. of Int. Particle Accel. Conf.* (Busan) 1287–1290
[2] Lapierre A *et al* 2018 *Phy. Rev. Accel. Beams* **21** 053401
[3] Beebe E N, Alessi J G, Graham D, Kponou A, Pikin A, Prelec K, Ritter J, and Zajic V 2004 *Journal of Physics: Conference Series* **2** 164–173
[4] Alessi J G, Beebe E N, Gould O, Kponou A, Lockey R, Pikin A, Prelec K, Raparia D, Ritter J, and Snydstrup L 2005 *Proc. of Particle Accelerator Conf.* (Knoxville) 363-365
[5] Field Precision LLC, http://www.fieldp.com.
[6] Pikin A, Alessi J G, Beebe E N, Kponou A, and Prelec K 2006 *Rev. Sci. Instrum.* **77** 03A910
[7] Marrs R E 1999 *Nuclear Instruments and Methods in Physics Research B* **149** 182-194

[8] Penetrante B M, Bardsley J N, DeWitt D, Clark M, and Schneider D 1991 *Phys. Rev. A* **43** 4861.
[9] Beiersdorfer P 1997 *AIP Conference Proceedings* **389** 121.
[10] Lapierre A 2017 *Can. J. Phys.* **95** 361–369
[11] Baartman R 2015 *TRIUMF Beam Physics Note* TRI-BN-15-07
[12] Son H J, Lapierre A, Villari A C C, Henriques A, Knowles C, Crisp D, Nash S 2020 *Proc. of 24th Int. Workshop on ECR Ion Sources* (East Lansing)
[13] Pikin A 2006 *Brookhaven National Laboratory Report* C-A/AP note #245

Low Z gas mixing effects of Ar and Xe multicharged ions in electron cyclotron resonance ion source

K Tsuda, W Kubo, S Harisaki, M Anan, K Sato, I Owada T Maenaka and Y Kato

Division of Electrical, Electronic and Infocommunications Engineering, Graduate School of Engineering, Osaka University, 2-1 Yamada-oka, Suita-shi, Osaka 565-0871, Japan

k.tsuda@nf.eie.eng.osaka-u.ac.jp

Abstract. An electron cyclotron resonance (ECR) ion source (ECRIS) has wide applications such as cancer therapy and material synthesis. We have constructed a single ECRIS capable of producing various ions with low to high mass/charge ratio, and have been conducting various experiments. We focus on the efficient production of multicharged ions and aim to increase the yield of multicharged ions by new effects of electromagnetic waves in ECRIS. We investigated the possibility of selective heating to low Z ions under low Z gas mixing. Therefore, as a preliminary step of these experiments, the low Z gas mixing effect with rare gases are investigated. The optimum mixing ratio of the main gas Ar and the low Z gas He was investigated by varying the mixing ratio. The Ar multicharged ion beam current, impurity fraction and plasma parameters were measured. As a result, it was confirmed that the Ar multicharged ion yield was sufficient reproducibly increased by mixing He gas into Ar plasma. In addition, this from the measured plasma parameters is not an effect of the plasma parameters, but maybe a cooling effect.

1. Introduction

Ion beams using an electron cyclotron resonance (ECR) ion source (ECRIS) plasmas has been applied in various fields such as accelerator physics and heavy particle cancer therapy [1]. In general ECRIS, high yield of multicharged ion currents have been achieved by increasing the magnetic field strength and microwave frequency. In contrast, our group has been conducting to increase the yield of multicharged ions by introducing other electromagnetic waves without changing the magnetic field strength and microwave frequency. We focus on the gas mixing effect [2][3], which is known as an empirical method to increase multicharged ions in ECRIS's. In our previous studies, the increase of Ar multicharged ion yield has been confirmed by the upper hybrid resonance (UHR) [4] using 4-6 GHz X-mode microwaves, the dual ECR heating method [5] in which microwaves are fed from two antennas simultaneously. We are attempting to increase the multicharged ions using lower hybrid resonance (LHR) by introducing low frequency electromagnetic waves (radio frequency waves: RF waves). It has been suggested that gas mixing increases the multicharged ion due to the ion cooling effect. For this reason, we conducted experiments using Ar as main gas and He as low Z gas. We investigated the change in the multicharged ion yield, the plasma parameters, the electron density n_e and the electron temperature, T_e and impurity ratio due to gas mixing.

2. Experimental apparatus and procedure

2.1 Experimental apparatus

Figure 1 shows the top view of ECRIS. The vacuum chamber has diameter of 160 mm and length of 1074 mm. The magnetic field is formed by mirror coils (Coils A and B), an auxiliary coil (Coil C), and four permanent magnets (octupole magnets). The cartesian coordinate system (x,y,z) is used, with the origin at the center of the vacuum chamber. The chamber is evacuated by a diffusion pump and the beamline by two turbo monocular pumps to maintain a base pressure $10^{-5\sim-6}$Pa. Ar and He gases are introduced into the vacuum chamber by mass flow controllers, and the pressure is directly monitored by B-A gauges. The pressures of Ar and He are calibrated by using the specific sensitivity coefficients, respectively. The magnetic field is formed by coil currents I_A, I_B and I_C and octupole magnets, and the resonance point of the ECR shown near the mirror bottom in Figure 1. The dashed line indicates the magnitude of the magnetic field corresponding to the ECR point at microwave frequency 2.45 GHz (0.0875 T). The 2.45 GHz microwaves generated by the magnetron are emitted from the rod antenna (z=175 mm) installed at the top of the vacuum chamber. The microwaves can be switched between the continuous wave (CW) mode and the external mode. The external mode is used to introduce pulse-modulated microwaves corresponding to pulse signals transmitted from the function generator. In this experiment, we used mainly CW mode. The incident and reflected microwave power are defined by P_{in} and P_{ref}. In the present experiment, $I_{A,B}$ = 150A and P_{in} = 100W.

The plasma parameters, the electron density n_e and the electron temperature T_e were measured by Langmuir probe LP1 (z=-175mm) and LP2 (z=175mm) with cylindrical Mo electrodes (0.5mm in diameter), which can be moved horizontally in the range of x=0~50mm.

The ion beam is extracted into the beam line, which consists of an ion beam extractor and an Einzel lens EL1. The ion beam extractor consists of a plasma electrode PE (15mm in diameter) and intermediate electrodes E1 (13mm in diameter) and E2 (13mm in diameter). The voltages applied to the PE and E1 electrodes and EL1 are defined as V_{PE},V_{E1}, and V_{EL1} respectively. In normal operation, V_{PE} was applied 10 kV. V_{E1}, V_{EL1} are used to optimize the extraction of each ion species. The ion beams are mass and charge separated by a sector magnet. After adjusting the current value with a movable slit, the ion beam current is collected by a faraday cup (FC) and amplified by a log amplifier or a linear amplifier to obtain its charge state distribution (CSD).

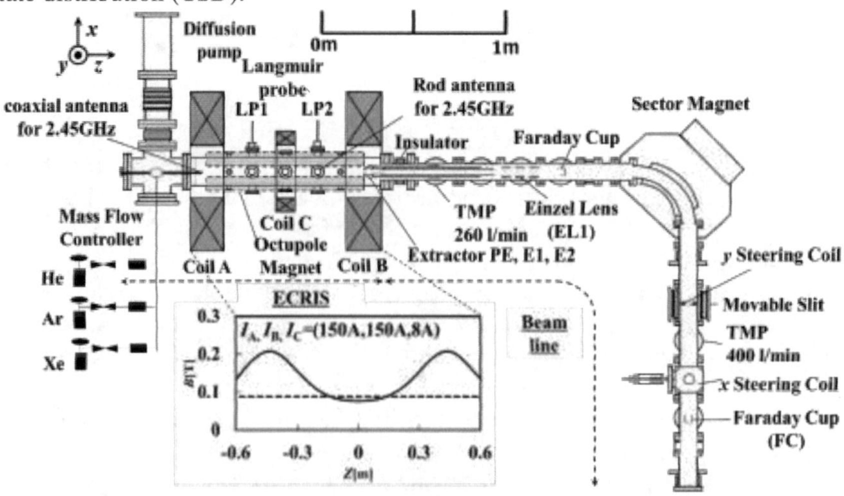

Figure 1. Top view of ECRIS

2.2 Experimental procedure

In the first step, the mixing ratio of He was varied with the total pressure kept constant, and the He mixing ratio that gives the maximum yield of Ar^{6+} ions were investigated. Then we surveyed the total pressure.

In the second step, the conditions (pressure, I_A, I_B and I_C, P_{in}) under which the Ar^{8+} ion yield is maximized without mixing He are investigated. After that, He is mixed under the previous conditions. We observed the change in Ar ion beam current and investigated the change in the multicharged ion yield, the plasma parameters and impurity ratio. The reproducibility was confirmed by repeated several times with and without He flow.

3. Experimental results

3.1 mixing He gas into Ar plasma under constant total pressure

Figures 2(a) and (b) show the dependence of the Ar ion beam on the He mixing ratio in mixing He into Ar plasma. The horizontal axis shows the He mixing rate, which is determined by the partial pressure of each gas. The total pressures are 3.0×10^{-4} Pa and 7.0×10^{-4} Pa, respectively. The extraction is optimized to maximize the Ar^{6+} current by adjusting V_{EI} and V_{ELI}. It is confirmed that the Ar^{6+} ion beam is maximized at the mixing ratio of about 30% at the total pressure of 3.0×10^{-4} Pa and about 40% at the total pressure of 7.0×10^{-4} Pa.

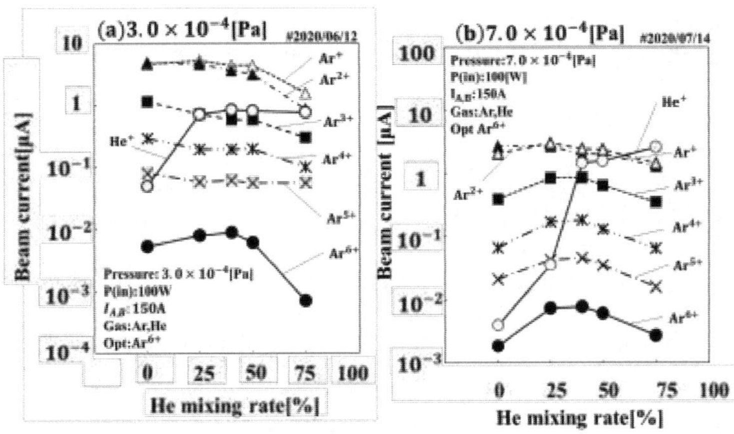

Figure 2. Dependence of Ar ion beam current on He mixing ratio under constant total pressure

3.2 Mixing He gas into Ar plasma under constant Ar partial pressure

Figure 3 shows the CSD for a typical Ar beam. The Ar operating pressure is 2.0×10^{-4}Pa. $I_{A,B}$ are 150A, P_{in} is 100W. These values are optimized to maximize the Ar^{8+} beam current. The blue line shows the CSD when the beam is extracted with pure Ar. Ar ions with charge states from 1 to 8 can be identified and then introduce He gas into Ar plasma. The red line shows the CSD obtained of the Ar ions beam in mixing He into Ar plasma. It is found that the beam current increases for $Ar^{3\text{-}8+}$. Especially, the Ar^{8+} beam current became about 5.5 times larger than that in the case of pure Ar.

Figure 4(a) shows the current values for each charge state for the CSD shown in Figure 3. The horizontal axis is the Ar ion charge state q and the vertical axis is the beam current. The experimental conditions are the same as those in Fig. 3. Each point shows the average value of the Ar^{q+} current in the cases of pure Ar and Ar/He mixing measured three times, and the error bars show the standard deviations. It can be seen that the mixing He into Ar plasma clearly and reproducibly increases the yield of Ar multicharged ions for $Ar^{3\text{-}8+}$ ion beam current. Figure 4(b) shows the Ar ion beam increase rate for each charge states. The horizontal axis shows the Ar ion charge state q and the vertical axis shows the increase rate of each Ar ion charge state in the case of pure Ar and Ar/He mixing. The experimental conditions are the same as in Fig. 3. It was confirmed that beam increase ratios were roughly proportional to the charge state q, especially those of Ar^{7+} and Ar^{8+} were remarkable large.

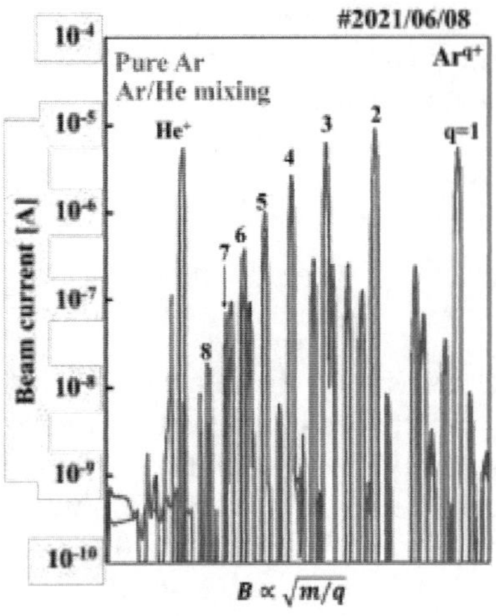

Figure . 3 Typical Ar ion beam CSD under constant Ar partial pressure

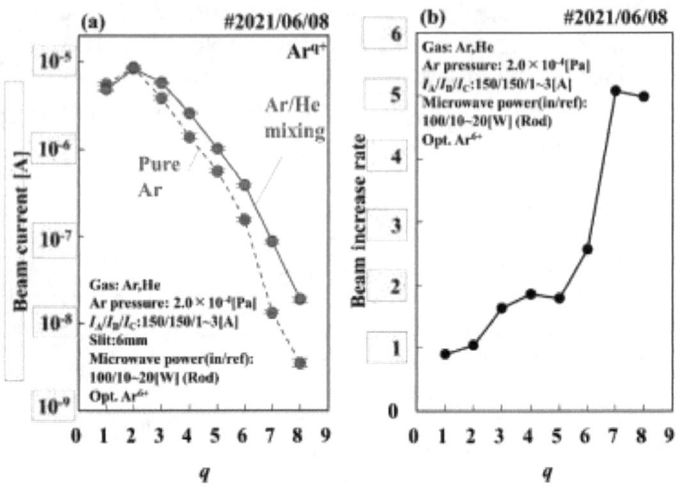

Figure 4. Effect of mixing He under constant Ar partial pressure (a) Ar ion beam current increase rate(b)

3.3 Probe measurement with/without gas mixing

Figure 5 shows the distributions of n_e and T_e in the cases of pure Ar and Ar/He mixing. The measurements were performed three times at each position using LP2, and the mean values and standard deviations are shown. For the points between x =10 and 40, reliable data could not be obtained due to the probe emission by the plasma.

The values of n_e are similar in the 2 cases for all the points. T_e was smaller for in the case of Ar/He mixing than the case of pure Ar at all positions. Therefore, it is considered that the increase in the multicharged ion beam yield is not due to the effect of plasma parameters, but rather to the cooling effect of the ions due to the gas mixing.

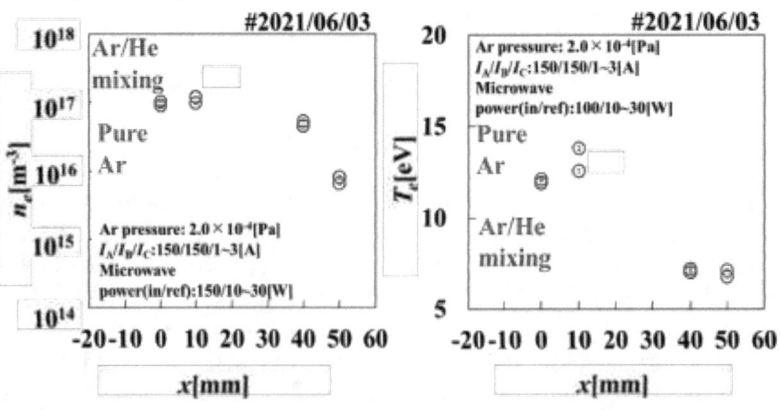

Figure 5. n_e, T_e in radial direction on pure Ar and Ar/He mixing

3.4 Effect of impurities with/without gas mixing

Figures 6 shows the Ar ion beam increase rates for each charge states. The three cases are shown in Fig. 6(a), (b), and (c). The experimental conditions are the same for (a), (b), and (c). The Ar operating pressure is 2.0×10^{-4}Pa, $I_{A,B}$ = 150A and P_{in} = 100W. The increase rates of Ar^{8+} are (a), (b), and (c) in the order of increasing, and the increase rates are 4.99, 3.18, and 1.50, respectively. The plasma discharge time just before each experiment is 10 hours in (a), 8 hours in (b), and 0 hours in (c) within 1 day.

The impurity ratio α_{imp} is the percentage of the beam current of all impurities except the Ar^{q+} and He^{+} ions current to the total ion current extracted. Figures 6. below shows the impurity ratio in each experiment in the CSD, where the vertical axis shows α_{imp}. In the case of (a) and (b), where the effect of He mixing was confirmed, it was about 5% for both pure Ar and Ar/He mixing, while in the case of (c), where the effect was not observed, it was about 10%.

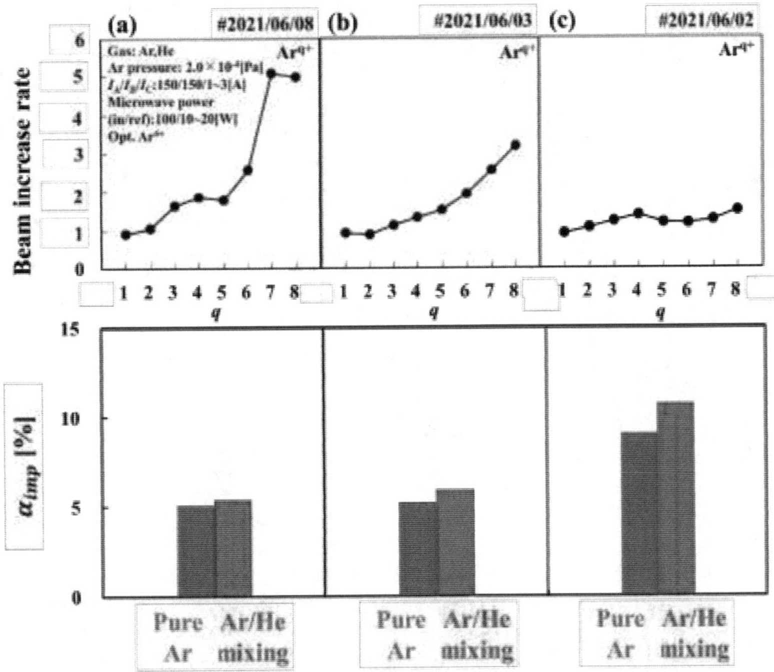

Figures 6. Effect of He mixing on Ar gas under the same conditions (a), (b), (c) impurity ratio on Ar gas under the same conditions (d)

4. Summary & future planning

Under the conditions where the $Ar^{6\sim8+}$ ion beam yield is maximized in the case of pure Ar, we have confirmed that the current value increases with sufficient reproducibility by mixing He. In order to obtain a sufficient gas mixing effect, the ratio of impurities must be less than around 5%. Since there is no significant change in n_e and T_e with and without mixing He to Ar plasma, it can be concluded that the increment of Ar multicharged ion currents with mixing He is not due to the change of plasma parameters, but maybe mainly due to cooling effects.

The emittance measurement will be carried out to estimate the ion temperature T_i with and without the low Z gas mixing in order to obtain for further confirmation of the selective heating of low Z ions. In addition, we will investigate the introduction of pulse-modulated microwaves, dual ECR heating, and more over low frequency electromagnetic waves under gas mixing.

References

[1] Kitagawa A, Fujita T, Muramatsu M, Sakamoto, Sakuma Y, Sasaki N, Sasano T, Takasugi W, Biri S, and Drentje A G 2010 in *proceedings of the 19th International Workshop on ECR Ion Sources (ECRIS'10)*, Grenoble, France, 23-26 August, http://www.jacow.org, pp. 20-22

[2] Drentje A G, 2003 *Rev. Sci. Instrum.* **74**, 2631

[3] Muramatsu M, Hojo S, Iwata Y, Katagiri K, Sakamoto Y, Takahashi N, Sasaki N, Fukushima K, Takahashi K, Suzuki T, Sasano T, Uchida T, Yoshida Y, Hagino S, Nishiokada T, Kato Y, Kitagawa A 2016 *Rev. Sci. Instrum.* **87**, 02C110

[4] Kato Y, Nishiokada T, Hamada K, Onishi K, Takeda T, Okumura K, Omori T, Kubo W, Ishihara M and Harisaki S 2020 *Rev. Sci. Instrum.* **91**, pp.013315-1-6.

[5] Kubo W, Harisaki S, Owada I, Sato K, Tsuda K and Kato Y 2021 *Rev. Sci. Instrum.* **92,** pp.043514-1-9.(online publication: 2021.4.2, https://doi.org/10.1063/5.0035631)

Experiments to improve the performance of the GTS-LHC ECR ion source

T Kövener, D Küchler, E Mahner, R Scrivens, A Baris, S Pfeiffer, L Mourier, W Vollenberg

CERN, CH-1217 Geneva, Switzerland

Richard.Scrivens@cern.ch

Abstract. The GTS-LHC 14.5 GHz ECR ion source provides the ion beam, which after acceleration in the ion injector complex, is injected into the LHC, as well as being sent to fixed target experiments at CERN. Two experiments have been performed on the source. The stainless steel plasma chamber has been sputter-coated with a rather thick layer of aluminium on the surface facing the plasma; and the lead micro-oven has been modified to avoid the build-up of lead-oxide on the oven outlet. Details of the changes will be given, and results of beam measurements will be shown, with particular attention on how the stability and oven-refill schedule is impacted by these changes.

1. Introduction

CERN's ion programme uses as a source an Electron Cyclotron Resonance Ion Source (ECR) constructed by CEA Grenoble in 2005 [1]. The source uses 14.5 GHz microwaves injected into a magnetic structure consisting of three normal conducting solenoids and a permanent hexapole magnet. Mostly lead ions are requested, hence the source uses a filament heated micro-oven to evaporate isotopically pure lead, as well as an oxygen support gas. The source uses the after-glow mode, where a 50 ms microwave heating pulse with a 10 Hz repetition rate is applied, after which a more than 1 ms long, high intensity pulse of highly charged lead ions is extracted. Ions are extracted with a three-electrode extraction system, with a total potential of 18.8 kV, and the Pb^{29+} is selected in a low energy spectrometer for acceleration through a 101 MHz RFQ, before further acceleration, stripping and injection into CERN Low Energy Ion Ring (LEIR). Further details on the source and previous upgrades can be found in [2], [3].

Two modifications to the source have been made recently and tested, in order to explore possible improvements in performance and long-term stability. These modifications and the results will be detailed in the following sections.

All the beam results presented in this report are for Pb^{29+}. As a source for an accelerator, the most important performance parameters are the beam intensity transported through the RFQ, and the shot-to-shot stability. The highest intensity through the RFQ does not always correspond to the highest beam intensity measured out of the source.

2. Plasma Chamber

The Linac3 source was originally designed with an aluminium (Al) plasma chamber, which has a length of 637 mm and an internal diameter of 78 mm. This was exchanged for a stainless steel (SS) version in 2010 [4], which led to an improvement in stability of the source, at the expense of needing a higher microwave power, and was less prone to damage. The lower stability of the aluminium chamber was suspected to be due to the lower rigidity of the chamber, which allowed very small leaks to develop in the O-ring vacuum sealing as the temperature varied.

To investigate if some of the benefits of lower microwave power can be recovered, an SS plasma chamber with a 20 μm thick coating of aluminium along the plasma facing surface, produced by dc magnetron sputtering in argon, denoted here as the Al/SS chamber, was tested.

The Al/SS chamber was installed for approximately 1500 hours of test running, and is compared with the beam performance reached with stainless steel chambers, at different times, in Table 1. All data from 2021 are taken with the beaked crucible oven (see next section). The beam intensity out of the RFQ is a critical performance indicator for Linac3 operation. However, this intensity is not recorded every pulse by the Faraday Cup. The source beam intensity measured with a transformer is also given, as well as the standard deviation of the intensity of all recorded pulses (averaged during the 200 μs pulse) over a 72 hour window, where pulses below 25% of the average are removed.

Concerning the Al/SS chamber test period, it is divided into two phases, the low-power (LP) phase where a microwave power of 400-600 W was sufficient for high intensity operation, and high-power (HP) phase where 1000-1500 W was required. In between these two phases, the source was very unstable. This was coincident to the microwave generator failing, which may have affected stability and lead to a 2 week pause in data. During the LP phase higher microwave powers did not lead to better source performance, whereas once the HP phase was reached, the source stability improved. Results for a new stainless steel chamber in 2021 are also given after about 200, 1000 and 2000 hours of operation and conditioning, where the first two periods correspond to the same periods after which the Al/SS chamber is assessed for LP and HP phases.

Once the Al/SS HP phase is reached, it is possible to have an intense stable beam. It was not possible to run the source in this test phase long enough to compare the performance of the source after 4000 hours of operation and regular optimization, as for the stainless steel chamber in 2018.

Figure 1 shows the beam intensity development over 300 hours, starting 200 hours after the source was ignited, for the stainless-steel, and Al/SS chambers.

Table 1. Beam performance for bare stainless steel (SS) and aluminium coated stainless steel (Al/SS) plasma chambers. Intensities given correspond to the best achieved value over approximately three days. Stdev is the standard deviation of measurements made over the assessment period of 72 hours, except (*) which was assessed over 36 hours only.

Plasma Chamber	Year	Start time for data analysis (Hours)	Source Beam Intensity Pb29+ (μA)		RFQ Beam Intensity Pb29+ (μA)
			mean	stdev	
SS	2018	4000	190	9	140
Al/SS (LP)	2021	200	190	14	140
Al/SS (HP)	2021	1000	175	10*	130
SS	2021	200	150	8	115
SS	2021	1000	150	8	115
SS	2021	2000	160	10	130

The Al coated stainless steel chamber has been analysed after dry cutting the chamber open, performing fast ion milling (with an added Pt protection layer) to expose a cross section of the material to a depth of ~50 μm and then performing secondary electron microscopy, allowing a cross-section of the chamber to be exposed, and analysed (see Figure 2 Left). Over most of the chamber surface, a layer containing Pb-O-C has been deposited on top of the Al coating. The depth of this Pb-O-C layer is variable, with a thickness up to 20 μm, with the thickest layer corresponding to the surface close to the Pb oven. The deposited Pb-O-C is mostly porous, and in all cases a thin intermediate layer of Al has formed within it. However, the three regions where the electron losses mostly occur, due to the hexapole magnetic field, are free of this Pb-O-C layer. At no location is the Al layer seen to be significantly removed. The analysis is still ongoing and will be fully reported in the future.

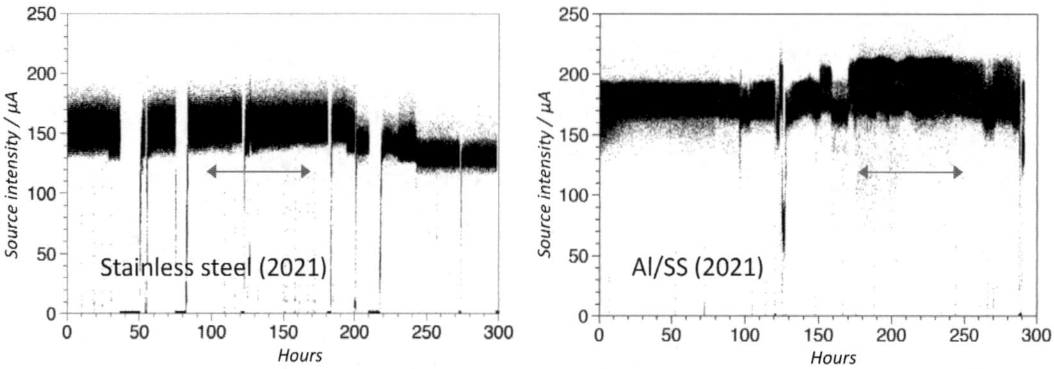

Figure 1. Source intensity of Pb^{29+} using Left: Stainless steel chamber; Right: Stainless steel chamber coated with aluminium. The arrow indicates the period used in Table 1 for the average and standard deviation of the intensity.

In summary, the Al coated plasma chamber initially provides a high intensity beam more quickly after start-up, although with poorer stability, but had a transition through a significant change in performance between 200 and 1000 hours, requiring a significant increase of the microwave power. The bare stainless steel plasma chamber starts with lower intensity, but after approximately 2000 hours its performance is very similar to the Al coated chamber. From this, we hypothesize that the Al surface initially delivers a high number of secondary electrons to the plasma, but as both chambers become coated with the Pb layer, the difference between them diminishes.

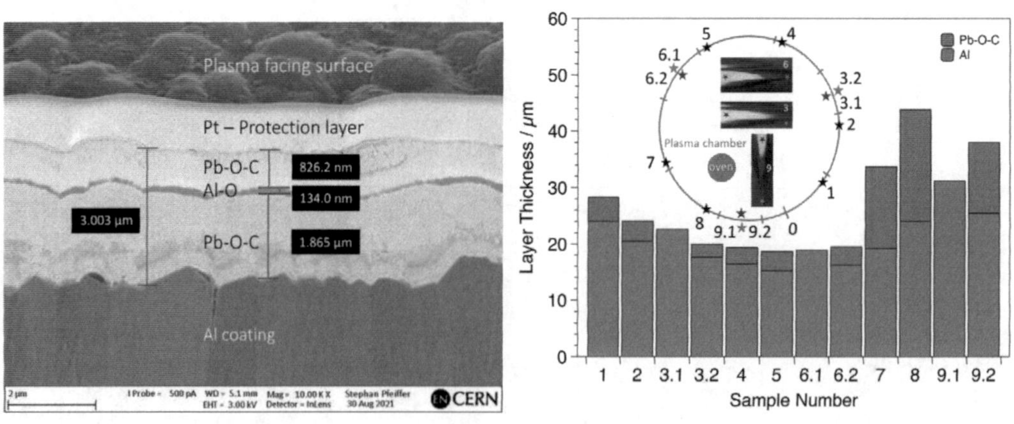

Figure 2. Left: Scanning Electron Microscope image of the Al coated plasma chamber surface, with the different surface layers marked. Right: Thickness of the Pb-O-C (red) and Al (blue) layers determined at 12 positions of the plasma chamber; samples #3, #6, and #9 were analysed at two different locations, i.e. inside visible "light" and "dark" layer areas (see insert). The oven position relative to the sample locations is marked in the insert.

3. Micro-oven

To evaporate lead into the plasma chamber of the ECR ion source a micro-oven containing an alumina crucible with the capacity of around 1.5 g of lead is used. A tantalum filament resistively heats this oven while a feedback loop stabilizes the applied heating power at a chosen level. During the operation of the source in the year 2018 and before, the applied power needed almost daily adjustment to maintain a stable beam output. After two weeks of uninterrupted operation, the oven needed a refill of the lead sample and a cleaning of its orifice, even though the crucible still contained around half of the initial lead sample.

A dedicated study to understand the reasons for the necessity of the frequent tuning and the failure mechanisms of the oven was conducted, including thermal simulations, gas flow simulations and measurements at an offline test stand [3], [5], [6]. Trials at the test stand showed the main failure mechanism to be the formation of a lead oxide blockage starting from lead condensate on the outer oven cover (also in [7]). The outcome of the study were proposals to prevent this blockage, with the least invasive being a modification of the oven crucible (see also [8]). By giving the crucible a prolongation at its orifice (for convenience called a beak), the Pb atoms no longer have a direct path towards the cover, while the crucible tip remains hot enough to prevent any condensation on itself. Figure 3 shows a thermal simulation of the conventional and the modified oven orifice. More details on the simulation model can be found in [6]. For a heating power of 8.5 W here the temperature of the crucible tip is 714 °C (without beak), 686 °C (with beak) while the outer cover only has 440 °C.

Testing with the modified crucibles at the GTS-LHC ion source started end of November 2020 and lasted all runtime of the source in 2021 so far. Here we present the experience during the longest possible operation of one lead filling with the new crucible design and compare it to the performance seen in 2018.

In order to operate the oven with the modified crucible the heating power was ramped up to 7 W. During the run the operator then adjusted the power as part of the source tuning to maintain a stable source output. In contrast to previous runs the refill of the oven was not pre-scheduled but it was operated until no further lead beam could be extracted from the source.

The longest oven fill lasted 59 days and upon inspection of the oven tip after the run no lead oxide blockage had formed. In this configuration it was possible to operate the source until the lead filling was completely consumed. Throughout the run the heating power was adjusted several times, usually in steps of 0.1 W leading to a final power of 9.1 W.

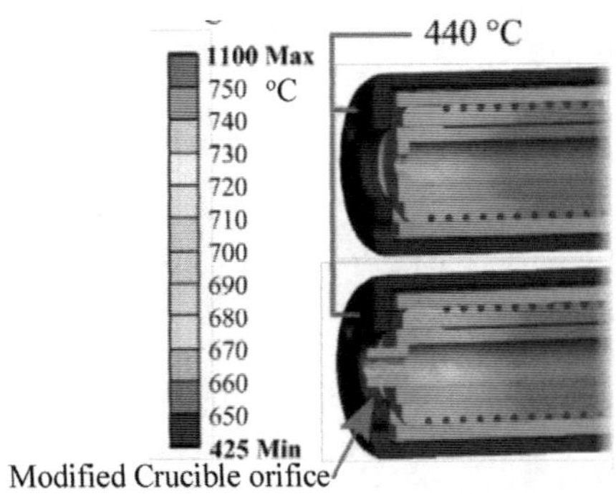

Figure 3. Simulations of the thermal profile of the micro-oven with Ansys [9]. The lower model includes the modified crucible. In both cases a heating power of 8.5 W is simulated by setting the filament temperature to 1100 °C.

Figure 4 shows the oven power and the Pb^{54+} beam current of each Linac3 pulse measured in the beam current transformer ITF.BCT25 at the end of Linac3 during the presented run, together with the same data during several refills of 2018 in a comparable timeframe (11.10.2018 - 11.12.2018), also spanning over the heavy ion run of the LHC. The red line shows the beam current requirement from the subsequent accelerator LEIR which is 30 μA of Pb^{54+}. It can be seen that Linac3 could provide the required beam output in 2021 for more than 30 days even though the delivery of beam to LEIR only started after the oven already ran for 17 days. During this time, the beam current did not drift significantly which can partly be attributed to a stable oven performance.

This experience differs from the previous operation in 2016 and 2018 where it was regularly necessary to ramp up the heating power up to 20 W during the source operation. The longest operation of one oven filling in 2018 lasted 16.8 days and on average a fill was operated for 14 days. Note that in 2018 the refills were pre-scheduled and the oven was not run as long as possible.

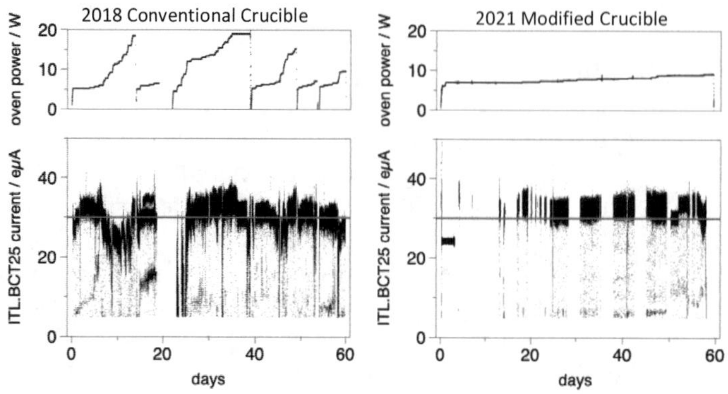

Figure 4. Plots of the oven power and the beam current measured in beam current transformer ITF.BCT25. Left: Selected timeframe of 2018 also spanning over the heavy ion run of the LHC, with six oven refills. Right: The longest run so far with the modified crucible. The red line indicates the intensity target.

In conclusion, the modified crucible successfully prevented the formation of a blockage at the oven orifice as it was expected from the previous test stand measurements. With no blockage forming, the oven could be operated until the lead sample was consumed (consumption rate 1 mg/h, in 2018 it was on average for the reference period 1.5 mg/h), which reduces the number of necessary refills or cleanings. This is promising especially considering the LHC heavy ion run that usually lasts four weeks and could be performed without having to include the downtime of an oven refill. Additionally, the oven needs less tuning as there is no closing orifice that needs to be compensated by higher vapour pressures.

References

[1] Dumas L, Hill C, Hitz D, Küchler D, Mastrostefano C, O'Neil M and Scrivens R 2006 Operation of the GTS-LHC Source for the hadron injector at CERN, *Proc. 17th Int. Workshop on ECR Ion Sources and Their Applications,* Lanzhou, China, p51

[2] Küchler D, O'Neil M and Scrivens R 2008 First experience with the operation of the GTS-LHC ion source at 18GHz, *Proc. 18th Int. Workshop on ECR Ion Sources ECRIS'08,* Chicago, Illinois

[3] Toivanen V *et al*, 2016 Recent developments with the GTS-LHC ECR ion source at CERN, *Proc. 17th Int. Workshop on ECR Ion Sources and Their Applications,* Busan, Korea, WEAO01

[4] Küchler D, Bellodi G, Lombardi A, O'Neil M, Scrivens R, Stafford-Haworth J and Thomae R 2012 Operation experience with the GTS-LHC ion source and future developments of the CERN Ion Injector, *Proc. of ECRIS2012,* Sydney, Australia, TUPP10

[5] Fichera C, Carra F, Küchler D and Toivanen V 2018 Numerical study of the thermal performance of the CERN Linac3 ion source miniature oven, *Nuclear Inst. and Methods in Physics Research, A* **901** p21

[6] Koevener T, Study of the micro oven at the Linac3 heavy ion source, Thesis, Staats- und Universitätsbibliothek Hamburg Carl von Ossietzky, Hamburg (2020), https://ediss2.sub.uni-hamburg.de/handle/ediss/8493?mode=full

[7] Koevener T, Küchler D and Toivanen V, 2018 Study of the lead evaporation from the oven of the GTS-LHC ion source, *Proc. ECRIS2018,* Catania, Italy, TUP21

[8] Koevener T, Küchler D and Toivanen V, 2020 Lead evaporation instabilities and failure mechanisms of the micro oven at the GTS-LHC ECR ion source at CERN, *Proc. ICIS2019, Lanzhou, China, Rev. of Sci. Instrum.* **91** 013320

[9] The official Ansys simulation tools website. https://www.ansys.com/

Progress on the development of key technologies for the fourth generation ECR ion source FECR

L Sun[1,2,b], W. Lu[1], H. W. Zhao[1,2,a], J. W. Guo[1], W. Wu[1], B. M. Wu[1], X. J. Ou[1], E. M. Mei[1], S. J. Zheng[1], X. Z. Zhang[1], L. B. Li[1], L. Zhu[1], C. J. Xin[1], M. Z. Guan[1], Y. Q. Chen[1], X. D. Wang[1]

[1]Institute of Modern Physics, Chinese Academy of Sciences, Lanzhou, China
[2]School of Nuclear Science and Technology, University of Chinese Academy of Sciences, Beijing, China

[a]zhaohw@impcas.ac.cn, [b]sunlt@impcas.ac.cn

Abstract. A 4th generation ECR (Electron Cyclotron Resonance) ion source FECR (First 4th generation ECR ion source) is under development at IMP. Aiming to be operated with the microwave power of 20 kW at 45 GHz, FECR is equipped with a fully superconducting Nb_3Sn magnet and conventional parts durable for high power ECR plasma heating and optimum for intense beam production and extraction. Breakthroughs on the Nb_3Sn superconducting magnet, high power density plasma chamber, high temperature oven, and so on have been successfully demonstrated recently. The prototype Nb_3Sn cold mass has been successfully tested. The plasma chamber with an innovative structure of micro-channel cooling structure has been tested which enables SECRAL-II ion source operated continuously for more than 1,100 hours with ~300 eμA Kr^{26+} for routine operation at the microwave power level of ~5 kW. A high temperature inductive heating oven has also been developed and used for intense uranium beam production. This talk will present the recent progress on the development of key technologies for the 4th generation ECR ion source.

1. Introduction

Heavy ion accelerators are the most powerful tools in modern nuclear physics research. Driven by the needs of large scale heavy ion accelerators, such as FRIB at MSU, RIBF in RIKEN, HIRFL at IMP and so on, ECR ion sources have been continuously supported and improved. The recently started national basic science research project called HIAF or High Intensity heavy ion Accelerator Facility [1], which is composed by high performance Superconducting ECR ion sources (SECR), SRF (Superconducting Radio Frequency) linac (iLinac), BRing (Booster synchrotron Ring), HFRS (High energy FRagment Separator) and SRing (Spectrometer synchrotron Ring), will accelerate and accumulate 1×10^{11} ppp U^{35+} of 800 MeV/u in BRing (figure 1). The injector accelerator iLinac is designed to work in two modes, i.e. the injector for BRing that needs to deliver pulsed U^{35+} beam with a peak beam intensity of ~1 emA at the energy of 17 MeV/u, and the heavy ion beam driver for the low energy experimental terminals that needs to deliver CW beams with the $M/Q\leq5$, typically U^{46+}. Consequently, the warm front end part will be able to produce and deliver high intensity U^{35+} beam at a afterglow mode and high intensity U^{46+} at a CW mode. A 70% efficiency is estimated to transport the intense U^{35+} with high quality to iLinac, and as a result ~40 pμA U^{35+} is literally needed for routine operation if the peak performance of HIAF is desired. Given some margin, the ion source should be able to produce 50 pμA U^{35+} when fully optimized to guarantee the routine operation performance. For CW beam U^{46+}, it is said to be as high as possible. A typical beam intensity of 10 pμA (or 460 eμA) is proposed which is about one order higher than the existing beam record [2]. Table 1 gives the typical parameters of HIAF and the requirements to the ion source.

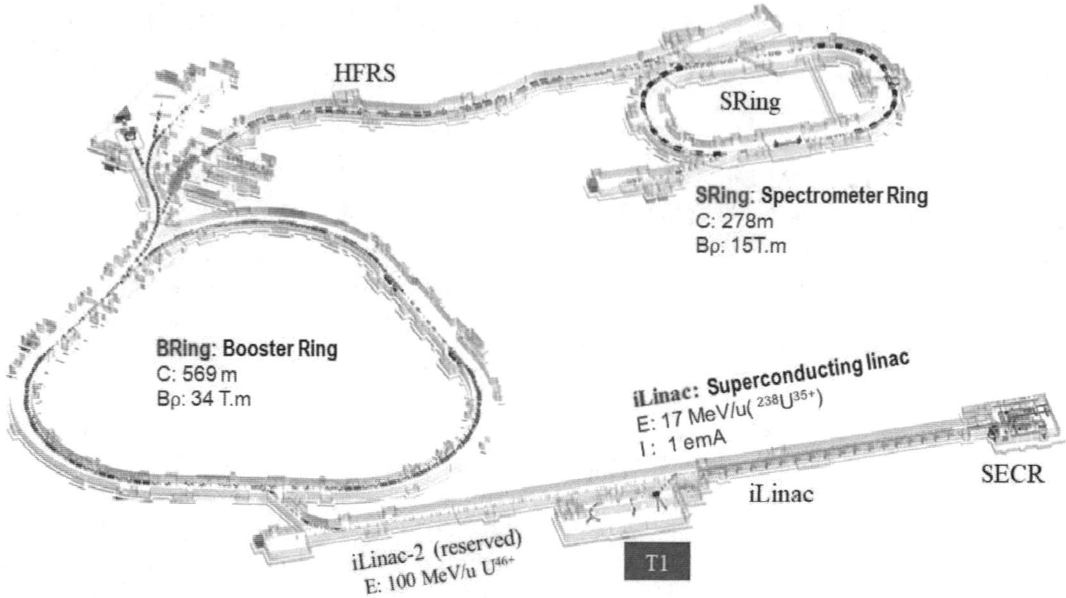

Figure 1. HIAF facility layout.

Table 1. Typical accelerator specifications of HIAF

Specs.	SECR	iLinac	BRing
Length / circumference (m)	/	114	569
Beam Species	H_2^+~U^{n+}	H_2^+~U^{n+}	H~U
Final energy of U (MeV/u)	0.014 (U^{35+})	17 (U^{35+})	835 (U^{35+})
Max. magnetic rigidity (T.m)	/	/	34
High Power Mode (pulsed)	50 pμA (U^{35+})	28 pμA (U^{35+})	10^{11} ppp (U^{35+})
High Energy Mode (CW)	>10 pμA (U^{46+})	>6 pμA (U^{46+})	/
Operation mode	DC	CW	/
	Pulsed 0.2~2 ms/3 Hz	Pulsed 0.2~2 ms/3 Hz	fast ramping (12 T/s, 3 Hz)
Emittance or Acceptance (H/V, π·mm·mrad, dp/p)	≤0.2 (n.rms)	5 / 5	200/100, 0.5%

The 3rd generation sources are regarded as the iconic machines of modern ECR ion sources which have been built since 20 years ago. Heated by the microwave of 24~28 GHz frequency and 10 kW maximum power, these sources such as VENUS in LBNL, SCECRIS in RIKEN, SuSI at MSU, SECRAL at IMP and so on, can produce (or potentially) U^{3x+} ion beams of the intensity 10~15 pμA which is more than 10 times the records set 20 years ago. But to meet the needs of HIAF, the performance still needs to be boosted by a factor of 3~4. As ECR ion source development is mainly governed by the scaling of $I_q \propto \omega_{ecr}^2$, the straightforward technical path is to develop a 4th generation machine. FECR ion source, which is one of the key components of the LEAF (Low Energy heavy ion Accelerator Facility) project [3], will develop the first 4th generation ECR ion source in the world. This project was literally started in 2015 from draft. As moving from a 3rd generation ECR ion source such as SECRAL to FECR, it is not just a scaling of microwave frequency or magnetic field strength, many challenges could be foreseen as already indicated during the development of the 3rd generation ECR ion source, such as the Nb_3Sn superconducting magnet, high power high frequency microwave coupling, high power ECR heating and long-term reliability in terms of the plasma chamber cooling and so on. Those issues will be detailed in this paper. Production of refractory metallic ion beams and high yield long pulse (>2 ms) ion beam production via afterglow mode are also big challenges to FECR project. We have made fundamental progresses recently, but as they are also general topics for other high performance ECR ion sources, these topics will not be discussed in this paper.

2. FECR cold mass development

In year 2015~2016, we were mainly in the discussion of the technical paths, for instance, the traditional sextupole inward of solenoids configuration or the innovative SECRAL reversed one to make the cold mass structure, Nb_3Sn cable or single strand to wind the coils, and so on. In year 2016~2017, we were working closely with the ATAP Group in LBNL on the cold mass structure design of FECR [4]. During the time of 2016~2018, we were mainly working on the prototype sextupole coil with single strand. From 2017 to 2019, the FECR team were mainly working on a 1/2 size prototype cold mass. In 2020, we have demonstrated the 1/2 prototype cold mass with more than 80% of the design currents.

Table 2. Critical parameters of FECR Magnet

Spec.	Unit	FECR			
B_{max} in conductor	T	~11.8			
Conductor	/	RRP Nb_3Sn			
Warmbore ID	mm	Ø162			
Mirror Fields	T	6.4/3.5			
B_{rad}@chamber inner wall	T	≥3.2			
Mirror Length	mm	500			
Nominal operation currents	A	Inj	Mid	Extr	Sext
		663	372	600	709
Gross weight	t	6.1			
Stored Energy	MJ	1.6			
Cooling Power@4.3 K	W	≥12.0			

FECR magnet is mainly composed by a fully Nb_3Sn superconductor cold mass and a cryostat integrated with 7 cryocoolers to recondense the evaporated helium gas. FECR is designed to be operated at the optimum microwave frequency of 45 GHz, and Table 2 gives the critical parameters of FECR magnet. As FECR magnet is the first in the world using Nb_3Sn superconductor to provide min-B field for dense plasma confinement, many challenges and uncertainties will exist during the magnet development. Compared to the NbTi technology widely used with existing 3rd generation ECR ion sources, Nb_3Sn technology features complicated procedures to make coils, wire brittleness, stress sensitive, stringent quench protection requirement and so on, other than high J_c. To demonstrate the feasibility and also lower the technical risk, we decided to make it in three phases, i.e. Phase I to prototype the coils and test technologies, Phase II to prototype a half-length cold mass and the related technologies, and finally the Phase III to develop the full-sized cold mass that will be put into a cryostat for operation.

2.1. Structure design

The coil system is based on the traditional sextupole-in-solenoid configuration (see figure 2), which allows minimizing the magnetic field in the sextupole coil conductor by placing it close to the plasma chamber. The axial field confinement is provided by solenoid coils consisting of an injection, middle and extraction solenoid. This coil configuration is used in present-day superconducting ECR magnets - including VENUS at LBNL [5] and SuSI at Michigan State University [6]. However, the higher confinement field required for 45 GHz operation and the transition from Niobium-Titanium (NbTi) to Niobium-Tin (Nb3Sn) superconductor introduce new, significant challenges, particularly in terms of the stress management of the brittle Nb3Sn superconductor. Special tooling and fixtures are required for handling and reacting Nb3Sn coils after winding. Also, strict limits on the maximum stress and strain need to be respected to prevent conductor degradation and a magnet structure has to prevent any conductor motion while minimizing the radial gap between the sextupole and solenoids.

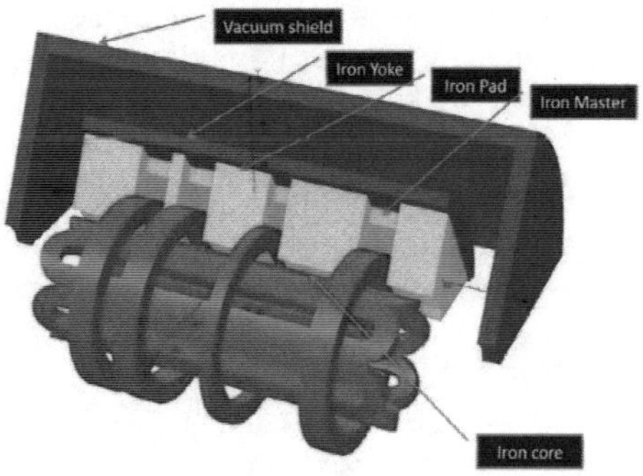

Figure 2. FECR magnet configuration.

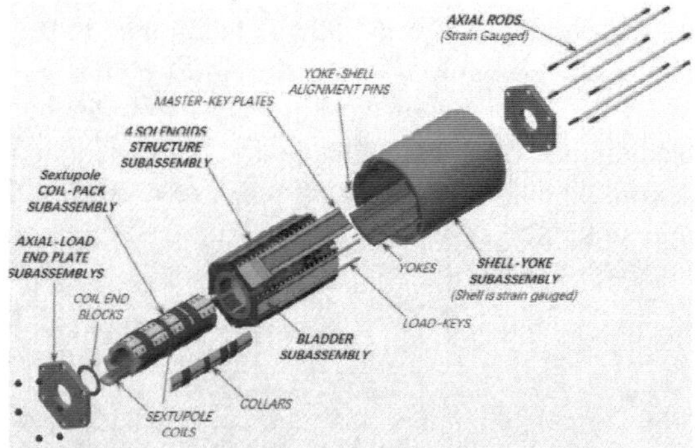

Figure 3. Exploded view of FECR magnet structure.

The new support structure is based on an aluminum shell surrounding the coils and iron yoke and is pretensioned using water-pressurized bladders and interference keys. This concept was originally developed for the LBNL high field dipole program [7-9] and adopted by LARP [10]. However, the conventional application of this approach to ECR magnets requires significant modifications from the preloading of accelerator magnets, due to the nesting of sextupole and solenoid coils, and the complex Lorentz force configuration introduced by their combined fields and currents. In a previous design study for a 56 GHz ECR [11], a two-stage system was adopted, with the sextupole pre-loaded by the solenoid assembly, and the solenoid assembly pre-loaded by the aluminum shell (figure 3). For the 45 GHz design, we developed a different approach where the aluminum shell supports the sextupole through longitudinally segmented loading pads placed in between solenoids (figure 4). These sextupole loading pads directly bear against thin curved, full length "collar" plates that are bolted around the coils. This approach required splitting the longer injection solenoid into two sub-coils with an axial gap, in order to reduce a long, unsupported longitudinal span between sextupole pads. Each of the solenoids are individually wound around a stainless steel bobbin and are radially pre-loaded by tensioned aluminum banding. The solenoids are separated and positioned longitudinally, over the sextupole coil-pack by a series of stainless steel "intercostal" elements. The aluminum shell provides the final preload through a second set of loading pads located in-between the pads of the first set. Longitudinal pre-load is applied to both sextupole and solenoid sub-assemblies by aluminum rods and end plates. Results of the 2D analysis showed that the proposed concept of the support structure is capable of sufficiently preloading the sextupole coil. After the cool-down, the model showed a maximum stress of approximately 143 MPa in the inner radius/pole corner areas, which is much lower than the design before the optimization giving a maximum stress more than 180 MPa that is too dangerous for Nb_3Sn coil (figure 5).

2.2. FECR coils

FECR magnetic field is supported by 4 solenoids, 6 sextupole coils and iron yokes as shown in figure 2. 4 solenoids are used to provide the mirror fields of 6.4 T and 3.4 T as the axial maxima at the injection and extraction sides respectively. The injection solenoid is split into 2 identical ones, each of them same as the extraction solenoid, which is for the consideration of quench protection [12], cost-efficiency and Nb_3Sn wire unit length. For prototyping, one extraction solenoid has been made. Sextupole coils are designed with the conventional configuration that has been used in the ion sources such as VENUS, and SuSI [5, 6]. Each sextupole coil is wound with one single Nb_3Sn strand. For prototyping, one ~1/2 length coil was made first, and after its success a full length one has been recently developed and tested accordingly.

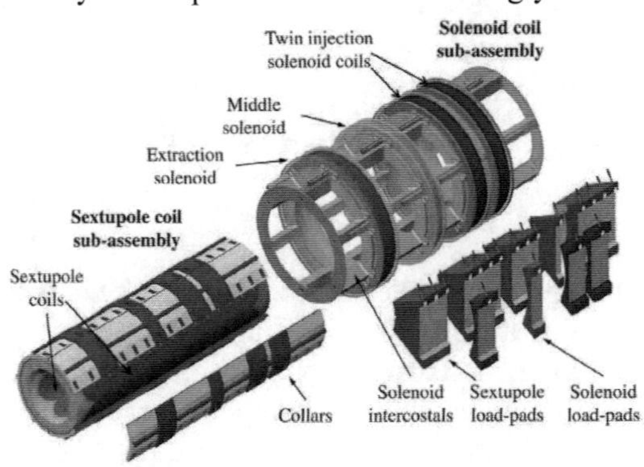

Figure 4. FECR coils assembly.

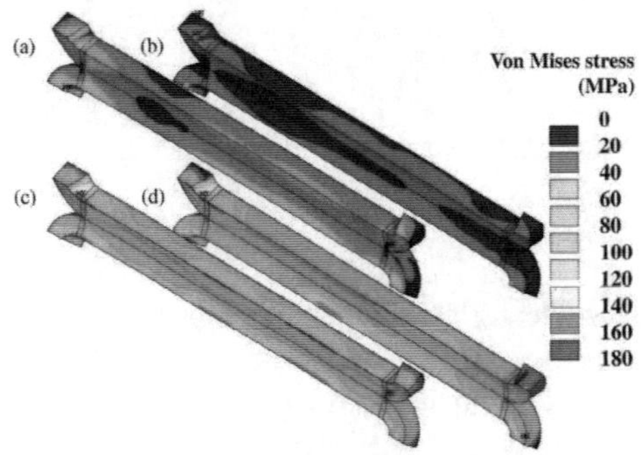

Figure 5. Von Mises stress in the sextupole during (a) the bladder operation, (b) the room temperature pre-load, (c) cool-down and (d) excitation.

2.2.1. Solenoid.

Compared to traditional solenoids, FECR solenoids have special requirements from active quench protection design. Consequently, it typically has 3 joints, i.e. 2 Nb_3Sn-NbTi joints for input and output current, one Nb_3Sn-Nb_3Sn joint between inner and outer layers that is reserved for symmetry Quench Protection System (QPS) instrumentation. The main risks in such a Nb_3Sn solenoid fabrication are High Voltage (HV) insulation and joint reliability. During the energy extraction through an active quench protection, 1 kV peak volage is expected in the circuit. As the solenoid is working in helium, according to Paschen discharge scaling the solenoid should be hi-potted up to 3 kV to guarantee its safe operation inside helium dewar which is a very critical requirement. Additionally, the turn-to-turn and layer-to layer insulation should also pass the impulse test up to 2.0 kV. FECR extraction solenoid design is given in figure 6. The solenoid bobbin is made of 316 L stainless steel (SS) with the inner sides facing the Nb_3Sn wire coated with Al_2O_3 and reinforced with laminated mica. After coil winding, heat treatment, decarbonization, the coil will be put into a mould for Vacuum Pressure Impregnation (VPI) with CTD-101K epoxy that is for the preliminary forming and HV insulation buildup. After the solenoid is unloaded from the impregnation mould, it will be reinforced with 0.8×1.6 mm^2 aluminium strip for 28 mm to provide additional prestress to the solenoid against the hoop stress during operation. As the cold mass setup has stringent tolerance requirement on all components, the prepared solenoid will be eventually put into another forming mould to complete a second VPI to refine the external shape with uniform outer radius and a tolerance better than ±100 μm. Once the solenoid is ready, a Quality Analysis (QA) is needed, such as the resistance at room temperature, hi-pot test, impulse test, mechanical check with Gage FaroArm. As the designed operation current is 600 A for the extraction solenoid, we tested the prototype solenoid up to 600 A only as the quench protection system was not fully validated and this was also the safety limit of the hoop-stress with the external aluminium prestress banding. Fortunately, this solenoid was tested to 600 A without any quenches. During the test, we also triggered several mandatory external quench protection and the solenoid insulation passed the shock of resultant energy extraction high voltages.

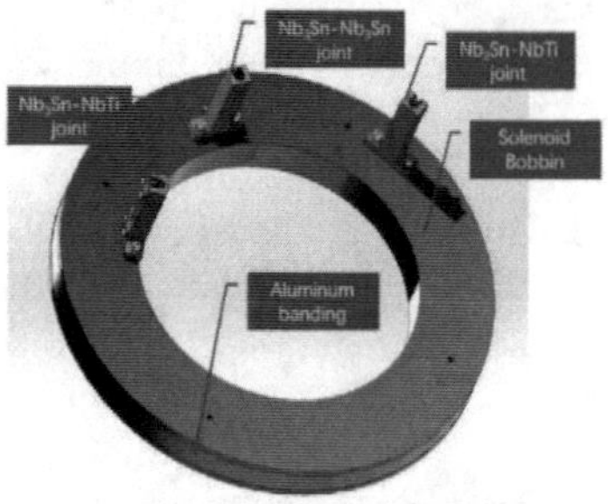

Figure 6. FECR extraction solenoid design.

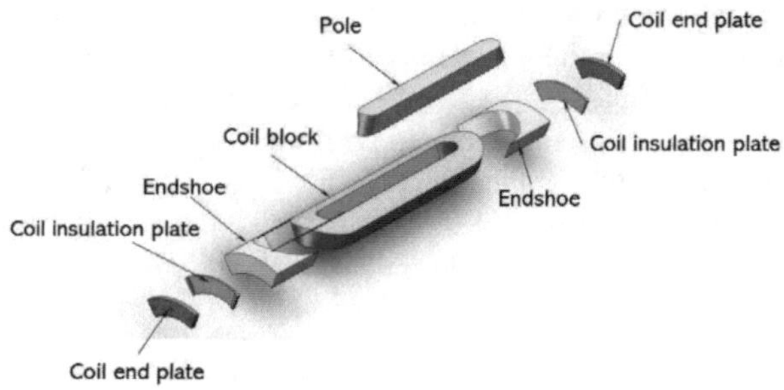

Figure 7. FECR sextupole coil structure.

2.2.2. *Sextupole Coil.*

For sextupole coil development, it also follows the similar recipe of solenoid fabrication. A wind and react approach is also used in the fabrication of FECR sextupole coils. There are many successful examples of high field solenoids with Nb_3Sn wire, but in terms of sextupole, no much successful information could be the references. The main challenges lie in the precise winding of more than 800 turns single strand in a saddle-like configuration. To have high standard electrical insulation (3~5 kV) of the coil, no damage in the S-glass braided material is allowed during the winding. After successful winding, wound coil will be sent to a programmable oven for reaction. To achieve high quality of electrical insulation, decarbonization process is mandatory. Finally, the coil will be put in a tooling for VPI to form the final configuration and more importantly to strengthen the coil and improve the electrical insulation property. To meet the needs of precise installation with bladder and key technology, the 937.4 mm (860 mm coil + end shoes + insulation plates + end plates) length sextupole coils should be eventually prepared within a tolerance of ±100 μm. As there are too much risk in the sextupole coil fabrication, we prototyped it with a half-length coil which had a half length in the straight bar of the saddle coil while remaining the two ends identical to a full-sized one which made a coil length of 537.4 mm (937.4 mm for full-size coil). The design plot of a half-length prototype sextupole coil is illustrated in figure 7.

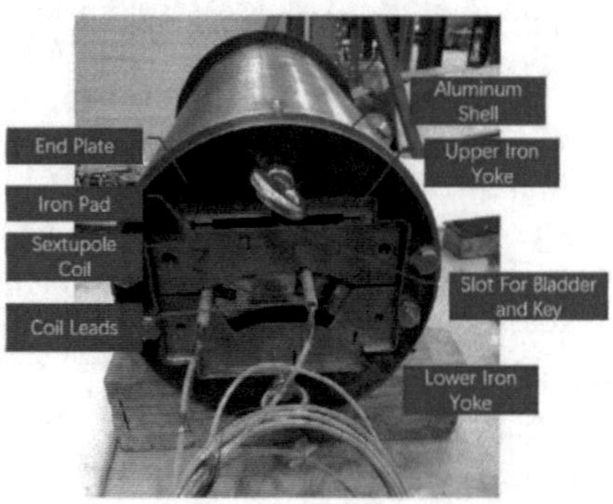

Figure 8. Half-length sextupole in the mirror structure prepared for cold test.

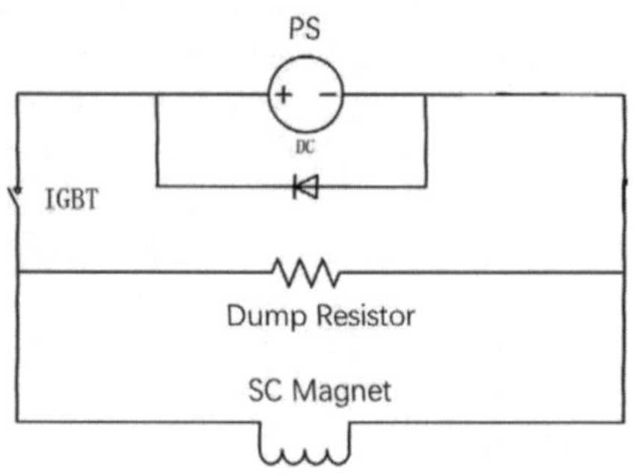

Figure 9. The active quench power supply system schematic design.

Not like solenoids fabricated with some aluminum banding for pretension, sextupole coil is fully exposed without external confinement. Therefore, to do training test, dedicated tooling is needed. We designed and fabricated so call mirror structures to test our half-length and full-sized sextupole coils respectively [13]. Our mirror structure is mainly composed of iron pad, iron yoke and aluminum shell. During the installation, a pretension system using bladder & keys and piston technology was applied, which utilized strain gauges to monitor the whole process. A 3D model was set up to do the finite element analysis (FEA) with ANYSYS. The simulation results were used to guide the mirror structure fabrication, assembly, coil installation, and current test at 4.2 K. Figure 8 is a picture of finished half-length sextupole mirror structure setup ready for cold test. According to our test, the coil was energized to 800 A that was ~113% of the design current of FECR sextupole coil. This test was lately found quite risky as the symmetric or asymmetric QPS can't efficiently protect the coil in case of any coil quench at high currents that need quick quench response in 20 ms or less. Therefore, when a full length sextupole coil was ready, we only tested it to 600 A that was within the safety margin of reliable quench detection.

2.3. Quench protection
Quench protection of the proposed 45 GHz design is considerably more challenging than in previous Niobium-Titanium (NbTi) ECR magnets [5]. First, the higher energy density stored in the conductor due to the higher magnetic field increases the potential for damage. Second, the normal-zone propagation velocities in the low-field regions are reduced due to higher thermal margin to quench. Third, the stabilizer fraction in Nb_3Sn internal tin wire is limited to lower range than NbTi wire, which increases the stabilizer current density. These challenges have been successfully addressed for Nb_3Sn accelerator magnets. However, accelerators magnets can benefit from high current, multi-strand cables to decrease the inductance. In the case of ECR magnets, a low current design based on a single wire is strongly preferred for compatibility with the cryogenic and current supply system. This significantly increases the magnets' self inductances and decreases the turn-to-turn normal-zone propagation speed. Simulations indicate that self-protection is not a viable option for any of the four magnets. Even in the case of the shortest middle solenoid, estimated highest temperature inside the coil reaches around 400 K, which is higher than the 350 K limit. The other magnets will reach much higher temperatures leading to extensive damage [12]. Thus, an active system is required, which detects the quench, switches off the power supply, and quickly discharges the magnet transport current (figure 9). The baseline powering and quench protection strategy features an independent power supply and energy-extraction system for each of the five magnets, i.e. the two injection solenoid, middle solenoid, extraction solenoid and sextupole coils.

2.4. half-length prototype
There will be complicated forces between the coils. For a 3rd generation ECR ion source, it has already been a big challenge. For FECR, this interaction forces will be more than doubled. Efficient pretension and clamping of all the coils will be the highest priority in the structure design. FECR cold mass design also adopted the state-of-the-art bladder and keys technique. To validate the cold mass design of FECR, a 1/2 length prototype has been proposed and developed based on the full-sized cold mass design [14, 15]. This prototype is mainly composed by 6 sextupole coils and 2 axial solenoids (figure 10), which can demonstrate the Nb_3Sn coil fabrication technique, cold mass structure, quench protection and cold mass integration technologies. By the end of August 2020, the prototype cold mass has been developed and the prepared prototype cold mass is shown in figure 11.

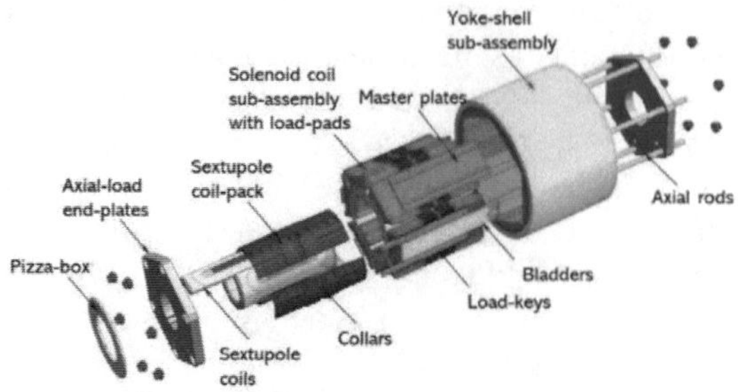

Figure 10. Explosive view of the half-length prototype cold mass.

Figure 11. The finished half-length cold mass ready for cold test.

2.4.1. half-length cold mass test.

Figure 12. The burnt lead connecting the copper lead and the sextupole coil.

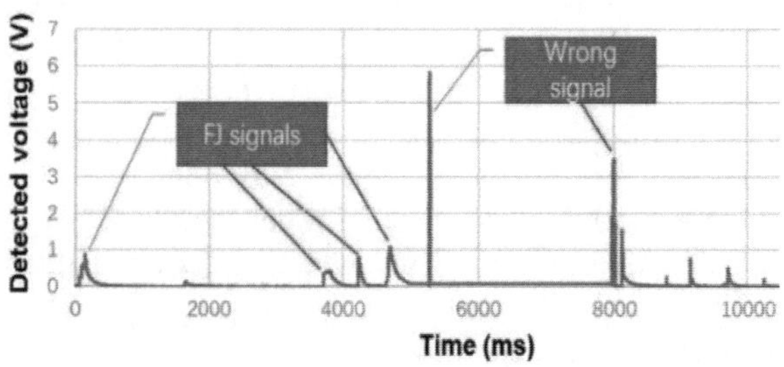

Figure 13. Flux Jump (FJ) signal captured by the quench detection system.

The cold mass test is much more complicated compared to that of one unit of solenoid or sextupole coil. The main challenges lie in the power supplies, quench detection and protection, high voltage issues during quench protection, and the most troublesome one Flux Jump (FJ) interference. During the first round of test, we were overwhelmed by so many unexpected problems, as a result we didn't ramp all the coils together. This round of test ended with a false triggered quench interfered by FJ signal when ramping all sextupole coils up to 671 A (~95% of the design current). In the second round of test, we upgraded the power supplies and added FJ filter code. We saw obvious improvement during the test. We were able to test sextupole coils and solenoid individually or all together. But unfortunately, as we were still not fully prepared for FJ challenge. When energizing sextupole coils to 681 A (~96% of the design current), a real quench happened but the signal was interfered by FJ signals and this quench was detected and triggered an external energy extraction about 100 ms after the quench time. This caused a hot spot inside the quench coil estimated to be more than 500 K that is devastating for a superconducting coil. We believe this accident had caused irreversible damage to one of the sextupole coils that was observed to show performance degradation in later test. After this round, our test was still strongly interfered by FJ problems and HV discharge issues in the PQS. Those two issues were efficiently handled during the third round of test. But then we noticed that two out of the six sextupole coils were not working properly although we can energize the cold mass to ~75% of the nominal operation currents. We decided to dissemble the cold mass with the assistance of bladder and keys technology after its warm-up. The sextupole coils were seen having moved azimuthally that indicated insufficient radial clamping stemming from improper assembly strategy despite the strain gauge on the aluminium shell was working as expected. With the assistance of pressure sensitive film and refined assembly strategy, we would have been able to complete the 4th and 5th round tests not for the severe damage to the current leads of the cold mass during a FJ triggered quench. IGBT (Insulated Gate Bipolar Transistor) of one of the power supply controllers failed to function in response to quench signal that caused power dumped into the leads that were totally burnt by the stored energy (figure 12). The exact reason is still unclear, but it is a strong indication that the power supply system needs to be further upgraded to be reliable for long-term operation. The cold mass has been repaired, and another round of test is scheduled to have systematic evaluation of this cold mass and the auxiliary systems.

2.4.2. Flux Jump issues.

When a high, but below the critical current density, current is run through a superconductor, a force is exerted on the fluxons within the superconductor. If these forces are great enough, the fluxons could potentially shift, causing the material to become locally resistive [16]. This is called a flux jump that normally happens with internal tin Nb_3Sn superconductors when they operate at low fields and high currents. Figure 13 shows the monitored FJ signals in the quench detection system. The typical voltage spikes can go from tens mV to several volts that have strong interference to normal quench signal that is typically of tens of mV. What makes things worse is that some of FJ peaks are very similar to real quench signal with regards to the voltage evolution over time. For operation safety and reliability, we developed a filter to bypass most of the interfering FJ signals, applied some quench judgement tricks to overcome those quench-alike FJ spikes. Even so, false triggered quench happened occasionally though. More efficient and reliable quench judgment algorithm or scheme is needed. We must mention that superconducting ECR ion source magnet coils had strong inhomogeneous field distribution throughout its volume as shown in figure 14. For FECR, when the magnet is energized from 10% to 100% of its design currents, there are always low field regions down to 0.1 T in contrast to the high field regions up to 11.8 T, which indicates that FJ issues will accompany the entire ramping process from low currents to 100% of the design currents.

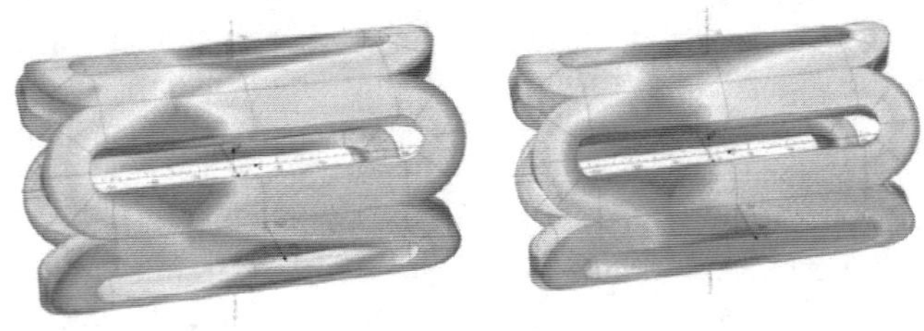

Figure 14. Magnetic fields on the sextupole coils at different currents: 10% (left) and 100% (right) designed currents.

2.5. Full-sized cold mass

As the half-length cold mass has demonstrated that we can reach closely to the 80% design goal, we plan to move on with the full-sized cold mass. For this moment, all the full-sized sextupole coils have been fabricated with very good mechanical tolerances and sufficient HV insulation. 6 of out of the 9 pieces (3 pieces for spare) have been put together and integrated into the sextupole subassembly (figure 15) with the assistance of Gage FaroArm and pressure sensitive film. All the auxiliary sub-systems have also been in place prepared for final assembly. Some challenges occurred when we proceeded with mass production of the axial solenoids. HV insulation problems in the solenoids stopped us. We are working closely with the company to solve this problem. A recently fabricated solenoid seems to be working properly up to 1.5 kV (goal 2.5~3.0 kV) without HV break-downs. After this, we will be back on track to prepare the final assembly of the full-sized cold mass.

Figure 15. Engineer making structure mapping with Gage FaroArm.

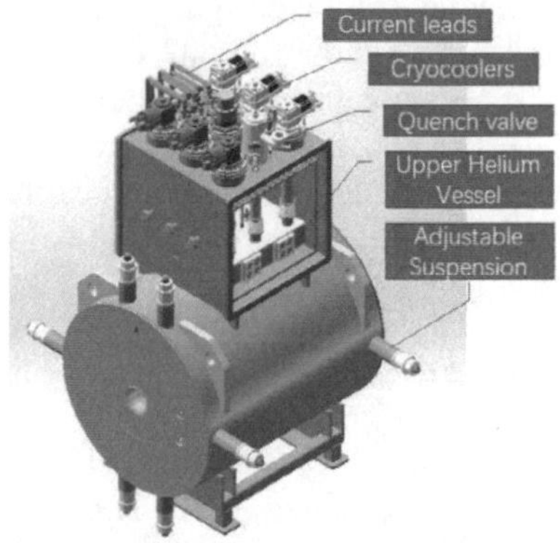

Figure 16. FECR cryogenic system.

3. FECR cryostat

Dynamic heat load to the 4.2 K cryogenic system is mainly caused by the high energy (>200 keV) electron bremsstrahlung radiation to the cold mass as most of the existing 3rd generation ECR ion sources have used 1~2 mm Ta shielding that could easily prevent the X-ray of <200 keV from penetrating into the cryogenic system. 3rd generation ECR ion sources have been designed with a total 4.2 K cooling capacity of 5~10 W. FECR will experience much higher microwave power operation and the resultant dynamic heat load to the cryostat. Therefore, a much higher dynamic heat load tolerance is desired.

This cryogenic system (Fig.16) will provide the LHe operation condition for the 3.5-ton cold mass. The critical issues are the precise alignment of the cold mass to the warm bore, high-capacity dynamic heat load, and instrumentations and feedthroughs to ensure cold mass safe energizing and operation. Cold alignment with the assistance of on-line field mapping is enabled in the cryostat design. To have high enough cooling power at 4.2 K and ensure safe operation at high magnet currents, a combination of different types of cryocoolers has been made in FECR cryostat design. The choice of the cryocoolers is specified as given in Table 3. A combination of 1 CH-110 and 3 RDE-418D4 coolers from Sumitomo Inc., and 3 KDE422 coolers from CSIC Pride (Nanjing) Cryogenic Technology Co., Ltd. has been adopted. One challenge of FECR cryostat instrumentation system is the HV breakdown voltage. As more than 1.0 kV HV (with varistor dump resistor) will be applied to the coils during quench energy extraction, all the feedthroughs on the helium dewar need to pass 3.0 kV HV hi-pot test in air that has been demonstrated recently in the company.

The experimental results from SECRAL-II indicate the dynamic heat load could reach 0.6~1.0 W/kW at 24~28 GHz when B_{min} was set in the range of 0.6~0.7 T (~0.7 of B_{ecr}). It has been accepted by the community that high temperature electron spectra temperature T_s increases linearly with B_{min} before the cyclotron instability is triggered at B_{min}/B_{ecr}~0.8 [17]. For FECR, it will be operated at the B_{min} of 0.9~1.1 T. Scaled by the B_{min} value, 1~2 W/kW dynamic heat load is predicted for FECR which means the cryogenic system will have a total dynamic heat load of 20~40 W during CW operation, which is much beyond the ~12 W design value. Before we can find better solutions, one compromised plan is to lower the operation duty factor by pulsing the microwave power or operate FECR within a reasonable B_{min} range when > 10 kW CW operation is desired.

Table 3: Cryocoolers for FECR

Model	1st Stage	2nd Stage	Quantity
CH-110	~130 W@50 K	N/A	1
KDE-422	~20 W@50 K	≥2.2 W@4.2 K	3

RDE-418D4	~42 W@50 K	≥1.8 W@4.2 K	3
Total	~316 W@50 K	≥12 W@4.2 K	/

4. Microwave system

As 45 GHz microwave has a wavelength of ~6.7 mm, transmission and coupling of high power microwave to ECR plasma may have the potential risk of arcing inside the circular waveguide. Figure 17 is schematic drawing of the microwave transmission system for FECR. The output power from the gyrotron is TEM_{00} mode. After a Matching Optics Unit (MOU), microwave will be transmitted through a 1.25 m length quasi-optical transmission system composed by several mirrors that serves as both transmission line and HV DC-break to connect the ion source floated on high potential and the microwave generator at ground potential, and the overall transmission efficiency is estimated to be better than 98%. Microwave power coupling to ECR plasma will be made through a TEM_{00} to TE_{01} mode convertor with an estimated transmission efficiency of ~97%. Quasi-optical transmission in air from the generator to the microwave window, and then converted to an oversized waveguide after the microwave window is proposed and successfully tested at IMP on the SECRAL-II test bench [18]. This has basically solved the problem of high power transmission and coupling to ECR ion source. However, the efficient coupling to an ECR plasma in terms of highly charged ion production remains unclear, which needs the real test with FECR. Additionally, in recent years several laboratories have demonstrated that multi-frequency heating is essential to stabilize the plasma when the ion source is fed with high power microwave for highly charged ion beam production. For FECR, it is still hard to make a choice in the auxiliary plasma heating frequency. As a baseline, 45 GHz/20 kW + 28 GHz/10 kW will be applied. Tentatively, a 32.9 GHz/10 kW gyrotron microwave generator is also proposed to have a better match of the operation conditions at 45 GHz.

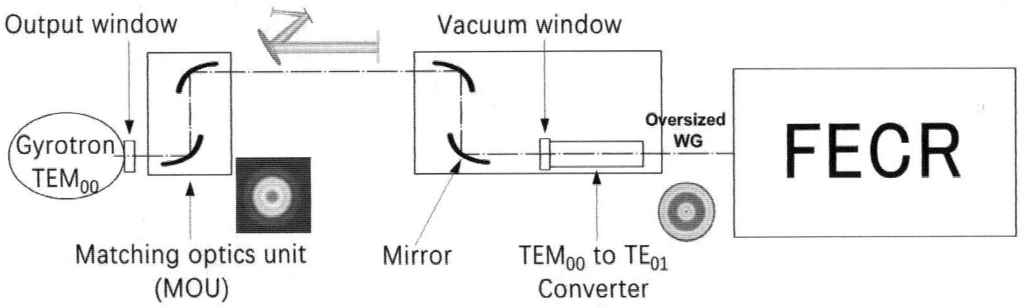

Figure 17. The microwave transmission and coupling scheme for FECR ion source.

5. Plasma heating cooling

More than 20 kW microwave power will be eventually injected into FECR plasma chamber, and this power will be mostly dumped into the components housing the hot dense plasma. Power dump is mostly through radiation and plasma loss that is governed by the B-min field configuration. The plasma flux mark on the surrounding components could provide some information on power dump distribution, i.e. the triangles on both ends and 6 plasma marks on the plasma chamber wall. Consequently, we can assume 10~20% power goes to the two ends mad 70~80% power goes to the plasma chamber [19]. Bulky materials and efficient cooling could be made to the design of the structure at the ion source plasma two ends. 10~20% power means efficient cooling of 1~2 kW at each end that could be removed by direct water cooling instead of conduction cooling used with most of the existing ECR ion sources.

Plasma chamber inner wall cooling is more challenging. Many laboratories had the experience of damaged plasma chambers by localized overheating caused by escaped plasma flux. ECR ion source min-B field is a supervision of axial mirror field and radial sextupole field. The radial component of the axial mirror field will cancel out some radial field that will result in some lower total field region on the plasma chamber wall where we can find the plasma loss. And there will be totally 3 lowest |B| points evenly distributed on the chamber wall separated by 120° at source injection and extraction sides. According to the estimation by T. Thuillier in his paper [19], the peak power sink on the inner wall of plasma chamber can reach to 1.25 MW/cm^2 for 1 kW 28 GHz microwave heating that will cause instant boiling of the cooling water at the weak |B| area on the plasma chamber at high power level. Created bubbles will prevent further cooling with flowing water and plasma chamber damage will happen when chamber material starts to degrade at high temperature. Normally, this damage is an instant process once the heated microwave power reaches the critical value of the chamber (magnetic field configuration and plasma stability also have influences). At 45 GHz, we might face much higher plasma flux density per kW microwave power. Most of the existing plasma chamber can barely take long-time operation at the power level of 10 kW. It is a big challenge for FECR to work at 20 kW/45 GHz.

A new plasma chamber structure design based on the micro-channel cooling concept [20] has been recently proposed at IMP. Compared to traditional double-wall cooling chamber, the new concept utilizes the strong heat exchange property of the micro-channel structure positioned at the localized hot spot on the inner chamber wall, which enables much lower temperature in both the plasma chamber wall and the cooling water flowing through the channel. An aluminum plasma chamber design based on this concept had been fabricated and tested in 2020 with SECRAL-II source. Up to a total power of 11 kW (9.3 kW@24 GHz and 1.7 kW@18 GHz) microwave power heating, this plasma chamber was validated to be durable and reliable. FECR plasma chamber with the inner diameter of Ø140 mm has been recently fabricated. It is expected to be able to handle the efficient cooling when high microwave power up to 25 kW is launched into FECR. Table 4 lists the main parameters of FECR plasma chamber in comparison of SECRAL-II plasma chamber designed with the same concept.

Table 4. Key parameters of FECR plasma chamber in comparison with SECRAL-II new chamber and conventional chamber.

Key Chamber Parameters	FECR	SECRAL-II New	SECRAL-II Conventional
Technology	Micro-channel	Micro-channel	Double-wall
Max. Power (kW)	25	12	6~8
Max. Localized Power Density (MW/m^2)	20	10	~5
Chamber ID (mm)	Ø140	Ø125	Ø125
Chamber OD (mm)	Ø156	Ø136	Ø136
Length (mm)	1225	887	887
Water pressure (bar)	10	8.9	6
Total water flow (l/min)	>50	>13	>20
Max. flow velocity (m/s)	25.5	11.0	1.8

6. Summary

FECR project was officially started in 2015. After 6 years continuous development, we have tackled most of the envisioned and new challenges. We have made fundamental progresses in FECR magnet development, 45 GHz microwave coupling, plasma chamber cooling and so on. Most of FECR manet components are now ready for the final assembly and test. Conventional parts fabrication based on the experience and understanding of the existing 3rd generation ECR ion sources have been mostly prepared. They will be put together for off-line test before FECR magnet is in position. According to the fabrication plan, FECR magnet might be ready in early 2022, and the first beam is expected at the middle of 2022.

7. Acknowledgement

We acknowledge the fruitful collaboration work on FECR cold mass design between ATAP/LBNL and IMP. The intense R&D work on Nb_3Sn prototyping made with IMP and XSMT colleagues is also warmly acknowledged. We are also grateful that so many people from our community and outside of our community have given us suggestive advice. We thank all the commercial vendors who have been supporting us in this challenging research. This work was supported by NSFC with contract No. 11427904 and 12025506.

References

[1] Yang J C *et al* 2013 *Nucl. Instrum. Meth. B* **317** 263-265.
[2] Lu W Sun L T Qian C Li L B Guo J W Huang W Zhang X Z Zhao H W 2019 *Rev. Sci. Instrum.* **90** 113318.
[3] Sun L Zhao H W Lu W Guo J W Yang Y Jia H Lu L Wu W 2020 *X-Ray Spectrometry* **49** 47-53.
[4] Sabbi G et al 2017 Design of the Nb_3Sn Superconducting Magnet System for the 45 GHz ECR Ion Source at IMP (LBNL Report 2001036).
[5] Taylor C Caspi S Leitner M Lundgren S Lyneis C Wutte D Wang S T Chen J Y 2000 *IEEE Trans. Appl. Supercond.* **10**(1) 224.
[6] DeKamp J C Zavodszky P A Arend B Hitchcock S Moskalik J Ottarson J and Zeller A F 2007 *IEEE Trans. Appl. Supercond.* **17**(2) 1217.
[7] Caspi S et al 2001 *IEEE Trans. Appl. Supercond.* **11**(1) 2272.
[8] Ferracin P et al 2006 *IEEE Trans. Appl. Supercond.* **16**(2) 378.
[9] Cheng D W et al 2013 *IEEE Trans. Appl. Supercond.* **23**(3) 4002504.
[10] Ferracin P et al 2014 *IEEE Trans. Appl. Supercond.* **24**(3) 4002306.
[11] Prestemon S et al 2009 *Trans. Appl. Supercond.***19**(3) 1336.
[12] Ravaioli E *et al* 2018 *IEEE Trans. Appl. Supercond.* **28**(3) 4700906.
[13] Zhu L Wu W Yu S R Sun L T Chen Y Mei E M Ni D S Du Z Y 2020 *Cryogenics* **112** 103192.
[14] Juchno M et al 2018 *IEEE Trans. Appl. Supercond.* **28**(3) 4602806.
[15] Wu B M Wang X Z Guan M Z Xin C J Wu W Chen Y Q Sun L T and Zhao H W *2020 IEEE Trans. Appl. Supercond.* **30**(4) 4002905.
[16] Donnelly C 2007 Flux Jump Analysis on Superconducting Nb_3Sn Accelerator Magnets (FNAL Work Report) p 10.
[17] Li J B et al 2020 *Plasma Phys. Control. Fusion* **62** 095015.
[18] Guo J W et al 2018 *AIP Conference Proceedings* **2011** 090001.
[19] Thuillier T Angot J Benitez J Y Hodgkinson A Lyneis C M Todd D S and Xie D Z 2016 *Rev. Sci. Instrum.* **87** 02A736.
[20] Naqiuddin N H Saw L H Yew M C Yusof F Ng T C Yew M K 2018 *Renewable and Sustainable Energy Reviews* **82** 901.

Journal of Physics: Conference Series **2244** (2022) 012022 doi:10.1088/1742-6596/2244/1/012022

Statistical analysis of operational data of the GTS-LHC ion source

D Küchler[1], **M Mihailescu**[2]

[1] BE Department, CERN, CH-1217 Geneva, Switzerland

[2] Rheinische Friedrich-Wilhelms-Universität Bonn, D-53113 Bonn, Germany

E-mail: `detlef.kuchler@cern.ch`

Abstract. The GTS-LHC ECR ion source delivers ion species for the physics programme of the CERN accelerator complex. Stability and reproducibility are key parameters for a successful operation. Therefore the source requires regularly intervention by an expert. The integrated functionality of data logging in the control system of the CERN accelerator complex allows to store all relevant control values and beam properties for the different machines. For the years 2015, 2016 and 2018, the data for the lead ion operation of the GTS-LHC ion source were extracted from the logging and treated with various statistical methods to identify recurring patterns in the operational settings of the source. If such patterns exist, they may be used in the future for the control of the source by semi-automatic feedback loops. The main study performed was based on a cluster analysis using the Optigrid algorithm, which is well suited for problems with a high number of parameters. The algorithm and selected results considering some operational aspects and ideas on further studies are presented.

1. Introduction

The GTS-LHC ion source [1] at Linac3 is the starting point of the heavy ion accelerator chain of the CERN accelerator complex. The performance and the reliability of the source have a direct impact on the performance and reliability of this chain.

The setting up and tuning of the source depends on a few experienced operators, but still suffers from some variability depending on experience. The GTS-LHC operation support (GHOST) software presently automates some simple procedures for source operation, and in an attempt to expand its role a study of the source operation has been made in terms of the settings applied by the operators, with a possible view to automating this process to help to improve the source stability and reliability and to reduce the necessity of the intervention of an operator or source expert.

In this paper the operational parameters of the source for the years 2015, 2016 and 2018 were analysed to find common settings or recurring patterns in the settings which could be used for software based feedback loops to control the source. For this analysis the source settings were treated with a clustering algorithm to try to identify commonly used setting giving good source performance.

In the first part of the paper the mathematical background of the algorithm and the procedures to prepare the data are explained. In the second part the results will be presented. At the end some conclusions concerning the possible use for GHOST are drawn and suggestions of further developments are given.

The settings and acquisitions used in this analysis are the most commonly adjusted settings (source solenoid currents, microwave power, gas injection, oven power and bias disk voltage). The acquisitions are intensities measured with the beam current transformers along the linac (one directly after the charge separation and one at the end of the linac) and the current of the high voltage power supply of the source.

The settings of the two ovens were not used in the clustering itself, but were included in the analysis of the resulting clusters. Since the need for ramping up the oven during one lead cycle is already well known, the oven power was excluded from the clustering input parameters to reduce the dimensionality and increase interpretability of the results.

2. The clustering algorithm

Given a set of parameters, referred to as data point, clustering is the process of grouping together data points that are *similar*. Clustering is an unsupervised learning method, i.e. it has to discover unknown classes in the data. For a general overview on clustering see for example [2].

The goal is to find similar settings of the source using clustering. So one will consider two points as similar if they represent the same setting of the source up to some variation that accounts for small changes or noise. The whole data pipeline process is illustrated in Figure 1.

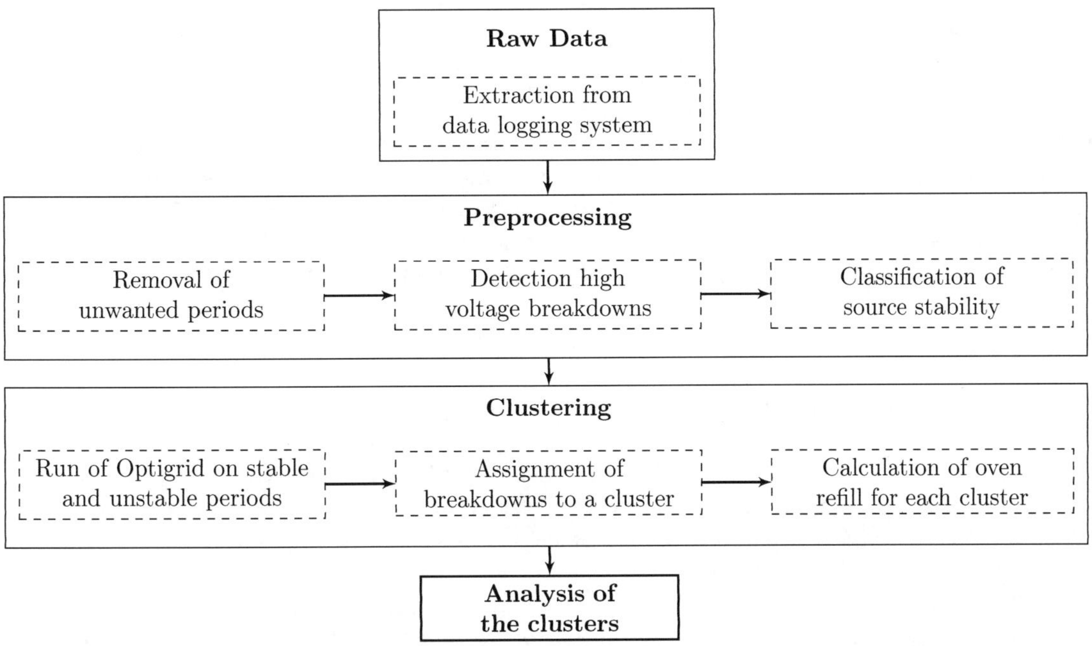

Figure 1: Flowchart of the different steps of the data analysis.

When it comes to the actual clustering, there is a wide range of algorithms to chose from, and while there exist several measures to grade clustering results (see [2]) it is generally not possible to say which algorithm will be the better one, as this depends on the actual use case and definition of similarity [3]. In the presented case clustering for unsupervised learning is used, meaning that the output classes are not known a priori. Hence the results have to be evaluated to see if the determined clusters fit the requirements.

For this analysis the Optigrid [4] algorithm developed by Hinneburg and Keim was selected. This algorithm was developed to solve the problem of clustering in high dimensional data (known as the "Curse of Dimensionality", see e.g. [5]).

The Optigrid algorithm partitions the input space into a grid, where each cell represents one cluster. Each cluster is described by its median and standard deviation in percent of mean, to capture the centre and dispersion of a cluster. In the present case, the two values can be interpreted as the setting that is represented by the cluster, and the variability of the setting, i.e. whether it was almost constantly on one value or if there were fluctuations.

After each point is assigned to one cluster, it is possible to calculate attributes for each cluster, e.g. its duration and the number of high voltage breakdowns while it was active. A duration can be assigned to each data point by taking the time delta to the next observation. Then, the duration of a cluster is defined as the sum of durations of all points that constitute the cluster.

3. Selected results of the cluster analysis and discussion of the data

The resulting data were analysed concerning the cluster structure, in respect to the operation and the oven refills. In addition the source and linac performance was investigated.

The first observation is that the calculation delivers a vast number of clusters which cannot easily combined to bigger clusters without losing information. But it can also be seen that if one orders the clusters by their duration, the size of the clusters drop quickly (see Figure 2). There are few large clusters and many small ones.

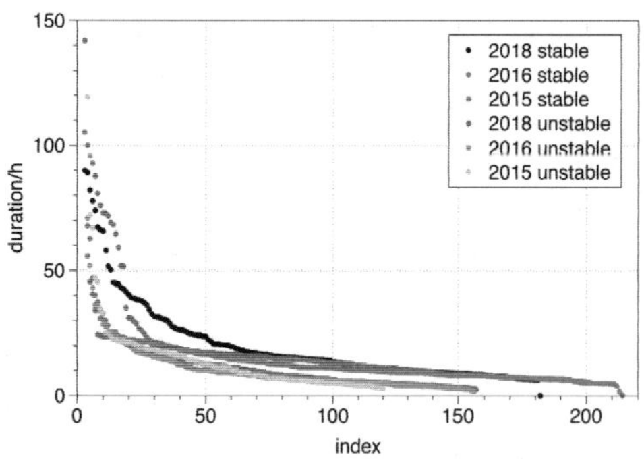

Figure 2: Cluster duration for the different years and operation modes ordered by the cluster size (lower index = larger cluster). Stability was measured by defining threshold in beam intensity and variation in time windows of several hours.

There are only small differences between the years and the operation modes. The stable operation mode in 2016 seems to have the highest number of long clusters. The longest one has a duration of 142 hours. But this is still short compared to the yearly operational period of around 6000 hours or compared to the time between two oven refills (two weeks = 336 h). The duration accumulated per cluster is for most of the cases less than 30 hours. Compared to the yearly operational period this shows that the settings are varied quite often. There is no dominant setting.

A detailed look at the internal distribution of the clusters in time also reveals that some clusters are continuous in time, while other are split up into many small fragments.

The RF power was chosen to give an impression of the distribution of the medians of the clusters. The Figure 3 shows the medians of the clusters for RF power of the stable period in 2016. The clusters are ordered by size. The lower index represents the larger cluster. One can see that the used settings do not accumulate around a certain value but only lie in a (relatively broad) range.

In the Figure 4 the duration of the clusters is plotted in relation to the time after the last oven refill for the stable operation periods in 2016. With the colour the number of the fill in the respective year is coded. There does not appear to be a relation between duration and time after oven refill.

The first week (t<168 h) after an oven refill the source needs normally very little tuning. So one could expect to find longer clusters in this time periods. But large clusters can be found

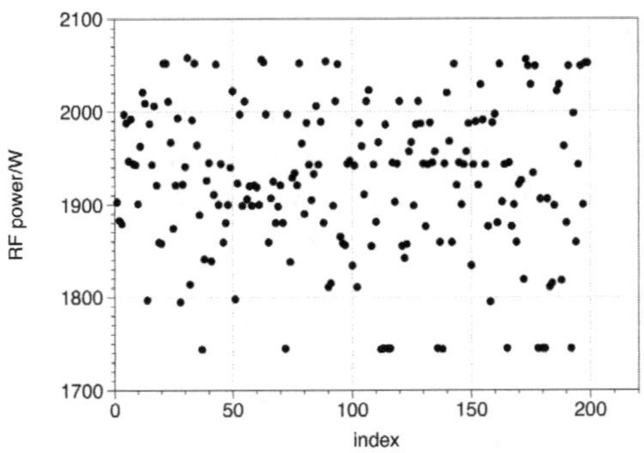

Figure 3: Plot of the medians of the RF power of the different clusters for the stable operation periods in 2016. The clusters are order by size (lower index = bigger cluster).

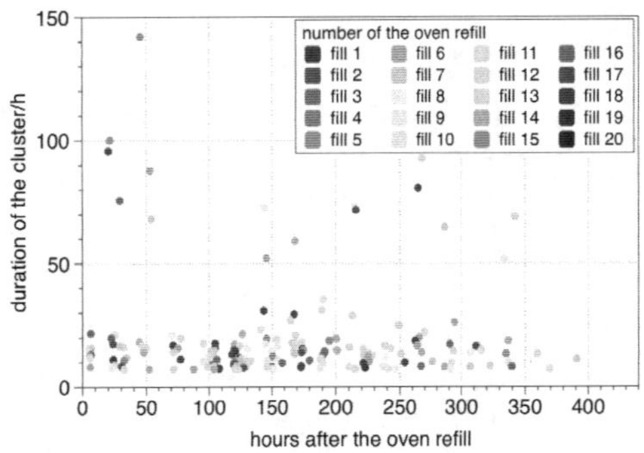

Figure 4: Plot of the duration of the clusters in relation to the time after the last oven refill for the stable operation in 2016.

at any time of the oven running period. No moment is preferred for longer clusters. In the second week (t>168 h) after an oven refill the source needs to be tuned quite often to keep the intensity and the stability. So one would expect to see more short clusters there. This could not be confirmed by looking only at the cluster duration.

After an oven refill there is normally only little tuning on the source. So one may have the expectation that the two time periods before and after an oven refill fall into the same cluster. But also this expectation could not be confirmed.

The collected data allow beside the cluster analysis to have a general overview over the source and linac performance.

Figure 5 shows the ion intensity of the Pb^{54+} beam at the end of the linac for the stable periods of the three different years. Stability was measured by defining threshold in beam intensity and variation in time windows of several hours.

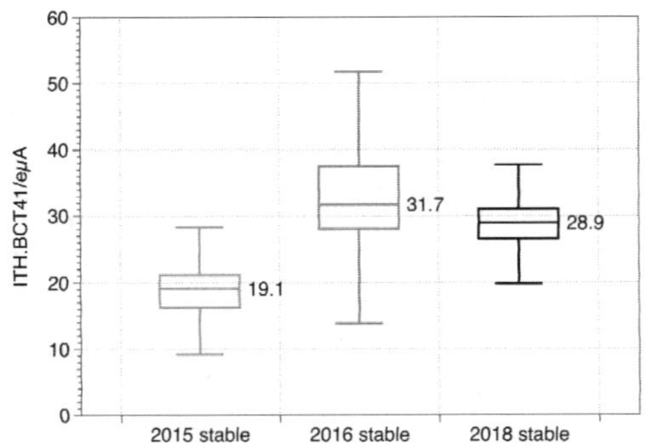

Figure 5: Box plots of the ion intensity of the Pb^{54+} ion beam measured in the beam current transformer at the end of the linac for the stable periods of the three different years. Outliers are not plotted.

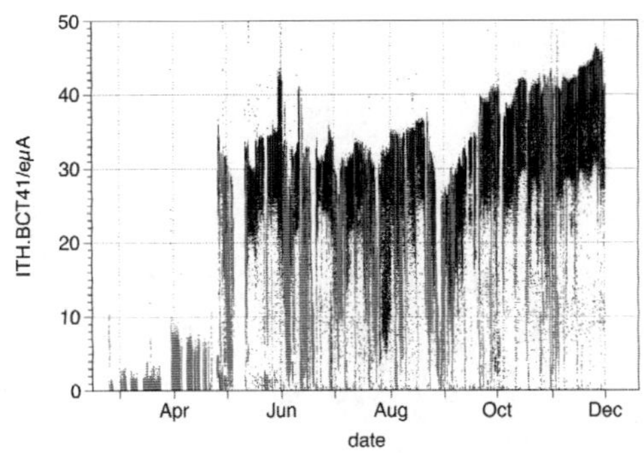

Figure 6: Ion beam current in the beam current transformer at the end of the linac for stable (black) and unstable (red) beam periods in 2016.

There are clear differences between the years visible. The increase from 2015 to 2016 was the result of the modification of the low energy beam transport [6]. During 2016 the best average intensity could be delivered. In 2018 the intensity was lower due to problems with the stripper foils.

The very high values (currents >40 eµA) for the year 2016 are an artefact, as for degrading stripper foils the charge states Pb^{53+} and Pb^{55+} can sneak through the filtering and so the integral measured current becomes higher. For the full explanation of this effect see [7].

In the Figure 6 the ion beam intensity at the exit of the linac is plotted for 2016. Stable periods are marked in black, unstable periods are marked in red. If one compares the integral of stable with the unstable periods per year one get 62.8% for 2015, 70% for 2016 and 68% for 2018. One can see that in 2016 there were long periods of stable beam operations, in particular during the month of November, when the 2016 heavy ion run took place (see Figure 6).

4. Conclusions and outlook

This present cluster analysis confirmed the qualitative experience of the operators that there is no obvious recurring pattern in the source settings. Neither between the years and nor within the years.

All source settings are within clear boundaries, but there are no sets of settings that can be use to reliably obtain either stable or unstable operation. None of the found clusters can be used directly to create one or several GHOST modules which can run the source at an automatised or semi-automatised mode.

More sophisticated time series analysis tools, e.g. LSTM neural networks, could potentially discover more complex relations, and could in the future be used for (mid- or long-term) beam quality prediction.

At the moment in all these studies the source is treated as a black box. Additional information about the physical processes inside of the box would help to improve the understanding of the source behaviour and the operation. Optical emission spectroscopy is being investigated as a potential source of addition diagnostics of the ion source plasma, and could lead to additional observation parameters on which to diagnose and predict source performance. This spectroscopy can be performed parasitically. The method was already demonstrated in the context of oven operation [8]. A corresponding study is in preparation.

References

[1] Dumas L, Hill C, Hitz D, Küchler D, Mastrostefano C, O'Neil M and Scrivens R 2007 Operation of the GTS-LHC Source for the Hadron Injector at CERN Tech. Rep. LHC-PROJECT-Report-985 CERN Geneva URL `http://cds.cern.ch/record/1019180`

[2] Rokach L and Maimon O 2005 *Data Mining and Knowledge Discovery Handbook* 321–352

[3] Gelbard R, Goldman O and Spiegler I 2007 *Data & Knowledge Engineering* **63**(1) 155–166

[4] Hinneburg A and Keim D A 1999 *Proceedings of the 25th International Conference on Very Large Databases* 506–517

[5] Verleysen M and François D 2005 *Computational Intelligence and Bioinspired Systems. IWANN 2005.* **3512** 758–770

[6] Toivanen V *et al.* 2016 *Proc. of 22. International Workshop on ECR Ion Sources (ECRIS'16), Busan, Korea, August 28 - September 1, 2016* (Geneva, Switzerland: JACoW) pp 50–54 https://doi.org/10.18429/JACoW-ECRIS2016-WEAO01 URL `http://jacow.org/ecris2016/papers/weao01.pdf`

[7] Bellodi G, Bertolo S, Kuchler D, Mahner E, Scrivens R and Wenander F J C 2020 Stripper foil investigations at Linac3 Tech. Rep. CERN-ACC-NOTE-2020-0016 URL `http://cds.cern.ch/record/2713678`

[8] Maimone F, Mäder J, Lang R, Patchakui P, Tinschert K and Hollinger R 2019 *Review of Scientific Instruments* **90** 123108

Journal of Physics: Conference Series **2244** (2022) 012023 doi:10.1088/1742-6596/2244/1/012023

Numerical Design of RF Antennas for Ion Cyclotron Resonance Heating in ECRIS Environment

G. S. Mauro[1*], G. Torrisi[1], A. Pidatella[1], A. Galatà[2] and D. Mascali[1]

1 Istituto Nazionale di Fisica Nucleare - Laboratori Nazionali del Sud (INFN-LNS), Via S. Sofia 62, 95123 Catania, Italy.

2 Istituto Nazionale di Fisica Nucleare - Laboratori Nazionali di Legnaro (INFN -LNL), Viale dell'Università, 2, 35020 Legnaro, Italy.

E-mail: `mauro@lns.infn.it`

Abstract. In this paper we present the numerical design and simulation of RF antennas to be employed in Ion Cyclotron Resonance Heating (ICRH) systems working in ECRIS environment. A 3D full-wave numerical model, based on the coupling between COMSOL FEM solution of Maxwell equations and the MATLAB-computed non-homogeneous plasma dielectric tensor, has been employed in order to study the performances of several ICRH antennas. Results in terms of S-parameters, on-axis electric field and RF absorbed power inside the plasma chamber have been obtained and compared between the chosen antenna geometries. The presented study will permit to better understand the fundamental aspects of ion dynamics in ECRISs as well as allowing the design of a proper matching network between the RF amplifier and the antenna, necessary to cope with the plasma properties' fast variations. Further ion kinetic simulations are ongoing.

1. Introduction and Motivation

Ion cyclotron resonance heating (ICRH) is a well-known method used for various applications such as: plasma-heating in, e.g., fusion [1], ICR isotope separation [2] and ion thruster plasma devices [3]. Hereinafter, we present the numerical simulation of three selected types of strap antennas (simple-loop, Nagoya III and helical) employed for ICRH and working inside a plasma chamber, that is in Electron Cyclotron Resonance (ECR) environment. The *slow* ion cyclotron wave excitation is radiated by left-handed circular polarization (LHCP) by the strap antennas [4, 5] or, more in general, by induction coils generating long-wavelength ion-cyclotron waves that propagates along the magnetic field lines from high-to-low magnetic field. Near the ICR region [6] (where $\omega_{RF} = \omega_{ci} = 2\pi f_{ci} = QeB/m_i$, ion charge Qe, mass m_i, and ion cyclotron frequency f_{ci}), wavelength becomes shorter, wave energy is damped and transferred to ions. Our model of ICRH, as a possible additional selective heating of the plasma ion component in ECRIS, is based on a three-dimensional (3D) full-wave Finite Element Method (FEM)-based calculation, performed in COMSOL. The paper is organized as follows. In section 2, the three types of strap antennas, working in ECR Ion Sources (ECRIS) environment, are presented, simulated in vacuum-filled plasma chamber and compared, for every single case, in terms of: a) values of incident power necessary to obtain a coupled power equal to 2 kW, and b) electric field intensity along the plasma chamber axis. As a second analysis step, power absorption and axial electric field have been obtained considering a non-uniform *cold* plasma, bounded by the metallic resonant cavity and immersed in the 3D magnetic configuration typical of ECRIS, obtained through a MATLAB script working in conjunction with the COMSOL simulation. Finally, in section 3, the effect of particle motion for the the simple-loop and helical antennas has been evaluated and performances have been compared.

2. RF antenna design in ECRIS environment

The presented setup consists of a cylindrical plasma chamber of radius $R = 140$ mm and length $L = 700$ mm. Inside the cavity a loop-type antenna, fed by a coaxial connector, is placed for ICRH purpose. Three types of antenna has been chosen for the study: a simple loop, a Nagoya type III and a helical, as visible in Fig. 1.

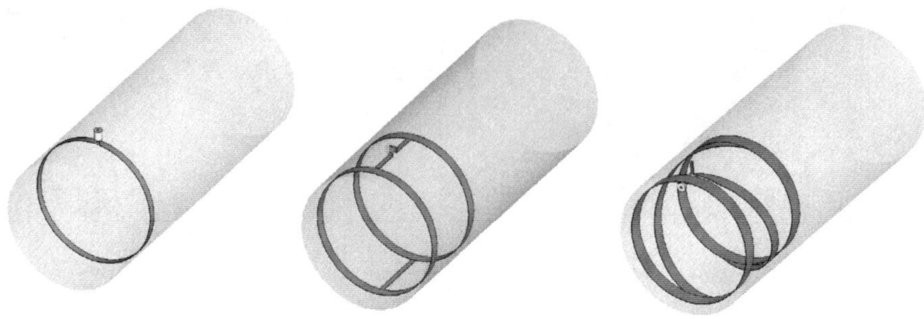

Figure 1: From left to right: simple-loop, Nagoya type III and helical antenna. The antennas have been placed inside a cylindrical cavity of radius $R = 140$ mm and length $L = 700$ mm.

The three structures have been first studied and compared in terms of required RF input power by fixing the value of coupled power to 2 kW, as typical in ECRIS setups. We used COMSOL to simulate the three structures, consisting of an empty cylindrical cavity with conducting walls (steel in our case, with electrical conductivity equal to 7.69×10^6 S/m), and a copper-made antenna. We consider the typical magnetic field configuration of the ECRIS, the so-called *minimum-B* configuration (where the value of the magnetic field has a minimum in the centre of the plasma chamber and from there it increases in all directions). A value at the electron resonance of the magnetic field, $\mathrm{B_{ECR}} = 0.64$ T has been considered (at 18 GHz operative frequency), and $^{16}O^{4+}$ has been used as a reference ion species. The implemented density profile ($n_e = n_i \sim 10^{18}$ m^{-3}), shown in Fig. 2 plotted along the plasma chamber axis, comes from modeling on plasma dynamics [7]. At a 3-dimensional level, the plasma is concentrated inside the ECR resonance surface. Finally, we imposed a value of the magnetic field $\mathrm{B_{ICR}} = 0.9\mathrm{B_{ECR}}$ at the ion resonance, in order to obtain an ICR surface nested in the ECR one. The ion cyclotron frequency can thus be calculated as

$$f_{ci} = 15.2\ [MHz]\ \frac{Q}{A}\ B_{ICR}\ [T] \tag{1}$$

where $\frac{Q}{A}$ is the charge-to-mass ratio for the selected ion species. The antenna operating frequency has been set equal to $f_{ci} = 2.2$ MHz. Defining $P_{coupled}$ the power coupled to the whole structure (cavity plus antenna), $P_{incident}$ the power at the coaxial connector input and $P_{reflected}$ the power reflected back at the coaxial connector input, we can write

$$P_{coupled} = P_{incident} - P_{reflected} \Rightarrow P_{coupled} = P_{incident} - |S_{11}|^2 P_{incident} \tag{2}$$

The procedure consists of two steps for each antenna. In the first step, the $|S_{11}|$ parameter is calculated from a simulation with an arbitrarily imposed incident power at the coaxial port.

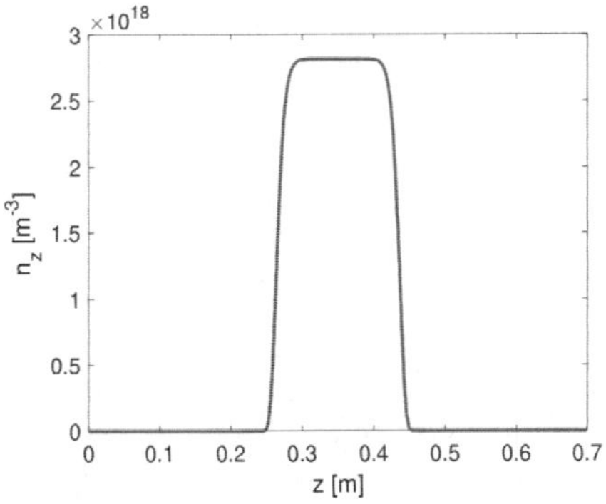

Figure 2: Implemented ion density profile plotted along the plasma chamber axis (z-axis).

Then, by substituting the obtained $|S_{11}|$ value and the desired coupled power, $P_{coupled} = 2$ kW in (2), the correct value for $P_{incident}$ can be calculated, and a second simulation can be performed with the new power value imposed at the coaxial port. Note that the obtained $P_{incident}$ values has only numerical significance: in fact, the real setup will include a proper matching network, placed between the RF generator and the antenna coaxial input.

The electric field along the cavity axis generated by the three antennas inside the vacuum-filled cavity is shown in Fig. 3 (left). For each antenna the plot also indicates the position of (from left to right): a) center of the first loop, b) the antenna center and c) the center of the last loop. For the simple loop antenna the loop center and the antenna center coincide.

In the vacuum-filled plasma chamber case, the field intensity results higher in the vicinity of the antenna (that for all the three cases is placed 'safely' distant from the region where the plasma is created) but, due to the antenna operating frequency ($f_0 = 2.2$ MHz), the wavelength of the generated wave results higher than the wavelength for which the ECR source plasma chamber has been built (usually it has a diameter equal to several operating microwave lengths in order to achieve good coupling efficiency). This means that, in vacuum, the antenna-generated wave results evanescent and attenuates as one moves away from the antenna(s).

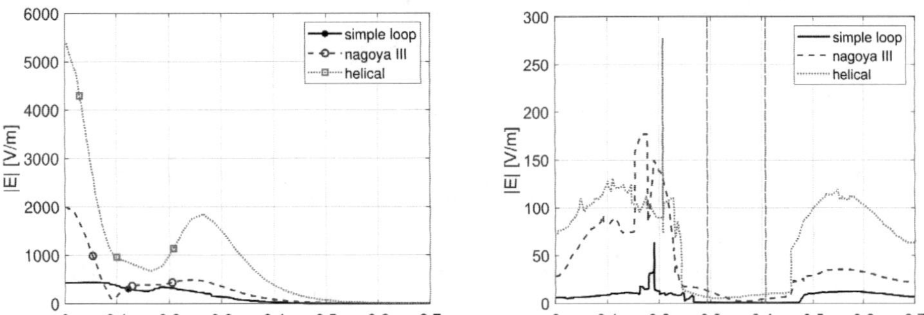

Figure 3: Electric field along the z-axis of (left) *vacuum cavity* vs. (right) *plasma-filled cavity* generated by the three antennas. For the plasma-filled cavity case, the ICR positions along the chamber axis are indicated with green vertical lines.

The next step has been the introduction of an non-homogeneous lossy anisotropic plasma medium in the COMSOL simulations, obtained through the stationary-PIC code typically used for electrons, and now extended to include ion properties. The full expression of the plasma dielectric tensor $\bar{\bar{\epsilon}}$ [8] has been derived in the *cold plasma approximation* in the presence of the 3D confining magnetostatic field and it has been generalized by including both the electron and ion species terms, providing a more comprehensive picture of the electromagnetic (EM) field. The *minimum-B* magnetic field profile has been derived by the classical ECRIS scaling laws considering an operating frequency f_{RF} = 18 GHz. Taking in mind the absence of the matching network in the current simulations, we calculated the S_{11} value at the input port for each antenna, obtaining −0.0377, −0.0409 and −0.135 dB respectively for the simple-loop, the Nagoya III and the helical antennas. This can be considered as a numerical evidence of a better wave-to-plasma coupling when the helical antenna is employed (less reflected power).

The electric field along the cavity axis generated by the three antennas in presence of the anisotropic plasma, is shown in Fig. 3 (right). In the plasma-filled plasma chamber case, since the plasma has a very high relative dielectric constant, the generated wave propagates inside the lossy medium with a much shorter wavelength, λ_p, with respect to the one in free-space, λ_0, namely for a larger distance, being then almost totally absorbed inside the ICR region ($z \simeq [0.3, 0.4]$ m). Even when interacting with a lossy medium, the field irradiated by the helical antenna has a larger magnitude if compared with the other two antennas.

3. ICRF antennas impact on ion dynamics

A further step forward has been done, by studying how the propagating EM field previously calculated affects the particle's motion, and in particular the ion dynamics in plasma. According to EM results in cavity, including the plasma dielectric tensor effects, two cases have been chosen as exemplary scenarios of ICRF antennas for ECRIS, i. e., the simple loop and the helical antennas. We have then started numerical simulations, by using a code to describe ICR stationary plasmas, by joining results from the FEM solver in COMSOL, and the particle mover code written in MATLAB [9–11]. The 3D EM field generated by the antennas is considered in the routine which follows the evolution of $^{16}O^{4+}$ ions for a simulation time of few ms, with an integration time step $\tau = 10^{-10}$ s. A 3D magnetostatic field, typical of an ECRIS, has been also included in the simulation. Local single-ion densities have been computed through a density accumulation of simultaneously-evolving ions' paths onto a 3D numerical grid (cell size ~ 2 mm^3), together with their energies (all ions started with an energy of 1 eV). Fig. 4 show results for density accumulation [a. u.] (top-line) and energy accumulation [eV] (bottom-line) for both simple-loop (left) and helical (right) ICRH antennas. Plots represent projection maps along the z-axis of the mentioned 3D data. Despite the ion density profile inside the cavity maintains a similar structure for both the cases, the helical antenna provides a larger ion energy accumulation than that of the simple loop one. This could be considered as a preliminary numerical evidence for a better electromagnetic coupling of the helical antenna. It is worth to be notice that these results have to be considered just as indicative and preliminary. They are helpful in order to advance in the design of the antenna on the basis of the energy transfer to the ions. At the present stage the output energy density seems to be exceptionally high with respect to other implementations of ion heating in different devices, at comparable input power. This could be due to the assumption of collisionless ion population. Further investigations are needed, via additional ion kinetics simulations including collisions, that are now in progress, in order to provide reliable estimates of expected ion temperature.

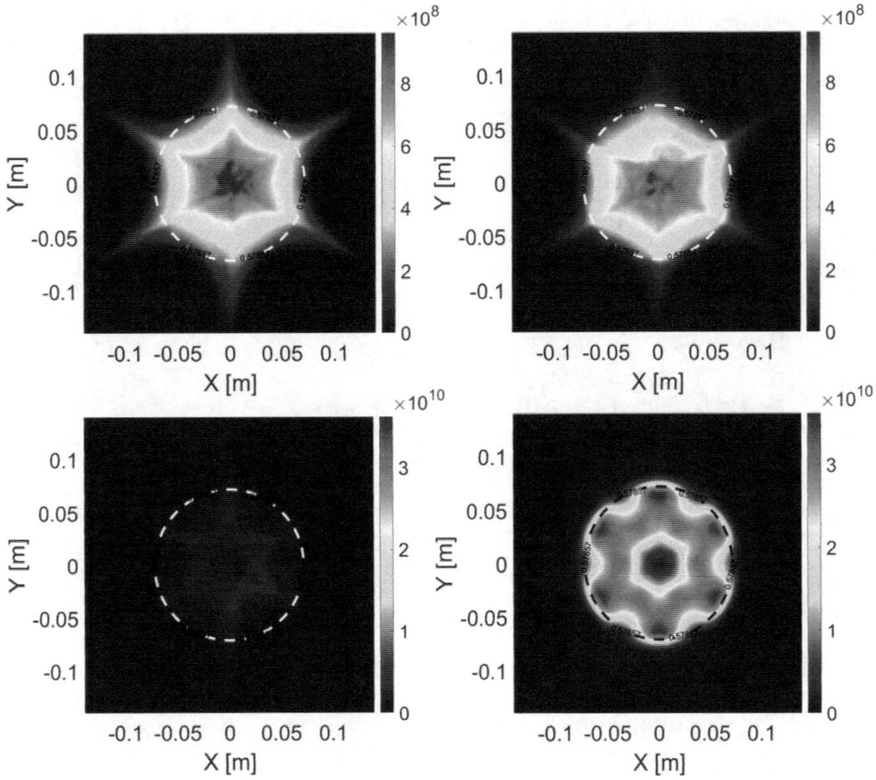

Figure 4: Projection maps along the cavity z-axis of particle numerical density (top) and energy (bottom) in [eV], resulting from kinetics simulation of ions $^{16}O^{4+}$ under the presence of cavity-coupled EM field as computed for simple loop (left) and helical (right) antennas. Dashed-lines indicate the ICR surface in the simulation domain.

4. Conclusion and perspectives

In this paper, the numerical design and simulation of RF antennas employable in ICRH systems working in ECRIS environment has been presented. The simulations are based on the coupling between the full-wave COMSOL simulations and the MATLAB-computed non-homogeneous plasma dielectric tensor. The performance of three loop antenna types have been studied in terms of S-parameters and on-axis electric field values, showing in each case an advantage of the helical antenna with respect to the other two models. Finally, the effect of ion dynamics in plasma has been studied for the simple-loop and helical antenna, showing also in this case a clear advantage of the helical antenna in increasing the ion energy. Further step forwards will be done in the next future by designing and introducing a matching network in order to minimize the reflected power at the antenna coaxial port, thus increasing the performances of the system.

Acknowledgement

The authors wish to thank the 3th Nat. Comm. of INFN, under the PANDORA_Gr3 project, for the financial support.

References

[1] J. Adam, "Review of tokamak plasma heating by wave damping in the ion cyclotron range of frequency," *Plasma Physics and Controlled Fusion*, vol. 29, no. 4, pp. 443–472, 1987.

[2] E. P. Potanin, "Analysis of the separation of gadolinium isotopes by the icr method," *Plasma Physics Reports*, vol. 34, no. 2, pp. 121–127, Feb 2008. [Online]. Available: https://doi.org/10.1134/S1063780X08020049
[3] A. V. Ilin, F. C. Díaz, J. P. Squire, and M. D. Carter, "Plasma heating simulation in the vasimr system," 2005.
[4] J. F. Caneses, B. D. Blackwell, and P. Piotrowicz, "Helicon antenna radiation patterns in a high-density hydrogen linear plasma device," *Physics of Plasmas*, vol. 24, no. 11, p. 113513, 2017.
[5] Y. Yasaka, Y. Hayashi, H. Takeno, and S. Nakamoto, "Bi-directional excitation of radio frequency waves using a helical antenna in non-uniform plasmas towards a compact magnetoplasma thruster," *Open Journal of Applied Science*, vol. 04, no. 13, 2014.
[6] J. L. Kline, E. E. Scime, P. A. Keiter, M. M. Balkey, and R. F. Boivin, "Ion heating in the helix helicon plasma source," *Physics of Plasmas*, vol. 6, no. 12, pp. 4767–4772, 1999.
[7] D. Mascali, L. Neri, L. Celona, G. Castro, G. Torrisi, S. Gammino, G. Sorbello, and G. Ciavola, "A double-layer based model of ion confinement in electron cyclotron resonance ion source," *Review of Scientific Instruments*, vol. 85, no. 2, p. 02A511, 2014.
[8] G. Torrisi, D. Mascali, G. Sorbello, L. Neri, L. Celona, G. Castro, T. Isernia, and S. Gammino, "Full-wave fem simulations of electromagnetic waves in strongly magnetized non-homogeneous plasma," *Journal of Electromagnetic Waves and Applications*, vol. 28, no. 9, pp. 1085–1099, 2014.
[9] D. Mascali, G. Torrisi, L. Neri, G. Sorbello, G. Castro, L. Celona, and S. Gammino, *Eur. Phys. J. D.*, vol. 69, no. 27, 2015.
[10] A. Galatà, C. S. Gallo, D. Mascali, and G. Torrisi, e-print arXiv:1912.01988, 2019.
[11] G. Torrisi, D. Mascali, G. Sorbello, L. Neri, L. Celona, G. Castro, T. Isernia, and S. Gammino, *Journal of Electromagnetic Waves and Applications*, vol. 28, no. 9, 2014.

High Performance 18 GHz ECR Ion Sources Development

C Qian[1,2], L T Sun[1,2], W Lu[1], Z H Jia[1], T J Yang[1,2], L Zhu[1], S J Zheng[1,2], X D Wang[1], X Fang[1], L B Li[1], J W Guo[1], J D Ma[1], H Wang[1], Y C Feng[1], X Z Zhang[1] and H W Zhao[1,2]

[1]Institute of Modern Physics, Chinese Academy of Sciences, Lanzhou 730000, China
[2]School of Nuclear Science and Technology, University of Chinese Academy of Sciences, Beijing 100049, China

E.mail: qianc@impcas.ac.cn, sunlt@impcas.ac.cn, luwang@impcas.ac.cn,

Abstract. At present, ECR ion sources are developing towards higher frequency, higher magnetic field technology. However, 18 GHz is the highest optimal operation microwave frequency for room temperature ECR ion sources, which can meet the needs of most of the existing heavy ion facilities. After the success of the 18 GHz ECR ion source LECR4, we developed the upgraded version source LECR5 aiming for higher beam intensity and higher charge state ions. With a higher radial field, bigger plasma chamber volume, longer mirror length, and flexible B_{min} field, promising results have been made at the power level of ~2.1 kW@18 GHz, for instance, 81 eμA Bi^{32+}, and 22 eμA Bi^{41+}. This ion source has been recently used for the heavy ion facility SESRI (Space Environment Simulation Research Infrastructure) as the pre-injector ion source. High beam intensities and reasonable beam qualities have been demonstrated during the test platform. Inspired by this outcome, a hybrid 18 GHz ion source called HECRAL has been proposed aiming for the similar performance of SECRAL at 18 GHz. This paper will present the recent update of the LECR5 ion source commissioning. The design and preliminary results of the HECRAL ion source will be described.

1. Introduction

R Geller and his colleagues first used the ECRIS (Electron Cyclotron Resonance Ion Source) to produce multiply charged ion beams. Due to superior performance gradually replaced other ion sources and is widely used for accelerators, atomic physics research, and industrial application [1, 2]. Over the last four decades, the min-B confinement magnetic configuration of the ECRIS was realized by superposing an axial magnetic mirror field and a radial multipole field. The ion beam intensity extracted from the ion source with the square increase of the magnetic trap is also summarized [3]. The performance of ECR ion sources in terms of charge state and current has been dramatically improved by using ever higher microwave frequencies for ECR heating of the plasma and corresponding higher magnetic fields. The design of modern ECR ion sources is based on semi-empirical magnetic field scaling laws [3] and can be divided into the following categories: a. RP ECRIS: Room temperature coils and Permanent magnet; b. PP ECRIS: All Permanent magnets; c. SP ECRIS: Superconducting coils and Permanent magnet; d. SS ECRIS: All Superconducting coils. The comparison diagram of the operating frequency of these ECR ion sources is shown in Figure 1. According to the peak intensity of the extracted charged ion beam, ECRIS has been developed to the 4th generation of 60 GHz. The 1st – 3rd generation ECR ion sources respectively operate at frequencies below 10 GHz, 10 – 18 GHz, and above 18 – 28 GHz. The 4th generation ECR ion source has only one FECR designed to operate at 45 GHz [4], which is still under construction, and its third generation ion source with high performance are SECRAL, SECRAL2 [4], and VENUS [5]. As can be seen from Figure 1, the 18 GHz ECR ion source is the limit of room temperature magnet technology, and higher frequency ECRIS must rely on superconducting technology. ECR ion sources operating below 18 GHz can meet the needs of most heavy ion accelerators by using multi frequency heating and afterglow technology. We have developed room temperature ECR ion sources from 10 GHz to 18 GHz on the IMP ECR ion source group [6, 7]. At present, LECR5 (Lanzhou Electron Cyclotron Resonance ion source No.5) upgraded LECR4 operating 18 GHz has also been successfully constructed and has been fully commissioned to meet the requirements of the SESRI (Space Environment Simulation Research Infrastructure) project [8]. Based on reducing power consumption and improving performance, a new hybrid superconducting ECR ion source named HECRAL (Hybrid superconducting Electron Cyclotron Resonance ion source with Advance in Lanzhou) was proposed for design in 2019. It was successfully constructed and commissioned in two years. The design parameters of LECR5 and HECRAL and other high performance ECR ion sources of the same type, GTS [9], HISSI [10], and AISHa [11], are shown in Table 1. 18 GHz is an optimal frequency for RP ECRIS and SP ECRIS to make a strong one with high performance and reasonable cost. Higher magnetic field and higher beam intensities result in a larger emittance, and magnets with all superconducting technology bring higher costs and risks.

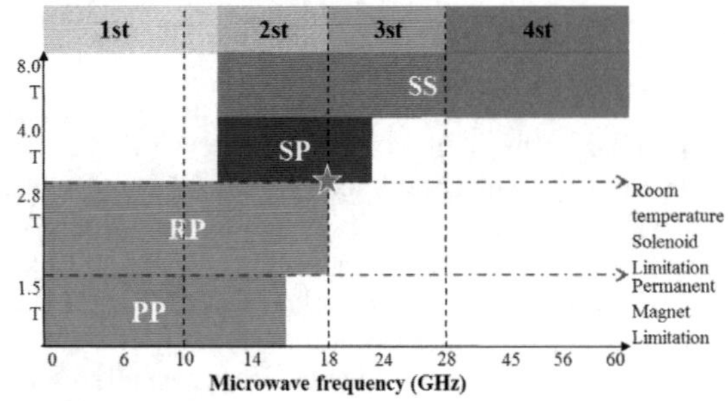

Figure 1. Classification and development of ECR ion sources.

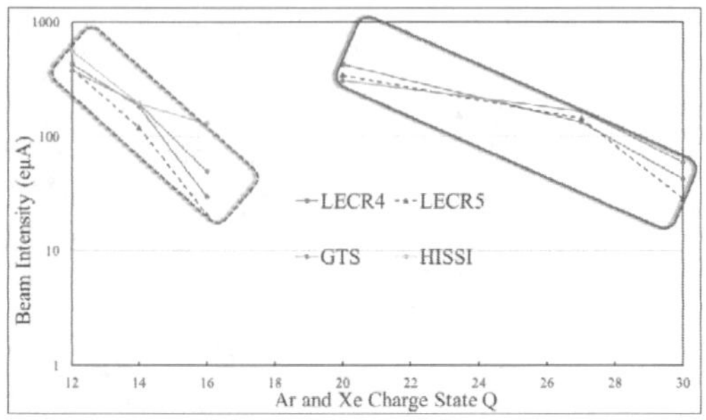

Figure 2. Comparison of highly charged ion beams production of argon and xenon on some high performance RP ECR ion sources.

Table 1. Design parameters of high performance 18 GHz ECR ion sources.

	LECR4 (RP/IMP)	LECR5 (RP/IMP)	GTS (RP/CEA)	HISSI (RP/JYFL)	HECRAL (SP/IMP)	AISHa (SP/INFN)
f (GHz)	18	18	18	18	18 - 24	18 - 21
P (kW)	~2.0	~2.0	~2.0	~3.0	~4.0	~3.0
B_{inj}/B_{ext} (T)	2.4/1.3	2.5/1.4	2.5/1.4	2.8/1.3	3.4/1.7	2.7/1.6
B_{min} (T)	0.4	~0.4	~0.4	~0.4	~0.5	~0.4
L_{mirror} (mm)	~280	~340	300	400	420	450
PC ID (mm)	76	80	80	100	100	92
B_r (T)	1.0	1.2	1.2	1.4	1.4	1.3

2. Production of Intensity highly charged ion beams with LECR5

2.1. Commissioning gaseous ion beams at 18 GHz

The LECR5 ion source and beam extraction focusing solenoid were installed on the SECRAL2 beam transmission line platform. The test bench includes the 90° double focusing bending magnet and beam diagnostic system. It was commissioned from stable beam production with oxygen and argon in August 2019. After one week of optimization, some good results were produced, such as 450 eμA of O^{7+} and 260 eμA Ar^{12+} at double frequency heating (14.5 + 18 GHz) [8]. As the control system and electromagnetic compatibility testing, it was commissioned with oxygen again in February 2020. Intense ion beam 2.12 emA O^{6+} and 485 eμA O^{7+} were produced at total power of ~1.5 kW. The typical drain current in these tunings is about 10 emA for intense O^{6+}. The working pressure is about 2×10^{-6} mbar at injection and 8×10^{-7} mbar at extraction. These ion beam results show it is a high performance RP 18 GHz ECR ion source. The LECR5 ion source needs to provide ion beams, such as H_2^+, $^{4}He^{2+}$, $^{84}Kr^{18+}$, $^{209}Bi^{32+}$ and commissioned with a single frequency of 18 GHz, and maximum a power of 2.5 kW [8]. General conditions were used, such as stainless steel plasma chamber, aluminum biased disk, and aluminum plasma electrode with 8 mm aperture. After two months of optimization, it can meet the requirements of the SESRI project. Because of the long-term operation of the ion source and providing a wide variety of ion beams for that project. Optimizing of typical charge states of gaseous ion beams was commissioned to research the performance of the LECR5 ion source, for instance, 385 eμA of Ar^{12+}, 121 eμA of Ar^{14+}, 21 eμA of Ar^{16+}, 345 eμA of Xe^{20+}, 145 eμA of Xe^{27+}, 29 eμA of Xe^{30+} were produced at a power of ~2 kW and extraction voltage 28 kV. The ion beam results compared with other high performance ion sources are shown in Figure 2. These result higher than the project requirements can ensure the stable and reliable operation of the ion source. LECR5 ion source has belonged to the level of the high performance 18 GHz ECR ion source.

2.2. Commissioning metal ion beams at 14.5+18 GHz

Several classic technologies were incorporated to explore the performance of the LECR5 ion source, such as double frequency heating that two off-axis WR62 waveguides make it possible to feed microwave power up to 3 kW, plasma chamber made of aluminum that good cooling condition for injection and extraction components. High microwave power density and high secondary electron emissivity result in an outstanding performance for highly charged ion beam production. By optimizing conditions, highly charged bismuth ion beams had been obtained with a resistance oven method at total power of ~2.1 kW and extraction voltage 25 kV. The micro resistance oven provided a maximum of ~1400 ℃ was developed at IMP for metallic ion beam production [12]. By directly evaporating bismuth, some highly charged bismuth ion beams have been produced successfully. After more than one week of commissioning for high power, especially above ion beams with 40 charge state were produced when optimizing all technical conditions. Table 2 compares the optimal beam results of LECR5 with SECRAL at 18 GHz [13]. Figure 3 gives the typical spectrum when $^{209}Bi^{32+}$ optimized at 18 GHz power of 1.8 kW and 14.5 GHz power of 0.3 kW and the typical magnetic field distribution B_{inj} = 2.5 T, B_{min} = 0.48 T, B_{ext} = 1.25 T. Figure 4 shows the normalized RMS emittance of 81 eμA of $^{209}Bi^{32+}$ (drain current 3.68 emA) is 0.09 π mm mrad in the x direction and 0.10 π mm mrad in the y direction. For refractory metals, a mini inductive heating oven provided a maximum of ~2000 ℃ has been fabricated and tested in 2020. Some high charge state calcium beams such as 52 eμA of $^{40}Ca^{16+}$ and 12 eμA of $^{40}Ca^{18+}$ have been produced successfully [14]. The performance of LECR5 can produce high current highly charged heavy ion beams is based on the optimized magnetic field and plasma volume.

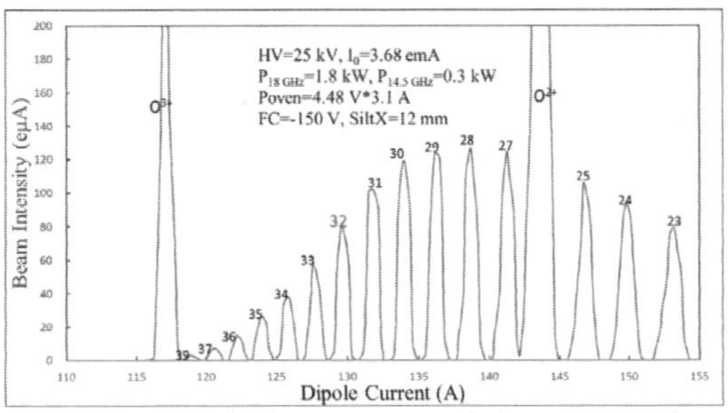

Figure 3. Typical spectrum of the LECR5 ion source optimizing for $^{209}Bi^{32+}$.

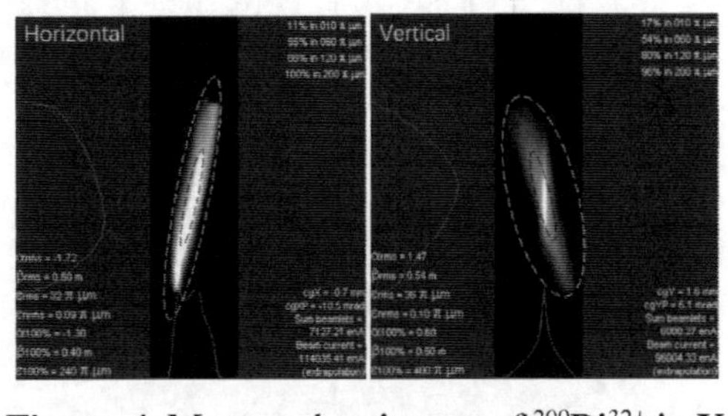

Figure 4. Measured emittance of $^{209}Bi^{32+}$ in X (horizontal) and Y (vertical) directions on theLECR5 ion source by an Allison-type emittance scanner.

Table 2. Comparison of optimal Bismuth ion beam results of LECR5 and SECRAL at 18 GHz.

	Charge State	LECR5 (eμA)	SECRAL (eμA)
^{209}Bi	30+	119	191
	31+	101	150
	32+	81	
	41+	22	22
	45+	12.5	15
	50+	3.8	1.5

3. Design and Preliminary results of HECRAL

3.1. The design of HECRAL

At present, the performance of SP ECR ion sources is worse than that of RP 18 ECR ion sources, for instance, 140 eμA Ar^{12+} with AISHa [11], 200 eμA Ar^{12+} and 100 eμA Ar^{14+} with PKISIS [15]. The higher magnetic field does not reflect the corresponding performance. The most important motivation of HECRAL development is to study the SP ECRIS and operate at 18 GHz and higher frequency. Besides, that development aimed to reduce the coils' power consumption from ~200 kW to ~15 kW and realize the cooling mode of zero volatilization through liquid helium immersion. The mechanical layout is present in Figure 5. The superconducting magnet with the warm hole of 260 mm and hexapole magnet with $B_r > 1.4$ T at the plasma chamber wall were designed separately and then assembled together, which is easy to upgrade hexapole magnet. Table 1 gives the design parameters. The assembled source body is about 864 mm, length 876 mm in diameter external size for a total weight of about 1.5 tons. Due to the long resonance length and in order to have a better tuning of the field gradient at resonance and of the minimum-B field at the midplane, a set of four coils made of NbTi is proposed, both central coils being supplied with a counter current as compared to the other two coils located at the injection and extraction sides. The axial magnetic field distribution was designed with reference to magnetic field of the SECRAL operating at 18 GHz [16].
The hexapole permanent magnet composed of NdFeB is an optimized Halbach structure, which special care was devoted to the volume size and demagnetization factors. Too large corresponding too large warm hole will increase the difficulty of producing more than 3 T by superconducting coils, and the leakage magnetic field will affect the electrical equipment at the injection and extraction ends. Dimensions of the hexapolar structure are: ID= 108 mm, OD = 232 mm, length = 420 mm. Because of the reverse magnetic field of the inner parts of the magnet and because of the strong opposite radial field created by coils and of the large axial field, some parts of the hexapole need to be built with magnets having a very strong coercitive force. Then three different magnet qualities were used, and the demagnetization risk was analyzed by software Opera-3D [17]. The calculated radial field is 1.45 T at the plasma chamber wall, which agrees with the magnetic field scaling laws at 18 GHz. The hexapole magnet has been successfully constructed, and B_r on the plasma chamber wall reached about 1.41 T in 2020, and it has been assembled with the superconducting magnet and successfully commissioned at 18 GHz in 2021. It is expected that the axial magnetic field can be increased to operate at a frequency of 24 GHz in 2022.

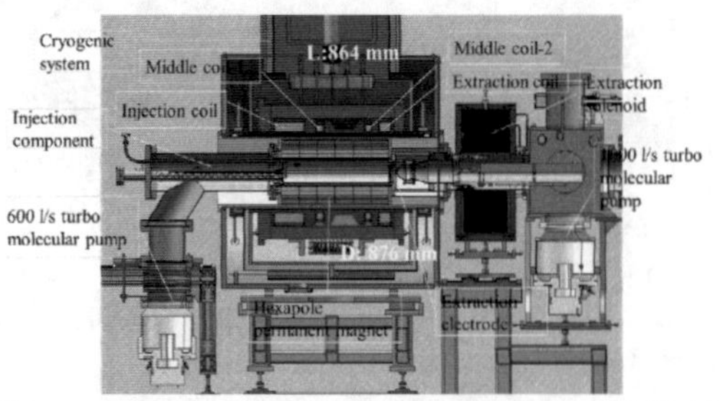

Figure 5. The assembled magnet of the HECRAL ion source shows four coils, a hexapole permanent magnet, injection and extraction components.

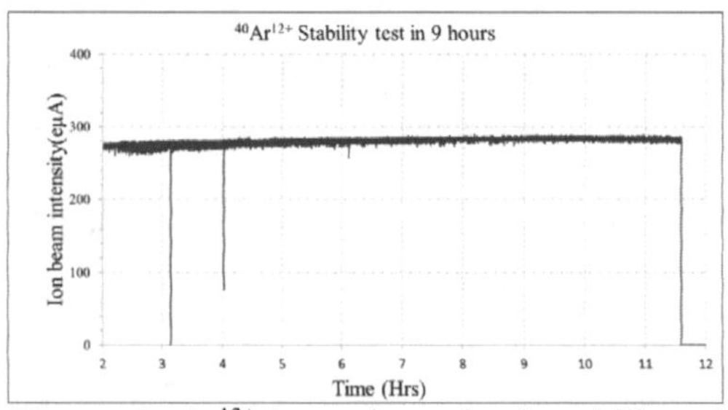

Figure 6. Ar^{12+} beam intensity instability of the HECRAL ion source during the source conditioning at 1.5 kW.

3.2. Commissioning preliminary results at 18 GHz

The HECRAL ion source and beam extraction focusing solenoid were installed on the SECRAL2 beam transmission line platform in 2021, when the LECR5 ion source system was containerized and transported to the SESRI project. The injection component of the ion source is designed with two off-axis WR62 waveguides and aluminum biased disk. Only one 18 GHz klystron microwave generator was limited to around 2.1 kW. Some classic technologies were incorporated to improve the performance of HECRAL, such as aluminum plasma chamber, aluminum plasma electrode with 8 mm aperture. After more than one week of baking and optimizing, it was commissioned for 1.61 emA O^{6+}, and 496 eμA O^{7+} were produced at a power of ~1.7 kW and extraction voltage 25 kV. Better than LECR4 and LECR5, 350 eμA Ar^{12+}, 45 eμA Ar^{14+} and 45 eμA Ar^{16+} were produced a power of ~2.0 kW and extraction voltage 25 kV. It will be used for routine operations next year. Therefore, it is necessary to know the performance of producing highly charged heavy ion beams. Figure 6 gives the instability of Ar^{12+} around 280 eμA is less than 5% in 9 hours of the test. Figure 7 shows the typical spectrum when $^{129}Xe^{27+}$ optimized and the typical magnetic field distribution B_{inj} = 2.7 T, B_{min} = 0.46 T, B_{ext} = 1.27 T. Figure 8 shows the normalized RMS emittance of 205 eμA of $^{209}Xe^{27+}$ (drain current 4.32 emA) is 0.09 π mm mrad in the x direction and 0.05 π mm mrad in the y direction. These results are superior to other SP ion sources.

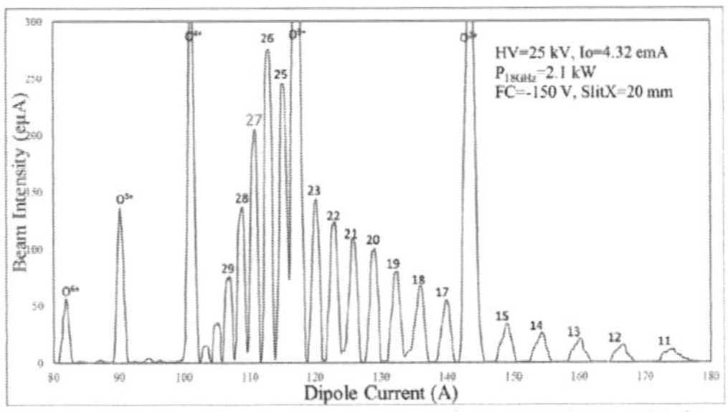

Figure 7. Typical spectrum of the HECRAL ion source optimizing for $^{129}Xe^{27+}$.

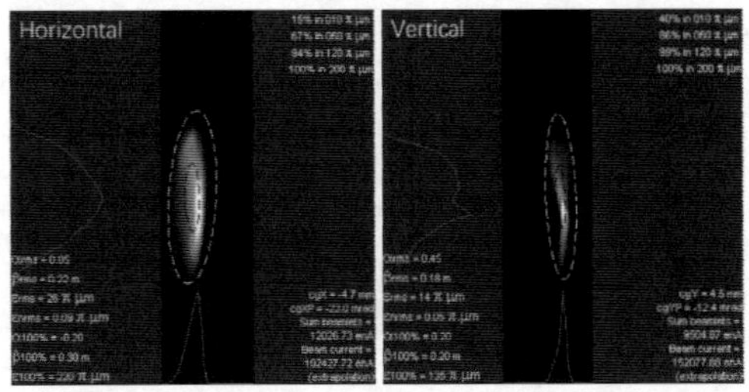

Figure 8. Measured emittance of $^{129}Xe^{27+}$ in X (horizontal) and Y (vertical) directions on the HECRAL ion source by an Allison-type emittance scanner.

4. Conclusion and Prospect

A high performance 18 GHz RP ECR ion source and a higher performance 18 GHz SP ECR ion source were successfully constructed in IMP. Some outstanding results of medium-high-charge ion beams have been produced. LECR5 can produce a higher charged heavy ion beam close to SECRAL operated at 18 GHz. HECRAL can reach the level of high performance 18 GHz ECR ion source. The emittance of the ion beam is reasonable at ~1.2 T extraction magnetic field. The optimization process of design parameters is as following the mirror length from 280 mm to 420 mm, plasma chamber diameter from 76 mm to 100 mm, the radial magnetic field from ~1.1 T to ~1.4 T. The 18 GHz ECR ion source can produce high current highly charged ion beams by using a higher radial magnetic field and bigger plasma volume when the flexible middle minimum magnetic field for magnetic mirror and resonance length. From single frequency heating to double frequency heating and higher microwave power, the performance of the 18 GHz ECR ion source will be improved in advance. Dual-frequency heating and around 3 kW microwave power will be used for LECR5 and HECRAL, and the HECRAL ion source will operate at 24 GHz with around 4 kW.

Acknowledgements

This project is supported by the National Natural Science Foundation of China (Grant No. 11875300) and National Twelfth Five-Year Plan of China.

References

[1] Arianer J and R Geller 1981 Annual Review of Nuclear and Particle Science 31: 19-51.
[2] Geller R 1990 Annual Review of Nuclear and Particle Science 40(1): 15-44.
[3] Hitz D 2006 Advances In Imaging And Electron Physics, Vol 144 144: 48-59.
[4] Zhao H W et al 2018 Review of Scientific Instruments 89(5).
[5] Leitner D et al 2006 Review of Scientific Instruments 77(3).

[6] Sun L T et al 2005 Nuclear Instruments & Methods in Physics Research Section B-Beam Interactions with Materials and Atoms 235: 524-529.
[7] Lu W et al 2015 Review of Scientific Instruments 86(4).
[8] Qian C et al 2020 Review of Scientific Instruments 91(2).
[9] Hitz D et al 2004 Review of Scientific Instruments 75(5): 1403-1406.
[10] Koivisto H et al 2020 Review of Scientific Instruments 91(2).
[11] Celona L et al 2019 Review of Scientific Instruments 90(11).
[12] Cao Y et al 2007 High Energy Physics and Nuclear Physics-Chinese Edition 31: 70-74.
[13] Zhao H W et al 2017 Physical Review Accelerators and Beams 20(9).
[14] Lu W et al 2021 Review of Scientific Instruments 92(3).
[15] Gaubert G et al 2012 Review of Scientific Instruments 83(2).
[16] Zhao H W et al 2006 Review of Scientific Instruments 77(3).
[17] http://www.operafea.cn/

19th International Conference on Ion Sources – ICIS2021 IOP Publishing
Journal of Physics: Conference Series **2244** (2022) 012025 doi:10.1088/1742-6596/2244/1/012025

The AISHa ion source at INFN-LNS

G. Castro[1], L. Celona[1], F. Chines[1], G. Costanzo[4], S. Gammino[1], O. Leonardi[1], L. Malferrari[2], A. Massara[1], C. Maugeri[4], L. Neri[1], F. Odorici[2], S. Passarello[1], R. Reitano[3], F. Russo[4], D. Siliato[1]

[1] Istituto Nazionale di Fisica Nucleare-Laboratori Nazionali del Sud, via S. Sofia 62, 95123 Catania, Italy

[2] Istituto Nazionale di Fisica Nucleare-Sezione di Bologna, Viale C. Berti Pichat , 6/2, 40127 Bologna, Italy

[3] Università degli studi di Catania, Dipartimento di Fisica e Astronomia, via S. Sofia 64, 95123 Catania, Italy

[4] CNAO, Pavia, Str. Campeggi 53, 27100 Pavia, Italy

Castrog@lns.infn.it

Abstract. The Advanced Ion Source for Hadrontherapy (AISHa) is an ECR ion source operating at 18 GHz, developed with the aim of producing high intensity and low emittance highly charged ion beams for hadrontherapy purposes. Due to its unique peculiarities, AISHa is a suitable choice for industrial and scientific applications. In the framework of the INSpIRIT and IRPT projects, in collaboration with Centro Nazionale di Adroterapia Oncologica (CNAO), new candidates for cancer treatment (including metal ion beams) are being developed. Moreover, within the IONS experiment, AISHa will be the test-bench for the development of an innovative active plasma chamber designed to increase plasma confinement by changing plasma fluxes. OES technique will be also used to refine techniques of non-invasive plasma diagnostics. Finally, a dedicated setup is under realization to provide impinging beams and detection systems for target production in nuclear physics experiments.

1. Introduction

The AISHa ion source has been designed to generate high brightness multiply charged ion beams for hadrontherapy and nuclear physics applications. AISHa is a compact electron cyclotron resonance ion source (ECRIS) [1] whose hybrid magnetic system consists of a permanent Halbach-type hexapole magnet and a set of independently energized superconducting coils. This allows to achieve high performances in a cost-effective way. AISHa has been already commissioned at 18 GHz operating frequency (see Ref. [2] for more information) and, in spring 2022, a second klystron amplifier operating in the 21 GHz band-1.5 kW, will be coupled to the source to carry out a two frequency heating mechanism, i.e. 21 + 18 GHz.

This final configuration will insert AISHa in an intermediate step between second generation ECRIS and third generation ECRIS as shown from the Golovanivsky plot (figure 1) [1].

The plasma chamber,able to hold a maximum power rate of 2 kW, the three electrodes extraction system and the waveguide DC break has been designed to permit reliable operation up to 50 kV.

The low energy beam transport (LEBT) consists of a focusing solenoid placed downstream of the source, a 90° bending dipole for ion selection, and two diagnostic boxes (fig. 2). The dipole supports the beam focusing by means of a double edge. Each diagnostic box consists of a Faraday Cup (FC), two beam wire scanners, and four slits. A beam viewer diagnostic system is being developed and its commissioning is foreseen in summer 2022.

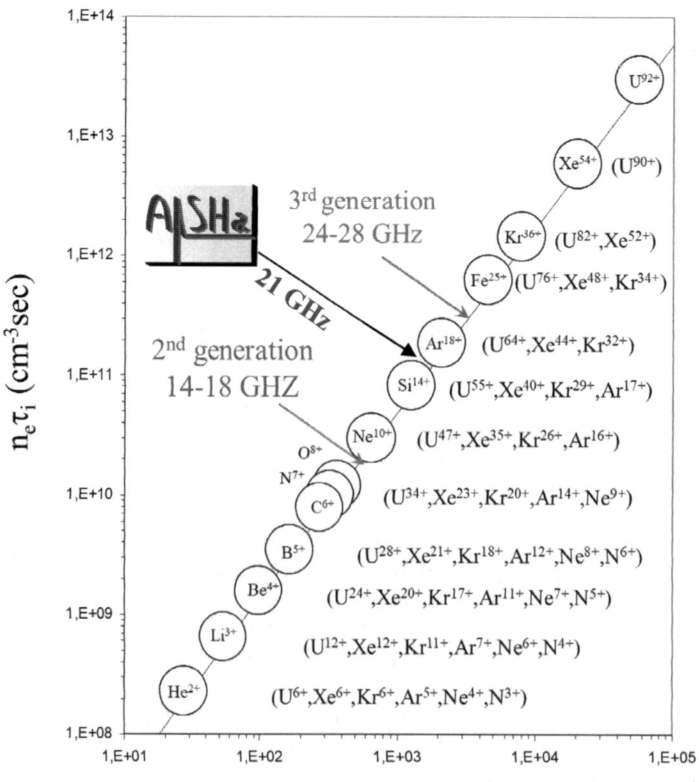

Figure 1. Position of the AISHa Ion source in the Golovanivsky diagram [1]

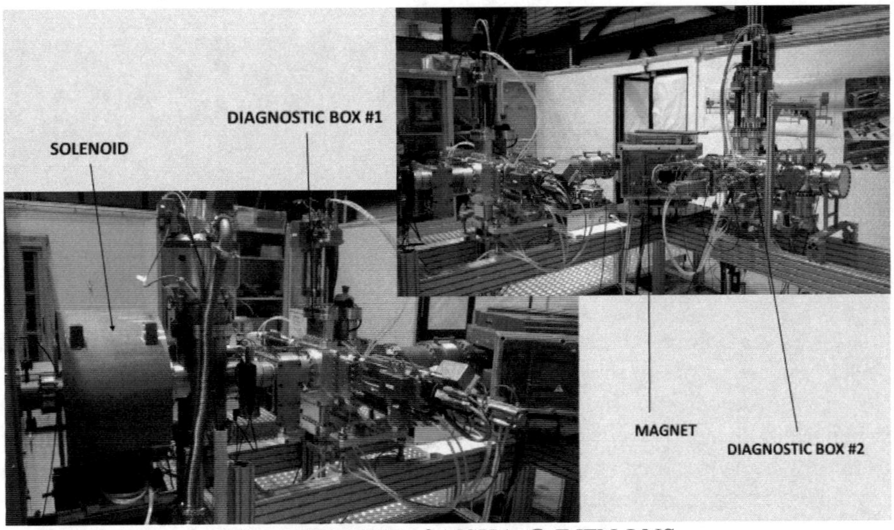

Figure 2. LEBT of AISHA @ INFN-LNS

2. Experimental activities

Peculiarities and characteristics of the AISHa ion source make it a suitable choice not only for medical purposes but also for industrial and scientific applications. In the following, the main activities that involve the AISHa ion source and the relative challenges are presented.

AISHa will be mainly used as a testbench to develop instrument and technologies to apply to the ECRIS for daily use.

The R&D program is funded by the Fifth National Commission of INFN under the IONS experiment and aims to the following main goals:

1) To project, develop and test an active plasma chamber able to decrease, by an opportune biasing, the radial ion losses induced by the Simon diffusion mechanism [3].
2) To develop a non - invasive diagnostics tool, based on optical Emission Spectroscopy to investigate the Electron Energy Distribution Function and ion densities in plasma during ECRIS operations.
3) Investigate wave to plasma coupling at 21 GHz.

The physical principle that has driven the R&D of the so called "active plasma chamber" is the well-known Simon diffusion mechanism: in a magnetized plasma, diffusion is anisotropic, electrons are mainly lost along the magnetic field while ion are mainly lost across the magnetic field. Therefore, the typical magnetic configuration of ECRIS causes electrons to be mainly lost axially and ions to be mainly lost in lateral walls [4, 5]. The reduction of electron and/or ion loss fluxes should improve plasma confinement, with positive consequences on ECRIS performances.

According to this principle, a negative bias is usually placed in the injection flange of ECRIS to reduce electron losses and improve the production of highly charged ion (HCI) beams [6]. At the same time, it was clearly proven that positively biased lateral electrodes decrease the ionic losses leading to higher HCI currents [7]. However, several technological limitations forbidden the daily use of this technology until now. The IONS project proposes to use a set of aluminium tails inserted into a support chamber. Each sector is insulated from the support to make possible a positive differentiated polarization. Each sector will then be connected to ground by means of an I-V measurement circuit, permitting the measurement of the plasma leakage currents based on an axial segmentation and thus enabling the optimization of the wall polarization to minimize radial losses and to optimize the plasma parameters. Figure 3 shows a lateral and front view of the system of tails together with a detail.

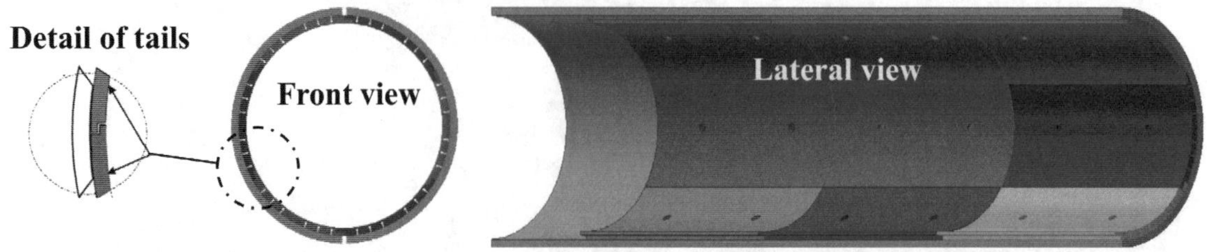

Figure 3. Lateral and front view of the aluminum ensemble of tails and a detail of the tails overlap.

The first prototype is foreseen to be tested in the second half of 2022 and in the same period we planned to start the operations of the Optical Emission Spectroscopy (OES) setup.

OES [8] has already demonstrated to be a reliable tool to investigate plasma parameters of a Hydrogen plasma discharge in a non-invasive way [9]. Moreover, opportune 0-dimensional model can be used to relate plasma parameters to beam parameters [10]. The goal of the IONS project is to relate the emission lines of some test element (such as hydrogen, Helium and Neon) to the plasma parameters. This result can be achieved by using databases concerning the emission characteristic lines in the optical domain for any charge state as a function of density and temperature. The CHIANTI database [11] has been identified as the most complete and appropriate for this purpose. Figure 4 shows a sketch of the envisaged optical coupling system which focus the light emitted by a 1.6° cone of view to an optical fiber by means of a system of collimators and lenses. Finally, the optical fiber leads the light towards a 15.000 resolving power ($\lambda/\Delta\lambda$) monocromator operating in the wavelength domain 350-700 nm.

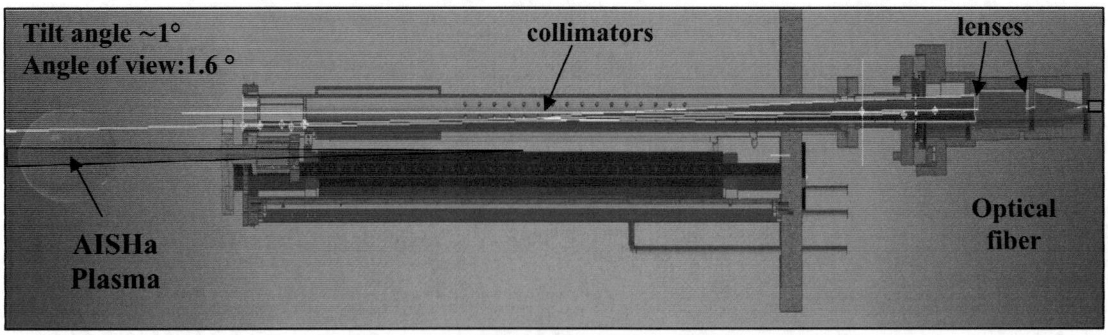

Figure 4. sketch of the optical coupling system that will be installed in the AISHa ion source for plasma diagnostics via OES.

The availability of high intensity beams and the scientific interest in precision nuclear physics experiments pushes the requests for demanding targets, able to resist at high intensity ion fluxes while maintaining its characteristics

In this framework, AISHa will be the test-bench for a feasibility study concerning the production of solid noble gas targets by implantation.

In particular, test will focus on some isotopes that are candidates for the Neutrinoless double beta decay, such as ^{136}Xe [12]. Implantation should take place in a surface quite small (5-10 mm^2) on a material able to resist to the high beam powers (up to 10 kW), such as Highly Oriented Pyrolitic Graphite (HOPG). High implantation uniformity is required to ensure the energy resolution during nuclear physics experiments. For this purpose, a rotating target support has been implemented. A sketch of the target support within the diagnostics box is shown in figure 5.

Figure 5. Sketch of the target support within the diagnostics box

3. Conclusion and Perspectives

The AISHa source is a LHe free compact ECR ion source adapt to work in hospital environment as well as for research purposes. Its use at INFN-LNS will enable some R&D activities playing a key role in the optimization of the ECR on sources for daily operation. Moreover, in the framework of the INSpIRIT project, funded by the Lombardy Region, a copy of the AISHa ion source is going to be installed at the Centro Nazionale di Adroterapia Oncologica (CNAO) in Pavia (Italy). The installation of a third ion source at CNAO will increase the beam availability both in terms of intensity and of ion species that can be produced. Helium and Lithium ions will be studied for their better lateral dose distribution with respect to protons, Oxygen is also considered due to its energy deposition characteristics [13]. Finally Iron beams will be also developed for space radiation research.

Acknowledgments
Authors wish to acknowledge the financial support of the 5th committee of INFN and the support of Lombardia Region program through POR FESR 2014-2020.

References
[1] Geller R, *Electron Cyclotron Resonance Ion Source and ECR Plasmas* 1996 (Institute of Physics, Philadelphia).
[2] Celona L et al., 2019 *Rev. Sci. Instrum.* **90** 113316
[3] Simon A, 1955 *Phys. Rev.* **98** 317
[4] Wolters U, Nadzeyka A, Meyer D, Wiesemann K 2002 *Rev. Sci. Instrum.* **73** 516#
[5] Castro G, Mascali D, Miracoli R, Celona L, Gammino S, 2019 *Rev. Sci. Instrum.* **90** 023301
[6] Gammino S, Sijbring J and Drentje A G, 1992 *Rev. Sci. Instrum.* **63** 2872
[7] Drentje A G, Kitagawa A, Uchida T, Rácz R, Biri S, 2014 *Rev. Sci. Instrum.* **85** 02A921
[8] U. Fantz, 2006 *Plasma Sources Sci. Technol.* **15** S137
[9] Mazzaglia M, Castro G, Mascali D, Celona L, Neri L, Torrisi G, Gammino S 2019 *Phys. Rev. AB* **22** 053401
[10] Castro G 2020 *Phys. Rev. AB* **23** 093402
[11] https://www.chiantidatabase.org/
[12] Cappuzzello et al., 2018 Eur. Phys. J. A **54** 72
[13] Mein S et al., 2018 Scientific Reports **8** 14829

A picosecond laser-based ion source for injection into a high capacity EBIS in both accumulation and single pulsed modes

S. Kondrashev[1], E, Beebe[1], T. Kanesue[1], M. Okamura[1], and R. Scott[2]

[1]Brookhaven National Laboratory, Upton, NY 11973, USA
[2]Argonne National Laboratory, Argonne, IL 60439, USA

skondrashev@bnl.gov

Abstract. Ps-lasers have advantages for generation of low charge state ions compared to ns-lasers because the influence of heat conductivity on a solid target is negligible in the case of ps-laser ablation for laser pulse durations shorter than 10 ps. By using a laser with high rep-rate, it is possible to produce quasi continuous 1+ ion beams for periods up to tens of milliseconds, making it possible to take advantage of the ability of the EBIS to trap 1+ ions in accumulation injection mode. We studied the properties of Al, Ti, Cu, Nb, and Ta plasmas generated by a ps-laser with 1.27 mJ energy within an 8 ps pulse to investigate feasibility and specify parameters of a laser ion source for RHIC EBIS using accumulation injection mode. It is shown that a both accumulation and single pulsed injection modes are accessible with a single ion source geometry and single injection line, providing the most attractive option for an ion source for external injection into RHIC EBIS trap based on a ps-laser.

1. Introduction

A high current Electron Beam Ion Source (EBIS) has been routinely supplying highly charged ion beams of elements from Helium to Uranium to the Relativistic Heavy Ion Collider (RHIC) and to the NASA Space Radiation Laboratory (NSRL) since it's commissioning in 2010 [1]. RHIC EBIS operates primarily as a charge bleeder of 1+ ions injected from external ion sources (neutral gas injection can be utilized when the need arises). A Laser Ion Source (LIS) provides 1+ ions for most species of solid materials in a high current single pulse ("fast") injection mode [2], and two Hollow Cathode Ion Sources (HCIS) provide 1+ ions of both gaseous and solid species in a lower current accumulation ("slow") injection mode [3]. Since the EBIS acts as a charge state multiplier, the stability and reliability of the external ion sources is as important as it is for the EBIS itself.

1.1. "Fast" and "slow" ion injection modes into EBIS

"Fast" and "slow" external ion injection into the EBIS trap are illustrated in Figs. 1 and 2. In the "fast" injection mode, (Fig. 1) the potential of the trap exit barrier is lowered to accept an incoming beam of 1+ ions and then raised to trap the ions after the time for one "round-trip" worth of 1+ ions. In the "fast" mode, efficiencies often more than 50% can be achieved. For the RHIC EBIS trap with a length of 1.7 m, the typical "round-trip" time of these ions is in the range of 100-300 μs for heavy elements of periodic table. In the "slow" injection mode (Fig. 2), the potential of the trap exit barrier is set just below the potential of the external ion source potential. If an injected ion is ionized further by the electron beam before exiting the trap region, it will remain in the trap as it is captured by the potential barrier. The "slow" injection mode efficiency is much lower compared to "fast" injection mode, typically <5%, but

Journal of Physics: Conference Series **2244** (2022) 012026 doi:10.1088/1742-6596/2244/1/012026

in the "slow" mode one can inject and accumulate ions into the EBIS trap for tens of ms which is advantageous in some cases.

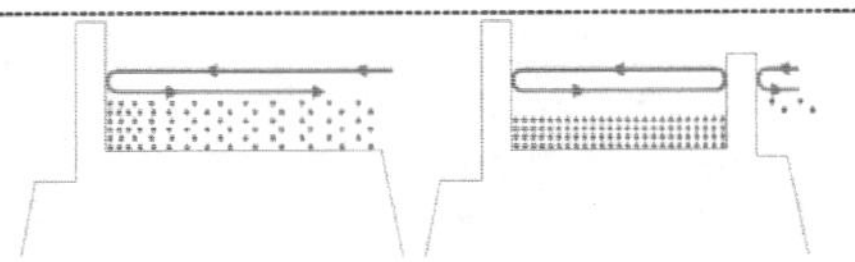

Figure 1. "Fast" ion injection mode into the EBIS trap: the potential of the trap exit barrier is lowered to accept an incoming beam of 1+ ions (left panel) and then raised to trap the ions after the time for one "round-trip" worth of 1+ ions (right panel). Red dots indicate injected 1+ ions and red lines schematically represent their trajectories.

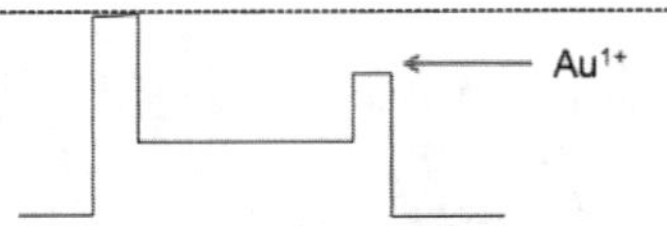

Figure 2. "Slow" ion injection mode into the EBIS trap: energy of injected 1+ ions is just slightly above of the potential of the trap exit barrier, so injected 1+ ions can enter the trap. If an injected ion is ionized further by the electron beam before exiting the trap region, it will not be able to escape past the trap barrier since the applied potential will now be almost twice as high as necessary for trapping the 2+ ion, comparing to the 1+ ion.

1.2. Advantages of ps-laser based ion source of 1+ ions for injection into EBIS

The focus of this paper is on practical aspects of ion generation by a ps-laser as a source of 1+ ions for external injection into the EBIS trap. There are at least two advantages of ps-laser generated plasma compared to the ns-laser generated plasma currently in use. First there can be higher efficiency of target material utilization resulting in lower rate of target material consumption and accompanied by a lower vacuum pressure increase. The physical picture of solid target ablation in the picosecond range qualitatively differs from nanosecond laser ablation [4]. Due to the very brief laser pulse, the influence of heat conductivity in the solid target is almost negligible in the case of ps-laser ablation, compared to ns-laser ablation which is dominated by heat conductivity. As a result, total number of atoms ablated from the target per laser pulse is significantly lower in the case of ps-laser. This can be important for generation of isotopically pure ion beams from expensive isotopically enriched targets, and also for interruption-free long-term operation even with conventional targets, because the required periods between target exchange can be significantly longer. A second and perhaps a more significant advantage of ps-laser ablation is that quasi continuous beams of singly charged ions can be generated to take advantage of the ability of the EBIS to accumulate ions in the "slow" injection capture mode. This injection mode is quite feasible because ps-lasers with pulse energy up to 10 mJ and rep-rate up to 10 kHz have recently become commercially available. The slow injection mode of beams from laser ion source has following attractive features:

- the required ion beam current is much lower than for "fast" injection mode
- low current ion beams are easier to transport and inject into the EBIS trap because space charge effects are less severe
- the EBIS pulse to pulse variation of highly charged ions can be improved since the ion pulse to pulse variations for the ps-laser plasma are averaged over a long train of ion pulses accepted by the EBIS
- an isotope separator can be placed downstream of laser ion source to select a single isotope ion beam generated from the targets with natural isotope abundance. This allows the EBIS to accumulate and be filled to a higher capacity with the isotope of interest
- laser ablation provides an intensity advantage compared with the HCIS for species such as Zr, which are difficult to sputter.

Recently it was shown with a HCIS in our test laboratory, that an isotope separator works quite well in the range of ion beam currents typical for "slow" injection mode into EBIS trap [5].

Journal of Physics: Conference Series **2244** (2022) 012026 doi:10.1088/1742-6596/2244/1/012026

2. Experimental setup

A 3D model of target irradiation and the diagnostics chamber specifically designed and built for ion yield measurements is shown in Fig. 3.

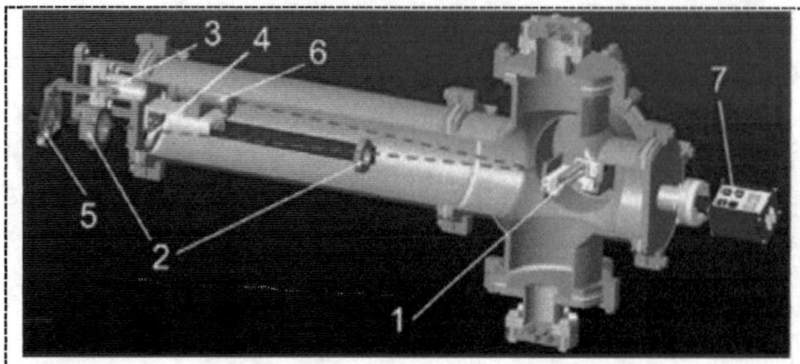

Figure 3. 3D model of the target irradiation and diagnostics chamber: 1—3D target positioner, 2—focusing lens holders, 3—CCD camera, 4—laser input window, 5—fast laser beam shutter, 6—Faraday cup, and 7—vacuum gauge. The laser path is shown by the orange dashed line and the plasma expansion axis is indicated by the blue dashed line.

We have used a ps-laser, which is available at Argonne National Laboratory (ANL) and was previously employed for material ablation into an ECR ion source [6]. This laser is commercially available from Passat Inc. and has the following parameters: wavelength—1064 nm, pulse width—8 ps, pulse energy—up to 5 mJ/pulse, repetition rate—up to 400 Hz. The maximum average laser pulse energy of this laser at the time of our experiment was found to be 1.27 mJ with standard deviation 0.06 mJ in a single pulse mode for all measurements described below. A laser beam with diameter of about 10 mm has been focused onto the target surface by lens with 300 mm focal distance mounted inside vacuum chamber. Irradiation angle between laser beam axis and the target normal was equal to 3°. A Faraday cup (FC) with input aperture of 10 mm is placed normally to the target at 57 cm distance and it is equipped with a highly transparent meshed electrode which can be biased up to –1 kV for ion extraction from plasma and for suppression of secondary electrons from the cup. A turbomolecular pump with a 450 l/s pumping speed is attached to the chamber to maintain the residual gas pressure inside the chamber at $2 \cdot 10^{-7}$ Torr. We use 3 ultra-high vacuum, high precision (±50 nm), long travel linear PPS-20 piezo positioner stages (MICRONIX USA) for 3D target positioning. It provides a 51 mm travel range in the horizontal and vertical directions and a 26 mm travel range in the lens focal depth direction. Five metal targets (Al, Ti, Cu, Nb, and Ta), each with dimensions 10 mm by 50 mm, were mounted on the target holder, which was attached to the 3D target positioner.

3. Experimental results

Parameters of ion yield were measured for both stationary targets and continuously slowly moving targets. The later option was used to minimize influence of target crater formation on ion yield parameters. Both options are of practical interest for implementation either in "fast" or "slow" ion injection modes into EBIS.

3.1. Stationary targets

3.1.1. Time dependences of ion current and total ion charge. FC signals for Ti, Cu, Nb and Ta targets and for first five laser shots onto the same target spot are presented in Fig. 4 (a, b, c, d). FC load resistor – 1 kOhm and each curve is averaged over 3 laser shots. One can see that time dependencies of ion current for the first and the second laser shots onto the same target spot are different from ion current time dependencies for subsequent shots for all target elements. There is one exception - ion yield for only the first laser shot was different from subsequent shots for the Ti target (Fig. 4 (a)). Further studies revealed that ion current pulse shape does not further change for at least about 100 subsequent laser shots onto the same target spot after the third shot.

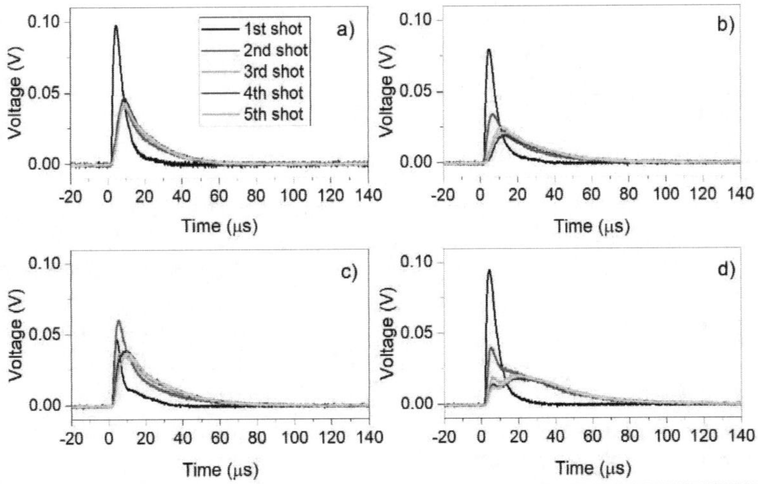

Figure. 4 (a, b, c, d). FC signals for Ti (a), Cu (b), Nb (c) and Ta (d) targets and for first five laser shots onto the same target spot (each curve is averaged over 3 laser shots).

3.1.2. Ion energy distributions. Time dependencies of ion current measured by the FC can be transformed into ion charge energy distributions using following equations:

$$\frac{dQ}{dE} = \frac{dQ}{dt} \cdot \frac{dt}{dE} = I(t) \cdot \left(\frac{dE}{dt}\right)^{-1} = I(t) \cdot \frac{t^3}{M \cdot L^2} \quad (1)$$

$$E(t) = \frac{1}{2}MV^2 = \frac{1}{2}M\frac{L^2}{t^2} \quad (2)$$

where I is ion current, t is time, E and M are ion energy and mass, Q is charge carried by ion flux, L is FC to target distance.

Ion energy distributions for all 5 target elements are presented in Fig. 5 (a, b, c, d, e) for the first 5 laser shots onto the same target spot. As one can see, ion energy distributions for all target elements have a plateau or a slight peak in the energy range of 10 - 100 eV and long low-intensity energy tail up to tens of keV.

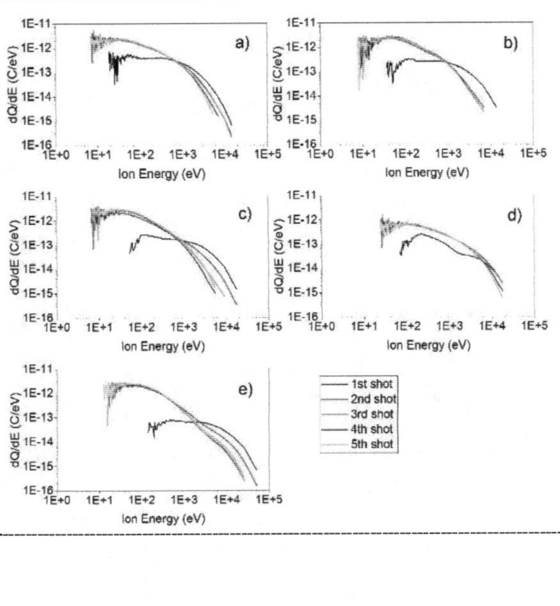

Figure 5 (a, b, c, d, e). Ion energy distributions for all 5 target elements for the first 5 laser shots onto the same target spot (laser pulse energy 1.27 mJ: a – Al target, b – Ti target, c – Cu target, d – Nb target, e – Ta target).

3.1.3. Ion Dependences of ion yield on number of laser shots onto the same target spot. Dependences of the total charge of registered ions on number of laser shots onto the same spot of stationary Al target for various laser pulse energies and rep-rates are summarized in Fig. 6.

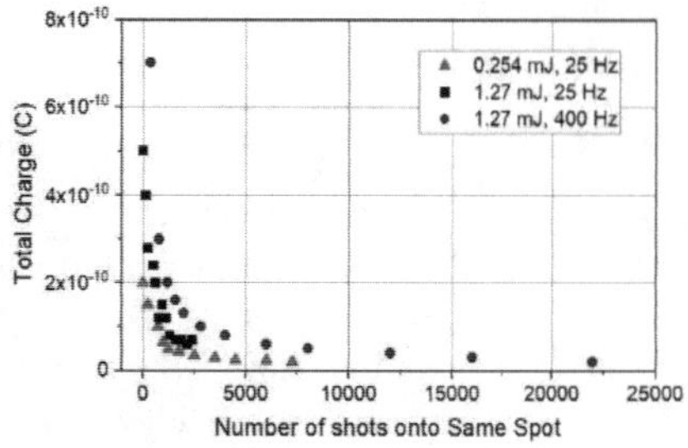

Figure 6. Dependences of total charge of registered ions on number of laser shots onto the same spot of an Al target for various laser pulse energies and rep-rates (blue circles – 1.27 mJ, 400 Hz, black squares – 1.27 mJ, 25 Hz, red triangles – 0.254 mJ, 25 Hz).

The dependences of the total charge of registered ions on number of laser shots onto the same target spot for stationary Ti, Cu, Nb and Ta targets are very similar to one for Al target.

3.2. Moving targets

3.2.1. Time dependences of ion current. Time dependences of ion current for different target translation speeds, for 1.27 mJ laser pulse energy and 400 Hz laser rep-rate, and for all five target elements are presented in Fig. 7 (a, b, c, d, e). Each curve was averaged over 128 laser shots.

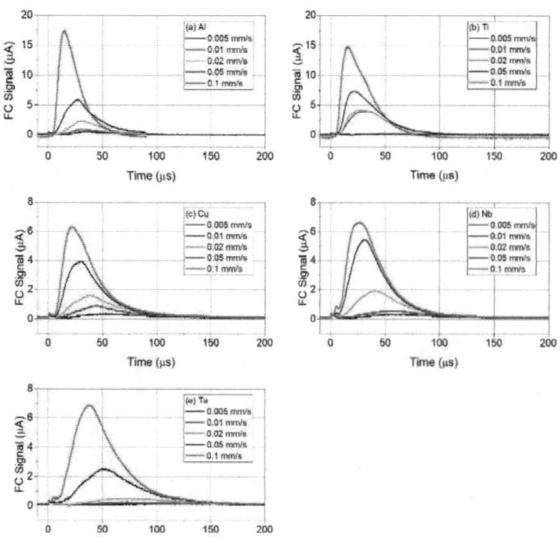

Figure 7 (a, b, c, d, e). FC signals for different target translation speeds for: (a) Al, (b) Ti, (c) Cu, (d) Nb and (e) Ta target elements A1.27 mJ laser pulse energy with a 400 Hz laser pulse rep-rate was used and each curve was averaged over 128 laser shots.

One can see that the amplitude of the ion current drops significantly and the arrival time of the peak current increases with decreasing target translation speed. Such dependence can be explained by formation of deeper craters for slower target translation speeds.

3.2.2. Ion energy distributions. Ion energy distributions are presented in Fig. 8 (a, b, c, d, e) for five target translation speeds plotted on separated graphs for each of the 5 target elements used. All measurements were performed for laser pulse energy of 1.27 mJ and 400 Hz rep-rate. The ion energy distributions resulting from the third laser shot onto the same target spot from stationary targets are displayed on each target element's graph as well. The ion energy distributions have very similar tendencies for all five target elements. The maximums of the distributions decrease, and high energy tails shrink in the distributions with decreasing target translation speed. The maximum of the ion energy distribution resulting from irradiation of the stationary targets is the highest compared to moving targets for Cu and Ta target elements. Distributions for stationary target have enhanced high energy tails compared to moving target for all target elements.

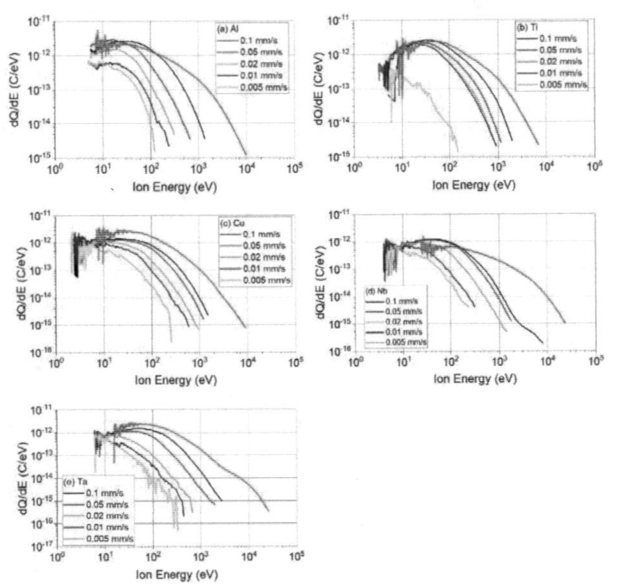

Figure 8 (a, b, c, d, e). Ion energy distributions for five target translation speeds and for five target elements (1.27 mJ laser pulse energy, 400 Hz laser rep-rate): a – Al target, b – Ti target, c – Cu target, d – Nb target, e – Ta target. The ion energy distributions resulting from the third laser shot onto the same target spot from stationary targets are displayed on each target element's graph in pink colour as well.

3.3. Ion yield for moving and stationary targets

Ion current amplitudes, pulse durations (FWHM) and total number of ions registered by FC per single laser shot for all target elements and for stationary and moving targets are summarized in Table 1.

Table 1. Ion current amplitudes (*I*), pulse durations (FWHM) (Δt) and total number of ions registered by the FC (*N*) per laser shot for all target elements for both stationary and moving targets with translation speed of 0.05 mm/s.

Element		**Al**	**Ti**	**Cu**	**Nb**	**Ta**
Stationary target	***I*, μA**	48	43	25	37	20
	***Δt*, μs**	11.5	15.5	20.4	20.2	42.1
	N	4.8×10^9	5.5×10^9	3.5×10^9	5.8×10^9	6.1×10^9
Moving target	***I*, μA**	6	7	4	5	2
	***Δt*, μs**	20	30	40	30	60
	N	$\approx 10^9$	$\approx 1.5\times10^9$	$\approx 10^9$	$\approx 10^9$	$\approx 10^9$

All data were obtained using a FC with the 10 mm entrance aperture diameter placed at 57 cm distance from target surface. Numbers for stationary target are taken for fifth laser shot onto the same target spot. All data for moving target are the data for target translation speed of 0.05 mm/s. For the estimation of the number of ions in both cases, it was assumed that most of the ions are singly charged because the laser power density at the target surface is purposely made relatively low. A small fraction of the registered ions could be in higher charge states affecting the accuracy of such estimations.

4. Specification of ion source parameters for external ion injection into RHIC EBIS

Using data summarized in Table 1, we can specify the main parameters of an ion source which will meet the requirements for external ion injection of singly charged ions into EBIS. There are scaling equations which are universal for laser ablated plasma dominated by hydrodynamic expansion. These scaling equations were first formulated and applied to laser ion source specification and design in [7] about 25 years ago. If the distance from the target surface to the ion extraction electrode (*L*) is significantly larger than diameter of extraction electrode aperture (*d*), and the time-of-flight of ions from target surface to extraction electrode is significantly longer than laser pulse duration, then the ion pulse length (Δt), number of ions (*N*) and ion current amplitude (*I*) are scaled with *L* and *d* according to following equations [7]:

$$\Delta t \sim L \qquad (3)$$

$$N \sim d^2/L^2 \qquad (4)$$

$$I \sim d^2/L^3 \qquad (5)$$

Once these parameters are measured at a given distance from the target using a given extraction aperture, they can be re-calculated to any distance and extraction aperture diameter using the above scaling equations to achieve the required ion pulse length, number of ions and ion current amplitude.

As was discussed in introduction section above, there are two different modes of external ion injection into EBIS trap – "fast" and "slow" injection modes. Typically, ~80 μA within ~100 μs (about 5×10^{10} ions total) and ~2 μA within ~30 ms (about 4×10^{11} ions total) of singly charged ions are required for "fast" and "slow" injection modes into RHIC EBIS trap, respectively. Below, we specify a geometry of a ps-laser-based ion source which will meet the requirements for external ion injection into RHIC EBIS for both the "fast" and "slow" injection modes, using the results of measurements (Table 1) and scaling equations (3) – (5).

For discussion here, we use 10 kHz laser rep-rate, which is commercially available. About 300 laser shots will hit target within a typical "slow" injection mode 30 ms time interval using a ps-laser with 10 kHz rep-rate. About 5×10^9 ions per laser shot for a stationary target and about 10^9 ions per laser shot for moving target with 0.05 mm/s speed have been measured at 57 cm distance into 1 cm aperture for all target elements (Table 1). Then, the total number of ions generated within a 30 ms time interval will be about 1.6×10^{12} and 3.3×10^{11} in this geometry for stationary and moving targets, respectively. One can see that both target irradiation options (moving and stationary target) are acceptable for "slow" injection mode into RHIC EBIS trap as the required number of singly charged ions is about 4×10^{11} within a 30 ms ion pulse. An ion extraction electrode with aperture diameter of 2.5 cm placed at 120 cm from the target surface appears to be an optimal source geometry: it provides about 5×10^{11} extracted ions for a moving target with the current amplitude reduced by a factor ~2 for both stationary and moving targets. Operation with a stationary target is preferable for the "fast" injection mode because it provides about five times higher ion yield than the case with a moving target. However, a 240 cm target-to-extraction electrode distance is needed to achieve the required ion pulse length of ~100 μs. The number of extracted ions at this distance into a 3 cm aperture according to scaling equation (4) is given by:

$$N_{f\ 240\ cm}^{3\ cm} \sim 5\times10^9 \cdot \frac{\left({}^{3\ cm}/_{1\ cm}\right)^2}{\left({}^{240\ cm}/_{57\ cm}\right)^2} \sim 3\times10^9 \qquad (6)$$

One can see that factor of ~17 is needed because about 5×10^{10} ions are typically required for "fast" ion injection mode into the RHIC EBIS trap. There are several possible ways to significantly increase number of extracted ions from the ps-laser ion source. It has been demonstrated that a solenoidal magnetic field can be used to guide a laser ablated plasma to the extraction aperture electrode thereby enhancing number of extracted ions for the same ion pulse duration by a factor of 10 [8]. One can expect that ion yield from ps-laser ablation plasma will also increase with higher laser pulse energies. So, an enhancement factor can be achieved by using more powerful laser of 5-10 mJ for the ion source, compared to the 1.27 mJ per pulse used in these tests. Another possible gain in ion yield can be achieved by using a focusing lens with longer focal length. The laser power density at the target surface will be lower in this case and pulse length of the extracted ion beam is expected to be longer for the same distance from the target to the extraction electrode. As a result, shorter target to extraction electrode distance will be needed to meet the required ion beam pulse length, and ion yield will be significantly higher according to equation (4). An additional gain of ion yield in this case can be expected due to larger focal spot size at the target surface for focusing lens with longer focal length. This larger focal spot size may also be necessary to keep the power density on the target below the threshold for ionization to multiple charge states. Based on above discussion, one can conclude that an ion source based on ps-laser for "fast" injection into the RHIC EBIS is also feasible.

A very attractive option would be the possibility to provide either "slow" or "fast" injection modes with single source geometry and single injection line. It would make ion injection system very flexible because for the generation of intense beams of highly charged ions by EBIS there may be certain advantages for different elements of periodic table either to use the "fast" or "slow" external injection modes. In general, the "slow" mode is advantageous for 1+ injected ion beam transport and stability of extracted from EBIS beam of highly charged ions, whereas the "fast" mode is advantageous when only short confinement times are needed to produce the EBIS high charge state ion of interest. The following is an estimate of the total number of ions extracted in a 30 ms ion beam pulse required during a "slow" injection mode, using the source geometry found to be feasible for "fast" injection mode, i.e., 240 cm target-to-extraction electrode distance and 3 cm diameter of extraction aperture. Using scaling equation (4), one can find that total numbers of ions will have following values for stationary (7) and moving (8) targets:

$$N_{s\,nm\,240\,cm}^{3\,cm} \sim 5 \times 10^{9} \cdot \frac{\left({}^{3\,cm}/{}_{1\,cm}\right)^{2}}{\left({}^{240\,cm}/{}_{57\,cm}\right)^{2}} \cdot 330 \sim 10^{12} \qquad (7)$$

$$N_{s\,m\,240\,cm}^{3\,cm} \sim 10^{9} \cdot \frac{\left({}^{3\,cm}/{}_{1\,cm}\right)^{2}}{\left({}^{240\,cm}/{}_{57\,cm}\right)^{2}} \cdot 330 \sim 2 \times 10^{11} \qquad (8)$$

The number of extracted ions for stationary target is well above the requirement for the "slow" ion injection mode (about 4×10^{11} ions), and the factor of ~ 2 missing for the moving target can certainly be compensated using the guiding solenoidal magnetic field technique [8]. Therefore, both the "slow" and "fast" injection modes can be provided using a single ps-laser ion source geometry and a single ion injection line.

5. Acknowledgment

This work was supported by the US Department of Energy under contract number DE-SC0012704 and by the National Aeronautics and Space Administration.
The authors are grateful to J. Ritter, R. Schoepfer and S. Picataggio for excellent technical support.

References

[1] Beebe E, Alessi J, Binello S, Kanesue T, McCafferty D, Morris J, Okamura M, Pikin A, Ritter J and Schoepfer R 2014 *Proc. Int. Symp. on Electron Beam Ion Sources and Traps (East Lansing)* (AIP Conference Proceedings 1640, 5) p 5
[2] Kanesue T *et al* 2014 *Proc. Int. Particle Accelerator Conf. (Dresden)* (JACOW) WEOAB01
[3] Alessi J, Beebe E, Carlson C, McCafferty D, Pikin A and Ritter J 2014 *Rev. Sci. Instrum.* **85(2)** 02C107
[4] Chichkov B, Momma C, Nolte S, Alvensleben A and Tünnermann A 1996 *Appl. Phys. A* **63** 109
[5] Kondrashev S, Beebe E, Coe B, McCafferty D and Ritter J 2017 *Proc. Int. Conf. on Ion Sources (Geneva)* (AIP Conference Proceedings 2011) 080005
[6] Bauder W *et al* 2015 *Nucl. Instrum. and Meth. in Phys. Res. B* **361** 465
[7] Sharkov B and Kondrashev S 1996 *Proc. European Particle Accelerator Conf. (Barcelona)* (JACOW) p 1550
[8] Ikeda S, Kumaki M, Kanesue T and Okamura M 2016 *Rev. Sci. Instrum.* **87(2)** 02A915

Journal of Physics: Conference Series **2244** (2022) 012027 doi:10.1088/1742-6596/2244/1/012027

Self-consistent Monte Carlo model for ECRIS plasma simulation

J A Méndez-Giono[1,2], T Minea[2,3], T Thuillier[1], A Revel[2]

[1]UGA/CNRS-IN2P3/LPSC - 53, rue des Martyrs - 38026 Grenoble, France
[2]LPGP - UMR 8578 CNRS - Bat. 210, rue Henri Becquerel - 91405 Orsay, France
[3]Université Paris-Saclay - 91190 Gif-sur-Yvette, France

E-mail: mendez@lpsc.in2p3.fr

Abstract. A self-consistent iterative Monte Carlo model to simulate electron cyclotron resonance ion source (ECRIS) plasma is presented. It computes the species' spatial and energy distribution in the whole plasma chamber in a three-dimensional mesh. A number of electrons and ions are propagated independently considering the static magnetic field, injected microwave field and local electrical potential field. The species trajectories populate the mesh allowing to compute their local density and velocity. Each species is pushed until it undergoes a destructive collision or after a fixed time limit. After each propagation phase, the local plasma potential and the heating electromagnetic microwave field are updated. This process is then iterated until convergence of species distributions and fields is reached. This method is intended to be a faster alternative to other methods to characterise the species distributions in the plasma for a specified ECRIS design and aid with their conception. The model and software development status are presented, along with prospects.

1. Introduction

The simulation of electron cyclotron ion sources (ECRIS) is of great interest for their design and development of such devices. This self consistent 3D Monte-Carlo method for simulating ECR plasma is intended to be a fast and reliable tool for predicting the plasma species' distributions. A Monte-Carlo method was chosen as it allows for easy parallelization of the code and a PIC simulation is not viable given the large volume simulated and predicted scale of the plasma's Debye length (λ_d), which constrains the simulation resolution. A similar, while not identical, method has been studied and is also under development by a team at INFN, Italy [1, 2, 3].

1.1. The PHOENIX V2 ECRIS

Helium plasma in the PHOENIX V2 ECR ion source [4] is simulated as a proof of concept. Helium beam production is chosen given that it has a maximum of two ionic species, further reducing computation requirements for the tests. The PHOENIX V2 ion source was developed at "Laboratoire de Physique Subatomique & Cosmologie" (LPSC), Grenoble. Then it was in operation until 2019 at the SPIRAL 2 accelerator at the "Grand Accelerateur National d'Ions Lourds" (GANIL), Caen. This ion source is now back at LPSC, where experimental measurements of its extracted ion beam will be performed, and compared with simulation.

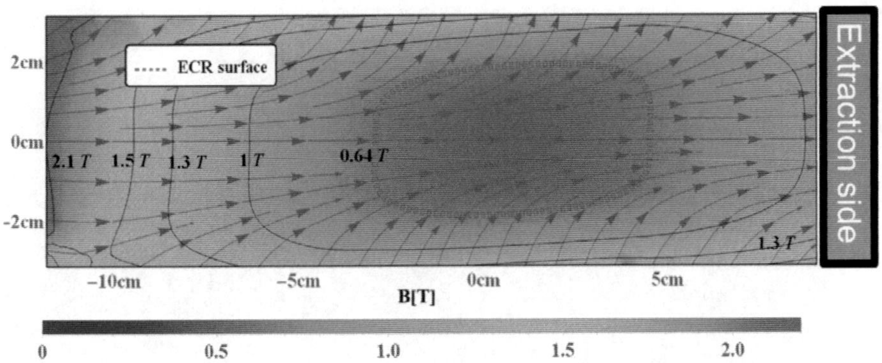

Figure 1: Minimum-B confining magnetic field of the PHOENIX V2 ECRIS

Table 1: Main characteristics of the PHOENIX V2 ECRIS

Dimensions	$L204mm \times \o 63mm$
Operation frequency	$18\,GHz$
Maximum extraction potential	$60\,kV$
$\{B_{inj}, B_{min}, B_{ECR}, B_{ext}\}$	$\{2.1\,T, 0.4\,T, 0.64\,T, 1.3\,T\}$

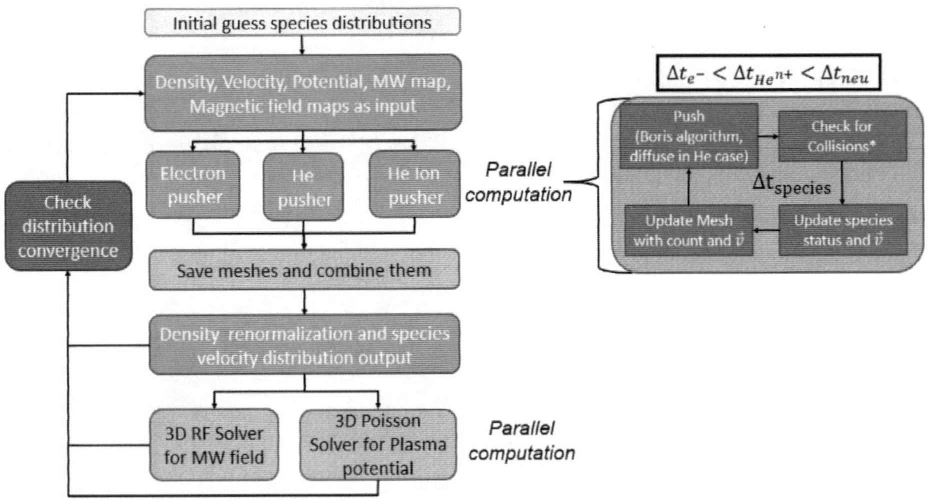

Figure 2: Flowchart of the ECR helium plasma Monte-Carlo simulation

Axial electron confinement is provided by solenoids and radial confinement by a permanent magnet hexapole. This provides a magnetic field configuration called "Minimum-B", see figure 1, and its main characteristics are listed in table 1 [4].

2. Simulation Overview

In its current state this code propagates both ionic species of helium and electrons using the Boris algorithm, the standard for magnetised plasma simulations [5, 6]. Given that simulated instances of the plasma species are propagated independently, this code lends itself very well to parallel computation. This can be done by simply running multiple instances of the executable

Table 2: Relevance of considered collision processes for the propagation of each plasma species

Collision processes considered*	**Energy range [eV]**	**e$^-$**	**He$^+$**	**He^{2+}**	**He**
e$^-$ impact ionization of He	24.5874<E	yes	no	no	no
e$^-$ impact ionization of He$^+$	54.4178<E	yes	yes	no	no
e$^-$ impact excitation of He 1s2	19.8196<E	yes	no	no	no
e$^-$ impact excitation of He$^+$ 1s	40.8<E	yes	no	no	no
e$^-$ radiative recombination with He$^+$	0<E	yes	yes	no	no
e$^-$ radiative recombination with He^{2+}	0<E	yes	no	yes	no
Single charge exchange He$^+$+He	0<E	no	yes	no	yes
Single charge exchange He^{2+}+He	0<E	no	no	yes	yes
Double charge exchange He^{2+}+He	0<E	no	no	yes	yes

which run sequentially on one thread each and then combining their results.

It takes maps of the source's magnetic fields as well of the injected microwaves as input. The latter is produced by an independent simulation using the commercial physics simulation software Comsol Multiphysics, which is a work in progress that in its current state is yet to achieve convergence when coupled to the electron distributions due to unmet hardware requirements. The microwave mode in vacuum is currently being provided instead. The source's magnets are modelled using the Radia software package developed at ESRF, Grenoble, which includes a Boundary Integral method solver for their magnetic fields [7].

The Plasma chamber is meshed using an array of cells delimited in a skew manner in the cylinder's circular cross section. This reduces memory requirements, as it allows for the meshing of a $2\pi/3$ section given the expected plasma symmetry with exact internal boundaries. This results from the hexapolar symmetry of the magnetic field and the validity of this approximation is confirmed by experimental observation of the source, specifically marks on the plasma chamber walls, the extraction electrode and the bias disk. The cells of the mesh store the density and a velocity distribution for each of the plasma species. The mesh from the previous iteration is saved in memory and given as input to the next. At the propagation phase at each time-step a species count is updated in the cell corresponding to the current position, this serves as a proxy for the local density and is re-normalised at the conclusion of each higher order loop. The species velocity is also binned in the corresponding cell in order to update its distribution.

2.1. Collision handling

Coulomb scattering and inelastic collisions are handled independently. Coulomb collisions are handled by an adapted Takizuka-Abe method[8, 9]. Inelastic collisions in turn are handled by the null-collision method [10, 11]. The original version of this algorithms, groups the simulated particle instances in pairs. This being impossible in this model, a collision partner is generated by randomly sampling the density and velocity distributions from the previous mesh, which remains static during each propagation loop. This creates the need to run a first propagation loop in non-collisional mode to populate the first mesh or define an initial distribution to sample from, the former option is currently being used.

3. Preliminary Results and Discussion

3.1. Collision Validation

Non-collisional propagation of electron in the plasma chamber lets one identify two distinct populations, confined and non-confined. Predictions based of the reflection rate of the magnetic mirror predict around 70% of electrons initialised in a random direction will be confined. This

Journal of Physics: Conference Series **2244** (2022) 012027 doi:10.1088/1742-6596/2244/1/012027

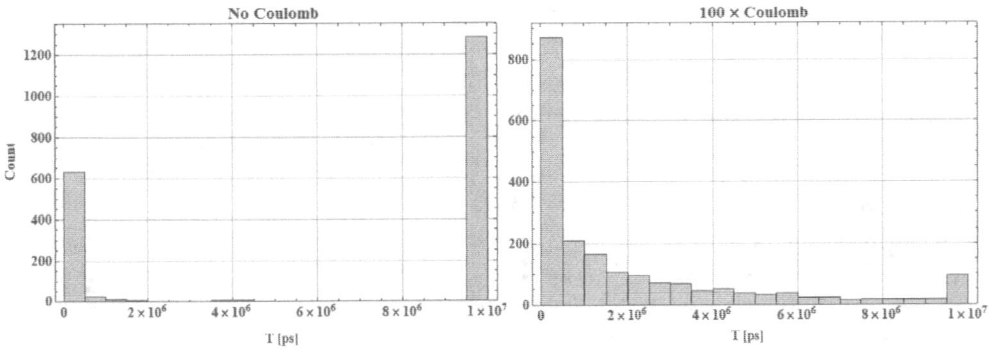

Figure 3: Comparisson of electron confinement times with and without considering coulomb collisions for $10^7\,ps$ of maximum propagation time. The strength of Coulomb collisions is synthetically increased by a factor of 100 in order to more easily observe their effects

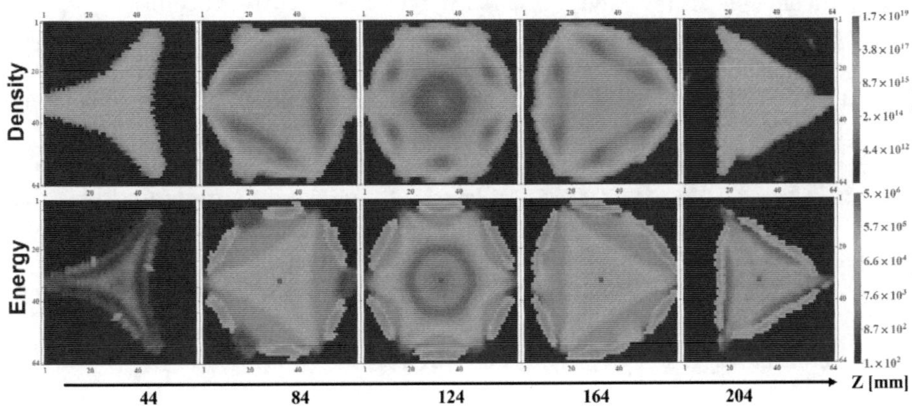

Figure 4: Transversal cuts of the preliminary electron density distribution (up) and energy distribution (down) for the collisional propagation of 12000 electrons with microwave heating for a $0.1ms$ maximum confinement time, Z is the position along the plasma chamber's axis.

is corroborated by simulation results, see figure 3 (left).

After the introduction of coulomb scattering, synthetically increased in strength to observe its effect in a shorter propagation, we see both population blend together towards a tail-like distribution with less electrons remaining confined for the $\sim 10\mu s$. Adjusting for the increased Coulomb strength, this suggests a $\sim 1ms$ maximum confinement time, consistent with expectation, see figure 3 (right).

The rates of inelastic collisions given by the implemented null-collision algorithm were computed at set positions in the plasma chamber and compared with expectation ($\nu = n_t \sigma u$). Good agreement was seen after a propagation time of the order of milliseconds, with a relative error below 1% in most cases. The exception being very infrequent processes, like in the case of radiative recombination at an energy of 1keV, where a considerable statistical error is seen ($\sim 50\%$), consistent with Poisson's distribution statistical uncertainties associated to rare events.

3.2. Electron Distributions

Qualitative examination of the electron density and energy distributions points to behaviour consistent with expectation. Electrons initialised at a temperature of 1keV are seen to heat up to up to 1MeV in some regions, while stay at around the tens of keV in others, remaining largely confined. A halo of high electron energy can be observed to surround the ECR surface.

High electron density is contained by the ECR region, which can point to a localised dip in the plasma potential and is consistent with some prediction. This dip is thought to play an important role in ionic species confinement. This is due to theoretical predictions taking only magnetic confinement into account fail to explain the long ionic confinement required for the high charged states produced [12].

A hotspot of well confined electrons with diverging energy along the plasma chamber's axis is seen, see figure 4. This phenomenon is being investigated as it seems to be a simulation artifact. One possibility being looked at is of it being a product of the microwave modelling and could go away with further iteration of this module.

4. Conclusions and Prospects

This Monte-Carlo model is so far promising for providing a relatively light-weight framework for this type of plasma simulation. After evaluation of the collision and microwave heating implementation, it is seen to have good qualitative agreement with expectation.

Charge exchange between neutral and ionic species is thought to play an important role on the charge distribution of the extracted beam. The implementation of this process is the next step in development. Then the higher order loop calling the propagation phase, as well as the microwave plasma coupling and Poisson equation solver for the plasma potential will be automated. This is a step towards consolidation of this frame-work and would be a good ease of use feature to have for future tests. After all the parts are completed a move towards obtaining higher statistics is required. A refining of the mesh towards the order of the Debye length ($\lambda_d \sim 10\mu m$) is the next step, alongside the parallel propagation of a larger number of simulated particle instances. Validation of this simulation results can then be performed by comparing with experimental data. Beam emittance and current measurements for a helium beam produced by the PHOENIX V2 ECRIS are expected to be done before the end of 2021.

References

[1] Neri L, Mascali D, Celona L, Gammino S and Ciavola G 2012 The Review of scientific instruments **83** 02A330

[2] Mascali D, Torrisi G, Neri L, Sorbello G, Castro G, Celona L and Gammino S 2015 The European Physical Journal D **69**

[3] Neri L, Castro G, Torrisi G, Galatà A, Mascali D, Celona L and Gammino S 2016 Review of Scientific Instruments **87** 02A505 (*Preprint* `https://doi.org/10.1063/1.4934616`) URL `https://doi.org/10.1063/1.4934616`

[4] Thuillier T, Angot J, Lamy T, Marie-Jeanne M, Peaucelle C, Barue C, Canet C, Dupuis M, Leherissier P, Lemagnen F, Maunoury L, Osmond B and Spädtke P 2012 Proc. 20th Int. Workshop on ECR Ion SOurce (ECRIS'12) (Sydney, Australia) pp 117, 120 URL `https://accelconf.web.cern.ch/ECRIS2012/papers/wezo03.pdf`

[5] Ripperda B, Bacchini F, Teunissen J, Xia C, Porth O, Sironi L, Lapenta G and Keppens R 2018 ApJS **235** 21 (*Preprint* `1710.09164`)

[6] Qin H, Zhang S, Xiao J, Liu J, Sun Y and Tang W M 2013 Physics of Plasmas **20** 084503 (*Preprint* `https://doi.org/10.1063/1.4818428`) URL `https://doi.org/10.1063/1.4818428`

[7] Chubar O, Elleaume P and Chavanne J 1997 Radia magnetostatics software Online URL `https://www.esrf.fr/Accelerators/Groups/InsertionDevices/Software/Radia`

[8] Takizuka T and Abe H 1977 Journal of Computational Physics **25** 205–219 ISSN 0021-9991 URL `https://www.sciencedirect.com/science/article/pii/0021999177900997`

[9] Wang C, Lin T, Calfish R, Cohen B and Dimits A 2008 J. Comp. Phys. **227** 4308–

[10] Koura K 1986 The Physics of Fluids **29** 3509–3511 (*Preprint* `https://aip.scitation.org/doi/pdf/10.1063/1.865826`) URL `https://aip.scitation.org/doi/abs/10.1063/1.865826`

[11] Levko D, Krasik Y, Vekselman V and Haber I 2013 Physics of Plasmas **20** 083512

[12] Chauvin N 2000 La transformation d'état de charge $1+/n+$ pour l'accélération des ions radioactifs Theses Université Joseph-Fourier - Grenoble I URL `https://tel.archives-ouvertes.fr/tel-00001315`

CANREB EBIS commissioning at TRIUMF

B.E. Schultz[1], C.R.J. Charles[1], M. Cavenaile[1,2], R. Kanungo[2], F. Ames[1] and O. Kester[1]

[1] TRIUMF, 4004 Wesbrook Mall, Vancouver, BC V6T 2A3, Canada
[2] Saint Mary's University, Halifax, NS B3H 3C3, Canada

E-mail: bschultz@triumf.ca

Abstract. The Canadian Rare Isotope facility with Electron Beam ion source (CANREB) is an essential part of the Advanced Rare IsotopE Laboratory (ARIEL) presently under construction at TRIUMF. CANREB was recently commissioned and can accept stable or rare isotope beams from a variety of ion sources, delivering high purity beams of highly charged ions (HCI) to experiments. The injected beams are bunched using a radiofrequency quadrupole cooler-buncher and energy adjusted using a pulsed drift tube for injection into the electron beam ion source (EBIS) charge state breeder. The EBIS is designed for a maximum electron beam current of 500 mA and a maximum magnetic field of 6 T. Ions with energies up to 14 keV can be injected and HCI with mass-to-charge (A/q) ratios $3 \leq A/q \leq 7$ can be charge bred and extracted. The HCI are A/q-selected using a Nier-type spectrometer before being transported to the ISAC linac for post-acceleration. Results from CANREB beam commissioning with focus on the EBIS will be presented.

1. Introduction

TRIUMF houses a rich nuclear and particle physics program centered around the production of rare isotope beams (RIB) [1]. At the heart of TRIUMF is the main cyclotron which strips H^- ions to produce proton beams at energies up to 520 MeV and currents up to 120 μA. For RIB production 480 MeV protons (< 100 μA) are transported to the Isotope Separator and ACcelerator (ISAC) facility and impinge on targets comprised of U, Th, Ta, Nb, or Si. Reactions products formed primarily through fission and spallation are ionized using surface, laser, or plasma discharge ion sources. Ions are extracted at energies < 60 keV and can be transported directly to a variety of low energy experiments, or to high energy experiments following post-acceleration. For the latter, ions are accelerated through the ISAC multi-stage linac with an energy acceptance of 2.04 keV/nucleon, which limits the mass-to-charge ratio to $A/q \leq 30$ for a 60 keV beam energy. Post-acceleration of heavier ions requires charge breeding prior to injection, with $3 \leq A/q \leq 7$. Currently, highly charged ions (HCI) are generated using an electron cyclotron resonance ion source (ECRIS) installed in ISAC, with efficiencies up to a few percent [2]. The main disadvantage of this source is beam contamination due to high residual gas pressure and sputtering from the plasma chamber. This generates isobaric contaminants which can be several orders of magnitude more intense than the species of interest. In addition, the maximum achievable charge state of HCI extracted from the ECRIS is limited, such that secondary stripping in the linac is necessary to reach the required A/q, and thus reducing the overall efficiency.

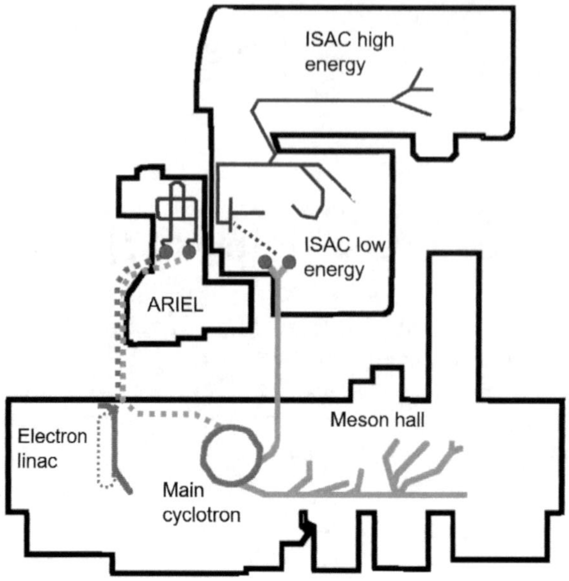

Figure 1. Overview of TRIUMF. Blue lines show existing beamlines in ARIEL and ISAC. Orange lines indicate existing proton beamlines from main cyclotron. Dashed green and orange lines show future driver beamlines for electrons and protons, respectively.

As part of the Advanced Rare IsotopE Laboratory (ARIEL) project currently in development at TRIUMF, the CANadian Rare isotope facility with Electron Beam ion source (CANREB) has been constructed for charge state breeding of ions for post-acceleration. CANREB utilizes an electron beam ion source (EBIS) to generate HCI. Unlike the ECRIS, the EBIS is designed for ultrahigh vacuum operation which greatly reduces the background contamination. The EBIS can also reach higher charge states, which will mitigate the need for a second stripping. Combined with the higher potential charge breeding efficiency of the EBIS, this will facilitate delivery of cleaner and more intense post-accelerated RIB to ISAC experiments.

2. ARIEL overview

ARIEL is currently under construction at TRIUMF, and is comprised of multiple facilities. Figure 1 shows an overview of the existing TRIUMF facility and future ARIEL installations.

2.1. Target stations for RIB production

ARIEL will add two additional target stations for RIB production, which can be operated in parallel with one of the ISAC targets. The first target station will use an electron driver beam incident on a high-Z converter material to generate gamma rays [3]. The gamma rays pass through a uranium carbide target producing radioactive nuclides through photofission. This mechanism favors the production of neutron-rich species while simultaneously reducing the production of neutron-deficient contaminants, as compared to the proton-induced fission currently employed at ISAC. The second target station will be similar to the existing ISAC stations, using an additional proton driver beam (up to 100 μA, 50 kW) extracted from the north side of the main cyclotron. The ARIEL electron and proton target stations are scheduled to come on-line in 2023 and 2026, respectively.

Both proton driver beams can be run in parallel with the electron driver, allowing for simultaneous operation of up to three target stations for RIB production. This will provide greater reliability and flexibility in the type and length of experiments that can be scheduled, as well as increase the overall yearly RIB hours by a factor of 2–3.

2.2. Electron linac

The electron linac (e-linac) [4] will be used primarily as the driver for the ARIEL photofission target. The e-linac front-end contains a thermionic cathode generating up to 10 mA of electron current at 300 keV energy. Electron extraction is RF-modulated for efficient coupling into the linac. The electrons are accelerated through a 10 MeV injector cryomodule followed by a 20 MeV cryomodule, giving a final energy of 30 MeV and up to 300 kW of power on target. The e-linac has been commissioned [5] and the ramp-up to higher powers is in progress. In the future, the e-linac facility can be upgraded to include a second cryomodule and a recirculation section to allow for increased acceleration up to 75 MeV. Alternatively, it could act as a user facility driven by a THz free electron laser, in parallel with the RIB driver beam.

2.3. CANREB

CANREB is located in the ARIEL experimental building adjacent to ISAC (Figure 1) and is split into two main parts. The lower level contains the ARIEL mass separator room, which houses the CANREB high resolution separator (HRS) which is designed to achieve a resolving power of 20,000 [6]. The HRS can accept beam from one ARIEL target station, which is then transported to the ground floor of ARIEL through a vertical beamline. In parallel, beam from the other target station can be sent directly upstairs through a second vertical beamline. If additional beam purification is not required, the HRS can be bypassed altogether. The HRS is currently being commissioned off-line.

The mass separator room also contains the ARIEL test ion source (ATIS) which is a small surface ion source used for commissioning tests [7]. It produces alkali beams with currents up to 20 nA at a beam energy up to 60 keV. The ATIS beam can be transported through the HRS for commissioning and/or upstairs for charge breeding tests.

The charge breeding part of CANREB is located on the ground floor of the ARIEL building (Figure 2). Beams can be accepted from ISAC or ARIEL ion sources. Ions are injected continuously at energies up to 60 keV into the ARIEL radiofrequency quadrupole cooler buncher (ARQB) which is used to cool and bunch the ion beam [8]. The ions are confined using a combination of electrostatic and RF fields (f_{RF} = 3 MHz and V_{RF} = 1.2 kV_{pp}). A helium buffer gas ($\approx$ 30 mTorr) cools the ions which are accumulated for up to 10 ms. Ion bunches are extracted at up to 100 Hz, giving $\approx 6 \times 10^6$ ions per bunch for the design current of 100 pA. The extracted ion bunches have a FWHM of 1 μs for efficient coupling into a pulsed drift tube (PDT) and EBIS downstream of the ARQB. The PDT is rapidly (< 500 ns) switched between high voltage and ground to reduce the ion beam energy for matching into the charge breeder. Beam is injected and extracted from the EBIS through a series of matching optics which are switched between injection and extraction values.

The CANREB EBIS was designed and constructed at the Max Planck Institute for Nuclear Physics in Heidelberg, Germany [9]. It can accept ion beams with energies up to 14 keV and $\sim 10^7$ particles per bunch, which is limited by the space charge limit of the ARQB. The ions are confined electrostatically using a series of electrodes in the center of a pair of superconducting Helmholtz coils generating a magnetic field up to 6 T. A barium dispenser cathode generates an intense electron beam (E_e < 15 keV, I_e < 500 mA) which is compressed by the magnetic field and transported through the charge breeding region to an electron collector. The central drift tubes are cooled to $\approx$ 4 K to act as a getter to reduce the residual gas pressure and the extracted ion background. Ions are charge bred for up to 10 ms (at 100 Hz) before being extracted and

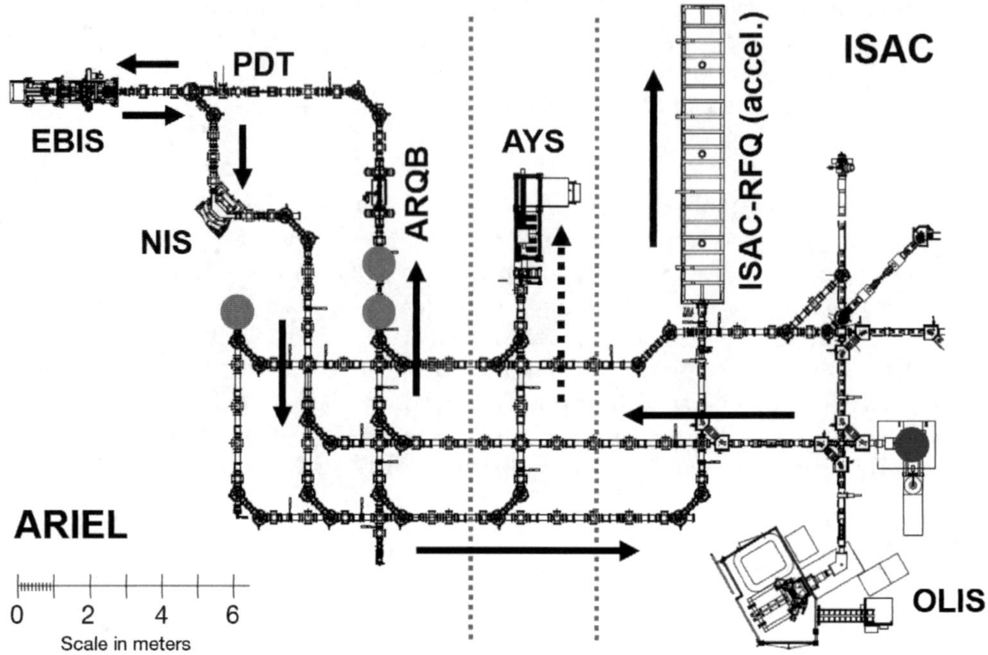

Figure 2. Layout of CANREB charge breeding facility. See text for acronym definitions. Solid black arrows indicate path of RIB from ISAC target stations (blue circle) through CANREB and into linac. Charge bred RIB from the EBIS can be sent to the ARIEL yield station (dashed black arrow) for yield measurement. Red circles indicate injection points from ARIEL mass separator room through vertical beamlines.

Table 1. List of ion sources and isotopes used for CANREB commissioning tests.

Source	Species
ATIS	^{39}K, ^{133}Cs
OLIS	^{39}K, ^{36,40}Ar, ^{85,87}Rb, ^{86}Kr
ISAC	^{59,63}Cu, ^{80}Rb, ^{124}Cs, ^{124m}In, ^{142}Cs

transported through a Nier-type spectrometer (NIS) for charge state separation ($m/dm \approx 300$). Charge bred RIB are then transported to the ARIEL yield station (AYS) for yield measurements or to ISAC for injection into the linac.

3. CANREB commissioning results

A number of isotopes were used for CANREB commissioning, listed in Table 1. The ATIS provided alkali ion beams up to 30 keV in energy. From ISAC, stable isotopes were injected from the off-line ion source (OLIS) [10] (< 20 keV) and radioactive ions were injected from the ISAC target stations (< 20 keV).

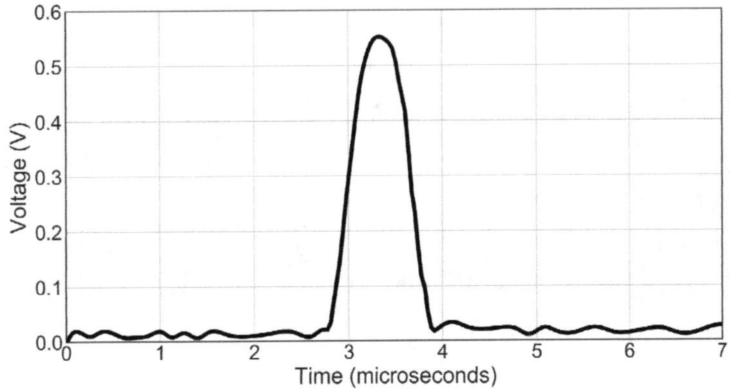

Figure 3. Pulse shape of ion bunch extracted from ARQB at 20 keV after 10 ms of cooling and accumulation. Ions were extracted by opening the trap for 5 μs. Pulse was measured downstream on a fast Faraday cup and output on an oscilloscope.

3.1. ARQB and PDT

The ARQB was commissioned using ion beams from the ATIS and OLIS at energies up to 30 keV and currents up to 500 pA. Beam was transported through the machine in both direct (i.e. no accumulation) and bunched modes. The direct mode efficiency was typically $> 90\%$ for all beams. The bunched mode efficiency was more sensitive to ion mass and current. For commissioning, the RF system was limited to a frequency of 3 MHz which was found to be optimal for confinement of masses $A > 40$. For this mass range, and with currents up to ≈ 200 pA, the bunched mode efficiency was typically $> 90\%$ and $\approx 100\%$ was achievable. For masses $A < 40$ the efficiency was reduced, but $> 80\%$ was still achievable. A 6 MHz circuit is currently under development to extend the mass range to lower values. For currents > 200 pA and up to 500 pA the efficiency was also reduced due to space charge limitations in the accumulation region. In the highest-current case, the maximum efficiency dropped to $\approx 60\%$.

The time profile of the ion bunches extracted from the buncher can be measured using a fast Faraday cup installed one meter downstream. The signal from the cup passes through a pre-amplifier and is recorded on an oscilloscope. Figure 3 shows a typical bunch profile. Ions were extracted by opening the trap for 5 μs at a repetition rate of 100 Hz, corresponding to a 10 ms cooling and accumulation time. The injected beam intensity was ≈ 130 pA with a bunching efficiency of 100%, giving $\approx 8 \times 10^6$ ions per bunch. The FWHM width of ≈ 1 μs is within the design specifications of the buncher.

Ions extracted from the ARQB were transported through the PDT for energy reduction. The PDT is powered using a 60 kV high voltage power supply connected to a 65 kV push-pull switch (Behlke Model HTS 651-10-GSM). The PDT was commissioned to a voltage of 54 kV, which was switched to ground potential in ≈ 500 ns. Transmission of ions through the PDT is optimized using two sets of quadrupole triplets and was typically $\approx 100\%$.

3.2. EBIS

For EBIS commissioning the injected ion beam energy was limited to < 7 keV due to EBIS drift tube high voltage issues, which are currently under investigation. This necessitated operation at a lower magnetic field of 1 T and a reduced electron beam current of < 50 mA.

An A/q spectrum for charge-bred ^{63}Cu (stable) extracted from the ISAC target station is plotted in Figure 4 (top, black line). The EBIS was operating with $B = 1$ T, $E_e = 4.4$ keV,

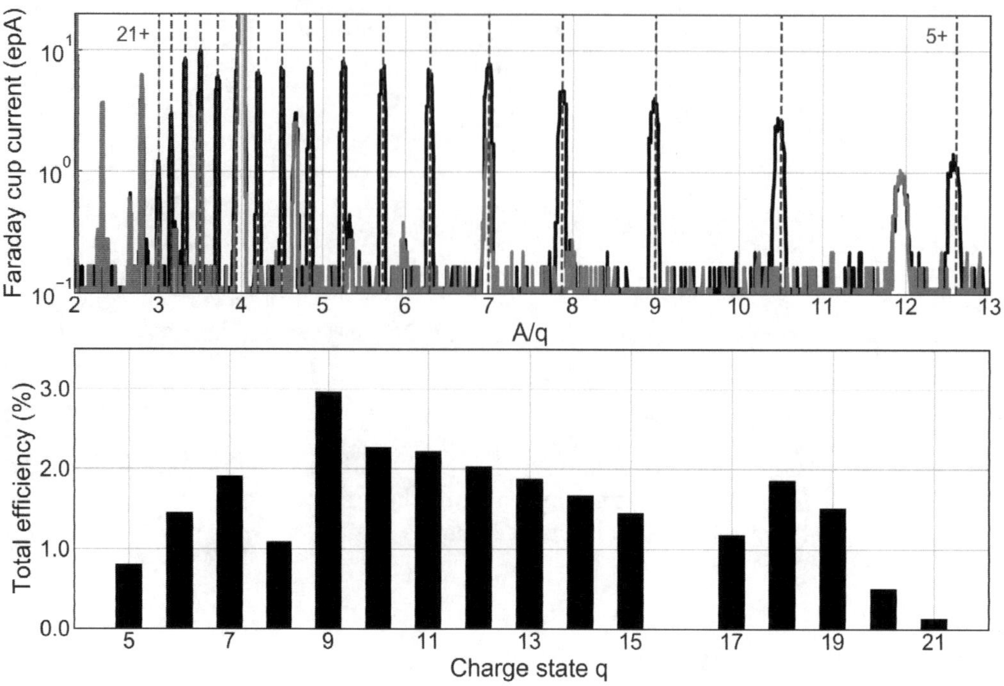

Figure 4. ^{63}Cu charge bred in the EBIS. $I_e = 30$ mA, $E_e = 4.4$ keV, $t_{CB} = 8$ ms, ion beam energy = 6.3 keV×q. (Top) A/q spectra for cases where Cu is injected into the EBIS (black) and blocked (red). Blue dashed lines indicate position of Cu charge states from $q = 5 - 21+$. (Bottom) Charge state distribution for states plotted in mass spectrum. Efficiency is calculated from the Cu beam current injected into the EBIS (≈ 30 pA).

$I_e = 30$ mA, charge breeding time $t_{CB} = 8$ ms, and an extracted ion beam energy of 6.3 keV×q. The spectrum was recorded by varying the magnetic field of the NIS up to 1150 G and measuring the beam current on a downstream Faraday cup. Charge states up to $q = 21+$ were observed for these settings. The red plot is EBIS background where the Cu beam was blocked at the source.

The charge breeding efficiency was estimated by comparing the intensity of each Cu charge state to the injected beam current measured on a Faraday cup in front of the EBIS (≈ 30 pA or $\approx 2 \times 10^6$ particles per bunch). The resulting charge state distribution is plotted in Figure 4 (bottom). The distribution peaked at $q = 9+$ with an efficiency of $\approx 3\%$, which was consistent with charge breeding results using other ion species. For the EBIS settings listed above (and assuming a residual magnetic field at the cathode of ≈ 1 mT) an upper limit on the electron beam current density is estimated to be 100 A/cm^2, which would be consistent with the observed peak charge state. The broad distribution could be attributed to poor overlap of the injected ion bunch with the electron beam. In addition, it is possible that the trap was not fully emptied between pulses. The form of the spectrum requires more investigation. To meet the final energy requirements following post-acceleration, the lower-efficiency state of $q = 19+$ was selected.

Typical background spectra from the ISAC ECRIS (black) and the EBIS (blue) are plotted in Figure 5. The EBIS spectrum was measured with $I_e = 30$ mA. For most A/q values the EBIS background is < 1 pA and several orders of magnitude less intense than that of the ECRIS. The exceptions are $A/q = 1, 2$ (not plotted) and $A/q = 4$, which are due to H^+, He^{2+}, and He^+, respectively. These species are not frozen out on the cryogenic drift tubes and are readily

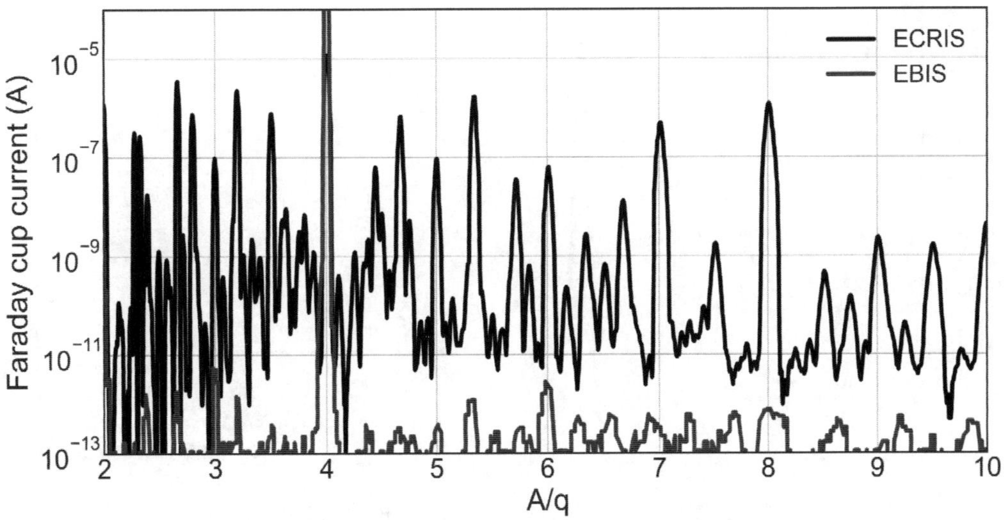

Figure 5. Typical background spectra extracted from ISAC ECRIS (black) and EBIS (blue). EBIS spectrum was measured with $I_e = 30$ mA. The EBIS peak at $A/q = 4$ saturated the Faraday cup; the actual current was $\approx 10^{-8}$ A.

ionized and extracted. Diffusion of He gas from the ARQB contributes to the helium peaks in the spectrum. The intensity of He^+ peak ($\approx$ 10 nA in this scan) is orders of magnitude more intense than most extracted charge-bred RIB will be, but this can be mitigated by selecting higher or lower A/q values for beam delivery to experiments.

Radioactive beam from ISAC was charge bred in the EBIS and used for commissioning of the ARIEL yield station, which is similar in design and capabilities to the ISAC yield station [11]. Surface-ionized isotopes (Table 1) were extracted from a uranium carbide target and injected into CANREB at 18 keV. Laser-ionized copper isotopes were also charge bred and used for first tests of post-acceleration of CANREB beams. ^{59}Cu ($t_{1/2} = 81.5$ s) was charge bred to $q = 19+$ ($A/q = 3.1$) with an efficiency of $\approx 1\%$ and accelerated through the linac to an energy of 9.0 MeV/nucleon. The ions were transported to the ISAC charged particle reaction spectroscopy (IRIS) [12] facility where the energy-loss spectrum (Figure 6) was recorded using an ionization chamber. The average beam intensity was $\approx$ 300 particles per second. The ^{59}Cu is clearly visible, along with a small peak from a lower-mass contaminant. Detector pile-up is present at higher energy loss, and results from the high instantaneous intensity of the ion pulse from the EBIS. Mitigation of this pile-up will require implementation of pulse-stretching during EBIS extraction, which is currently under development.

4. Conclusion

As part of the ARIEL project at TRIUMF, CANREB has been built to charge breed RIB for post-acceleration to experiments in ISAC. A variety of stable and radioactive species up to $A = 142$ have been injected into CANREB at energies up to 30 keV. Ions were cooled and bunched in the ARQB with efficiencies typically $> 90\%$ and energy matched into the EBIS using a pulsed drift tube. Ions were charge bred for up to 8 ms, achieving efficiencies in the peak of the charge state distribution of $\approx 2\%$ with charge states as high as $q = 21+$ in Cu. The yields of charge bred radioactive ions were measured using the ARIEL yield station, demonstrating efficiencies comparable to those obtained using stable isotopes.

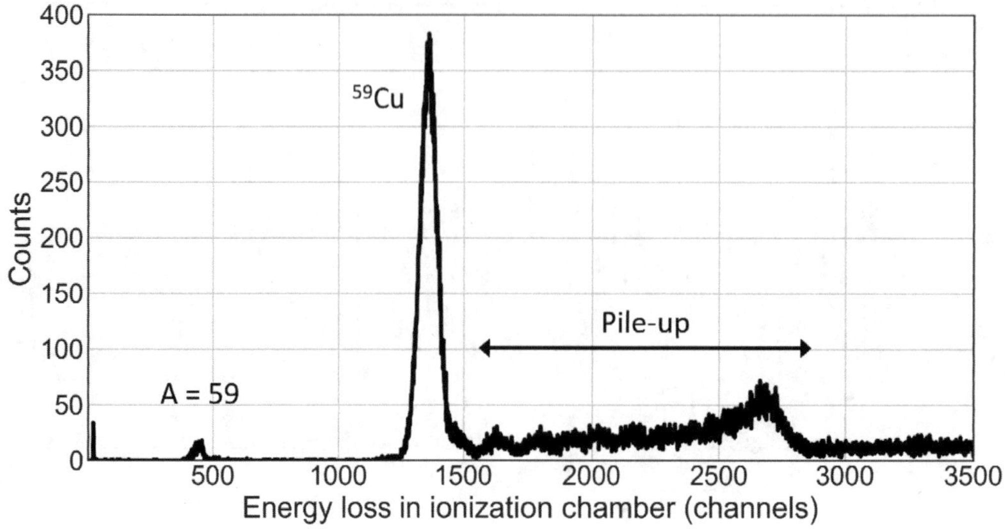

Figure 6. Energy loss spectrum obtained in IRIS ionization chamber with $^{59}Cu^{19+}$ ($A/q = 3.1$) accelerated to an energy of 9.0 MeV per nucleon. Average beam intensity at IRIS was ≈ 300 particles per second.

EBIS performance during commissioning was limited due to stability issues relating to high voltage discharge in the EBIS drift tube assembly. This limited the ion beam energy to < 7 keV and required a reduction in magnetic field and electron beam current. While sufficient for commissioning, these issues need to be resolved for beam delivery to experiments. The high voltage issues are currently being investigated using simulations and a repair plan is being formulated. Following resolution of these issues, the EBIS will be brought back to operational status in preparation for beam delivery in 2022.

Acknowledgments

TRIUMF is funded by the National Research Council of Canada (NRC). The CANREB project is funded by the Canadian Foundation for Innovation (CFI) in collaboration with Saint Mary's University and the University of Manitoba.

References

[1] 2014 *ISAC and ARIEL: The TRIUMF Radioactive Beam Facilities and the Scientific Program* ed J Dilling *et al* (Dordrecht: Springer)
[2] Ames F, Baarthman R, Bricault P and Jayamanna K 2014 *Hyperfine Interact.* **225** 63–67
[3] Egoriti L, Cervantes M, Day Goodacre T and Gottberg A 2020 *Nucl. Instrum. Methods Phys. Res.* B **463** 232-236
[4] Koscielniak S 2014 *Hyperfine Interact.* **225** 263–273
[5] Marchetto M *et al* 2015 *Proc. 6th International Particle Accelerator Conference* (Richmond, VA: JACoW) p 2444–2449
[6] Marchetto M *et al* 2020 *Nucl. Instrum. Methods Phys. Res.* B **463** 227–231
[7] Saminathan S, Ames F, Baartman R, Marchetto M, Lailey O and Mahon A 2019 *Rev. Sci. Instrum.* **90** 123302
[8] Barquest B R, Bale J C, Dilling J, Gwinner G, Kanungo R, Krücken R and Pearson M R 2016 *Nucl. Instrum. Methods Phys. Res.* B **376** 207–210
[9] Blessenohl M A *et al* 2018 *Rev. Sci. Instrum.* **89** 052401
[10] Jayamanna K *et al* 2008 *Rev. Sci. Instrum.* **79** 02C711

[11] Kunz P, Andreoiu C, Bricault P, Dombsky M, Lassen J, Teigelhöfer, Heggen H and Wong F 2014 *Rev. Sci. Instrum.* **85** 053305
[12] Kanungo R 2014 *Hyperfine Interact.* **225** 235–240

Impurities Reduction Conditionings to Recover Best Beam Quality of J-PARC Cesiated RF-Driven H⁻ Ion Source with New Parts Exposed to Plasma

Akira UENO, Kiyonori OHKOSHI, Kiyoshi IKEGAMI, and Hidetomo OGURI

J-PARC Center, Tokai-mura, Naka-gun, Ibaraki-ken 319-1195, Japan
akira.ueno@j-parc.jp

Abstract. For the J-PARC operations, only three plasma chambers (PCHs) of #7, #8 and #9PCHs among ten PCHs have been used since the transverse emittances of the beam produced with them are more superior than those with others for unknown reasons. However, the #7PCH beam emittances were enlarged by 16 % with a brand-new plasma electrode temperature control plate (PETCP) to solve the air leak at the VCR vacuum fitting. The impurities from the new parts exposed to the plasma seemed to cause the degradation. The beam with almost the best emittances was reproduced by #4PCH with a new PETCP after a new impurities reduction conditionings for 48 hours, in which the impurities produced by the plasma were reduced as small as possible by scanning both of 2-MHz-RF power and axial magnetic field in #4PCH.

1. Introduction

The J-PARC cesiated RF-driven H^- ion source (IS) is successfully suppling about 58 mA beam with a duty factor of 1.25 % (0.5 ms × 25 Hz) for the ordinary J-PARC LINAC 50-mA operations and about 72 mA beam for the 60-mA acceleration studies during a few days at each end of the operation [1] of about 3 months by using the internal-RF-antenna developed at the Spallation Neutron Source (SNS) [2]. For the J-PARC operations, only three plasma chambers (PCHs) of #7, #8 and #9PCHs among ten PCHs have been used since the transverse emittances of the beams produced with them are more superior than those with others for unknown reasons. However, the #7PCH beam emittances were enlarged by 16 % with a brand-new plasma electrode (PE) temperature control plate (PETCP) to solve the air leak at the VCR vacuum fitting [3]. The impurities from the new parts exposed to the plasma seemed to cause the degradation. The results of the impurities reduction conditionings (IRCs) performed on #4PCH with a brand-new PETCP are presented. Since the most likely parts, from which the impurities were released by the plasma bombardments and/or heating for a very long duration, were the 99.7 % alumina tubes covering the PETCP stainless-steel (SS) tubes for the PE temperature control air flow inside, sapphire (purity 99.9 %) tubes were installed instead of the alumina tubes for #4PCH.

2. Experimental Setup, Impurities Reduction Conditionings (IRCs) and Beam Measurements

The cross-sectional view of the experimental setup of an IS test-stand, which consists of a SS PCH unitized from an end-flange to the PE, extraction and ground electrodes (EE and GE), an ejection angle correction magnet (EACM), a vacuum chamber for differential pumping by a 1500 L/sec turbo-molecular-pump (TMP) with an alumina insulator duct for the insulation of the terminal voltage (V_T) of 50~66 kV, a solenoid magnet (SM) and a vacuum chamber for beam monitors and a 500 L/sec TMP, is shown in Fig.1. An internal-RF-antenna, a cesium (Cs) injector composed of a Cs-reservoir, a remotely controlled Cs-valve and a Cs-tube, each of which is temperature controlled by using a thermocouple and a heating mantle attached to it, the PE made of molybdenum, a PETCP made of oxygen free copper (OFC) attached to the PE and an axial magnetic field (AMF) correction (AMFC) coil located around the downstream flange are installed to #4PCH. The AMFC coil was operated with coil currents between -14.25 A (-1311 AT) and 14.25 A (1311 AT) for the IRCs and of 0 A (0 AT) for all beam measurements presented in this paper to maximize the RF efficiency [4]. The PE temperature (T_{PE}) is controlled by air temperature control heater power. The IS was operated with a rather low H_2 flow rate of 17 SCCM for a rather large PE aperture of 9 mm by using the CW 30-MHz-RF plasma igniter with a 42 W power [4].

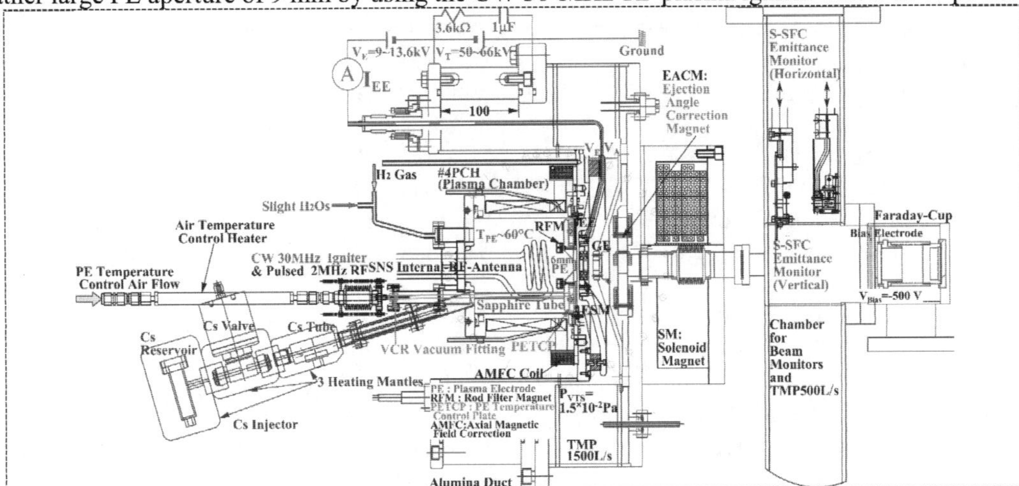

Figure 1. Cross-sectional view of experimental setup of J-PARC IS test-stand.

The H^- ion beam extracted by the extraction voltage (V_E) between the PE and the EE is accelerated by the acceleration voltage ($V_A = V_T - V_E$) between the EE and the GE. The ejected H^- ion beam is focused by the SM into the chamber for beam monitors. The horizontal and vertical emittances are measured by using two sets of movable slit (S) and movable slit with Faraday-cup (SFC) (S-SFC) emittance monitors installed in the chamber. The measured emittance is visualized by randomly plotting dots, whose number is proportional to the voltage signal and is normalized to make the total number of 400,000, in each mesh area defined by the moving steps (typically 0.2 mm and 2 mrad) and the positions of the S and the SFC [5]. The H^- ion beam intensity (I_{H^-}) is measured with a Faraday-cup attached on the downstream flange of the chamber, whose cup is grounded with a 50 Ω terminator. The I_{H^-} is calculated by converting the voltage on the terminator with a coefficient of 20 mA/V. The coil current of the SM (I_{SM}) was set to 300 A (42000 AT) for the IRCs and 350 or 390 A (49000 or 54600 AT) for the I_{H^-} of 72 or 110 mA, respectively, to avoid too small beam size causing the sputtering of the emittance monitor slits and Faraday-cup made of tungsten (W).

The following IRCs were performed to reproduce a 72-mA beam with 95 % beam transverse normalized rms emittances ($\varepsilon_{95\%nrmsx/y}$) of ~0.2/~0.24 πmm•mrad necessary for the J-PARC operations.
(1) The 1st chamber vacuum pressure (P_{V1}) lower than 1×10^{-5} Pa was waited with T_{PE} of 454 °C.
(2) On Day1, at first, $\varepsilon_{95\%nrmsx/y}$ of a 72-mA beam with a duty factor of 2.5 % (1 ms × 25 Hz) was measured as 0.239/0.260 π mm•mrad, after the shortest start-up procedure composed of 15-min plasma operations with P_{2MHz} of 21, 25 and 29 kW (after each operation P_{V1} lower than 1×10^{-5} Pa was waited) with T_{PE} of 454 °C, and the cesiation to produce a 72-mA beam with P_{2MHz} and T_{PE} of 29 kW and 50 °C. Since the $\varepsilon_{95\%nrmsx/y}$ doesn't satisfy the J-PARC requirements, the IRCs were started. Next, 15-min plasma operations with P_{2MHz} of 33.4, 37.6, 41.8, 45.9 and 50 kW with the 2.5 % duty factor, T_{PE} of 454 °C, Cs-valve open and Cs-reservoir temperature (T_R) of 100 °C (after each operation P_{V1} lower than 1×10^{-5} Pa was waited).

(3) On Day2, 4 and 5, the P_{2MHz} and AMFC scanning plasma operations with P_{2MHz} between 14.7 and 50 kW, AMFC between -1311 and 1311 AT, V_E and H^- ion beam energy (W_{H^-}) of 4 kV and 52.5 keV and a duty factor of 5 % (1 ms × 25 Hz) as shown in Figs. 2(a), (b) and (c) are performed, respectively. The axial magnetic field (AMF) inside of #4PCH produced with the SM fringe field {I_{SM} = 300 A (42000 AT)} and AMFC between -1311 and 1311 AT were calculated with a two-dimensional magnetostatic Poisson code [6] as shown in Fig. 3. As shown in Fig. 2(b), the I_{EE} was minimized for the AMFC of about 200 AT, with which the absolute value of the AMF (|AMF|) around the antenna was minimized as shown as the AMFC of -252 AT of Fig. 3.
(4) On Day3, at first $\varepsilon_{95\%nrmsx/y}$ of a 72-mA beam with a duty factor of 5 % was measured as 0.225/0.253 πmm•mrad, which was larger than the requirements. Next the remaining P_{2MHz} and AMFC scanning operations were performed.
(5) On Day6, after the remaining P_{2MHz} and AMFC scanning operations, it was confirmed that the 72-mA beam production with W_{H^-} of 52.5 keV was impossible probably due to the impurities on the PE.
(6) On Day7, as the final IRCs, the P_{2MHz} and AMFC scanning plasma operations with P_{2MHz} between 22.1 and 29.2 kW, AMFC between -1109 and -303 AT, which were the ranges used in the J-PARC operations, the for V_E and H^- ion beam energy (W_{H^-}) of 4 kV and 52.5 keV and a duty factor of 5 % were performed as shown in Fig. 2(d). The operation parameters for the above IRCs are listed in the 1st row of Table 1.
(6) On Day8 and 9, there were the full maintenance of #4PCH, in which the inner surfaces were cleaned up and the antenna was replaced to brand-new one, and the vacuum pumping as same as (1).
(7) On Day10, $\varepsilon_{95\%nrmsx/y}$ of a 110-mA beam with a duty factor of 2.5 % was measured as 0.219/0.262 π mm•mrad with P_{2MHz} of 36.2 kW, after the shortest start-up procedure composed of 15-min plasma operations with P_{2MHz} of 21, 25, 29, 33 and 37 kW (after each operation P_{V1} lower than 1×10^{-5} Pa was waited) with T_{PE} of 454 °C, and the cesiation to produce a 110-mA beam with P_{2MHz} and T_{PE} of 37 kW and 50 °C, respectively. The emittances are the smallest ones measured for 110-mA beams. On the remaining time of Day10 and on Day11, 72- and 110-mA operations were performed to attain the stationary beams with T_R of about 130 °C and Cs-valve open.
(8) On Day12, $\varepsilon_{95\%nrmsx/y}$ of 72- and 110-mA beams were reproducibly measured as 0.204/0.236 and 0.220/0.263 πmm•mrad, respectively. The parameters of the 72- and 110-mA beam operations are listed in the 2nd and 3rd rows of Table 1. The 72-mA beam satisfies the requirements for the J-PARC operations.

Table 1. Parameters for impurities reduction conditioning (IRC) and 72- and 110-mA operations.

	IRC	I_{H^-}=72 mA	I_{H^-}=110 mA
H_2 gas flow rate	17 SCCM	17 SCCM	17 SCCM
CW 30-MHz-RF igniter power	42 W	42 W	42 W
2-MHz-RF duty factor ~ Beam duty factor	5% (1 ms×50 Hz)	5% (1 ms×50 Hz)	4.5% (1 ms×45 Hz) *For 5 mA average limited by radiation safety permission.
Beam energy (W_{H^-} = V_T)	52.5 keV	52.5 keV	65 keV
Extraction Voltage (V_E)	4 kV	10.4 kV	13.25 kV
2-MHz-RF power (P_{2MHz})	16~50 kW	19.3 kW	28.6 kW
Extrac. Electrode Current (I_{EE})	-	27 mA	59 mA
RF power efficiency (I_{H^-}/P_{2MHz})	-	3.73 mA/ kW	3.85 mA/kW
I_{H^-} density at PE (ϕ_{PE}=9mm)	-	1245 A/m^2	1729 A/m^2
$\varepsilon_{95\%nrmsx/y}$ (πmm•mrad)	-	0.204/0.236	0.220/0.263
Solenoid magnet current	300 A (42000AT)	350 A (49000AT)	390 A(54600 AT)
AMFC coil current	-167.84~167.84 A (-1315~1315 AT)	0 A (0 AT)	0 A (0 AT)

The measured 100 % beam distributions in horizontal phase-planes of 72-mA beams without and after the IRCs are shown with red dots in Figs. 4(a) and 5(a), respectively. The 1.5 πmm•mrad ellipses fitting 95 % beam and backward traced to the GE downstream surface with Trace2D (T2D) [7] are also shown with blue and red lines in Figs. 4(a) and 5(a), respectively. The $\varepsilon_{95\%nrmsx}$ are evaluated as 0.237 and 0.204 πmm•mrad, respectively. The measured relationships between the horizontal normalized emittances (ε_{nx}) and the beam fractions (f_{bx}) included in the ellipses fitting the 95 % beam are shown with blue lines in Figs. 4(b) and 5(b), respectively. The f_{bx} included in the RFQ injection beam distribution with the normalized rms emittances of 0.25 πmm•mrad commonly used in the PARMTEQ (PQ) simulation [8] are plotted with red lines as the functions of ε_{nx} in Fig. 4(b) and 5(b), respectively.

In Fig. 6, the measured waveforms of the V_E (a), V_T (b), forward and reflected 2-MHz-RF voltages (V_{2MHzF} and V_{2MHzR}) detected with -60 dB directional couplers (c) and I_H- and I_{EE} (d) of one beam pulse of a 110-mA beam are shown. The RF efficiency and I_{EE} of 3.83 mA/kW and 59 mA for the 110-mA operation are also much improved from 2.64 mA/kW and 135 mA with #4PCH without IRCs [9].

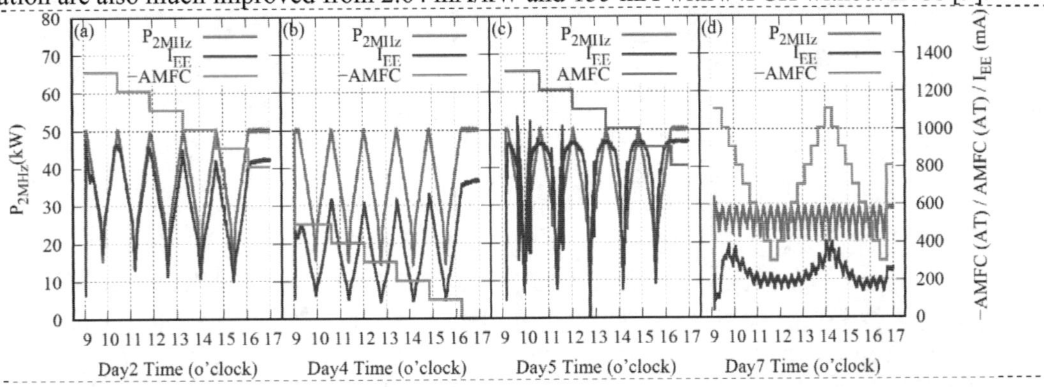

Figure 2. Trends of P_{2MHz}, I_{EE} and AMFC in IRCs on Day2 (a), Day4 (b), Day5 (c) and Day7 (d).

3. Conclusions

The third PCH indispensable for the J-PARC operations as the backup providing against accidents was restored by performing the IRCs for 48 hours, in which the impurities produced by the plasma were reduced as small as possible by scanning both of 2-MHz-RF power and axial magnetic field, on #4PCH with a brand-new PETCP. Not only $\varepsilon_{95\%nrmsx/y}$ of 72-mA beam satisfied the J-PARC requirements, but also RF efficiencies and I_{EE} of both of 72- and 110-mA beams were much improved. After the IRCs, the full maintenance of #4PCH, in which the inner surfaces were cleaned up and the antenna was replaced to brand-new one, was mandatory to recover the H^- ion production efficiency decreased due to the impurities heavier than H^- ions around the PE surface. The H^- ions momentum spreads should be increased by the collisions with them. The sapphire tubes instead of the alumina tubes should contributed to reduce the time of the IRCs.

Acknowledgment

The authors express their sincere thanks to Dr. Martin P. Stockli of the SNS and the SNS ion-source group members for their support to purchase antennas and their information on the SNS H^- ion-source.

References

[1] A. Ueno, Rev. Sci. Instrum. 91, 033312 (2020).
[2] M. P. Stockli, et. al., Rev. Sci. Instrum. 89, 052202 (2018).
[3] See http://www.swagelok.com/downloads/WebCatalogs/EN/MS-01-24.pdf
[4] A. Ueno, New J. Phys. 19 (2017) 015004.
[5] A. Ueno, et. al., AIP Conf. Proc. 1655, 030009 (2015).
[6] See http://laacg.lanl.gov/laacg/services/download_sf.phtml#ps6

[7] K. R. Crandall and D. P. Rusthoi, TRACE3D Documentation, LA-UR-97-886, May 1997.
[8] K. R. Crandall and T. P. Wangler, AIP Conf. Proc. 177, 22 (1988).
[9] A. Ueno, et. al., AIP Conf. Proc. 2373, 040002 (2021)

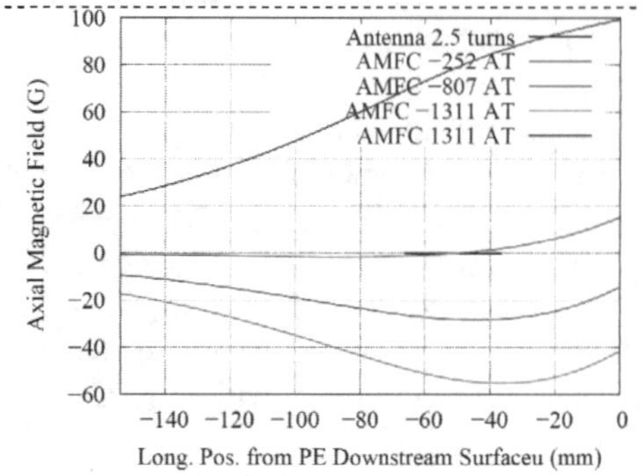

Figure 3. Axial magnetic field calculated with Poisson code and antenna position in #4PCH.

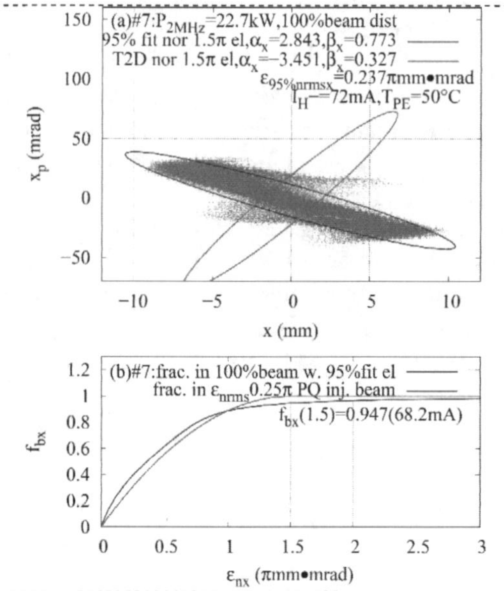

Figure 4. 100 % beam distribution in horizontal phase-plane and fitted ellipses (a) and beam fractions in ε_{nx} (b) of 72-mA beam without IRCs.

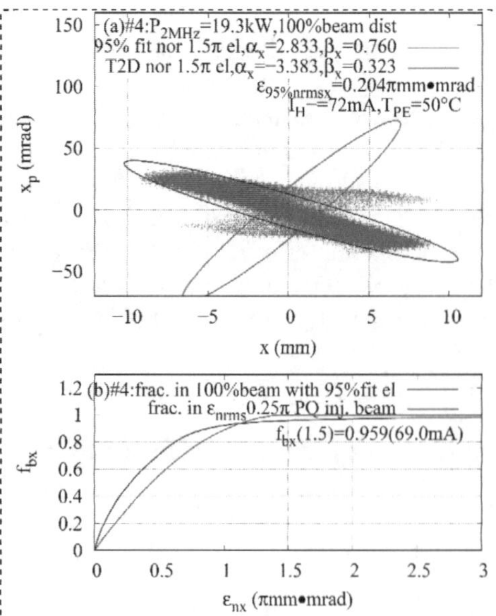

Figure 5. 100 % beam distribution in horizontal phase-plane and fitted ellipses (a) and beam fractions in ε_{nx} (b) of 72-mA beam after IRCs.

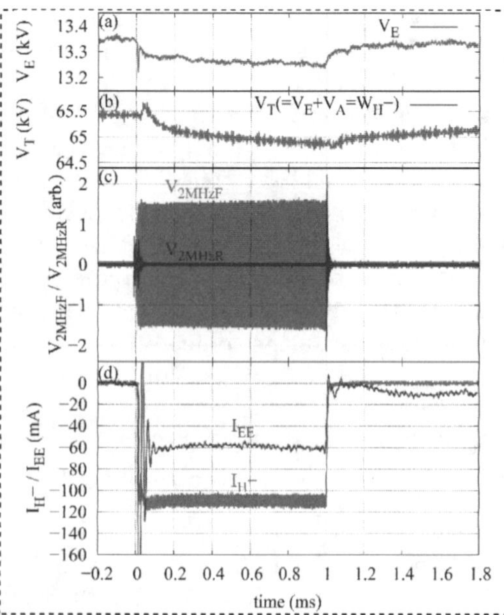

Figure 6. Waveforms of V_E (a), V_T (b), V_{2MHzF} and V_{2MHzR} (c) and I_{H^-} and I_{EE} (d) of 110-mA beam.

Journal of Physics: Conference Series **2244** (2022) 012030 doi:10.1088/1742-6596/2244/1/012030

Predictive fluid model for self-consistent description of inductive RF coupling in powerful negative hydrogen ion sources

D Zielke[1], S Briefi[1,2], S Lishev[3] and U Fantz[1,2]

[1] Max-Planck-Institut für Plasmaphysik, Boltzmannstr. 2, 85748 Garching, GER
[2] AG Experimentelle Plasmaphysik, Universität Augsburg, 86135 Augsburg, GER
[3] Faculty of Physics, Sofia University, 1164 Sofia, BG

E-mail: dominikus.zielke@ipp.mpg.de

Abstract. RF-driven negative hydrogen ion sources are typically operated at low frequencies around 1 MHz, gas pressures around or below 1 Pa and large power densities up to 10 $\mathrm{Wcm^{-3}}$. Owing to these conditions as well as the current discharge geometries and antenna layouts, the RF power coupling is far from optimized, i.e. only a fraction η of the power delivered by the generator is absorbed by the plasma. This considerably limits the performance and reliability of RF-driven ion sources. To study the bidirectional RF power coupling a self-consistent fluid model is introduced. Taking into account the interplay between the nonlinear RF Lorentz force and the electron viscosity (usually neglected in state-of-the-art fluid models) a steady state solution is obtained, where the trends reflect the experimental data. Solutions calculated in hydrogen but with increased ion masses indicate that the latter are responsible for the systematically increased η, which is observed experimentally when deuterium instead of hydrogen is used as feed gas.

1. Introduction

Optimizing the RF power coupling in inductively driven radio frequency (RF) ion sources is a worthwhile task, since it can lead to significant improvement of the *RF power transfer efficiency* [1–6]

$$\eta = \frac{P_{\mathrm{pl}}}{P_{\mathrm{RF}}} = \frac{R_{\mathrm{pl}}}{R_{\mathrm{pl}} + R_{\mathrm{net}}}, \tag{1}$$

where P_{pl} and P_{RF} denote the power absorbed by the plasma and the total power delivered by the RF generator, respectively. Since only a negligible part of the electromagnetic energy is radiated away, the power is either lost in the RF network or it is deposited in the plasma. In the former, Joule heating of the RF coil produces undesired heat or eddy currents are driven e.g. in a Faraday shield (which is often present to suppress capacitive coupling) or in conducting support structures surrounding the discharge. These losses are conveniently combined in the network resistance R_{net}, whereas the plasma resistance R_{pl} quantifies the power absorbed by the plasma, i.e. $P_{\mathrm{pl}} = \frac{1}{2} R_{\mathrm{pl}} I_0^2$, where I_0 is the RF coil current amplitude. If η is increased, P_{pl} can be held constant but at the same time the total delivered power P_{RF} can be decreased. This in turns relaxes requirements on the RF power generator (i.e. a cheaper model with smaller power output and space required can be used) as well as on the voltage holding and cooling of the various components of the RF network, such as RF coil, Faraday shield (if present) and matching system. RF ion sources are complex systems, wherefore several external parameters affect η. These are e.g. type of gas, filling pressure, power and driving frequency. Furthermore, different magnetic fields are typically present to increase the plasma confinement (cusp field) or to decrease the plasma density and temperature in specific regions (filter field). To study how and why the various parameters affect η, a numerical fluid model is developed. The challenges in simulating an RF ion source discharge are manifold. The used generator frequency f_0 is typically in the single digit MHz range, wherefore large RF magnetic fields are needed for producing induced electric RF fields comparable to the ones at higher frequencies of 13.56 MHz, which is the industry standard. Because of the high applied powers of up to 100 kW, the produced RF magnetic fields are even further elevated to values well above 100 G. To describe the RF coupling in this regime, an electron momentum balance is needed, where the RF Lorentz force has to be retained. However, solely retaining the Lorentz force leads to a compression of the plasma due to the DC component of the RF Lorentz force (ponderomotive effect). In state-of-the-art numerical models, this effect is so pronounced at the large RF magnetic fields in RF ion sources, that no steady state solutions are obtained [7]. It was recently found that the electron viscosity in the momentum balance mitigates the compression by the RF Lorentz force. In this way, self-consistent steady state numerical solutions are obtained for the first time. Further information about the self-consistent fluid model such as the used simulation domain, collisional processes and solved equations as well as boundary conditions can be found in [8].

2. Investigation of RF power coupling mechanism in RF ion sources

Experimental data from discharges of the ITER prototype RF negative ion source is used for the validation of the model. Typical operation is at a low filling pressure of 0.3 Pa, a driving frequency of 1 MHz and at large RF powers up to 75 kW [6]. The numerically obtained solutions are compared to values which are experimentally obtained from the plasma production region (driver). The comparison is exemparily performed in hydrogen as a variation of the total delivered power P_{RF} at a fixed filling pressure of $p_{\mathrm{fill}} = 0.3\,\mathrm{Pa}$. Shown in figure 1 are the RF coil current amplitude I_0 as well as the ratio of the electron density in the heating zone (in the vicinity of the radial wall) to the one in the driver center, $n_{\mathrm{e,h}}/n_{\mathrm{e,c}}$. As evident from the three trends, the model captures the trends of both quantities, and the absolute values are within or close to the experimental error bars, when the RF Lorentz force as well as the viscosity are retained. Note that the numerically obtained absolute values of the RF coil current are around 20% too small (compared to the experimentally obtained ones), when the RF Lorentz force and the viscosity are neglected. In this case the electron density ratio is overestimated by a factor of roughly two due to the missing compression effect. If only the RF Lorentz force is retained and the viscosity is neglected, the plasma compression by the RF Lorentz force is so strong, that no steady state solution develops (not shown in figure 1).

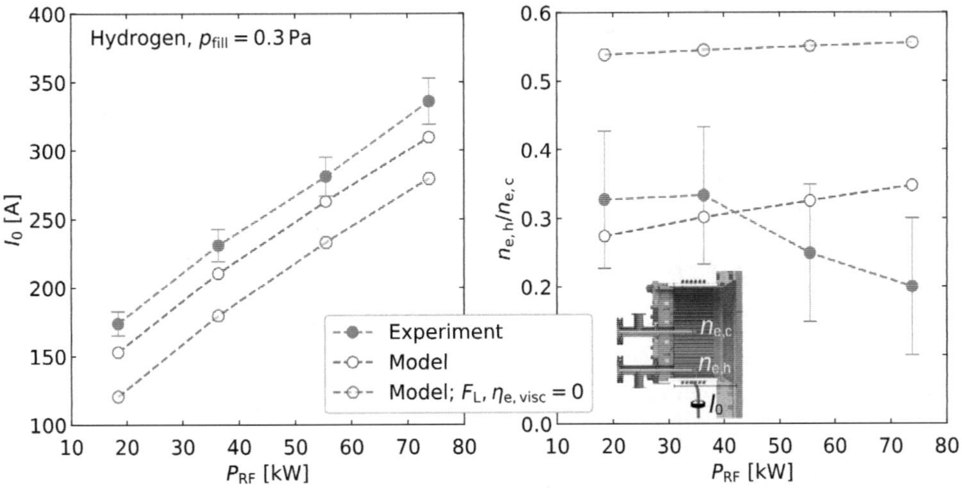

Figure 1. Validation of the self-consistent RF power coupling description of the model. Shown are the experimentally (full symbols) and numerically obtained (open symbols) RF current amplitude I_0 as well as the ratio of the electron density in the plasma heating zone to the one in the driver center, $n_{\text{e,h}}/n_{\text{e,c}}$. The total delivered power P_{RF} is varied from 20 to 75 kW. 'Model' denotes values calculated by the model, where the RF Lorentz force and the electron viscosity are retained. The points denoted by 'Model; $F_{\text{L}}, \eta_{\text{e,visc}} = 0$' are obtained by neglecting the RF Lorentz force as well as the electron viscosity. Figure adapted from [8].

3. Validation studies in hydrogen and deuterium

Negative RF ion sources used in fusion science are operated in deuterium as well. Hence it is interesting to study how the different isotopes affect the plasma parameters and the RF power coupling. Figure 2 shows at a fixed total delivered power of $P_{\text{RF}} = 60$ kW the RF power transfer efficiency η, the central electron temperature $T_{\text{e,c}}$ and density $n_{\text{e,c}}$ as well as the volume averaged atomic to molecular ratio $n_{\text{H}}/n_{\text{H}_2}$. In hydrogen the modeled trends of η and the plasma parameters $T_{\text{e,c}}$ and $n_{\text{e,c}}$ scale as the experimental ones. As expected from 0D global models, the electron temperature decreases, whereas the electron density increases with increasing pressure [9]. However the behavior of the RF power transfer efficiency η cannot be explained in this way. But the self-consistent power coupling model can be used to study the underlying physics effects. As demonstrated in figure 2, the trend of η agrees well with the experimentally observed one, i.e. the plasma's capability to absorb more RF power increases mostly because of the increasing electron density. Note that the calculated atomic to molecular ratios are by a factor of roughly two to four larger than the ones obtained by optical emission spectroscopy and collisional radiative modeling (CRM) [10]. The deviations are possibly caused by uncertainties in the input data (both in the numerical model and in the CRM evaluation) regarding atomic recombination coefficients of the wall material as well as opacity effects.

As a first step towards a model for description of deuterium, the masses of all heavy species from the validated hydrogen model, i.e. H, H_2, H^+, H_2^+ and H_3^+ are doubled, whereas everything else (such as e.g. collisional cross-sections) is left unchanged. This affects the flux of the ions to the discharge walls, which is in essence controlled by the Bohm speed $u_{\text{i}} = (\text{e}T_{\text{e}}/m_{\text{i}})^{1/2}$. For deuterium, the doubled ion masses yield a smaller flux towards the walls, when compared to hydrogen at the same pressure. A consequence of the decreased wall losses is a larger plasma density. The increased number of electrons in the volume available for the RF fields to couple to

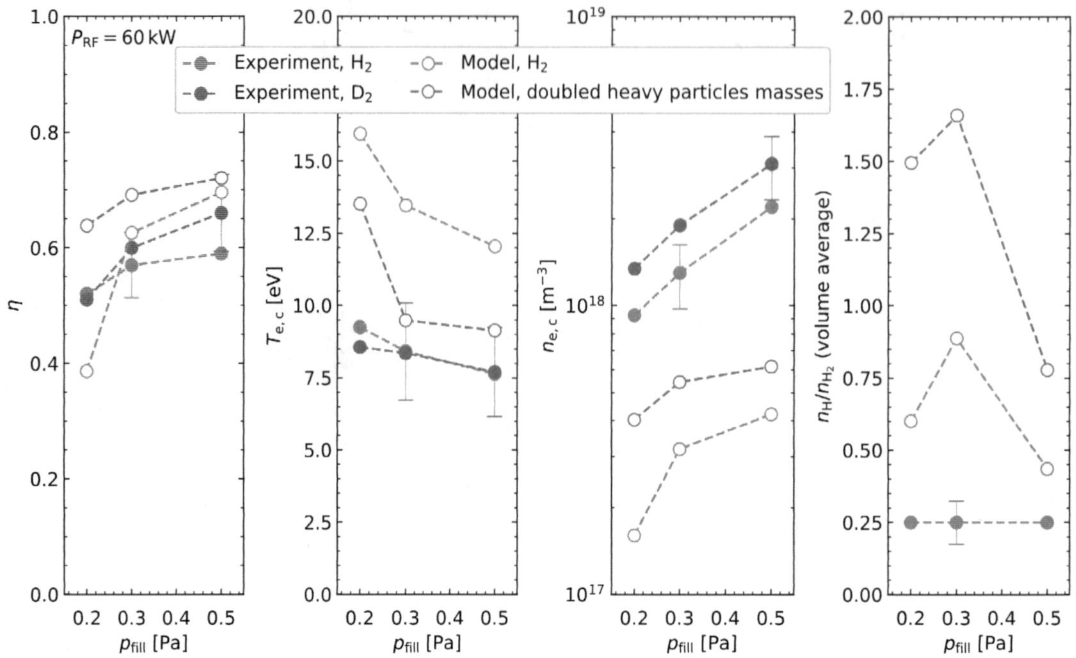

Figure 2. Comparison of experimentally (full symbols) and numerically (open symbols) obtained electrical and plasma parameters for hydrogen and deuterium at $P_{RF} = 60\,kW$ as a function of the filling pressure p_{fill}. Shown are the RF power transfer efficiency η, the electron temperature $T_{e,c}$ and density $n_{e,c}$ in the center of the driver as well as the volume averaged atomic fraction n_H/n_{H_2}.

consequently lead to an increased RF power transfer efficiency in deuterium. Interestingly, in the case of the increased mass, the model predicts that also the atomic to molecular ratio n_H/n_{H_2} will be systematically increased. The reason for this increase is a larger atomic production rate, since the latter is proportional to the electron density.

4. Conclusion

A mechanism for the RF power coupling in the RF negative ion sources operating regime (low applied RF around 1 MHz, high power densities up to $10\,Wcm^{-3}$ and low pressures around or below 1 Pa) has been validated with experimental measurements. Here it has been shown that the RF power coupling can be described by an electron momentum balance, where the RF Lorentz force as well as the viscosity are retained. The latter mitigates the ponderomotive effect caused by the RF Lorentz force. In this way steady state solutions are obtained for the first time. The numerical solutions scale as in the experiment. Moreover, the absolute values of the measured RF coil current and the intensity of the plasma compression match at almost all investigated operation points.

Further validation activities regarding the isotopic effect have been started by simply doubling the mass of all heavy species. This leads to a slight increase in the RF power transfer efficiency calculated by the model, which is in agreement with the experimental measurements. A plausible explanation for the increased η are the decreased plasma losses in the model that go along with larger plasma densities. However, increasing the ion mass can only be considered as a first step towards a comprehensive deuterium model. Here also cross sections for the elastic and inelastic collision processes as well as for various processes at the walls have to be exchanged.

The validated model can be applied to determine and optimize the external parameters that have the largest positive effect on the RF power coupling in RF ion sources. This is shown exemplarily in the contribution by S. Briefi et al. to this conference.

References

[1] J. Hopwood. "Planar RF induction plasma coupling efficiency". In: *Plasma Sources Sci. Technol.* 3.4 (1994), p. 460. URL: http://stacks.iop.org/0963-0252/3/i=4/a=002.

[2] V. A. Godyak et al. "Experimental setup and electrical characteristics of an inductively coupled plasma". In: *Journal of Applied Physics* 85.2 (1999), pp. 703–712. DOI: 10.1063/1.369150. URL: https://doi.org/10.1063/1.369150.

[3] V. A. Godyak. "Ferromagnetic enhanced inductive plasma sources". In: *Journal of Physics D: Applied Physics* 46.28 (June 2013), p. 283001. DOI: 10.1088/0022-3727/46/28/283001. URL: https://doi.org/10.1088/0022-3727/46/28/283001.

[4] E. A. Kralkina et al. "RF power absorption by plasma of a low-pressure inductive discharge". In: *Plasma Sources Science and Technology* 25.1 (Jan. 2016), p. 015016. DOI: 10.1088/0963-0252/25/1/015016. URL: https://doi.org/10.1088%2F0963-0252%2F25%2F1%2F015016.

[5] V. I. Kolobov and V. A. Godyak. "Inductively coupled plasmas at low driving frequencies". In: *Plasma Sources Science and Technology* 26.7 (July 2017), p. 075013. DOI: 10.1088/1361-6595/aa7584. URL: https://doi.org/10.1088%2F1361-6595%2Faa7584.

[6] D. Zielke et al. "RF power transfer efficiency and plasma parameters of low pressure high power ICPs". In: *Journal of Physics D: Applied Physics* 54.15 (Feb. 2021), p. 155202. DOI: 10.1088/1361-6463/abd8ee. URL: https://doi.org/10.1088/1361-6463/abd8ee.

[7] G. J. M. Hagelaar et al. "Model of an inductively coupled negative ion source: I. General model description". In: *Plasma Sources Sci. Technol.* 20.1 (2011), p. 015001. DOI: 10.1088/0963-0252/20/1/015001. URL: https://doi.org/10.1088%2F0963-0252%2F20%2F1%2F015001.

[8] D. Zielke. "Development of a predictive self-consistent fluid model for optimizing inductive RF coupling of powerful negative hydrogen ion sources". PhD thesis. Augsburg University, 2021. URL: https://nbn-resolving.org/urn:nbn:de:bvb:384-opus4-894170.

[9] M. A. Lieberman and A. J. Lichtenberg. *Principles of Plasma Discharges and Materials Processing.* Second Edition. John Wiley & Sons, Inc., Hoboken, New Jersey, 2005.

[10] S. Briefi et al. "Spectroscopic Investigations of the Ion Source at BATMAN Upgrade". In: *AIP Conference Proceedings* 2052 (2018), p. 040005.

This work has been carried out within the framework of the EUROfusion Consortium and has received funding from the Euratom research and training programme 2014-2018 and 2019-2020 under grant agreement No 633053. The views and opinions expressed herein do not necessarily reflect those of the European Commission.

Caesium Balance of the ISIS H^- Penning Ion Source

O Tarvainen, D Faircloth, S Lawrie, T Sarmento, R Abel, J Macgregor, C Cahill, T Stanley, M Whitehead, T Wood

STFC ISIS Pulsed Spallation Neutron and Muon Facility, Rutherford Appleton Laboratory, Harwell, OX11 0QX UK

E-mail: olli.tarvainen@stfc.ac.uk

Abstract.

We have developed a model predicting the equilibrium caesium coverage work function of the cathode (in eV) and the expected H^- beam current (in arb. units) of the ISIS Penning ion source. The model is based on semi-empirical expressions for the cathode work function, negative ion surface ionization yield, and caesium adsorption and desorption rates. We compare the model predictions to experimental data with 760 µs discharge and 250 µs beam pulses. It is concluded that the ISIS Penning ion source operates near the optimum cathode work function in a wide range of caesium pressures and cathode temperatures. The model implies that long pulse operation of the source requires elevated Cs oven temperature and improved cathode cooling.

1. Introduction

The ISIS H^- Penning ion source [1] is a cesiated pulsed DC discharge surface plasma source routinely delivering 55 mA beam of negative hydrogen ions (H^-) in 250 µs pulses for neutron and muon physics at 50 Hz pulse repetition rate. The H^- ions are produced on the caesiated molybdenum (Mo) cathode surfaces and are drawn into the discharge volume where they are believed to undergo resonant charge exchange (see e.g. Ref. [2]) with slow neutral ground state hydrogen atoms, i.e. $H^-_{\text{fast}} + H(1s)_{\text{slow}} \rightarrow H(1s)_{\text{fast}} + H^-_{\text{slow}}$. The "second generation" slow H^- ions are then extracted from the discharge volume through a 0.6 mm by 10 mm rectangular slit.

The extraction of square 250 µs H^- beam pulses with $< 5\%$ change at constant discharge current (plasma density) implies that the cathode surface conditions, relevant for the H^- production, remain practically unchanged during the 750 µs discharge pulse. Considering that a significant fraction of the several kW discharge power is deposited onto the $2 \times 49\,\text{mm}^2$ cathode surfaces, which results in a temperature fluctuation of approximately 100 °C at 50 Hz [3], it is far from obvious that this should be the case. To explain the stability of the H^- output, we have developed a model predicting the equilibrium caesium coverage work function of the cathode (in eV) and the expected H^- beam current (in arb. units). The model is based on semi-empirical expressions for the cathode work function, negative ion surface ionization yield, and Cs adsorption and desorption rates at specific volumetric Cs density and cathode surface temperature.

2. Description of the H^- Surface Yield and Caesium Balance Model

The negative ionisation probability p_- and the corresponding yield Y of H^- ions formed by kinetic surface ionization (resonant tunneling) has been described by Rasser [4]. According to the quantum mechanical "amplitude model"

$$p_- \propto Y \simeq \frac{2}{\pi} \exp\left(-\frac{\pi(\phi(\theta) - S)}{2av_{||}}\right), \tag{1}$$

where the work function ϕ depends on the Cs-coverage θ measured in monolayers (one monolayer equals to 4.8×10^{14} Cs atoms/cm^2 [5]), S is the electron affinity of hydrogen (0.754 eV), $v_{||}$ the velocity of the emitted H^- ions parallel to the surface normal and a is a constant in the range of 2–5 $\times 10^{-5}$ eVs/m. According to Alton [6] the work function of an alkali metal covered metal surface as a function of the coverage (in fractional monolayers) can be estimated from

$$\phi(\theta) = \phi_0 + \frac{6\Delta\phi_m}{(3-\theta_m)\theta_m}\theta - \frac{3\Delta\phi_m(\theta_m + 1)}{(3-\theta_m)\theta_m^2}\theta^2 + \frac{2\Delta\phi_m}{(3-\theta_m)\theta_m^2}\theta^3 \tag{2}$$

where θ_m is the coverage corresponding to the minimum of the work function - for the majority of metals, including Mo, the value of θ_m is 0.5–0.7 [6, 7] - and $\Delta\phi_m \simeq -1.24[\phi_0 - \frac{1}{2}(I_A + E_A)]$ is the maximum change of the work function. Using the ionisation potential of $I_A = 3.894$ eV and electron affinity of $E_A = 0.4716$ eV for Cs we obtain $\Delta\phi_m \simeq 2.707 - 1.24\phi_0$ for caesiated metal surfaces and $\Delta\phi_m = -2.59$ eV for Mo with $\phi_0 = 4.27$ eV.

The equilibrium Cs coverage of the Mo cathode surfaces is determined by the Cs vapour pressure in the ion source volume, temperature of the Mo surface and the Cs density (coverage) of the surface. The adsorption flux Γ_a of Cs onto a surface in a closed system can be evaluated from the kinetic gas theory:

$$\Gamma_a = \frac{1}{4} n_{\mathrm{Cs}} v = \frac{p_{\mathrm{Cs}}}{\sqrt{2\pi k T_{eq} m_{\mathrm{Cs}}}}, \tag{3}$$

where v is the mean thermal speed of the Cs atoms and n_{Cs} their number density in the gas phase, p_{Cs} is the Cs vapour pressure in the discharge chamber, T_{eq} is the equilibrium temperature of the system and m_{Cs} is the atomic mass of Cs. The order-of-magnitude Cs pressure in the discharge volume can be estimated from two starting points. The Cs pressure range, estimated from the Cs vapour pressure curve at 145–190 °C oven temperature, is 1–10 Pa. We consider the vapour pressure corresponding to the given oven temperature range as an upper limit for the true Cs pressure in the discharge chamber. Long-term experience with the ion source, supported by quartz microbalance measurements, has shown that, on average, 5 g of Cs is consumed in 50 days. This corresponding Cs flux escaping the discharge volume via the extraction slit is approximately 1.2 µg/s. Assuming that the Cs equilibrium temperature, defined by the ion source wall temperature [3], is approximately 900 K, the Cs pressure corresponding to the above leakage flux is approximately 0.1 Pa. This is an order of magnitude lower than the Cs pressure estimated from the vapor pressure curve, and is therefore considered as a lower limit. Thus, we estimate the Cs pressure in the ISIS Penning ion source to be 0.1–10 Pa.

The desorption flux Γ_d of Cs from a fractional monolayer ($0 \leq \theta \leq 1$) on a metal substrate at temperature T_s depends strongly on the coverage θ (in monolayers), and can be expressed as

$$\Gamma_d = \Gamma_{d,0} \exp\left(\frac{-e(3.37 - 2.78\theta)}{kT_s}\right), \tag{4}$$

where $\Gamma_{d,0} = 7.5 \times 10^{26}$ atoms/cm^{2}s is a factor derived from experiments with tungsten [8]. According to Ref. [9] similar dependence is found for other metal substrates. 3D thermal simulations [3] reveal that the cathode surface temperature fluctuates by approximately 100 °C rising from 670 °C to 770 °C with the nominal discharge parameters (55 A, 70 V, 760 µs pulses at 50 Hz) during the plasma pulse. Taking into account some operational variation we estimate that the cathode surface temperature is 600–800 °C.

The time-dependent Cs coverage of the cathode is calculated by first equating the adsorption and desorption rates of Cs in the beginning of the discharge pulse, and then numerically estimating the temporal change of the cathode Cs coverage (and work function) in differential time increments dt. This is done by taking into account the adsorption rate at given Cs pressure (affected by the Cs released from the electrode surfcaces during the discharge pulse) and the time-varying cathode temperature and Cs coverage affecting the Cs desorption rate. The corresponding beam current in arb. units is calculated from the amplitude model assuming that the extracted beam current depends linearly on the surface ionisation yield.

3. Results

Figure 1 shows the Cs desorption flux as a function of the cathode surface temperature for various (fractional) monolayer coverages. The adsorption flux within the estimated Cs pressure range is represented by the shaded area in Fig. 1. The vertical lines in Fig. 1 depict the expected range of cathode temperature. The adsorption and desorption fluxes must be equal at some point of the 50 Hz thermal cycle or otherwise the Cs coverage of the cathode would evolve with time resulting in pulse-to-pulse drift of the H^- current. This corresponds to 0.50–0.66 monolayer Cs coverage for 0.1–10 Pa Cs pressure. At lower cathode temperature the equilibrium Cs coverage shifts towards a full monolayer. At higher cathode temperature the equilibrium coverage is lower, i.e. below the optimum for H^- production. For Fig. 1 it was assumed that the Cs gas temperature corresponds to the average wall temperature of approximately 600 °C [3].

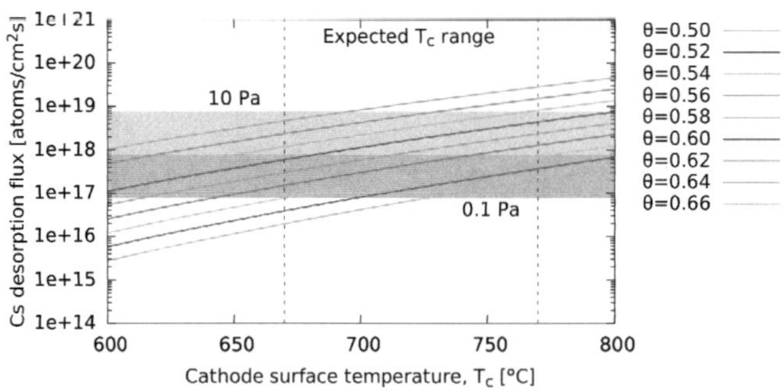

Figure 1: The desorption flux of Cs as a function of the cathode surface temperature for various Cs coverages, θ. The vertical lines indicate the expected range of the cathode temperature T_c. The shaded areas represent the range of adsorption flux for the expected Cs pressure range estimated from the oven temperature (1–10 Pa) and from the Cs consumption (0.1–1 Pa).

Figure 2(a) shows the simulated cathode surface temperature (from [3]) and the predicted equilibrium Cs coverage over the 760 µs discharge pulses at 50 Hz in 0.1–10 Pa Cs pressure. Figure 2(b) shows the predicted temporal variation of the cathode work function (in eV) and the H^- beam current (in arb. units). The beam currents are normalised to the beam current in the beginning of the 250 µs extraction pulse with 10 Pa Cs pressure. The beam current prediction assumes that (i) the optimum Cs coverage of the Mo cathode is 0.6 monolayers [7], (ii) the parameter a in Eq. 1 is 3.5×10^{-5} eVs/m [4], and (iii) the H^- escape velocity at the surface corresponds to a typical sputtering energy of 10 eV [10]. The model implies that the ~100 °C temperature rise of the cathode surface during the discharge pulse affects the Cs coverage by 0.05–0.07 monolayers, which translates to less than 0.02 eV change of the work function during the beam pulse at 0.1–10 Pa Cs pressure. However, the yield of negative ions is expected to be approximately 10-15 % lower at 0.1 Pa Cs pressure in comparison to 1–10 Pa range. Both predictions agree with experiments: (a) the observed droop of the H^- beam is less than 5 %

over a wide range of Cs oven temperatures [11], and (b) the H^- beam current extracted from the standard ISIS Penning ion source in the nominal short pulse mode increases by 20 % when the Cs oven temperature is increased from 140 °C to 180 °C [12] that , which is expected to change the Cs pressure from lower than 0.1 Pa to higher than 10 Pa.

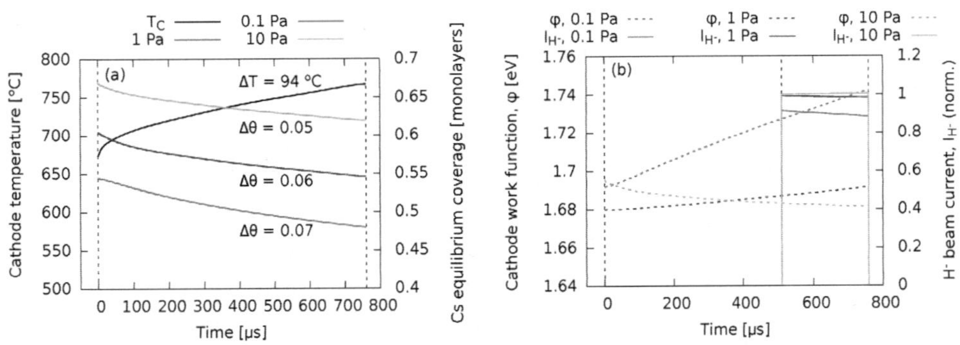

Figure 2: (a) The variation of the cathode surface temperature and equilibrium Cs coverage; (b) The predicted variation of the cathode work function and H^- beam current.

Figure 3(a) shows that the cathode temperature can vary ±100 °C around the nominal range of 680–770 °C without significant effect on the cathode work function or extracted beam current. The example is calculated with Cs pressure of 10 Pa (lower pressures yielding similar result). To validate the model prediction we used the database of the ISIS ion source fleet and identified two ion sources operating on the ion source development rig (prior to installation on ISIS) with "hot" and "cold" cathode, i.e. with T_c (reading) of 454 °C and T_c (reading) of 266 °C, and compared their beam currents and beam pulse shapes. The variation between ion sources is caused by small differences in thermal conductance from the cathode to the water-cooled mounting flange and, to some degree, the variation of thermal contact of the thermocouple measuring the cathode temperature. Despite the related uncertainties and the predicted 200–400 °C difference between the thermocouple reading and the actual cathode surface temperature [3] it is understood that these two sources represent the range of cathode surface temperatures in Fig. 3(a) and are labelled as "hot" and "cold" in Fig. 3(b) showing a comparison of their experimental and predicted beam current pulses (normalised). For short pulses the model prediction of more than 200 °C cathode temperature window for stable H^- yield agrees well with the experiment.

4. Discussion

The estimated Cs coverage of the cathode surface and its implications on H^- production in short pulses fit the experimental observations very well. High enough Cs pressure is required to sustain the Cs-H_2 discharge, whereas the cathode surface temperature needs to be sufficiently high and well-controlled to reach near optimum Cs coverage for H^- ion production. This could appear "too good to be true", but it must be kept in mind that stable operation of the Penning ion sources is possible only in a specific window of various parameters – it is the high surface temperature that allows operating the cathode near the optimum Cs coverage to minimise the work function and, hence, maximise the H^- ion production. Overall, the range of experimentally found parameters matches the range where the above calculations best make sense.

The model supported by thermal simulations [3] reproduces trends observed in long pulse operation of the ISIS Penning source. A significant droop of the beam current due to cathode surface temperature rise is expected for 2.2 ms beam pulses at 50 Hz, required for the Front End Test Stand [13]. Furthermore, the Cs balance model predicts that the beam pulse length of the standard source can be extended to 1 ms at 50 Hz or 2 ms at 25 Hz by increasing the Cs oven

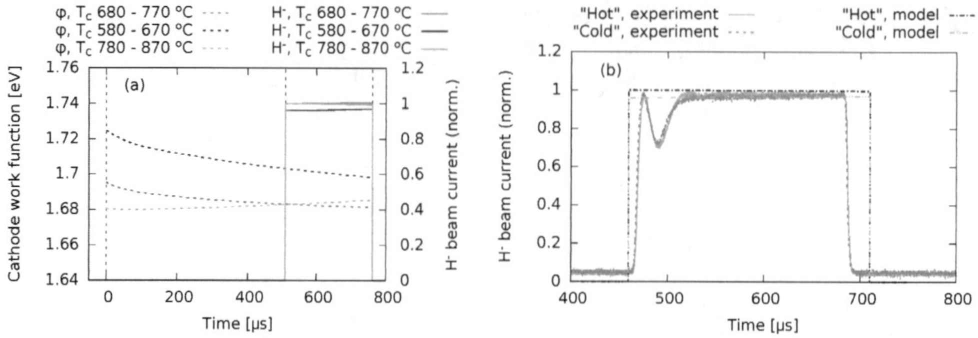

Figure 3: (a) The predicted cathode work function and H^- beam current for different cathode temperature ranges. (b) The (normalised) experimental H^- beam currents of "hot" and "cold" cathode sources, and the beam pulses predicted by the model with 200 °C cathode temperature difference. The model results are normalised with respect to the maximum predicted current of the "hot" cathode source. The experimental data are normalised with respect to the maximum measured beam current of 48.6 mA achieved with the "hot" cathode source. The irregular shape of the experimental pulses is due to extraction voltage overshoot and beam transport mismatch.

temperature by 10–30 °C above the nominal range of 150–160 °C. All of the above predictions match very well with the experiments [12]. The beam current droop is partially compensated by recirculation of Cs during the discharge pulse. At the equilibrium condition (at 680–770 °C cathode temperature) the number of Cs atoms adsorbed onto the cathode surface is expected to be $3.17–2.91\times10^{14}$ i.e. approximately 0.26×10^{14} atoms per pulse are released. At 10 Pa Cs pressure the number of atoms in the discharge 105 mm^3 volume is $0.87\cdot10^{14}$, i.e. the volumetric Cs density is increased by almost 40 % during the pulse, which mitigates the beam current droop in long pulse operation. The long pulse beam current droop can be eliminated only by reducing the rise of cathode surface temperature during the discharge pulse, which can be achieved with the so-called 2X Penning source under development at ISIS [11]. We plan to use the developed Cs model for source development towards longer pulses for FETS and the potential ESSνSB project [14].

References

[1] R. Sidlow, P. J. S. Barratt, A. P. Letchford, M. Perkins and C. W. Planner, Proc. EPAC'96, THP084L, p. 1525, (1996), jacow.org
[2] M. A. Huels, R. L. Champion, L.D. Doverspike et al., Phys. Rev. A 41, 4809 (1990).
[3] D. C. Faircloth, S. R. Lawrie, H. Pereira Da Costa, and V. Dudnikov, AIP Conf. Proc. 1655, 030013 (2015).
[4] B. Rasser, J.N.M. van Wunnik, J. Los, Surf. Sci., 118 (1982), p. 697.
[5] J.B. Taylor and I. Langmuir, Phys. Rev., 51 (1937), p. 753.
[6] G.D. Alton, Surf. Sci., 175 (1986), p. 226.
[7] P.W. van Amersfoort, *Formation of negative ions on a metal surface*, Thesis, FOM-Institute, Amsterdam, the Netherlands, 1987.
[8] P. W. Van Amersfoort, Y. Tong, and E. H. A. Granneman, J. Appl. Phys. 58 (1985), 2317.
[9] R. Gutser, *Experiments and Simulations for the Dynamics of Cesium in Negative Hydrogen Ion Sources for ITER N-NBI*, doctoral thesis, University of Augsburg, IPP report 4/291, (2011).
[10] J. Sherman, et al., AIP Conf. Proc., 763, (2005).
[11] D. C. Faircloth, S. R. Lawrie, J. Sherman, P. Wise, M. O. Whitehead, T. Wood, and T. Sarmento, AIP Conf. Proc. 2011, 050028 (2018).
[12] D. Faircloth, S. Lawrie, C. Gabor, A. Letchford, M. Whitehead, T. Wood, and M. Perkins, Rev. Sci. Instrum. 83, 02A701 (2012).
[13] A. Letchford et al., Proc. IPAC'15, THPF105, jacow.org.
[14] T. Ekelof and M. Dracos, Proc. Neutrino Oscillation Workshop PoS (NOW2018), 2019.

Journal of Physics: Conference Series **2244** (2022) 012032 doi:10.1088/1742-6596/2244/1/012032

RF power transfer efficiency of H⁻ ion sources: fluid modeling of accelerator source geometries

S Briefi, D Zielke, U Fantz

Max-Planck-Institut für Plasmaphysik, Boltzmannstr. 2, 85748 Garching, Germany

E-mail: stefan.briefi@ipp.mpg.de

Abstract. The plasma in RF driven negative hydrogen ion sources is sustained via inductive coupling with large power levels of up to 100 kW and low frequencies around 1 MHz. This leads to currents of around 100 A flowing over the RF coil and corresponding voltages in the kV range. The associated risk of arcing limits the reliability of the whole ion source. The required power level can be reduced via optimizing the RF power transfer efficiency η, which is typically only in the range of 50 to 60% for H^- sources used for neutral beam injection systems. In order to study the optimization of η systematically, a self-consistent fluid model has been set up and successfully validated with experimental measurements at the BATMAN Upgrade test bed. For H^- sources applied at particle accelerators, no experimental measurements of η are available so far. In order to gain a first insight into the RF power transfer efficiency of these sources, exemplary simulations were carried out with the fluid code. The simulated plasma parameters are in good agreement with results from OES measurements. η shows an increasing trend with larger source radius and a virtually constant value with increasing RF power. For benchmarking these first results, dedicated measurements at an accelerator source setup are inevitable.

1. Introduction

The plasma of RF driven negative hydrogen ion sources is generated via inductive coupling in a cylindrical vessel. High RF powers of up to 100 kW are typically applied both for sources used in neutral beam injection (NBI) systems [1] as well as for particle accelerator applications [2]. This high power together with the low RF frequency of 1 or 2 MHz leads to high currents in the range of hundreds of Amps flowing over the RF coil and voltages in the kV range. In general, this poses the risk of arcing especially at the RF coil [3, 4, 5] what limits the reliability of the whole ion source setup. Not all the RF power P_{RF} supplied by the generator is coupled to the plasma. Typically, a significant share is lost in the RF circuit due to Joule heating or via Eddy currents in metallic components of the source setup (such as a Faraday screen or metallic support structures in the vicinity of the RF coil) [6, 7] what makes active cooling of these components necessary. The effect is quantified by the RF power transfer efficiency η:

$$\eta = \frac{P_{plasma}}{P_{RF}}, \tag{1}$$

where P_{plasma} denotes the amount of RF power which is absorbed by the plasma. In general, an improvement of η offers the possibility to retain P_{plasma} while reducing P_{RF}. This would result one the one hand in a lower required RF generator output power, relaxing the voltage holding demands on the RF circuit. On the other hand, the cooling requirements of the source components are lowered when η is improved due to a better source design leading to less Eddy currents. However, the RF power transfer efficiency is influenced in a non-trivial way by many operational and design parameters such as gas pressure, RF power level, driving frequency, and source geometry. A dedicated measurement campaign at the BATMAN Upgrade test bed revealed, that η is only in the range between 50 and 60% for typical operational conditions of NBI H^- ion sources providing a substantial optimization potential [8].

In order to enable a systematic study and optimization of the RF power transfer efficiency of H^- ion sources, a dedicated 2D fluid model exploiting cylindrical symmetry has been set up and successfully validated with experimental measurements at the BATMAN Upgrade test bed [9]. However, due to the strong differences in source design and operational parameters between NBI and accelerator H^- ion sources (filling pressure, source diameter, ...), the obtained results cannot be directly transferred to accelerator sources. In addition, measurements of η are not available for these sources up to now. In order to gain a first insight, dedicated model runs were performed for a geometry and parameters being exemplary for accelerator sources.

2. Description of the fluid model

In order to simulate the RF power coupling of H^- ion sources self-consistently, two important challenges need to be overcome: First, the low driving frequency together with the high power levels leads to large RF magnetic fields in the range of 100 G. Hence, a correct description of the power coupling requires the consideration of the RF Lorentz force in the electron momentum balance leading to a compression of the plasma due to the ponderomotive effect. In addition, the counteracting electron viscosity must also be included in the equations as otherwise no steady state solution of the model can be obtained [9]. Second, the effect of neutral depletion has a large impact on the results. Therefore, the assumption of a uniform neutral gas background, which is often made cannot be applied. Instead, Navier-Stokes equations for both H and H_2 including the energy transport have to be solved. With this approach, a successful model validation could be carried out (a detailed description of the model can be found in [9]).

As input parameters, the fluid model requires the source geometry, filling pressure, RF frequency, generator power and the real part of the RF network impedance R_{net}. The latter can be obtained either from measurements or via 3D electromagnetic simulations of the setup. Unfortunately, for accelerator sources neither measurements nor simulations of the impedance are available. Therefore, the resistivity of the RF network has to be estimated. For the BATMAN Upgrade test bed, measurements showed that $R_{net} = 0.6\ \Omega$, but 74% of this resistivity is caused by eddy currents in the Faraday screen [8]. As no Faraday screen is typically present in the accelerator sources, the network resistivity can be reduced accordingly. Furthermore, these sources usually operate at an RF frequency of 2 MHz instead of the 1 MHz used for NBI sources. As approximation, R_{net} can be scaled up by a factor $\sqrt{f}$ to consider the impact of the frequency-dependent skin effect. In summary, this results in $R_{net} \approx 0.2\ \Omega$ which has been used for the investigations presented in this paper. It should be noted, that the absolute value of η determined by the model strongly depends on the absolute value of the network resistivity. Therefore, it is highly desirable to perform either measurements or simulations at an accelerator source setup.

3. Modeling results

Figure 1 shows the 2D distribution of the electron density and temperature calculated for a filling pressure of 3 Pa H_2 ($P_{RF} = 20$ kW, $f = 2$ MHz). As geometry, a cylindrical setup with a length of 13.8 cm and a diameter of 4.8 cm has exemplarily been chosen. In general, these parameters reflect the ones being typical for CERN's Linac4 ion source [10]. External static magnetic fields like cusp or filter fields have not been included in the simulation. Concerning the electron density, a maximum with an absolute value of 2.7×10^{19} m^{-3} in the discharge

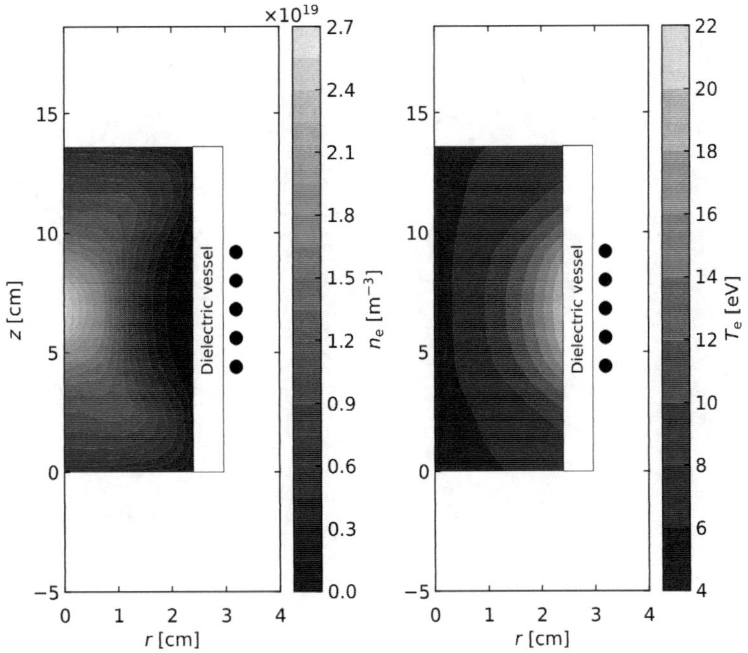

Figure 1. Electron density and temperature calculated with the fluid code [9] for a filling pressure of 3 Pa H_2 and an RF generator power of 20 kW using the geometry of the Linac4 ion source.

vessel center is evident. Close to the RF coil, the density is lower as the plasma is pushed away from the region close to the RF coil due to the RF-averaged Lorentz force. In contrast, the electron temperature has the highest value in this area and drops from 22 eV to about 4 eV in the discharge center. These results are in good agreement with the ones obtained from optical emission spectroscopy (OES) measurements at the Linac4 test stand [10, 11].

The corresponding RF power transfer efficiency is shown in Figure 2 for a variation of the source radius as well as the generator power. In general, η is in the range between 80 and 90% which is much higher than the ones obtained for NBI H^- ion source. However, it must be strongly emphasized that the absolute value of η depends on the value of R_{net}, which could only be estimated as described in the last section. Increasing the source radius results in a higher power transfer efficiency. This is somewhat unintuitive, as one would expect from classical collisional heating a power deposition only within the first one or two centimeters under the antenna. As the source radius of 2.4 cm is already larger than this RF skin depth this would result in a constant value of η with increasing radius. However, the induced RF current can diffuse much more inside the discharge due to the viscosity of the electrons [9]. Therefore, with increasing radius more electrons can be heated resulting in an increasing RF power transfer efficiency with higher radius.

In contrast, η is almost unaffected for a variation of the RF generator power. However, this behavior is the result of a complex interplay of several effects influencing the RF power transfer efficiency: a higher power leads to a higher current flowing over the RF coil and therefore to lager RF fields in the plasma. This leads to a higher electron density but also to an increased impact of the RF Lorentz force affecting the plasma distribution. In turn, this results in a different coupling between the RF fields and the plasma. [9] Furthermore, a higher power also leads to

Journal of Physics: Conference Series **2244** (2022) 012032 doi:10.1088/1742-6596/2244/1/012032

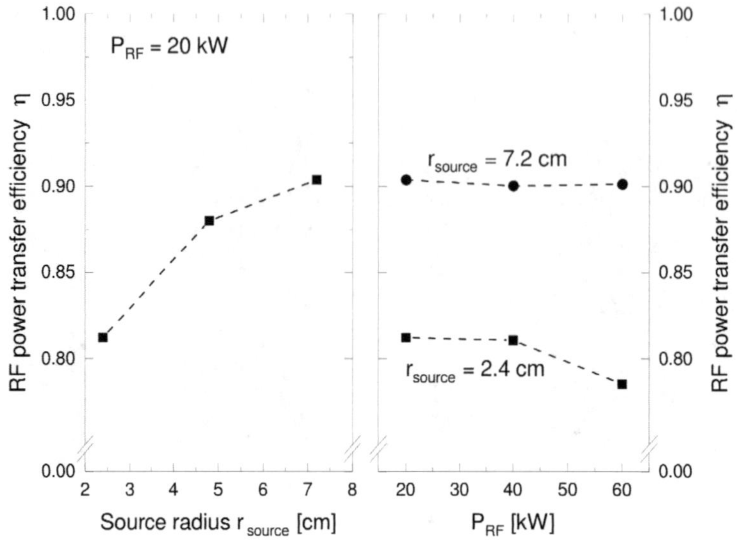

Figure 2. RF power transfer efficiency η for varying source radius at fixed $P_{RF} = 20$ kW and for varying P_{RF} at fixed source radius.

an increased neutral depletion, what lowers the electron density in the central discharge region again affecting the plasma distribution. In the model, all effects are considered self-consistently resulting in an almost constant value of the RF power transfer efficiency with varying P_{RF} for the investigated parameter range.

In order to demonstrate the possible impact of a different value of R_{net} present in the real setup of an accelerator ion source, calculations of the RF power transfer efficiency were carried out for $0.1\ \Omega < R_{net} < 0.5\ \Omega$ (see Figure 3). It can be seen that η drops from values just below 90% down to about 65% with increasing network resistivity highlighting the relevance of a correct determination of R_{net}.

4. Conclusion

Investigations at NBI H^- ion sources revealed that about 40 to 50% of the RF generator output power is not coupled to the plasma but lost as heat in metallic components of the setup. Reducing this share, i. e. improving the RF power transfer efficiency η, will result in a strongly reduced demand on the RF generator and a higher reliability of the source setup due to the reduced arcing probability. Transferring these results to H^- sources applied at accelerators is very difficult due to the different geometrical and operational parameters. Therefore, exemplary simulations were carried out with a fluid model describing the RF power coupling self-consistently. The real part of the RF network impedance is required as input for the model but unfortunately neither measurements nor simulations of R_{net} are available for accelerator sources. Scaling R_{net} from the measured value at the BATMAN Upgrade ion source to the typical accelerator ion source parameters yields $R_{net} \approx 0.2\ \Omega$. This results in modeled RF power transfer efficiencies between 80 and 90%. However, in order to benchmark the application of the fluid model to accelerator source setups, a dedicated measurement campaign at such a source is inevitable.

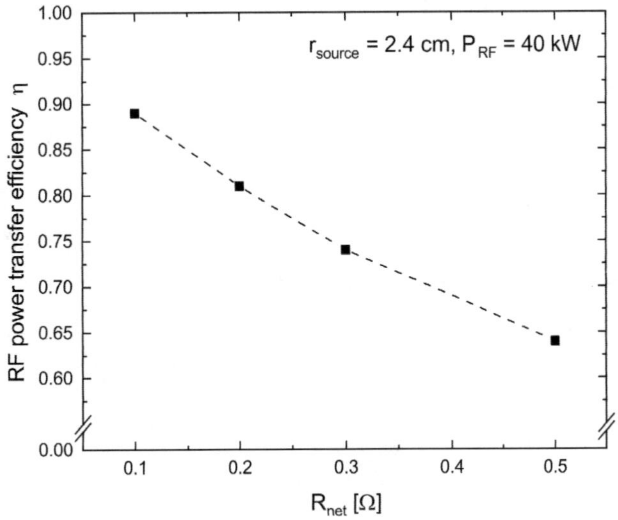

Figure 3. Modeled RF power transfer efficiency η for varying R_{net} at fixed $P_{RF} = 40$ kW and $r_{source} = 2.4$ cm.

References

[1] Heinemann B, Fantz U, Kraus W, Schiesko L, Wimmer C, Wünderlich D, Bonomo F, Fröschle M, Nocentini R and Riedl R 2017 *New Journal of Physics* **19** 015001

[2] Lettry J, Aguglia D, Alessi J, Andersson P, Bertolo S, Briefi S, Butterworth A, Coutron Y, Dallocchio A, David N, Chaudet E, Faircloth D, Fantz U, Fink D A, Garlasche M, Grudiev A, Guida R, Hansen J, Haase M, Hatayama A, Jones A, Koszar I, Lallement J B, Lombardi A M, Machado C, Mastrostefano C, Mathot S, Mattei S, Moyret P, Nisbet D, Nishida K, O'Neil M, Paoluzzi M, Scrivens R, Shibata T, Steyaert D, Thaus N and Voulgarakis G 2016 *Review of Scientific Instruments* **87** 02B139

[3] Mattei S, Nishida K, Onai M, Lettry J, Tran M Q and Hatayama A 2017 *AIP Conference Proceedings* **1869** 030018

[4] Recchia M, Maistrello A, Bigi M, Marcuzzi D and Gaio E 2018 *AIP Conference Proceedings* **2052** 040010

[5] Kraus W, Wünderlich D, Fantz U, Heinemann B, Bonomo F and Riedl R 2018 *Review of Scientific Instruments* **89** 052102

[6] Hopwood J 1994 *Plasma Sources Science and Technology* **3** 460–464

[7] Godyak V A, Piejak R B and Alexandrovich B M 1999 *Journal of Applied Physics* **85** 703–712

[8] Zielke D, Briefi S and Fantz U 2021 *Journal of Physics D: Applied Physics* **54** 155202

[9] Zielke D 2021 "Development of a self-consistent fluid model for optimizing inductive RF coupling of powerful negative hydrogen ion sources" PhD thesis. Augsburg University

[10] Briefi S, Mattei S, Rauner D, Lettry J, Tran M Q and Fantz U 2017 *New Journal of Physics* **19** 105006

[11] Briefi S, Mattei S, Lettry J and Fantz U 2017 *AIP Conference Proceedings* **1869** 030016

Acknowledgments

This work has been carried out within the framework of the EUROfusion Consortium and has received funding from the Euratom research and training programme 2014-2018 and 2019-2020 under grant agreement No 633053. The views and opinions expressed herein do not necessarily reflect those of the European Commission.

Plasma commissioning in a high power external RF-coil volume-type H^- ion source

S Lawrie, R Abel, D Faircloth, T Sarmento, J Speed and O Tarvainen

ISIS Pulsed Spallation Neutron and Muon Facility, Rutherford Appleton Laboratory, Harwell Campus, Oxfordshire OX11 0QX, UK

E-mail: `scott.lawrie@stfc.ac.uk`

Abstract. A high power, high duty cycle, negative hydrogen ion source is in development at ISIS. It will operate in pure volume-production mode and is driven by a $5\frac{1}{2}$-turn RF-coil mounted external to the plasma chamber. A solid-state amplifier with a maximum output of 100 kW in 50 Hz, 1 ms pulses delivers RF power to the coil via an impedance-matching network. The amplifier has a relatively wide bandwidth, able to deliver full power from 1.8-4.0 MHz. This flexibility allowed straightforward commissioning of the matching network into an inductively-coupled plasma. Striking of the pulsed plasma is facilitated by a compact microwave ignition gun, requiring only 10 W of power at 2.45 GHz to deliver 1 mA seed pulses of electrons. Experiments have shown that it is vital to encapsulate the RF-coil properly to mitigate high voltage sparking. In addition, the location of the coil relative to the ion source's permanent magnets has a critical effect on the ease of plasma ignition. The result of commissioning work is that a full duty-factor (50 Hz, 800 μs) plasma has been achieved at nominal operating power (30 kW) and detailed optical studies have begun.

1. Introduction

The ISIS pulsed spallation facility [1] at the Rutherford Appleton Laboratory (RAL) begins with a cesiated Penning-type surface-plasma H^- ion source [2]. The limited cathode lifetime means the ion source is replaced every three weeks. Moreover, poor beam transport efficiency along the linac means only 25 mA of beam is delivered to the synchrotron from the 55 mA created by the ion source. A project is underway to overhaul the 'pre-injector' section of the linac with a medium energy beam transport (MEBT) [3]. This will improve beam efficiency such that only 35 mA will be needed from the ion source. This beam current is within reach of a state-of-the-art, non-cesiated RF-driven H^- ion source. The benefit of an RF ion source (RFIS) utilising an external RF-coil is the lack of internal wear, meaning replacement will be much less frequent; bringing large benefits to overall facility operation and reliability. The RFIS being developed at RAL is based primarily on the designs at SNS [4] and CERN Linac4 [5], with some components inspired by J-Parc [6] and CSNS [7]. A major difficulty in operating a high duty-factor RFIS is thermal management. Significant simulation and engineering work resulted in a robust design with several novel features [3; 8]. This paper discusses the RFIS plasma commissioning.

2. RF Amplifier and Matching

A custom high power, high duty-factor, air-cooled, solid-state, low frequency RF amplifier was purchased from R&K [9]. Based around 128 LD-MOS devices, it benefits from a modular design with enough operational overhead to continue delivering full power with four out of 32 front-end modules inoperable. It has a relatively large bandwidth and gain, so can deliver 100 kW, 50 Hz, 1 ms RF pulses anywhere between 1.8 and 4.0 MHz. Due to the inductive and changeable nature of the plasma load, it was designed specifically to cope with high reflections and transitory high voltage breakdowns. As such it is very robust and flexible, yet extremely easy to use.

The amplifier outputs to a coaxial transmission line isolation transformer such that power can be delivered to the RFIS floating at -36 kV for beam production. After the transformer is an air-cooled matching box consisting of a series capacitor and inductor. Together with a $5\frac{1}{2}$-turn RF-coil surrounding the plasma chamber, this forms a resonant circuit to match the 50 Ω amplifier output to the ≈4 Ω plasma load. The inductance was tuned by hand initially to fix the resonance to around 3 MHz, but no active or variable components are necessary to set the frequency thereafter. Instead, the resonance is measured on a vector network analyser (VNA) after any mechanical changes to the RFIS. The amplifier's input low-level RF (LLRF) signal is then set to the measured frequency. As well as the inherent cost-effectiveness and simplicity of this system, having an adjustable frequency gives the benefit of tuning the LLRF to achieve the lowest reflected power during plasma operation. Typically, the LLRF is set to around 2.87 MHz in vacuum, with the frequency ramped linearly over 5 μs to around 2.93 MHz when the plasma pulse is ignited; i.e. the plasma shifts the resonance by around 50 kHz. The LLRF is extremely flexible, being able to deliver arbitrary RF pulse shapes. Ultimately, this will be used to tailor the output H^- beam current pulse to suit the accelerator requirements.

3. Major Components

A labelled photo of the RF ion source operating at 50 Hz is shown in Fig. 1. To achieve full power and full duty cycle, improvements were made to several key components, compared to the design discussed in [3].

Figure 1. The RF ion source installed and operating at full duty cycle. Visible components include filter magnets (a), plasma light visible through translucent plasma chamber and cooling jacket (b), main support housing (c), RF coil (d), cooling jacket connection manifold (e) and ignition gun (f).

3.1. RF-Coil

The coil consists of a $5\frac{1}{2}$-turn, water cooled solenoidal conductor encapsulated in dolomite resin. The inner diameter of the coil is reduced to fit as close to the plasma chamber as possible to minimise the required RF power. The outer diameter did have C-shaped ferrite blocks encapsulated with the conductor. The ferrites were meant to reduce RF leakage which would couple to nearby metallic components; instead concentrating the power into the plasma region. However it was found that the ferrites proximate to the conductor would float to a sufficient voltage to spark to the nearby metallic housing. The ferrites were removed, however sparking still occurred at higher RF power. It was discovered that the high permittivity resin compressed the electric field across the small air gap between the coil and housing, again leading to breakdowns. To short out the small air gap, the next coil had a thin copper mesh encapsulated and exposed partially on its outer diameter, such that grub screws could provide electrical continuity to the nearby housing, ensuring good grounding. Finally, the last source of sparks was the copper legs which protrude out of the coil and connect onto the matching box. The high field on the 'hot' leg caused field emission of electrons which would propagate across the surface of the resin to the housing. Machining the slots through which the legs exit the housing even larger – to several times the air breakdown distance – and adding large rounded smooth edges did not help prevent the breakdowns. The problem was solved by using successively longer layers of heat shrink tubing to cover the exposed legs to prevent surface tracking. These improvements have resulted in the RF-coil being able to operate spark-free up to 70 kW.

3.2. Compact Microwave Ignition Gun

Each 50 Hz plasma pulse must be ignited reliably. Introducing gas puffs to increase the pressure temporarily toward the Paschen breakdown limit is unfeasible for a large plasma chamber at high repetition rate. Equally, we did not want complicated multiple RF frequencies, nor power-inefficient plasma- or spark-guns. Instead, we use a compact microwave ignition gun. Developed at LPSC Grenoble for ion beam applications [10], this device has been modified to serve as an external gun to seed the plasma chamber with electrons for plasma ignition. The gun uses just 10 W of 2.45 GHz microwave power, coupled into a quarter-wave resonant cavity. Permanent magnets generate the necessary field to incite electron cyclotron resonance (ECR) at the cavity tip. The small ECR zone in the hydrogen plasma, combined with a pulsed bias voltage of $\approx$400 V, is sufficient to extract up to 1 mA of electrons. The gun has its own hydrogen feed separate to that of the main plasma chamber, such that the seed electrons and H^- production may be optimised independently. We typically use 10 and 8 ml/min H_2 flow into the ignition gun and plasma chamber, respectively. The gun is simple to operate, has indefinite lifetime, does not require cooling and ensures reliable ignition every pulse.

3.3. Magnets

The magnetic filter field, used to reduce the electron temperature and enhance H^- production, is generated with a pair of permanent magnets located external to the plasma chamber and main housing. The original design had the filter magnets integral to the octopole cusp arrangement for plasma confinement. The measures taken to reduce RF-coil sparking necessitated removal of the cusp magnets for the time being, to leave just the filter magnets. Measurements at CERN indicated that the cusp field may not in fact be necessary for optimal H^- production [11], so the cusp magnet arrangement will be re-visited during beam commissioning.

The pulsed current of seed electrons from the ignition gun were measured on the plasma electrode and was shown to be removed completely by the filter field; thus demonstrating its effectiveness for H^- production. On the downside, significantly more power than expected (around 50 kW) was required to ignite the pulsed plasma, because so few seed electrons made it to the plasma forming region. Therefore, the RF-coil was moved back toward the gun such that the RF field no longer coincided with the peak filter field. In this location, the plasma ignites reliably with 7 kW of power. It remains to be seen whether the RF-coil (and hence peak plasma density) being moved away from the outlet aperture affects the extracted beam current. If so, an alternative filter magnet solution has been manufactured which will reduce the field overlap between RF-coil and filter when the coil is re-positioned closer to the outlet.

3.4. Cooling Jacket
The most challenging part of the RFIS is the plasma chamber cooling. The plasma chamber is a plain cylindrical tube of pure aluminium nitride, with smooth faces at either end for vacuum sealing. No features such as cooling channels or mounting holes are machined into it, so as to avoid any stress or fracture points. As such, the chamber is cooled with an external 'jacket' which fits snugly on its external diameter. To prevent the RF-coil coupling power to the jacket, it must be non-conductive. Combined with the complex geometry of the cooling channels, this lends itself to 3D-printing techniques using glass-reinforced toughened Nylon-12. The difficulties then lie in ensuring the plastic cooling jacket does not melt, achieving excellent vacuum & water seals and attaching the water hoses securely. True to the 'rapid prototyping' ethos, several iterations were made until all requirements were met. At full power and duty cycle, the cooling jacket reaches a steady temperature of less than 30 °C, which is far below its softening point. Long-term full power tests are underway to ensure its reliability.

4. Preliminary Plasma Optical Measurements
Through the ignition gun body are two lines of sight into the plasma chamber for optical emission spectroscopy. A variety of optical instruments are available at RAL and a detailed experimental campaign will follow. As a first test, we used a range of optical filters and silicon photomultipliers to take fast photometry measurements. Figure 2 shows the H_α^{Balmer}, H_β^{Balmer} and $H_2^{Fulcher}$ emission during a plasma pulse. It can be seen that there is an initial burst of molecular emission at plasma ignition, before the atomic emission builds up over 30 μs. Thereafter, all optical emissions continue to increase throughout the pulse, regardless of the shape of the forward or reflected RF power; with the molecular emissions increasing at a faster rate.

The Fulcher-band emission arises from electron impact excitation from the molecular ground state up to triplet d-states. The d-state decays to the a-state, emitting the Fulcher-α ro-vibrational band. Thereafter the triplet a-state can only decay to the un-bound b-state, thereby

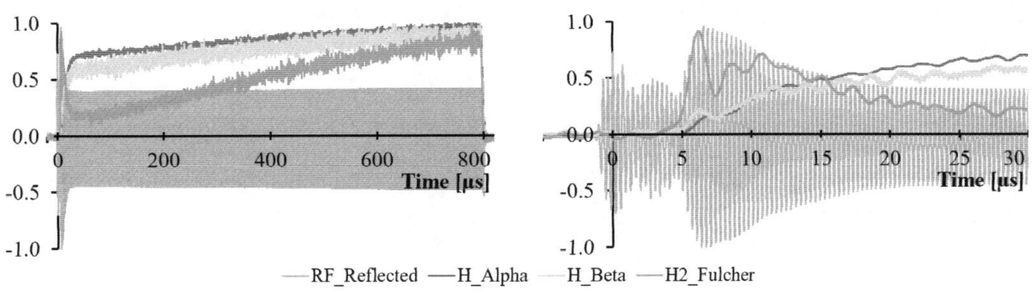

Figure 2. Optical emissions for a range of bandpass filters throughout the entire RF pulse (left) and zoomed into the first microseconds after plasma ignition (right). Amplitudes scaled to the maximum of each dataset to emphasise relative changes. Input RF frequency optimised for minimum reflected power. The first 5 μs of the RF pulse is programmed such that the frequency ramps from the resonant conditions in vacuum, to the resonance with plasma. This is why the optical emissions are delayed until the plasma builds up over time.

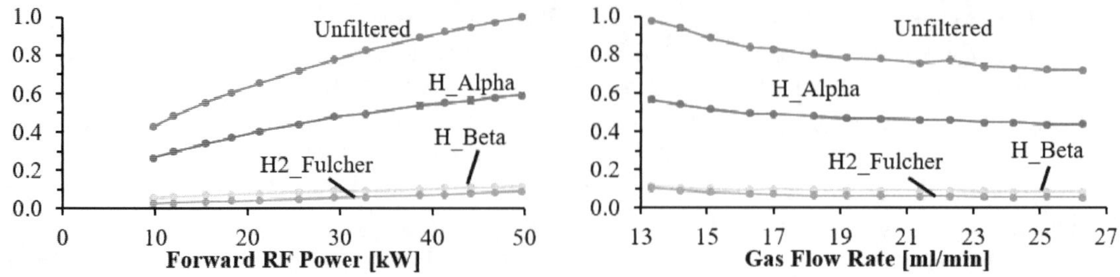

Figure 3. Variation of the optical emissions with forward RF power (left) and with total hydrogen gas flow rate (right), on an arbitrary scale. Note that the optical filters have different transmissions and the total, unfiltered channel was attenuated significantly, deliberately. Therefore the channel amplitudes should not be compared directly with each other.

forcing the molecule to auto-dissociate [12]. Therefore, Fulcher-band emission is proportional to the molecular dissociation rate in the plasma. Figure 3 shows that both atomic and molecular emissions increase with RF power, since more energy is available for electrons to generate excitation and dissociation. On the other hand, emissions decrease with hydrogen flow because the increased gas density results in more low energy inelastic collisions; so the electrons have on average less energy to drive the neutral particles into the required excited states.

5. Conclusion

A new external RF-coil ion source plasma is operational at ISIS. The RF amplifier, RF-coil, gas delivery, ignition gun and cooling jacket all perform nominally to sustain a stable 30 kW plasma at 4% duty factor (50 Hz, 800 μs). Operation is much more simple than the legacy Penning-source in use at present on the facility; with push-button start and no ramping of power or repetition rate necessary. A full optical measurement campaign is underway whilst long-term soak-testing begins. The next step is to finish assembly and installation of personnel protection systems and high voltage power supplies in order to extract an H^- ion beam.

References

[1] Thomason J 2019 *Nucl. Instrum. and Meth. A* **917** 61
[2] Thomason J W G and Sidlow R 2000 *Proc. of the European Particle Accelerator Conf.*
[3] Lawrie S et al 2019 *Rev. Sci. Instrum.* **90** 103310
[4] Welton R F et al 2021 *AIP Conf. Proc.* **2373** 070004
[5] Lettry J et al 2018 *AIP. Conf. Proc.* **2052** 050008
[6] Ueno A et al 2021 *AIP Conf. Proc.* **2373** 040002
[7] Li H et al 2021 *Nucl. Instrum. and Meth. A* **996** 165149
[8] Tarvainen O et al 2018 *AIP. Conf. Proc.* **1655** 050005
[9] R&K Company Limited URL `https://rk-microwave.com/`
[10] Sortais P et al 2010 *Rev. Sci. Instrum.* **81** 02B314
[11] Briefi S, Mattei S, Lettry J and Fantz U 2017 *AIP Conf. Proc.* **1869** 030016
[12] Farley D R, Stotler D P, Lundberg D P and Cohen S A 2011 *J. of Quantitative Spec. and Rad. Transfer* **112**(5) 800–819

Journal of Physics: Conference Series **2244** (2022) 012034 doi:10.1088/1742-6596/2244/1/012034

Development of a surface plasma method for negative ion beams production

Vadim Dudnikov[a)],
Muons, Inc. 552 N. Batavia Ave. Batavia, IL 60510, USA
a) Corresponding author: dvg43@yahoo.com

Abstract. By now, the fluxes of accelerated negative ions of steel an indispensable tool for accelerators, plasma physics, solid state physics, chemistry, biology, medicine, important areas of new technique and technology. It arouses increased interest to the development of intensive sources of these particles.
The cesiation effect - a significant increase in the emission of negative ions from the discharge with a simultaneous decrease in the flux of accompanying electrons below the current of negative ions after adding a small amount of cesium or other substances with a low ionization potential to the discharge was discovered 50 years ago, on July 1, 1971 in the Budker Institute of Nuclear Physics (BINP), Novosibirsk, Russia.
The cesiation effect has opened the door to high-intensity, high-brightness negative ion beams production. The efficiency of generating beams of negative ions was significantly increased due to the invention at the BINP of geometric focusing of generated negative ions. Based on the cesiation effect a surface-plasma method for producing beams of negative ions was developed at the BINP, which provided the creation of surface plasma sources of negative ions for accelerators with a current of hundreds of mA and sources for Controlled Nuclear Fusion with a current of tens of A. (the previous intensity record was 5 mA).
Surface plasma sources are being developed all over the world in England, France, Germany, Switzerland, Italy, Japan, Korea, China, India, USA, Canada, Sweden, Spain, New Zealand, Russia, Ukraine, Finland with investments in billions of USD. Specialists in surface plasma sources are trained at the Plasma Physics Department of Novosibirsk State University.

1. Introduction

Soviet academician N. Semenov proposed in 1945 to suppress the explosions of nuclear bombs by irradiating them with neutrons [1]. A large accelerator development program was established in the USSR. V. Teplyakov has invented Radio Frequency Quadrupole (RFQ, Teplyakov Accelerator) within the framework of this program [2]. A similar proposal was put forward by R. Wilson in 1953 [3].
After experimental development of charge exchange injection with the conversion of negative ion into neutrals [4] (my Ph.D. thesis and doctoral dissertation by G.I. Dimov) [5,6], director of Institute of Nuclear Physics (Novosibirsk, Russia) G. Budker suggested using high-energy neutral beams in space to influence space objects (inspecting satellites for the presence of nuclear materials, suppressing nuclear explosions ...)
In the Korolev Space Central Design Burro (already without Korolev), after the closure of the Soviet lunar program, unused funds remained, and in 1969 a contract was signed with the INP to develop a neutral injector for an ion beam current of H^- 10 mA for 2 Mrub (at that time in USSR, Novosibirsk, a nine-story building for 286 apartments was cost 1 Mrub , 1 \$ = 0.86 rubles). Strategic Defense Initiative with using neutral particle beams was established in USA in 1976 [7]. Then 3 groups were organized in INP, which were engaged in different methods of negative ions beams obtaining. For charge exchange injection into the Dimov's lab. charge-exchange and plasma sources of negative ions (of the Ehlers type) were developed. In the 1st year of Novosibirsk State University I read Gaponov's

textbook "Electronics" and then reread it. In it, I found a mention of the secondary emission of negative ions, little known at that time. I began to study secondary ion-ion emission in more detail. There was a lot of data, but they were very contradictory and everywhere the probabilities of secondary emission was very small, and the theoretical foundations of this phenomenon were absent. It was known that the deposition of alkali metals increases the coefficient of secondary ion-electron emission.
The works of Ayukhanov of 1961 and of Kron of 1962 were discovered, in which it was shown that when deposited alkali metals, the coefficient of secondary emission of negative ions also increases, but the resulting currents of H^- ions were at a sub-microampere level.
I suggested that Dimov take up the secondary emission method for obtaining negative ions. He enthusiastically accepted this proposal and a group was organized to obtain H^- ions by bombarding a surface with cesium ions under his leadership. I took an active part in this work. A stand was set up and rather soon H^- beams with currents up to -2.5 mA were obtained [8], but the brightness of these beams was very low and the service life of these devices was very short. G. Roslyakov's group was engaged in charge exchange sources [9]. My student Y. Belchenko and I worked on plasma sources. The groups worked intensively, but they were far from reaching the required parameters. Budker gathered weekly meetings where various proposals were discussed, but no solutions were found. By the end of the contract, everyone despaired of the possibility of obtaining the necessary currents and went on vacation for the summer. Yura Belchenko left with a building team in Bilibino in Kolyma (to build the Bilibino nuclear power plant) and took our laboratory assistant with him. The mechanics also went on vacation.

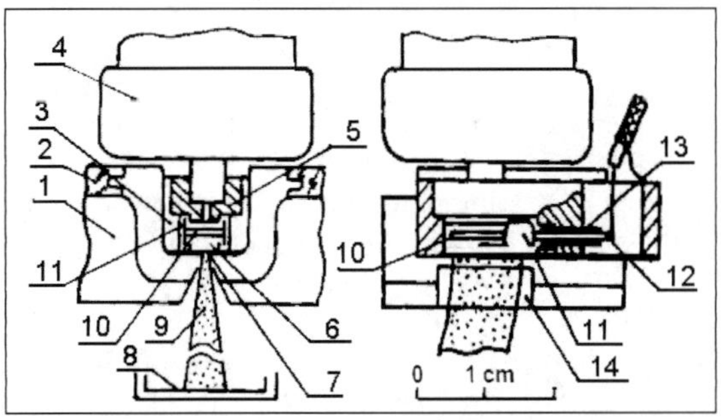

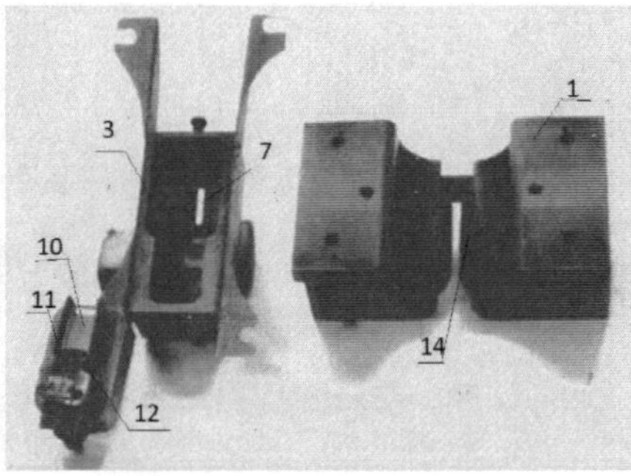

FIGURE 1. Schematic of planotron plasma configuration for producing a negative ion beam.
1 – electromagnet pole tips, 2 – high-voltage insulators, 3 – body of plasma chamber, 4 – pulsed gas inlet valve, 5 – anode insert, 6 – screen of emission slit, 7 – emission slit, 8 – collector, 9 – negative ion beam, 10 – central cathode plate, 11 – cathode side shields, 12 – cathode holders, 13 – cathode insulators, 14 – extractor plates.
FIGURE 2. Photograph of the first planotron with numeration of parts.

2. Discovery of cesiation effect

I stayed alone for the summer and continued to work with a plasma source with a planotron geometry. It yielded up to 4.5 mA H^- in 1 ms pulses, but with an electron current to the extractor 30 times higher, a discharge current of 100 A, and a discharge voltage of 600 V (a power of 60 kW put into a device with a volume of 1 cm^3). Taking into account the acquired "negative" experience gained after a number of trial experiments, the design of a new source was brought to the form depicted schematically in Fig. 1. The body of the plasma cell 3 was fixed on insulators 2 of plexiglass in the gap between the pole pieces 1. The plates of extraction electrode 14 were welded to special protrusions from the pole pieces 1, creating a magnetic field in the high-voltage gap (but not a Penning trap configuration). A pair of pole-pieces with the source was installed between the grounded poles of the electromagnet. A plasma cell with planotron configuration is formed by the cathode, consisting of the central plate of cathode 10 and cathode side-shields 11, and a cathode-enclosing anode formed by parts of the plasma cell body 3 and anode insert 5. A cathode made of 0.2 mm thick molibdenum foil was attached to tantalum current leads 12 passing through the wall of anode insert 5 and insulated from it by ceramic tubes 13. The volume of the plasma cell was minimized as much as possible. Gaps between the cathode and the anode, in which a discharge should not burn, were reduced to 1 mm. Hydrogen was supplied to the plasma cell through a short channel by a pulsed electromagnetic valve 4. Emission slit 7 with dimensions 0.5 x 10 mm^2 oriented across the magnetic field was cut in the thin-walled body of the plasma cell. From the discharge region, particles could pass to the emission slit through the gap between the anode projections 6 shielding the emission slit from the dense, high-current plasma. A photograph of the first planotron is shown in Fig. 2.

The old power supply systems were used for pulsed hydrogen gas injection, plasma ignition and support of the discharge, and ion extraction. A beam collector 8 was installed to monitor the beam current 9. The body of the plasma cell was held at the negative-polarity extraction voltage, and the collector was held at low voltage.

On July 1, 1971, I fixed on the anode of the planotron a tablet of cesium chromate with 1 mg of released cesium and turned on the discharge. The emission slit was shielded and an ion current of 1.5 mA was recorded on the collector. After several minutes of operation, a current surge of up to 3 mA appeared at the end of the pulse. After optimizing the gas supply, the current at the end of the pulse increased to 4 mA, but after 20 minutes, the current surge disappeared and the current to the collector became 1 mA again.

Deciding that the current surge was associated with the release of cesium, I placed several tablets on the cathode, which was heated more strongly, and covered them with a nickel mesh.

In this configuration, the collector current quickly increased to 12.5 mA, and after optimizing the gas supply and discharge, a rectangular pulse of 15 mA was obtained.

The discharge voltage dropped from 600 V to 100 V. After that, within a week, various configurations of discharges were tested, it was verified that these are mainly H^- ions, and the currents of electrons and heavy ions to the collector are small [10]. After that, I left everything in working order and went on vacation to my village Gunda in Buryatia.

After my return, the source worked as it should, the H^- current was quickly increased to 100 mA, by optimization of design and mode of operation, and then to 300 mA, with emission current density up to 4 A/cm^2, then to 0.9 A from a source the size of a cigarette lighter [11].

Under the contract with the Central Design Bureau, we successfully reported and concluded 2 new contracts of 5 Mrub each for the manufacture of a surface plasma sources (SPS) for 100 mA for the linear accelerator of the meson factory of Institute of Nuclear Research (Moscow) and for the Institute High Energy Physics (Protvino).

Then the results with cesiation were recognized as deep secret and were not allowed to be published. But many delegations from the Soviet Union (Central Design Bureau, NIIEFA, Kurchatov Institute, Institute of Nuclear Research, Institute of High Energy Physics) and from the USA began to visit INP. Budker allowed the SPS to be shown to high-ranking US visitors as objects for bidding. International publication [11] was permeated only in 1974.

One practical result of development of high brightness negative ion source with cesiation is accepting of the charge-exchange injection in circular accelerators for a routine operation [12]. Now SPS are "Sources of life" for gigantic accelerators complexes as SNS, FNAL, LANSE, BNL, ISIS, DESY, KEK, IHEP, LHC and an efficiency and reliability of these sources operation are determined a productivity of these big collaborations. Many results of the high energy physics were discovered with using of negative ion sources. Development of high brightness H^- sources was stimulated by first success of high current proton beam accumulation with using a charge-exchange injection [5] and supported by interest of "Star War" [7]. A Beam Experiment Abroad Rockets (BEAR experiment), conducted by LANL and Grumman, Corp [13] was cost 749 M$. A recent circumstance was the reason of difficulties and long delay of first publications, but nonofficial communication was relative fast. Until 1971 main attention was concentrated on development the charge-exchange negative ion sources, because there was no hope to extract from the plasma directly more than 5 mA of H^-. On July 1, 1971 at the Institute of Nuclear Physic (INP), Novosibirsk, significant enhancement of negative ion emission from the magnetron (planotron) plasma source followed by introducing small cesium admixture to the gas discharge was observed by the author [10].

This observation, considered in book [14], was further developed and understood as a principle new method of negative ion production. In the patent application [10] it was stated: "a method of negative ion production in gas discharges, comprising adding to the discharge an admixture of substance with a low ionization potential such as cesium, for example, for enhance a negative ion formation".

3. Development of surface plasma method of negative ion beams production

Further development of SPS was conducted by cooperation Belchenko, Dimov, Dudnikov (BDD) [15 ,16]. Cesiation effect-increase of negative ion emission after adding into discharge subtances with low ionization potential was explained by M. Kishinevskii [17]. Probability of escaping as negative ions for reflected and sputtered particle increased significantly after decreasing of a work function of surface after adsorbing the cesium. The first high brightness SPS for accelerator has been developed by author [18]. The Semiplanatron SPS with an efficient geometrical focusing has been developed by author at 1977 [19]. Further R&D of high brightness SPS was conducted in cooperation with Derevyankin. Very fast development and adaptation of SPS has been start in many USA laboratories, in Europe and in Japan, and International Symposiums for Production and Neutralization of Negative Ions and Beams has been established [20]. Now the Surface Plasma Method of negative ion production and SPS are considered in many books (see for example books [14,21,22]). Good reviews of SPS for accelerators were presented in reports of Peters [23]. Development of high current SPS (tens of Amperes) for thermonuclear plasma heating is in progress and used in large Tokamaks and Stellarators [24]. Production of polarized negative ions by charge- exchange with a slow negative ion in SPS has been proposed by author and has been realized with a good success [25].

In R&D of heavy negative ion SPS for technology application a good success has been reached [26] but it is still necessary to further improve the DC SPS for heavy negative ion production to meet the broad requirements of many industrial applications. DC SPS for long time operation with accelerators has been developed [27]. Cesiated SPS with RF discharges for accelerators [28] and for Neutral Beam Injectors (NBI) [29,30] were improved recently significantly.

The efficiency of negative ion formation depends very much on the catalytic property of the surface, mainly the work function. For enhanced negative ion formation in the SPS, an admixture of substances with low ionization energy, such as alkaline or alkaline earth elements or compounds are used. Most efficient is the addition of cesium. Still, the surface work function and catalytic properties of the surface for negative ion formation depend on many parameters, including surface-cesium concentration, and admixtures of other compounds, such as oxides, halides, nitrides, and surface temperature. At the same time, the extracted current (which includes electrons and back-accelerated positive ions in addition to the H^- ions measured by the Faraday cup) decreased significantly.

The highest brightness can be reached only with noiseless operation. The level of discharge noise (hash) depends on many parameters. For stable discharge, the surface properties should be favourable for stable conditions and the frequency of electron scattering by plasma particles should be higher than the Larmor frequency. Discharge noise can be suppressed by a decrease of magnetic field and by an increase of the gas or Cs density.

Development of surface plasma sources with cesiation for neutral beam injection into fusion plasma trap with beam current more then 70 A and energy 1 MeV was established [31] with a cost of ~1 billion of USD. Early publication for SPS is collected in review report of Wells (Rand corporation) [32].

4. **Summary**

Since the discovery of the cesiation effect and the development of the surface plasma mechanism of negative ion formation with cesiation, many modifications of the SPS have been developed and many improvements have been introduced. Now SPSs have become a highly reliable technology.

5. **References**

[1] Matyushkin V, *Everyday life of Arzamas-16*, Molodaya Gvardia, 2007 (in Russian).

[2] Kapchinskii M and Teplyakov V. A., Prib. Tekh. Eksp., No. 4 (1970) p 19 and No. 4 (1970) p. 17.

[3] Silvan S. Schweber, R. Wilson Defending against nuclear weapons: A 1950s proposal, Physics Today 60(4), 36 (2007); https://doi.org/10.1063/1.2731971

[4] Dimov GI, Dudnikov VG, Cross sections for stripping of-1-MeV negative hydrogen ions in certain gases, Soviet physics technical physics-USSR 11 (7), 919 (1967)

[5] Budker GI, Dimov GI and Dudnikov VG, in Proc. of the Intern. Symp. on Electron and Positron Storage Ring, France, Sakley, 1966, Rep. VIII, 6.1
[6] Budker GI, Dimov GI and Dudnikov VG, "Experiments on producing intensive proton beams by means of the method of charge-exchange injection", Soviet Atomic Energy **22**(5), 441 (1967).
[7] Robinson C, Aviation Week&Space Tech., p.42, Oct., 1978; Rev. Mod. Phys., 59(3), Part II (1987).
[8] Bender ED, Kishinevskii M, Morozov II. Proc. of Second International Symposium on Production and Neutralization of Negative Hydrogen Ions and Beams, Brookhaven, NY. Preprint BNL 51304, 60 (1980).
[9] Dimov GI, Roslyakov GV, ZTF, 42,1186 (1972).
[10] Dudnikov V, Method of negative ion obtaining, USSR Patent 411542, 10/III. 1972; http://www.findpatent.ru/patent/41/411542.html, Dudnikov, V.G., Technique for producing negative ions, https://inis.iaea.org/search/search.aspx?orig_q=RN:9355182
[11] Belchenko YI, Dimov GI, Dudnikov VG, Powerful injector of neutrals with a surface-plasma source of negative ions, Nucl. Fusion, v. 14, no. 1, pp. 113-114 (1974).
[12] Dudnikov V G "Charge exchange injection into accelerators and storage rings", *Phys. Usp.* 62 405–412 (2019)
[13] https://accelconf.web.cern.ch/l90/papers/th454.pdf
[14] Dudnikov Vadim , *Development and Applications of Negative Ion Sources*, Springer, 2019..
[15] Belchenko Yu. I., Dimov GI, and Dudnikov VG, Proc. Symp. Production and Neutralization of Negative Hydrogen Ions and Beams, Brookhaven, 1977 (BNL, Upton, NY, 1977), BNL- 50727, p. 79.
[16] Belchenko Yu. and Dudnikov V., "Surface Negative Ion Production in Ion Sources", in Production and Application of Light Negative Ions, 4rd European Workshop, W. Graham, Editor, Belfast, 1991, pp. 47–66.
[17] Kishinevskii ME, Zh. Tekh. Fiz. 48, 1281 (1978); translated in Sov. Phys. - Tech. Phys. 20, 799 (1975).
[18] Dudnikov V, Proc. 4th All-Union Conf. on Charged Part. Accel., Moscow, 1974,V.1, p.323.
[19] Dudnikov, Belchenko Yu, Journal de Physique, v. 40, p. 477, (1979); Preprint, INP 78-95, Novosibirsk,1978; Dudnikov V., Fiksel G., Journal de Physique, v. 40, p. 479 (1979).
[20] Schmidt CW, Prod. Neutralizat. of Negative Ions and Beams, 8th Int. Symp. AIP 1-56396-773-5.(1998).
[21] Dudnikov VG, *Negative Ion Sources*, NGU, Novosibirsk, 2018.
[22] Zhang H, *Ion Sources,* Springer (1999)
[23] Peters J, LINAC' 98, Chicago, 1998; Peters J, Rev. Sci.Instrum.,71(2),1069(2000); Peters J, EPAC'2000.
[24] Kojima A et al., Rev. Sci. Instrum, **81**, 02B112, (2010).
[25] Belov AS et al., Nucl. Instrum. Methods Phys. Res., Sect. A 333, 256 (1993).
[26] Dudnikov Vadim and Farrell J. Paul, "Compact surface plasma sources for heavy negative ion production", Eev. Sci. Instrum., V. 75, N. 5 (2004).
[27] Bashkeev A and Dudnikov V, AIP Conf. Proc. No 210, p. 329 (1990); Belchenko Yu., Sanin A and Ivanov A, AIP Conf. Proc. No. 1097 (American Institute of Physics, Melville, NY, 2009), pp. 214–222.
[28] Welton RF, Dudnikov VG, Gawne KR, Han BX, Murray SN, Pennisi TR, at al, H^- radio frequency source development at the Spallation Neutron Source, Rev. Sci. Instrum. 83 (2), 02A725 (2012).
[29] Kraus W, Fantz U, and Franzen P, Rev. Sci. Instrum., **81**, 02B110 (2010).
[30] Belchenko Yu. I., Ivanov A. A.,. Sanin A. L, and Sotnikov O. Z., Development of surface-plasma negative ions sources at the Budker Institute of nuclear physics, AIP Conference Proceedings 2052, 030006 (2018); https://doi.org/10.1063/1.5083734
[31] ITER Neutral Beam Test Facility - 2017;https://www.youtube.com/watch?v=DKOHJYxK15o
[32] Wells N, The Development of High-Intensity Negative Ion Sources and Beams in the USSR, Rand corporation, 1981. http://www.rand.org/content/dam/rand/pubs/reports/2007/R2816.pdf

Journal of Physics: Conference Series **2244** (2022) 012035 doi:10.1088/1742-6596/2244/1/012035

Numerical analysis of isotope effect in NIFS negative ion source

Ryo Kato[1, a)], Kazuo Hoshino[1], Haruhisa Nakano[2], Takanori Shibata[3], Kenji Miyamoto[4], Kengo Iwanaka[1], Katsuya Hayashi[1], Akiyoshi Hatayama[1]

[1]Graduate School of Science and Technology, Keio University, Yokohama, Japan

[2]National Institute for Fusion Science, National Institutes of Natural Science, Toki, Japan

[3]Accelarator Laboratory, KEK, Tsukuba, Japan

[4]Graduate School of Education, Naruto University of Education, Naruto, Japan

[a)]Corresponding author: kato@ppl.appi.keio.ac.jp

Abstract. Result of hydrogen (H) and deuterium (D) experiments done by NIFS research and development negative ion source (RNIS) demonstrated that the co-extracted electron current with the negative ions and the electron density in the driver region in the D experiment have been around three times higher than that in the H experiment. To investigate mechanism of this difference, electron transport simulation using 3D kinetic particle tracking model KEIO-MARC code has been modified and applied to analysis of the isotope effect in the NIFS negative ion source. Simulation result suggests that impacts of isotope effects of sheath potential drop, coulomb collisions, and some reactions of ground state molecules and ions on the electron density is not large to explain the experimental result of the increase in the electron density in the plasma.

1. Introduction

Negative deuterium ion beam is utilized for Neutral Beam Injection (NBI), heating equipment of nuclear fusion plasma. Development of the deuterium negative ion source is crucial to achieve negative deuterium ion beam which is required to maintain high temperature fusion plasma with large volume. For the development of the deuterium NBI system, the difference of the beam characteristics between the negative hydrogen ion (H^-) and the negative deuterium ion (D^-) has been investigated by the device of National Institute of Fusion Science (NIFS), NIFS-Research and development Negative Ion Source (NIFS-RNIS) [1]. In this experiment, the electron density in the D plasma discharge has been three times larger than that in H plasma discharge. Additionally, in the extraction region, the electron density in the D plasma case was also three times larger than that in H plasma. The result demonstrates that the isotope effects of H and D plasma cause difference in electron density in the ion source plasma, which leads to the difference of the co-extracted electron current. Therefore, it is necessary to identify causes which lead to the isotope effects observed in the experiments, for example, the difference in the electron density in H and D plasma discharges.

In this paper, to analyze isotope effects in the ion source, electron transport simulation in the ion source is carried out using 3D kinetic particle tracking model KEIO-MARC code [2,3]. First, the KEIO-MARC code has been modified for the application to the NIFS-RNIS. It is confirmed that it can reproduce basic characteristics, such as electron flow by magnetic drift, filter effect, etc. Then, the isotope effects via sheath potential drop, coulomb collision, and some reactions of ground state molecules and ions are analyzed.

2. Simulation Model

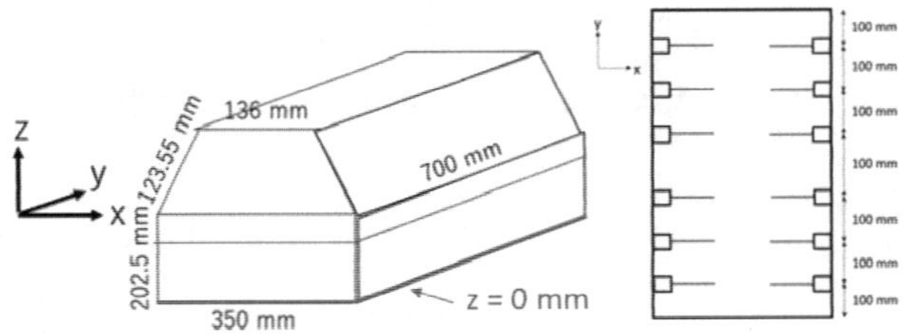

Fig 2.1 Schematic view of geometry. Twelve filaments are located at the surface of $z = 142$ mm [6]

Table 2.1 Input parameters of background plasma

Particles	Density [m^{-3}]	Temperature [K]
H_2	1.43×10^{19}	774
H^0	1.43×10^{18}	774
H^+	4×10^{17}	457
H_2^+	1×10^{17}	457
H_3^+	5×10^{17}	457

In the Kinetic modeling of Electrons in the IOn source plasmas by the Multi-cusp ARC-discharge (KEIO-MARC) code [2,3], equation of motion for electron is solved in three dimensions by the multiple Benemann-Boris solver [4]. By assuming quasi-neutral plasma in the chamber, electric field is assumed to be zero for the transport. The sheath potential at the chamber wall is modeled based on the standard formula [5]. When the test particle reaches the wall, electrons which have higher energy than the sheath potential are lost, and others are reflected. The model geometry in the KEIO-MARC code has been modified for 3D geometry of NIFS-RNIS shown in Fig. 2.1 and the real 3D magnetic configuration has been taken into account. Also, as shown in Fig. 2.1, 12 filaments are located at $z = 142$ mm. In the model, electron-electron and electron-ion coulomb collisions are treated by the Binary Collision Method [7]. Furthermore, reactions between the electron and the hydrogen species ($H, H_2, H^+, H_2^+,$ and H_3^+) are modeled by the Null Collision Method [8]. The density and temperature of the hydrogen species are given as input parameters, which are shown in Table. 2.1. Estimated parameters from measurements and other simulations are used in Table 2.1 for the sake of initial test calculation. The space inside chamber is divided into 646,800 cells with the size of $5 \text{ mm} \times 10.3 \text{ mm} \times 2 \text{ mm}$. The electron energy distribution function (EEDF) is calculated for each spatial cell. Other input parameters, such as arc discharge power, arc voltage, and H_2 gas pressure, are given as 83 kW, 82 V, and 0.3 Pa, respectively. The initial energy of primary electrons emitted from the filaments is set to be the same as the arc discharge voltage. Secondary electrons are produced by the ionization and traced as the same as the primary electrons.

3. Numerical Result

To NIFS-RNIS application, KEIO-MARC code has been modified and basic characteristics in the NIFS-RNIS is confirmed. Figure 3.1 shows the EEDF averaged at $z = 100$ mm plane. The peak at 82 eV corresponds with arc voltage. Figure 3.2 shows the spatial profile of electron density in x-y

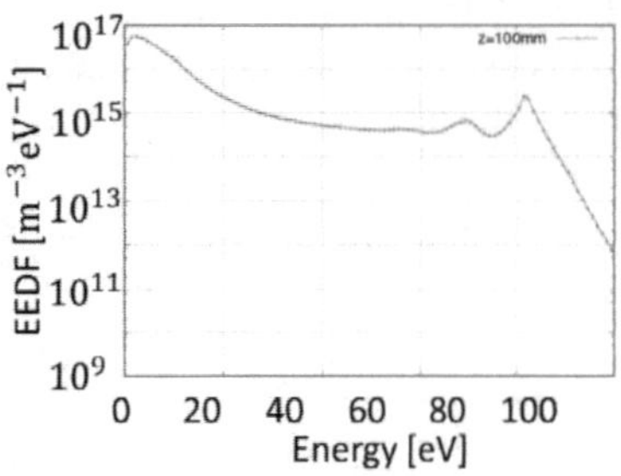

Fig.3.1 Electron Energy Distribution Function (EEDF) for the electrons in the $z = 100$ mm plane.

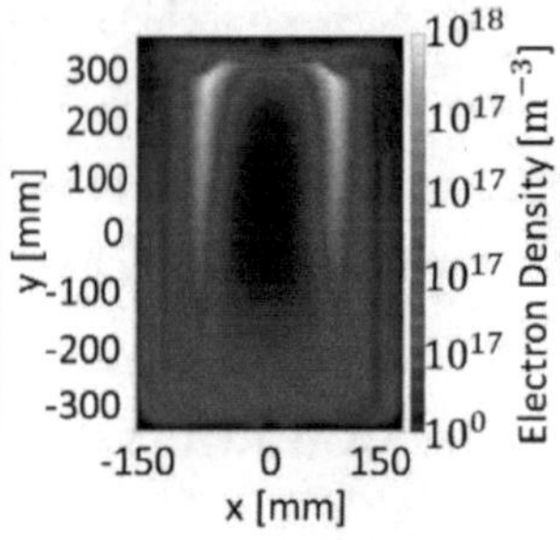

Fig 3.2 n_e distributions in *x*-*y* plane at $z = 142$ mm

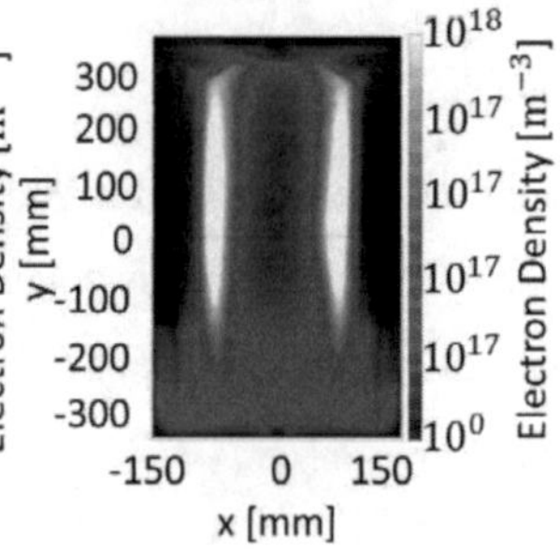

Fig 3.3 n_e distributions in *x*-*y* plane at $z = 142$ mm after control

plane at $z = 142$ mm (the same *x*-*y* plane as filament). Electrons tend to drift toward $+y$ direction due to the magnetic drift. This electron drift leads non-uniform beam. Therefore, in NIFS-RNIS experiments, arc current to each filament is adjusted to make the beam more uniform. In this study, the similar measure is implemented in the KEIO-MARC simulation, and the electron density profile has been improved as shown in Fig. 3.3. Figure 3.4 compares the ratio of electron production processes in the simulation. As electron production processes, primary electrons from 12 filaments and secondary electrons by 4 reactions shown in Table 3.1 are implemented in KEIO-MARC code. It is shown that injection from filament and non-dissociative ionization are the dominant electron production processes. Figure 3.5 shows the ratio of electron reduction processes. KEIO-MARC code takes into account loss at the wall, extraction from the extraction hole, and 3 reactions shown in Table

Table 3.1 Production reactions

Nondissociative Ionization	$e + H_2(X^1\Sigma_g^+) \rightarrow H_2^+ + 2e$	[10]
H(1s) Ionization	$e + H(1s) \rightarrow H^+ + 2e$	[11]
H_2^+Dissociation 1	$e + H_2^+ \rightarrow H^+ + H^+ + 2e$	[10]
Dissociative Ionization	$e + H_2(X^1\Sigma_g^+; v = 0) \rightarrow H^+ + H(1s) + 2e$	[9]

Table 3.2 Reduction reactions

Dissociative Attachment	$e + H_2(v) \rightarrow H + H^-$	[10]
H_2^+ Dissociation 3	$e + H_2^+ \rightarrow H(1s) + H^*$	[10]
H_3^+ Dissociative Recombination	$e + H_3^+ \rightarrow 3H(1s)$	[10]

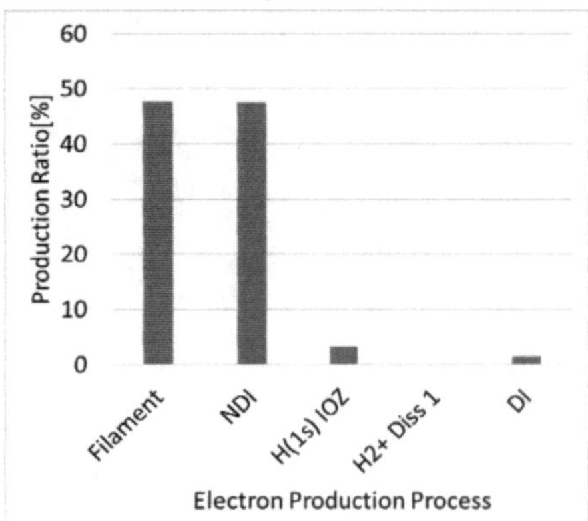

Fig 3.4 Composition ratio of electron production process

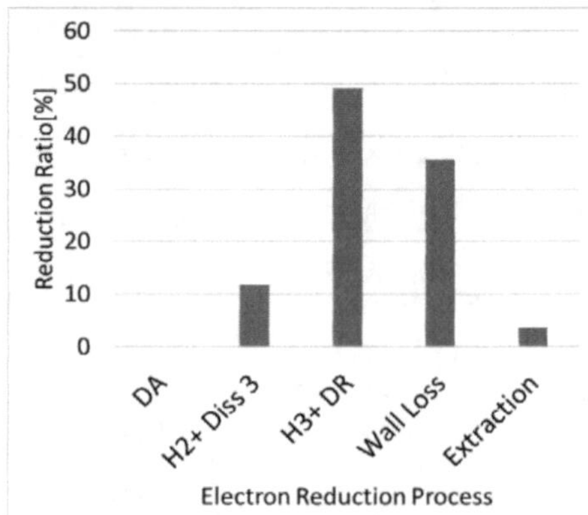

Fig 3.5 Composition ratio of electron reduction process

3.2. It is shown that H_3^+ dissociative recombination is the dominant electron reduction process, followed by loss at the wall.

After confirming similar n_e and T_e by calculation of updated KEIO-MARC code, in this study, isotope effects on sheath potential drop, coulomb collision, and 6 reactions are separately analyzed below.

For analysis of the isotope effect by the sheath potential drop, the mass of H^+ ion m_i in the following standard formula of sheath potential [5] is replaced to the mass of D^+ ion:

$$\Phi_{sh} = \frac{k_B T_e}{2e} \ln\left(\frac{m_i}{2\pi m_e}\right) \tag{3.1}$$

where Φ_{sh} is the sheath potential, k_B is Boltzmann constant, T_e is the electron temperature, e is charge of electron, m_e is the mass of electron. Due to the isotope effect on sheath potential drop, the ratio of electron density averaged in the chamber increased by 4.41% compared to hydrogen condition, and the ratio of electron temperature decreased by 0.43%. From Eq. 3.1, the difference of sheath potential would only be around 12%. Therefore, electron density is not expected to significantly increase as observed in the experiment.

For Coulomb collision, the mass of collision partner, H^+ ion, is replaced to the mass of D^+ ion. By changing the ion mass in the Coulomb collision, diffusion process, especially in the filter region, may change, and then change of the density is expected. However, simulation result shows that changes of the electron density and temperature are only +0.05% and +0.24%, respectively.

Table 3.3 Isotope effect on reactions

Reaction		Ref	Electron Density Ratio	Electron Temperature Ratio
DI	$e + H_2(X^1\Sigma_g^+; v = 0) \rightarrow e + H^+ + H(1s)$	[9]	99.81%	100.09%
Nondiss IOZ	$e + H_2(X^1\Sigma_g^+) \rightarrow e + H_2^+ + e$	[10]	99.40%	100.94%
DA	$e + H_2(v) \rightarrow H + H^-$	[10]	99.96%	100.02%
eV①	$e + H_2(X^1\Sigma_g^+; v = 0) \rightarrow e + H_2(X^1\Sigma_g^+; v = 1)$	[9]	97.38%	97.66%
eV②	$e + H_2(X^1\Sigma_g^+; v = 0) \rightarrow e + H_2(X^1\Sigma_g^+; v = 2)$	[9]	100.25%	99.01%
e Excitation	$e + H_2(X^1\Sigma_g^+) \rightarrow e + H_2^*(B^1\Sigma_u^+)$	[10]	105.07%	100.51%

The isotope effect, i.e., effect of cross section and mass on the molecular processes, shown in Table 3.3, is evaluated. Cross section and mass of collision partner for each reaction is replaced one by one, to those of deuterium from hydrogen. In KEIO-MARC, the atomic and molecular processes more than 20 are considered. However, here, important 6 reactions such as dissociative ionization, non-dissociative ionization, dissociative attachment, eV reaction from $v = 0$ to $v = 1$, eV reaction from $v = 0$ to $v = 2$, and electron excitation, are evaluated. The result for change of the total amount for each reaction is also summarized in Table 3.3. The results are shown as a percentage to total amount of the reaction (in the chamber) for the reference case of H+ ion. Molecular processes are changed by the partner cross section and mass, but its impact on the electron density is very small.

4. Discussion and Future Works

As shown in results above, increase of electron density caused by the isotope effects investigated in this paper are significantly small compared with the experimental data. Other factors for the isotope effects needs to be considered to make clear the isotope effect which leads to the increase in electron density observed in D+ plasma experiments. For example, difference of energy loss through each reaction, isotope effect of other reactions, such as H_2 vibrational excitation, possibly affect the electron density. In the present KEIO-MARC code, the effect of local electric field and its effect on the transport process are neglected. More specifically, effects such as the Short-Circuit effect [12] and/or the effect of the ambipolar electric field on the plasma transport across and along the magnetic field may also be important.

In addition, difference of vibrationally excited levels between H_2 and D_2 can have an important role, which is not considered in the present KEIO-MARC model. Using zero-dimensional particle balance model, the isotope effect, especially focused on the atomic and molecular processes has been investigated [13]. The results show the increase of the electron density by 65%. One of the reasons for increase in the electron density is difference of vibrationally excited levels of molecules (H: 0 ~ 14, D: 0 ~ 20). In the D case, the ionization channel number via vibrationally excited levels increases and therefore the effective ionization rate is enhanced.

Acknowledgements

This work is performed with the support and under the auspices of the NIFS Collaboration Research Program (NIFS21KLER110).

Reference

[1] H. Nakano, *et al.*, Jpn. J. Appl. Phys, **59** SHHC09 (2020).

[2] T. Shibata, *et al.*, J, Appl, Phys, **114** 143301 (2013).

[3] A. Hatayama, *et al.*, New J. Phys., **20** 065001 (2018).

[4] S. Zenitani, T. N. Kato, Comput. Phys. Commun., **247** 106954 (2020).

[5] R. Terasaki, I. Fujino, A. Hatayama, T. Mizuno, and T. Inoue, Rev. Sci. Instrum. 81 02A703 (2010).

[6] H. Nakano, private communication

[7] K. Nanbu, S. Igarashi, and Y. Watanabe, *Proc. Soviet Union-Japan Symp. Comput. Fluid Dynamics*, (Computing Center of the USSR Academy of Sciences, 1989), Vol.2, pp.126-132

[8] G. A. Emmert, *et al.*, Phys. Fluids **23**, 803 (1980).

[9] R. Celiberto, *et al.*, Atom. Data and Nucl. Data Tables **77**, 161 (2001).

[10] R. K. Janev, D. Reiter and U. Samm, Julich Report No. JUEL-4105 (2003).

[11] L. E. Johnson, Astrophys. J. **174**, 227 (1972).

[12] A.Simon, Phys. Rev., **98** 317 (1955).

[13] T. Shibata, *et al.*, **68** in this conference

Beam Formation Studies on the CERN IS03b H⁻ Source

J Lettry[1], A Vnuchenko[1], S Bertolo[1], C Mastrostefano[1], M O'Neil[1], F di Lorenzo[1], Y Coutron[1], D Steyeart[1], B Riffaud[1], J Thiboud[1], R Guida[1], K Kapusniak[1], C Charvet[1], B Teissandier[1], P Moyret[1], F Roncarolo[1], S Bart Pedersen[1], M Duraffourg[1], C Vuitton[1], U Fantz[2], S Mochalskyy[2], D Wunderlich[2], M Lindqvist[2], N den Harder[2], A Mimo[2], S Briefi[2], A Hurlbatt[2], T Kalvas[3], T Minea[4], A Revel[4]

[1] CERN, Esplanade des Particules 1, P.O. Box, 1211 Geneva 23, Switzerland
[2] Max-Planck-Institut für Plasmaphysik, Boltzmannstr. 2, D-85748 Garching, Germany
[3] Department of Physics, P.O. Box 35 (YFL), 40014 University of Jyväskylä, Finland
[4] Laboratoire de Physique des Gaz et des Plasmas, UMR 8578 CNRS, Université Paris-Saclay, Bat. 210, rue Henri Becquerel, 91405 Orsay, France

Jacques.lettry@cern.ch

Abstract. An H⁻ ion source is being operated at the new 160 MeV linear injector (Linac4) of the CERN accelerator complex. The source's plasma is of the Radio Frequency Inductively Coupled Plasma type (RF-ICP), without magnetic cusp and runs with Cs-loss compensation [1]. Vertical downward oriented filter- and electron dump-dipolar magnetic fields expand over the plasma chamber, beam-formation, beam-extraction and electron dump regions and generate horizontal asymmetry and beam angular deflection partially compensated by mechanical alignment of the front-end. The H⁻ beam is generated via volume and caesiated plasma surface modes, the latter inducing a radial asymmetry characterized by an increased current density close to the plasma electrode surface [2]. Asymmetries affecting the meniscus shape, or its current density have to be simulated via 3D Particle In Cell Monte Carlo (PIC-MC) solvers, such as the Orsay Negative Ion eXtraction code (ONIX) [3]. Validation of these simulations require dedicated measurements. This contribution describes experimental methods newly implemented at CERN's ion source test stand and initial results for Optical and Beam Emission Spectroscopy (OES, BES), emittance and beam profile measurements. In a later stage, the gathered data sets can be compared to source plasma parameters and extracted beam parameters from PIC-MC simulations, once coupled to the Ion Beam Simulation (IBSimu) [4] beam transport code. The experimental parameter space includes RF-power, density of neutrals, position of the RF coil and extraction field. Beams of H⁻, D⁻ and protons were produced; examples of measured data are presented in this contribution.

1. Introduction

H⁻ sources for Linac4 are designed to generate a specified beam current within RFQ-acceptance. The beam optics is iteratively optimized with IBSimu runs of variable electrodes and co-extracted electron beam dump (e-dump) geometries, extraction field, and e-dump magnetic field. IBSimu simulates electron and ion beams, encompasses an electron impact energy density map essential for e-dump design and enables tracking of secondary electrons. The plasma parameters, the total ion and electron currents extracted through the meniscus input to IBSimu were used to design the IS03b H⁻ source operated at Linac4 and presented in figure 1.

RF-ICP heating of the plasma was simulated with the 2.5D PIC-MC code NINJA [5] and validated by optical emission spectroscopy (OES) and photometry [6,7]. However, high-resolution 3D PIC-MC simulation is mandatory to resolve the physics in the vicinity of the meniscus, plasma sheath and caesiated plasma electrode. Neutral Beam Injector (NBI) for fusion pioneering work [8,9] brought deep insight into the processes resulting from combined volume and plasma surface H^- ion production and identified halos and the negative potential well at the origin of the massive reduction of co-extracted electrons observed under Ceasiated Surface Plasma (CSP) production mode. Initial ONIX runs [10] on the Lina4 source illustrated the phase space differences between volume and surface emitted ions. When compared to NBI-sources applications, radial boundary conditions in ONIX must be modified to match single aperture conditions, furthermore, the larger plasma density of Linac4 sources (few 10^{18} m^{-3}) is challenging. Extension of ONIX to single aperture is required to simulate the Linac4 source, preliminary low-density results are presented in [11].

The dual-origin of H^- ions, the filter-field induced asymmetry and the expected radial variation of ion current-density along the meniscus are likely detrimental to beam properties, i.e. presence of beam halos. Beam formation studies around the meniscus region is the link between RF-ICP-plasma heating and beam transport that determines the beam properties (position and velocity). Ions and electrons extracted from ONIX simulations a few mm after the meniscus, will be injected into IBSimu and transported to beam diagnostics locations. This contribution presents experimental methods to refine plasma parameters input to ONIX, to complement the Linac4 test-stand slit-grid beam emittance diagnostics with a new 2D beam profile monitor and to measure BES based angular distribution [12]. Setups and illustrative examples of the results are presented in the following sections.

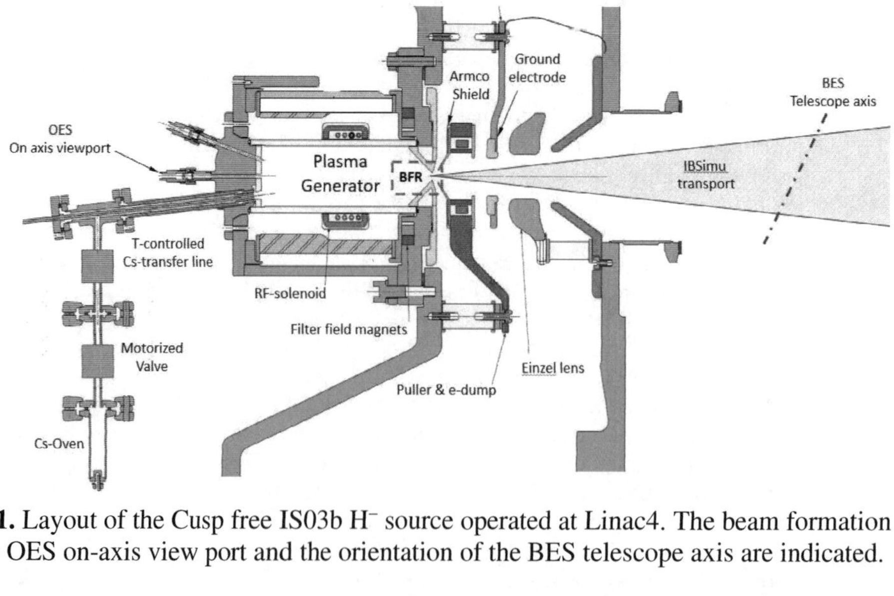

Figure 1. Layout of the Cusp free IS03b H^- source operated at Linac4. The beam formation region, the OES on-axis view port and the orientation of the BES telescope axis are indicated.

2. Plasma electrode prototype and Optical emission spectroscopy

A new plasma electrode (PE) geometry was produced (PE75) to provide simple metallic radial boundary conditions at the periphery of the beam formation region (BFR) of the ONIX simulation region. Black anodized (20 μm) Al-mockups of the prototype and standard plasma electrode (PE45) are produced and equipped with sets of mirrors and focusing lens to an optical fiber. The IS03 plasma chamber equipped with plasma electrode mockups can be operated without high voltage, the collisional radiative analysis of the light captured via this view port shall provide insight to plasma parameters, at the location of the meniscus. BFR-plasma will be compared to ONIX simulations with no extraction field. The new plasma electrode geometry and OES-mockups are presented in figure 2, the light is collected at 7 and 3.5 mm (PE75-mock-up) and 4 mm (PE45 mock-up) upstream from the PE-aperture.

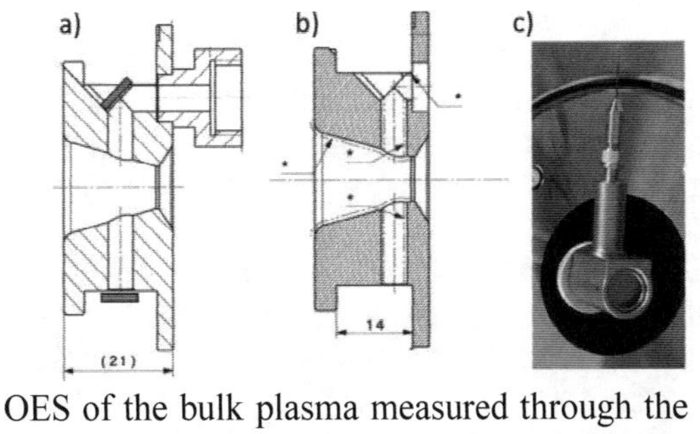

Figure 2. OES-mock-ups of the PE75 plasma electrode, 5 mm diameter view port are located: a) 7 mm and b) 3.4 mm upstream the plasma electrode aperture. Flat and 45 deg. mirrors illustrated in a) guide the plasma light to a lens focussing it onto an optical fibre c). The 20 μm black anodized surfaces are indicated with a star in b). The angle of the PE75 inner conus is 30 deg.

OES of the bulk plasma measured through the on-axis view ports of the plasma generator provides a reference point to previous work [5,6]. For each beam setting, 15 wavelength spectra are recorded to cover the Fulcher band (~200 H_2 molecular transition lines) and 4 atomic Balmer lines $H_{\alpha-\gamma}$. After noise subtraction, the spectra are corrected to account for the spectrometer calibration. The leakage of light from the bulk plasma to the beam formation detector amounts to approximately 1%.

3. Beam emission spectroscopy

A fraction of the H^- ions propagating through the space charge compensation gas is neutralized, the excited atoms maintain direction and velocity and, when observed in the laboratory's rest frame, the wavelength of its H_α transition is Doppler shifted according to its velocity and angle of observation. The H_α spectrum of the light collected from a telescope oriented vertically at an angle of 62°-65° and pointing to the beam axis can therefore be converted into an angular distribution [12]. Due to our low duty factor, a few hours acquisition and a residual H_2 pressure of 2×10^{-2} Pa are necessary to obtain clean signals. The telescope viewing zone partially overlaps with the beam, extracting the BES signal from a simulated beam does not require any assumption, however, cylindrical symmetry is necessary to reconstruct a beam from an experimental BES angular distribution. BES angular distributions are presented in figure 3, effects of variating the extraction field (3.a) for a constant plasma heating power and of increasing beam intensity (3.b) under fixed extraction field on the beam angular distributions are shown, the difference between volume and CSP modes operated D^- is illustrated (3.c).

The beam halo is larger for the D^- cesiated surface production as a likely result of the fraction of the beam originating form the cesiated surface at the vicinity of the plasma electrode aperture, however this remains to be demonstrated via simulations of ONIX beam formation folded with IBSimu transport to the BES detection region.

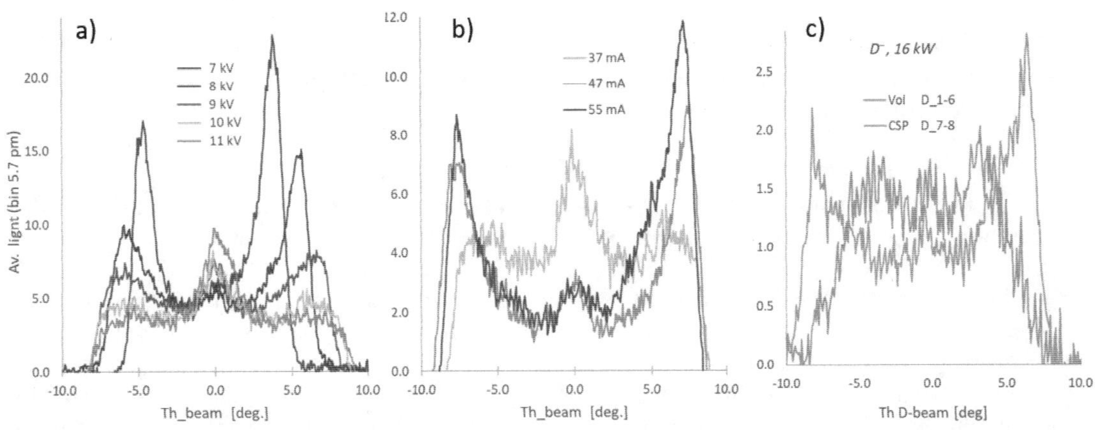

Figure 3. Beam emission spectroscopy angular distributions: a) CSP H^- beam at variable extraction field and b) CSP H^- beam at variable beam intensity, c) D^- beam in volume and CSP production modes.

4. Beam Emittance and profile

The Slit Grid emittance meter operated at the ion source test stand delivers velocities and positions integrated along parallel slit and wires. Assumptions on the beam profile are mandatory to reconstruct the beam phase space. The active surface 30×30 mm^2 is smaller than the beam and focusing is required to match the beam to the detector. The detection of positive charges on the wires is achieved with a bias voltage extracting secondary electrons, however, a negative signal is observed around each emittance measurement and neglected in the beam analysis. Operating the slit and wires in perpendicular orientation provides insight to the beam current density distribution, as the beam is converging, its image on the grids (located 200 mm downstream) is smaller than on the slit. Beam profile at the slit plane is reconstructed from two drift corrected scans of the horizontal (resp. vertical) slit across the beam monitored via vertical (resp. horizontal) grid. In figure 4 the profile of a 40 mA H^- beam is presented, the reconstructed picture is neither cylindrical nor axis symmetric, the observed shape cannot be extracted from emittance scans and shows a strong current density inhomogeneity. The larger radial current density is partially due to optics non-linearities, our long-term goal is to understand whether part of it originates from the beam formation process. Beam profiles can be compared to simulations. Future beam phase space reconstruction in view of beam transport (usually based on emittances) should account for the observed inhomogeneous current density.

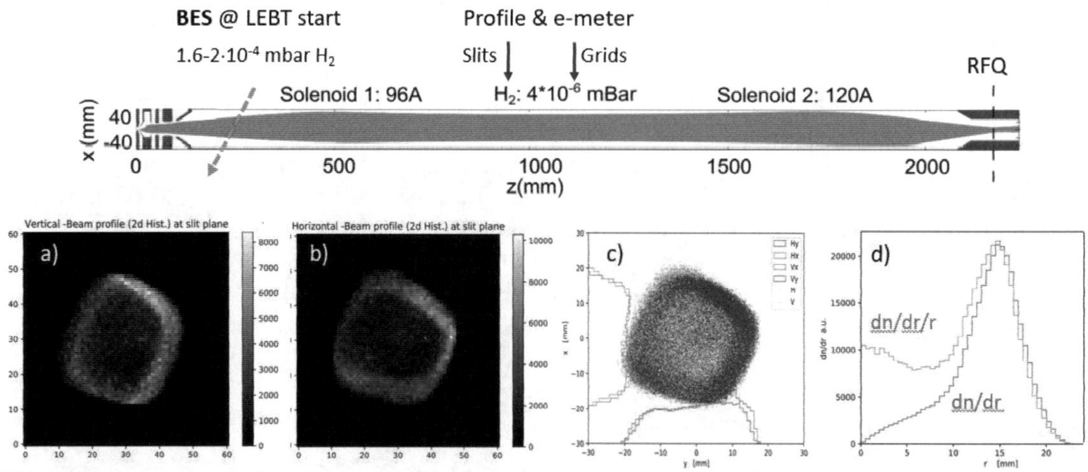

Figure 4. Beam simulation of Linac4's low energy beam transport section [13], the location of the BES telescope and of the slit-grid profile and emittance meter are indicated. H^- Beam profile at the slit plane measured via a) vertical- and b) horizontal-slits scans. c) scatter plots and projection histograms on horizontal and vertical axis and d) average radial beam current and current density distributions.

5. Conclusion and outlook

Volume and CSP production modes coexist in caesiated H^- sources. For permanent magnet filter field configuration, the plasma components' temperatures and densities, ion-current and ion to co-extracted electron ratio scale to the plasma RF-power; their contribution to the beam position and velocity distribution is complex and often reduced to a set of IBSimu parameters fitted to measurements. Furthermore, immediately after extraction through the meniscus, beam optics effects may be predominant at high current (i.e. non-linearities far from beam axis) and are difficult to disentangle from BFR induced asymmetries. Comparing volume and CSP production modes for identical plasma density is complexified, as co-extracted electrons drop by a factor around 20 while H^- beam intensity is doubled. Beam profile now provides a mean to cross check and possibly improve previous beam transport simulations. Plasma densities above 10^{18} m^{-3} are expected in Linac4 plasma generator and currently beyond reach of available superclusters. Simulation of the beam formation region without direct coupling to NINJA relies on complex OES calibration and production of dedicated prototypes for which radial boundary conditions are well defined.

BES shows a clear response to plasma and extraction field parameters, the low duty factor of the source implies multi-hours acquisition and, in the future, should be performed along horizontal and vertical axis to highlight asymmetries. Plasma electrode geometry will be developed to improve the match to ONIX's simulation domain. XRF based Cs-coverage measurement method is developed [14], a calibration down to 0.01 μg enables to measure Cs-coverage, averaged on the inner surface of the plasma electrode, down to sub monolayer thickness. An experimentally demanding correlation between H^- beam yield, electron suppression and Cs-coverage would provide insight to CSP core physics.

References

[1] Lettry J, Bertolo S, Fantz U, Guida R, Kapusniak K, Di Lorenzo F, Machado C, Mastrostefano C, Minea T, O'Neil M, Neupert H, Noll D, Steyaert D, and Thaus N *AIP Conference Proceedings* 2052, 050008 (2018); https://doi.org/10.1063/1.5083762

[2] Ichikawa M, Kojima A, Chitarin G, Agostinetti P, Aprile D, Baltador C, Barbisan M, Delogu R, Hiratsuka J, Marconato N, Nishikiori R, Pimazzoni A, Sartori E, Serianni G, Tobari H, Umeda N, Veltri P, Watanabe K, Yoshida M, Antoni V and Kashiwagi M *AIP Conference Proceedings* 1869, 030024 (2017); https://doi.org/10.1063/1.4995744

[3] Mochalskyy S, *Doctoral thesis*, Université Paris Sud - Paris XI, 2011, theses.fr – Serhiy Mochalskyy , Modeling of the negative ion extraction from a hydrogen plasma source : application to ITER neutral beam injector

[4] Kalvas T, Tarvainen O, Ropponen T, *et al*, IBSimu, *Rev. Sci. Instrum.* 81, 02B703, (2010).

[5] Mattei S, Nishida K, Onai M, Lettry J, Tran M Q and Hatayama A *J. Comput. Phys.* (2017), http://dx.doi.org/10.1016/j.jcp.2017.09.015.

[6] Briefi S, Mattei S, Lettry J, and Fantz U NIBS Oxford, 2016, *AIP conference proceedings* 1869, 030016.

[7] Briefi S, Mattei S, Rauner D, Lettry J, Tran M Q and Fantz U 2017 *New J. Phys*. 19 105006

[8] Montellano I M, Wünderlich D, Mochalskyy S and Fantz U *J. Phys. D: Appl. Phys.* 52 (2019) 235202 (15pp), https://doi.org/10.1088/1361-6463/ab0f44

[9] Miyamoto K, Okuda S and Hatayama A 2012 Appl. Phys. Lett. 100 233507

[10] Mochalskyy S, Lettry J and Minea T 2016 *New J. Phys*. 18 085011

[11] Vnuchenko A, Lettry J, Mochalskyy S, Wünderlich D, Fantz U, Revel A, Minea T, *Simulation of Beam Formation in the CERN Negative Ion Source for the Linac4 Accelerator*, These proceedings ICIS-Victoria 2021.

[12] Hurlbatt A 2019 *J. Phys. D, Appl. Phys*. 10.1088 1361-6463 ab6145.

[13] Fink D A, Kalvas T, Lettry J, Midttun Ø and Noll D *Nuclear Inst. and Methods in Physics Research*, A 904 (2018) 179–187.

[14] Ory J, Charvet C and Teissandier B https://edms.cern.ch/document/2469324/1, https://edms.cern.ch/document/1555439/2

Design and first operations of a ECR based He^- source at INFN-LNS

M. Mazzaglia, A. S. Amato, G. Battiato, G. Calabrese, L. Celona, L. G. Cosentino, L. Dantoni, S. Gammino, G. Manno, S. Marletta, A. Massara, A. Maugeri, G. S. Mauro, A. Miraglia, E. Naselli, S. Passarello, A. Pidatella, M. Piscopo, A. D. Russo, A. Seminara, D. Siliato, G. Torrisi, D. Tudisco, G. Vecchio, S. Vinciguerra and D. Mascali

INFN-Laboratori Nazionali del Sud, via S. Sofia 24, 95123 Catania, Italy

A. Galatá

INFN Laboratori Nazionali di Legnaro, Viale dell'Universitá 2, 35020 Legnaro (PD), Italy

E-mail: maria.mazzaglia@lns.infn.it

Abstract. A new source for the TANDEM accelerator of LNS has been designed and installed. It is called NESTOR (Noble Elements Source for acceleraTORs) and consists of an ultra-compact ECR microwave discharge type ion source [1] operating around 6 GHz and up to 40 W of RF power, provided by a solid state power amplifier, coupled to a Li-Charge Exchange Cell (Li-CEC). It is engineered for the production of a wide range of 1^+ and/or 1^- ion beams from gaseous elements, in particular for noble gases. This work presents the characterization of the primary source and first operations of the whole setup on the HV platform (injector) of the Tandem. The He^+ beams have been formerly characterized in terms of current, beam shape (by BaF_2 beam viewers) and emittance (by the three-gradients method). Measurements have been carried out varying pressure, microwave frequency and RF power. Then, the source has been moved to the HV platform, coupled to the Li-CEC for first operations running in gas-exchange mode. Activities are ongoing to optimize beam transport towards the Tandem.

1. Introduction

An ECR source of MDIS (Microwave Discharge Ion Sources) type coupled to charge exchange cells may represent a remarkable innovation in the field of sources used for the production of negative ion beams. Some attempts have been done by ECR groups in the recent years [2, 3, 4]. Unlike duoplasmatron or RF sources, which coupled to charge exchange cells are able to produce only light ion beams, ECR sources allow the production of a wide range of (both light and heavy) ion beams, in particular beams of noble element ions. Moreover, ECR sources are versatile and reliable and they are typically cheaper and more compact than the others kind of sources above mentioned. The robust downsizing of the MDIS (guaranteed by the properly designed frequencies and magnetic fields) allow both extreme compactness and the optimization the coupling of electromagnetic energy to the plasma, and therefore the optimization of the overall efficiency of the system. This kind of systems can be really useful to broaden the portfolio of beams available at nuclear physics facilities. In particular, they are quite attractive for Tandem accelerators, which are largely used worldwide for performing nuclear physics and nuclear astrophysics experiments. The demand of, among the others, He^- beams is steadily increasing since these nuclides are responsible of a number of nuclear reactions of astrophysical interest (see for example [5, 6, 7]). In addition, intense beams of noble gases are also used in applied research and industry, e.g. for implantation in semiconductors.

2. Source design

The NESTOR (Noble Elements Source for acceleraTORs) setup is based on an ultra-compact ECR ion source (whose total size, including ancillaries such as the high power isolator/circulator system, is of 30x30x30 cm^3) coupled to a Li-based Charge Exchange Cell (hereinafter, CEC). Figure 1 shows the configuration of the ECR ion source and CEC. In details, the source includes:

- the plasma chamber (56 mm diameter and 61 mm length);
- the permanent magnet providing in-plasma electron cyclotron resonance and a weak plasma confinement ($B_{max} = 0.28T$);
- the primary beam extraction system (20 kV);
- the frequency variable RF generator, operating in the range 5.85 - 6.425 GHz, consisting of a solid state power amplifier, with 40 W of maximum power;
- the microwave injection line (waveguide);
- a water-based cooling system (both of the plasma chamber and charge exchange cell);
- the gas injection system and pipes (He, hitherto).

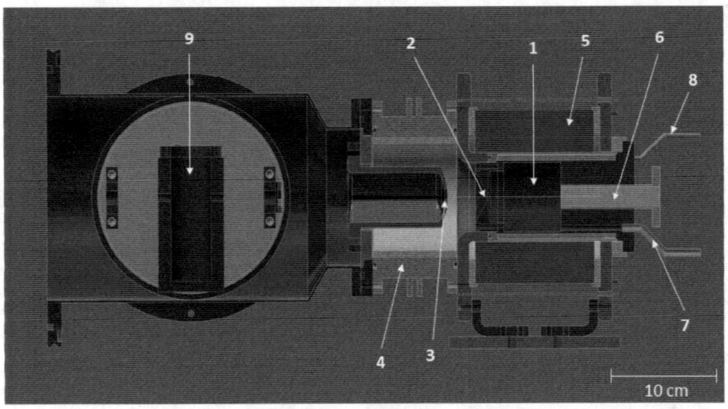

Figure 1. NESTOR sketch, including the ECR ion source and CEC: 1-plasma chamber, 2-plasma electrode, 3-ground electrode, 4-insulation column, 5-permanent magnet, 6-waveguide, 7-gas injection, 8-water cooling, 9-CEC.

The CEC has been manufactured by the High Voltage Engineering Company. The positive ion beams are produced by the ECR source and then pass through the CEC which contains high temperature (≈ 600 °C) Lithium vapours. The incident positive ions can likely pick up electrons by means of multi-several scattering processes with Li atoms, whereby a fraction of them emerges with a net negative charge. Overall charge exchange efficiency is expected in the order of 1-1.5 % [8].

3. Installation setups

Figure 2 shows the general layout of the primary source during the commissioning, when it was installed on a dedicated testbench. Also the beamline components are shown in the picture (they were only used for primary beam characterization, but are not installed on the HV platform of the TANDEM). In the beamline two Faraday Cups, two CCD cameras associated with two beam viewers and one quadrupole were installed. The diagnostics has been temporarily used to characterize the primary source in terms of current, beam shape and to perform some approximated estimations of the beam emittance by using the so-called three-gradients method [9].

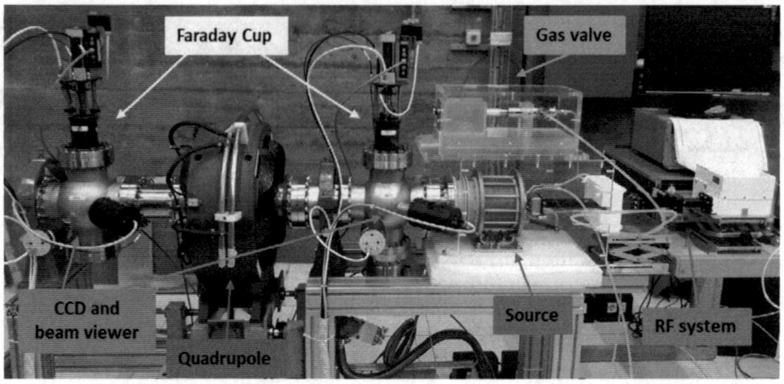

Figure 2. ECR ion source and beam line layout.

4. Primary source commissioning

The beam commissioning of the primary source started in May 2020. NESTOR was operated at 20 kV of extraction voltage, with a beam extraction hole of 6 mm in radius. The characterization of the He^+ beam has been done by means of a Faraday Cup and a system consisting of a quadrupole associated with a beam viewer in order to perform the emittance measurements by the mentioned three-gradients method.

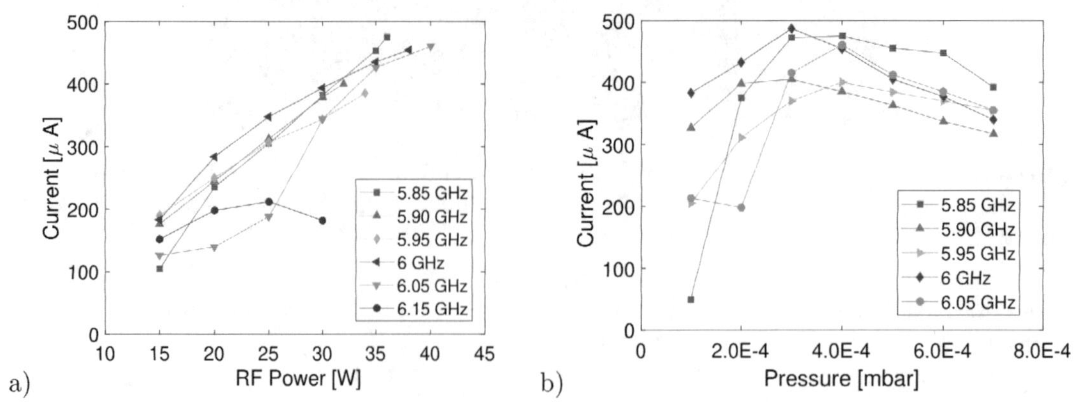

Figure 3. (a) Extracted current as a function of RF power and frequency at constant pressure ($4 \cdot 10^{-4}$ mbar). (b) Extracted current as a function of pressure and frequency at constant RF power (38 W).

Measurements have been carried out in different experimental conditions, i.e. varying pressure, microwave frequency and power. Figure 3 (a) shows the extracted current, in μA units, as a function of RF power (at step of 5 W) and frequency at constant pressure ($4 \cdot 10^{-4}$ mbar). The plot displays a clear increase of current vs. RF power, as expected, without appreciable saturation. The only saturated curve was the one corresponding to 6.15 GHz. In this latter configuration, was not possible to operate SSA at RF power larger than 30 W, due to a very bad matching. We attempted to calculate the linear correlation coefficient R for evaluating the degree of correlation of current versus the RF power, for six different values of frequency, from 5.85 GHz to 6.15 GHz at steps of 0.05 GHz. The plot shows that during the power scaling the data were very well linearly correlated with almost all the frequencies (very high correlation with $0.97 < R < 1$, p-value < 0.0014). The best performance in terms of output currents were obtained at 6 GHz of input microwave frequency.

a)

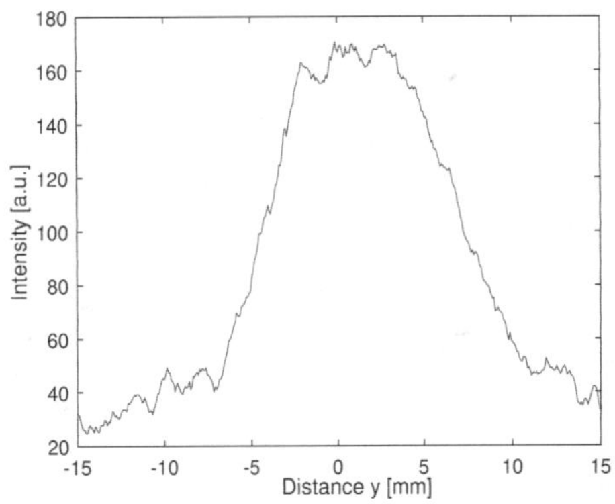

b)

Figure 4. a) Beam image. b) Beam profile.

The extracted current as a function of pressure and frequency at constant RF power (38 W) is shown in figure 3 (b). It can be observed that the extracted current increased remarkably at the lower pressures, then saturating and even slightly decreasing. The optimum was reached in the range $3 \div 4 \cdot 10^{-4}$ mbar. These data demonstrate that there are margins of improvements in terms of absolute output currents, operating at slightly higher pressure but with higher RF power as well. However, current levels observed already in this configuration are deemed to be very promising for the routine operations, in light of the expected, aforementioned charge exchange efficiencies. Hence we opted to maintain a so compact solid state power amplifier system, despite the low power, which was however suitable to reach a maximum measured current of 487 μA at 6 GHz of microwave frequency. A further, finer tuning of the frequency and pressure, with a power reaching the limit of the amplifier (40 W), allowed to reach 1 mA of current during the successive tests on the high voltage platform of the Tandem.

After the investigation of the main source performance scalings with power and pressure, we analyzed the beam shape and overall properties. Figure 4 (a) shows the beam image obtained by BaF_2 beam viewers at 6 GHz of microwave frequency and 35 W of microwave power. A tablet of BaF_2 powder has been realized at the INFN-LNS chemical laboratory and deposited through evaporation on a 40 mm $\times$ 40 mm square aluminum plate 1 mm thick. A series of holes of 1 mm diameter, 5 mm apart from each other, has been drilled on the plate before evaporation, in order to have a reference frame. The beam profile is shown in figure 4 (b).

Beam images were taken especially to evaluate the matching in the real space of the primary beam with the CEC. The latter, has an inner diameter of about 6 mm. Beam images and profiles where thus measured exactly where the CEC input hole is planned to be positioned in the final, operative setup. The observed beam size FWHM value was around 12 mm. Considering this value, compared with the absolute value of the primary current, moderate losses of beam halos were considered to be tolerable, and we opted to avoid any beam optics elements in between the ECR source and the CEC, in order to keep the overall system as much compact as short as possible. Despite some inefficiency, this solution allowed to keep really compact the overall setup, matching with the actual size of the high voltage platform. We then attempted to estimate the emittance by the three-gradients method. Results are reported in Table 1.

Beam intensity [μA]	ϵ_x[π mm mrad]	ϵ_y[π mm mrad]
480	14.13	15.07
420	13.24	14.05
345	11.82	13.30
260	11.26	11.89

Table 1. Emittances calculated with the three-gradient method.

Measurements of the emittance in the horizontal and vertical planes were done with quadrupole currents ranging from 0.25-6 A for four different beam intensities. Absolute values were considered to be compliant with design parameters and the target transport efficiency.

5. Status of NESTOR at INFN-LNS and ongoing activities

NESTOR has been installed in early September 2020 on the 450 kV platform for the production of noble gases negative ions for the Tandem Accelerator. Figure 5 shows the sketch of NESTOR installation and beamline.

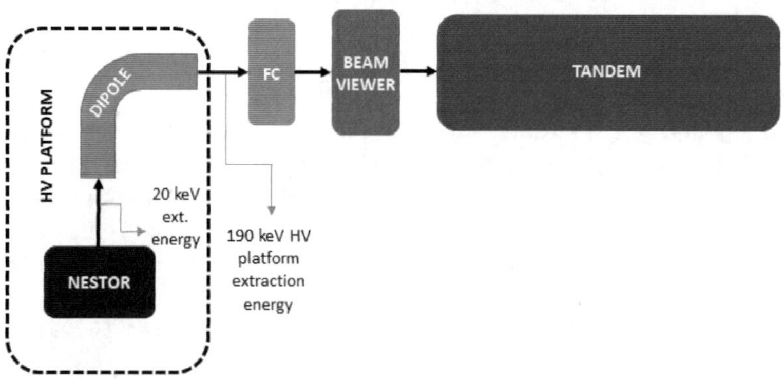

Figure 5. Sketch of NESTOR installation and beamline, including the HV platform and beam diagnostics.

Preliminary commissioning tests were carried out in September-October 2020. First He^- beams were extracted and transported till the last FC before the Tandem accelerator. Measured current was of only few nA, while the target is to get a range of 100 nA-1μA. The main criticality was found to be in the beam transport system. The beam optics elements remained exactly the same of the sputtering sources setup, in place of which NESTOR has been installed. This is of course requiring a major optimization due to the different type of beams and the slight changes operated in the beamline settings, and, especially, dimensions and sizes. To address this task, beam optics simulations are ongoing, by considering for the electrostatic focusing the *Einzel lens* placed inside the beam transport line. The challenging goal is to have a flexible and versatile setup, able to swap and replace the sources each others, depending on the type of beams required, in short times and without significative worsening of the overall efficiencies. Figure 6 a) shows the CST Studio Suite 3D model of the NESTOR setup, that is currently

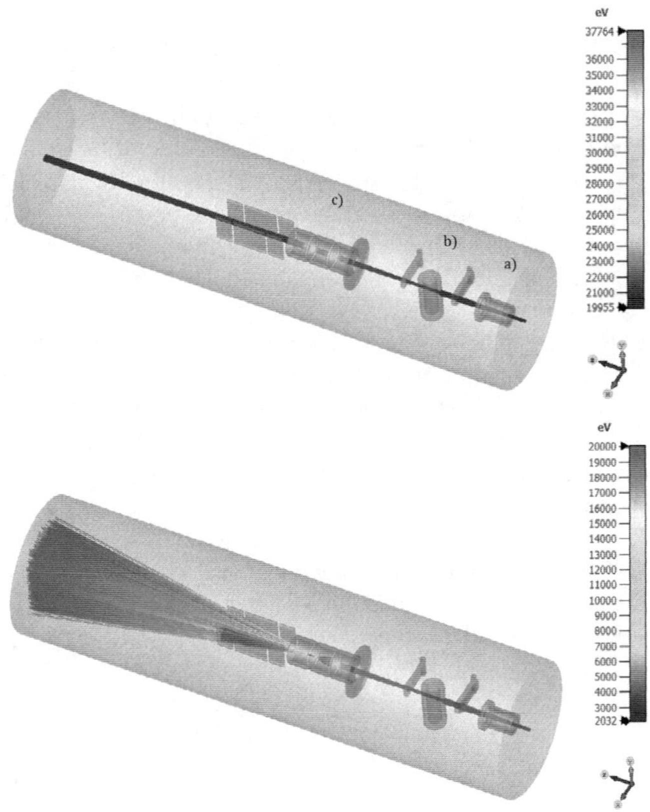

a)

b)

Figure 6. a) He^- beam dynamics including focusing effects of an Einzel lens at 20 kV. b) He^+ beam dynamics including focusing effects of an Einzel lens at 20 kV.

employed for particle tracing simulations. In particular, from right to left, we have: a) the extraction aperture, with diameter equal to 6 mm, from which the particles are launched, b) the charge-exchange cell, that in this simulation acts only as a collimator, c) the three Einzel lenses (along with the suppressor plate), with 20 kV potential set at the central electrode, the other two being grounded and d) the set of steerer plates, which are set to off in the simulation. The simulation domain, in which the particles travel, is a vacuum box with radius and length sufficiently large to observe the particle trajectory behavior. The beam initial energy has been set equal to 20 keV. From the simulation it can be seen that, in the case of negative charged beam and positive potential at the Einzel lens central electrode, the lenses have almost no effect on the beam, which is transported from the start to the end of the computational cylinder with beam losses only when it encounters 'stop-walls', i.e. when it passes inside holes with acceptance lower than the one of the beam itself. However, from figure 6 b), it can be seen that one needs to be careful if a positively charged beam is employed. In fact in the latter case the effect of the lenses on the beam in terms of focal point generation is much more pronounced and the central electrode potential needs to be properly adjusted in order to avoid beam blow-up. The HV platform undergone to major maintenance in the first half of 2021 (vacuum and control system, especially). Further tests (including acceleration through the Tandem) will run in early

2022, lasting until the major shutdown of LNS facilities due to Cyclotron upgrade (POT-LNS project). According to the simulations results, lens electrodes configuration (in terms of both sizes and applied voltages) will be optimized.

6. Conclusion

In this paper we have reported about the design, construction and first experimental test of the NESTOR source of negative noble elements. It has been conceived to be the new source for the TANDEM accelerator of the LNS, in addition to the two already operative negative sputtering sources. NESTOR will provide negative ion beams of especially gaseous elements, giving the unique opportunity to produce beams of noble gases, namely He^-, which has been requested by several user groups with the aim of making nuclear astrophysics experiments. It has been designed to be compact, reliable, easy for maintenance, and able to operate over long runtimes on the order of weeks without significative stops and optimization operations. Since the source has been designed to operate on the already existing high voltage platform hosting the other two sputtering-type sources, for sake of available room it has been coupled to the available beam optics. However, after the technique validation that has been done in late 2020, the source is now undergoing to a robust phase of further commissioning, including the adaptation/optimization of the beam optics, especially the Einzel lenses, based on CST numerical simulations. On this last purpose, a preliminary study has been presented in this paper, and it will be carried out more in details in the coming months. Before the installation on the platform, the primary ECR source was characterized in terms of current, beam shape and emittance for different values of the pressure, microwave frequency and RF power, providing very promising results in terms of total output currents, beam stability and quality. It is worth to be noticing that the primary source represents a significative advancement in the field of MDIS-type devices, since it is based on solid state power amplifier-technology in the range of 6 GHz.

References

[1] S. Gammino et al., *Review on high current 2.45 GHz electron cyclotron resonance sources (invited)*, Review of Scientific Instruments 81, 02B313 (2010).

[2] J. Angot et al., *Development of a new compact 5.8 GHz ECR Ion Source*, Proc. of Int. Work. on ECR Ion Source (ECRIS'16), WEPP09 (2016).

[3] M. Bacal et al., *Negative ion sources*, J. Appl. Phys. 129, 221101 (2021)

[4] P. Sortais et al., *Ultracompact/Ultralow power electron cyclotron resonance ion source for multipurpose applications*, Proc. 13th Int. Conf. on Ion Sources (ICIS13), 02B314 (2010).

[5] R. Sparta et al., *Direct and Indirect Measurements for a Better Understanding of the Primordial Nucleosynthesis*, Front. Astron. Space Sci., (2020), https://doi.org/10.3389/fspas.2020.00009

[6] C. Spitaleri et al., *Astrophysics studies with the Trojan Horse Method*, The European Physical Journal A 55, 161 (2019).

[7] P. Descouvemont, *Nuclear Reactions of Astrophysical Interest*, Front. Astron. Space Sci., (2020) https://doi.org/10.3389/fspas.2020.560149.

[8] L. Calcagnile, *SCIRES-IT (SCIentific RESearch and Information Technology)*, Vol 1, Issue 1 (2011), 79-112

[9] T. Olsson et al., HIE-ISOLDE-PROJECT-Note-0016 (2012).

Journal of Physics: Conference Series **2244** (2022) 012038 doi:10.1088/1742-6596/2244/1/012038

H^- beam emittance analysis in a multicusp ion source

A George[1,3], T Kalvas[2], S Melanson[3], M Dehnel[3] and N G R Broderick[1]

[1]Department of Physics, University of Auckland, Private Bag 92019, Auckland, New Zealand.
[2]Department of Physics, University of Jyväskylä, P.O. Box 35, 40014, Finland.
[3]D-Pace, 625 Front St 305, Nelson, BC V1L 4B6, Canada.

E-mail: a.george@auckland.ac.nz

Abstract. Emittance of the ion beam extracted from an ion source is dependent on the initial focusing action at the plasma sheath. The properties of the plasma sheath is further dependent on the local electric fields and charge densities around the sheath. Experiments are conducted for creating different sets of conditions around the plasma sheath in an H^- multicusp filament ion source and the resulting emittance of the extracted H^- ion beam is measured. Variation of beam emittance under different plasma densities, electrode voltages and gas flows are analysed.

1. INTRODUCTION

Emittance is one of the most important quality parameters representing the extent of the spread of particles in an ion beam in phase space. Emittance of the ion beam, soon after the beam formation from the plasma, is determined by the initial focusing action at the plasma sheath. This plasma sheath is the boundary layer between the quasi neutral plasma and the extraction region where mostly negative charges are present. The shape and location of the plasma sheath is determined by properties of the plasma, geometry of the electrodes around the plasma and electric potential on the electrodes. At optimal shape of the plasma sheath, the produced beam emittance has a minimum value. Depending on the properties of the extraction system and beam transport, there might be other effects which contribute to the emittance. For example, the beam might suffer from aberrations in the extraction resulting in emittance growth or collimation which can decrease both the emittance and the measured current. Nevertheless, in a suitable extraction system, the emittance minimum should be observable in conditions where these other effects are avoided or at least minimized. In the current paper, different factors affecting the H^- beam emittance from D-Pace's TRIUMF licensed multicusp ion source are presented [1].

2. DETAILS OF THE ION SOURCE AND EXTRACTION SYSTEM

Section view of the filament ion source is presented in Fig. 1. Plasma is sustained inside the plasma chamber via thermionic emission from electrically heated Ta filaments. The filaments are biased at -120 V with respect to the plasma chamber and the corresponding current measured on the filament power supply is the arc current. H^- ions are generated inside the plasma mostly through volume production methods [2]. The plasma chamber is surrounded by Sm_2Co_{17} permanent magnets that form cusp fields, which confines the electrons in the plasma. The extraction system consists of the plasma electrode, the extraction electrode and the ground electrode [3]. The plasma and extraction electrodes are biased positive with respect to the plasma chamber for negative ion extraction. The co-extracted electrons are deflected on to the extraction electrode, before reaching the ground electrode, by the perpendicular magnetic field created by permanent magnets in the extraction electrode and is measured as the electron

Journal of Physics: Conference Series **2244** (2022) 012038 doi:10.1088/1742-6596/2244/1/012038

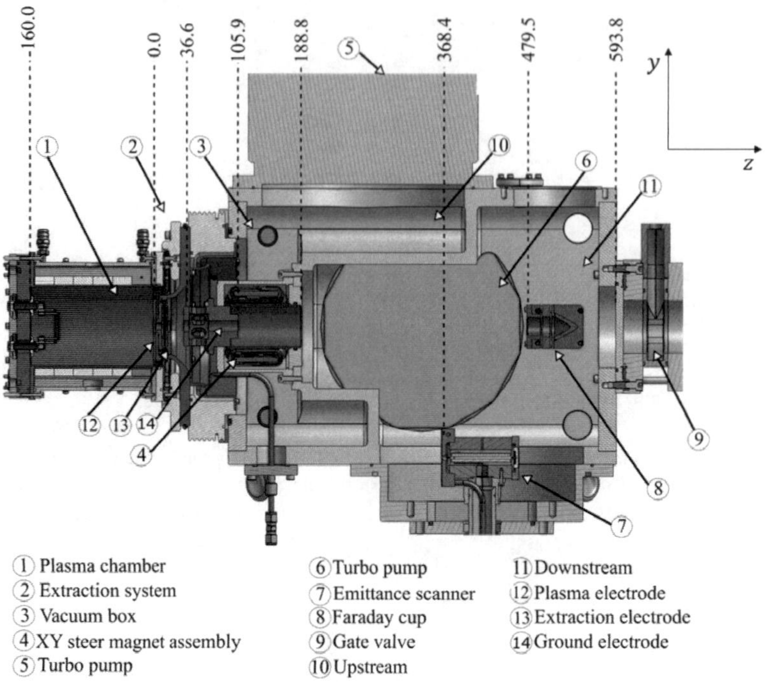

Figure 1: Section view of D-Pace's TRIUMF licensed multicusp filament ion source and extraction system. Ion beam is extracted in the z direction.

current, I_e. The plasma chamber and the plasma and extraction electrodes are biased at a negative potential (−30 kV) with respect to the ground electrode by the bias power supply. The current measured on the bias power supply, I_{bias}, represents the net charge flow from the ion source to the ground electrode region. Beam emittance is measured at about 368 mm downstream from plasma electrode, by D-Pace ES4 emittance scanner. This is an Allison-type emittance scanner which measures the beam intensity as a function of position (y) and angle (y′) simultaneously. Background electronic noise in the emittance data measurement is eliminated by discarding the beam intensity values below 4% of the maximum value. The Faraday cup is located at about 480 mm from the plasma electrode and can measure current (I_{FC}) in a beam of size up to 55 mm diameter. Since I_{FC} values are affected by the beam emittance, I_{bias} variable is used for characterizing the emittance variation in the current set of experiments. I_{FC} values are having a linear dependence on the measured I_{bias} values in the ion source.

3. EXPERIMENTAL RESULTS AND DISCUSSIONS

3.1. Effect of gas flow on beam emittance

A set of experiments were performed at 30 keV beam energy to study the influence of H_2 gas flows on the beam emittance values from the filament ion source. In order to isolate the influence of the gas flows, the experiments were conducted at constant values of electrode potentials (V_{plasma} = 5 V, V_{ext} = 3.5 kV). Gas flows were varied from 8 sccm (standard cubic centimeter per minute) to 17 sccm for constant arc currents. This corresponds to about 10 mTorr to 20 mTorr of pressure inside the plasma chamber. The increase in H^- bias current (I_{bias}) extracted from the plasma is obtained by increasing the arc current and hence the plasma density, as shown in Fig 2(A). The beam emittance values obtained for the different conditions created in the experiment are shown in Fig. 2(B). As can be seen from the figure, the curve passes through an emittance minimum value at a certain I_{bias} value, for all gas flows. The minimum corresponds to an optimum plasma sheath shape [4]. The emittance grows to either side of the minimum,

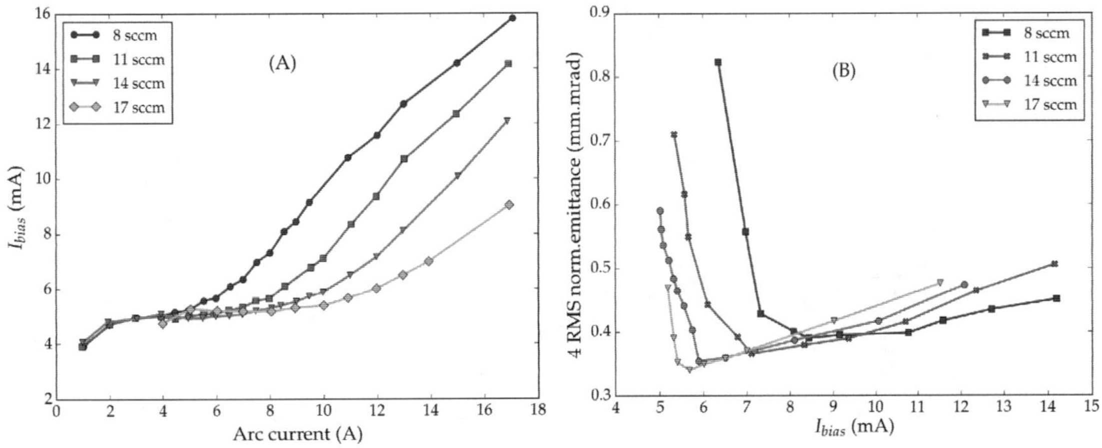

Figure 2: (A) Variation in I_{bias} and (B) beam emittance values at different arc currents and gas flows.

indicating the plasma sheath becoming more convex or concave from the optimum shape. The emittance minimum drifts to lower bias current values as the gas flow increases. This is contrary to the behaviour observed in some other filament ion sources, as mentioned in [4]. It can also be seen that the shift in emittance minimum decreases as gas flow values increases. The variation in the co-extracted electron current (I_e) obtained at the extraction electrode for the different arc currents and gas flows is shown in Fig. 3(A). As can be seen from this figure and Fig 2(A), both I_{bias} and I_e decrease as the gas flow increases. These currents correspond to the charged particle density in the regions close to the plasma sheath. The decrease in the extracted currents with higher gas flows suggests a decrease in the electron and negative ion charge densities near the plasma sheath. This could lead to the observed variations in the beam emittance values. The dependence of the emittance on gas flow points to the fact that the plasma sheath dynamics is influenced by the electron and negative ion charge densities near the sheath. Fig 2(B) considers only the negative ion charge density through the I_{bias} values. In order to include the effect of electron densities, we can define an effective current at the plasma sheath, I_{eff}, which corresponds to the sum of currents due to H^- ions and electrons [4].

$$I_{\text{eff}} = R_{ic} I_{\text{bias}} + R_{ec} I_e \sqrt{m_e / m_H}, \tag{1}$$

where m_e and m_H represent the mass of electron and H^- ion. Here R_{ic} and R_{ec} represents the negative ion density correction factor and electron density correction factor. These correction factors represent the transport properties of negative ions and electrons from the plasma sheath to the electrodes. A value of 1 for the correction factors means that the charge densities measured on the electrodes are equal to the charge densities near the sheath. The $R_{ic} = R_{ec}$ scenario suggests that the electrons and ions behave in the same manner in the plasma sheath and there are no factors that affect only the ions or electrons. The emittance values in Fig. 2(B) can now be plotted in terms of the effective current, I_{eff} in (1), such that the emittance minimum occurs at the same value in x-axis by choosing the appropriate correction factor values. This is shown in Fig. 3(B). The values used for the correction factors are $R_{ic} = 1$ and $R_{ec} = 1, 6, 11$ and 14, for the different gas flows. $R_{ec} > 1$ values suggest that the electron density in the plasma sheath region is higher than the observed current density corresponding to the co-extracted electron current at the extraction electrode. This is an anticipated result, as part of the electrons near the sheath will never get extracted. These electrons would get accelerated from the sheath, but they bend due to the magnetic fields and hit the plasma electrode. Hence the real electron to ion ratio at the sheath would be different from the values obtained by measuring currents at the electrodes in the experiments. It can also be seen that the value of R_{ec} increases with gas flow. This can be understood in terms of the

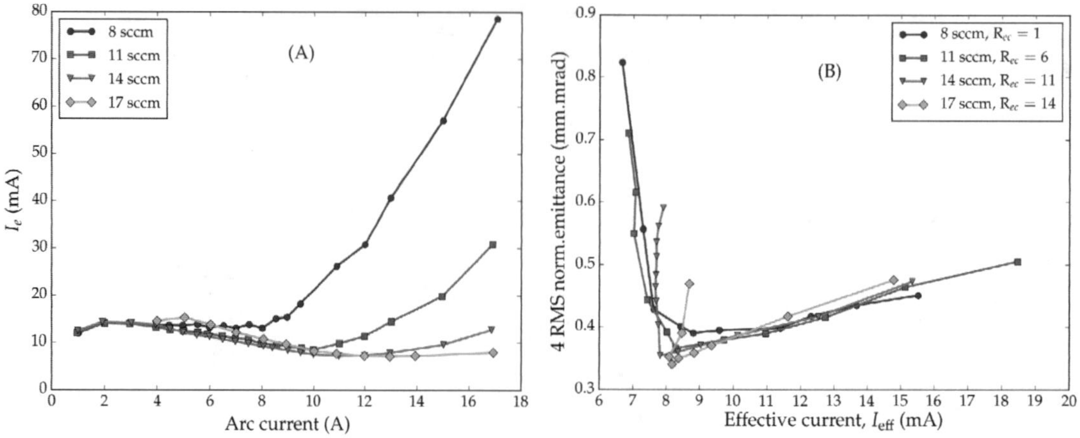

Figure 3: (A) Variation in co-extracted electron current (I_e) for different arc currents and gas flows. (B) Emittance values (as in Fig 2(B)) in terms of the effective current, using correction factors.

effect of pressure on electron transport and collisions in the plasma [4]. But such high R_{ec} values are difficult to comprehend as it represents very high electron density at the plasma sheath. Another set of correction factors which can produce similar results is when R_{ec} = 1 and the ion density correction factor R_{ic} =0.65, 0.8, 0.95 and 1.0 for the different gas flows. This indicates that the bias currents recorded at lower gas flows consist of charges other than the negative ions. In all cases, $R_{ec} \neq R_{ic}$ for different gas flows indicate that the emittance is dependent on the density of the type of particle species, ions or electrons, near the plasma sheath and not only on the total charge density.

3.2. Effect of extraction electrode voltage on beam emittance.

The variation in beam emittance values for 3 different extraction electrode voltages is shown in Fig. 4(A), for fixed values of gas flow (15 sccm) and plasma electrode voltage (5 V). This ensures that the variation in emittance values along a curve is only due to the variation in plasma densities from changes in arc currents. These graphs also reveal an emittance minimum for all the experiments. Fig. 4(A) also shows a decrease of emittance at higher bias currents (> 13 mA) for 3 kV extraction electrode voltage curve. The current at the Faraday cup (I_{FC}) was also found to be decreasing at this point even though I_{bias} continued to increase. The emittance scanner scans a region of about 50 mm length in the direction perpendicular to the beam propagation. Simulations reveal that the beam is more divergent due to space charge effects, leading to the loss of charged particles to regions outside the emittance measurement and hence a decrease in emittance values [3]. Also, there are more particles hitting the extraction electrodes, leading to beam collimation. Furthermore the emittance curve drifts to slightly lower values as the extraction electrode voltage increases. This is mostly seen when the plasma density increases from the emittance minimum. The curvature of the plasma sheath might play a role in this behaviour. Increase in the extraction voltage causes higher positive electric field near the plasma electrode, pushing the positive ions towards the plasma in the ion source. This causes the boundary of the quasi neutral plasma also to be shifted further towards the plasma, resulting in a more concave plasma sheath and this could lead to lower emittance values. Another factor that could contribute to the lower emittance values at high extraction voltages is the reduced space charge effects. Higher beam velocities experience reduced space charge forces and this could reduce the beam spread.

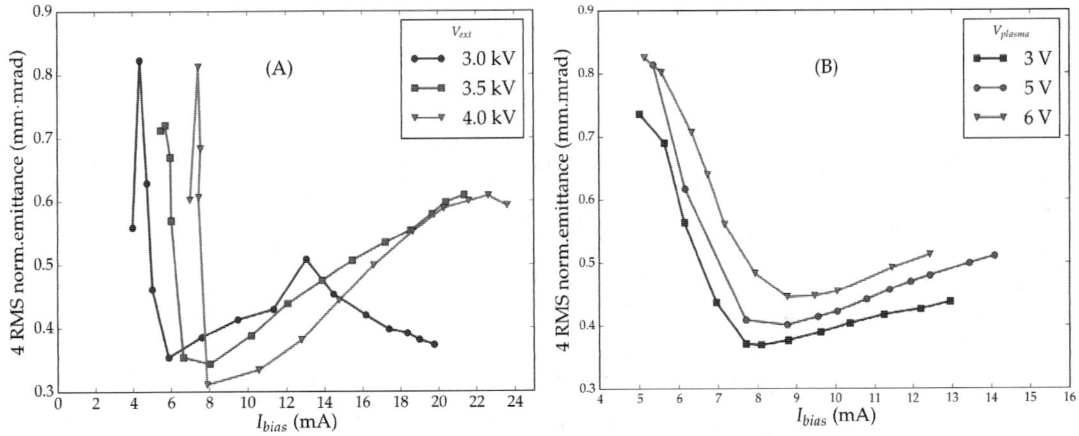

Figure 4: Beam emittance values for (A) different extraction and (B) plasma electrode voltages.

3.3. Effect of plasma electrode voltage on beam emittance

Experiments were performed to understand the influence of plasma electrode voltages on the beam emittance, using a fixed H_2 gas flow (8 sccm) and extraction electrode voltage (3.5 kV). The plasma electrode current increased proportional to the electrode voltage for fixed arc currents. This was accompanied by a decrease in the I_{bias} and I_e values. It suggests that increasing plasma electrode voltage depletes the corresponding charge densities at the plasma sheath, since the gas flow is constant. The effect of plasma electrode voltages on the beam emittance is shown in Fig. 4(B). As seen in the figure, the beam emittance minimum does not shift appreciably in terms of bias current. But, in general, the beam emittance values increase as the plasma voltage increase, for a fixed bias current. This could be attributed to the higher total charge densities needed for attaining a particular I_{bias} at high plasma electrode voltages. Furthermore, variations in the plasma electrode voltage could also affect the electric fields in the plasma sheath, leading to the trends in the figure.

4. CONCLUSIONS

The current paper reports the results of the experiments conducted to study the influence of different factors on the H^- beam emittance from D-Pace's TRIUMF licensed filament ion source. Different conditions of gas flows, extraction and plasma electrode potentials and arc currents were studied. The possible relation between the emittance results and the plasma sheath behaviour is also discussed. The results indicate that the H^- and co-extracted electron currents measured at the electrodes might not be a true representation of the charge densities at the plasma sheath, and the dependency between these parameters needs to be established. The study also suggests differences between the transport of negative ions and electrons from the plasma sheath to extraction.

References

[1] T Kuo et al. "On the development of a 15 mA direct current H- multicusp source". In: *Review of scientific instruments* 67.3 (1996), pp. 1314–1316.

[2] M Bacal. "Physics aspects of negative ion sources". In: *Nuclear Fusion* 46.6 (2006), S250.

[3] Anand George et al. "A study of the optical effect of plasma sheath in a negative ion source using IBSIMU code". In: vol. 91. 1. AIP Publishing LLC, 2020, p. 013306.

[4] Taneli Kalvas and J Lettry. "Deviation of H- beam extraction simulation model". In: *AIP Conference Proceedings*. Vol. 2052. 1. AIP Publishing. 2018, p. 050007.

Study on influence on plasma under current-generated magnetic filter using Langmuir probe in HUST negative ion source

Jinghan Yang, Dong Li, Peng Chen, Zengshan Li, Yubo Zhang, Dezhi Chen

State Key Laboratory of Advanced Electromagnetic Engineering and Technology, School of Electrical and Electronic Engineering, Huazhong University of Science and Technology, Wuhan 430074 Hubei, China

Corresponding author: lidonghust@hust.edu.cn

Abstract. The current-generated magnetic filter is widely used in negative ion sources to bring down electron temperature and optimize source operation. The effect of the current-generated magnetic filter on plasma generation and transport is important. The axial distribution of cesium-free plasma parameters under different magnetic filter configurations was measured by a movable Langmuir probe at HUST negative ion source. Abnormal changes of plasma parameters are discovered at the driver exit. When the magnetic filter field penetrates into the driver, plasma density inside the driver increase, and the high-density high-temperature region of plasma expands outside of the driver. And the radiofrequency power coupling is also influenced by the magnetic filter. These experimental phenomena indicate that a new magnetic filter configuration is required to optimize source operation.

1. Introduction

The negative ion source is an important part of the neutral beam injector (NBI) for fusion devices, such as ITER and CFETR. Negative ions are extracted from the ion source and accelerated to hundreds of keV energy. The production of negative ions comes from volume and surface processes. The ion source needs to achieve a higher negative ion production rate to fulfill the ITER beam density requirement[1], 300A/m^2(H-) or 200A/m^2(D-). Cesium evaporation to the plasma grid (PG) reduces the surface work function, enhances the production of negative ions in the vicinity of the extraction apertures. However negative ions can be destroyed easily by high-energy electrons. Thus, a magnetic filter is introduced to reduce the electron energy near the PG to minimize the destruction of negative ions from hot electrons. Electrons are highly magnetized due to their large charge-to-mass ratio, while ions are nearly not affected. Magnetized electrons experience more collisions in the path to PG, thus their energy is lost to other heavy particles. In general consideration electron temperature around the PG should be controlled to ~1 eV. BATMAN source in Max Planck Institute for Plasma Physics (IPP) used a set of permanent magnets to create a Gaussian-like magnetic field[2]. In the upgrade of BATMAN, an ITER-like magnetic filter is generated by a current flowing through the PG. And good results are obtained from the experiment[3]. Consorzio RFX also conducts studies on the influence of the current-generated magnetic filter in the NIO1 source[4]. Other devices use return conductors to change magnetic field in the driver, we chose not to set the return conductors, let transverse magnetic field exist in the driver. This offers a simple scenario for further study on plasma ignition and maintain difficulty under magnetic field influence.

To provide assist in NBI beam optimization in CFETR, Huazhong University of Science and Technology (HUST) NBI Team build a compact radio frequency negative ion source for studies of experiment conditions influence on plasma and for extraction beam optimization. Experiment operation is currently working on the ignition of hydrogen plasma and extraction without cesium. This paper investigates the influence of MF on RF power coupling and whether the MF is enough for cooling electron temperature. The HUST negative ion source will be detailed in the next section.

2. Experiment setup

2.1. *HUST negative ion source*

The HUST negative ion source contains one driver, expansion chamber, and extraction grids, as shown in Figure.1. The driver is a 160 mm long alumina ceramics barrel (R = 150 mm) surrounded by RF coils. An RF power source that can deliver a maximum 1MHz 25 kW power is connected to the driver through a matching network. A Faraday shield is installed in the driver. A 230 mm long cylindrical expansion chamber is equipped with the ion source to use current-generated magnetic filter. The inner of the expansion chamber is implanted with water cooling tubes for long-time operation. The extraction structure consists of plasma grid (PG), extraction grid (EG), and ground grid (GG). Total 5x5 extraction apertures are designed in the extraction system. A bias plate connected to the chamber is surrounding apertures. Bias power supply can offset the PG voltage from the chamber by approximate plasma potential (20V~40V). The source can work in the range of filling pressure from 0.3 Pa to 1.1 Pa.

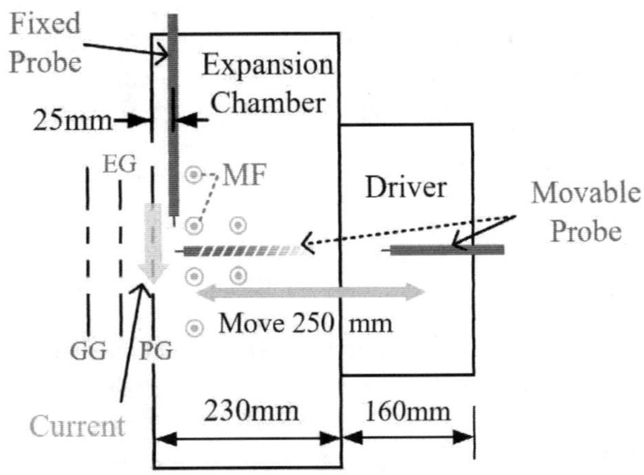

Figure 1. HUST negative ion source.

2.2. *Magnetic filter configuration*

The magnetic filter is produced by current flowing through PG. The equipped current power source can output a maximum of 4 kA current. This current is delivered by feed-through to the busbar and then to the PG shown in Figure 2. In order to study the influence of magnetic filter penetration into the driver, return current wire is not arranged as in BATMAN Upgrade[5]. In terms of our magnetic filter configuration, the FEM simulation results and Hall probe measurement results of the transverse magnetic field are shown in Figure 3. There are several mT of magnetic field in the driver position, this could lead to influence on RF power coupling and plasma generation. Despite the mechanism of this influence is unclear, we exhibit the experiment results of MF influence in section 3 for further study.

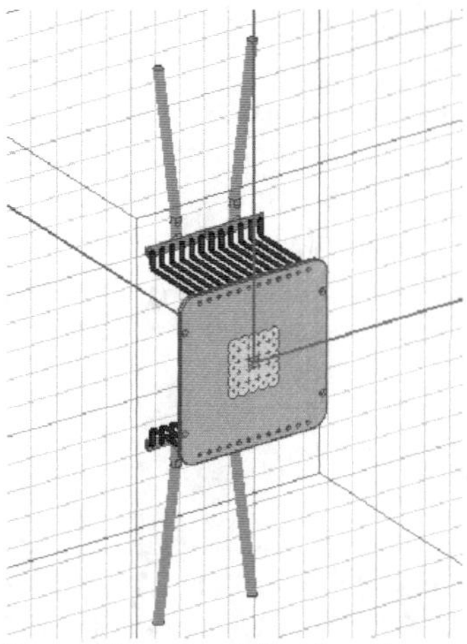

Figure 2. current flowing through PG

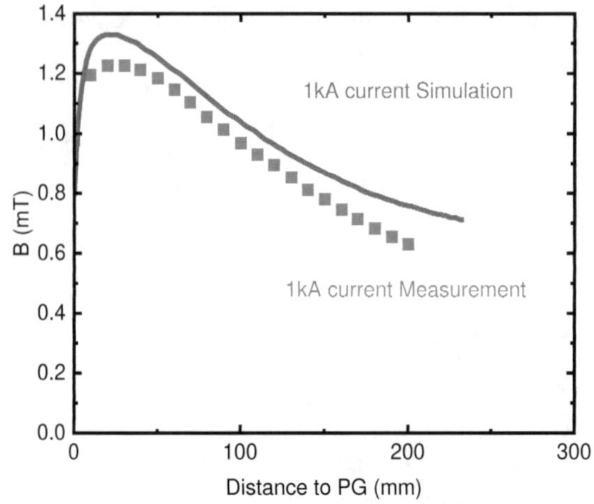

Figure 3. Profile of magnetic filter generated by 1kA current.

2.3. Langmuir probe arrangement

We use two cylindrical single Langmuir probes, one is placing fixed downstream of the source, the other one is a movable probe inserting from the backplate of the driver as illustrated in Figure 1. The fixed probe is inserted transversely into the chamber, therefore it is parallel to the magnetic filter field. The movable one is perpendicular to the magnetic field. Under our experimental condition, the maximal magnetic flux density is 5mT at 4kA current flowing through the PG, in this field strength range it is experimentally proved that the direction of the probe tip to the magnetic field has negligible effect on the measurement of the plasma[6]. So we can expect that both probes would obtain the same measurement results at the same position. The fixed probe is intended to verify the influence of movable one on plasma. We measured plasma parameters using fixed probe in the same condition while the movable probe is located in different positions. Our experiment results show that the movable probe does not have a noticeable influence on plasma parameters measured by the fixed probe. The experiment data is not list here due to limitation of space.

The fixed Langmuir probe tip is 25 mm from the PG surface. And the movable probe can move 250 mm, cover the range from the PG surface vicinity to the driver center. These two probes use the same size tungsten tip with radius and length of 0.05 mm and 5 mm respectively. The probe tip radius is small enough compared to electron cyclone radius, which is about 0.5 mm, so the magnetic field does not affect Langmuir probe analysis[7].

3. Influence of magnetic filter

As mentioned by L Schiesko[7], plasma might not sustain under high magnetic field influence, this is because of electron trajectory distortion by magnetic field[8]. We tried two experiment sequences, one is to build the magnetic field first, then start the ignition of plasma. The other one is to ignite plasma first and wait until it is stable, then enable the current-generated magnetic filter. The results turned out that magnetic field existence in the drive has a negative effect on plasma ignition. With starting magnetic filter first condition, the driver is harder to start the plasma compared to no magnetic filter condition. In detail, we set RF power at 6kW and pressure at 0.6Pa, the driver would not properly ignite plasma when the magnetic field in the driver exceeds about 1mT. The ignition difficulty is suspect to have relation with unusual electron trajectory in the driver, and this idea is being studied.

3.1. Influence on power delivery

RF power coming from power source transport through coaxial line to the matching circuit. Matching circuit matches inductive load of RF coils to 50 Ohm load. A directional coupler is connected in transmission line as a power meter where RF incident power (P_{inc}) and reflected power (P_{refl}) signals are obtained from. P_{refl} would be zero if in perfect match. If match circuit works badly, efficiency goes down and power source takes unnecessary damage. Different power conditions are observed with different magnetic field cases shown in Figure 4. All experiments are conducted at the filling pressure of 0.6 Pa. When no magnetic field is applied, P_{inc} is 6.2 kW, P_{refl} is 0.2 kW. After the plasma is stable, we apply various currents to the PG generating magnetic filter in the ion source chamber. P_{inc} and P_{refl} increase to 6.4 kW and 0.25 kW at 1000A (0.5mT at the driver). P_{inc} and P_{refl} increase to 7.2 kW and 0.4 kW at 2000A (1mT at the driver). P_{inc} shift to 7 kW, and P_{refl} increase to 0.6 kW at 3000A (1.5mT at the driver). It is suspect that the power shift has relation to plasma parameters changes in the driver. The magnetic field that penetrates the driver affects plasma transportation, thus parameters of plasma inside the driver change. Dielectric parameters of plasma inside the RF coils change causing coils inductance to change. Our device uses fixed match circuit which can only fit pre-tuned inductance parameters. Therefore RF power transmission shifts to different conditions. Diagnostic results of plasma in the driver are detailed in the next section.

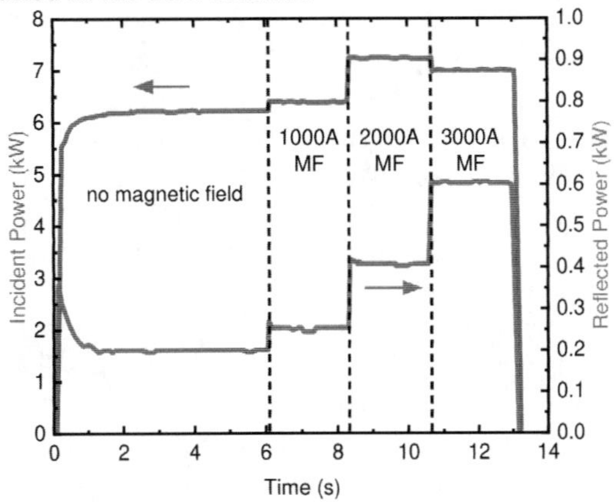

Figure 4. Incident RF power and reflected RF power during a shot under different MF.

3.2. Influence on plasma axial distribution

Plasma parameters axial distribution from the PG vicinity to the driver measured by Langmuir probe is demonstrated in Figure 5. With no magnetic field applied, electron temperature drops from 5.5eV in the driver to 3.2eV around PG.

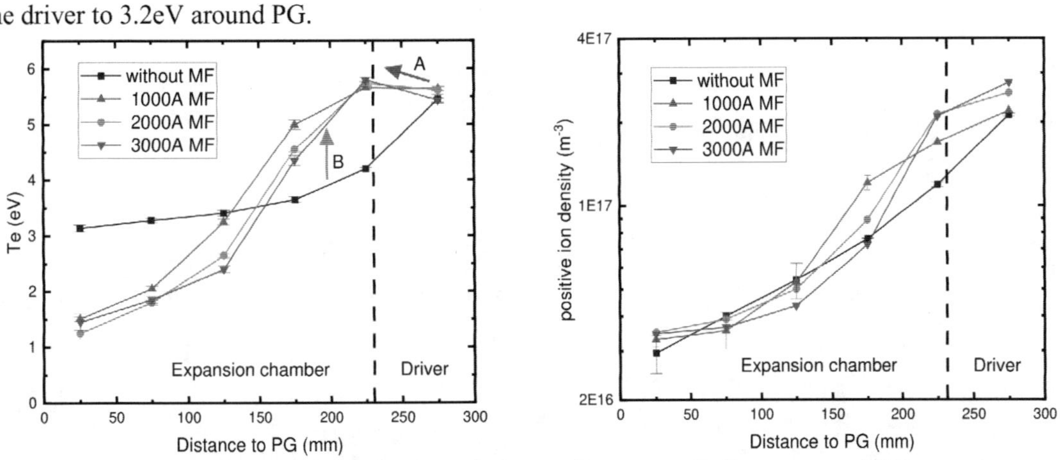

Figure 5. Axial distribution of plasma. Pressure = 0.6Pa, power = 6kW.
Electron temperature distribution (left); Plasma density distribution (right).

After different cases of magnetic filter applied, electron temperature around the PG undergoes a more significant cooldown to about 1.5eV, while the electron density has a slight raise around PG. In general consideration, electron temperature outside the driver would be low than it in the driver, and electron temperature would be lower with MF applied. Yet at 3000A case, we noticed an abnormal phenomenon at the driver exit that electron temperature is higher than it in the driver (mark A in Figure 5 left). Another strange phenomenon is that electron temperature with MF is higher than it without MF in the middle of expansion chamber (mark B in Figure 5 left). Similar results were observed in reference[9], we are currently studying mechanism behind.

Plasma density at the driver exit exhibits an increment compared with no magnetic filter case, because plasma transportation is slightly suppressed. In general, current-generated magnetic filter have two effects on plasma axial distribution. First, it cools down electron temperature at PG vicinity, under our experimental conditions 1000A-current magnetic field is sufficient for cooling electron temperature. Second, it expands high-temperature-high-density plasma core area from the driver center to the driver exit. Plasma parameters variation inside the driver is suspect to cause matching point shift in the matching circuit. This is to be further studied in the future work.

4. Conclusion

This paper has run through a brief introduction to HUST negative ion source research project. We monitored RF incident power and reflected power under different MFs. It is shown that transmit condition of RF power changes related to magnetic field penetration in the driver. Source operation requires a self-adjusting match circuit to follow the matching variation.

We presented the measurement of plasma parameters axial distribution under different magnetic filter cases. In the driver area, high-temperature-high-density plasma core region expands from the center point of the driver to the outside of the driver exit. As for electron temperature near the PG, 1000A current-generated MF is good enough to cool down the temperature to around 1.5eV at our experiment condition. Higher magnetic field, on the contrary, may increase electron temperature slightly. These results are obtained for further beam extraction experiment study, the extracted beam condition is standard for evaluation of MF. In addition, magnetic field existence in the driver area makes ignition and maintenance of plasma more difficult. Optimization of magnetic filter should take these factors into consideration.

Acknowledgement

This work was supported by the National Key R&D Program of China [Grant No.2017YFE0300105], the National Natural Science Foundation of China [Grant No. 11775088].

References

[1] Serianni G, *et al.* 2019 *Fusion Eng. Des.* **146** 2539–46
[2] Fantz U, Schiesko L and Wünderlich D 2014 *Plasma Sources Sci. Technol.* **23** 044002
[3] Kraus W, Schiesko L, Bonomo F, Fantz U, Heinemann B and Hurlbatt A 2018 *AIP Conf. Proc.* **2052** 040004
[4] Cavenago M, *et al.* 2018 *AIP Conf. Proc.* **2052** 040013
[5] Schiesko L, Wimmer C and Fantz U 2018 *AIP Conf. Proc.* **2052** 040006
[6] Schiesko L, Franzen P and Fantz U 2012 *Plasma Sources Sci. Technol.* **21** 065007
[7] Schiesko L, McNeely P, Franzen P, Fantz U, and the NNBI Team 2012 *Plasma Phys. Control. Fusion* **54** 105002
[8] Serianni G, *et al.* 2020 *Rev. Sci. Instrum.* **91** 023510
[9] Veltri P, Buonincontri L, Sartori E, Cervaro V, Fincato M, Laterza B, Moro G, Cavenago M and Serianni G 2018 *AIP Conf. Proc.* **2011** 050011

Study of negative ion beam emittance characteristic using 3D PIC-MCC simulation

K. Miyamoto[1], K. Nagaoka[2, 3], A. Hatayama[4], K. Hoshino[4], H. Nakano[2,5], T. Shibata[6], K. Tsumori[2,5]

[1]Naruto University of Education, Naruto 772-8502, Japan
[2]National Institute for Fusion Science, NINS, Oroshi-cho, Toki 509-5292, Japan
[3]Graduate School of Science, Nagoya University, Furo-cho, Chikusa-ku, Nagoya 464-8602, Japan
[4]Faculty of Science and Technology, Keio University, Yokohama 223-8522, Japan
[5]SOKENDAI (The Graduate University for Advanced Studies), 322-6 Oroshi-cho, Toki 509-5292, Japan
[6]Accelerator Laboratory, KEK, Tsukuba 300-3256, Japan

[+]Corresponding author: kmiyamot@naruto-u.ac.jp

Abstract. It is reported from the measurement of a negative ion beamlet emittance that the characteristic three-Gaussian components are observed in the emittance diagram. The negative ion trajectories are calculated using 3D PIC-MCC simulation in order to clarify the origin of this complicated phase space structure and underlying physical mechanism. The characteristic three-Gaussian components can be reproduced in the emittance diagram from the simulation result. It is verified that the Gaussian components are caused by the negative ions extracted from the different plasma meniscus region, that is, the central region or the region near the edges of the meniscus.

1. Introduction

The negative ion sources are used for a wide range of applications such as neutral beam injection (NBI) for heating and current drive of magnetically confined fusion plasma [1, 2], intense H^- injection into accelerators for elementary particle physics [3, 4], and compact acclerators for medical field [5]. It is essential for the negative ion source to produce a negative ion beam with good optics. For example, a beam e-folding divergence angle of 3 ~ 7 mrad is required in the design of ITER-NBI [6, 7]. If the beam e-divergence divergence angle is large, the negative ion beam will impinge on the acceleration grid and other components during beam acceleration and transport, and then causes the heat loads, loss of the negative ion beam current, and break down of voltage holding due to the resultant secondary electrons. Therefore, the negative ion beam optics has been studied intensively in view of the development of the negative ion sources.

In the measurement of a negative ion beamlet emittance at the NIFS test-bed, that the characteristic three-Gaussian components were observed in the emittance diagram [8]. These components can be identified as a combination of three beam components with different beam axes. In order to clarify the origin of this complicated phase space structure and underlying physical mechanism, potential structure and particle dynamics including the negative ion trajectories have been calculated self-consistently using 3D PIC-MCC simulation. In this article the PIC-MCC simulation model is

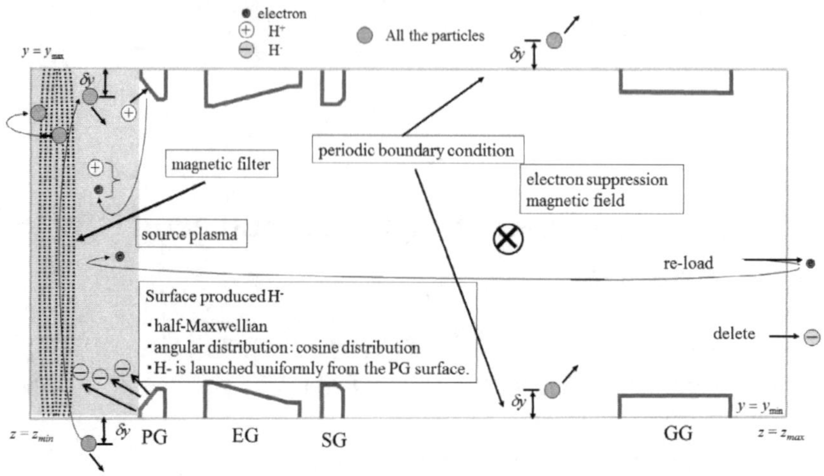

Figure 1 A schematic view of the simulation model.

Table 1. Main physical parameters in the simulation.

Physical quantity	Value
Electron density	1.8×10^{17} m^{-3}
Electron temperature	0.8 eV
H^+ ion temperature	0.3 eV
H^- ion temperature	0.1 eV (volume produced) 1 eV (surface produced)
H_2 molecular density	1.88×10^{19} m^{-3} (at 0.3 Pa)
H_2 molecular temperature	0.1 eV
H atomic temperature	0.3 eV
n_{H2} / n_H	0.5

described in the section 2. The simulation results are compared with the experimental results of ref. [8] in the section 3, and the section 4 is a summary.

2. Simulation model

The motion of charged particles (H^+ ions, H^- ions, and electrons) is solved in their self-consistent electric field using the PIC-MCC method. The region from the source plasma up to the accelerator is modeled for a single aperture as shown in Fig. 1. A magnetic filter and an electron suppression magnetic field are taken into account. The x-axis is taken to be parallel to the magnetic filter while the y-axis is taken to be parallel to the electron suppression magnetic field. The direction of the negative ion extraction is taken as the z-axis. The NIFS negative ion source in Ref. [8] is modelled, i.e., the extractor and accelerator consist of the four grids, namely a plasma grid (PG), an extraction grid (EG), an electron suppression grid (SG), and a grounded grid (GG). The main physical parameters in the source plasma such as an electron temperature are listed in Table 1. These values are obtained from the experimental data [8, 9]. Electrons, H^+ ions, volume produced H^- ions are assumed to be launched as Maxwellian distributions with the temperatures of 0.8 eV, 0.3 eV, 0.1 eV, respectively. The simulations start with the source plasma containing the electrons, the H^+ ions, and the volume produced H^- ions.

After the steady states are obtained, the surface produced H^- ions are launched uniformly from the PG surface, thus, the angular distribution of the surface produced H^- ions is initially a cosine distribution. The surface produced H^- ions are assumed to be the half-Maxwellian distribution with the temperature of 1.0 eV. In the present simulation, the constant number of H^- super-particles is launched

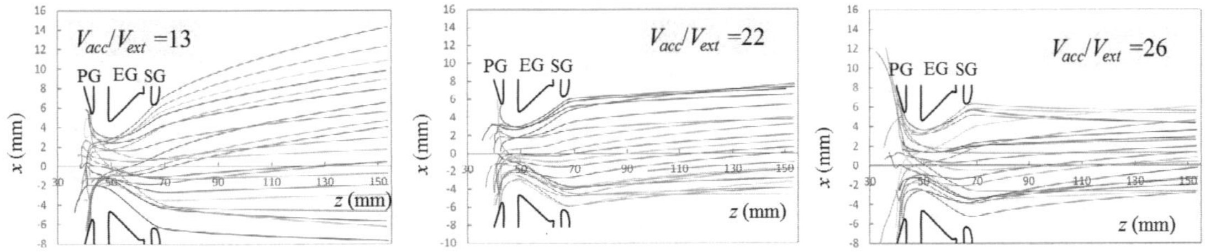

Figure 2 The negative ion beam trajectories for V_{acc}/V_{ext} = 13, 22, 26.

every time step (4800 H^- super- particles per time step). The negative ion super-particles moved out of $z = z_{max}$ are removed to obtain the steady state.

As for the energy relaxation processes of the surface produced H^- ions, the Coulomb collision with the H^+ ions and the charge exchange with the H atoms are taken into account. A binary collision model (BCM) by the Monte Carlo method [10] is applied to the Coulomb collision with the H^+ ions. As for the charge exchange collision, an electron moves from the surface produced H^- ion (fast ions) to a slow H atom, which then leaves behind a slow scattered H^- ion. This slow H^- ion is scattered isotopically: the scattering angles Θ and Φ are randomly chosen as $\Theta = \pi U_1$ and $\Phi = \pi U_2$ (U_1, U_2,: a isotopically random number), respectively. The speed of the scattered H^- ion is chosen to match the H atom velocity distribution (Maxwellian). On the basis of experimental result [9], the temperature of the H atom is set to be 0.3 eV. Whether the charge exchange collision occurs or not is determined by using the null collision model [11].

3. Simulation results

The negative ion beam trajectories for V_{acc}/V_{ext} =13, 22, 26 are shown in Fig. 2. One can find that negative ion beam is varying from the diverging beam to the converging beam as the voltage ratio V_{acc}/V_{ext} increases. This tendency is due to the lens effect of the focusing electric field.

It is pointed out from the experimental result that the Gaussian components are caused by the negative ions extracted from the different plasma meniscus region, that is, the central region or the region near the edges of the meniscus. In order to verify this indication, the relation between the location of negative ion extraction on the plasma meniscus and the emittance diagram is numerically investigated. The contour plot of the two dimensional profiles of the H^+ ion density at V_{acc}/V_{ext} = 13 is shown in Fig. 3(a). The contour with $n_{H+} = 0$ clearly defines the plasma meniscus. The H^+ ions cannot cross this boundary and return back towards the driver region, while the H^- ions cross this boundary and are extracted. In Fig. 3(a), the plasma meniscus is divided into three regions, namely, an upper edge of the meniscus (region 1), the central region (region 2), and a lower edge of the meniscus

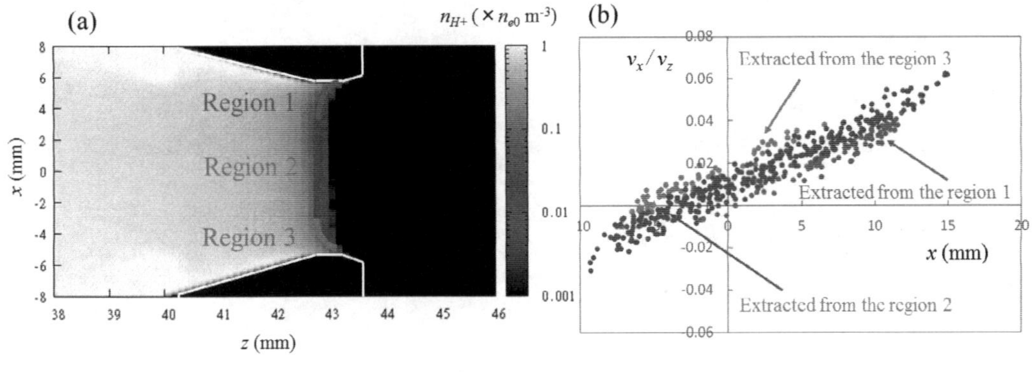

Figure 3 (a) The contour plot of the two dimensional profiles of the H^+ ion density at V_{acc}/V_{ext} = 13 and (b) the plots of negative ions extracted from the regions 1 ~ 3 in the emittance diagram.

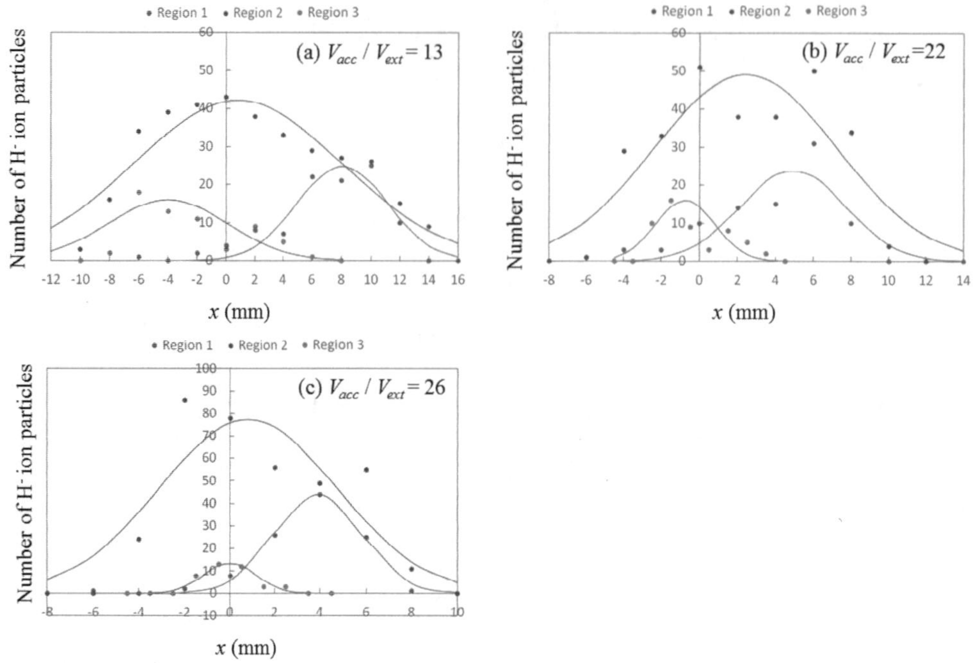

Figure 4 Spatial profiles of the negative ion beam component extracted from the meniscus with Gaussian fitting for (a) V_{acc}/V_{ext} =13, (b) V_{acc}/V_{ext} =22, and (c) V_{acc}/V_{ext} =26.

(region 3). The plots in the emittance diagram are classified into three categories, i.e., the plots of negative ions extracted from the regions 1 ~ 3 as shown in Fig. 3 (b). Similarly the plots in the emittance diagram are classified into three categories for the voltage ratio of V_{acc}/V_{ext} = 16, 22, 26.

Spatial profiles of the negative ion beam component extracted from the meniscus regions 1 ~ 3 are shown in Fig. 4. The characteristic three-Gaussian components in the emittance diagram in ref. [8] can be reproduced. These simulation results show that the H^- ions coming from different meniscus regions can be mapped to different areas in the emittance diagram, allowing to interpret the three-Gaussian components in Ref. 8 to be originated from different meniscus regions.

Dependence of the spatial profiles of the negative ion beam component extracted from the meniscus region 1 and 3 on the voltage ratio V_{acc}/V_{ext} is shown in Fig. 5. The peak of the spatial profile negative ion beam component extracted from the meniscus region 1 shifts to the minus x direction with increase of the voltage ratio V_{acc}/V_{ext}, while that from the region 3 shifts to the plus x direction with increase of the voltage ratio V_{acc}/V_{ext}. This opposite direction of the peak shift corresponds to a negative ion beam

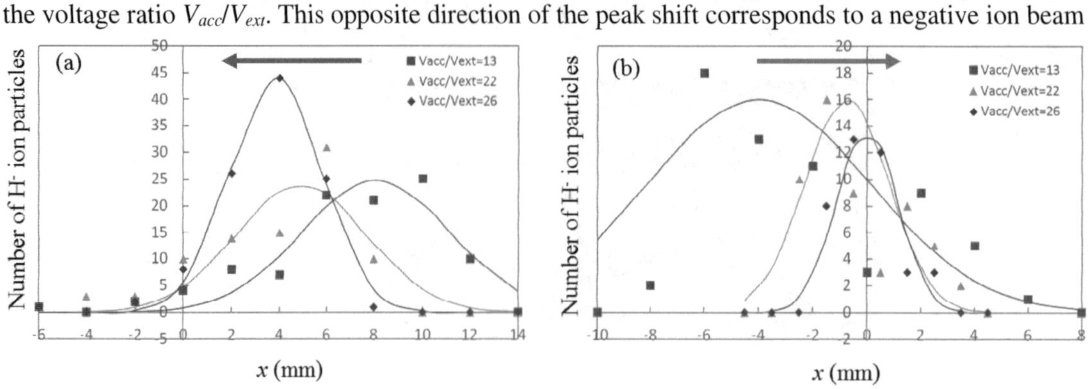

Figure 5 Dependence of the spatial profiles of the negative ion beam component extracted from the meniscus (a) region 1 and (b) region 3 on the voltage ratio V_{acc}/V_{ext}.

Table 2. A number of the negative ion beam trajectories extracted from the meniscus regions 1 ~ 3.

	$V_{acc}/V_{ext.} = 13$	$V_{acc}/V_{ext.} = 22$	$V_{acc}/V_{ext.} = 26$
Total number of the negative ion beam trajectories	515	421	505
A number of the negative ion beam trajectories extracted from the region 1	100 (0.19)	87 (0.21)	107 (0.21)
A number of the negative ion beam trajectories extracted from the region 2	353 (0.69)	278 (0.66)	359 (0.71)
A number of the negative ion beam trajectories extracted from the region 3	62 (0.12)	56 (0.13)	39 (0.08)

changing from a diverging beam to a converging beam.

The number of negative ion beam trajectories extracted from the meniscus regions 1 ~ 3 is compared in Table 2. The values in parenthesis correspond to the fractions of particles from the three regions with respect to the total number of trajectories. It is shown that the number of the negative ion trajectories extracted from the upper meniscus region (region 1) is not the same as that extracted from the lower meniscus (region 3), which reflects that the negative ion density spatial profile is not symmetric with the axis.

4. Summary

From the emittance measurement, it is pointed out that the characteristic three Gaussian components in the emittance diagram are related to the location of negative ion extraction on the plasma meniscus. The 3D PIC-MCC simulation is carried out to verify tis experimental result. The simulation results are summarized as follows:

1) The negative ion beam changes from a diverging beam to a converging beam due to an electrostatic lens effect as the voltage ratio V_{acc}/V_{ext} increases.
2) The H^- ions coming from different meniscus regions can be mapped to different areas in the emittance diagram, allowing to interpret the three-Gaussian components to be originated from different meniscus regions.
3) The peaks of the negative ion beam components extracted from the upper and lower meniscus region shift in the opposite direction each other as the voltage ratio V_{acc}/V_{ext} increases, which corresponds that the negative ion beam changes from the diverging beam to the converging beam.

These results agrees well with the experimental result qualitatively.

Acknowledgement

This work was supported by KAKENHI 18KK0080. This work is performed with the support and under the auspices of the NIFS Collaboration Research Program (NIFS17KLER061).

References

[1] K. Tsumori et al., Rev. Sci. Instrum. **87**, 02B936 (2016).
[2] B. Heinemann et al., New J. Phys. **19**, 015001 (2017).
[3] J. Lettry et al., Rev. Sci. Instrum. **85**, 02B122 (2014).
[4] H. Oguri et al., Rev. Sci. Instrum. **87**, 02B138 (2016).
[5] Y. Etoh et al., Rev. Sci. Instrum. **87**, 02B135 (2016).
[6] R. S. Hemsworth et al., New J. Phys. **19**, 025005 (2017).
[7] M. J. Singh et al., New J. Phys. **19**, 055004 (2017).
[8] Y. Haba et al., New J. Phys. **22**, 023017 (2020).
[9] H. Nakano et al., AIP Conf. Proc. 1655, **020018** (2015).
[10] T. Takizuka and H. Abe, J. Comput. Phys. **25**, 205 (1977)
[11] V. Vahedi, M. Surendra, Comput. Phys. Commun., **87**, 179 (1995).

Soundness evaluation of J-PARC RF ion source after 5-month continuous operation

T Shibata[1,*], K Ohkoshi[2], K Shinto[2], K Nanmo[1], K Ikegami[1] and H Oguri[2]

[1] J-PARC/KEK, 203-1 Shirakata, Tokai, 319-1106 Japan
[2] J-PARC/JAEA, 2-4 Shirakata, Tokai, 319-1195 Japan

*Corresponding author: takanori.shibata@kek.jp

Abstract. In the J-PARC user operation from Nov. 2020 – Apr. 2021, continuous operation of J-PARC Radio Frequency (RF) negative hydrogen ion (H^-) source up to 3,651 hours (5 months) has been achieved. The ion source was operated with the output H^- current of 60 mA, the duty factor (for plasma generation) 2% and the input RF power up to 30 kW. After the operation, phase space diagrams at the Radio Frequency Quadrupole linac (RFQ) entrance were measured by the emittance monitor at the ion source test stand (IS-TS) under the same operation condition as in the J-PARC Linac. Comparison of the phase spaces and the beam emittances between the ion sources in the present and the previous operations shows slight difference. From the direct observation of the antenna coil, no exhaustion or the decrease in the thickness of the enamel coating of the coil have been confirmed. The results indicate the possibility of the next goal of the long-run up to 7 months, which is the same as the full duration of the J-PARC user operation in 1 year.

1. Introduction

Operation of the Radio Frequency (RF) ion source started in J-PARC Linac from 2014 as part of the linac intensity upgrade project. The negative hydrogen ion (H^-) peak beam current extracted from the ion source has been increased from 34 mA to 60 mA for the user operation. On the other hand, beam instability due to the ion source replacement is one of the issues. By replacing the ion source, the beam phase space of the extracted beam becomes unstable for several days due to the reconstruction of the cesiated surface and the variation of the mixed H_2O or oxygen at the moment of ion source installation. In order to solve the issue, extension of the ion source life-time is effective. The present J-PARC user operation duration which takes place between each summer maintenance shut down is around 7 months excluding several short shut downs. To extend the life time of the RF ion source to this 7 month is a goal for improving beam stability in the J-PARC. From the beginning of the RF ion source operation in 2014, the duration of the continuous operation has been extended step by step. The latest status of the long-run operation, the emittance measurement at the test-stand and the direct observation of the antenna coil condition after the operation are given in this manuscript.

2. J-PARC RF ion source

Figure 1 shows the schematic drawing of the J-PARC RF ion source. The detail of the ion source equipment is shown in Refs. [1,2]. The internal antenna coil, which was developed in SNS [3], is installed inside the ion source chamber. The antenna coil is a copper pipe which is coated by enamel for the insulation between the pipe and the plasma. The antenna coils delivered from the factory are checked by the original criterion in J-PARC. Antenna coils with crack, pin-hole or irregular thickness of the enamel coating are omitted. In the J-PARC user operation in these two years, the peak H^- beam current extracted from the ion source is 60 mA. The duty factor of plasma generation is 2% (800 μs pulse width and 25 Hz repetition). In the recent ion source condition, the input 2 MHz RF power to the ion source is around 30 kW.

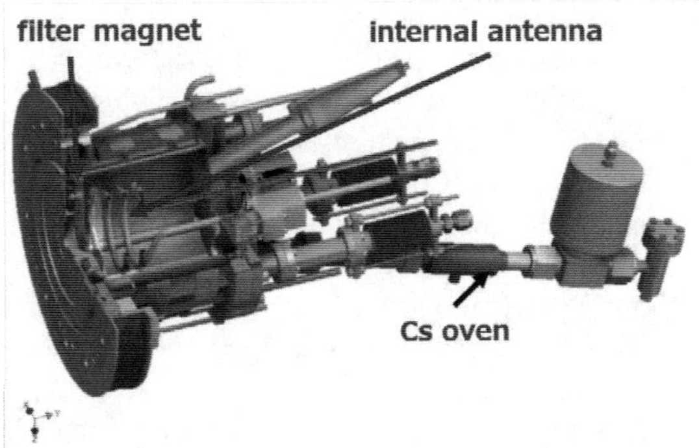

Figure 1. Schematic drawing of the J-PARC RF negative hydrogen ion source.

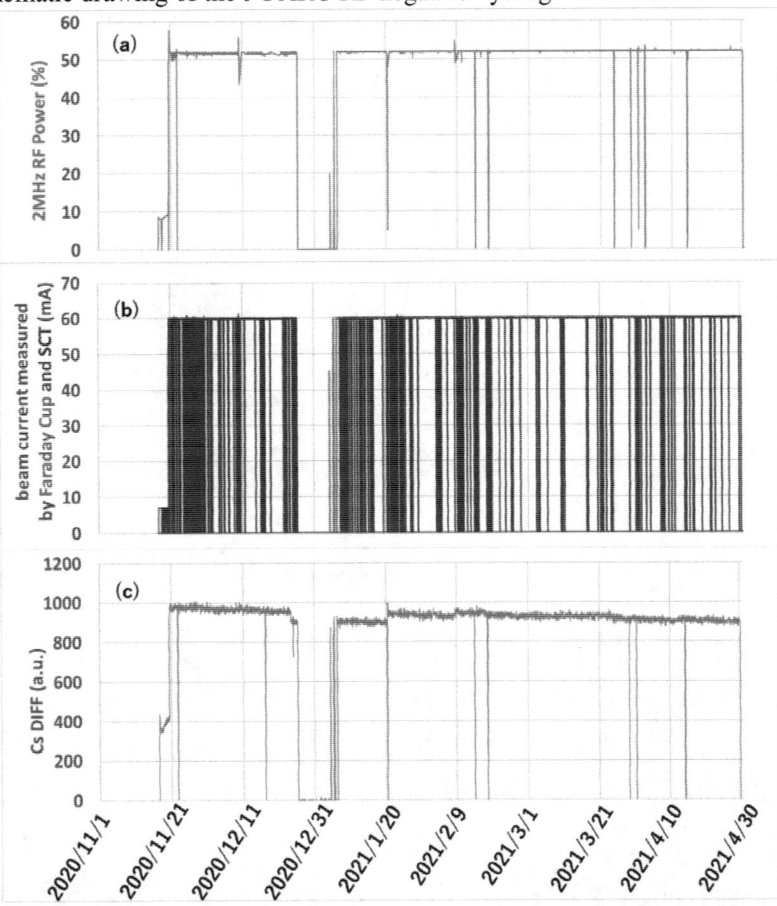

Figure 2. Trend of the ion source operation parameters in the latest long-run (RUN#86) J-PARC user operation. (a) The 2 MHz RF power is given in the occupancy rate of the amplifier. The value is around 52 %, which corresponds to the RF power up to 30 kW. (b) The beam current is measured by Faraday Cup (FC) or Slow Current Transformer (SCT) in the LEBT. (c) The Cs (cesium) DIFF is the line intensity of the Cs^+ emission (852 nm) difference from the base intensity in arbitral units.

3. Long-term continuous operation of J-PARC RF ion source

Figure 2 shows the trend graph of the ion source parameters in the latest long-run user operation (RUN#86) from Nov. 2020 – Apr. 2021. The continuous operation time of the ion source is 3,651 hours (5 months) with the same operation conditions in the previous RUNs. The peak beam current is kept stable up to 60 mA at the Low Energy Beam Transport (LEBT) line which is located downstream of the ion source exit. The fraction of the beam current is 0.3 mA around the peak current. The beam current is measured with the Slow Current Transformer (SCT) which pass through the beam from the ion source to the linac cavities. When a failure takes place in accelerator cavities or RF amplifiers, the trigger of the ion source operation is shifted from the triggers of the RF cavities for safety. Also, the Faraday-Cup (FC) is inserted instead of the SCT to prevent the beam transport to the linac. The ion source is kept in operation during these moments. The beam current value becomes zero at the moment of this trigger shift which are shown as the spikes of the beam current from 60 mA to 0 in Fig.2. There were several short stop durations (0.5 – 2 hours) of the ion source which was not due to the ion source failure (e.g., due to earthquake). At the end of year duration, the ion source operation was paused for a week. The same ion source was equipped before and after this duration.

Figure 3 is the history of the continuous operation time and the beam current increase of the RF ion source (from 2014). The continuous operation duration was extended from 2,445 hours in the RUN#84 before to the present duration by a factor of around 1.5. The ion source life-time is mainly decided by

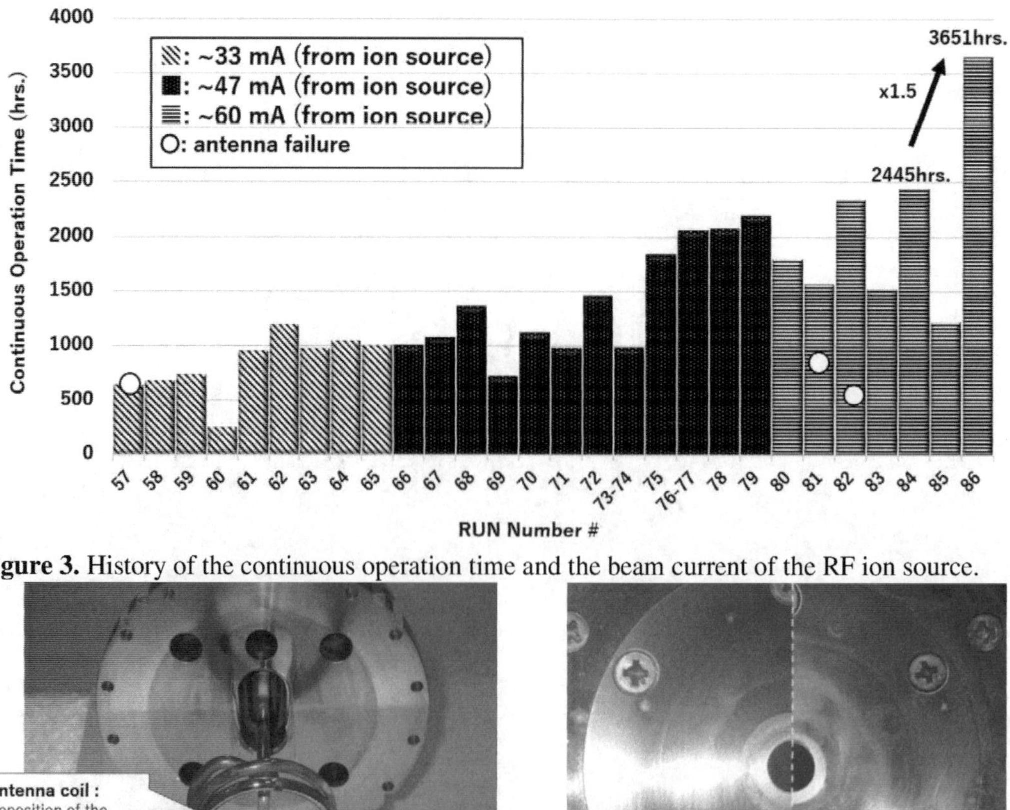

Figure 3. History of the continuous operation time and the beam current of the RF ion source.

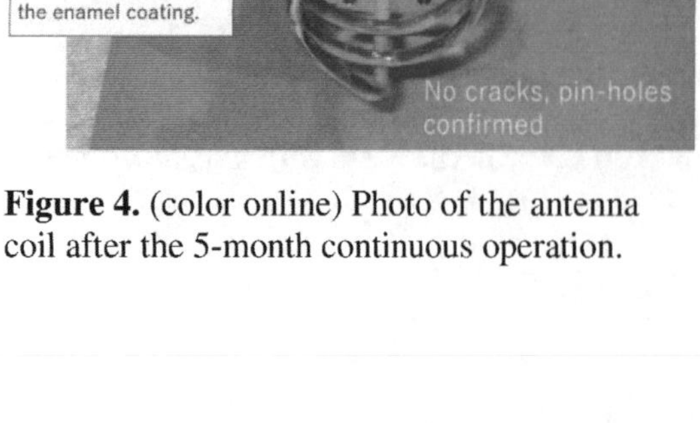

Figure 4. (color online) Photo of the antenna coil after the 5-month continuous operation.

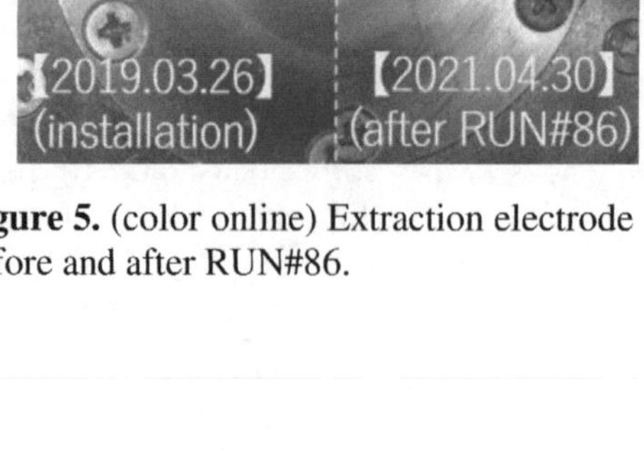

Figure 5. (color online) Extraction electrode before and after RUN#86.

the antenna coil. Figure 4 shows the photo of the antenna coil after RUN#86. Although there was deposition of the metal component which is seen as gloss on the enamel coating, no cracks or pin-holes of the coating itself was confirmed. Thickness of the enamel coating was also measured along the coil. The average thickness of the coating is up to 5.92 ± 0.15 mm before the operation while the thickness is 5.94 ± 0.14 mm after the operation. The difference is smaller than the measurement error.

Figure 5 shows the extraction electrode before and after RUN#86. Although slight damage was observed, no difference in the aperture shape was confirmed. The results indicate that no beam emittance increase takes place during the 5 months due to deformation of the meniscus.

4. Emittance measurement after long-term operation

By sustaining the beam emittance in a constant value, the stable beam acceleration can be obtained. However, the emittance can vary in the long-term operation due to exhaustion of the ion source components. For example, deformation of the electrodes directly changes the initial divergence angle of the extracted beam. On the other hand, exhaustion of the RF antenna leads to shift or non-uniformity of the plasma spatial distribution via reduction of the plasma ignition efficiency. These effects result in the gradual variation of the beam emittance at the RFQ entrance and the resultant transmission ratio. It is confirmed in the Ion Source Test Stand (IS-TS) that the beam emittance is maintained after every user operation which has different durations (for 1 – 5 months). After the user operation, the ion source was installed to the IS-TS equipped with a double-slit EM at the position corresponds to the RFQ entrance in the linac as in figure 6. The detail configuration of the IS-TS is explained in Ref. [4,5]. Resolution of the EM is 0.2 mm and 2 mrad in the real space and the angle axes. In the present measurement, beam current fractions at each phase space mesh point are averaged over the beam extraction pulse width 500

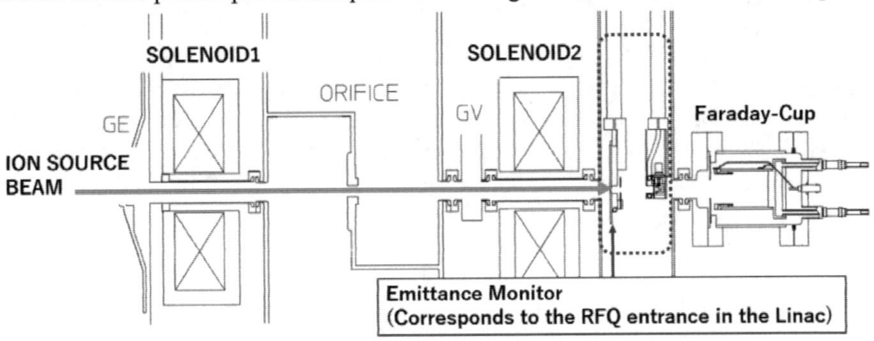

Figure 6. A schematic drawing of the IS-TS and the double-slit emittance monitor.

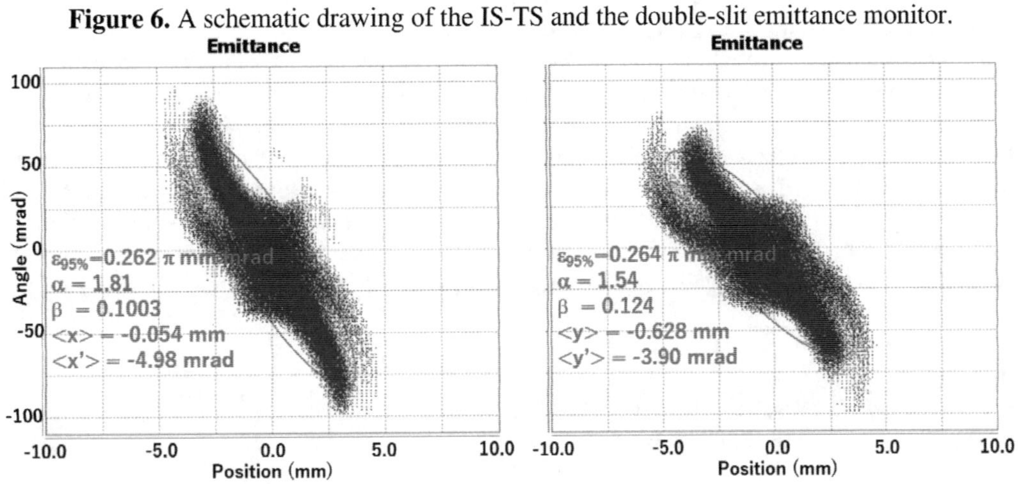

Figure 7. Phase space diagrams of the beam extracted from the ion source after 3,651 hours continuous operation. The horizontal (X) phase space is shown in the left and the vertical (Y) one is shown in the right. The emittance is normalized RMS with 95 % Gaussian.

Table 1. Recent history of the beam emittance of the ion source operated in the J-PARC long-run.

RUN # and period	RUN#83: Nov. 2019 – Dec. 2020	RUN#84: Jan. 2020 – Apr. 2020	RUN#85: May. 2020 – Jul. 2020	**RUN#86: Nov. 2020 – Apr. 2021**
Gas flow rate	23.5 sccm	22.0 sccm	21.0 sccm	**21.0 sccm**
2MHz RF power input	25.5 kW	31.2 kW	27.2 kW	**27.0 kW**
emittance	0.244 / 0.244 π mm mrad	0.277 / 0.273 π mm mrad	0.284 / 0.269 π mm mrad	**0.262/0.264 π mm mrad**

μs. The bias voltage for secondary electron suppression is 500 V.

Figure 7 shows the phase space diagrams of the H^- beam in horizontal (X) and vertical (Y) directions which are perpendicular to the direction of the beam acceleration. For the calculation of the RMS emittance in the analysis software, positions of the beam particles are given randomly according to the beam current fractions at each phase space point. The normalized RMS emittance in X and Y directions are calculated as

$$\varepsilon_{\mathrm{RMS},95\%} = 4\sqrt{\langle u^2\rangle\langle u'^2\rangle - \langle uu'\rangle^2}$$

from the beam phase space positions (u, u') in the real space and the angle. In the figure, the normalized emittance is shown in the units of π mm mrad.

The operation condition of the IS-TS is the same as in J-PARC Linac in RUN#86, which optimizes the RFQ transmission rate. In these settings, the phase spaces and the RMS emittances in X and Y directions matches well. History of the beam emittance in the recent long-run operations measured in the same manners is shown in table 1. The average beam emittance in these two years is 0.262 – 0.266 π mm mrad. The emittance of the ion source operated in the RUN#86 is 0.262 and 0.264 π mm mrad in X and Y directions, respectively. This result shows that the beam emittance is kept stable even after the 5-month continuous operation.

5. Summary

For achieving the RF ion source without replacing during the J-PARC user operation period, sequences of the long-run experiments have been performed. In the user operation from Nov. 2020 – Apr. 2021, the continuous operation of the ion source was up to 3,651 hours. The emittance measurement at the IS-TS shows that the no emittance increase has been observed. The normalized RMS emittance is 0.262 and 0.264 π mm mrad, which is same as the average emittance in these 2-year operation. From the direct investigation of the antenna coil, no crack, pin-holes and the thickness variation of the enamel coating has been observed. Also, no deformation of the extraction aperture has been confirmed. From these results, ability of J-PARC RF ion source for the 5-month continuous operation with the peak current up to 60 mA, the duty factor of 2 % and the injection power up to 30 kW has been confirmed. The next goal is to achieve 7-month long-run, which corresponds to the full duration of J-PARC user operation in 1 year.

References

[1] H. Oguri, *et al.*, JPS Conf. Proc. **33**, 011008 (2021).
[2] K. Shinto, *et al.*, AIP Conf. Proc. **2052**, 050002 (2018).
[3] M. P. Stockli, et al., Rev. Sci. Instrum. **81**, 02A729 (2010).
[4] T. Shibata, *et al.*, AIP Conf. Proc. **2373**, 050002 (2021).
[5] T. Shibata, *et al.*, JPS Conf. Proc. **33**, 011009 (2021).

Journal of Physics: Conference Series **2244** (2022) 012042 doi:10.1088/1742-6596/2244/1/012042

Simulation of Beam Formation in the CERN Negative Ion Source for the Linac4 Accelerator

A. Vnuchenko[1], J. Lettry[1], S. Mochalskyy[2], D. Wünderlich[2], U. Fantz[2], A. Revel[3], T. Minea[3]

[1] CERN, Esplanade des Particules 1, P.O. Box, 1211 Geneva 23, Switzerland
[2] Max-Planck-Institut für Plasmaphysik, Boltzmannstr. 2, D-85748 Garching, Germany
[3] Laboratoire de Physique des Gaz et des Plasmas, UMR 8578 CNRS, Université Paris-Saclay, Bat. 210, rue Henri Becquerel, 91405 Orsay, France

anna.vnuchenko@cern.ch

Abstract. Linac4 is the negative hydrogen ion (H^-) injector of the CERN accelerator complex. Modelling of the beam formation is essential for optimizing the current and emittance of the H^- ion source. We exploited the 3D PIC-Monte Carlo ONIX (Orsay Negative Ion eXtraction) code for studying H^- beam formation processes in caesiated negative ion sources. The various geometries of the IS03 prototypes have been implemented into ONIX. The code, designed for neutral injector multi-aperture sources for fusion has been adapted to match the single-aperture extraction region of the Linac4 H^- source. A plasma electrode designed to ensure radial metallic boundary conditions was produced and tested. The simulation results of the beam formation region at low plasma density to validate the functionality of the modified ONIX version are presented.

1. Introduction

The Radio Frequency Inductively Coupled Plasma (RF-ICP) H^- source prototype (IS03b) is being operated on CERN's Linac4 [1-3]. This caesiated molybdenum-surface plasma source produces H^- through the volume and surface production mechanisms [4–6]. Details of the plasma generator and operation under Cs-loss compensation can be found in reference [7].

The initial formation of the beam determines the design of the required beam optics. The properties of the ion beams after their extraction from the ion source strongly affect ion losses during the beam transport in Low Energy Beam Transport Lines (LEBT) and efficiency of particle injection into accelerators. Modelling the beam phase space around the meniscus region (positions and velocities of each ion and electron) shall provide proper input to beam transport engineering, minimization of emittance growth and beam losses. The current densities, shape and the position of the meniscus are key parameters for the beam formation.

The physics of the meniscus is complex due to the presence of surface produced negative ions and the three-dimensional magnetic field topology. The magnetic field in the beam extraction system is formed to deflect and dump the co-extracted electrons into the puller-dump electrode. The filter and dump fields are generated by pairs of permanent dipole magnets located around the plasma and puller-dump electrodes (PE, PD). The filter field strength has to ensure a low electron energy in the beam formation region upstream of the PE aperture with minimal influence on H^- trajectories. Therefore, simulations of the self-consistent plasma in beam formation region are mandatory to understand the processes of negative ion extraction.

2. Simulation model

Negative ion beam formation and plasma in the vicinity of the PE extraction aperture have been simulated with the ONIX code [8-11], used to the modelling of the extraction of negative hydrogen ions and co-extracted electrons in negative ion multi-aperture sources for ITER's Neutral Beam Injector (NBI). ONIX has been modified and adapted to match extraction region of the Linac4 H^- source [12].

Boundary conditions of the simulation volume in directions orthogonal to the beam axis were assumed non-periodic in order to define the single-aperture extraction region of the Linac4 H^- source. In this case, all plasma particles that strike the boundaries in y and z directions and left side wall are reinjected in the "bulk" plasma, following certain rules. Sketches of the simulation region are shown in figure 1. Standard plasma electrode (PE45) and a new plasma electrode geometry (PE75) have been implemented to the code. PE75 was produced to provide simple metallic radial boundary conditions at the periphery of the beam formation region of the ONIX simulation region. The extraction potential of 10 kV is applied to the right boundary of the domain, the remaining domain boundary potentials are assumed to be zero. The initial bulk plasma covers the first 8 mm of the simulation domain, the bulk plasma composition and energy distributions were taken from previous measurements [13].

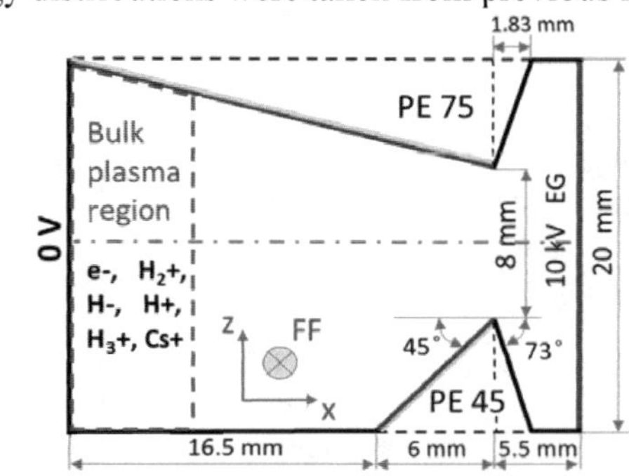

Figure 1. Schematic view of the simulation domains used in the ONIX code (x–z mid plane) for modelling beam formation of IS03b PE45 bottom and PE75 top. The filter field (FF) orientation is indicated. The bore diameter of PE is 8 mm.

3. Result and discussions

The simulation has been performed on a CERN cluster using 20 CPUs (total 360 cores) with a meshing of 420×310×310 PIC nodes and a time step 5×10^{-12} s. The cell size is 65 μm, slightly larger than the Debye length ($\lambda_D \approx 41$ μm). These numerical parameters are sufficient for PIC stability requirements for a plasma density of 10^{16} m^{-3} in the bulk region. The H^- plasma surface production mode has been simulated assuming a constant homogenous H^- emission rate of 550 Am^{-2} from the caesiated Mo-plasma electrode surface. This yield was theoretically estimated for the fusion negative ion sources [14].

3.1. Simulation of PE45

In the initial state, the particles are uniformly distributed in the bulk plasma region. The plasma expands into the simulated domain due to the extraction potential, the populations of the plasma migrate according to their energy to mass ratios. The self-consistent meniscus is formed in the vicinity of the plasma electrode aperture. The meniscus stabilizes only once all plasma populations expanded into the beam formation region. The meniscus position and the depth of its curvature play an important role for the beam formation since they define the velocity and angle of trajectory of each extracted particle. The density maps of electrons, positive ions and H^- in the (x-y) and (x-z) planes of the PE45 simulation domain are shown in figure 2, after steady state was reached. The electron density decrease in the x direction is caused by the magnetic field. The H^- density distribution is different in the (x-y) and (x-z) planes. The reason is that the filter field is predominant in the y direction and limits the electron flow, that influences the distribution of the positive charged species.

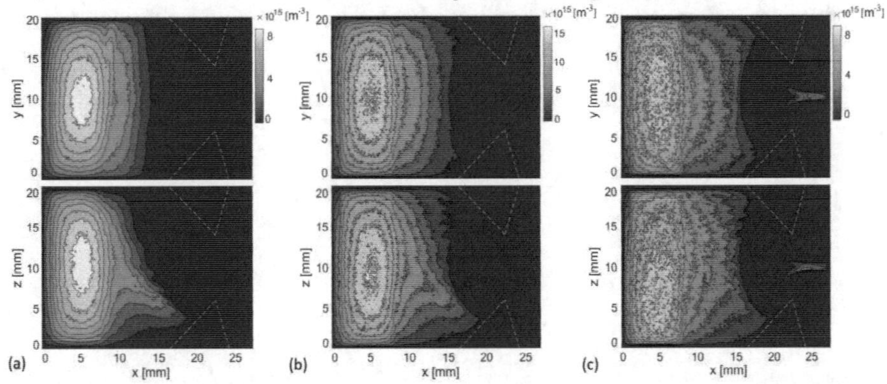

Figure 2. Density maps of electrons (a), positively charged particles (H^+, H^{2+}, H^{3+}) (b) and H^- (c) in the (x-y) (top) plane and (x-z) (bottom) plane where filter field generated asymmetry is clearly visible.

The particle is considered extracted when crossing the right boundary of the simulation domain. Figure 3a shows the evolution of the H^- and co-extracted electron beam currents during simulation. Rapid increase of electron current is observed at the beginning of the simulation. After this transitory phase, the electron and H^- currents stabilize, and the system evolves asymptotically towards a quasi-steady state. The value of both currents, for co-extracted electrons and H^-, is about 9 mA. The H^- current includes volume and surface production modes. The e/H current ratio is about 1, that corresponds to typical values for a well-caesiated source. The current is significantly lower than the experimental values, this is due to the low plasma density used as input for simulations to fulfil the stability criteria. The extracted H^- and co-extracted electron currents calculated for reduced plasma densities, provide the first microscopic information on the kinetics of plasma species and the beam formation in this type of plasma source.

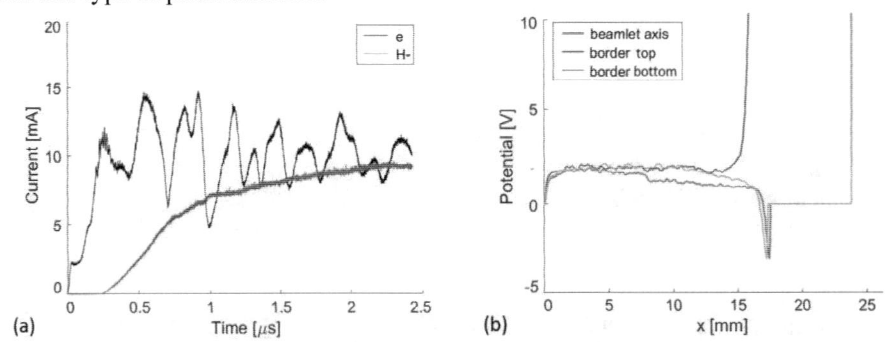

Figure 3. Time evolution of the extracted H^- current (red line) and co-extracted electron (black line) current for the PE45 system using an extraction voltage of 10 kV and a magnetic field of 11 mT (a); potential profile along the simulation domain (b).

The potential along the plasma region has an almost constant value of about 2 V, which is in agreement with the sheath potential expected for a plasma with the electron temperature of 1 eV [13]. Close to the meniscus the potential increases due to the applied extraction potential of 10 kV. The electrostatic potential profile along a beamlet axis and close to the border of the calculation domain is shown in figure 3b. The plasma density drastically drops close to the PE aperture and Debye length condition is completely fulfilled in the region of the plasma sheath including the meniscus. The depth of the plasma sheath is about 2 V near the PE aperture.

The influence of the magnetic field strength on the extracted current was considered additionally, see figure 4. The co-extracted electron current increases significantly with reducing of magnetic field, while the value of the proton yield remains unchanged. This demonstrates that the magnetic field of 11–13 mT meets the requirements for reducing the co-extracted electron current in the experiment.

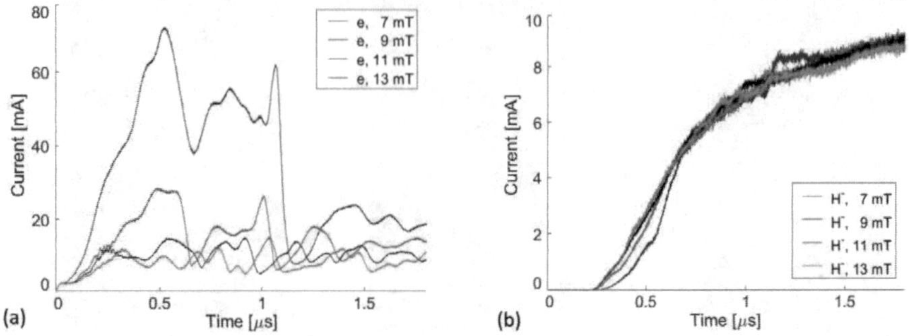

Figure 4. Time evolution of co-extracted electron (a) and the extracted H^- (b) current for the PE45 system simulated with different magnetic field.

The density profiles of electrons, negative and positive ions along a beamline and H^- profile at extraction aperture depend on their production origin in x-y and x-z plane are shown in figure 5. The simulation indicates that the majority of extracted H^- are created at the conical part of the PE. At the same time, it is observed that the production of H^- from the surface of PE is uniform, while the particles from the bulk plasma region are strongly influenced by the magnetic field. This distribution of H^- extracted from simulation domain indicates the appearance of a halo around the main beam.

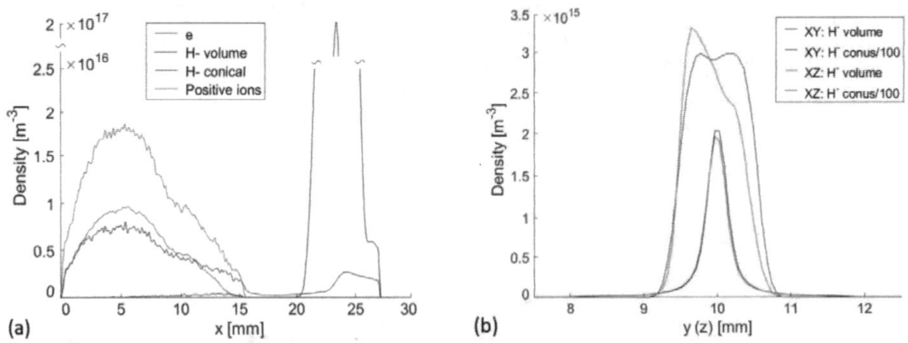

Figure 5. Density profile of electrons, negative and positive ions along a beamline (a) and H^- profile at extraction aperture (b).

3.2. Simulation of PE75

Similar simulations were carried out for the PE75 with the same settings. For this case, a significant decrease in H^- current was observed, see figure 6. The value of H^- current is about 4.5 mA, that is two times lower than for PE45 geometry. The e/H current ratio is below 1.

Comparative analysis of the density of positive and negative charged particles was performed with and without an applied extraction potential, see figure 7. The non-uniform density profiles of positive and negative ion are observed at extraction potential of 10 kV, but with significant lower impact of magnetic field on the meniscus. The negative ions produced in the surface of the plasma electrode are transported more uniformly due to the large emission surface over the entire domain. For the simulation without the applied voltage, the density distribution of the particles is completely different. Negative particles remain in the emission area (figure 7d), while positive particles are more evenly distributed over the entire domain (figure 7c).

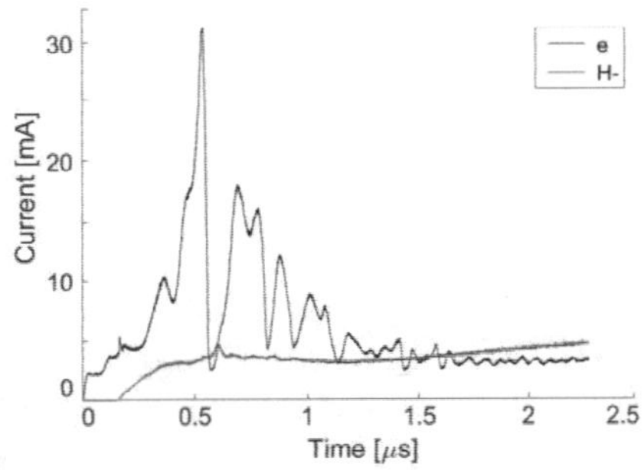

Figure 6. Time evolution of the extracted H^- and co-extracted electron current for the PE75 system.

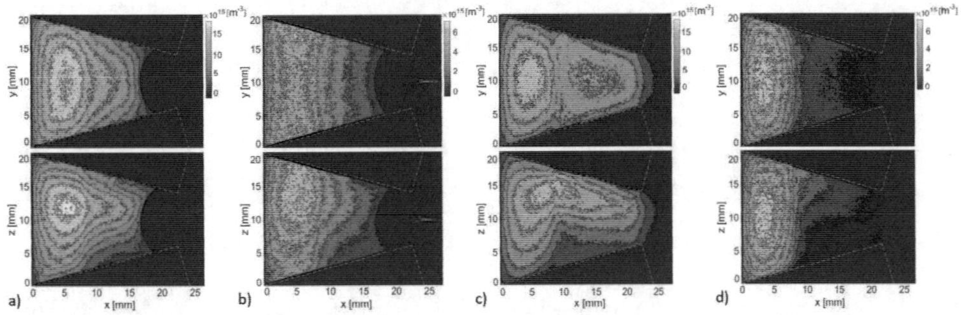

Figure 7. Density maps of positively charged particles (H^+, H^{2+}, H^{3+}) and H^- using an extraction voltage of 10 kV (a – b) and 0 V (c – d) in the (x-y) (top) plane and (x-z) (bottom)

4. Conclusion outlook

A modified version of ONIX code with a single extraction aperture and non-periodic boundary conditions to model the beam formation region of the Linac4 ion source was successfully tested on a CERN's computing cluster using low plasma density. The non-uniform profile of the current density was observed in the simulations of two type of plasma electrodes due to the influence of the magnetic field. As a result, the meniscus shape is strongly curved and the beam extracted from the peripheral region is uniform, that could lead to the formation of so-called beam halo. Relying on the demonstrated result, simulation of higher plasma densities with different species energy distributions and proportions are planned to be performed with initial conditions refinements base on analysis from OES system installed around the plasma generator at the Linac4 ion source test stand [13].

References

[1] Arnaudon L et al 2006 Linac4 Technical Design Report Rep. CERN-AB-2006-084 ABP/RF
[2] Lettry J et al 2017 AIP Conf. Proc. 1869 030002
[3] Noll D et al 2019 Proc. Conf. IPAC19 (Melbourne, Australia) 1090-1093
[4] Bacal M, Hatayama A and Peters J 2005 Volume production negative hydrogen ion sources IEEE Trans. Plasma Sci. 33 1845-1871
[5] Bel'chenko, Yu Dimov, G Dudnikov V and Ivanov A 1973 Dokl. Akad. Nauk SSSR 213 1283
[6] Bacal M and Wada M 2015 Appl. Phys. Rev. 2 021305
[7] Lettry J et al 2018 AIP Conf. Proc. 2052 050008
[8] Mochalskyy S et al 2013 AIP Conf. Proc. 1515 31-40
[9] Mochalskyy S, Lettry J and Minea T 2016 New Journal Phys. 18 085011
[10] Revel A et al 2017 J. Appl. Phys. 122 103302
[11] Wünderlich D et al 2021 Journal of Applied Physics 130, 053303
[12] Vnuchenko A et al 2021 Proc. Conf. IPAC21 (Brazil, South America)
[13] Briefi S, Fink D, Mattei S, Lettry J and Fantz U 2016 Rev. Sci. Instrum. 87 02B104
[14] Wünderlich D, Gutser R and Fantz U 2009 Plasma Sources Sci. Technol. 18 045031

Beam instability in the vicinity of beam extraction region of negative ion source

K. Nagaoka[1,2], R. Nakamoto[1,3], T. Sasaki[3], T. Hamajima[2], H. Nakano[1,4], K. Ikeda[1], Y. Fujiwara[1,5], M. Osakabe[1,5] Y. Takeiri[1], and K. Tsumori[1,5]

[1]National Institute for Fusion Science, National Institutes of Natural Sciences, Toki, Gifu 509-5292, Japan
[2]Graduate School of Science, Nagoya University, Nagoya, Aichi 464-8603, Japan
[3]Graduate School of Engineering, Nagaoka University of Technology, Nagaoka, Niigata 940-2188, Japan
[4]Graduate School of Engineering, Nagoya University, Nagoya, Aichi 464-8603, Japan
[5]The Graduate University for Advanced Studies, SOKENDAI, Toki, Gifu 509-5292, Japan

Corresponding author: nagaoka@nifs.ac.jp

Abstract. Beam instability in the presheath region of negative ion beam extraction is investigated in theoretically and experimentally. The linear stability analysis shows that the beam instability is unstable due to coupling between positive ion flow and negative ion flow. On the other hand, no clear activity can be seen in the experiment in the frequency range predicted by the theory. The beam instability in the presheath region of negative ion beam extraction may not cause the degradation of the beam focusing because of collisional damping and/or Landau damping.

1. Introduction

Negative ion beam technique is applied to plasma experiment for fusion research, particle physics experiment using huge accelerators, Boron neutron capture therapy etc. The negative ion beam focusing is a common issue for all applications. For example, the negative ion beam focusing is one of the most important issues for development of RF negative ion source for ITER [1-2]. Recently, an oscillation of negative ion beamlet with the same frequency as the RF frequency was observed by the emittance measurement with high-time resolution in the RF negative ion source for J-PARC [3]. In the filament-arc type negative ion source, the beamlet was also observed to responses to the plasma density oscillation in the vicinity of the plasma grid [4]. These experimental observations suggest the importance of the meniscus stability for the negative ion beam focusing. Moreover, it was found that the beam instability (two-stream instability) may cause density fluctuation in the presheath region [5-6].

Therefore, the beam instability was investigated theoretically and experimentally in this study. The linear stability analysis is discussed with the parameter regime of negative ion source plasma in section 2. The experimental observation of the fluctuation of the negative ion beamlet is presented in section 3, and summary is given in section 4.

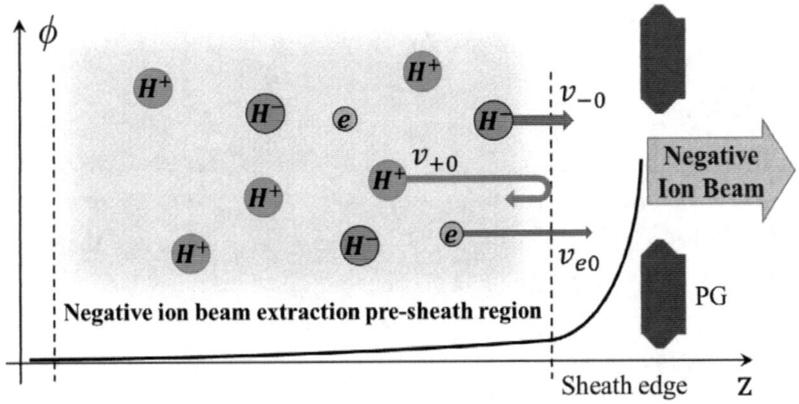

Fig. 1 Schematic of the pre-sheath region of negative ion extraction boundary.

2. Linear stability analysis of beam instability

The negative ion source plasma composed with positive ions, negative ions and electrons is considered here. When negative ion beam is extracted from the source plasma, the pre-sheath (finite electric field region) penetrates deeply from the plasma grid (PG) aperture toward the plasma (see Fig. 1). The linear stability analysis for the negative ion source plasma is described here. The set of equations are equation of motion and equation of continuity given by

$$\frac{\partial v_j}{\partial t} + v_j \frac{\partial v_j}{\partial z} = -\frac{e_j}{m_j} E, \tag{1-1}$$

$$\frac{\partial n_j}{\partial t} + \frac{\partial}{\partial z}(n_j v_j) = 0, \tag{1-2}$$

where v_j, e_j, m_j, n_j and E are velocity, charge, mass, density of the j particle species and electric field, respectively. The subscript $j = +$ for positive ion, $-$ for negative ion and e for electron. We consider streaming negative ions, streaming electrons and fixed positive ions, then linearization is given by

$$v_+ = v_{+1} \quad (2\text{-}1) \qquad n_+ = n_{+0} + n_{+1} \quad (3\text{-}1)$$

$$v_- = v_{-0} + v_{-1} \quad (2\text{-}2) \qquad n_- = n_{-0} + n_{-1} \quad (3\text{-}2)$$

$$v_e = v_{e0} + v_{e1} \quad (2\text{-}3) \qquad n_e = n_{e0} + n_{e1} \quad (3\text{-}3)$$

where the equilibrium components are indicated by a subscript "0" and the perturbation components are indicated by a subscript "1". The charge neutrality is kept for the equilibrium density component ($n_{+0} = n_{-0} + n_{e0}$). The Poisson's equation is given by

$$\frac{\partial E_1}{\partial z} = \frac{e}{\varepsilon_0}(n_{+1} - n_{-1} - n_{e1}) \tag{4}$$

Using Fourier transformation and plane wave approximation, the dispersion relation is given by

$$\Omega_{pi}{}^2 \left\{ \frac{1}{\omega^2} + \frac{1-\alpha}{(\omega - kv_{-0})^2} + \frac{m_+}{m_e}\frac{\alpha}{(\omega - kv_{e0})^2} \right\} = 1 \tag{5}$$

where $\Omega_{\mathrm{pi}} = \sqrt{n_{+0}e^2/\varepsilon_0 m_+ e}$ is ion plasma frequency, and $\alpha = n_{e0}/n_{+0}$ is the electron density normalized by positive ion density. Three terms in the left-hand side of the eq. (5) represent positive ion contribution, negative ion contribution, and electron contribution, respectively. Low frequency mode is our interest, thus, the third term is neglected. In order to investigate the properties of the solution of the dispersion relation, we define a function as

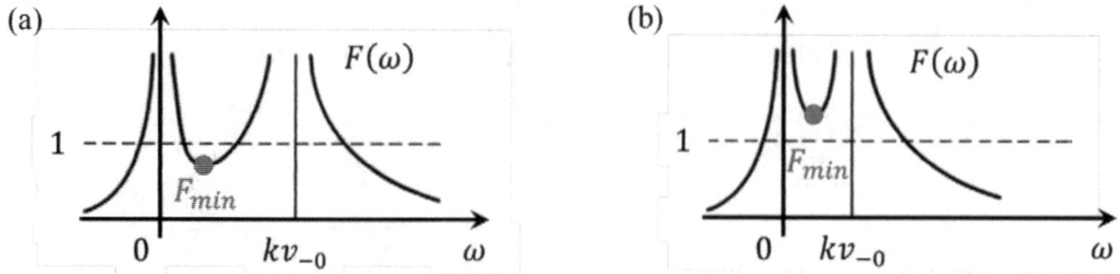

Fig. 2 The function $F(\omega)$ has four real roots is shown in (a) and two real roots and two imaginary roots is show in (b).

$$F(\omega) \equiv \Omega_{pi}{}^2\left\{\frac{1}{\omega^2}+\frac{1-\alpha}{(\omega-kv_{-0})^2}\right\}=1 \tag{6}$$

When eq. (6) has complex solutions, the mode become unstable. It can be evaluated by whether extremity value is larger than unity or not, which is shown in Fig. 2. The condition that the mode is unstable is given by

$$F_{min}=F(\omega_m)=\frac{\Omega_{pi}{}^2}{(kv_{-0})^2}(1+\beta)^3>1 \tag{7}$$

where $\omega_m$is the root of $\partial F/\partial\omega=0$ and $\beta=\sqrt[3]{1-\alpha}$. The negative ion velocity is written by mean of the electric potential ϕ; ($v_{-0}=\sqrt{2e\phi/m}$), then this inequality can be solved for the electrical potential as

$$0<\phi<\frac{\Omega_{pi}{}^2 m_-}{2ek^2}(1+\beta)^3 \tag{8}$$

The stability boundary in the ϕ and k space is shown in Fig. 3 (a). The parameter regime of the negative ion source plasmas (NIFS-RNIS : National Institute for Fusion Science - Research and development Negative Ion Source) is in the unstable region. The frequency of the beam instability is given by

$$\omega_{Re}=\omega_m=\frac{k}{1+\beta}\sqrt{\frac{2e\phi}{m_-}} \tag{9}$$

The predicted frequency range is the order of 10^5-10^6 Hz, which is shown in Fig. 3 (b). The linear stability analysis of the beam instability shows that the mode is unstable in the presheath region of negative ion beam extraction in negative ion source (NIFS-RNIS).

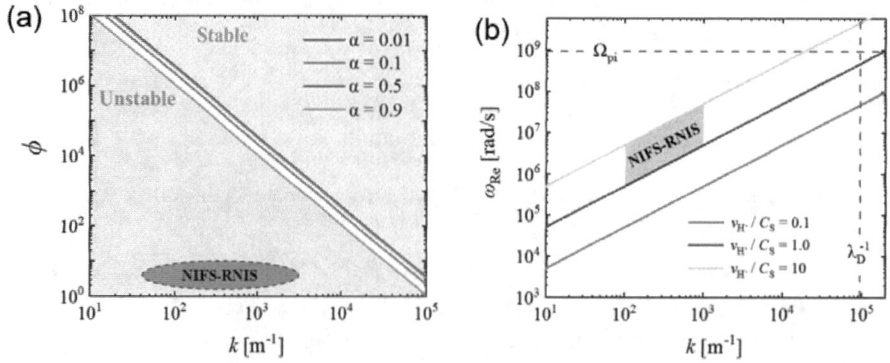

Fig. 3 (a) The stability boundary in p and k space. (b) The frequency of the beam instability. The parameter regime of NIFS-RNIS is also shown.

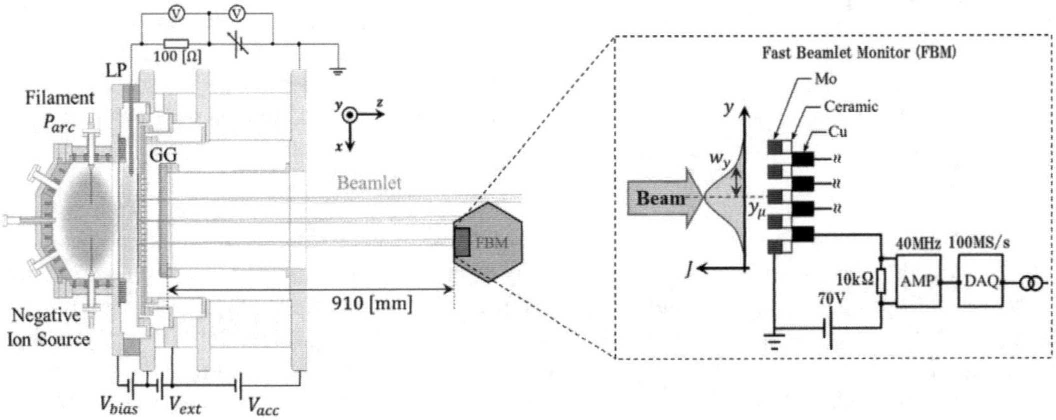

Fig. 4 Schematic of the experimental setup. Langmuir probe (LP) and fast beamlet monitor are used for fluctuation measurements in negative ion source plasma and the beamlet.

3. Experiment

The experiment to confirm the linear stability analysis for the beam instability in the presheath of negative ion beam extraction was conducted at NIFS-NBTS (Negative ion Beam Test Stand at National Institute for Fusion Science). The NIFS-RNIS, which is a filament-arc type negative ion source with one third scaled the negative ion source for the LHD (Large Helical Device), was used. The schematic of this experiment is shown in Fig. 4. The density fluctuation in the vicinity of the PG is measured by a Langmuir probe (LP) [7]. The location of the probe measurement is 12 mm from the PG, and at the almost center of the aperture, where the presheath is formed when the extraction voltage is applied to the second grid (extraction grid). The voltage applied to the LP is swept from -40 V to +40 V with repetition frequency of 20 Hz. The probe signal is digitized with 50 MS/s. The beamlet profile was measured by fast beamlet monitor (FBM), which is a 32ch Faraday cup array with the time resolution of 40 MHz [8]. The secondary electron emission current is reflected with the bias voltage to the collectors. The FBM is inserted at the 910 mm downstream from the NIFS-RNIS. The beamlet to which LP is applied is different from that where FBM measurement is carried out to avoid the disturbance to beamlet formation / beam focusing.

The parameters of this experiment are 45 – 75 kW of arc power, beam extraction voltage of 3 kV and acceleration voltage of 30 - 43 kV. The typical example of this experiment is shown in Fig. 5. The probe voltage is swept up in time and the jump down at $t = 0.045$ s and $t = 0.095$ s. The frequency

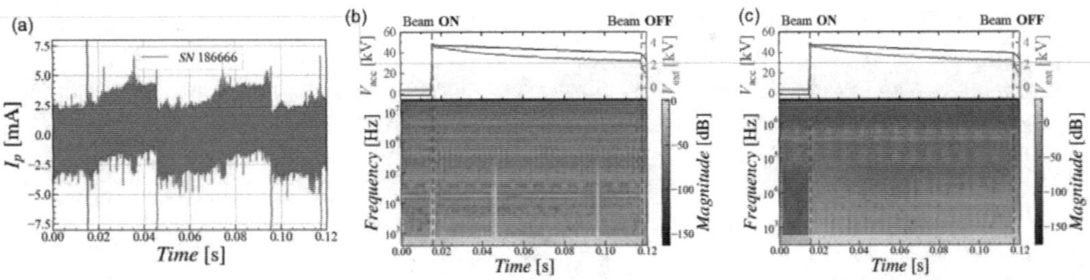

Fig. 5 (a) Time evolution of probe current. The probe voltage is swept in time and jumps at $t =$ 0.045 s and $t = 0.095$ s. (b) The time evolutions of beam extraction and acceleration voltage and the frequency spectra for the probe current in the source plasma. (c) The time evolutions of beam extraction and acceleration voltage and the frequency spectra for beamlet current density measured by FBM.

spectra of probe current and negative ion beam current measured by the FBM are shown in Fig. 5 (b) and (c), respectively. One can see no clear activities, which is excited only when beam extraction turns on, in the frequency range predicted by the linear stability analysis. Therefore, it is found that the beam instability in the presheath region of negative ion beam extraction does not drive density fluctuation and does not cause the degradation of the negative ion beam focusing.

4. Concluding remarks

The negative ion beam focusing is a common issue in the development of negative ion source for a variety of application. The beam instability in the presheath of negative ion beam extraction was investigated in this study for negative ion source for fusion research. The linear stability analysis shows the beam instability become unstable due to the positive ion – negative ion coupling. The frequency of the unstable mode is 10^5-10^6 Hz. However, the activity was not observed in the experiment carried out at NIFS-NBTS. The collisional damping is a candidate to explain the experimental observation because the ion-ion collision frequency is close to the growth rate of the beam instability. Landau damping is also considered, but the precise ion velocity distribution function measurement is necessary to evaluation of the damping rate. It should be noted that the beam instability does not degrade the beam focusing of the negative ion source for LHD, and the situation could be the same for ITER because the parameter regime in the vicinity of the plasma grid of RF type negative ion source for ITER is almost identical to that for LHD.

Acknowledgments

The authors would like to thank the technical staffs of the LHD-NBI group for supporting the experiment. This research is supported by the NIFS (NIFS20KLER103) and JSPS KAKENHI (17H03002 and 18KK0080).

References

[1] B Heinemann, Ursel Fantz, W Kraus, L Schiesko, Christian Wimmer, Dirk Wünderlich, F Bonomo, et al., New Journal of Physics 2017, **19**, 015001.
[2] M Barbisan, F Bonomo, Ursel Fantz, and Dirk Wünderlich, Plasma Physics and Controlled Fusion 2017, **59**, 055017.
[3] T. Shibata, K. Shinto, M. Wada, H. Oguri, K. Ikegami, K. Ohkoshi, and K. Nanmo, AIP Conference Proceedings 2021 **2373**, 050002.
[4] Y. Haba, K. Nagaoka, K. Tsumori, M. Kisaki, K. Takahashi, H. Nakano, K. Ikeda, S. Yoshimura, and M. Osakab, Japanese Journal of Applied Physics 2020, **59**, SHHA01.
[5] B. Scheiner, S.D. Baalrud, B.T. Yee, M.M. Hopkins, and E.V. Barnat, Physics of Plasmas 2015, **22**, 123520.
[6] R. Hood, S.D. Baalrud, R.L. Merlino, and F. Skiff, Physics of Plasmas 2020, **27**, 053509.
[7] K. Tsumori and M. Wada 2021 Appl. Phys. Rev. **8**, 021314
[8] Y. Haba, K. Nagaoka, K. Tsumori, M. Kisaki, H. Nakano, K. Ikeda, Y. Fujiwara,S. Kamio, S. Yoshimura, and M. Osakab, Review of Scientific Instruments 2018, **89**, 123303.

H^- ion source operational performance and latest development at the Spallation Neutron Source

Baoxi Han, Martin Stockli, Robert Welton, Chip Piller, Syd Murray Jr., Terry Pennisi, Chris Stinson, and Sang-Ho Kim

Spallation Neutron Source, Oak Ridge National Laboratory, Oak Ridge, TN 37831, USA

Corresponding author: hanb@ornl.gov

Abstract. The U.S. Spallation Neutron Source (SNS) at Oak Ridge National Laboratory is the world's most powerful accelerator-driven pulsed neutron source. An H^- injector feeds the accelerator with high current (>50 mA), time structured (1ms, 60 Hz) H^- beam. The injector consists of an RF-driven, Cs-enhanced H^- ion source and an electrostatic low energy beam transport section. In the recent three run cycles, an H^- ion source operated ~4 months for each run with excellent availability near 99.9%. The presently ongoing run is also on track to finish another ~4 months run with similar high availability. Lately, we have tested a solid-state RF system for the ion source to replace the existing vacuum-tube type RF amplifier for the sake of improving the ease of operation and maintenance. Progress has also been made on the continued development of an external antenna RF H^- ion source with a focus on improvement to its plasma ignition scheme.

1. Introduction

The U.S. Spallation Neutron Source (SNS) at Oak Ridge National Laboratory is the world's highest power accelerator-based pulsed neutron facility. Typically, the SNS accelerator delivers ~35GW of instantaneous power proton pulses to a mercury target for a duration of ~700-800 ns at 60 Hz (average power ~1.4 MW) producing intense, short-pulse neutrons for scientific research for a broad range of disciplines, such as physics, chemistry, biology, and materials science. The accelerator system of the SNS consists of a 65 keV H^- injector, a 2.5 MeV RFQ, a 1-GeV linac chain (DTL-CCL-SCL), and a proton accumulator ring. The accumulator ring requires ~35 mA, time-structured (60 Hz, 1.0-ms with ~200-300 ns chopped for every ~1 μs) H^- beam delivered from the linac. The H^- injector produces and delivers the beam current with the required time structure to the RFQ accelerator. Depending on the operation power of the RFQ, the injector is required to deliver beam current varying from ~40 mA to ~60 mA to achieve ~35 mA in the linac. The H^- injector includes an RF-driven, Cs-enhanced, multi-cusp H^- ion source and a compact, two-lens electrostatic Low Energy Beam Transport (LEBT) system. Figure 1 shows a schematic view of the H^- injector coupled to the RFQ accelerator. The ion source plasma is driven by a 2-MHz RF through a 2.5-turn porcelain-coated, water-cooled, copper-tube antenna which is placed internally in the ion source chamber. Typically, the 2-MHz RF is operated at 60 Hz with 1 ms pulse length with 40-65 kW to generate the required H^- beam. To facilitate fast and reliable ignition of the 2-MHz high-density pulsed plasma, a continuous, low power (typically ~300W) 13.56-MHz RF maintains a low-density background plasma in the chamber. A solid reaction cesium dispenser system (Cs_2CrO_4 + Zr, Al) is used for ion source cesiation. A Mo cone attached to the

---

This manuscript has been authored by UT-Battelle, LLC, under contract DE-AC05-00OR22725 with the US Department of Energy (DOE). The publisher acknowledges the US government license to provide public access under the DOE Public Access Plan (http://energy.gov/downloads/doe-public-access-plan).

cesium dispenser and fixed near the ion source outlet aperture enhances H^- yield through negative ion conversion on its cesiated surface. The co-extracted electrons are separated from the H^- beam and dumped on the e-dump electrode by a transverse magnetic field generated in the ion source extraction region by an array of permanent magnets that are embedded in the outlet electrode. The effect of the transverse magnetic field on the H^- beam can be compensated by mechanically tilting and offsetting the ion source against the LEBT beam axis. The second lens electrode of the LEBT in front of the RFQ entrance is split into electrically isolated four quadrants to allow application of individual voltages for beam steering and chopping. The LEBT output / RFQ input beam currents can be measured on the chopper target by deflecting the entire beam onto it and draining it through a resistor circuit. The beam output from the RFQ is continuously monitored with a beam current monitor toroid (BCM). Besides the H^- injector on the SNS accelerator system, we also operate two similar H^- injectors on test facilities for ion source and LEBT development, and for new RFQ commissioning and various beam studies [1-5].

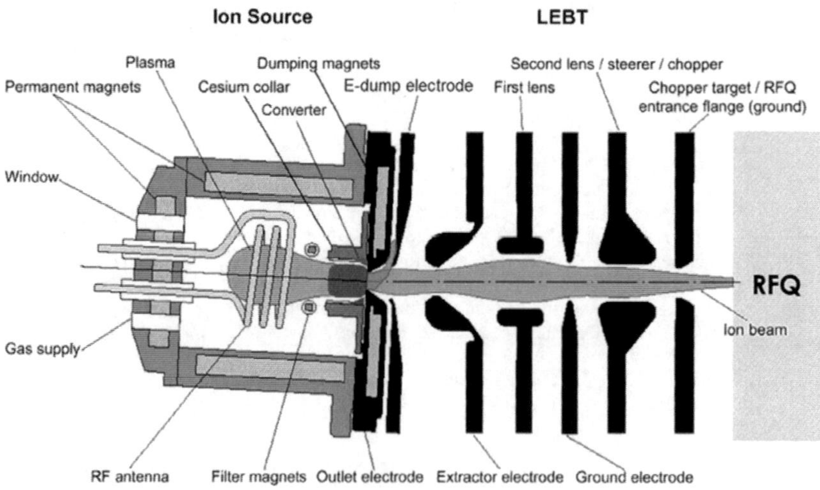

Figure 1. A schematic view of the SNS H^- injector coupled to the RFQ entrance

2. Recent operational performance of the H^- ion source on the SNS accelerator

The SNS H^- injector performance has been improved over the years both in beam current capability and operational reliability [2]. The ion source and LEBT are individual assemblies, they can be removed and refurbished as needed after a certain service duration. A LEBT can typically operate for a year without significant degradation of voltage holding capabilities. The service duration of an ion source has recently been extended to ~4 months covering a whole run-cycle of the SNS neutron production operation without changing ion source. A newly installed ion source undergoes a startup process that includes outgassing and sputter-cleaning the plasma chamber, ion source cesiation and beam tuning [6]. Typically, a single dose of cesiation with <30 mg cesium covers several months of ion source operation with high beam current (>50 mA) [4]. As summarized in Table 1, in the most recently completed 3 run cycles, the ion source was operated ~4 months for each run with ~54-56 mA beam current delivered to the RFQ. The ion source availability for each run was near 99.9%. The presently ongoing run which was started in early July and scheduled to end in late October 2021 is on track to accomplish another ~4 months run with an excellent availability of ~99.9%. Figure 2 shows strip-charts of several selected parameters of the ion source and accelerator in these long runs. The co-extracted electron current generally increases over time. If the electron current becomes excessive it causes the e-dump voltage to sag below its set point value. In those cases, the electron current was reduced and the e-dump voltage was stabilized by elevating the cesium collar temperature (by ~10-15 ^{0}C) as shown in the dashed ellipses on the figure 2. The H^- beam current shows slight decay over several months long runs. The beam power on target is regulated via beam current and/or beam chopping fraction in front of the RFQ.

Table 1. Summary table of H^- ion source operational performance

Start date	End date	Ion source ID#	Days of operation	Ion source availability	Date of LEBT output / RFQ input beam measured	RFQ input beam (mA)	RFQ field setting (arb.u.)	RFQ output beam (mA)
3/12/2020	7/16/2020	#2	116[a]	99.88%	3/16/2020	~54	0.370	34
					4/10/2020	~56	0.370	34.5
					7/15/2020	~54	0.370	32.5
7/24/2020	11/30/2020	#2	129	99.93%	8/7/2020	~55	0.373	34
					9/22/2020	~54	0.381	35
					10/27/2020	~54	0.381	34.5
1/29/2021	6/1/2021	#2	120[b]	99.94%	2/1/2021	~54	0.381	34
					5/13/2021	~54	0.397	35.5
7/2/2021	10/23/2021	#2	113[c]		7/4/2021	~54	0.397	34
					8/30/2021	~54	0.409	34.5

[a)] During 3/16-3/27, 2020, the ion source was switched off while waiting for other systems of the accelerator to be ready for operation following maintenance, these 11 days were excluded.
[b)] During 5/4-5/7, 2021, the source was switched off while waiting for a target issue to be resolved, these 3 days were excluded.
[c)] As is currently scheduled.

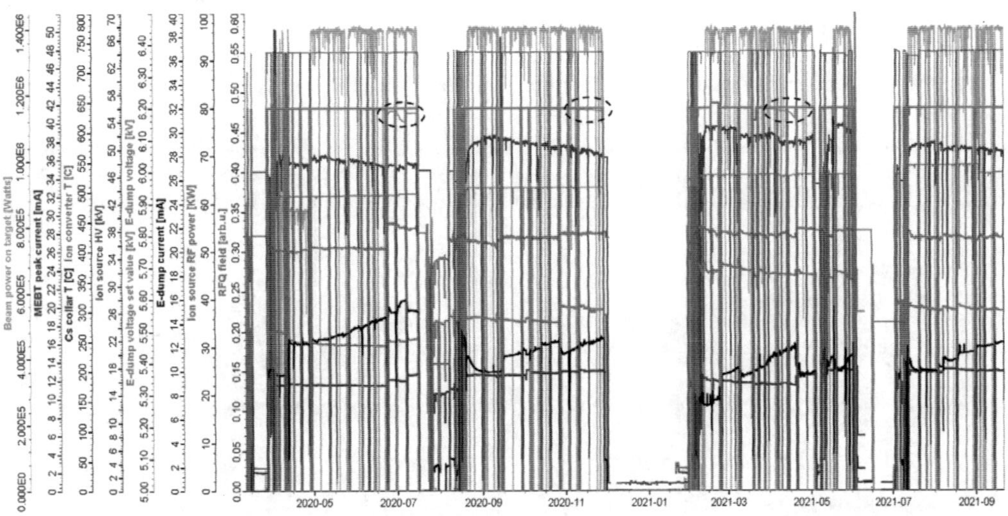

Figure 2. Strip-charts of some parameters of the ion source and accelerator for ~4 months long runs

3. Latest development activities of H⁻ ion sources

Latest development activities on H⁻ ion source and infrastructure include testing of a new solid-state, modular 2-MHz RF system on the accelerator injector and continued development of external antenna RF H⁻ ion source on an ion source test stand.

3.1. Tests of a new solid-state 2-MHz RF system

It has been in our plan to replace the present vacuum-tube type 2-MHz RF amplifier with a solid-state, modular RF system for the ion source on the accelerator injector. Based on the experience with TOMCO Technologies solid state 2-MHz RF amplifier modules operated on the H⁻ injectors on test facilities, we set up a new system on the accelerator injector with newer version of the same TOMCO modules. The solid-state RF system provides several advantages over the vacuum-tube amplifier system: 1) it can turn on and immediately produce the desired RF output power level while the vacuum-tube system requires warm-up with delicate matching tunes for different power levels; 2) it is modular chassis based and is easier to maintain; 3) it features fault detection for high reflected power, which protects the system in situations like plasma outages. The new system can provide up to 120 kW of total output power by combining 8 units of 15 kW RF modules. A series of tests on the ion source covering different power levels 0-70 kW, different pulse lengths (0.5 ms and 1.0 ms) and different repetition rates (30 Hz to 60 Hz) were carried out. Fault protection during a plasma outage and uninterrupted operation in an event of failure of individual module(s) were also verified. The reduced power with missing module(s) can be compensated for the typical operational power range of the ion source by driving up the RF control set point. The only issue we had in the tests was the system's tolerance to ion source high voltage sparking even though the RF system was operated on ground and isolated from the high voltage platform through an isolation transformer. Measures including spark guards for surge protection will be installed for the next test opportunity in a long maintenance outage of the accelerator.

3.2. Continued development of external antenna RF H⁻ ion source

The external antenna source development was started in the early years of SNS when the insulation coating of the internal antenna had frequent failures. Now, we continue to develop the external antenna source as a contingency ion source for the SNS though the internal antenna quality had been dramatically improved to a lifetime of 4 months and beyond. Figure 3 shows the external antenna source with the internal antenna source showing beside. The external antenna source uses the same outlet electrode assembly and electron dumping electrode as the internal antenna source and is compatible with the three H⁻ injector infrastructure. The plasma chamber is made of AlN ceramic, and the high-density pulsed plasma is driven by an externally placed antenna using the same 2-MHz RF system as the internal antenna source. Initial trials applying continuous 13.56-MHz RF to the same antenna to generate background starter plasma were not successful in the earlier development.

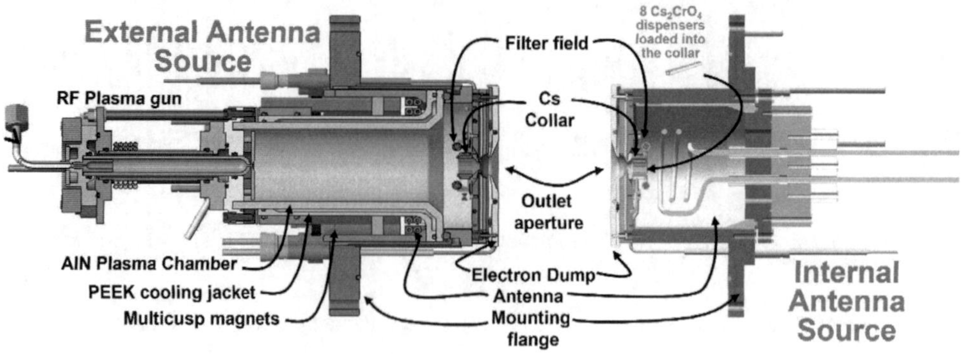

Figure 3. The external antenna ion source vs. the internal antenna ion source

A particle injection gun attached to the back of the main plasma chamber was utilized to assist the ignition of the pulsed plasma. A DC, 800 V electron gun had limited lifetime of ~2 weeks due to cathode sputtering or surface degradation. An RF assisted DC gun allowed lower voltage to start with, but the voltage had to be increased over time still beyond the cathode sputter threshold [5]. Recently, a pulsed DC gun (60Hz, 500 μs) was successfully tested with ~2 kV for ~90 days on a test bench. Testing the external antenna source with this pulsed DC gun is currently underway.

We also revisited the idea of direct generation of background starter plasma in the main chamber by applying a continuous, low power RF on the same external antenna. In this effort, by extensive optimization of RF matching, we were able to create background plasma with 13.56-MHz RF and facilitated the ignition of 2-MHz high-density pulsed plasma. Figure 4 (a) and (b) show the plasma emission spectra of background plasma generated with ~350 W 13.56-MHz RF and high-density plasma created with ~24 kW 2-MHz pulsed RF superimposed on the 13.56-MHz RF. The issue we had encountered with this trial was interference from the 2-MHz RF on the 13.56-MHz circuit, which was much stronger than what was seen with the internal antenna source and damaged the 13.56-MHz RF supply. A filter circuit has been designed to protect the 13.56-MHz system from the high power 2-MHz RF. Further testing is scheduled.

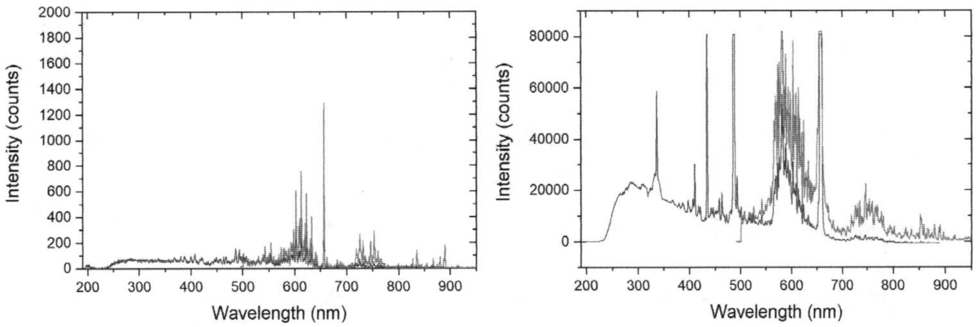

Figure 4. External antenna ion source plasma emission spectra of background plasma (a) and high-density pulsed plasma (b)

References

[1] R. Keller, R. Thomae, M. Stockli, R. Welton, "Design, Operational Experiences and Beam Results Obtained with the SNS H^- Ion Source and LEBT at Berkeley Lab" in AIP Conf. Proc. 639, 47 (2002).

[2] M.P. Stockli, R.F. Welton, and B.X. Han, "Record productions establish RF-driven sources as the standard for generating high-duty-factor, high-current H^- beams for accelerators", in Rev. Sci. Instrum. 89, 052202 (2018).

[3] B. X. Han, R. F. Welton, S. N. Murray, T. R. Pennisi, M. Santana, C. M. Stinson, and M. P. Stockli, "Refined beam measurements on the SNS H^- injector" in AIP Conf. Proc. 1869, 030014-1 (2017).

[4] B. X. Han, M. P. Stockli, R. F. Welton, S. N. Murray, T. R. Pennisi, C. M. Stinson, and S. M. Cousineau, "Recent performance of the SNS H^- ion source with a record long run" in AIP Conf. Proc. 2373, 040004 (2021).

[5] R. F. Welton, M. P. Stockli, B. X. Han, S. N. Murray, T. R. Pennisi, C. Stinson, A. Aleksandrov, C. Piller, Y. Kang, O. Tarvainen,, "H^- ion source research and development at the Oak Ridge National Laboratory" in AIP Conf. Proc. 2373, 070004 (2021).

[6] B. X. Han, M. P. Stockli, R. F. Welton, S. N. Murray Jr., T. R. Pennisi, and M. Santana, "Application of optical emission spectroscopy for the SNS H^- ion source plasma studies", in AIP Conf. Proc. 1655, 030003 (2015).

Negative hydrogen ion sources for particle accelerators: Sustainability issues and recent improvements in long-term operations

Robert Welton[1], Dan Bollinger[2], Morgan Dehnel[3], Ilija Draganic[4], Dan Faircloth[5], Baoxi Han[1], Jacques Lettry[6], Martin Stockli[1], Olli Tarvainen[5], Akira Ueno[7]

[1]Spallation Neutron Source, Oak Ridge National Laboratory, Oak Ridge, TN 37831, USA
[2]Fermi National Accelerator Laboratory, Batavia, IL 60510, USA
[3]D-Pace, Nelson, British Columbia, Canada
[4]Los Alamos Neutron Science Center, Los Alamos National Laboratory, Los Alamos, NM 87545, USA
[5]UKRI-STFC-ISIS Pulsed Spallation Neutron and Muon Facility, Rutherford Appleton Laboratory, Harwell, UK
[6]CERN, 1211 Geneva, Switzerland
[7]J-PARC Center, Tokai-mura, Naka-gun, Ibaraki-ken 319-1195, Japan

Corresponding author: welton@ornl.gov

Abstract. High brightness, negative hydrogen ion sources are used extensively in scientific facilities operating worldwide. Negative hydrogen beams have become the preferred means of filling circular accelerators and storage rings as well as enabling efficient extraction from cyclotrons. Several well-known facilities now have considerable experience with operating a variety of sources such as RF-, filament-, magnetron- and Penning-type H^- ion sources. These facilities include the US Spallation Neutron Source (SNS), Japan Proton Accelerator Research Complex (J-PARC), Rutherford Appleton Laboratory (RAL-ISIS), Los Alamos Neutron Science Center (LANSCE), Fermi National Accelerator Laboratory (FNAL), Brookhaven National Laboratory (BNL), numerous installations of D-Pace (licenced by TRIUMF) ion sources used mainly on cyclotrons and, most recently, the CERN-LINAC-4 injector. This report first summarizes the current performance of these ion sources in routine, daily operations with attention toward source service-periods and availability metrics. Sustainability issues encountered at each facility are also reported and categorized to identify areas of common concern and key issues. Recent ion source improvements to address these issues are also discussed as well as plans for meeting future facility upgrade requirements.

1. Introduction

Large accelerator complexes have supported scientific user facilities which have had an enormous societal impact spanning many decades and enabling the work of 10's of thousands of scientific users worldwide [1]. Many of these facilities employ synchrotrons, storage rings, cyclotrons and tandem accelerators which depend on long-term operation with stable, highly available beams of negative hydrogen ions from high-brightness ion sources [2]. H^- ions are utilized because H^- and H^+ beams can be conveniently merged into a single beam by passage through an orthogonal magnetic field and, after merging, H^- can be converted efficiently into H^+ by charge exchange stripping. This ability makes H^- ideally suited to fill synchrotrons and storage rings by allowing both the injected beam and the circulating beam to pass through the same charge exchange device allowing injection with minimal emittance growth. Similarly, in cyclotrons, H^- ions can be extracted with near 100% using a stripping foil rather than using magnetic kickers and a septums which are employed in H^+ machines. In tandem accelerators, H^- can be converted to H^+ in a high voltage terminal thereby doubling the beam energy for a given terminal voltage. It should also be noted that H^- ions are capable of being simultaneously accelerated along with H^+ ions in linear RF accelerators by utilizing the 180 degree phase shift of the alternating direction of the applied RF electric field. Accelerated H^- ions can also efficiently form neutral beams for applications such as fusion energy but those ion sources tend to be much larger devices with multi-aperture extraction and are not included in this study.

The high brightness H^- ion sources which serve the facilities discussed here deliver beams from single apertures ranging in intensity from <1 to ~120 mA, pulse lengths of ~100 μs to Continuous (CW), repetition rates of <1 Hz to CW and generally have a normalized-RMS emittance of around 0.25 π mm mrad. Ion source service cycles range from a few weeks to many months of steady operation between maintenance interventions (source replacement / refurbishment). Generally, the length of each facilities service cycle has been optimized over time to deliver maximum source availability (minimum unplanned downtime). This report first summarizes the current performance of these ion sources in routine, daily operations with attention toward source service-periods and availability metrics. Sustainability issues encountered at each facility are also reported and categorized to identify areas of common concern and key issues. Recent ion source improvements to address these issues are also discussed as well as plans for meeting future facility upgrade requirements.

The facilities which participated in this study include Oak Ridge National Laboratory, the Spallation Neutron Source (SNS) which is a Megawatt (MW)-class, pulsed-neutron facility; Japanese Proton Accelerator Research Complex (J-PARC), a multi-purpose, MW-class, neutron, meson and neutrino facility; Fermi National Accelerator Laboratory (FNAL), a TeV-class, high energy physics and neutrino facility; Los Alamos National Laboratory, Neutron Science Center (LANSCE), a MW-class, multipurpose facility; Rutherford Appleton Laboratory, a Pulsed Spallation Neutron and Muon Facility (ISIS), a neutron and muon facility; the European Organization for Nuclear Research (CERN), LINAC-4 injector, a TeV-class high energy physics multipurpose laboratory; Brookhaven National Laboratory (BNL) and D-Pace, a company who provides their TRIUMF-licensed H^- source to many facilities worldwide. This list of participating facilities is by no means fully inclusive but should be largely representative of the field.

The following definitions will be used throughout this work; *Issues*: recent H^- ion source problems or concerns as defined by system experts in long-term facility operations or those anticipated in future facility upgrades; *Improvements:* recent developments, planned or implemented, which address sustainability issues; *Availability:* Beam time delivered by the source (in hours) / Beam time requested from source (in hours) per production run period, this definition only accounts for beam downtime attributed to the ion source.

2. The SNS Cs-enhanced, RF-driven, multicusp H^- ion source

The SNS RF ion source routinely produces ~55 mA H^- ion pulses, 1 ms in duration with a 60 Hz repetition rate [3]. Over time, this source and the Low Energy Beam Transport (LEBT) coupling it to an RFQ, have been improved to now have a >99.9% availability during a typical 4-month operational period [4]. Fig. 1 shows the operational SNS ion source. Some operational concerns reported include: (1) *Beam persistence*: sources have a slight beam decay months after the initial cesiation requiring minor adjustments to the time structure of the accelerated H^- beam filling the SNS storage ring to maintain full power on target. (2) *Consistency between ion sources*: presently only one of the four SNS production source bodies show minimal beam decay compared to that from the spare ion sources.

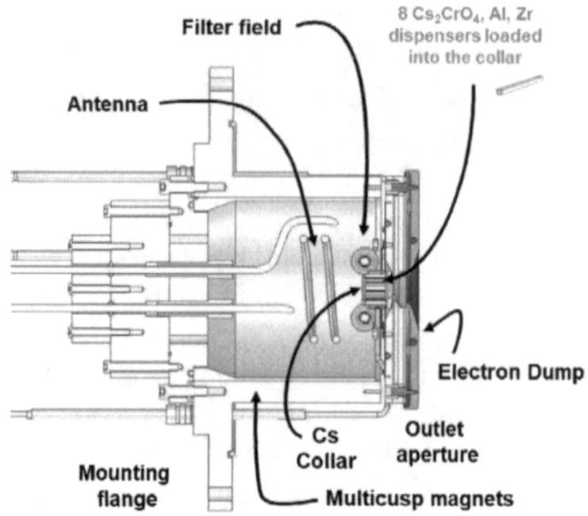

Figure 1. The operational internal-antenna SNS ion source

(3) *Co-extracted electron management*: the amount of these electrons can slowly grow over time which can cause the electron dump power supply to sag which can necessitate a second source cesiation. Cesiation is usually performed only once during startup by briefly heating the Cs-chromate dispensers. (4) *Beam intensity*: future SNS upgrade projects will require ~30% more beam out of the RFQ which the operational SNS ion source can provide if the RFQ performs as expected with an 80-90% transmission. More beam current, however, will be required if the RFQ underperforms this metric. (5) *Source lifetime*: the current internal-antenna source lifetime certainly exceeds the 4-month operational periods as it has not experienced an antenna failure in many years due to improved antenna manufacture and selection process [5]. We do, however, still recognize this as a risk since the antennas are manufactured by a single vender using a proprietary porcelain coating process [6].

Over the years, many improvements have been made to both the SNS internal [3] and external [7] antenna ion sources and are too extensive to discuss here. Looking forward, we are trying to minimize minor physical differences between the production ion source and spare sources. Towards this end, we hope to improve the thermal contact between the Cs-dispenser cartridges in the spare ion sources to be the same as that in the main SNS production source. We also plan to explore incremental improvements in beam intensity by testing slightly larger outlet apertures as well as improving electron management by implementing filter circuits and improved power supplies. Development of the external-antenna ion source also continues as a contingency source for the SNS should any issues with the internal antennas reemerge [7]. Development of this source now focuses on standardizing build quality of the source and improving the reliability of plasma ignition [4].

3. The J-PARC Cs-enhanced, RF-driven, multicusp H^- ion source

The J-PARC RF ion source routinely produces ~60 mA H^- ion pulses, 0.8 ms in duration with a 25 Hz repetition rate [8]. Fig. 2 shows the operational J-PARC ion source. Over time this source has been improved to now have near ~100 % availability as demonstrated during the last 5-month operational period [9]. Overall, the source operates well with the only reported operational issue being a couple of *antenna failures* in 2019 which have been attributed to degradation and issues with virtual and actual leaks in the forward O-ring on the plasma chamber and VCR vacuum seals on the gas feed line (see Fig. 2). The issues were resolved by replacing the VCR fitting with a welded structure, reducing the size of the O-ring groove and venting trapped air-pockets on the vacuum side of the seal [10]. Like many other facilities, J-PARC is also working towards achieving consistent performance across the available pool of ion source bodies [11]. From November 2020 to April 2021, the J-PARC RF-driven H^- ion source (IS) successfully produced the H^- beam of about 58 mA and 1.25 % duty-factor to J-PARC delivering a LINAC beam current of 50 mA with no ion source maintenance for ~152 operational days.

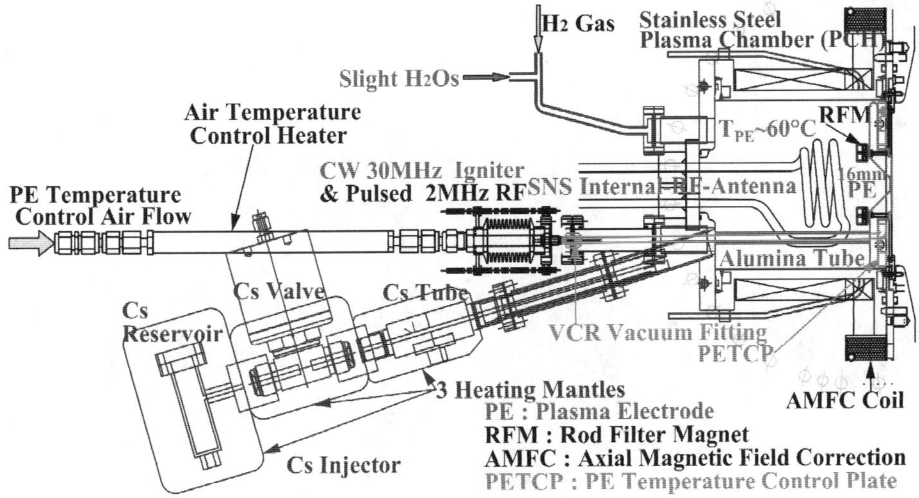

Figure 2. J-Parc operational internal-antenna source.

The sustainability of the J-PARC source was realized by successive improvements of the RF efficiency and transverse emittances and the reduction of impurities in the plasma. By the superior RF efficiency of about 2.6 mA/kW (making 72 mA H^- beam with only 28 kW of 2 MHz RF power) [12], the longer lifetime of the internal-RF-antenna, which was developed by the SNS [5], is expected. Also, the excellent transverse emittances suitable for the J-PARC RFQ, which were measured on the test stand, enabled the excellent J-PARC RFQ transmission efficiency of 94.3% (67.9 mA acceleration for the injected 72 mA H^- beam) [12]. This source performance was enabled by the following improvements [8]; (1) 45 ° tapered Plasma Electrode (PE) with 16 mm thickness to significantly increase I_{H-}, (2) continuous-wave igniter plasma driven by 50 W, 30 MHz RF to reduce hydrogen pressure in plasma-chamber by 50 % and beam loss in the Low Energy Beam-Transport (LEBT) by 12 % compared with that by popularly used 300 W 13.56 MHz-RF, (3) operation with low PE temperature of ~60 °C to significantly reduce the transverse normalized rms emittances ($\varepsilon_{95\%}$) and (4) slight water (H_2O) feed in hydrogen line to avoid $\varepsilon_{95\%}$ growth and divergence angle expansion. It should also be noted that during beam development periods on the J-PARC accelerator and on the test stand the ion source was shown to be capable of meeting the near-term, J-PARC RCS upgrade LINAC beam current requirement of ~60mA, 0.6 ms, 25 Hz for 1.44 MW of beam power on target.

4. The FNAL cesiated magnetron H^- ion source

Overall, the ion source runs very well and routinely produces ~60 mA H^- ion pulses, 0.2 ms in duration, with a 15 Hz repetition rate. Fig. 3 shows a simplified schematic and installation view of the FNAL ion source. FNAL had been using Magnetron ion sources to supply the LINAC with H^- ions since 1997. From then until 2012 Cockcroft-Walton (C-W) accelerators were used as the injector accelerator. During this time the magnetron sources had a lifetime of ~3 months. In 2012 the C-W injectors were replaced with a Radio Frequency Quadrupole (RFQ) accelerator. At the same time the magnetrons were changed to a design used by Brookhaven National Lab (BNL). The lifetime of the sources is now greater than 9 months with a ~99.8% availability as demonstrated during the last 9-month operational period [13].

Some of the current operational issues for the FNAL sources come in 3 categories: *functional, procedural and equipment.*

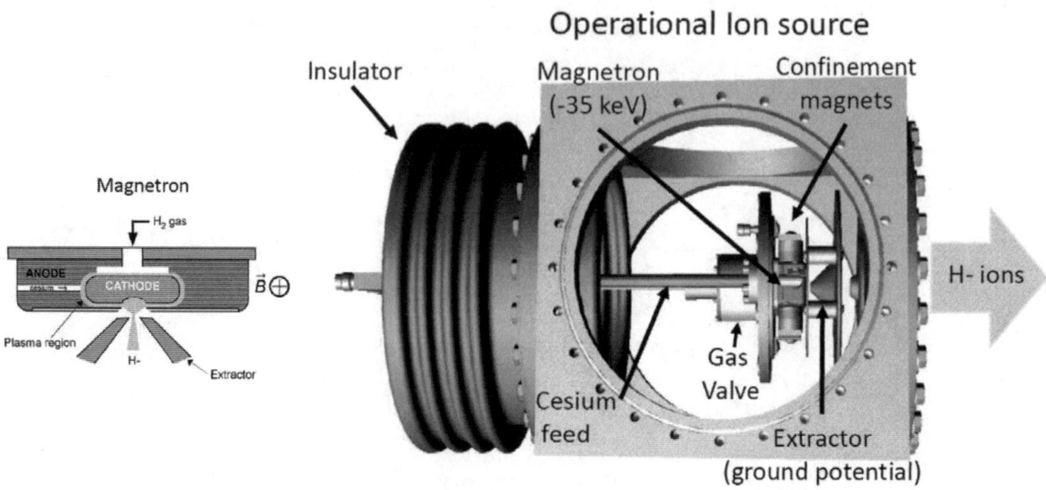

Figure 3. FNAL H⁻ ion source schematic (left) and actual installation (right).

The major issues are maintaining the required beam intensity and quality over the run cycle time of 9 months. We do have Proportional–Integral–Derivative (PID) loops to keep the gas pressure and cesium oven temperature mostly constant, but there is still frequent adjustment of source parameters to keep the sources running well. The major procedural issue is the annual ion source refurbishment and rebuild consistency. Finally, equipment issues include the arc pulser not terminating the end of the arc pulse when the source impedance is too high, inconsistent pulsed solenoid (electromagnetic) gas valves, and extractor pulser cables which fail twice per run on average.

The future of the FNAL complex will use the Proton Improvement Plan II (PIPII) super conducting LINAC. The front end of that accelerator was built as a test stand to test equipment that will be used in the final installation. The source used was a 10 mA DC H^- source (D-Pace) with a pulsed extractor. With a lifetime of around 300 hours, the source is not ideal for accelerator operations. FNAL have been studying a magnetron source and will soon be studying a Penning ion source to meet the initial PIP II parameters [14]. To date, the magnetron source seems capable of meeting the current and duty-factor requirements (~2mA, 550μs, 20 Hz in the LINAC), but beam quality studies still need to be performed.

5. The LANL surface converter H^- ion source

Over the past three decades at LANSCE, the negative ion beams supplied to our various user programs have been generated using the surface-converter negative ion source. The source now routinely produces ~16 mA H^- ion pulses, 0.84 ms in duration with a 120 Hz repetition rate, ~98.5% availability and ~4-week service period. Fig. 4 shows a side-view photograph of the interior of the source with the major components labeled. In the past several years and under high duty-factors of 5% and 10% (60/120 Hz x 840 μs), several issues with the converter ion source were observed and attempts made to resolve them. The issues included: (1) modest H^- current in range 12 – 14 mA (no electrons), (2) limited source lifetimes, i.e., 4 weeks or shorter with filaments failures, (3) long rise-time of the arc discharge (<180 μs) that reduces the usable H^- pulse length, (3) beam instability associated with frequent 80 kV arc downs (every few minutes), (4) larger-than-expected transverse, normalized, RMS emittances, > 0.2 π mm mrad, (5) 4 days of beam downtime for a source exchange, full Cs transfer to achieve production beam-intensity levels, (6) excessive high energy (voltage) droops over the beam pulse (~1-2 keV), (7) significant equipment failures due to numerous HV arc breakdowns, (8) low beam stability and low charge per pulse to all H^- user facilities.

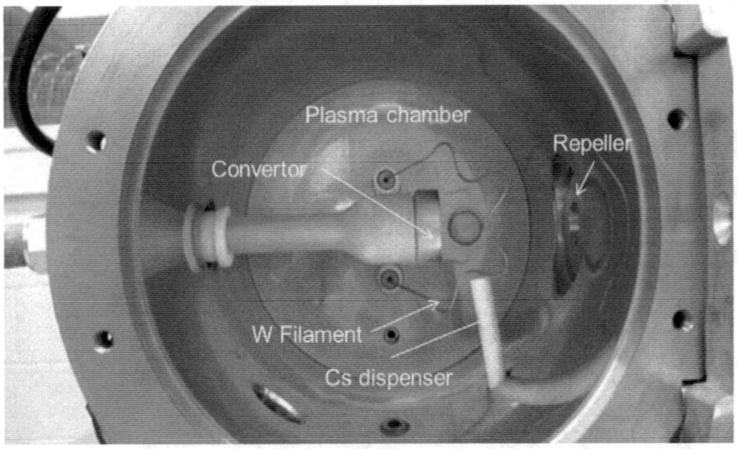

Figure 4. LANL surface conversion source.

In recent beam production cycles, however, incremental *performance improvements* were achieved for the negative ion injector, linear accelerator, and proton storage ring including: (1) average production currents now range from 15 - 16.7 mA, (2) Extracted peak beam currents in the range 17 - 18.5 mA during beam development times, (3) demonstrated W-filament lifetimes of 5 weeks (DF=10%) and 6 weeks (DF=5%), (4) Improved H^- beam stability with 3-5 hours between 80 kV arc downs, (5) reduced high voltage droops now < 500V over the full macro pulse, (6) average current injected into PSR over <120 μA, (7) no major equipment failures due to reduced HV breakdowns.

Several critical *ion source preparation improvements* were also made that enhanced the negative ion source performance. These included: (1) improved converter-head cooling, (2) installation of a new 80 kV high voltage regulator, (3) improved converter electrode alignment, (4) improved processing of the Cs reservoir, valve cleaning and Cs loading procedures, (5) adoption of a lower vacuum leak rate of ion source vacuum assembly at processing stand and in the H^- dome, (6) implementing a new method of filament outgassing and preparation with monitoring outgassed molecules, (7) use of converter quartz insulation tube with rougher surfaces, and (8) careful HV column electrode assembly and using SF_6 gas in Plexiglas jacket.

Further improved beam performance was realized by employing a new ion source *operational tuning procedure*. This procedure was applied in recent beam production cycles and features:

1) Lower arc voltage 160 - 170 V, which reduces plasma sputtering, increases filament lifetime, and improves arc discharge rise time
2) Higher arc discharge currents of 40-55 A to increase the extracted H^- beam currents to the range of 16-18 mA
3) Lower Cs oven temperature <180 C, which reduces plasma sputtering of the hot filaments, increases filament lifetime, minimize number of HV breakdowns, but increases arc discharge rise time over 180 μsec
4) Convertor voltage peak centered at ~ -250 V: Peaking a source at lower voltage (up to -200 V) will lower extracted-beam transverse emittances thereby reducing spill along LINAC and beam lines
5) Optimized H_2 gas flow (in the range from 2 - 3 sccm): The right amount of injected gas will increase beam output, decrease arc time response, and reduce number of HV arcs.

Finally, to dramatically improve beam performance for the H^- users and overcome the modest output current limitation of our present H^- ion source, a collaboration between SNS-ORNL and LANSCE-LBNL to build and install an SNS RF negative ion source at LANSCE is in progress [15].

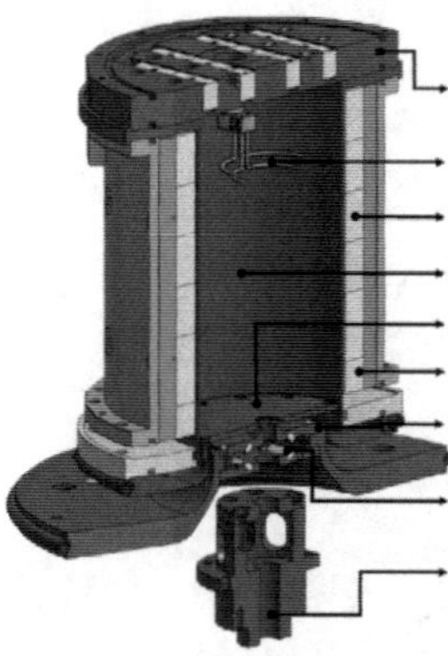

Table A: Filament Life-Time Data for TRIUMF DC Volume-Cusp, Filament-Powered H^- Source

Filament Type	Institute/Company	Average DC H^- Current (mA)	Cumulative Runtime at Changeout (hrs)	Filament Life-Time (mA•hrs)
(a)	TRIUMF TR30 Cyclotron	4.5	600	2700
(b)	INER TR30/15 Cyclotron	0.9	224	202
(b)	TRIUMF TR13 Cyclotron	0.4	~130	~52
(c)	TRIUMF TR30 Cyclotron	6	840	5040
(c)	Fermilab	6.5	840 Maximum 591 Average 391 Minimum	5460 3842 2542
(d)	TRIUMF TR13 Cyclotron	0.4	1200 (estimated, run in progress)	~480

Figure 5. The D-Pace (TRIUMF) filament-driven, multicusp source shown with the 4x semi-circular Tantalum filament (φ = 1.6 mm) structure (c). The table contains filament lifetime data for various structural filament types (a,b,c,d), see text. [21-27].

6. The D-Pace (TRIUMF) filament-driven, multicusp H^- Ion Source

D-Pace licensed the 15 mA TRIUMF Direct Current (DC), volume-cusp, filament-powered, H^- source in 2001 [16]. A cross sectional model of the source is shown in Fig. 5. This source technology is primarily used in three areas:

a) Commercial axial injection type cyclotrons for medical radioisotope production by way of companies/institutes such as IBA, SHI, THALES, ACSI, KAERI, INER [17]
b) Boron Neutron Capture Therapy (BNCT) applications for the case of tandem accelerators or cyclotrons in companies/institutes such as TAE, SHI, CNEA, KIRAMS, BINP [18,19]
c) National laboratories for various applications i.e., TRIUMF, Fermilab, KIRAMS, KAERI, CNEA, INER, BINP.

Since these sources are often used by end-clients in confidential business environments, it has been difficult to garner sustainability information that would be useful for the ion source community. For example, filament lifetime data, the key sustainability issue for the TRIUMF DC H^-/D^- source utilized in commercial and research environments is sparse. The authors of this paper are thankful for contributors [21-27] for the table in Fig. 5 containing lifetime data for the following filament configurations: (a) 1 Tungsten arch-shaped filament segment (φ = 2.4 mm), (b) 2 semi-circular Tantalum filament segments (φ = 2 mm), (c) 4 semi-circular Tantalum filament segments (φ = 1.6 mm), and (d) 1 pigtail Tungsten alloy filament segment (φ = 3 mm) [20]. Recent source improvements include implementing single machined Ta filament holder which was superior to earlier Elkonite structures which could fail after months of service [26]. Efforts to develop and RF driven version of this source are ongoing.

7. The CERN-LINAC-4 RF H^- ion source

The IS03b H^- source is shown in Fig. 6 and is the first element of CERN's new 160 MeV injector (LINAC-4) that started operation in September 2020. The ion source provides 35 mA H^- pulses of 0.8 ms duration at 0.83 Hz, and LINAC 4 delivers 0.6 ms, 27 mA pulses at 160 MeV which is sufficient to produce all CERN beams. To date the source has operated with >99.9% availability on LINAC-4 for ~6 months experiencing only minor preventive maintenance of the gas pulsing valve (see below). The design of all ion source components aims at utmost sustainability via systematic testing, preventive maintenance and identifying source components which could be potentially lifetime limiting. Three fully tested

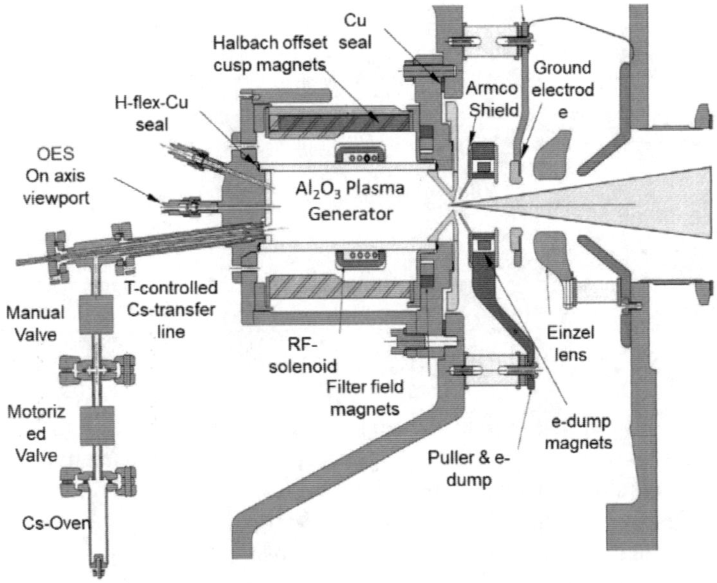

Figure 6. The CERN-LINAC-4 RF H^- ion source.

spare ion sources are ready to be installed in the event of operational source hardware failure, in addition the test stand's source front end can be used as a spare. HV-sparks induced damages are related to capacitive stored energy; therefore, the HV-system is designed to minimize HV-cable length, plasma and puller electrodes are connected to the secondary of standardized and stabilized 700 μs pulse generators equipped with protection resistors and current detection monitoring [28]. Currently, plasma- or puller- electrodes do not show signs of wear and their voltage stability is within the specified 1% [28].

Vacuum is an important aspect of the sustainability of all ion sources and, therefore minimizing the presence of impurities to protect the very reactive Cesium layer deposited on the Mo-plasma electrode is essential. Therefore, all components of the ion source and LEBT are UHV compatible featuring bakeable metallic seals eliminating the need for O-rings and their associated gas permeation issues. Dedicated Helicoflex / Cu seals surround the cylindric Al_2O_3 plasma chamber with Ti rings brazed on the ceramic to ensure tightness. The large Al_2O_3 main insulator is brazed, equipped with a bellows and employs Al seals which are made at CERN. And all other seals are standard con-flat Cu rings.

A 5-turn solenoid of silver-coated 4 mm diameter Cu-tube capacitively (for ignition) and inductively (after plasma formation) couples RF power into the ion source plasma. 3D printed epoxy spacers, vacuum molded around the solenoid are used to prevent arcing of the antenna structure by equalizing turn spacing and reducing the possibility of air ionization. Multicusp magnetic plasma confinement is well known to improve plasma densities in small laboratory plasmas for a given amount of input power. However, in our case, simulation and experiment showed that our external antenna inductive coupling is more than twice as efficient without cusp [29,30]. Therefore, the source is now operated without the cusp which halves the RF-power and avoids the e-loss lines that were visible on the inner surface of the ceramic plasma chamber.

Automated operations were necessary to adjust cesiation, hydrogen injection, RF pulse shape and frequency [31] to follow changes in the surface production yield from the source. A fail-safe control system ensures a constant, very low flux of cesium that stabilizes operational parameters while keeping the oven at low temperature (60°C). The source now operates under cesium loss compensation [30] and a simple RF-driven, H^- current regulation feedback control system ensures the required stability over time. Every beam interruption is recorded, its origin identified and attributed to sub-systems;

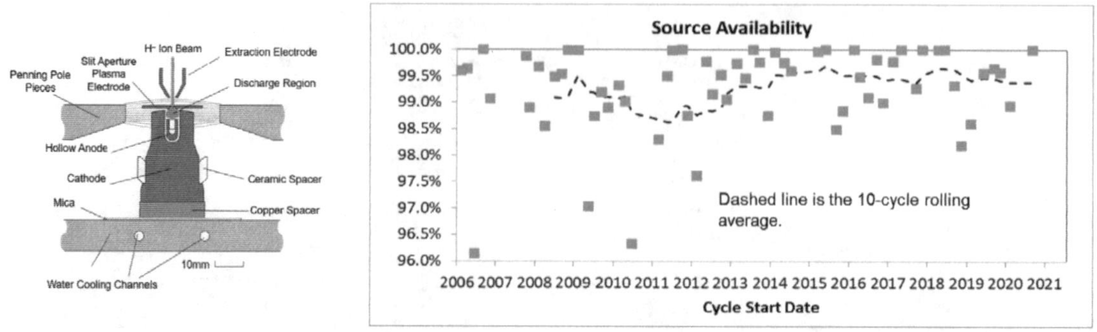

Figure 7. Schematic diagram of the ICIS H⁻ ion source and the 15-year source availability data.

the analysis of these failure statistics is the key to prioritize improvement. This led to studying insertion of a manual valve between the source and electromagnetic pulsed hydrogen valve, which is known to evolve and would occasionally fail, allowing the source to remain under vacuum during valve replacement. This would dramatically reduce the H_2 valve replacement time [32]. H^- pulse of 35 mA extracted from the source require typically 20 kW of RF-power while the RF system was tested up to 90 kW. Under cesium loss compensation the electron to ion ratio remains below 4, the co-extracted electrons current is below 150 mA while the e-dump was tested to 3A; these considerable operating margins will provide room for the improvement required to meet LINAC4's 45 mA out of the RFQ goal set by the High-Luminosity LHC project [33]. Preliminary test stand results from a new extraction system and an RFQ mask suggest that modified extraction optics may meet this requirement at similar duty-factor [34]. Beam sustainability (availability and reliability) are essential to respect the impressive investment of the 14000 physicists' experiments on CERN site.

8. The ISIS Penning H^- ion source

The ISIS Penning Surface Plasma Source, shown in Fig. 7, is a pulsed DC discharge surface plasma source, which produces 48-55 mA of H^- current pulsed at 50 Hz repetition rate [35]. The discharge pulse length is typically set to 700 µs (3.5 % plasma duty factor) and the extraction pulse length is modulated from 220 to 270 µs (1.1-1.35 % beam duty factor) to compensate for beam variations while injecting a constant charge per pulse into a rapid cycling synchrotron. The source lifetime varies between 5 and 40 days, with scheduled changes often taking place at 14-21 days and requiring a fleet of ten nominally identical sources to serve ISIS.

The main factor limiting the source lifetime is electrode sputtering by cesium ions and formation of macroscopic molybdenum debris electrical shorting the anode and cathode or causing a partial blockage of the outlet aperture. Most of the sputtering damage is believed to occur during the first 100-200 µs of the discharge pulse when so-called breakdown oscillations result in drastic fluctuations of the discharge voltage and current. The breakdown oscillations can be suppressed by improving the discharge power supply and its auxiliary circuits, which presumably leads to reduced sputtering damage [36]. The formation of the macroscopic flakes is thought to be related to the thermal cycling of the cathode and hydrogen embrittlement of the metal. The surface temperature of the cathode varies by approximately 100 C during each discharge pulse, which causes a strong local temperature gradient within the cathode material [37] and exposes it to the formation of stress layers. Maintaining a favorable cesium balance on the cathode surfaces is only possible in a certain temperature window, which prevents mitigating the flake formation by more aggressive cooling of the cathode.

Another sustainability issue of the ISIS Penning ion source is the gradual increase of the co-extracted electron current as the source ages with the fastest increase occurring in the first 5 days of operation. To counter-act the electron current drift and to avoid operating the extraction power supply at its limit the source is tuned throughout its lifetime by very slowly adjusting the cesium oven temperature, hydrogen flow rate and air cooling.

The ISIS Penning source availability for each user cycle for the last 15 years is shown in Fig. 7. The dashed black line is a 10-cycle rolling average. The average source availability for that period was 99.3%. As the ISIS ion source is fundamentally lifetime-limited by electrode damage and requires a specific operational window to run trouble-free, the ISIS Low Energy Beams Group has started a project to replace the Penning source with a cesium-free, RF ion source aiming to produce 28-30 mA H^- pulses. The reduction of the beam current is enabled by simultaneous improvements to the LEBT and matching of the beam into the drift tube linac by constructing a new medium energy beam transport section. The commissioning progress of the RF ion source is reported elsewhere in these proceedings [38]. The long-term (> 10 years) plan of ISIS is to upgrade the existing facility into a MW-level spallation neutron source. The requirements for the ion source depend on the accelerator design, which has not been completed. It is expected that the injector will need to deliver 57-120 mA of H^- out of the RFQ in 600–1000 μs pulses at 50 Hz with 0.15-0.25 π mm mrad rms-emittance. Both, the Penning source and the (cesiated) RF source are being considered for now.

9. The BNL Magnetron H^- ion source

Recently the BNL-RHIC-LINAC injector was upgraded to include 3 ion sources (2 magnetron and 1 optically polarized source). In normal accelerator operation an H^- magnetron ion source routinely produces ~120 mA H^- ion pulses, 0.65 ms in duration with a 7 Hz repetition rate for ~3-month service periods with excellent availability achieved after the upgrade [39]. In this configuration, one source delivers beam while the others remain in standby should a source fail minimizing injector downtime. The recent injector upgrade involved several sustainability improvements to the ion source systems which were incorporated into the last BNL production run:

a) Improved heating system for the source and Cs supply to ensure steady Cs flow and minimum consumption while keeping the source body 160-180C range (unable to be heated by arc power alone)
b) Improved power supplies reducing footprint and achieving better arc stability with reduced noise.
c) Improved ceramic supports for the ion optics: reducing Cs and W deposition on insulators and consequent spark frequencies.
d) Significant improvements were also made to the pulsed gas valve

Due to the high gas and pulsed power efficiency (~100mA/kW) of this source, work is ongoing to increase the pulse length to 1 ms to support the BNL isotope program with encouraging initial results. Also encouraging is the low co-extracted electron current and low average operating power of the source (e.g. ~100W at 50Hz, 1 ms) [40].

Source Function	SNS	J-PARC	FNAL	LANSCE	D-Pace	CERN	ISIS	Impact Score
Beam persistence (long-term drifts in current, emittance)	x		x					2
Beam Intensity (beam current, transmission and margin)	x			x		x		3
Beam stability (arc events, pulse ignition)			x	x			x	3
Intrinsic Source Lifetime (cathodes, filaments and antenna)			x	x	x		x	4
Electron management	x					x	x	3
Build/source consistency	x	x	x					3

Cs System	SNS	J-PARC	FNAL	LANSCE	D-Pace	CERN	ISIS	Impact Score
Cs system	x		x	x				3
RF system	x	x						2
Gas feed system			x			x		2
DC power sources	x							1
Pulsed power sources			x					1
Cathode / filament structure			x	x	x		x	4
Ion /electron extraction structure	x						x	2

Table I. Operational & anticipated issues reported: grouped by source function and components.

10. Summary and Outlook

Overall, H^- ion sources for accelerators operate very well with most having an availability > 99% with service cycles ranging from a few weeks to greater than 9 months. The high availabilities have resulted from numerous and continuous source improvements and judicious selection of ion source service periods. These efforts have enabled decades of high-impact science which have advanced the frontier of human knowledge considerably in many areas. Now, many of these facilities are looking toward upgrade projects which will require higher beam current / duty-factors and most seem well-positioned to meet their near-term upgrade goals.

Ion source experts from each facility also shared their 'behind the scenes' concerns and detailed recent improvements which were summarized in the previous sections. Table I, provides a quick reference to issues reported which are grouped by source function and source component to help identify key and common areas of concern. Not surprisingly, issues associated with filaments and cathodes were still most notable in the survey results with most of these facilities now actively pursuing the development alternative RF sources which have simultaneous high-availability, long-service cycles and proven scalability to higher duty-factors. Concerns about co-extracted electrons and Cs management were also clearly voiced in the survey.

Acknowledgements

A special thank you to Anatoli Zelenski who contributed information about the BNL ion source on short notice and Kiersten Ruisard for serving as a technical reviewer on this and other publications. This manuscript has been co-authored by UT-Battelle, LLC, under contract DE-AC05-00OR22725 with the US Department of Energy (DOE). This research was supported by the DOE Office of

Science, Basic Energy Science, Scientific User Facilities. The US government retains and the publisher, by accepting the article for publication, acknowledges that the US government retains a nonexclusive, paid-up, irrevocable, worldwide license to publish or reproduce the published form of this manuscript, or allow others to do so, for US government purposes. DOE will provide public access to these results of federally sponsored research in accordance with the DOE Public Access Plan (http://energy.gov/downloads/doe-public-access-plan)

References

[1] Vladimir Shiltsev, "Particle beams behind physics discoveries", Physics Today 73, Issue 4, 32 (2020)

[2] D. Faircloth and S. Lawie, "An overview of negative ion sources for accelerators", New J. Phys. 20, 025007 (2018)

[3] M.P. Stockli, R.F. Welton, and B.X. Han, "Record productions establish RF-driven sources as the standard for generating high-duty-factor, high-current H^- beams for accelerators", in Rev. Sci. Instrum. 89, 052202 (2018).

[4] B. Han, M. Stockli, R. Welton, C. Piller, S. Murray Jr., T. Pennisi, C. Stinson, and S. Kim, "H-ion source operational performance and latest development at the Spallation Neutron Source", these proceedings.

[5] R. Welton, V. Dudnikov, B. Han, S. Murray, T. Pennisi, C. Pillar, M. Santana, M. Stockli, and M. Turvey, "Improvements to the internal and external antenna H^- ion sorces at the SNS", Rev. Sci. Instrum. 85, 02B135 (2014)

[6] Cherokee Porcelain Enamel Corporation, 2717 Independence Lane, Knoxville, TN 37914, USA.

[7] R. F. Welton, M. P. Stockli, B. X. Han, S. N. Murray, T. R. Pennisi, C. Stinson, A. Aleksandrov, C. Piller, Y. Kang, O. Tarvainen,, "H^- ion source research and development at the Oak Ridge National Laboratory" in AIP Conf. Proc. 2373, 070004 (2021).

[8] A. Ueno, "Cesiated surface H^- ion source: optimization studies", New J. Phys. 19 015004 (2017)

[9] T. Shibata, K. Ohkoshi, K. Shinto, K. Nanmo, K. Ikegami and H. Oguri, "Emittance measurement of J-PARC RF ion source after 5-month continuous operation", these procedings

[10] A. Ueno, K. Ohkoshi, K. Ikegami, A. Takagi and H. Oguri, "Emittance Improvements of cesiated RF-driven H^- ion source to enable 60 mA operation of high-energy and high-intensity LINACs by plasma impurity controls", AIP Conf. Proc. 1869, 030011 (2017).

[11] A. Ueno, K. Ohkoshi, K. Ikegami and H. Oguri, "Impurities reduction conditioning to Recover Best Beam Quality of J-PARC Cesiated RF-Driven H^- Ion Source with New Parts Exposed to Plasma", these procedings.

[12] A. Ueno, "J-PARC H^- ion source and space-charge neutralized LEBT for 100-mA high energy and high duty factor LINACs", Rev. Sci. Instrum. 91, 033312 (2020)

[13] D. Bollinger, "Installation and commissioning of the new Fermi National Accelerator Laboratory H^- Magnetron", Rev. Sci. Instrum. 85, 02B121 (2014)

[14] S. Holmes, P. Derwent, V. Lebedev, S. Mishra, D. Mitchell, V. Yakovlev, "PIP-II status and stratagy", 6th Internaitonal Particle Accelerator Conference, Richmond, VA, USA, THPF116 (2015)

[15] M. Stockli, B. Han, M. Clemmer, S. Cousineau, A. Justice, Y. Kang, S. Murray Jr., T. Pennisi, C. Piller, C. Stinson, R. Welton, I. Draganic, Y. Batygin, R. Garnett, D. Kleinjan, J. Medina, J. Montross, G. Rouleau, V. Dudnikov, "Upgrading the LANSCE Accelerator with a SNS RF-driven H^- Ion Source", Review Scientific Instruments 91, 013321(2020)

[16] T. Kuo et al., "On the development of a 15 mA direct current H^- multicusp source", Rev. Sci. Instrum 67 1314 (1996)

[17] M. Dehnel et al. "An ion source upgrade for an axial injection based commercial cyclotron", NIMB, 241, Issues 1-4, 896 (2005)

[18] J. Ma, "Tandem electrostatic accelerators for BNCT", 13th Int'l. Conf. on App. of Accel. in

Research and Industry, Denton, 201 (1994)
[19] T. Mitsumoto et al, “Cyclotron-based neutron source for BNCT”, AIP Conf. Proc. Vol. 1525 (1), 319 (2013)
[20] D. Prevost et al. “New ion source filament for prolonged ion source operation on a medical cyclotron”, Instruments 3, 5 (2019)
[21] B. Milton et al. “Commissioning and first operation of a 500 μA, 30 MeV, H^- cyclotron: The TR30”, Particle Accelerator Conference (PAC), San Francisco, 67 (1991)
[22] T. Duh, INER, Private Communication, August 2021.
[23] D. Prevost, TRIUMF, Private Communication, August 2021.
[24] G. Cojocaru and J. Lofvendahl, “Operational experience at the intensity limit in compact cyclotrons”, Proc. of the 20th Int'l Conf. on Cyclotrons and their Applications, Vancouver, 432, (2013)
[25] L.R. Prost, Fermilab, Private Communication, July 2021.
[26] M. Bacal, K. Maeshiro, S. Masaki, M. Wada, “Performance of tantalum as plasma electrode material in negative hydrogen ion sources”, Plasma Sources Sci. Technol. 30, 075014 (2021)
[27] T. Stewart, D-Pace, Private Communication, July 2021.
[28] D. Aguglia, “Design of a System of High Voltage Pulsed Power Converters for CERN's Linac4 H^- Ion Source”, CERN-ACC-2014-0345 (2014).
[29] S. Briefi, S. Mattei, J. Lettry, and U. Fantz, “Influence of the cusp field on the plasma parameters of the Linac4 H^- ion source”, NIBS-2016, Oxford, NIBS Oxford, AIP conference proceedings 1869, 030016 (2016)
[30] J. Lettry, S. Bertolo, U. Fantz, R. Guida, K. Kapusniak, F. Di Lorenzo, C. Machado, C. Mastrostefano, T. Minea, M. O'Neil, H. Neupert, D. Noll, D. Steyaert, and N. Thaus, Linac4 H– source R&D: “Cusp free ICP and magnetron discharge”, AIP Conference Proceedings 2052, 050008 (2018)
[31] G. Voulgarakis, J. Lettry, B. Lefort and V. J. Correia Costa, “Autopilot regulation for the Linac4 H^- Ion source”, NIBS-2016, Oxford, AIP conference proceedings 1869, 030012 (2016)
[32] C. Mastrostefano, S. Bertolo, M. O'Neil and F. di Lorenzo, CERN, Private Comminucaiton, August 2021
[33] High-Luminosity LHC | CERN (home.cern)
[34] F. Wenander, “CERN hadron sources: status and innovation overview”, these proceedings
[35] R. Sidlow, P. Barratt, A. Letchford, M. Perkins and C. Planner, Proc., “Operational experience of penning H^- ion sources at ISIS”, EPAC'96, THP084L, 1525, (1996), jacow.org
[36] O. Tarvainen, R. Abel, S. Lawrie, D. Faircloth, J. Macgregor, T. Sarmento and T. Wood, “Breakdown oscillations of the ISIS H– Penning ion-source hydrogen–cesium discharge”, IEEE Transactions on Plasma Science, 47, 11, (2019).
[37] D. Faircloth, S. Lawrie, H. Pereira Da Costa, and V. Dudnikov, “Operational and theoretical temperature considerations in a Penning surface plasma source”, AIP Conf. Proc. 1655, 030013 (2015).
[38] S. R. Lawrie, “External RF-coil H^- ion source plasma commissioning”, in these proceedings.
[39] A. Zelenski, G. Atoian, T. Lehn, D. Raparia and J. Ritter, “High-intensity polarized and un-polarized sources and injector developments at BNL linac”, AIP Conference Proceedings 2373, 070003 (2021)
[40] A. Zelenski, a private communication 2021

Comparison of ion source plasma responses to extraction grid bias between hydrogen and deuterium operations in NIFS-RNIS

H Nakano[1,2], S Masaki[3], E Rattanawongnara[3], K Tsumori[1,3], K Ikeda[1], T Hamajima[4], K Nagaoka[1,4] and M Osakabe[1,3]

[1]National Institute for Fusion Science, Natural Institutes of Natural Sciences, Toki, 509-5292, Japan
[2]Graduate School of Engineering, Nagoya University, Nagoya, 464-8603, Japan
[3]The Graduate University for Advanced Studies, SOKENDAI, Toki, 509-5292, Japan
[4]Graduate School of Science, Nagoya University, Nagoya, 464-8603, Japan

E-mail: nakano@nifs.ac.jp

Abstract. An isotope effect of negative ion motion in plasma near a plasma grid (PG), which is a plasma-beam boundary grid, has been investigated in a negative hydrogen-ion source with the surface production process of the negative ion on the PG surface. Negative deuterium-ion (D^-) density was higher than negative hydrogen-ion (H^-) density in the condition without extraction-grid bias (V_{EGB}) and approached the H^- density in the condition with V_{EGB}. Thus, the D^- density responded stronger than the H^- density to the V_{EGB}. The density of the negative hydrogen-ion isotopes and the density response to the V_{EGB} have been organized into respective identical trends by momenta of isotopes emitting from the PG. The Larmor motion can be a dominant mechanism of the negative ion transport from the PG to the plasma.

1. Introduction

High ion particle flux to a boundary between an ion source plasma and a beam, so-called a meniscus, from the plasma is one of the important factors to obtain an intense ion beam current in ion sources. While high plasma density directly contributes to the high ion particle flux as extracted ions as the beam especially in positive sources, in negative hydrogen-ion source with the dominant negative-ion process of surface production, the negative hydrogen-ion yield on the production surface and efficient negative ion transport from the production surface to the meniscus should be also considered to increase the beam current. The elementary process of surface production is known. The negative ion transport has not been fully elucidated even by previous studies [1-3] and should be clarified more.

An isotope effect on the negative hydrogen ion source is one of the attractive phenomena to elucidate especially toward the development of a high-power negative-deuterium-ion source for neutral beam injector on fusion device. Most previous studies in large-scaled sources for the neutral beam report global differences by the isotopes such as co-extracted electron with a negative ion beam, global beam homogeneity, and so on [3-6]. Elucidation of the isotope effect on the negative ion transport from the production surface to the meniscus can support a deeper understanding of the global differences between operations with hydrogen and deuterium and can help further improvement of the negative deuterium ion source for fusion.

To investigate the negative ion transport from the negative ion production surface to the meniscus and its isotope effect in the negative hydrogen ion source, negative ion density in the vicinity of the surface of a plasma grid (PG), which is a plasma-beam boundary electrode, has been measured at the Research and development of Negative Ion Source (NIFS-RNIS) on Neutral Beam Test Stand (NIFS-NBTS). An experimental setup is described in section 2, experimental results on the isotope effect on the negative ion density are shown in section 3, and interpretations of the experimental results with the isotope effect and the negative ion transport are discussed in section 4. Finally, section 5 summarizes this study.

2. Experimental setup

A schematic of the NIFS-RNIS, which is the same type and a half-size of negative ion source for neutral beam injector on a Large Helical Device, is shown in Fig. 1. A discharge chamber size of the NIFS-RNIS is 350 mm width in the x-direction, 700 mm height in the y-direction and 263 mm depth in the z-direction. An ion source plasma generated by filament-driven arc discharge with a typical arc voltage of 80 V expands from the driver region to the beam extraction region across the transverse filter magnetic field. The plasma density in the vicinity of the PG is the order of 10^{17} m^{-3} and electron temperature reduces around 1 eV to suppress negative ion destruction by high-energy electrons. Cesium is seeded from the backplate, which is the most upstream part of the arc chamber, to produce the negative ion on the PG surface. An accelerator consists of the PG, extraction grid (EG), and grounded grid (GG). The thickness of the PG is 3.5 mm and the gap between the PG and EG is 5.5 mm. High voltage applications between the PG and the EG, and between EG and GG in deuterium operation are limited at the NIFS-NBTS. To observe the plasma responses to the electric field through the PG apertures, extraction grid bias (V_{EGB}) of hundreds of volts is applied to the gap between the PG and the EG in hydrogen and deuterium operations.

Plasma parameters and line averaged negative ion density in the vicinity of the PG are measured with the Langmuir probe (LP) and Cavity RingDown technique (CRD) [7], respectively. Optical Emission Spectroscopy (OES) is used for monitoring whether unexpected variation occurs. These diagnostics are installed at a bias insulator with a thickness of 52 mm between the discharge chamber and the PG mounting flanges. The LP is positioned at a horizontal center in the upper half side, vertical center of neighboring PG apertures, and 14 mm from the PG. A CRD measure line is aligned along the vertical center of the neighboring PG apertures, 17 mm from the PG, and symmetric position to the vertical

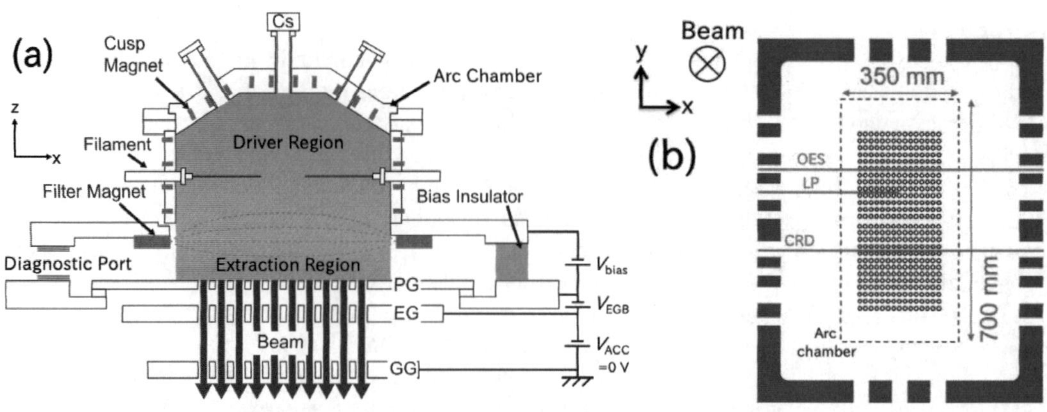

Figure 1. (a) Cross-sectional schematic view of the NIFS-RNIS parallel to filter magnetic field. The number of the aperture in the acceleration grid system is reduced. (b) Diagnostics arrangement in the bias insulator with the plasma grid aperture positions and the arc-chamber size perpendicular to the beam direction. OES, LP, and CRD indicate optical emission spectroscopy, Langmuir Probe, and Cavity RingDown technique, respectively.

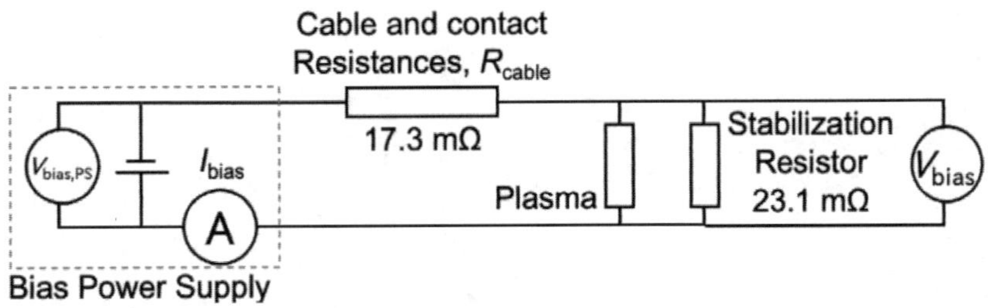

Figure 2. Bias circuit with resistors of cable and contact, plasma, and stabilization.

center of the ion source from the position of the LP. A line of sight of the OES is 19 mm from the PG on the upper half side.

A bias voltage is applied between the arc chamber and the PG. A maximum output voltage from the bias power supply is 10 V. While the bias voltage is normally applied as the PG is positive, reversed polarity bias can be also applied by exchanging feed-through connections to the ion source. The output bias voltage from the bias power supply ($V_{bias,\,PS}$) is directly stored in the data server. In the present system, the stabilization resistor with a resistance of 23.1 mΩ is connected between the PG mounted flange and the discharge chamber to stabilize applied bias voltage (Fig. 2). Plasma resistance changes by the plasma parameters and is normally not less than one order higher than the resistance of the stabilization resistor. The voltage drop due to the cable and contact resistance R_{cable}, which is 17.3 mΩ, from the power supply to the ion source is not able to be ignored because the R_{cable} is comparable to the combined resistance of the plasma and stabilization resistors. The effective bias voltage (V_{bias}) between the discharge chamber and the PG is corrected by $V_{bias} = V_{bias,\,PS} - I_{bias} R_{cable}$, where I_{bias} is output current from the bias power supply and directly stored in the data server.

3. Experimental results

Figure 3 shows time traces of the line-averaged negative hydrogen-ion (H^-) and deuterium-ion (D^-) densities in the same operation condition where the filling pressure is 0.3 Pa, the V_{bias} is 0.8 V, and the average discharge power is 60 kW during the V_{EGB} of 110 V applied phase from zero to one second. The H^- and the D^- densities ramp up with the discharge power. The D^- density is higher than H^- density, as

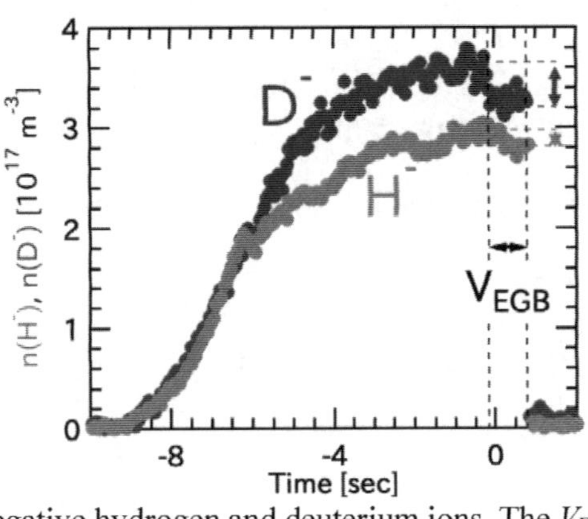

Figure 3. Time traces of negative hydrogen and deuterium ions. The V_{EGB} is applied the gap between the PG and the EG from zero to one second.

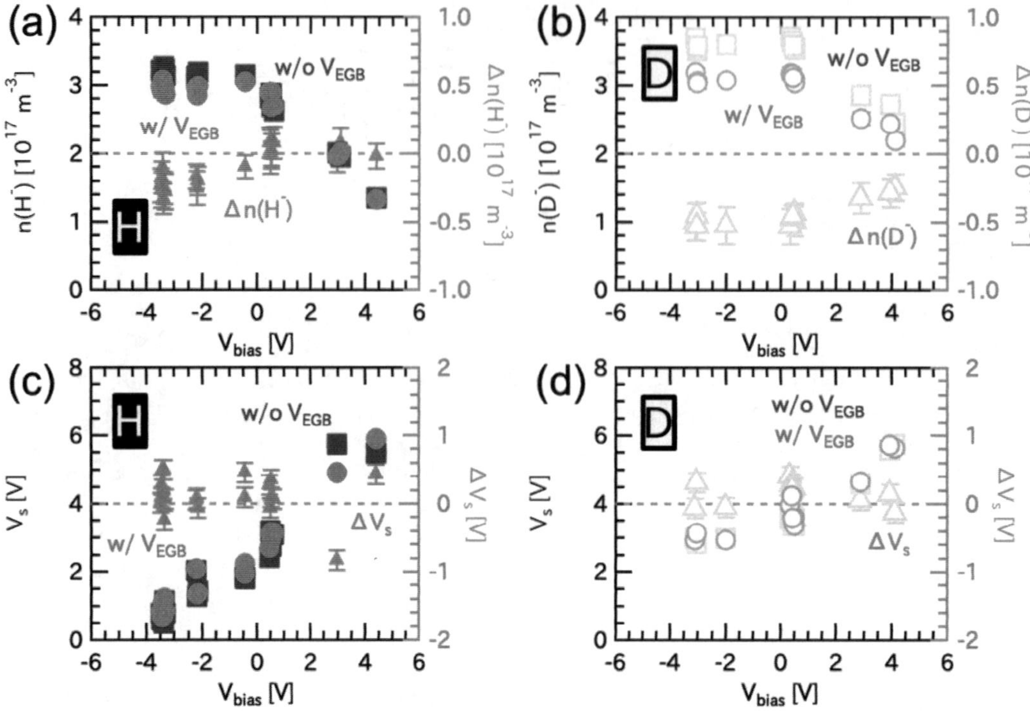

Figure 4. Negative ion density and plasma space potential with (the reds circles) and without (the blues squares) the V_{EGB} and their responses (the greens squares) to the V_{EGB} in the hydrogen (dark-colored closed symbols) and deuterium (light-colored open symbols) operations as a function of the bias voltage between the arc chamber and the PG.

is shown in Ref. 8, but approaches H^- density during V_{EGB} applied phase. Consequently, a larger response of the D^- density, which decreases by 9 %, to the V_{EGB} than that of the H^- density, which decreases by 5 %, has been observed. Thus, there are isotope effects in the response of the negative ion density to the V_{EGB} as well as the absolute line-averaged negative ion density.

Dependences of the H^- and D^- densities, plasma space potential V_s, and their responses to the V_{EGB} on the V_{bias} have been investigated with the discharge power of 60 kW (Fig. 4). The V_{bias} has been varied from -4 V to +4 V by changing the bias-power-supply output voltage from -10 V to +10 V. The H^- and the D^- densities do not change much in low V_{bias} and decrease in high V_{bias} in the conditions with and without the V_{EGB}. This result is similar to Ref. 1 in hydrogen operation without V_{EGB}. The D^- density is higher than the H^- density in the whole bias range without the V_{EGB}, while the H^- and the D^- densities are equivalent in V_{EGB} application phase. Consequently, the D^- density response is larger than that of the H^- in the whole V_{bias} range. The V_s monotonically increases with the bias voltage in both the hydrogen and deuterium operations and is higher in deuterium operation in low V_{bias}. A clear V_s difference between conditions with and without the V_{EGB} cannot be distinguished.

4. Discussion

Since the negative-ion density measured with the CRD increases by cesium seeding, the surface-produced negative ions emitted from the PG reach the CRD measure line. To explore the causes for the

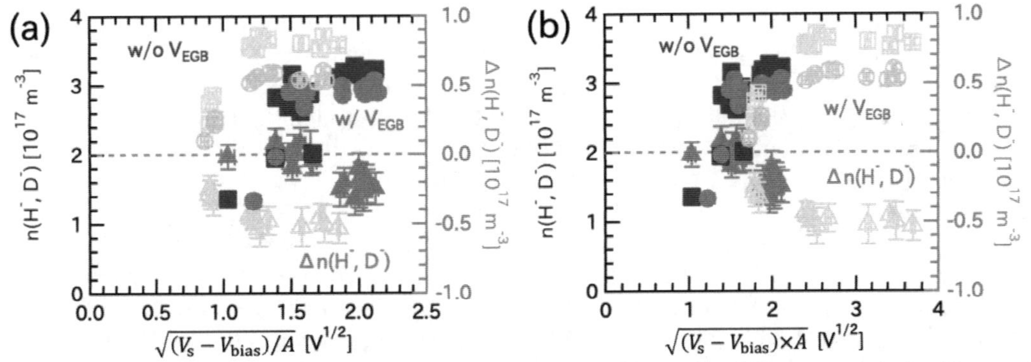

Figure 5. Negative ion density with (the reds circles) and without (the blues squares) the V_{EGB} and its response (the greens squares) to the V_{EGB} in hydrogen (dark-colored closed symbols) and deuterium (light-colored open symbols) as functions of (a) emission speed and (b) momentum of negative ion emission from the PG.

isotope effect shown in Fig. 4, the speed of the surface-produced negative ions perpendicular to the PG is considered. Assuming an energy source for the motion perpendicular to the PG is the only potential difference between the V_s and V_{bias}, the perpendicular speed is proportional to a square root of the potential difference divided by mass number (A), i.e., $\sqrt{(V_s - V_{bias})/A}$. Figure 5(a) represents the negative ion density and its response to the V_{EGB} as a function of the value proportional to the perpendicular speed. The perpendicular speed is not able to explain the isotope effect because it is difficult to identify the H^- group with the D^- group in Fig. 5(a). A momentum component perpendicular to the PG is proportional to a square root of the potential difference times A, i.e., $\sqrt{(V_s - V_{bias}) \times A}$. Figure 5(b) expresses the negative ion density and its response to the V_{EGB} as a function of the value proportional to the momentum component perpendicular to the PG. The H^- density is involved in identical trends of the D^- density in respective conditions with and without the V_{EGB}. The H^- density response to the V_{EGB} is also an extension of the D^- density response trend. One of the possible interpretations for a relation between the isotope effect and the momentum is that the Larmor motion, whose radius is proportional to the momentum, is a dominant mechanism for the surface-produced negative ion in the vicinity of the PG. There is a magnetic field component perpendicular to the PG, such as the filter field, whose typical strength is 10 mT in the vicinity of the PG. The density increase with the momentum in a small momentum range indicates the larger number of negative ions with larger momentum can statistically reach the CRD measure line. The Larmor radius is 20 mm when $\sqrt{(V_s - V_{bias}) \times A}$ equals 1.4. This radius is comparable to the distance of the CRD measure line from the PG. Saturation of the negative ion density in a large momentum range results from the surface-produced negative ions mostly reaching the CRD measure line regardless of the momentum. A difference between plasma equilibriums with and without the V_{EGB} causes the negative-ion density response, whose variation for the momentum seems to follow the negative-ion density. However, compared with the negative ion density, the negative-ion density response becomes zero in larger momentum. This is not explainable in present. The time trace of the H^- in Fig. 1 has been obtained in the condition with the $\sqrt{(V_s - V_{bias}) \times A}$ of approximately 1.5 where the negative-ion density response is close to zero. By contrast, the time trace of the D- in Fig. 1 has been obtained in the condition with the $\sqrt{(V_s - V_{bias}) \times A}$ of approximately 2.5 where the negative-ion density response is saturated. Therefore, the H^- density reduction by the V_{EGB} may be lower than the D^- density reduction. Further study is necessary to verify more correctly these possible interpretations and to elucidate deeply the isotope effect.

5. Summary

The experiment has been conducted to investigate the isotope effect on the surface-produced negative-ion transport in the vicinity of the PG at the NIFS-RNIS on the NIFS-NBTS. Higher D^- density than H^-

density was observed and the D^- density reduction by the V_{EGB} of110 V, which is applied to the gap between the PG and the EG, is larger than the H^- density reduction. The negative ion density and its response to the V_{EGB} were measured by changing the V_{bias}, which is applied between the arc chamber and the PG, from -4 V to +4 V. The H^- and the D^- densities keep at a high level in low V_{bias} and decrease with V_{bias} in high V_{bias}. Although the H^- density response to the V_{EGB} increases with V_{bias} decrease in low V_{bias} and disappear in high V_{bias}, the D^- density response to the V_{EGB} keeps constant in the low V_{bias} and weakens with the V_{bias} in high V_{bias}. The D^- density without the V_{EGB} and the D^- density response to the V_{EGB} of the D^- are higher levels than the H^- density without the V_{EGB} and the H^- density response. The negative ion density regardless of the isotopes is able to be grouped in respective conditions with and without the V_{EGB} by the negative-ion momentum component perpendicular to the PG. The density response to the V_{EGB} is also able to be organized by the momentum. This suggests that the Larmor motion is one of the candidates for the dominant mechanism concerning the negative ion transport from the PG to the plasma in the vicinity of the PG.

Acknowledgments
This work was supported by NIFS Research Programs ULRR702 and KLER110, and JSPS KAKENHI Grant Numbers JP19H01883.

References

[1] Nakano H *et al.* 2013 *AIP Conf. Proc.* **1515** 237
[2] S. Geng *et al.* 2017 *Fusion Eng. Des.* **123** 481
[3] S. Nishioka *et al.* 2018 *J. Appl. Phys.* **123** p 063302
[4] H. Nakano *et al.* 2020 *Jpn. J. Appl. Phys.* **59** SHHC09
[5] K. Ikeda *et al.* 2019 *Nuclear Fusion* **59** 076009
[6] U. Fantz *et al.* 2020 *Fusion Eng. Des.* **156** 111609
[7] H. Nakano *et al.* 2016 *J. Instrum.* **11** C03018
[8] S. Masaki *et al.*, 2020 *Rev. Sci. Instrum.* **91** 013512

Application of Thomson scattering system toward direct measurement of extraction surface of the negative ion beam

J Hiratsuka[1*], K Tomita[2], M Kashiwagi[1], A Kojima[1], G Q Saquilayan[1], T Kaihori[1], M Ichikawa[1], M Kisaki[1], H Tobari[1] and K Watanabe[1]

[1] National Institutes for Quantum Science and Technology, 801-1 Mukoyama, Naka-shi, Ibaraki, 311-0193, Japan
[2] Hokkaido University, Kita 8, Nishi 5, Kita-ku, Sapporo, Hokkaido, 060-0808, Japan

[*]Corresponding author's e-mail: hiratsuka.junichi@qst.go.jp

Abstract. As a challenge to measure the extraction surface of the negative ion beam, a laser Thomson scattering system with high spatial resolution was applied to a weakly ionized plasma in a negative ion source for the first time. The first target was to measure electron density around extraction region because shape of the extraction surface can be estimated by electron spatial profile. However, stray light by injected laser, filament and plasma was too intense compared to signal from the low density region. To solve this, a triple grating spectroscopy and a multilayer black screen were introduced, and the laser spot size was minimized to reduce slit sizes of the spectroscopy. As a result, electron density was successfully measured accurately in low density region. From an error estimation, measurement of electron density 3×10^{16} m^{-3} with spatial resolution of 0.5 mm at extraction region can be realized with accumulation of signals from multiple laser shots. This system can contribute to the understanding of the mechanism of formation of the extraction surface, which is a long-term issue in the negative ion source.

1. Introduction

For the heating of fusion plasma such as in ITER and JT-60SA, negative ion beam acceleration with 1 MeV, 40 A for 3600 s, and 500 keV, 22 A for 100 s is required, respectively [1, 2]. An issue to realize such high power and long pulse beam is that the conventional simulation codes [3, 4, 5] of negative ion trajectory have not predicted heat loads on acceleration grids due to direct interception of negative ion observed in the experiments [6, 7]. Typically, divergence of the negative ions in the simulation is relatively lower than that in the experiment. This difference could be caused from an assumption of negative ion profile around an aperture of plasma grid (PG) to extract the negative ions. In the negative ion source, main source of negative ion production is a surface production on the PG with lower work function by cesium (Cs) seeding. The distribution of the negative ions produced on the PG is influenced by the PG configuration, namely non-uniform, but it is assumed to be uniform in the conventional simulation codes as assumed in the positive ion beam codes. The non-uniform negative ion profile could generate distorted electrical lens as the potential surface of the negative ion extraction, so-called meniscus in this study, and causes larger beam divergence.

This study tries to measure the meniscus directly by introducing a non-disturbance tool, laser Thomson scattering (LTS) diagnostics with excellent spatial resolution and high sensitivity, which has been developed to measure micro-structure of industrial plasmas with high dense electrons of $>10^{23}$ m^{-}

^{3} [8]. The LTS can measure electrons directly and also negative ions by measuring photo-detached electrons from the negative ions [9]. The measurement of the negative ion may be difficult due to relatively low density. However, it is considered that the electron density profile may give the prediction of the meniscus structure because the electrons are extracted with the negative ions from the aperture of the PG. Therefore, in this study, the LTS has been adjusted to measure the electron density with $10^{16} – 10^{18}$ m^{-3} at the special area with mm order in the large negative ion source for the first time.

2. Idea to investigate extraction surface by Thomson scattering measurement

Figure 1 shows a schematic view of the extraction region, in which negative ion is produced on the surface of a Cs seeded PG. Extraction voltage is applied between the extraction grid (EXG) and the PG, and the extraction electric field forms a meniscus, around an aperture of PG. Negative ions and electrons downstream of the meniscus are extracted/accelerated and having high velocity. From the equation of continuity, electron density must drop rapidly at the meniscus. Therefore, the shape of the meniscus can be investigated by measuring electron density profile around the aperture with incoherent LTS diagnostics. When Cs is seeded for negative ion production, Thomson scattering signal of background electron and photo-detached electron from negative ion by injected laser for LTS is measured. Even in that case, the meniscus can be found at the position of rapid drop of total density of background and photo-detached electron in the same way.

In order to design the LTS diagnostics, target values of electron density and spatial resolution of this measurement were considered. From an extrapolation of measured electron density profile with Langmuir probe in the negative ion source, electron density at the meniscus region can be assumed as $\sim 3 \times 10^{16}$ m^{-3}. To estimate required spatial resolution, meniscus shape of several cases with different beam perveance was calculated by the BEAMORBT code [10] as shown in figure 2, with beam current density of 220 A/m^2. Red line (V_{ext} = 6 kV) is the meniscus of best optics for the beam extraction, and a part of ion beam extracted from the meniscus shown with blue (V_{ext} = 7.3 kV) and green (V_{ext} = 5.3 kV) lines are intercepted by acceleration grid in the accelerator. To distinguish such differences, a target of spatial resolution of LTS diagnostics was determined as < 0.5 mm.

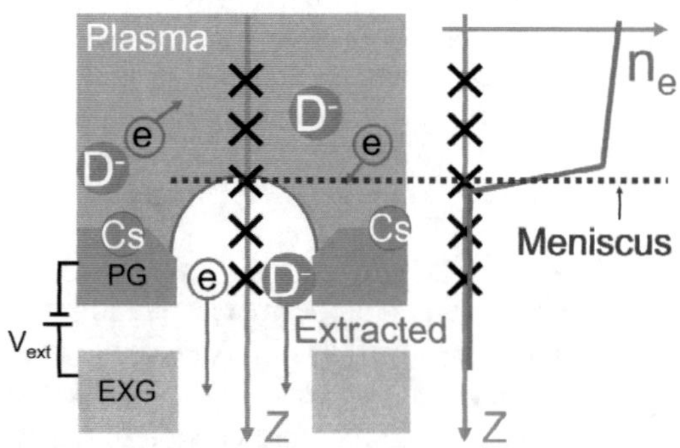

Figure 1. Schematic view of the extraction region in a negative ion source.

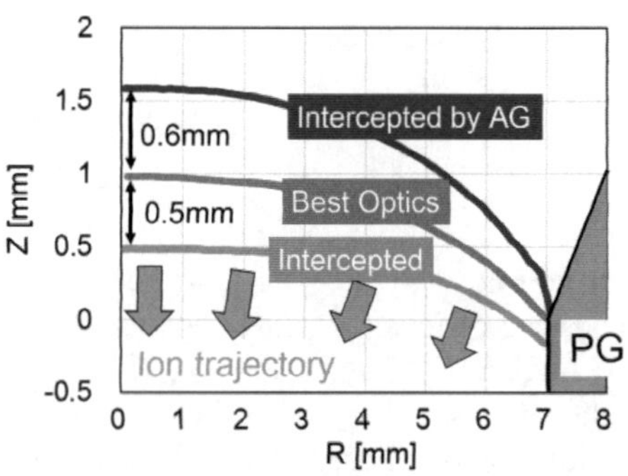

Figure 2. Equipotential surfaces simulated with Child-Lamgmuir law.

3. Design of Thomson scattering diagnostics in negative ion source

Figure 3 shows a schematic view of the LTS system set in a negative ion source. The LTS system was designed to detect enough photon number at extraction region assuming $n_e = 3 \times 10^{16}$ m^{-3} and spatial resolution $\Delta z = 0.5$mm. A high-power Q-switch pulsed Nd:YAG laser (wavelength: 532 nm, laser energy: 300 mJ, pulse width: 3~5 ns, 8×10^{17} photon) is focused on the extraction region. FWHM wavelength of the laser (< 0.46 nm) is much thinner than expected FWHM wavelength of Thomson scattering signal (2.0 nm) assuming T_e = 0.3 eV. The total number of photon of Thomson scattering light $N_{TS} = n_e \Delta z N_i r_e^2 = 96$ photon/shot, where laser photon number N_i, electron radius r_e. A friction of the backscattered (scattering angle: 130°) Thomson scattering light $N_{TS}\Delta\Omega = 0.58$ photon/shot is collected and detected in a spectrometer, where $\Delta\Omega$ is solid angle of measurement line of sight. Because the extraction region is surrounded by electrodes, filaments and inner wall of vacuum chamber, stray light of reflected incident laser from such metal structure is predicted to be much higher intensity than the Thomson scattering signal. Therefore, the Triple-Grating Spectrometer (TGS) [8], which can drastically suppress the stray light, was introduced. The TGS works as a strong notch filter to remove components of spectrum directly coming from the incident laser. TGS consists of an entrance slit (S1), an intermediate slit (S2), three diffraction gratings (G1-G3) and a reverse slit which is called as a 'laser-wavelength stop' (diameter D = 0.5 mm). After dispersed by the first grating, the light is focused by a lens (f_s = 250 mm) and only the central part of spectrum (532 nm) with intense stray light is cut by the laser-wavelength stop. Edge components of the stray light, which is not blocked by the laser-wavelength stop, is cut by the intermediate slit. Finally, dispersed light by the third grating is focused with a lens (f_l = 170 mm) and detected by a CCD camera. Because of the low intensity of Thomson scattering signal, a high-sensitivity and low-noise electron-multiplying intensified charge-coupled device (emICCD) camera (1024 × 1024 pixel, interval (dx) 13μm × 13μm, quantum efficiency at 532 nm ~ 49.6 %, electron-multiplying gain > 400) was selected. In order to calibrate spatial resolution, a ruler was temporally set on the extraction region and measured by the CCD camera. By the method, the spatial resolution on CCD camera was measured as 54 mm/pixel. In addition, because the width of spectrum of the Thomson scattering is thin due to low electron temperature in the extraction region, a fine grating (N = 2400 gratings/mm) was selected for G1-G3 to have thin width of the cut wavelength $D\cos\beta/Nf_s$ = 0.48 nm and a fine wavelength resolution $dx\cos\beta/Nf_l$ = 0.018 nm/pixel at CCD camera, where β = 55° is reflection angle at grating. When assuming transmission of collection optics η ~ 0.3, 500 shot accumulation gives 87 photon and error of signal can be estimated as $\sqrt{N_{TS}\Delta\Omega\eta}/N_{TS}\Delta\Omega\eta$ = 11 % by assuming detected number of photon follows Poisson distribution without any stray light.

Figure 4 shows the first signal of Thomson scattering from a Kamaboko negative ion source. Although the Gaussian spectrum around the laser wavelength of 532 nm was detected, noise was the same level with the signal due to background. Because the background was observed with plasma and without laser injection, the background noise was assumed to be caused by stray light of emission from filament and plasma. To block such stray light, an anodized multilayer black screen was designed and set before the collective window. The multilayer screen consists of a series of large and small apertures, and the difference of aperture size makes optical trap. Stray light of which path is out of alignment of the Thomson scattering light is trapped and absorbed in the optical trap. In addition, width of the slits (S1, S2) was reduced to 0.2 mm by reducing beam spot size of the injection laser. Focal length of the injection lens was tuned so that the laser spot size is minimized but laser power density at the entrance window is less than the damage threshold. When changing measurement point of different axial distance to the PG, the focal length was kept constant and position of the injection lens was moved. By these countermeasures, S/N ratio was improved as figure 6(a).

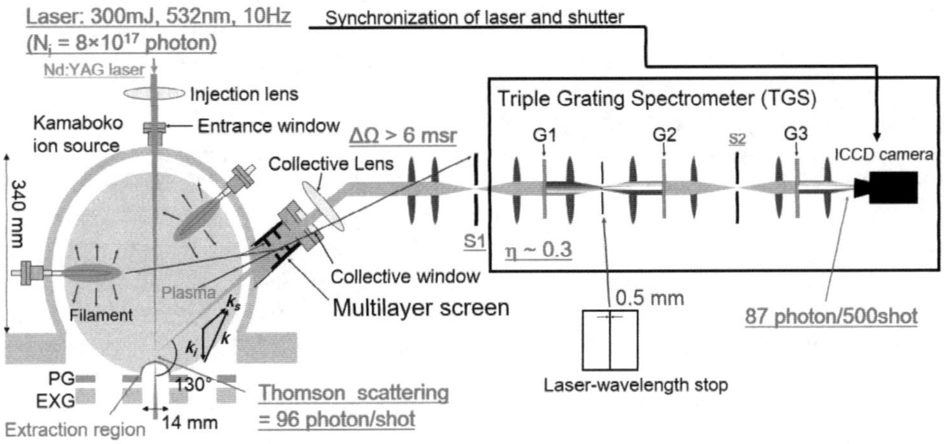

Figure 3. Design of LTS diagnostics in negative ion source.

4. Calibration result

In order to calibrate sensitivity and spatial and wavelength resolution, Rotational Raman Scattering (RRS) from nitrogen gas (200 Torr) was measured by the LTS diagnostics system [11]. Figure 5(a) shows 2D image of the measured spectrum of RRS at the center axis of the extraction aperture. Figure 5(b) shows spectrum of the RRS at 30 mm from the surface of PG. From the measurement, sensitivity was estimated as 0.36, which is enough to detect the Thomson scattering light by accumulation. As shown in figure 5(b), width of the cut wavelength = 0.46 nm and wavelength resolution = 0.018nm/pixel were measured as designed, which confirms the feasibility of the LTS measurement.

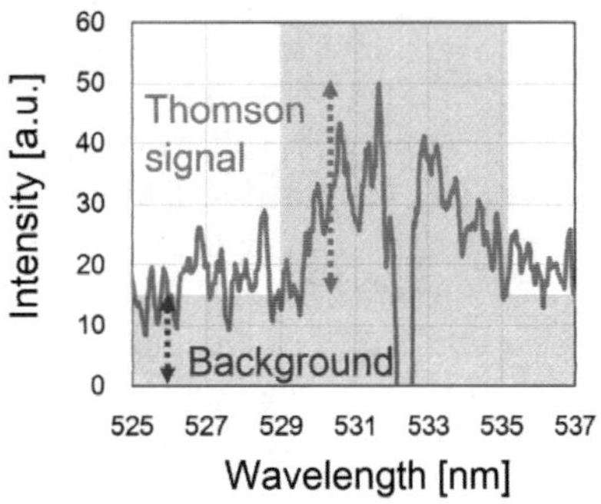

Figure 4. First signal of Thomson scattering measurement in negative ion.

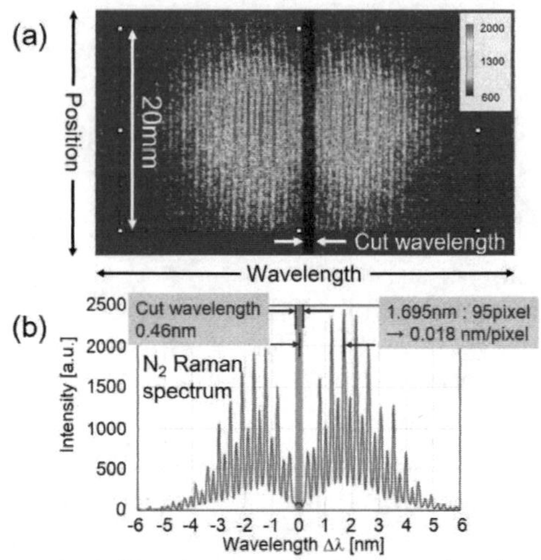

Figure 5. Calibration result with Rotational Raman Scattering from nitrogen gas (200 Torr).

5. Experimental result and estimation of achievable spatial resolution at extraction region

Figure 6 shows an accumulated Thomson scattering signal measured by the LTS system with 500 laser shot on the center axis of the PG aperture and at 30 mm from the surface of PG. A clear Gaussian spectrum of Thomson scattering was observed in figure 6(a) and (b), where the signal was integrated over 20 mm and 0.5 mm around the measurement point, respectively. Electron density and temperature estimated from the spectrum in figure 6(b) were calculated as $n_e = 1.2 \times 10^{18}$ m^{-3} and $T_e = 0.52$ eV with fitting error of $n_e = 16$ % from the spectrum. The fitting error was estimated as $\sqrt{\sum_i (y_i - f_i)^2/(N_s - 2)}$, where y_i , f_i , N_s are signal, fitting value, total number of signal, respectively, by assuming y_i distributes around f_i as the normal distribution. This measurement was repeated at 14 mm from the surface of PG with 3000 laser shot, and the results was $n_e = 3.9 \times 10^{17}$ m^{-3} and $T_e = 0.63$ eV with fitting error of $n_e = 31$ %. The measured values of electron density are in the similar order of magnitude as measured by Langmuir probe, though reliability of Thomson scattering is higher than Langmuir probe in the high magnetic filter field in negative ion source.

In order to estimate spatial resolution at extraction surface, relationship between electron density and integration length Δz along the laser path to achieve fitting error < 20 % was plotted in figure 7. Red dotted line is extrapolation from measurement result (accumulation of 500 laser shot) on an assumption that the number of detected photons follows Poisson distribution and the error is proportional to square root of the photon number. The target of electron density (3×10^{16} m^{-3}) and spatial resolution (0.5 mm) is on the green dotted line of which electron density is 24 times lower than the red dotted line. Therefore, accumulation of 500 shot × 24 = 12000 laser shot is necessary to measure electron density around extraction surface, which is acceptable because it corresponds to signal accumulation of 120 times of 10 second discharge with 10 Hz laser injection. This estimation confirms feasibility of detection of extraction surface even in such low-density region.

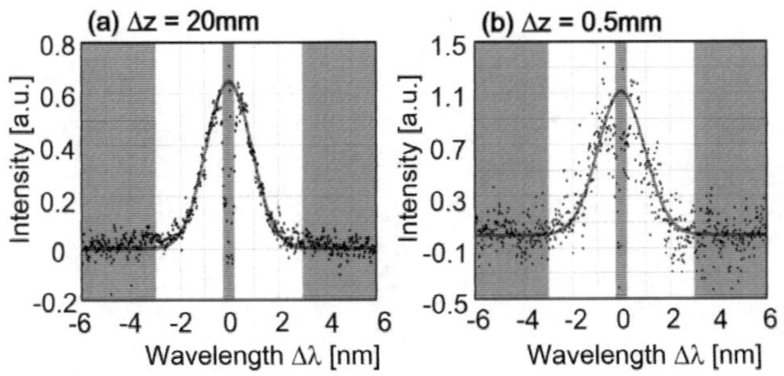

Figure 6. Thomson scattering signal at 30 mm from surface of PG.

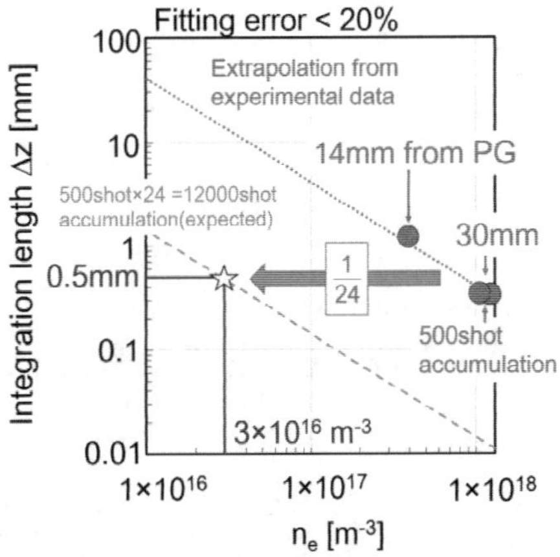

Figure 7. Relationship between electron density and spatial resolution of LTS.

6. Summary

Toward measurement of meniscus shape from rapid drop of electron density, Thomson scattering diagnostics in a negative ion source was developed. To achieve high spatial resolution in low density region, a LTS system with TGS was designed. In addition, stray light caused by emission from filament and plasma was suppressed by introducing multilayer screen and minimizing slit size in the spectroscopy. By calibration with RRS from nitrogen, high spatial resolution at camera (54 μm), high wavelength resolution (0.018 nm) and thin cut wavelength of TGS (0.46 nm) were confirmed. By applying the LTS system to negative ion source at 14 – 30 mm from PG, measurement of density of 3.9×10^{17} m^{-3} with high spatial resolution of 0.5 mm and fitting error of 31 % was achieved. By assuming the error follows Poisson distribution, electron density profile in the extraction region is expected to be measured by accumulating Thomson scattering signals of 12000 laser shots. The system contributes to understanding extraction region of negative ion source and design optimization of negative ion source.

References

[1] Doyle E J, Houlberg W A, Kamada Y, Mukhovatov V, Osborne T H, Polevoi A, Bateman G, Connor J W, Cordey J G, Fujita T *et al* 2007 *Nucl. Fusion* **47** S18
[2] Ishida S, Barabaschi P, Kamada Y and the JT-60SA Team 2011 *Nucl. Fusion* **50** 094018
[3] Miyamoto K, Nishioka S, Goto I, Hatayama A, Hanada M, Kojima A and Hiratsuka J 2016 *Rev. Sci. Instrum.* **87** 02B124
[4] Agostinetti P, Aprile D, Antoni V, Cavenago M, Chitarin G, de Esch H P L, De Lorenzi A, Fonnesu N. Gambetta G, Hemsworth R S *et al* 2016 *Nucl. Fusion* **56** 016015
[5] Kisaki M, Kojima A, Saquilayan G Q, Hiratsuka J *et al* Reverse trajectory analysis of the hydrogen negative ion beam in a prototype accelerator for ITER presented in ICIS2021
[6] Hiratsuka J, Hanada M, Kojima A, Umeda N, Kashiwagi M, Miyamoto K, Yoshida M, Nishikiori R, Ichikawa M, Watanabe K and Tobari H 2016 *Rev. Sci. Instrum.* **87** 02B137
[7] Denizeau S, Aprile D, Fubiani G, Taccogna F, Minelli P, Ichikawa M, Hiratsuka J, Kashiwagi M, Kojima A and Chitarin G 2020 *Plasma Sources Sci. Technol.* **29** 075012
[8] Tomita K, Sato Y, Nishikawa K, Uchino K, Yanagida T, Tomuro H, Wada Y, Kunishima M, Kodama T, Mizoguchi H, and Sunahara A 2015 *Appl. Phys. Express* **8** 126101
[9] Naitou H, Sakurai Y, Tauchi Y, Fukumasa O, Yagi M, Yamagata Y, Uchino K and Muraoka K 2004 Plasma Phys. Control. Fusion **46** 1217-1230
[10] Ohara Y 1976 *JAERI-M* 6757
[11] Gessel A F H, Carbone E A D, Bruggeman P J B and Mullen J J A M 2012 Plasma Sources Sci. Technol. **21** 015003

Probe measurement of an ECR hydrogen plasma facing the C12A7 electride surface

K. Bito[1*], M. Sasao[2] and M. Wada[1]

[1]Graduate School of Science and Engineering, Doshisha University, Kyoto 610-0321, Japan
[2]Office of R&D Promotion, Doshisha University, Kyoto, 602-8580, Japan

*Corresponding author's e-mail: ctwf0303@mail4.doshisha.ac.jp

Abstract. Plasma parameters of an ECR discharge excited in a compact ion source with a replaceable plasma electrode (PE) were measured with a Langmuir probe. The PE was installed to investigate how an electride PE changes the plasma parameters from those with a Mo PE. The effect upon the plasma parameters in front of the PE due to the material will be elucidated by changing the probe position axially along the beam extraction direction. Analyses of the measured probe characteristics revealed that electride PE has a higher electron density (n_e) and lower electron temperature (Te) than the Mo PE. When the PE was biased negatively, the ratio of the positive saturation current (I_i) to the negative saturation current (I_e+I_{H^-}) for the electride PE was about 1.6 times of the ratio for the Mo PE.

1. Introduction

Large-scale, high-power, long-pulse negative hydrogen (H^-) ion sources have been used as powerful plasma heating tools in magnetic confinement fusion programs to date due to their high neutralization efficiency. Cesium (Cs) is injected and covers the plasma electrode (PE) or plasma grid (PG) of an ion source for increasing the H^- and negative deuterium (D^-) ion current while suppressing co-extracted electron current [1]. However, Cs removed from PE or PG accumulates in the source and pollutes the source components requiring regular maintenance. This can cause a severe disadvantage for a future electrical power generation system as maintenance activities in an intense neutron / gamma ray irradiation environment are costly. To solve the problems, the programs of Cs free H^- / D^- ion source development are in progress [2].

The reason for realizing low coextracted current while keeping high current density H^- ion beam by Cs coverage on PE/PG is attributed to the surface production of H^- ions. Surface production forms H^- flux by forming H^- ions on low work function surfaces such as Cs-covered molybdenum surfaces ($\varphi \sim$ 1.5 eV) [3,4]. Therefore, the idea for enhancing surface formation is to find a suitable material with a low work function. Alternative materials for the use of Cs, such as carbon materials, are being studied and high H^- production yields from the surface of Cs-free plasma are expected [5,6]. Recently, a new inorganic material with low work function and high conductivity, C12A7 electride, has been studied as a material for PE/PG [7,8,9].

In the previous experiment, it was confirmed that by applying a low acceleration voltage, the ratio of the H^- ion current to the electron current extracted from the source was higher for an electride PE than a Mo PE [10]. In the plasma, the positive ion density (n_{i+}) and electron density (n_e) are nearly identical: $n_i = n_e$ in the absence of negative ions which can be confirmed by saturation currents of the opposite polarity. However, when substantial fractions of electrons are replaced by H- ions, the relationship ni+ > ne should appear on the ratio of the positive saturation current to the negative saturation current and the ratio of hydrogen ion positive saturation current to the negative saturation current is large because the contribution of negative ion density to the negative saturation current is small [11]. In this paper, we try to see the plasma parameters of electride PE with low work function and compare with those of Mo PE from current-voltage (*I-V*) characteristics of a Langmuir probe. Furthermore, some hint on H^- production efficiency is discussed by calculating the ratio of hydrogen ion positive saturation current (I_i) to the negative saturation current (I_e+I_{H}-).

2. Experimental setup

Figure 1 shows a schematic view of the experimental setup: (a) compact ECR ion source [12], and (b) enlarged view of the Langmuir probe. The device produces a hydrogen plasma with a magnetic field strength of over 87.5 mT on the axis of the Mo antenna. The magnetic field is generated by six permanent magnets, and 2.45 GHz microwave power excites the plasma to achieve ECR conditions where the axial magnetic field strength is equal to 87.5 mT. A 1.5 mm diameter Mo antenna fixed to the center conductor of a vacuum airtight N-type coaxial feedthrough is spirally wound to a length of 9.5 mm. The end of the antenna is fixed to a stainless-steel screw attached to the chamber and the microwave current through the antenna excites a stable plasma around the antenna. The locally generated plasma diffuses into a volume about 20 mm long and 50 mm in diameter facing the PE. PE can be electrically biased against the plasma source chamber. Two Wilson-sealed 6 mm diameter diagnostic ports are located to optically observe the plasma volume near the PE. Plasma ignition status was confirmed through these ports. A pair of additional permanent magnets installed near the PE, called vertical magnetic field magnets (VFMs), generates a magnetic field that plays the roles of the magnetic filter field and the electron deflection magnetic field. A Langmuir probe is mounted axially along the beam extraction direction, and the probe core wire has the 0. 3 mm diameter and 1 mm length exposed to a plasma. The Langmuir probe travels the distance from -1 to 7 mm relative to the PE surface of the smaller diameter opening.

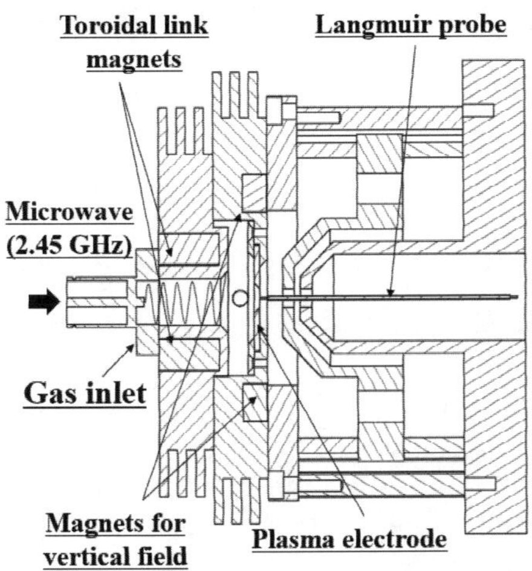

(a)

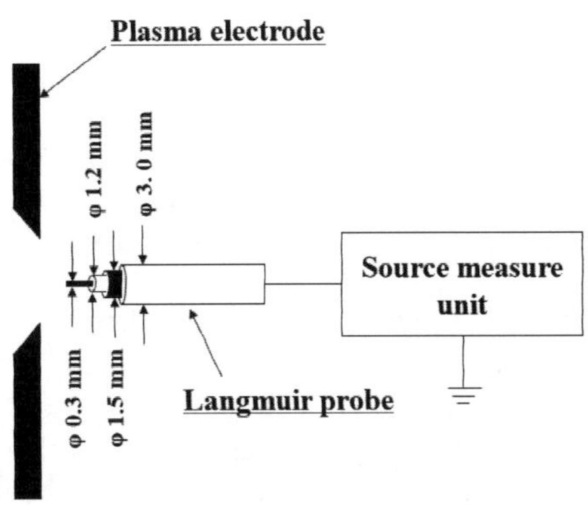

(b)

Figure 1. Schematic view of the experimental setup (a) compact ECR ion source, (b) enlarged view of Langmuir probe.

The PE has a diameter of 34 mm and a thickness of 2 mm, and the conical extraction hole has been extended from 6 mm to 10 mm in diameter (taper angle 45 °). The C12A7 electride, having a cage structure centered on free oxygen ions replaced by electrons, was heated by the supplier and chemically processed before shipment. However, because the surface is coated with an antioxidant, the sample must be reheated before use. Confirmation of the low work function surfaces condition was achieved by using a separate vacuum vessel to remove the membrane according to the reheating procedure instructions provided by the supplier and detecting the photoelectron current by the exposure of a He-Cd laser light (3.8 eV photon energy). After cooling and vacuum rupture, the chamber was pumped down again to confirm a low work function surface condition, then the chamber was moved and attached to the ion source.

3. Experimental results

3.1. Probe parameter analysis

Under the conditions of gas pressure 0.3 Pa, 0.6 Pa, 1.2 Pa, microwave input power 30 W and PE bias voltage of floating, the probe characteristics were measured when the probe distance was changed from -1 to 7 mm and the electron temperatures (T_e) were calculated. Figure 2 show the electron temperature characteristics against the probe distance for electride PE and Mo PE. Parameters are compared with (a) when VFM magnets are off, and (b) when the VFM magnets are on. The results shows that the plasma in front of the electride PE has a lower electron temperature than the plasma in front of the Mo PE. And it is also interesting to note that the electron temperature of electride PE and that of Mo PE increased between 5 mm to 7 mm probe distance by using a magnet for vertical magnetic field.

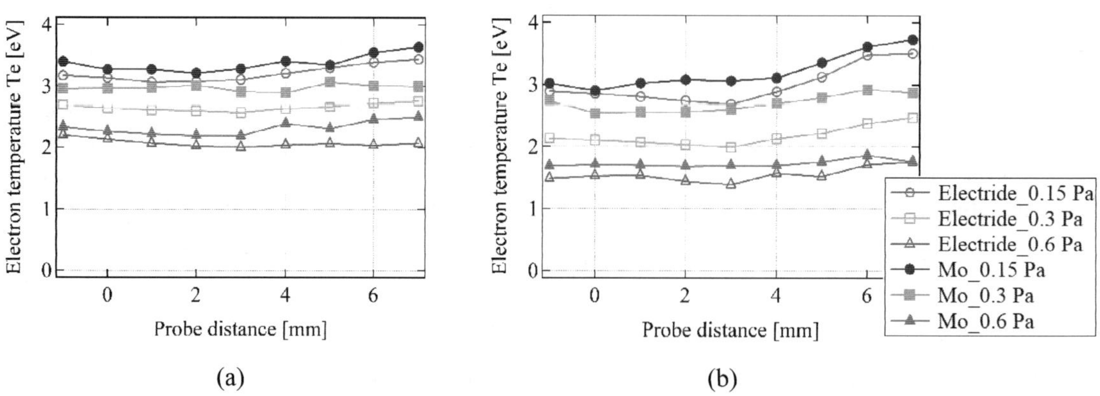

Figure 2. The electron temperature characteristics against the probe distance for the electride PE and the Mo PE (a) when VFM is detached, (b) when VFM is attached.

Figure 3 shows the electron density (n_e) characteristics against the probe distance for electride PE and Mo PE for identical discharge conditions with figure 2. Two magnetic field configurations are compared between (a) when VFM field is off and (b) when the VFM field is on. The results show that the electride PE produced higher density plasma than the Mo PE particularly at the lower H_2 pressure. The spatial distributions of electron density without the VFM field increased monotonically toward the plasma center. (Larger distance from the PE toward the ECR plasma region.) On the other hand, the distribution with the VFM field decreased from 6 to 7 mm for electride PE for two cases.

3.2. PE bias voltage characteristics

Under the conditions of H_2 0.3 Pa gas pressure, probe distance 7 mm from the end of the PE front surface, and the microwave input power 30 W, the PE bias was varied from -30 to 50 V. The ratios of positive ion saturation current, I_i, with respect to the negative saturation current, ($I_e + I_{H^-}$), for both PE materials, electride and Mo were measured and compared as shown in Figure 4 as functions of PE bias with respect to the chamber potential where the ECR plasma was excited. Here the contribution of the negative ion density to I_{es} would be small [11]. At above PE bias more positive than +40 V, both ratios for electride and Mo PE saturated at about 0.12, while in the negative bias region, we see the clear decrease of the rate corresponding to more negative ions, saturated at about 0.047 for electride and 0.075 for Mo PE, showing the higher negative ion concentration at the larger negative bias (the maximum difference in ratios should be around 0 V).

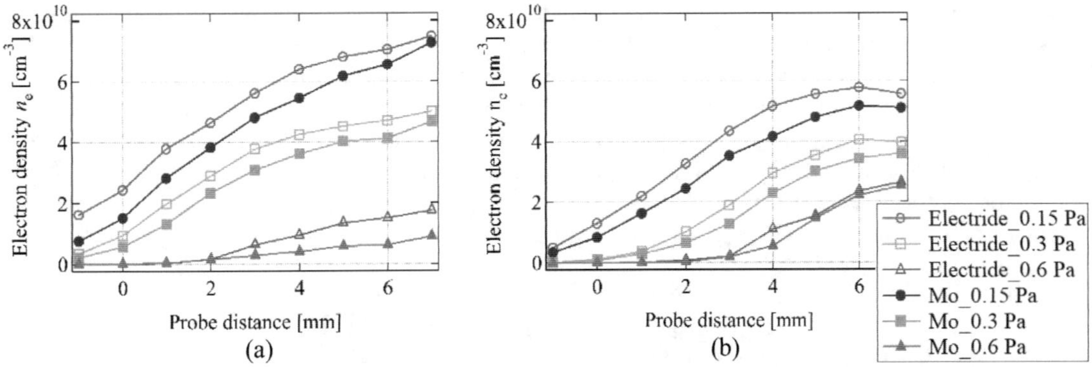

Figure 3. The electron temperature characteristics against the probe distance for the electride PE and the Mo PE (a) when VFM is detached, (b) when VFM is attached.

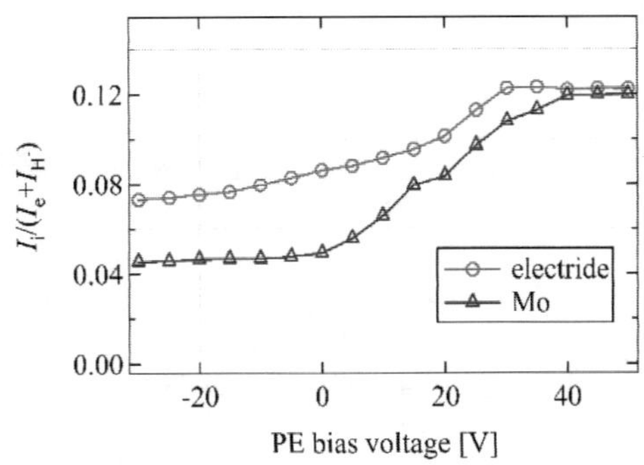

Figure 4. The ratio of positive ion saturation current to negative saturation current with respect to the PE bias voltage on the Mo PE and on the electride PE.

4. Discussion

4.1. Electride PE probe parameters

Positive ions are collected on the surface of electride PE by applying a negative bias voltage. Because electride is a material with a low work function, positive ions near PE are absorbed by the electride surface and combined with electrons existing inside the electride and reemitted into the plasma as negative hydrogen ion. Electride with a low work function emits electrons with low energy from the surface. Also considered is the higher yield of secondary electron emission from the electride surface that should also contribute to a higher electron density. Therefore, it is considered that the electron temperature is relatively lower, and the electron density is higher than when Mo PE is used.

As is pointed out in the paper by P. McNeely et al., 2008 Plasma Sources Sci. Technol. 18 014011, the probe characteristics would be affected by the negative ion distribution, particularly in the positive probe voltage region (the electron suppression region), and must be handled with care. In the probe characteristics measurement of this study, we did not analyse the transient regions, and the electron temperature was calculated within a narrow range of the probe voltage, where the probe characteristics were not affected by the influx of negative ions [13].

4.2. Electride PE surface condition

Figure 5 shows the Photograph of electride PE surface (a) a new sample before using plasma exposure, (b) after hydrogen plasma exposure for about 4 hours with -30 ~ 40 V bias. It was confirmed that the contact resistance of the insulating layer measured by the point contact method under the condition of a pressure of 10 Pa order or less so as not to damage the generation curtain was about 16 MΩ and that the other parts were about 180 kΩ, and that the insulating layer lost the electrical conductivity of the initial electride. It has been known that hydrogen plasma bombardment on an electride surface does not affect or diminish the work function in case of positive biasing [2,14]. Formation of the insulating layer, which can be removed easily by polish, was caused by deposition of metal impurity and the sputtering component from the side surface of the ion source in the vicinity and they are oxidized, because the distance from the nearest ECR antenna surface was only 10 mm. In future design of a H^- source, it should be noted this problem.

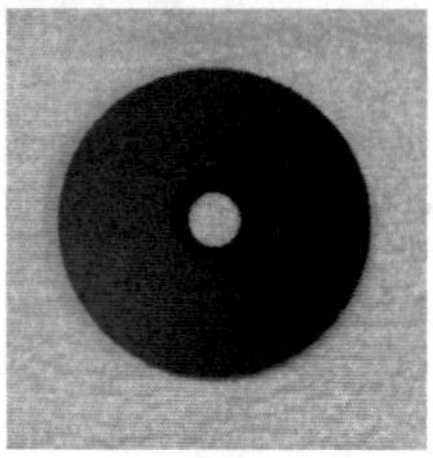

(a)

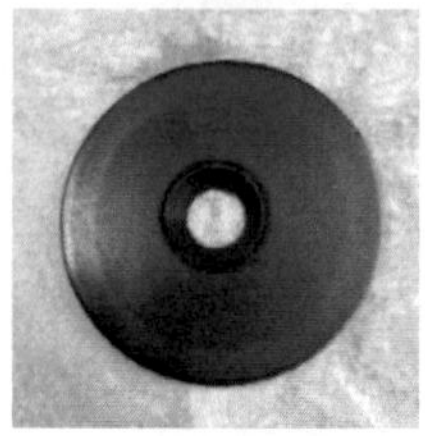

(b)

Figure 5. Photograph of electride PE surface (a) before plasma exposure, (b) after hydrogen plasma exposure for about 4 hours with -2 V bias.

5. Summary and conclusions

Using a compact ECR ion source experimental device [12], we researched PE materials suitable for Cs-free H^- sources. By investigating the influence of the PE bias voltage, the H^- ion generation efficiency was high in the electride, and it showed a great advantage especially when a negative voltage was applied. By examining the probe characteristics and comparing the plasma parameters between two materials, the electride showed stable low electron temperature and high electron density.

6. Acknowledgements

The authors would like to thank H. Hosono for the permission to use C12A7-electride. The C12A7-electride was supplied from AGC Inc., and technical support by Naomichi Miyakawa, Satoru Watanabe, and Kazuhiro Ito of AGC Inc. and deep discussion with them are greatly acknowledged. This work was supported by Grant-in-Aid for Scientific Research of JSPS, 19H01883, 17H03512.

References

1. Bacal M and Wada M 2015 *Appl. Phys. Rev.*, **2**, 0211305
2. Tacogna F *et al* 2021 *Eur. J. D* **75**, 227
3. Belchenko Yu *et al* 1974 *Nuclear Fusion*, **14**, 113-114
4. Wada M *et al* 1983 AIP Conf. Proc. **111**, 227
5. Kumar P *et al* 2011 *J. Phys. D :Appl.* Phys., **44**, 372002
6. Ahmad A *et al J. Phys. D : Appl.* Phys., **47**, 085201
7. Sasao M, Tanemura K, Wada M and Tsumori K 2017 AIP Conf. Proc. **1869** 020005
8. Sasao M et al 2018 Appl. Phys. Express **11** 066201
9. Kobayashi M, Sasao M, Kisaki M, Eguchi T and Wada M 2018 AIP Conf. Proc. 2052, 020003
10. K. Bito *et al*., AIP Conf. Proc. **2373,** 030006 (2021).
11. Amemiya A *et al* 2005 J. Plasma Fusion Res. **81** 482-525 (in Japanese)
12. T. Eguchi *et al*., Rev. Sci. Instrum. 91, 013508 (2020).
13. P. McNeely *et al.,* 2009 Plasma Sources Sci. Technol. **18** 014011
14. A. Heiler *et al*., AIP Conf. Proc. **2373,** 020004 (2021).

Journal of Physics: Conference Series **2244** (2022) 012049 doi:10.1088/1742-6596/2244/1/012049

Ion source developments at IPP: On the road towards achieving the ITER-NBI targets and preparing concepts for DEMO

U Fantz, S Briefi, M Fröschle, N den Harder, A Heiler, B Heinemann, A Hurlbatt, C Hopf, M Lindqvist, F Merk, A Mimo, R Nocentini, G Orozco, R Riedl, G Starnella, C Wimmer, D Wünderlich, D Yordanov, D Zielke

Max Planck-Institut für Plasmaphysik, Boltzmannstr. 2, 85748 Garching, Germany

E-mail: ursel.fantz@ipp.mpg.de

Abstract. The large ion source of ITER's neutral beam injection (NBI) systems (0.9 m×1.9 m) with 1280 apertures has to deliver 57 A D^- for 3600 s (286 A/m^2) and 66 A H^- for 1000 s (329 A/m^2). The RF ion source test facilities ELISE and BUG at IPP are aimed to demonstrate the ion source parameters, the homogeneity of large beams (up to 1 m×1 m) and to perform beam optic studies. While the ITER parameters could be demonstrated in hydrogen, the achievement in deuterium for long pulses is still pending due to the large fraction of co-extracted electrons, their temporal dynamics, and inhomogeneity in vertical direction, limiting the ion source performance. Biasing of the bias plate in the vicinity of the plasma grid improves the symmetry of the co-extracted electrons and contributes to its stabilzation being thus a promising alternative to potential rods formerly used to achive high performance. With the replacement of the high-voltage power supply at ELISE first 100 s steady state extraction is demonstrated increasing the relevance of the test facility for ITER and DEMO studies. For the latter, the IPP contributions focus on improvement of the RF coupling, the caesium management and conceptual studies of a beam driven plasma neutraliser as alternative to the gas neutraliser system.

1. Introduction

Neutral beam injection (NBI) systems for upcoming fusion devices like the international fusion experiment ITER being presently under construction and the DEMOnstration Power Plant DEMO under conceptional design within the framework of the EUROfusion Consortium rely on the generation of negative hydrogen ions, their electrostatic acceleration and neutralization before injecting the high energetic particles into the Tokamak. At ITER the injected power is 16.5 MW per beamline hosting two beamlines with the option to extend to three for sufficient heating and current drive. Considering the particle losses in the beam line and the neutralization efficiency of the gas neutralizer the ion current has to be 46 A H^- accelerated to 870 keV and 40 A D^- accelerated to 1 MeV [1,2]. According to the staged approach, hydrogen injection is planned for the Pre-Fusion Operation Phase II, starting in 2032 where the beam should be temporally stable for up to 1000 s. Fusion Power Operation starts in 2035, for which stable deuterium injection for 400 s is needed for the baseline scenario (Q = 10) and 3600 s for the advanced scenario (Q = 5). Regarding the ion current extracted with up to 10 kV from the ion source, 66 A H^- and 57 A D^- are needed to compensate the estimated stripping losses of 30% in the 7-grid accelerator. The extraction area of 0.2 m^2 is composed of 1280 beamlets with apertures of 14 mm in diameter arranged in 16 beamlet groups in a source area of roughly 0.9 m × 1.9 m. The extracted current density has to be uniform across the large area at values of 329 A/m^2 (H^-) and 286 A/m^2 (D^-)

with the goal to achieve a beam uniformity of better that 90%. The core of the accelerated beamlets (carrying ≈85 % of the power) needs to have a divergence of less than 7 mrad, while divergences of 15–30 mrad are allowed for a halo carrying the remaining ≈15 %. A beamlet misalignment of 2 mrad in the horizontal, and 4 mrad in the vertical direction is allowed as worst-case scenario.

The present EU-DEMO design concentrates on a pulsed machine for which 2 GW fusion power converted to 500 MW net power are envisaged over 2-hour phases [3]. A heating power of 50 MW is planned, where NBI is one of the options; current drive plays - if at all - a minor role. The general strategy is to rely as much as possible on the Return of Experience (RoX) from ITER. Hence the beam source of DEMO NBI will be based on the same ion source and accelerator concept, whereas for the neutralizer the option of a beam driven plasma neutralizer (BDPN) as alternative to the gas neutralizer is investigated [4]. The BDPN would relax the demands to the accelerated D^- current density as the neutralisation efficiency enhances.

For the ion source, which has to operate reliably at the low pressure of 0.3 Pa, the RF-driven concept (Inductively Coupled Plasma, ICP) has been chosen using a frequency of 1 MHz and a total power of up to 800 kW delivered by four RF generators. The modular concept is based on the prototype source developed at IPP [5]. At present, IPP is contributing with the RF ion source test facilities ELISE and Batman UpGrade (BUG) to the ion source development aimed to demonstrate the ion source parameters, the homogeneity of large beams (up to 1 m×1 m) and to perform beam optic studies. The paper focusses the recent investigations contributing to the achievement of the ITER NBI targets (section 2) and on the activities for the NBI for an EU-DEMO compromising the beam source and the BDPN (section 3).

2. Towards the ITERs' ion source targets with the test facilities ELISE and BUG

The prototype source is mounted at the test facility BUG [6] consisting of one RF-driver in which the plasma is generated and an expansion chamber such that it represents an 1/8 size-scale of the ITER source. A magnetic filter field is used to reduce the electron temperature and density by roughly an order of magnitude to 1 eV and 10^{17} m^{-3}, respectively close to the extraction system. The extraction area is about 0.01 m^2 and comprises with 14×5 apertures almost one beamlet group of the ITER arrangement (16×5 apertures). For extraction and acceleration three grids are used: the plasma facing grid (PG) with chamfered apertures (14 mm diameter), the extraction grid (EG) equipped with permanent magnets to deflect co-extracted electrons out of the beam, and the grounded grid (GG). The source is at high potential and a total voltage of up to 50 kV (15 A) is available provided by the HV power supply capable to operate in continuous wave mode. At present, BUG is still working in pulsed operation mode, typically 10 s plasma with a beam phase of 6 s; up to 100 s are possible at full RF power but longer pulses only at reduced power as some components (e.g. the plasma grid) are still under preparation for steady state operation at full source performance. Investigations at BUG are presently focused on beam optics studies and the understanding of beam properties for which the test facility is equipped with several beam diagnostic tools [6], extended recently by a retractable 1D-CFC beam target to measure the beam footprint from the backside with an IR camera [7].

The test facility ELISE is equipped with a half-size ITER source in order to bridge the gap towards the full- size ITER source. The size-scaling route is, for example, illustrated in [8]. The source consists of four RF-drivers two of each connected in series to one RF generator. The same width but only half the height of the source allows to host 8 beamlet groups in ITER arrangement. As at BUG, a three-grid system is used for extraction and acceleration of the negative ions with 60 kV total voltage available, however in pulsed mode only (10 s extraction blips with about 150 s in between). The ion source is capable to operate continuously and the beam is probed with those short beam blips. For continuous beams a new HV power supply is currently tested (see below). Steady state operation at full performance will be possible after replacing the current short pulse calorimeter with a properly cooled diagnostic calorimeter similar to the one successfully tested at BUG [9] and is presently under manufacturing. ELISE and BUG are equipped with a proper neutron shielding allowing for deuterium operation.

In the ion source concept for ITER and EU-DEMO the negative ions are produced by the surface conversion process for which caesium is injected in the ion source. Caesium is evaporated by ovens: one at the back plate of the ion source at BUG, two at the side walls (left and right) at ELISE and three (at the backplate) for the ITER source. The evaporation rate per oven is typically several milligrams per hour [8] and caesium deposits on surfaces from which it is re-distributed in the source by the plasma. To avoid cold spots and thus caesium accumulation in the source the surfaces are temperature controlled; typically the source walls around 40°C and the PG around 150°C. Both temperatures reflect the experimentally determined optimal temperatures for the RF sources, with the PG temperature in particular having a wide stable range (up to 200-250°C) before the electron currents increase and the ion currents decrease. The caesium dynamics in the ion source determines the temporal behavior and uniformity of the work function of the caesium layer on the PG and has thus consequences for the temporal stability of the extracted currents (dominantly on the co-extracted electrons) and the beam uniformity. In front of the plasma grid an electronegative plasma evolves with negative ion densities higher than electron densities. Subsequently, the co-extracted electron current reacts much more sensitive than the extracted ion current. In fact, the co-extracted electron current can limit the achievable source performance as the electron current should not exceed the ion current in order to keep the heat load at the EG at the tolerable value.

The tasks of ELISE focus on the achievement of ITERs' ion source targets with respect to extracted current densities stable for the required long pulse operation with the current of the co-extracted electrons lower than the ion current and the required beam uniformity. This comprises investigations on (i) the reduction, the symmetry and the temporal stability of co-extracted electrons, (ii) caesium management studies, (iii) studies on plasma uniformity and (iv) on the beam uniformity for which ELISE is well equipped with plasma and beam diagnostic tools. Furthermore, gaining experience in routine operation of such large sources and developing reliable operation scenarios as well as training of future NBI operators and scientist is among the tasks as well.

2.1. Ion source performance and its challenges

At ELISE, the ITER parameters could be demonstrated in hydrogen, i.e. more than 90% of the required extracted ion current for 1000 s was achieved in a reproducible series, limited only by technical constraints regarding the available RF power and the high voltage power supply [10]. For deuterium, more than 90 % are achieved as well but only for short pulses, i.e. 20 s plasma with 10 s beam extraction. Over 60 % could be demonstrated in long pulses up to one hour [8,11]. Extending the pulse length at high performance is limited by the higher co-extracted electron current than in hydrogen accompanied by a stronger inhomogeneity in vertical direction and stronger increase with time. The measures developed for hydrogen are obviously not effective enough in deuterium. Among them are the improved filter field configuration, i.e. in addition to the filter field created by a current of several kA flowing vertically through the PG (as in the ITER source) permanent magnets are mounted at the side walls to influence the field topology. Further reduction of the co-extracted electrons is achieved by implementing metallic rods in vertical direction to influence the particle fluxes by biasing these potential rods [8]. In the standard configuration (BUG, ELISE, ITER), a metallic plate (6 mm thick) parallel to the PG with openings for the beamlet groups is mounted in the source 7 mm above the PG being electrically connected to the source walls. The PG is positively biased against these zero potentials, with bias values above the floating potential of the plasma to attract the electrons to the PG resulting in a bias current of up to several 10 Amps and thus reducing the co-extracted electron current. With the potential rods (3 cm thick mounted at 15 mm distance from the PG), the electrons from the driver are (depending on their potential w.r.t. the plasma potential) attracted by the potential rods; the co-extracted electron current is reduced, the asymmetry improved as well as the temporal stability. In order to provide sufficient Cs flux to the PG in long pulses, the ion source was over-conditioned with caesium such that the very first beam blip of a long pulse has a reduced ion current due to a much too high caesium content. Details of these measures and the results are reported for example in [8,10,11]. Regarding the vertical asymmetry of the co-extracted electrons it is to mention that ELISE has the unique feature of measuring the electron current segment wise, i.e. currents are obtained for the top and the bottom half separately. Consequently, ELISE focused in the last two experimental campaigns on deuterium to identify proper measures.

The combination of magnetic filter field strength and potentials applied to (additional) surfaces was extensively investigated in short pulses (20 s plasma with 10 s beam). Figure 1 shows the extracted ion current and the currents of the co-extracted electrons on the top and the bottom segments of the EG for several combinations. At an RF power of 35 kW/driver and a fixed magnetic filter field of 3.8 mT, the bias plate was connected to the source (0 V case) and floating (Fig. 1 (a) and (b), respectively). The high asymmetry and the different time dependence of the co-extracted electron current observed for the zero-volt case becomes almost symmetric in the case of the floating bias plate. In both cases the PG was positively biased such that a bias current of 20 A is achieved (see [12] for more details). Please note that the comparison was possible only for reduced parameters (low RF power and an extraction voltage of 6 kV) as an increase would inevitably drastically increase the electron currents and an interlock would stop the pulse to protect the EG. The maximum available RF power of 75 kW/driver at an extraction voltage can be applied in case of a floating bias plate (Fig. 1 (d)) and by using the additional potential rods (part (c)). The potential of the floating bias plate is close to the plasma potential (the difference is a few Volts). In case of using the potential rods, they were electrically connected with the PG. For those both cases the highest performance can be achieved (the roughly 25 A correspond to a current density of about 250 A/m^2 and reflect the best short pulses obtained so far for deuterium, namely more than 90% of the ITER target). The floating bias plate allows also to reduce the strength of the magnetic filter field from 4 mT to 2.5 mT which improves the symmetry of the co-extracted electrons (even reverses it) which is interpret as a consequence of the reduced plasma drift. Furthermore, both segments show a similar increase with time and a smaller slope than in the case with potential rods, top segment. Hence, an adequate replace of the potential rods has been identified as their implementation for ITER's ion sources is not planned. Investigations towards the one-hour pulses in deuterium with floating bias plate have already started and will be continued. So far, similar behavior as with potential rods has been observed, electrons still increasing in time. Particularly the latter, requires measures to provide sufficient caesium to the PG and might be in a first approximation decoupled from the reduction and improvement of the symmetry of the co-extracted electrons. Regarding the beam parameters, the floating bias plate indicates a slightly improved divergence and improved symmetry of the beam power at the calorimeter.

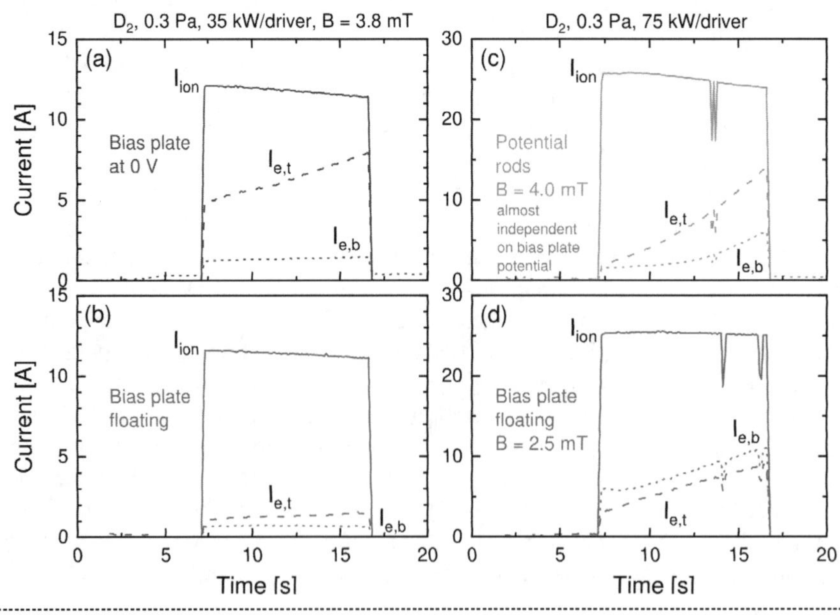

Figure 1. Time traces of the extracted current of negative ions (I_{ion}) and electrons for different potential configurations (see text). The electron current is plotted segment wise: for the top segment ($I_{e,t}$) and the bottom segment ($I_{e,b}$) of the EG at ELISE. All data is taken in deuterium at 0.3 Pa filling pressure. Note changes in the RF power and in the field strength of the magnetic filter field.

One of the challenges for long pulses is the caesium management, i.e. providing more caesium flux to the PG as the plasma removes the caesium from the grid [12]. The increase of the caesium evaporation is limited by breakdowns in the grid system due to caesium leakage. Furthermore, too much caesium changes the plasma parameters in the driver (even leading to an unstable behavior) and causes a reduction of the ion current as well most probably due to a reduction of the atomic density [10]. In the plasma, the dominant part of the caesium evaporated from the ovens at the side wall or at the back plate is ionized and may not overcome the potential barrier caused by the positive bias applied to the PG [13]. Therefore, alternative Cs evaporation techniques are under investigation (see section 3.1.). Further investigations with the particle tracking code CsFlow3D are ongoing, highlighting as well the role of caesium released from the back plate of the ion source by sputtering of caesium from the back streaming positive ions created in the first acceleration grid [13]. The effect on the co-extracted electrons and also the ion current is been seen clearly with the beam blips: the co-extracted electrons decrease during such a beam blip and the ion current increases slightly. [8,10,11].

The temporal dynamics of the co-extracted electrons, influenced also by the beam blips, motivated the change to a steady-state power supply for extraction and acceleration. The specifications of the selected power supply based on single power modules connected in series, are summarized in [10]. Installation started in 2020 and tests on a dummy load went smoothly. Recently, the power supply was commissioned at ELISE and Figure 2 shows the very first continuous wave extraction for 100 s. As the present calorimeter is not sufficiently cooled, the power density on the calorimeter has to be limited by operating the source at low RF power and with low extraction voltage. For extraction the power supply is specified to 12 kV,70 A and 50 kV, 35 A for acceleration. Tested in short pulses at ELISE so far are 11.5 kV, 45 A for extraction and 50 kV, 27 A for acceleration. First comparisons of steady state extraction with the previously used beam blip operation mode in view of ion source performance in long pulses are expected for the next experimental campaign.

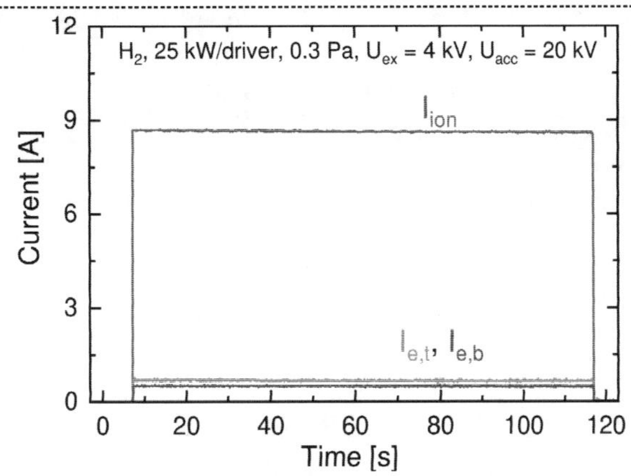

Figure 2. First 100 s extraction with the steady state power supply at ELISE. Plotted are the **e**xtracted current of negative ions and electrons (segment wise).

2.2. Beam optics investigations

The experiments at the BUG test facility concentrated on the investigations of the beam optics in terms of divergence of the beam composed of several beamlets (beamlet group) and one single beamlet as well as on beamlet deflection. An overview on the recent activities is presented by Wimmer et al. [15]. For these beam optics studies, the 14×5 beamlet arrangement was masked for the two rows in the centre to have a sufficient separation in two groups. For single beamlet investigations the centred aperture of one subgroup is kept open, allowing for a direct comparison with the results of a beamlet group (see [16]).

The electron deflection magnets in the EG create a small deflection of the negative ions which changes the horizontal direction row-by-row due to vertically alternating field direction of the cusp field. This deflection should be corrected in order to satisfy the beamlet misalignment requirements for which the ITER system uses additional correction magnets in the EG, the Asymmetric Deflection Compensation Magnets (ADCM). These are not installed in ELISE although the respective grooves are available in the EG and are not available at BUG in the standard configuration. As demonstrated by Hurlbatt et al. [16] at BUG, the zig-zag deflection is the reason for measuring a higher divergence for one beamlet group than for a single beamlet. In collaboration with the ITER Organization an extraction system with ADCM has been designed for BUG and first results are reported by den Harder et al. in [17] focussing on the single beamlet shape and deflection.

Figure 3 shows the comparison of a beamlet group for which the ADCM are installed (upper group) with the group without ADCM in the EG. The perveance is changed by varying the acceleration voltage at fixed extraction voltage. Figure 3 (a) represents the (perspective corrected) data measured at the 1D CFC calorimeter installed at about 86 cm distance from the GG. Figure 3 (b) shows corresponding simulations using IBSimu [18]. The lateral heat conduction in the 2 cm thick CFC title is mimicked by applying a 4.8 mm wide Gaussian to the beamlet. The circles represent the position of the beamlets. The simulations reflect very well the measurements. The best beam optics is obtained for the 32.5 kV/5 kV voltage ratio which is with 6.5 in the perveance minimum for this grid system. The grid half with the ADCM shows a robust compensation of the zig-zag reflection, also against other parameter variations. Further investigations, in particular on the beamlet shape and divergence, will follow.

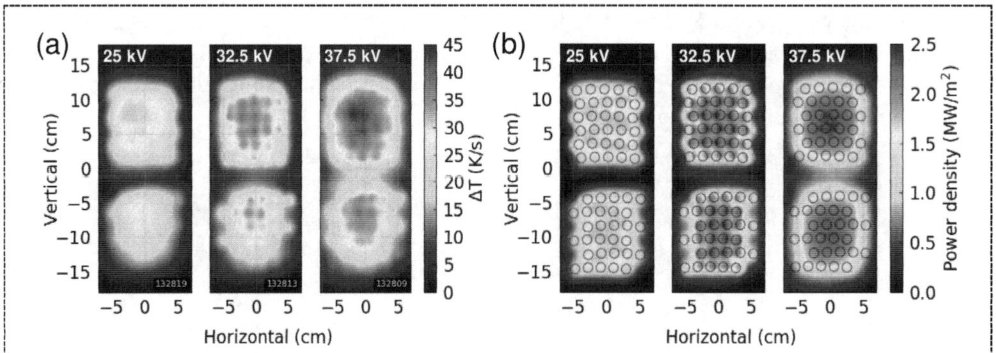

Figure 3. Beam profiles as measured by the 1D CFC calorimeter (a) and simulated power density profile (b) for variations of the acceleration voltage (25, 32.5 and 37.5 kV) at 5 kV extraction voltage. The measurements were taken at a source pressure of 0.6 Pa, RF power of 55 kW and an extracted ion current density of about 150 A/m^2. The upper beamlet group is compensated by the ADCM, the lower group is without the ADCM.

3. Contributions to the EU-DEMO neutral beam activities

Although the design of an NBI system for DEMO has been chosen to be similar to the system at ITER with the aim to maximize the RoX from ITER, some DEMO requirements are different: one of them is related to the pulse duration (7200 s for DEMO) leading to the need for an improved Cs management system. Other modifications like efficient RF power coupling, number of drivers and their arrangements for optimized plasma generation, magnetic filter field configuration or alternatives to generate a uniform plasma with proper plasma parameters in front of the extraction system, or grid systems with enhanced transparency and improved beam optics are being looked at. While the reference neutralizer is a gas neutralizer, the design allows the change to the BDPN if the concept is convincing, i.e. a prototype experiment is successful. A compilation of the beam line parameters, together with aspects of the conceptual design and the R&D activities performed within EUROfusion are compiled in [19].

3.1. Ion source

The replacement of ITER's four pairs of cylindrical drivers by four larger, racetrack-shaped drivers promises to simplify RF matching and to reduce RF power consumption for similar ion source performance [20]. The vertical stacking (Figure 4) maintains the ability to compensate vertical plasma asymmetries by adjusting the RF power to each driver. A test of two racetrack drivers on the half-size ITER source is planned at ELISE. At ELISE and BUG, the RF power transfer efficiency to the plasma is about 60% only which encourages studies for its optimization by self-consistent modelling of the power coupling using a fluid code [21, 22].

Figure 4. Concept for a beam source for DEMO based on an ITER size ion source but using 4 racetrack-shaped drivers.

Regarding the caesium management, the investigations needed for ITER will be extended by focusing on deuterium with the aim of demonstrating two hour pulses, both test facility are equipped with the proper pumping capability and neutron shielding. Additional investigations concern alternative caesium evaporation studies like the recently tested shower concept tested at BUG [8,15]. As the shower evaporates in the region close to the PG, a high flux of neutral caesium reaches the grid system. The concept will be followed in combination with the caesium ovens accompanied by further CsFlow3D studies. Designs for a shower at ELISE are in preparation. Already a concern for ITER, caesium consumption should be kept similar or even reduced. Proper measurement techniques and extrapolation for ITER relevant operations to DEMO scenarios are under investigations as well.

3.2. Beam driven plasma neutralizer (BDPN)

A BDPN relies on the ionization of the gas by the stripped electrons from the beam which distinguishes this concept from a plasma neutralizer in which the plasma is generated by additional concepts, e.g. an arc discharge. A neutralization efficiency of about 75% can be reached in comparison to the 55% achieved by a gas neutralizer, relaxing thus the demands on the accelerated ion current at same beam power or increasing the beam power from one beam line. Recent investigations on the efficiency of the BDPN revealed that there is a need to benchmark the simulations by experimental investigations [23]. Uncertainties concern the plasma confinement in the magnetic cusp field, the achievable electron density and thus ionization degree being together with the gas temperature the key parameters to be benchmarked. Studies showed that present day experimental facilities have either not the required beam power to ionize the gas sufficiently or are not available yet [23]. Therefore, a proof-of-principle setup is under consideration in which the stripped electrons of 42 A are mimicked by electrons generated by a filament at the required energy of 272 keV.

4. Summary and Outlook

The test facilities ELISE and BUG focus on the ion source performance to be achieved for ITER NBI. Challenges are the achievement of long pulses in deuterium due to the high co-extracted electron current accompanied by a strong asymmetry caused by the plasma drift. Additional potential rods improve the symmetry and reduce the strong increase with time. It is demonstrated in short pulses that biasing the bias plate instead of connecting the bias plate to source potential can replace the additional rods . Techniques need to be developed to stabilize further the increase of the co-extracted electrons for which the strategy is to provide sufficient flux of neutral caesium to the PG. The Cs management is also a concern for an NBI system for DEMO as stable ion currents for two hours must be provided reliably. Other points for DEMO concern the caesium consumption in deuterium and its reliable determination. Present day experiments need a two to three times higher caesium evaporation rate for deuterium that for hydrogen. The caesium re-distribution is influenced by the back streaming positive ions which sputter Cs from the back plate of the ion source, being beneficial in the operation region of ELISE in which the beam is probed by beam blips so far. Continuous wave extraction will be available for the next experimental campaign giving first insights into the temporal behavior of the co-extracted electrons in large sources. Racetrack-shaped drivers seem to be very promising in terms of simplifying the RF circuit, increasing its efficiency and reducing the demands on the RF generators and are thus under investigations for DEMO's ion sources. To relax the demands on the achievable ion current density or to increase the beam power of one beam line, modelling activities for a BDPN concept are carried out and the conceptual design of a proof-of-principle experiment is developed. Regarding the beam optics dedicated experiments are performed at BUG to study the zig-zag compensation by ADCM showing a robust compensation in the perveance optimum. Details on the beamlet shape and the divergence are an ongoing task, by analyzing results of single beamlet investigations for which the grid has been masked. Although DEMOs' ion source relies as much as possible on the RoX from ITER, grid modifications for increase transparency or improved beam optics are desirable and still acceptable at the present stage of the design.

Acknowledgments
This work has been carried out within the framework of the EUROfusion Consortium and has received funding from the Euratom research and training programme 2014-2018 and 2019-2020 under grant agreement No 633053. The views and opinions expressed herein do not necessarily reflect those of the European Commission.

The views and opinions expressed herein do not necessarily reflect those of the ITER Organization.

References

[1] Hemsworth R et al. 2017 *New J. Phys.* **19** 025005
[2] Singh M J, Boilson D, Polevoi A R, Oikawa T. and Mitteau R 2017 *New J. Phys.* **19** 055004
[3] Federici G et al. 2019 *Nucl. Fusion* **59** 066013
[4] Tran M Q et al. submitted to *Fus. Eng. Des.* 2021
[5] Speth E et al. 2006 *Nucl. Fusion* **46** S220
[6] Fantz U et al 2019 *Fus. Eng. Des.* **146** 212
[7] Orozco G et al submitted to *Fus. Eng. Des.* 2021
[8] Wünderlich D et al. 2021 *Nucl. Fusion* **61** 096023
[9] Nocentini R, Bonomo F, Heinemann B, Hurlbatt A and Mario I 2019 *Rev. Sci. Instrum.* **92** 023504
[10] Wünderlich D et al. 2019 *Nucl. Fusion* **59** 084001
[11] Fantz U et al. 2020 *Fus. Eng. Des.* **156** 111609
[12] Yordanov D et al., submitted to *J. Phys. Conf. Ser.,* ICIS conference 2021
[13] Cristofaro S, Friedl R and Fantz U 2020 *Plasma Res. Express* **2** 035009
[14] Mimo A, Wimmer C, Wünderlich D and Fantz U (2018) *AIP Conf. Proc.* **2052** 040009
[15] Wimmer C et al., submitted to *J. Phys. Conf. Ser.,* ICIS conference 2021
[16] Hurlbatt A et al. 2021 *AIP Advances* **11** 025330
[17] Den Harder et al., submitted to *J. Phys. Conf. Ser.,* ICIS conference 2021
[18] Den Harder et al. 2021 *Fus. Eng. Des.* **173** 112837
[19] Tran M Q et al. to be submitted to *Fus. Eng. Des.* 2021
[20] Wimmer et al. 2017 *AIP Conf. Proc.* **1869** 030021
[21] Zielke D, Rauner D, Briefi S, Lishev S and Fantz U 2021 *Plasma Sources Sci. Technol.* **30** 065011
[22] Briefi S, Zielke D and Fantz U submitted to *J. Phys. Conf. Ser.,* ICIS conference 2021
[23] Starnella G, Hopf C and Maya P N, submitted to *Nucl. Fusion* 2021

On the effect of biased surfaces in the vicinity of the large extraction area of the ELISE test facility

D Yordanov, D Wünderlich, C Wimmer and U Fantz
Max-Planck-Institut für Plasmaphysik, Boltzmannstr. 2, 85748 Garching, Germany
E-mail: dimitar.yordanov@ipp.mpg.de

Abstract. The ion source at the ELISE test facility is an intermediate step towards the operation of the ITER NBI source and it demonstrated fulfillment of the ITER NBI requirements for accelerated negative current density in hydrogen. However, in deuterium operation the co-extracted electron current density (j_e) is higher and much more unstable and limits the source performance. In the standard setup of ELISE, j_e is reduced by a positive potential applied to the plasma grid (PG) with respect to the source body and the bias plate (BP). To further reduce and stabilize j_e in deuterium, an alternative scheme by biasing the BP is investigated. Measurements of the BP currents and of the extracted currents, combined with probe diagnostics in the vicinity of the PG for short-pulse in deuterium operation are presented. Biasing the BP, leads to a change in the distribution of the plasma potential in front of the extraction area, affecting the charged particle fluxes towards the BP and the PG and a strong reduction of j_e.

1. Introduction

The Neutral Beam Injection (NBI) system for the upcoming fusion experiment ITER requires large negative ion sources [1, 2], which must be capable to deliver a negative ion current density (from 0.2 m^2 extraction area) of 230 A/m^2 up to 1000 s accelerated to 870 keV in hydrogen and 200 A/m^2 accelerated to 1 MeV up to 3600 s in deuterium (j_{ex}) with beam uniformity better than 90% and filling pressure of 0.3 Pa or less. For this purpose low-pressure and low-temperature plasma ion sources are used. The extracted negative ions are mainly produced by conversion of atoms on a caesiated low-work function surface [3] in the vicinity of the extraction region. To reduce negative ion destruction by electron stripping, the electron temperature is reduced to approximately 1 eV close to the extraction area. This is achieved by applying a magnetic filter field strong enough to magnetize the electrons, but not the ions. The magnetic filter field causes diamagnetic and $\mathbf{E} \times \mathbf{B}$ particle drifts [4], leading to a vertically asymmetric plasma distribution. Electrons are inevitably co-extracted with the negative ions from the source and collected prior to the acceleration stage by the second grid of the extraction system. In order to avoid damage to the grid by the heat load, a current density ratio of co-extracted electrons to negative ions of one at maximum is allowed. In deuterium operation, the co-extracted electron current density (j_e) is higher compared to hydrogen operation and j_e can strongly increase during pulses [5]. As a consequence in deuterium operation the RF power and/or the extraction voltage have to be reduced, resulting in a lower extracted negative ion current density, limiting the source performance.

An intermediate step towards the development and operation of the ITER source is the half-sized ion source (0.1 m^2 extraction area and 1 × 1 m source size) at the ELISE (Extraction from a

Large Ion Source Experiment) test facility at IPP Garching [6], which demonstrated in hydrogen ITER-relevant accelerated negative ion current densities over 1000 s [7]. In the standard setup of ELISE, the amount of co-extracted electrons are reduced by a positive bias applied to the plasma grid [8] (PG) – the first grid of the extraction system – with respect to the source body and the bias plate (BP) – a metal surface in front of the PG. In order to further reduce, symmetrize and stabilize the unwanted co-extracted electrons, alternative biasing technique by applying a combination of potentials to the PG and BP with respect to the source walls is investigated. Electrical measurements of the BP current, j_e and j_{ex}, combined with Langmuir probe diagnostics in the vicinity of the PG in deuterium discharge for short pulses are presented.

2. Experimental setup

Figure 1(a) schematically presents the ELISE ion source. The plasma is generated in the four drivers by RF coupling ($f = 1$ MHz and RF power (P_{RF}) up to 75 kW/driver) and expands from the drivers towards the extraction system in a common chamber. The expansion chamber is equipped with two ovens for Cs evaporation. The negative ions and the electrons are extracted by a three-component extraction system (640 apertures in 8 extraction beamlet groups): (1) plasma grid (PG) - in contact with the plasma -, (2) extraction grid (EG) - with embedded magnets for separation of the co-extracted electrons from the negative ion beam - and (3) grounded grid (GG) - for final acceleration of the beam. Besides, a bias plate (BP) with a window-frame structure is positioned 7 mm in front of the PG and covers the area between the beamlet groups. Each grid together with the BP are electrically insulated, allowing to bias them separately. Furthermore, the BP and EG are vertically divided into two parts allowing to measure the currents (I_{BP} and I_e, respectively) on the top and bottom segments. Typically, the PG is biased positively with respect to the source body and BP for initial suppression of the co-extracted electrons [8]. An extraction voltage of up to 10 kV is applied between the PG and EG and the total beam extraction voltage is up to 60 kV applied between the GG and the source. A horizontal magnetic filter field [9, 10] of a few millitesla, generated by a current flowing from insulated power supply through the PG (I_{PG}), is used to reduce the electron density and temperature in the extraction region, as well as j_e. For plasma diagnostics purposes the expansion chamber is equipped with diagnostic ports close to the PG. Two cylindrical, RF

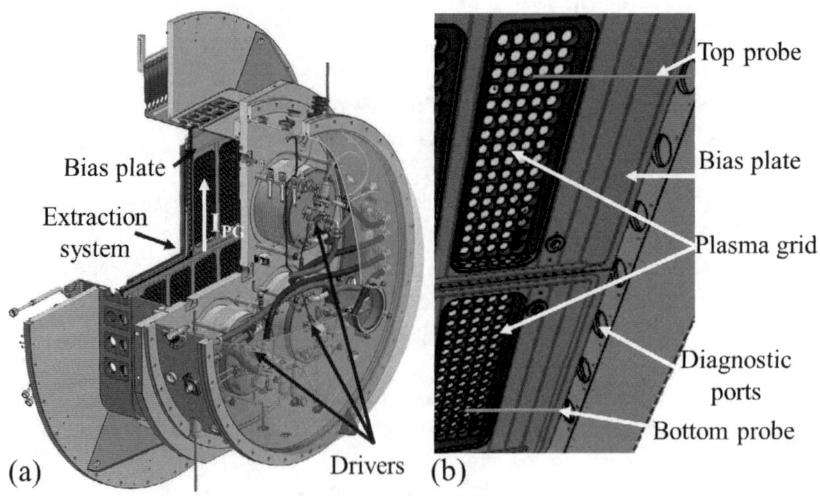

Figure 1. Schematic representation of the ELISE ion source (a) and a view close to the PG with the location of the Langmuir probes (b)

compensated Langmuir probes are mounted on the right side of the source, measuring the top and bottom (fig. 1(b)) positive ion density (n_+) and the plasma potential (Φ_{pl}). The tip (5 mm length and 150 μm radius) of both probes is fixed at 2 cm axial distance away from the PG. The top probe is located in the middle of the upper part close to the top beamlet group and the bottom one, respectively, in the lower part close to the bottom beamlet group.

3. Results and discussions

All measurements are carried out in deuterium with filling pressure at 0.3 Pa and for short pulses (20 s plasma time, including 9.5 s beam extraction phase). It is explicitly to mention that compared to other results the measurements were performed without the additional potential rods [11]. Two cases are considered: (i) standard configuration with the BP on source potential and (ii) floating BP. The potentials are taken with respect to the source body, which is defined as zero potential regardless the applied total (extraction and acceleration) voltage. A bias current (I_{bias}) is applied to the PG, defining the PG potential and in case (ii) current (I_{BP}) $\approx$ 0 is applied to the BP. The following parameters are discussed below: I_{BP} - measured with top ($I_{BP,t}$) and bottom ($I_{BP,b}$) segments of the BP - as well as Φ_{pl} and n_+ - measured with the top probe ($\Phi_{pl,t}$ and $n_{+,t}$) and bottom one ($\Phi_{pl,b}$ and $n_{+,b}$). To avoid high heat load on EG, the main results are for $P_{RF} = 35$ kW/driver and magnetic filter field current $I_{PG} = 3.6$ kA, as well as extraction (U_{ex}) and accelerating (U_{acc}) potentials 6 kV and 30 kV, respectively. The data points of the extracted currents (j_e and j_{ex}) below are averaged by the duration of the second part of the beam extraction phase.

In the standard set-up ($I_{bias} = 20$ A and $\Phi_{BP} = 0$ V) the plasma sheath on the whole BP is electron repelling since $\Phi_{BP} < \Phi_{pl}$ and the positive ions contribute the most to I_{BP} (fig. 2(b)). The co-extracted electron current density on top ($j_{e,t}$) is higher than bottom one ($j_{e,b}$) (fig. 2(a)) and the ratio of the total co-extracted electron ($j_e = j_{e,t} + j_{e,b}$) to negative ion current densities $j_e/j_{ex} \approx 0.75$. Due to **E** $\times$ **B** particle drifts cause an asymmetry of the plasma parameters, on the bottom side of PG $\Phi_{pl,b} \approx \Phi_{PG}$ (fig. 2(c)), the sheath is between electron repelling and electron attracting. On the top side $\Phi_{pl,t} < \Phi_{PG}$ defines electron attracting sheath [12]. A higher positive ion density is observed above the top segment (fig. 2(d)), resulting in $\approx$ 6 times

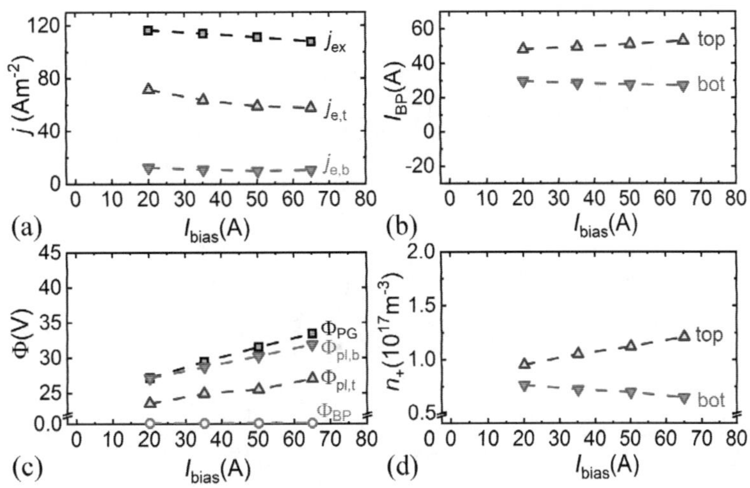

Figure 2. Extracted currents densities (a), BP currents (b), BP, PG and plasma potentials (c) and positive ion density (d) as a function of I_{bias} for $\Phi_{BP} = 0$ V; $P_{RF} = 35$ kW/driver and $I_{PG} = 3.6$ kA

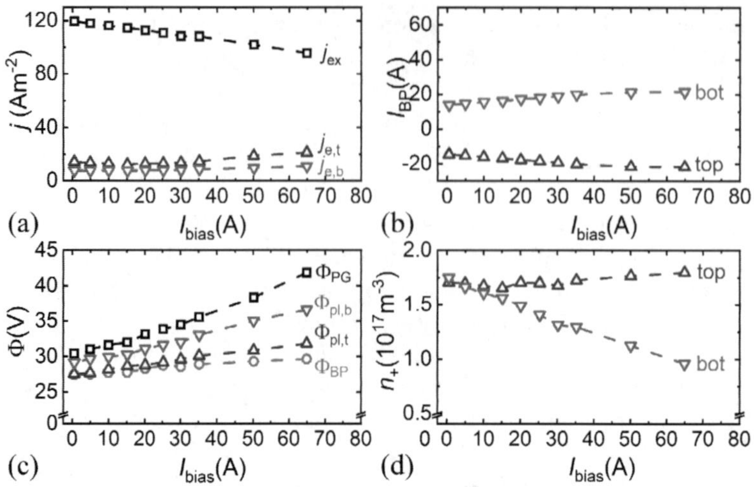

Figure 3. The same as figure 2, but for floating BP.

higher $j_{e,t}$ than $j_{e,b}$. With increase of I_{bias}: j_{ex} and j_e decrease and Φ_{pl} increases, as well as the asymmetry in the n_+. In addition, I_{BP} follows the same trend as n_+ - $I_{BP,t}$ and $n_{+,t}$ higher than $I_{BP,b}$ and $n_{+,b}$.

For the BP on floating potential and I_{bias} = 20 A, j_e is approximately 4 times reduced compared to the standard case, while j_{ex} is only slightly lower (fig. 3(a)), resulting in $j_e/j_{ex} \approx 0.18$. In addition, $j_{e,t}$ and $j_{e,b}$ become symmetrical. The floating BP shifts Φ_{PG} and Φ_{pl} to higher values and it requires zero net currents flowing through the top and bottom BP segments - $I_{BP} = I_{BP,t} + I_{BP,b} \approx 0$ A. These effects in a combination with the asymmetry of the plasma, I_{BP} on top and on bottom parts show different behaviours: $I_{BP,b}$ is positive and $I_{BP,t}$ is negative (fig. 3(b)). As $\Phi_{BP} \approx \Phi_{pl,t}$ (fig. 3(c)), a higher electron flux is provided towards the top BP segment, collecting more electrons than positive ions, causing the negative value of $I_{BP,t}$. Consequently, less electrons are co-extracted in the top part, leading to lower $j_{e,t}$. The sheath on the bottom part of BP remains electron repelling due to $\Phi_{BP} < \Phi_{pl,b}$ and it collects more positive ions than electrons and $I_{BP,b}$ is positive. The electron flux on bottom is unaffected by the BP potential and $j_{e,b}$ remains the same as in the standard set-up. The floating BP collects the majority of the electrons on the top, causing more than 5 time lower $j_{e,t}$ compared to the case $\Phi_{BP} = 0$ V (fig. 2(a)). The positive ion density increases as well in case of floating BP (fig. 3(d)). A working hypothesis is the influence of the recombining plasma [13],

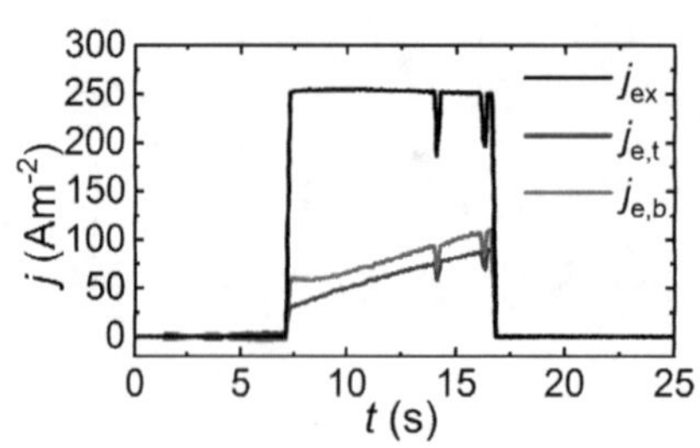

Figure 4. Time traces of the extracted currents for floating BP and maximum RF power; $P_{RF} = 75$ kW/driver and $I_{PG} = 2.2$ kA

which is present in front of the plasma grid and the decrease of the electrons leads to a lower recombination rate and respectively to higher n_+ and negative ion density in the plasma. An increase of I_{bias} at floating BP results in a asymmetry of n_+ and of Φ_{pl}. The increase of the potential difference between $\Phi_{pl,t}$ and Φ_{BP} results in an increase of $j_{e,t}$, whereas $j_{e,b}$ remains low and j_{ex} decreases slightly. $I_{BP,t}$ and $I_{BP,b}$ increase by absolute values, while the total I_{BP} remains approximately zero.

To study the effect of a floating BP in a high RF power scenario, the maximum RF power ($P_{RF} = 75\,\mathrm{kW/driver}$) was used with the following parameters: $I_{bias} = 1\,\mathrm{A}$, decreased magnetic filter field to a value of $I_{PG} = 2.2\,\mathrm{kA}$, higher $U_{ex} = 10.5\,\mathrm{kV}$ and $U_{acc} = 40\,\mathrm{kV}$ with an achieved averaged values of $j_{ex} = 256\,\mathrm{A/m^2}$ and $j_{e,t} = 76\,\mathrm{A/m^2}$ and $j_{e,b} = 97\,\mathrm{A/m^2}$. The time traces of j_e are almost symmetric and increase during the pulse, while j_{ex} stays constant (fig. 4). Compared to the standard case the strong increase in j_e with time is clearly reduced, but still present since the increase during the pulse is related to the caesium-redistribution caused by the plasma. Consequently, the floating BP gives a simple method to decrease the electron currents and to influence their symmetry. Investigations towards one-hour pulses are ongoing.

4. Summary and conclusions

One of the crucial points in large-scale negative ion sources for the ITER NBI system is reducing the co-extracted electrons since they can strongly limit the source performance, particularly in deuterium operation. The effect of the floating BP has been investigated at the ELISE ion source for medium RF power and short pulses in deuterium. The floating potential of the BP requires zero total net current. This provides together with the asymmetry of the plasma parameters in the vicinity of the extraction area different behaviour of the BP sheath on top and bottom. An electron repelling sheath remains on the bottom part of the BP, resulting in no changes in $j_{e,b}$. On the top segment of BP a higher electron flux is observed, resulting in negative $I_{BP,t}$ and in 5 times lower $j_{e,t}$ and $j_{e,t} \approx j_{e,b}$, while j_{ex} slightly decreases compared to the standard configuration, i.e. zero potential of the BP. Therefore, this method gives an outlook for a simple technique for reducing and symmetrizing the co-extracted electrons during the deuterium operation of the ITER NBI systems. Investigations are ongoing with the aim to check the applicability of this method to the one-hour pulses foreseen in the ITER standard operational scenario.

Acknowledgments

This work has been carried out within the framework of the EUROfusion Consortium and has received funding from the Euratom research and training programme 2014-2018 and 2019-2020 under grant agreement No 633053. The views and opinions expressed herein do not necessarily reflect those of the European Commission.

References

[1] Hemsworth R S and *et al* 2017 *New Journal of Physics* **19** 025005
[2] Singh M J, Boilson D, Polevoi A R, Oikawa T and Mitteau R 2017 *New Journal of Physics* **19** 055004
[3] Bacal M 2006 *Nuclear Fusion* **46** S250–S259
[4] Fantz U, Schiesko L and Wünderlich D 2014 *Plasma Sources Science and Technology* **23** 044002
[5] Wünderlich D and *et al* 2019 *Review of Scientific Instruments* **90** 113304
[6] Heinemann B and *et al* 2017 *New Journal of Physics* **19** 015001
[7] Wünderlich D and *et al* 2019 *Nuclear Fusion* **59** 084001
[8] Speth E and *et al* 2006 *Nuclear Fusion* **46** S220–S238
[9] Fantz U, Schiesko L and Wünderlich D 2014 *Plasma Sources Science and Technology* **23** 044002
[10] Fröschle M and *et al* 2013 *Fusion Engineering and Design* **88** 1015–1019
[11] Wimmer C, Mario I, Wünderlich D and Fantz U 2018 *AIP Conference Proceedings* **2052** 040003
[12] Wimmer C and Fantz U 2016 *Journal of Applied Physics* **120** 073301
[13] Wünderlich D, Fantz U, Franzen P, Riedl R and Bonomo F 2013 *Review of Scientific Instruments* **84** 093102

Overview of recent and upcoming activities at the BATMAN Upgrade test facility

C. Wimmer, S. Briefi, M. Fröschle, B. Heinemann, N. den Harder, A. Hurlbatt, A. Mimo, R. Nocentini, G. Orozco, U. Fantz

Max Planck Institut f. Plasmaphysik (IPP), Boltzmannstr. 2, 85748 Garching, Germany

E-mail: christian.wimmer@ipp.mpg.de

Abstract. The BATMAN Upgrade (BUG) test facility at IPP is contributing to the development of RF-driven H^- sources towards the ITER neutral beam injection and beyond. BUG is equipped with an 1/8 size of the ITER NBI ion source and thus is highly flexible for setup changes or diagnostic access. The present strategy for BUG is aligned along two paths: (i) BUG is continuing to be upgraded upgrading for long pulse operation (up to 1 hour, both in H and D) to identify measures for the stabilization of long pulses. For pulses > 100 s a cooling of the plasma grid (first grid of the extraction system) is required in order to avoid the dominance of thermal Cs desorption. A Cs evaporation concept with evaporation close to the extraction system is being tested, showing a stabilization of co-extracted electrons in 100 s H pulses. (ii) BUG is used for investigations of whole beam and beamlet optics for which the beam diagnostics have been massively upgraded. The beam diagnostics have been used to characterize a newly installed MITICA-like extraction system including asymmetric deflection compensation magnets.

1. Introduction

The Neutral Beam Injection (NBI) system of ITER will be based on large sources for negative hydrogen (and deuterium) ions, accelerated to an energy of up to 1 MeV. Two test facilities at IPP contribute to the development of the ITER NBI: BATMAN Upgrade (BUG) is a rather small (1/8 ITER source size) and thus very flexible test facility, whereas ELISE (1/2 ITER source size) has the main aim to demonstrate the required performance in a size-scaled source [1]. Full size ion sources are part of the Neutral Beam Test Facility at Consorzio RFX, Padova [2].

The sources rely on the plasma production in RF drivers (one in BATMAN Upgrade, eight in the full size sources) and H^- production by conversion of mainly hydrogen atoms produced by the driver plasma on the caesiated plasma grid, which is the first grid of the electrostatic extraction and acceleration system. There are strict requirements for the ITER source, among them are: extracted D^- current density $j_{D^-} = 286$ A/m^2 (or $j_{H^-} = 329$ A/m^2), a low amount of co-extracted electrons $j_e/j_{D^-} < 1$, a sufficient low beam divergence of the accelerated beam (core divergence carrying at least 85% of the power < 7 mrad), and these parameters kept stable for an operational time of up to one hour. One of the challenges for the ITER sources and possible sources beyond ITER is the long pulse stability of the ion sources: the co-extracted electrons increase in long pulses and limit the source performance particularly in deuterium operation [1]. This effect is attributed to the complex Cs dynamics inside the source, where in long pulses up to now no sufficient flux of Cs onto the convertor surface can be achieved.

BATMAN has been upgraded with an ITER-like extraction system in 2017 to BATMAN Upgrade [3], however, it has been still restricted in pulse lengths of ≈ 10 s. In order to investigate measures for the stabilization of pulses, BUG is currently being further upgraded for long pulse operation. In addition, BUG contributes to the negative ion source development with detailed investigations of the beam optics, for which the diagnostic capabilities have been massively enhanced in 2020.

2. BATMAN Upgrade

A schematic horizontal cut through BATMAN Upgrade is shown in Fig. 1. At present, the extraction system consists of three grids (Plasma Grid PG, Extraction Grid EG, and Grounded Grid GG); the formerly installed Repeller Grid [3] has been removed in 2021. A total high voltage of 50 kV is available from the HV power supply and up to -45 kV can be applied to the source. For beam optics studies, the PG is masked in order to isolate a single aperture in the top half, thus resulting in 36 open apertures (14 mm diameter). The PG consists of a frame and an insert; the latter can be flexibly exchanged allowing for changing the aperture geometry. The beam optics can be diagnosed with three diagnostics: Beam Emission Spectroscopy (BES) is available at two different axial positions with vertical resolution (BES1 5 LOS at 26 cm from the GG, and BES2 11 LOS 129 cm from the GG) [4], giving insight into the properties of the single beamlet and the group of beamlets formed in the bottom. A CFC tile calorimeter, which can be flipped into the beam, has been installed at the beginning of 2020 [5]. This diagnostics is mainly used to analyze single beamlet properties. The beam dump calorimeter has been replaced in mid of 2020 by a CW-capable one, which gives in addition a certain resolution of the beam power density [6]. Typical beam footprints on both calorimeters are shown in Fig. 1 as well. Whereas the CFC calorimeter results in a high resolution ($< 1\text{mm}^2$), the CW calorimeter gives a rather

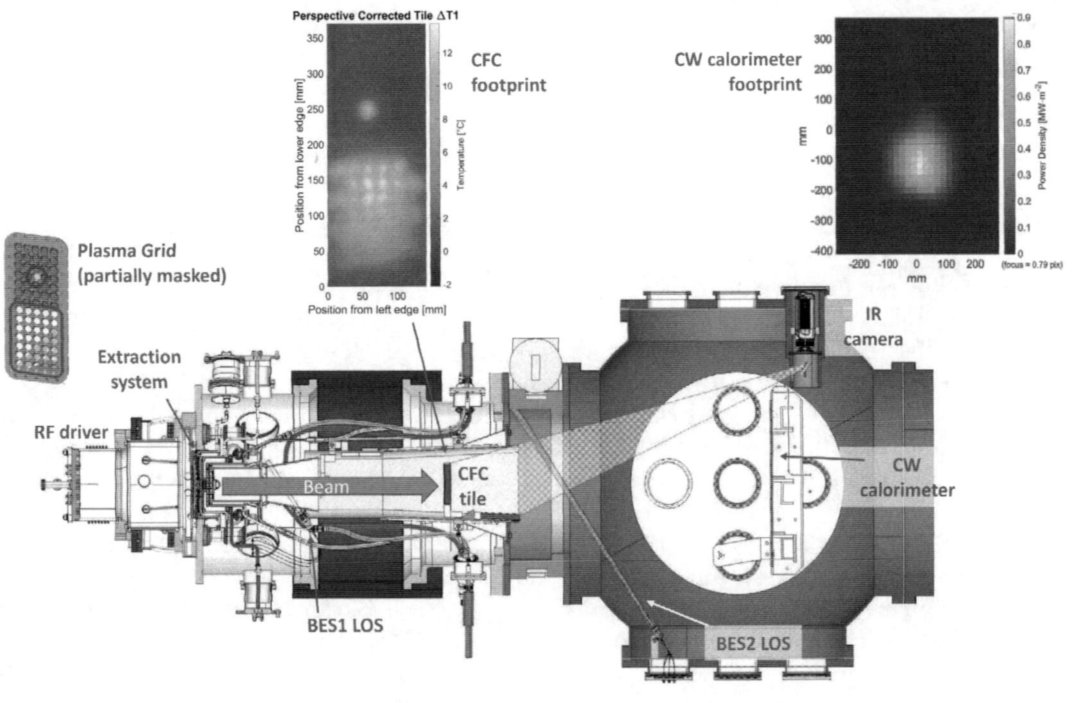

Figure 1. Horizontal cut through BATMAN Upgrade showing the beam diagnostics as well as typical beam footprints on the CFC and CW calorimeters.

broad beam overview (resolution $20 \times 20\text{mm}^2$).

A comparison of the beam divergence of a single beamlet with a group of beamlets revealed that the CFC and BES diagnostics measure a similar divergence for the single beamlet, whereas the measured divergence with BES is higher for the group of beamlets [7]. The reason behind this is the horizontal row-wise zig-zag deflection of the beamlets caused by electron deflection magnets mounted in the EG with alternating vertical polarity (zig-zag to be seen on the CFC footprint in Fig. 1); and since the BES diagnostics collects signal from a multitude of beamlets at the lower lines of sight, the resulting measured horizontal divergence is increased. A magnetic correction system of this deflection, pioneered by Consorzio RFX (Italy) [8], has been designed and installed at BUG recently in collaboration with the ITER Organization [9]. It could be shown that these magnets largely cancel out the row-wise zig-zag deflection [10]; the resulting BES divergence for the group of beamlets is thereby lowered.

The pulse length of BATMAN Upgrade has been limited mainly by three components: (1) The Ti getter pumps require frequent evaporation of fresh Ti in order to keep a stable pumping rate. These pumps have been replaced by a cryopump in 2019, allowing now for 2–3 h gas injection before regeneration. (2) The beam dump calorimeter had to be replaced (see above). (3) The heat-up of the not directly cooled PG (and in addition the so-called bias plate, which is a plate covering the PG outside the extraction area) by the plasma presently limits the pulse length. The design phase for an active cooling solution has started recently.

3. Influence of the PG temperature on the source performance

After the installation of the cryopump and the CW calorimeter at BUG, longer beam pulses (up to 2 minutes) have been carried out. Fig. 2 shows the increase of the bias plate and PG temperature (the latter measured both at the frame and the insert) during a 115 s pulse (60 kW RF power). Usually, both temperatures are set to 150 °C; the starting temperature is slightly higher because longer pulses have been performed already prior the shown pulse. The PG insert shows the highest temperature rise of 81 K, whereas the temperature rise of the frame is the lowest. The lower temperature rise of the PG frame is explained by the fact that it is covered from the plasma by the bias plate and is only heated indirectly by the insert, thus leading to a further temperature rise after the end of the pulse.

Technically, the PG can withstand temperatures of several hundred °C. However, results at the former MANITU test facility as well as at QST Japan revealed that an unstable temperature of source components can significantly influence the Cs dynamics inside the source [11, 12]. The influence of the PG temperature on the source performance (here: ratio of electrons to ions) has been studied in more detail at BUG. For this purpose, the PG has been actively heated by the heating wire installed in the PG frame to temperatures up to 450 °C and indirectly cooled (contact cooling) down again. In between, short pulses (10 s) have been carried out. In this way, the effect of long pulses on the source performance could be decoupled from the effect of higher PG temperatures, and the temperature of the frame and insert is almost similar in this scenario. The result for four different operational days in different experimental campaigns is shown in Fig. 3. The different starting value of the electron/ion ratio can not be compared, since the operational parameters were different in these campaigns. One observes that from a certain PG temperature onward (between 250 °C and 300 °C) the electron/ion ratio starts rising strongly. The likely reason for this is that the thermal desorption of Cs from the PG becomes dominant and thus no sufficient Cs layer on the PG can be established. The main change in the increasing electron/ion ratio comes from the electron current; the extracted ion current stays much more stable with a decrease of less than 15 %. After cooling down the PG again, a comparable performance can be reached (with a certain hysteresis). These results show that the PG temperature should not exceed ≈ 250 °C in long pulses and thus there is a need for active PG temperature control at BUG for pulses > 100 s.

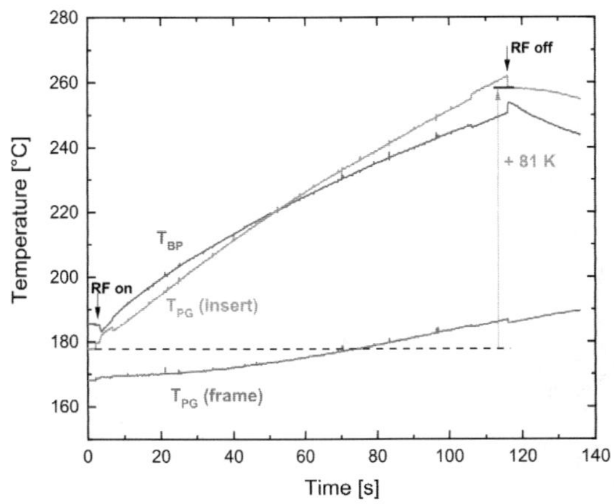

Figure 2. Temperature rise of the PG (insert and frame) and of the bias plate during a 115 s pulse with 60 kW RF power.

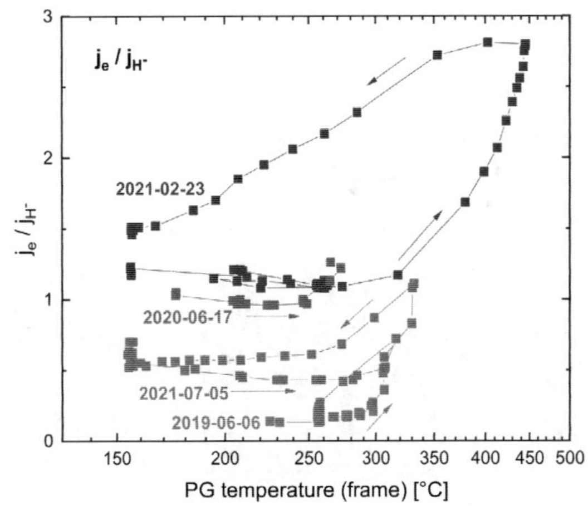

Figure 3. Ratio of co-extracted electrons to extracted ions in short pulses during a heat-up and cool-down phase of the PG as function of the PG frame temperature. Different colors indicate different experimental days in various campaigns.

4. Alternative Cs evaporation concepts

The normal Cs management in the source relies on evaporation of Cs from an oven mounted at the backside of the source and plasma redistribution inside the source. During plasma operation, a large fraction of Cs is ionized, which cannot penetrate the evolving electron-attracting plasma sheath formed at the positively biased PG [13]. The lack of Cs flux onto the PG is a likely explanation for the degrading source performance in long pulses [14]. An alternative Cs evaporation concept for stabilizing long pulses has been tested at BUG, a so-called Cs shower, i.e. a tube around the beamlet group containing 54 holes (0.4 mm diameter) directly evaporating Cs onto the PG. An oven with multiple evaporation openings has been formerly tested at IPR India [15]. The setup at BUG as well as two 100 s pulses (with and without additional Cs evaporation from the shower) are shown in Fig. 4. Short pulses have been carried out in between these long pulses in order to cool down the PG (both pulses had a comparable increase of the PG temperature as the pulse shown in Fig. 2). These first tests revealed promising results with a clear increase of the Cs emission close to the PG accompanied by a more stable co-extracted electron current when additional Cs is evaporated from the shower [1, 14]. Further tests (e.g. influence on the HV holding of the grid system, stabilization in long deuterium pulses) are foreseen at BUG.

5. Outlook and conclusions

The beam diagnostic capabilities have been massively upgraded at BUG in the last two years, now allowing for detailed studies of the beam optics of the prototype ion source. Knowledge has already been gained on the interpretation of measured beam divergences between a single beamlet and a group of beamlets, and the suppression of the row-wise zig-zag deflection of the beam by a magnetic correction system could be successfully characterized.

In parallel, BUG is being upgraded for long pulse operation. The high-speed vacuum pump and the beam dump calorimeter have already been replaced. The heat-up of the not cooled PG limits at present the pulse length to $\approx$ 100 s; an active cooling is foreseen for the future. The

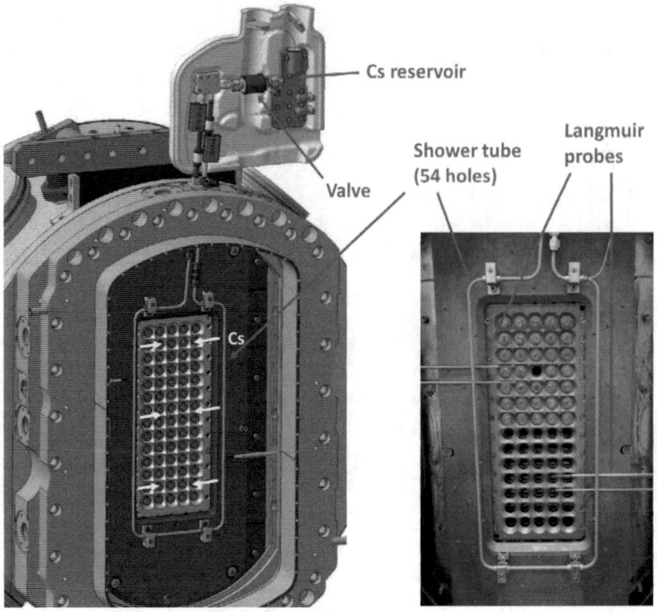

Figure 4. Cs shower as an alternative Cs evaporation concept, tested at BATMAN Upgrade. Left/center: CAD drawing and picture of the Cs shower installed at BUG. Right: Two 100 s pulses, both with permanent Cs evaporation from the standard oven and one with additional Cs evaporation from the shower (Cs emission close to the PG and co-extr. electron current).

possibility of long pulses allows measures for stabilizing long pulses to be investigated. The Cs shower as an alternative Cs evaporation concept close to the PG revealed promising first results in stabilizing pulses. Whereas BUG has been operated mainly in hydrogen in the last years, the focus will be set more on deuterium operation in the future, still with the task of beam optics investigations and finding measures for stabilizing pulses (the latter is more critical in deuterium operation).

Acknowledgments

This work has been carried out within the framework of the EUROfusion Consortium and has received funding from the Euratom research and training programme 2014-2018 and 2019-2020 under grant agreement No 633053. The views and opinions expressed herein do not necessarily reflect those of the European Commission.

References

[1] Wünderlich D *et al.* 2021 *Nuclear Fusion* **61** 096023
[2] Toigo V *et al.* 2017 *New Journal of Physics* **19** 085004
[3] Kraus W *et al.* 2018 *AIP Conference Proceedings* **2052** 040004
[4] Fantz U *et al.* 2019 *Fusion Engineering and Design* **146** 212–215
[5] Orozco G *et al.* 2021 *Fusion Engineering and Design* **165** 112225
[6] Nocentini R *et al.* 2021 *Review of Scientific Instruments* **92** 023504
[7] Hurlbatt A *et al.* 2021 *AIP Advances* **11** 025330
[8] Chitarin G *et al.* 2017 *AIP Conference Proceedings* **1869** 030026
[9] den Harder N *et al.* 2021 *Fusion Engineering and Design* **173** 112837 ISSN 0920-3796
[10] den Harder N *et al. submitted to Journal of Physics: Conf. Series, contribution to the ICIS conference 2021*
[11] Gutser R, Wünderlich D and Fantz U 2011 *Plasma Physics and Controlled Fusion* **53** 105014
[12] Ichikawa M *et al.* 2020 *Review of Scientific Instruments* **91** 023502
[13] Wimmer C and Fantz U 2016 *Journal of Applied Physics* **120** 073301
[14] Mimo A *et al. submitted to Journal of Physics: Conf. Series, contribution to the ICIS conference 2021*
[15] Bansal G *et al.* 2012 *Review of Scientific Instruments* **83** 02B118

The H^- multiaperture source NIO1: gas conditioning and first cesiations

M. Barbisan[1], M. Cavenago[2], R. S. Delogu[1], A. Pimazzoni[1], C. Poggi[1], M. Ugoletti[1], V. Variale[3], V. Antoni[1,4], D. Ravarotto[1], G. Serianni[1], C. Baltador[2], L. Franchin[1], A. Minarello[2], D. Martini[2], M. Maniero[1], R. Rizzieri[1], L. Romanato[1], F. Rossetto[1], F. Taccogna[3,4]

1 Consorzio RFX (CNR, ENEA, INFN, Universitá di Padova, Acciaierie Venete SpA), Cso Stati Uniti 4; 35127 Padua, IT

2 INFN-Laboratori Nazionali di Legnaro, Legnaro, IT

3 INFN-Sezione di Bari, Bari, IT

4 CNR-Istituto Scienza Tecnologia Plasmi, Milan, Padua, Bari, IT

E-mail: marco.barbisan@igi.cnr.it, cavenago@lnl.infn.it

Abstract. In NIO1, a compact H^- ion source installed at RFX operated in continuous mode, gas conditioning was needed to improve results in Cs-free regimes (up to a current density about 25 or 30 A/m^2). Installation of a cesium oven gave an additional increase of current, progressively limited also by a rapid overcesiation: we observed an average density about 50 A/m^2 (with a peak of 67 A/m^2). Obtained beam optics is briefly compared with simulations and previous results in the Cs-free regime in similar conditions (as source voltage, often set below or equal 11 kV). In both regimes, central beamlet apparent density is larger, possibly because of a narrow bias plate mask, now enlarged. Further improvements for Cs-based regime include a moderate oven reservoir temperature, careful tuning and a substantial increase of beam voltage, and power handling capability.

1. Introduction

The production of D^- ions (arranged in a matrix of low emittance beamlets) is needed for Neutral Beam Injectors (NBI) of fusion devices, with large physical and technical challenges[1, 2]. Radiofrequency (rf) ion sources (BATMAN, ELISE[3]) were preferred to arc sources (Kamaboko[4]) for expected durability. A large rf source (SPIDER[1], 1280 beamlets) aiming to a total 55 A D^- current is now installed in the NBTF[2] at Consorzio RFX, where, as a companion activity and as a training facility and to verify some systems and theory, a compact ion source (NIO1, Negative Ion Optimization 1; 9 beamlets) was installed in 2014, strictly limited to H^- ions (to avoid deuterium activation issues), see figure 1.a. In Cesium-based regimes, ion current is strongly boosted (factor 10) by a thin deposit (some monolayers) of cesium on the plasma Grid PG (see figure 1.c); for these sources, as a conventional goal, an ion current density $j_i \cong 300$ A/m^2 is required (which scales to $j_i \cong 200$ A/m^2 for D^-), with a small co-extracted electron current density j_e: goal is $R_j = j_e/j_i < 1$. Up to now NIO1 reached only $j_i \cong 67$ A/m^2 for the overcesiation issue, see section 3. In Cesium-free regime goals are about $j_i \cong 30$ A/m^2, and $R_j < 10$. Moreover, practically continuous and stable operation is required.

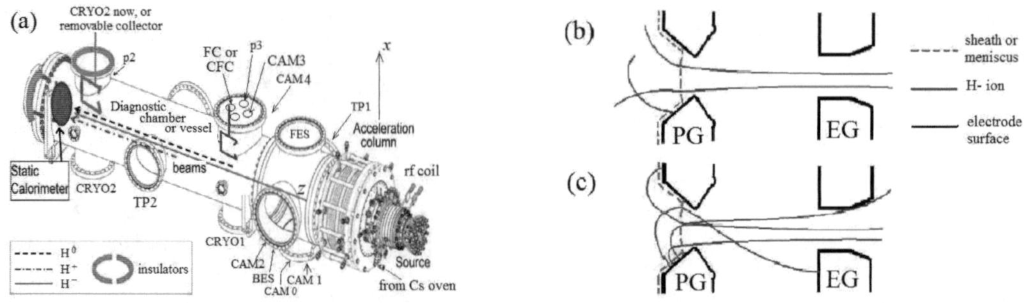

Figure 1. (a) NIO1 setup showing diagnostic chamber, with pressure gauge at p_2 and p_3, and pumps (TP and CRYO) and use of other ports; (b) sketch of beam extraction for volume produced H^- and (c) for surface produced H^-.

A NIO1 section[5] is shown in figure 2.a; let z be the beam axis, with origin at the extraction hole; x axis is vertical, which is the deflection direction of co-extracted electrons (due to permanent magnets PM embed in the extraction grid EG, with deflection field $\mathbf{B}^d$ in the yz plane[6]; the ion x-deflection compensation condition[7] is also satisfied). Inside the plasma $z < 0$ we have a static magnetic filter $\mathbf{B}^s$; after some optimization in 2017-2018, filter pattern is aligned with $\hat{x}$, to reduce interferences with $\mathbf{B}^d$; now $\mathbf{B}^s$ is the sum of 3 terms[5, 8], one due to inner PMs, another to easily removable PMs, and the last one due to adjustable current I_{pg}.

2. Setup, improvements and simulations

Apart from the dielectric rf window (in Pyrex), plasma chamber walls are connected to 3 different electric potentials: ϕ_{pg} is the plasma grid PG potential (see figure 2.b); ϕ_{pb} is Bias Plate potential; ϕ_{ww} is the potential of the other walls (the wider surface), covering the front and rear multipole; we define the main bias voltage $V_b = \phi_{pg} - \phi_{ww}$, the second bias voltage $V_{b2} = \phi_{pg} - \phi_{bp}$ and the related absolute current I_b and I_{b2}. Note also acceleration voltage $V_{ag} = -\phi_{pg}$ and the extraction voltage $V_e = \phi_{eg} - \phi_{pg}$, with ϕ_{eg} the EG potential. We have $j_i = I_H^-/A_e$ where $A_e = 410$ mm^2 is the area of extraction and ion current I_H^- is estimated by the corrected current I_a, that is the absolute current I_{ag} supplied by HV power supply AG, minus leakage due to known resistance R_t (including bleeder and water cooling): $I_a = I_{ag} - V_{ag}/R_t$, see figure 2.b. Moreover I_e is the absolute current of the Extraction Grid Power Supply (EGPS).

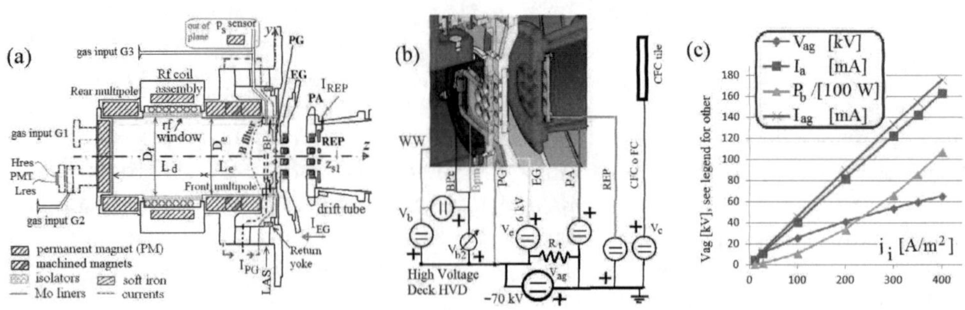

Figure 2. (a) NIO1 main section; note that High Resolution (Hres) and Low Resolution (Lres) spectrometer fibers are multiplexed a CF16 port, where also a second gas input G2 is provided; photomultiplier (PMT) is placed on a CF16 port below; (b) scheme of power supplies (see also figures 2 and 4.a in Ref. [9]); (c) scaling of voltage and power on electrodes EG and PA vs j_i.

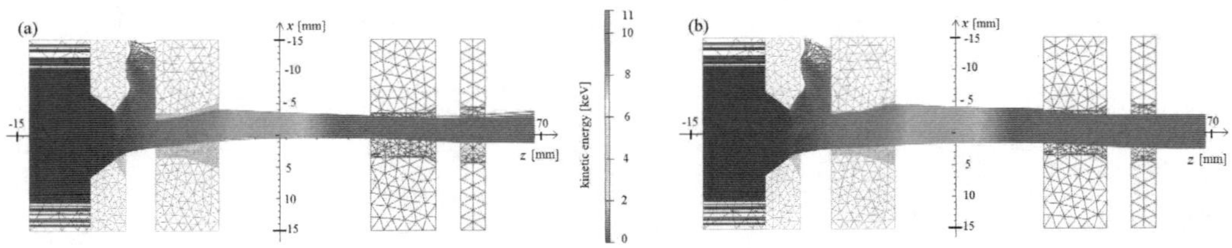

Figure 3. (a) xz projection of electrode wireframe and simulated beams (e^-, H^-) with $j_i = 25$ A/m^2 , $R_j = 5$, $V_{ag} = 11$ kV $V_e = 1.3$ kV; (b) as 'a', with $j_i = 35$ A/m^2 .

Four viewing position are now provided for lateral cameras, looking towards x (CAM1 and CAM3) or y (CAM2 and CAM4), using ports centered at $z = 0.515$ m (CAM1 and CAM2) or $z \cong 0.93$ m (CAM3 and CAM4). Emission produced by atomic processes due to H^--gas collisions is visible and easily imaged by these cameras; so luminosity is roughly proportional to line of sight (LOS) integral of beam times gas density; the streaming images of CAM1 and CAM2 are particular useful in optimizing beam optics.

Simple beam simulation helps to understand extracted beam limits; two class of beam optics simulation exists: (a) ray-tracing with an effective plasma model (including an ambipolar plasma potential), where H^- ions are launched at some plane in the plasma (called emitter) with guessed or selfconsistent initial kinetic energy K_i (of order eV_{b2}); for example in figure 3, $K_i \cong 3$ eV; similarly for electrons; by convention plasma meniscus is at +3 V (respect to emitter and PG); (b) kinetic codes[10], where details of H^- production (see figures 1b and 1c) are also modeled. Class (a) requires much less CPU time; both academic and commercial examples exists[11, 12, 13], and details of plasma and emission model are often critical; their advantage is that effect of change of electrode geometry or of extraction voltages on beam transmission can be rapidly seen; for example in figure 3 at $V_{ag} = 11$ kV see $j_i = 25$ A/m^2 and 35 A/m^2 cases: in the latter ion beam collides with EG, so that I_a decreases and I_e increases.

The simple simulations of figure 3 confirm that $V_{ag} \geq 11$ kV is needed, for good transport of $j_i \leq 30$ A/m^2 with $R_j = 5$. Co-extracted electrons are deflected by permanent magnets (PM) embedded in the extraction grid EG; scaling of power deposited on EG and of beam power P_b with j_i is shown in figure 2.(c).

The 2021 setup introduces minor ion source modifications, in the bias plate (BP) mask and in the oven control. As to BP, we noted that central beamlet apparent density resulted larger both in Cs-based and Cs-free regimes; one factor was the reduced opening in the bias plate mask

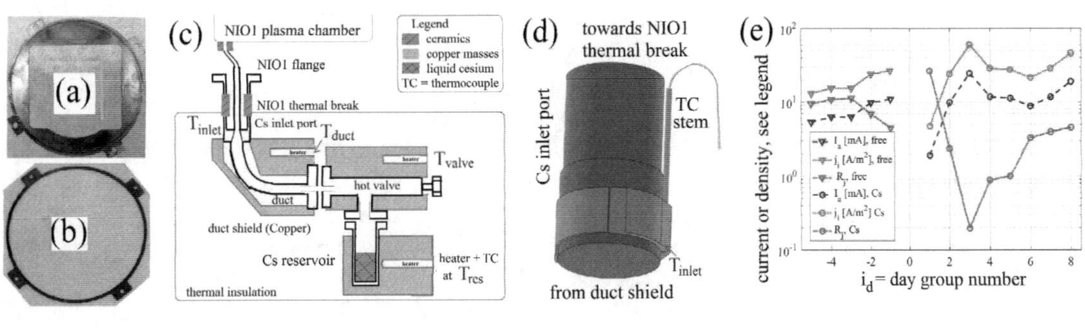

Figure 4. (a) old BP mask (b) new BP mask; (c) Scheme of Cs oven, heaters and TCs placement with detail: (d) 3D view of TC inlet clamp; (2) Comparison of 2019 Cs-free result ($i_d < 0$) and 2020 Cs-based results, where i_d enumerates the result group (see text).

(a 58 mm x 52 rectangle, see figure 4.a) now enlarged to a 79 mm diameter circle (figure 4.b), in order to increase plasma density n_e at lateral extraction holes. Other possible reasons are more difficult to correct: the filter field is lower at central hole (centered at $x = y = 0$) than it is at lateral holes ($|x|$ and/or $|y|$ about 14 mm), so more plasma arrives on axis; the expansion plasma diameter D_e has no extra margin over the rf driver diameter D_f (defined in figure 2.a), so peripheral plasma is further weakened. In other sources, expansion chamber section is often rectangular, anyway a proportion $D_e \cong 1.3 D_f$ is considered for next project (say NIO2).

The first experimental campaigns with Cs evaporation into NIO1 (performed in 2020) proved that the oven can produce a large quantity of Cs vapor, up to 10^{16} m^{-3} as measured from laser absorption spectroscopy (LAS[8], see figure 2.a) while optimal H^- production is expected about in the 10^{14} m^{-3} range. So temperatures at several points (including T_{inlet}, T_{res} and T_{duct}, see figure 4.c) should be controlled by thermocouples (TC) as carefully as possible, also in view of long equilibration times. Even assuming a good thermal insulation, the placement of TC matters: for example the inlet of Cs pipe into NIO1 vacuum chamber T_{inlet} is measured by a TC tightly pressed against pipe terminal, with a very difficult assembly due to limited space available; reservoir temperature T_{res} is measured by in the heating cartridge itself (with an error of about 10 K over true reservoir temperature as verified by other TC in a oven test-stand at LNL). A criterion to avoid overcesiation is that Cs vapor pressure should be (much) smaller than H_2 pressure. This suggests $T_{\text{res}} \cong 390$ K; a practical limit including errors may be $\check{T}_{\text{res}} \cong 400$ K. Moreover to avoid large deposit of Cs on the inlet pipe we plan $T_{\text{inlet}} \geq T_{\text{res}}$, with corrections for local thermal losses and T_{inlet} measurement errors (to be calibrated again). To improve T_{inlet} accuracy, a special clamp is being built, see figure 4.d; only a 1 mm outer diameter TC (with its junction insulated by MgO) may fit in the available space.

3. Selected Results

Up to 2019 NIO1 was operated to understand the physics of Cesium-free ion sources. The observation of a transient effect, in Cs-free regime and long term NIO1 operation, led to the development of stabilization techniques[5], including conditioning (with oxygen or other gases, as argon, nitrogen, and xenon), interleaved with days of use of H_2 as a process gas. Stabilization of transients gave more H^- current j_i and less plasma luminosity.

In Cs-free operation, we found an optimal $B_f^o \cong 11$ mT on axis, which rises two issues: 1) this large filter reduces $n_e^{pg} = n_e(x, y, 0)$ plasma density at PG (as shown by preliminary simulation and a measurement of heat deposited H_{pg} in PG, less than 30 W); 2) to agree with results $\check{B}_f^o$ of larger sources (as SPIDER), we must assume the scaling $B_f^o \propto D_f^{-1}$, so $\check{B}_f^o \cong 4$ mT. In Cs-based operation, we found $B_f^o \leq 7$ mT, which scales to $\check{B}_f^o \leq 2.5$ mT for H^- (not too much different from results of other sources).

Beam power may be large in Cs-based regimes (see figure 2.c), while in Cs-free operation, beam power is modest (100 W order). This beam can be stopped by an uncooled and fixed Carbon Fiber Composite (CFC) tile, biased to a small positive voltage V_c, where two other beam current measurement are taken; I_{cfc} from the power supply of V_c, and I_{cal} from the observed temperature rise/fall of the tile in beam on/off condition (the latter method is time consuming and requests to stop beam for a few minutes, which is unpractical during V_{ag} scans); measured values largely differ (for a factor 2), with the ordering: $I_{cal} < I_a < I_{cfc}$. In 2020 Cs-based experiment, control of Cs layer thickness failed (due to tests including a too large reservoir temperature $T_{res} \cong 500$ K, giving an overcesiated source condition; so performance was rapidly degrading and anyway limited to $j_i < 67$ A/m^2 , averaged on all beamlet, see figure 4.e). As apparent from figure 2.c, improvements to $j_i \cong 100$ A/m^2 can be tolerated by a fixed CFC. Further improvements will request: 1) more acceleration voltage V_{ag} as shown in figure 2.c; 2) a movable CFC (or a Faraday cup FC, removable as standard), normally retracted, so that beam is stopped by a water cooled end calorimeter (rated 8 kW) at $z = 2.04$ m.

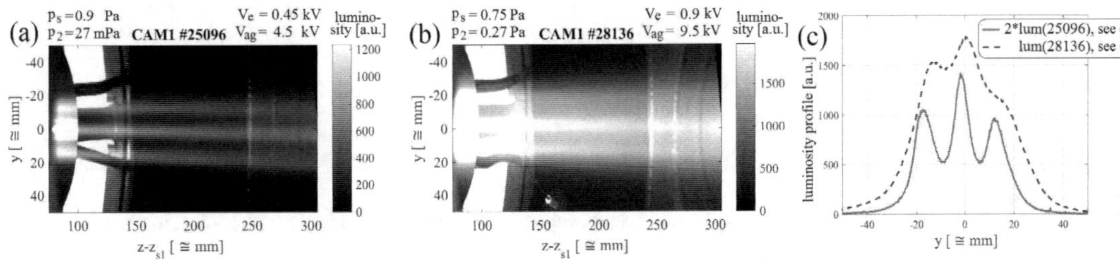

Figure 5. (a) Typical Cs-free optics, at cryopump on, $I_a = 1.6$ mA, before O_2 and Xe conditioning; (b) Cs-based optics, cryopump off, $I_a = 19$ mA, $I_e = 11$ mA; (c) profile comparison.

In figure 4.e results for I_a and R_j of selected groups of days for Cs-free and Cs-based regimes are compared; for example $i_d = -4$ is the result for 3 days of H^- following the day of Argon conditioning, $i_d = -3$ groups results after Nitrogen, and $i_d = -2, -1$ after Oxygen, and after Xenon[14]. For Cs-based results, $i_d = 1, \ldots, 6$ refers to initial days in a 2020 campaign, and $i_d = 7, 8$ to other days; for each quantity, the 4th best value is shown (to filter rare results out). Even so, a remarkable peak shows at $i_d = 3$ where $j_i > 60$ A/m^2 and $R_j = 0.2$ was obtained; but this condition was not lasting: simpler explanation is overcesiation (also mechanism similar to Cs-free transient should be investigated).

Other performances in term of I_a, source pressure and voltages in Cs-free[14] and Cs-based regimes[14, 15] were compared elsewhere. Typical beam images are shown in figure 5, with luminosity depending on gas pressure p_2 (practically uniform in the vessel) as labeled and on camera gain. In case 'a' optics was easily optimized, since source yield was stable and matched to voltage $V_{ag} < 6$ kV; for case 'b' less optimization were possible (in view of the larger current and progressing overcesiation). A deviation in y is apparent in 'a' related to the fact that $B_x \cong 11$ mT and $I_{pg} \cong 0.4$ kA were used; in 'b' this deviation is not visible, corresponding to $I_{pg} \cong 0$ used in Cs-based experiment, where a local filter $B_x \cong 7$ mT was provided by magnets only. In conclusion, this large effect of I_{pg} fringe field deserves further investigation, also in consideration that ion sources larger than NIO1 necessarily have to rely on I_{pg} only.

Acknowledgments
Work set up in collaboration and financial support of INFN (project INFN-E and Group 5, experiments Plasma4beam and Ion2neutral) and EUROFusion. This work within the framework of the EUROfusion Consortium has received funding from the Euratom research and training programme 2019-2020 under grant agreement No 633053. The views and opinions expressed herein do not necessarily reflect those of the European Commission.

[1] Serianni G. et al. 2020 *J. Vac. Sci. Technol.* B **91** 023510.
[2] Toigo V et al. 2019 *Nucl. Fusion* **59** 086058.
[3] Kraus W, Fantz U, Heinemann B, Franzen P, 2015, *Fus. Eng. Design* **91** 16
[4] Kojima A et al, 2010, *Rev. Sci. Instrum.* **81** 02B112.
[5] Cavenago M et al. 2020 *Rev. Sci. Instrum.* **91** 013316.
[6] Cavenago M and Veltri P 2014 *Plasma Sources Sci. Technol.* **23** 065024
[7] Chitarin G, Agostinetti P, Aprile D, Marconato N and Veltri P 2014 *Rev. Sci. Instrum.* **85** 02B317.
[8] Barbisan M., Pasqualotto R. and Rizzolo A. 2019 *Fus. Eng. Design* **146** 2707.
[9] Cavenago M et al., 2017 *AIP Conference Proceedings* **1869** 030007.
[10] Taccogna F. et al., 2016 *Rev. Sci. Instrum.* **87** 02B145.
[11] Spadtke P 2014 *Rev. Sci. Instrum.* **75** 1643.
[12] Cavenago M, Veltri P, Sattin F, Serianni G and Antoni V 2008 *IEEE Trans. on Plasma Science* **36** 1581.
[13] OPERA-3d, Cobham Co. Ltd., Vector Fields software (2003 version or higher).
[14] Taccogna, F., Bechu, S., Aanesland, A. et al. 2021 *Eur. Phys. J.* D **75**, 227.
[15] Cavenago M et al., in preparation.

ITER-HNB plasma and extraction grids in BUG: Characterization of magnetic deflection correction

Niek den Harder, Andrew Hurlbatt, Christian Wimmer, Guillermo Orozco, Riccardo Nocentini, Markus Fröschle, Bernd Heinemann and Ursel Fantz

Max-Planck-Institut für Plasmaphysik, Boltzmannstraße 2, D-85748 Garching bei München

E-mail: niek.den.harder@ipp.mpg.de

Abstract. The BATMAN Upgrade test facility aims to extract H^-/D^- at high current densities. The inevitably co-extracted electrons are deflected out of the beamlet by permanent magnets before acceleration. These magnets induce a row-wise zig-zag deflection of the beamlets. In the new MITICA-like extraction system, this deflection is compensated by additional deflection correction magnets. IBSimu modeling studies indicate that the deflection correction system not only acts on the horizontal beamlet deflection, but also causes aberrations to the vertical angular distribution. Dependencies of the beamlet position and size are studied in several parameter scans with the available diagnostic tools (CFC tile calorimetry, Beam Emission Spectroscopy). The horizontal beamlet deflection is robustly compensated within the systematic uncertainty of the experiment. The triangular beamlet shape predicted on basis of the magnetic field structure is visible in the CFC calorimetry.

1. Introduction

The ITER Neutral Beam Injection system is based on negative ions[1]. Coextracted electrons are inevitable for negative ion based NBI systems. To avoid wastefully accelerating the coextracted electrons to full energy, they are deflected out of the beamlet directly after extraction by CESM (Co-extracted Electron Suppression Magnets). The electron deflection creates a small deflection of the ions, which should be corrected in order to satisfy the beamlet misalignment requirements. The ITER Heating Neutral Beam magnetically corrects the beamlet angles with a second set of magnets which, together with the deflection magnets, form an asymmetric magnetic flux structure[2, 3]. These correction magnets are called ADCM (Asymmetric Deflection Compensation Magnets), which were first introduced in Reference [4]. In the framework of a research collaboration between Consorzio RFX and QST, magnetic deflection correction has been demonstrated successfully in the QST Negative Ion Test Stand (NITS), albeit under restrictive conditions: an acceleration potential below 10 kV, an extraction potential of approximately 1 kV, and thus current densities in the order of 10 A/m^2[5]. To obtain more experience with this technique, a magnetic deflection correction system has been designed in combination with a MITICA-like extraction geometry (MLE) for BATMAN-Upgrade (BUG) in collaboration with the ITER Organization[6]. BUG is a well diagnosed and highly versatile testbed for the extraction of negative ions[7, 8]. This contribution reports on the first CFC calorimetry measurements of the single beamlet shape and position for this new grid system[9].

2. Methods

A technical description of the BUG facility is given in [7], an overview of recent activities is found in [8]. The BUG-MLE grid geometry and magnet design are described in [6]. In the BUG ion source, an RF power (P_{RF}) generates a plasma which expands from the driver region onto the caesiated plasma grid, where negative ions are produced by surface conversion. A negative ion current density J_{ext} is extracted by an extraction potential U_{ext} that is applied between the plasma grid (PG) and extraction grid (EG). The plasma grid and extraction grid contain 5 x 14 apertures, the single aperture geometry is identical to the ITER-HNB PG and EG[3]. The beamlet is further accelerated by an acceleration potential U_{acc} applied between the EG and the grounded grid (GG). The CESM inside the extraction grid create a vertical magnetic field that horizontally deflects the electrons. The ADCM create an axially asymmetric field structure, so that there is no net average angle of the beamlet particles at the end of the grid system. The horizontal magnetic field component introduced by the ADCM defocuses one side of the beamlet, and focuses the other side of the beamlet, leading to a triangular beamlet with an increased vertical divergence[6].

IBSimu is used to calculate the ion-optics[10, 11]. For the divergence, the perpendicular temperature $T_\perp$ is a crucial input parameter since it adds $\sqrt{eT_\perp/(U_{ext}+U_{acc})}$; 1 eV was used.

During the experimental campaign, the BUG-MLE plasma grid was masked to isolate a single beamlet. Beam Emission Spectroscopy (BES) was used to diagnose the beamlet direction from the Doppler shift[12]. Due to constant systematic uncertainties in the angle of the optical heads and the high-voltage, the spectroscopically determined beamlet angle has an unknown offset and thus only gives a relative value. IR imaging of the backside of the CFC calorimeter (376 x 142 x 20 mm) diagnoses the spatial distribution of the beamlet at $l_{CFC} = 0.866$ m from the grounded grid[9]. A lens correction is applied to correct radial distortions; the corners of the visible part of the CFC are identified in the lens corrected IR data, and used for a perspective correction. Contrary to earlier studies, the measured and simulated beamlets are fitted with a double Gaussian instead of a single Gaussian[13]. The beamlet width $\sigma_{Beamlet}$ at the front of the CFC is the sum of the beamlet width σ_{GG} at the end of the grid system, and the broadening due to the divergence $\theta_{1/e}$ which amounts to $l_{CFC}\tan(\theta_{1/e})/\sqrt{2}$. The measured beamlet width σ_{CFC} additionally increases due to lateral heat conduction. This is taken into account for the simulated profiles by FEM modeling of the CFC tile. Measurements are reported after 0.5 seconds of beamtime, which corresponds to a lateral conduction $\sigma_{Cond.}$ of 4.8 mm on the CFC backside for a point source. The lateral heat conduction can also be approximated by a convolution of the simulated power density profile. The widths add geometrically: $\sigma_{CFC} = \sqrt{\sigma^2_{Beamlet} + \sigma^2_{Cond.}}$. Assuming the beamlet is central in the GG aperture, the measured position y_{CFC} can be converted to an angle by $\arctan(y_{CFC}/l_{CFC})$, which amounts to approximately 1.15 mrad/mm.

3. Results

The BUG-MLE grid system was extensively characterized with BES and CFC measurements of an isolated beamlet. Figure 1a shows a typical single beamlet CFC measurement. Operating parameters such as the potentials and the grid averaged extracted current density are indicated above the plot. The fitted double Gaussians are indicated by a solid (core) and dashed (halo) ellipse; both run from $-\sigma$ to $+\sigma$. The beamlet tends to be triangular, as predicted in the design paper[6]. This is especially clear in the fit residue, which systematically underestimates the measured data in the tips of the triangle. Still, the double Gaussian fits the measurement data reasonably well with an R^2 of 0.992, calculated with the area visible in the plot.

The average horizontal angle was used as BUG-MLE design criterion: the calculated average horizontal particle position is zero around a U_{acc}/U_{ext} ratio of 5.2. Figure 2a shows simulated angular distributions for several U_{acc} values at 5 kV U_{ext}. The average horizontal angle decreases with increasing U_{acc}, but since the simulated horizontal angular distribution is increasingly

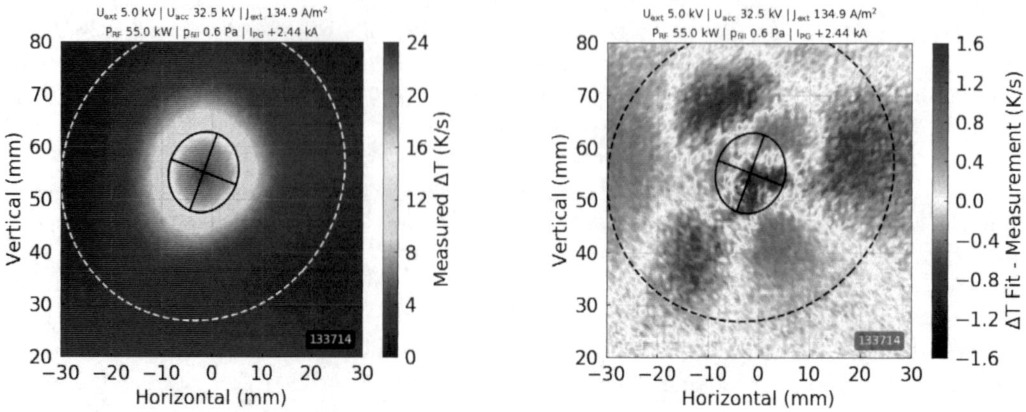

(a) Measured CFC profile

(b) Beamlet fit residue

Figure 1: The measured beamlet profile is fitted by a double Gaussian. The core size is indicated by solid lines, the halo is dashed; both run from $-\sigma$ to $+\sigma$. The beamlet tends to be triangular, the fit systematically underestimates the measured data in the tips of the triangle.

asymmetric, the peak of the distribution shows different behavior. The asymmetric angular distribution could complicate the determination of the beamlet angle/center in the experiment.

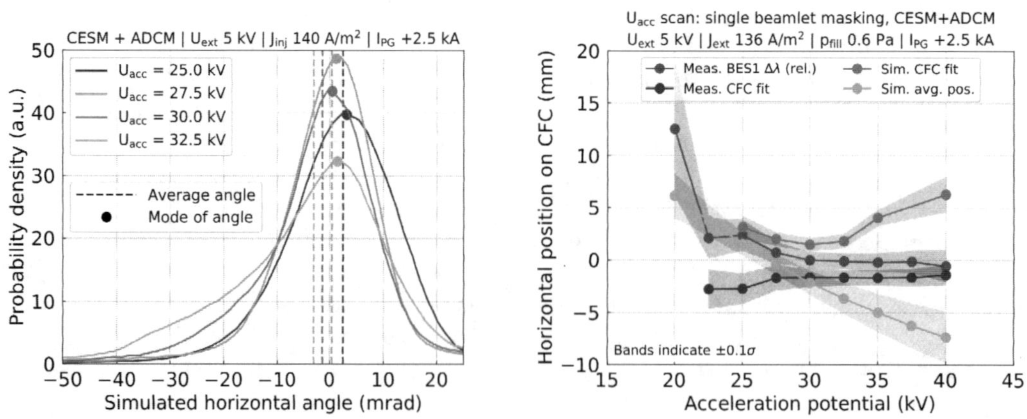

(a) Simulated angular distributions

(b) Measured and simulated beamlet positions

Figure 2: As shown in (a) the simulated angular distribution is asymmetric, which complicates the determination of the beamlet angle/center in the experiment. (b) shows the measured and simulated beamlet position on the CFC over the U_{acc} scan.

Figure 2b shows the measured and simulated horizontal beamlet position on the CFC. The measured beam emission Doppler shift is used to calculate a relative horizontal angle, which is converted to an equivalent displacement. Since the measurement is relative due to constant systematic uncertainties, the data point at 30 kV is arbitrarily set to zero. The Doppler shift displacement qualitatively matches the measured beamlet position on the CFC, but deviates at lower U_{acc}. Corresponding IBSimu calculations were performed, particles were tracked up to the CFC. FEM calculations were used to obtain the temperature profiles at the back of the CFC tile, which were fitted, resulting in the calculated horizontal fit positions; only fits with an R^2 above 0.95 are shown. The calculated average position on the CFC is shown as well, analogous to the calculated average particle angle. As expected based on the simulated angular distributions, the average position decreases over the scan, whereas the fitted location tends to positive y. In the perveance optimum, the deviation between measured and simulated horizontal position is approximately 2.5 mm, which is just within the systematic uncertainty. A likely reason for the discrepancy between measured and simulated position at the extremes of the scan is the asymmetry of the horizontal angular distribution. Further investigation is needed to clarify what parameter of the angular distribution should be tuned to optimize NBI transmission. The CFC measured horizontal beamlet position is robustly compensated throughout the scan. For reference, with the standard BUG grid system, the horizontal beamlet position varied between -10 and -15 mm for a similar scan.

Figure 3a shows the measured BES divergence, and the measured beamlet core size on the CFC for the acceleration potential scan. The size is converted to a divergence assuming that the beamlet has a zero width at the grounded grid ($\sigma_{GG} = 0$ mm), and that there is no broadening by the CFC tile ($\sigma_{Cond.} = 0$ mm). Neglecting both effects results in an upper limit of the divergence, taking the initial width and heat conduction into account will result in lower values. The modeled σ_{GG} decreases from 3.3 mm to 1.4 mm over the scan, and is 2 mm at a U_{acc} of 32.5 kV. The 6.9 mm minimum width converts to a divergence upper limit of 11.3 mrad; a 2 mm σ_{GG} would bring the divergence down to 8 mrad. Additionally including heat transfer would result in a 4.9 mrad divergence. The measured divergence and CFC widths match relatively well as shown before [13], although different assumptions on the beamlet size at the GG and heat transport would result in lower divergences as determined from the CFC widths. The modeled CFC widths are close to the measured values, but the optimum U_{acc}/U_{ext} ratio is lower in the simulations, similar to the BES measurements. The vertical width σ_z is larger than the horizontal width σ_y in measurement and simulations.

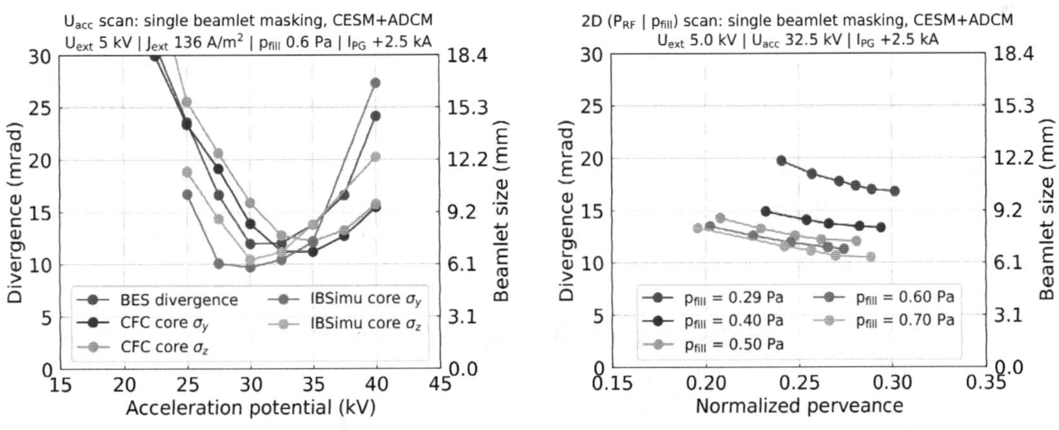

(a) Measured and simulated U_{acc} dependence (b) Measured p_{fill} dependence in RF power scans

Figure 3: (a) shows the measured (BES + CFC) and calculated (IBSimu) single beamlet divergence/size for a U_{acc} scan. (b) shows the CFC measured single beamlet size for a 2D scan of the RF power and the filling pressure.

Several other scans were made to investigate the deflection compensation. The measured horizontal beamlet position is not sensitive to the U_{acc}/U_{ext} ratio, the perveance around the divergence optimum, and is not influenced by the filling pressure. As expected on basis of the simulations, the beamlet width varies much more with the experimental parameters. One spurious dependence that is not explicitly accessible in the modeling is the source filling pressure. Figure 3b shows the measured beamlet size on the CFC as function of the perveance for scans in the RF power at various source filling pressures. Below a source filling pressure of 0.4 Pa, the measured beamlet size increases; the filling pressure dependence is weak above 0.4 Pa. A similar behaviour is seen for negative ion based NBI that used filamented arc sources, but the threshold filling pressure is lower, although it is difficult to compare amongst experiments since the location of the pressure measurement is not standardized[14]. Speculated causes for this dependence are an increase of the perpendicular temperature of the negative ions at low pressures, or a changed potential structure upstream of the plasma grid. Similar effects are well known for volume ion sources, but it is unclear how these finding translate to surface-production dominated negative ion sources[15].

4. Conclusions

The BUG-MLE grid system was experimentally characterized in BATMAN Upgrade. Single beamlet properties were studied in a masking configuration that leaves only a single beamlet open in the upper grid half, where the correction magnets were installed.

The simulated and measured horizontal position agree at the standard operating parameters ($U_{ext} = 5$ kV, $U_{acc} = 32.5$ kV). The measured horizontal position is not sensitive to the U_{acc}/U_{ext} ratio, the perveance around the divergence optimum, and is not influenced by the filling pressure. The measured horizontal positions are fully compensated within the positioning uncertainty in the experiment, independent of the total potential.

As predicted in the design[6], the beamlets appear triangular, partially as result of the correction magnets. The shape of the beamlet complicates the determination of the position and width in the measured and simulated data. The triangular beamlet shape highlights the need to study the angular distribution of the beamlet, rather than reducing it to a single number. The impact of the asymmetric angular distribution on the transmission is under investigation.

Acknowledgments

This work has been carried out within the framework of the EUROfusion Consortium and has received funding from the Euratom research and training programme 2014-2018 and 2019-2020 under grant agreement No 633053. The views and opinions expressed herein do not necessarily reflect those of the European Commission.

The views and opinions expressed herein do not necessarily reflect those of the ITER Organization.

References

[1] INTERNATIONAL ATOMIC ENERGY AGENCY 2002 *ITER Technical Basis* 24 (Vienna: IAEA)
[2] J. Mallinson 1973 *IEEE Transactions on Magnetics* **9** 678–682
[3] H. de Esch et al. 2015 *Nuclear Fusion* **55** 096001
[4] G. Chitarin et al. 2014 *Review of Scientific Instruments* **85** 02B317
[5] G. Chitarin et al. 2019 *Fusion Engineering and Design* **146** 792 – 795 ISSN 0920-3796 sI:SOFT-30
[6] N. den Harder et al. 2021 *Fusion Engineering and Design* **173** 112837 ISSN 0920-3796
[7] W. Kraus et al. 2018 *AIP Conference Proceedings* **2052** 040004
[8] C. Wimmer et al. 2021 *This conference*
[9] G. Orozco et al. 2021 *Fusion Engineering and Design* **165** 112225 ISSN 0920-3796
[10] T. Kalvas et al. 2010 *Review of Scientific Instruments* **81** 02B703
[11] T. Kalvas 2013 Ph.D. thesis University of Jyväskylä
[12] A. Hurlbatt et al. 2021 *AIP Conference Proceedings* **2373** 030004
[13] A. Hurlbatt et al. 2021 *AIP Advances* **11** 025330
[14] P. Veltri 2019 Final discussion 3rd EAC meeting
[15] M. Bacal et al. 1991 *Journal of Applied Physics* **70** 1212–1219

Techniques to widen the operational space of SPIDER Radio Frequency driven plasma source

R Casagrande[1,2], M Dan[1] and A Maistrello[1]

[1]Consorzio RFX, Corso Stati Uniti 4, I-35127 Padova, Italy
[2]Department of Applied Physics, Ghent University, 9000 Gent, Belgium

Corresponding author: riccardo.casagrande@igi.cnr.it

Abstract. SPIDER is the full-size prototype of the ion source of the ITER Neutral Beam Injectors, and is in operation since June 2018 as part of the ITER Neutral Beam Test Facility. The ion source is composed of 8 Radio Frequency drivers supplied by 4 RF oscillators, rated for operation at 1 MHz, for a total power of 800kW. This article presents a strategy for the selection of source biasing and of operating frequencies of the different RF generators to facilitate plasma initiation and to improve plasma stability. This operational strategy improved the plant reliability at high RF power in a broader range of operational conditions in SPIDER.

1. Introduction

SPIDER is the full-size prototype of the RF driven ion source for ITER NBI [1], [2]. SPIDER RF system is composed of 4 RF tetrode oscillators rated at 200 kW each, working in the frequency range 0.9-1.1 MHz. Each oscillator provides power to two RF drivers connected in series, which generate a plasma inside the beam source through inductive coupling. The layout of the RF drivers is a 4 by 2 grid, with each row defined as "segment". Negative ions are extracted from the plasma and accelerated through a series of electrostatic grids, while electrons are removed from the beam by deviating their trajectory on the second grid of the extraction system, by means of a magnetic filter field. SPIDER early operation has been characterized by several performance limitations, primarily related to the novelty of the ion source configuration. Plasma initiation is affected by the beam source filling pressure and the magnetic filter field, whereas in steady-state operation, limitations mainly come from RF system-related issues (frequency flip [3], RF stray currents [4], RF induced discharges outside the ion source [5], mutual coupling between RF circuits [4]). Some of these problems have been partially addressed, to improve SPIDER reliability: RF stray currents in the electrical system and mutual coupling between RF circuits were significantly reduced [4], and a mask was installed over the plasma grid [6] to reduce the vessel background pressure to nominal conditions. However, the frequency flip still represents a fundamental issue, thus an upgrade to solid-state amplifiers is foreseen during the 2021/2022 shutdown to overcome this limitation [7]. A campaign devoted to the study of caesium conditioning was planned before the shutdown; the technical requirement to perform an efficient caesiation of the ion source surfaces is a high reliability of the RF drivers, i.e. of the RF generators [8]. In addition, to explore the conditioning effectiveness in several operating conditions, reliability was required over a significant range of machine parameters (source filling pressure, Plasma Grid filter field, RF power). This paper describes the analysis of plasma start-up and plasma stability during steady-state operation, in correlation with machine settings. This analysis led to a strategy to define machine settings which facilitates plasma initiation and guarantees better RF load stability, resulting in higher reliability of SPIDER, and in particular of RF oscillators.

2. Parameters affecting beam source operation

The main parameters available to the beam source operator during experimentation are the following:

- Plasma Grid (PG) filter current: used to reduce electron temperature and density in the extraction region; it also influences plasma confinement inside the RF drivers;
- Filling pressure: the neutral gas pressure influences the electron density and temperature of the plasma generated inside the drivers;

- Biasing of the beam source: it allows the reduction of co-extracted electrons. The biasing can be carried out with 2 independent power supplies (ISBI and ISBP), polarizing the beam source body with respect to:
    - Bias Plate: the first grid facing the plasma expansion region, featuring large apertures
    - Plasma Grid: the second grid from the expansion region and the first one involved in the accelerations of the beamlets
- RF power and frequency: they can be controlled independently for the four couples of RF drivers;

All these parameters have an effect on plasma ignition and on the RF load, thus on the operation of the RF oscillators.

3. Plasma ignition

During SPIDER operation, we have found that beam source bias provides an additional way to control plasma ignition. At the output of the two polarizing units, large capacitor banks and pre-load resistors are installed (C_{BP}-R_{BP} and C_{BI}-R_{BI}), for voltage filtering and energy absorption from the load [9], respectively. Because of SPIDER configuration, the plasma in contact with driver walls, the beam source walls, the bias plate and the plasma grid has different density and temperature. For this reason, during the plasma ignition phase, electrons are lost to the driver and the beam source walls, leading to a current circulation in the biasing circuits and an equilibrium voltage is reached. The time constant of this transient is determined by the R-C elements at the output of the power supplies (Figure 1). For certain values of RF power, PG current and gas pressure, the excessive electron current drain prevents successful plasma ignition. Pre-charging the capacitors to the expected equilibrium voltage can therefore assist the plasma ignition, reducing the loss of electrons.

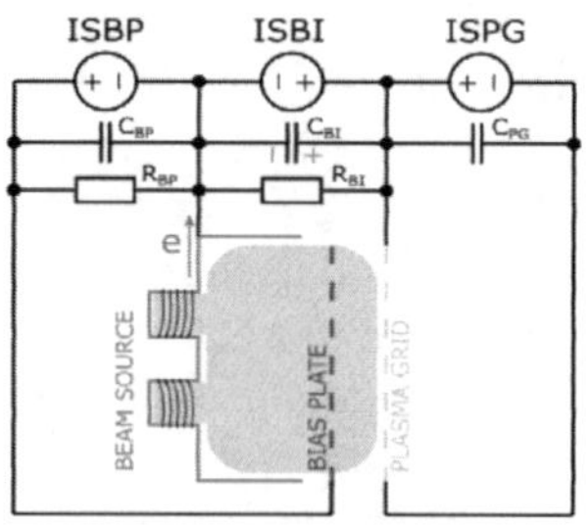

Figure 1. Schematics of the beam source biasing circuit. ISBP: Ion Source Bias Plate power supply, ISBI: Ion Source Biasing power supply, ISPG: Ion Source Plasma Grid power supply

4. Steady-state operation

4.1. Limits of the RF oscillators

SPIDER early operation has been limited, in part, by problems related to the RF system. Several issues have been addressed, as mentioned in the introduction, but some limitations are inherent to the vacuum tube technology used for the RF generators. In particular, the problems encountered during steady-state operation are:

A) Frequency flips. SPIDER RF oscillators consist of two tetrodes in push-pull configuration. A variable capacitor, from now on dubbed Cv, is used to tune the resonant frequency of a tank circuit, so as to define the operating frequency. The oscillator provides power to the RF load, which is composed of a capacitive matching network and the RF drivers. Ideally, the working frequency should be such that the impedance seen at the oscillator output is as close as possible to $50\boldsymbol{\Omega}$. Because of the connection between the two resonant circuits (tank circuit on-board the oscillator and RF load) through the transmission line, an abrupt change in the operating frequency happens when the working point approaches the minimum reflection coefficient. A forbidden frequency band in the operational space ensues, which prevents the transfer of the nominal oscillator power to the RF drivers.

B) Screen Grid Control Fault. The tetrode screen grid regulates the output power of the oscillator, by modulating the flow of electrons to the anode plate. The screen grid control fault is an alarm acting for the protection from overvoltage and overcurrent of the screen grid power supply. When large output power is required, the screen grid voltage is raised and it approaches the values of the anode voltage. In such conditions, a fraction of the secondary electrons emitted from the anode can reach the screen grid. This process reduces the tetrode output power, thus the control system tries to increase the screen grid voltage to compensate for it, leading to a positive feedback and eventually to the overload of the screen grid power supply. [9]

The forbidden frequency band, imposed by the frequency flip phenomenon, prevents the oscillator from operating in matched conditions. Therefore, even for 100 kW output power (half of the nominal value), the screen grid power supply is significantly stressed. When working at limit conditions, RF load instabilities can trigger the positive feedback described in 4.1B, leading to the screen grid control fault.

4.2. *Influence of machine parameters on RF load stability*

The parameters described in section 2. have an impact on plasma properties and stability, thus on the characteristics of the load seen by the RF generators. Figure 2 shows the plasma light ripple (right driver on segment 4) for shot 8322, where the aforementioned parameters were varied independently to observe their effect on the plasma light oscillations. The diagnostic system used for these studies is a plasma light detector [10]: for each RF driver, a photodiode collects the light emitted from the plasma, along a line of sight parallel to the driver axis. The signals are acquired with a sample rate of 30 kHz. In the pulses used for the analysis, the relevant parameters are varied at a rate much lower than the sampling frequency, or kept constant over large intervals. Thus, even though plasma oscillations at frequencies larger than the sampling rate are not fully resolved, their amplitude can be still evaluated by looking at the noise magnitude. From *Figure 2* we can observe that:

- Increasing the PG current over a threshold leads to significantly large plasma oscillations (in t_3 the oscillations are still contained, while in t_5 the threshold was surpassed);
- Biasing the beam source with respect to PG and Bias Plate slightly reduces plasma oscillations (t6). This effect is not clear due to a simultaneous increase of the plasma light average;
- In t8, increasing the relative frequency between the adjacent drivers reduced the ripple, while the average plasma light intensity remains constant over the interval (t7 and t8).

However, PG current, gas pressure, source biasing and RF power are bound to the experimental requirements, and typically cannot be varied freely during operation. The only free parameter available to minimize the RF load oscillations is then the operating frequency of the oscillators.

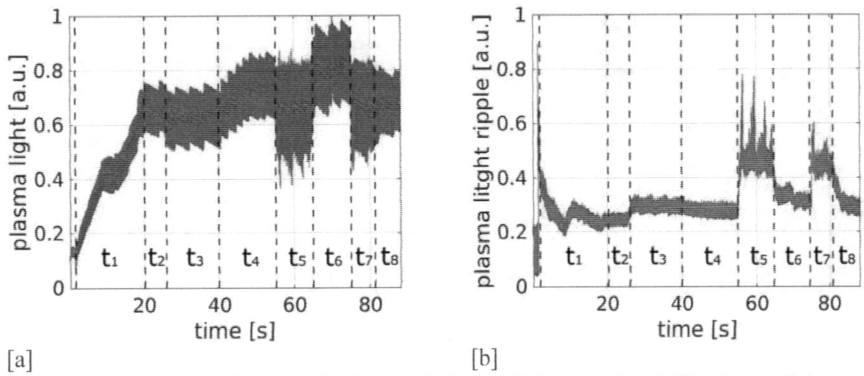

Figure 2. Dependency of plasma light oscillations ([a] plasma light raw signal, [b] plasma light normalized to its average) on various machine parameters.
t2: ramp-up of RF power to 80kW in segments 3 and 4; t2: PG current is raised step-wise to 1kA; t3: PG current raised step-wise to 1.3kA; t4: filling pressure linearly increased up to 110% of the initial value; t5: PG current raised step-wise to 1.7kA; t6: beam source biased with ISBI and ISBP; t7: ISBI and ISBP switched off; t8: relative frequency between operating oscillators varied from 9kHz to 11kHz

4.3. *Effect of the oscillator frequencies on operational stability*

Due to the mutual coupling between RF circuits, when oscillators operate at different frequencies the instantaneous power delivered to each RF load is subject to modulation. This results in oscillations, or beats, of the plasma density and temperature in the kHz range. In principle, beats can be avoided if the oscillators operate at the same frequency, but we observed experimentally that if RF generators are synchronized in frequency, the screen grid control fault is triggered when PG current is ramped-up to nominal values (1.5-2 kA). Prior to the caesium campaign, the strategy used to operate four oscillators simultaneously was based on the minimization of the reflection coefficient seen at the output of each unit: the oscillators are switched on in sequence, setting an initial Cv value which prevents frequency flip in vacuum. Then, the RF power is ramped-up and the Cv is simultaneously varied, until the lowest possible value prior to frequency flip in load conditions is achieved. Therefore, the operating conditions are defined by the frequency flip limits of each oscillator, which are not necessarily optimal in terms of low plasma oscillations. In addition, prior to the modification of the RF connections inside the vacuum vessel, the mutual coupling between adjacent and non-adjacent RF drivers was comparable. Thus, while operating with all four oscillators, minimizing the beat amplitudes without incurring in instabilities has proved virtually impossible. Lowering the mutual coupling between RF circuits [4] helped significantly with the reduction of beat amplitudes when operating with non-adjacent segments. This allowed us to develop a new strategy for the simultaneous operation of four oscillators: the RF generators powering two adjacent ion source segments are operated at the required conditions (PG current, filling pressure), and the Cv is varied during the pulse flat top to observe the effect on plasma light oscillations for operation at different relative frequencies (*Figure 3*). From this analysis, the operating frequencies for these two oscillators are defined, while avoiding frequency-locking and instabilities in its proximity. The other two oscillators are then added one at a time, still with varying Cv during the flat top to establish the optimal operating frequency.

With this strategy, the oscillators do not operate at the minimum achievable reflection coefficient, so the single RF generator cannot provide the maximum theoretical power. Nonetheless, thanks to the reduced oscillations of the RF load, operation at high power (100 kW per couple of drivers) is far more stable and less prone to trigger screen grid control faults. Therefore, through this expedient, it was possible to operate SPIDER oscillators up to a total RF power of 440 kW, with high reliability, during the first caesium campaign [11].

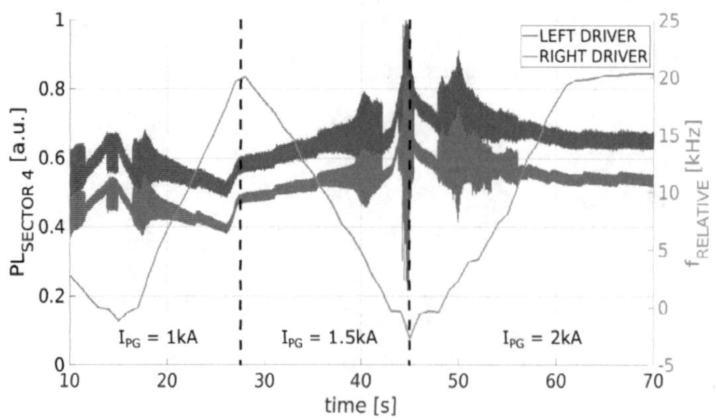

Figure 3. Effect of oscillators relative operating frequency on plasma light ripple, for three values of PG current and filling pressure of 0.3Pa. The operating frequencies are chosen so that the large plasma oscillations in proximity of the frequency locking interval are avoided, and at the same time the reflection coefficient for each oscillator is kept as low as possible

5. Conclusion

Thanks to two experimental observations, namely the charging transient of the capacitors installed at the output of the source biasing power supplies, and the effect of the relative difference between oscillators operating frequencies on plasma stability, it was possible to develop a strategy to operate SPIDER in a broad range of filling pressures, PG current and RF power with increased reliability. The performance of SPIDER during the 2021 campaign did not improve in terms of maximum output power from the oscillators, which still incur in the hard limit caused by frequency flips, but the overall reliability of the RF system was improved, allowing the simultaneous operation of four oscillators, up to 4x110 kW, with hundreds of pulses per day.

Acknowledgement
The views and opinions expressed herein do not necessarily reflect those of the ITER Organization. This work has been carried out within the framework of the EUROfusion consortium and has received funding from the Euratom research and training programme 2014-2018 and 2019-2020 under grant agreement No 633053. The views and opinions expressed herein do not necessarily reflect those of the European Commission.

References

[1] Sonato P et al. "The ITER full size plasma source device design." *Fusion Engineering and Design* 84.2-6 (2009): 269-274.
[2] Toigo V et al. "The PRIMA test facility: SPIDER and MITICA test-beds for ITER neutral beam injectors." *New Journal of Physics* 19.8 (2017): 085004.
[3] Gasparini F et al. "Investigation on stable operational regions for SPIDER RF oscillators." *Fusion Engineering and Design* 146 (2019): 2172-2175.
[4] Maistrello A et al. "Improvements in the SPIDER RF system." *Fusion Engineering and Design* 167 (2021): 112337.
[5] Dan M et al., “RF induced discharges in vacuum in the SPIDER experiment” *ISDEIV 2021*
[6] Pavei M et al. "SPIDER plasma grid masking for reducing gas conductance and pressure in the vacuum vessel." *Fusion Engineering and Design* 161 (2020): 112036.
[7] Zanotto L et al. “Radio frequency generators based on solid state amplifiers for the nbtf and iter projects”, submitted to SOFE2021 (2021).
[8] Wünderlich D et al. "Operation of Large RF Driven Negative Ion Sources for Fusion at Pressures below 0.3 Pa." *Plasma* 4.1 (2021): 172-182.
[9] Zamengo A et al. "Power supply system for large negative ion sources: Early operation experience on the SPIDER experiment." *Fusion Engineering and Design* 173 (2021): 112790.
[10] Pasqualotto R et al. "Plasma light detection in the SPIDER beam source." Fusion Engineering and Design 146 (2019): 709-713.
[11] Sartori E et al. “First Operations with Caesium of the Negative Ion Source SPIDER” *ICIS 2021, contribution 170*

Characterization of Hydrogen Plasma in an ECR based Large Volume Plasma Chamber

Shweta Sharma[1], D Sahu[1], R Narayanan[1], S Kar[1], M Bandyopadhyay[2], A Chakraborty[2], M J Singh[2], R D Tarey[1] and A Ganguli[1]
[1]Centre for Energy Studies, Indian Institute of Technology Delhi, New Delhi, India 110016
[2]Institute for Plasma Research, Gandhinagar, Gujarat, India 382428

E.mail: shwetasharma1102@gmail.com

Abstract. Hydrogen plasma characterization was carried out in a large volume (dia~ 1.0 m, h~ 1.0 m) plasma chamber to evaluate the efficacy of production of uniform, large-area H^- beam for fusion applications. Up to seven Compact ECR Plasma Sources (CEPS; Indian Patent #301583, Patentee: IIT Delhi) can be mounted on the top dome of the chamber (one in centre and six on a ~60 cm dia circle). Axially poled permanent ring magnets provide the magnetic field for each CEPS and the total field in the chamber is the combined field of all CEPS. Separate experiments were conducted with: (i) six CEPS with opposite polarity on adjacent magnets, (ii) seven CEPS with identical polarity on all the magnets, and (iii) a single CEPS at the center. In case (ii) the field lines repel while for (i) a cusp field is formed between adjacent sources. For case (iii), one obtains a uniform plasma density over 35 cm radius (n_e ~ 4 × 10^{10} cm^{-3}, T_e ~1–2 eV), ~ 60 cm downstream (from central source mouth) with 400W power at ~ 1-3 mTorr pressure, yielding ideal conditions for volume mode H^- production over a large area. For case (ii), however, plasma from each source flows along its individual field lines, without forming uniform plasma downstream. Whereas for case (i), because of cross-talk between adjacent sources, the system became unstable, giving oscillations in plasma formation and microwave reflected power. Hence, it appears that the single CEPS configuration is the most efficacious for large area H^- generation.

1. Introduction:

Plasma based large area negative hydrogen ion source is an essential part of Neutral Beam Injector (NBI) modules of thermonuclear fusion reactors like ITER, since H^-/D^- ion beams can be efficiently converted into high energy neutral H^0/D^0 beams (~ 1 MeV/u) for heating of fusion plasma **[1]**. The target H^- current that has to be delivered to the NBI of ITER is around 40 A which requires production of ~ 33 mA/cm^2 H^- current density over a grid area of approximately 2 m^2 **[2]**. This in turn demands production of high-density uniform hydrogen plasma of the order of 10^{12} cm^{-3} at operating pressures of 2-3 mTorr (0.27 – 0.4 Pa) **[3]** and electron temperature (T_e) ~ 1-2 eV, to avoid electron detachment of H^-. Further, plasma uniformity of 10% over the plasma grid is desired, which ensures a good quality beam extraction. Present negative ion sources use around 1 MW RF (1 MHz) inductive power to achieve the target plasma density **[4]**. Such high-power investments face several technical and economic challenges that have triggered the search for alternate technologies offering more efficient power coupling / absorption **[5]**. Microwaves based on Electron Cyclotron Resonance (ECR) absorption can be one such scheme **[6],** since in this scheme the microwave power is coupled very efficiently to the plasma electrons enabling production of high plasma densities at low pressures and relatively lower powers, encouraging utilization of ECR schemes for negative hydrogen ion sources. Tarey *et. al.* **[7]** have shown that *multiple Compact ECR Plasma Sources* (CEPS, patentee: Plasma Lab, IIT Delhi) **[8]** *can be configured in appropriate arrays to produce uniform, large area, high density plasma in argon at very modest power.* However, unlike argon, hydrogen is a molecular gas and as suggested by some recent experiments using both gases **[9]**, the behaviour of hydrogen plasma differs significantly from that of argon, emphasizing the need for investigation of hydrogen plasma production *using the CEPS-based source array configuration used for argon* **[7]** and also providing motivation for the present work.

The experimental system (dia: ~1.0 m, height ~ 1.0 m) utilized here is suitable for large area hydrogen plasma production. It has provision for attaching multiple CEPS on its top dome and side wall. Each CEPS has its permanent ring magnets for producing within its interior the required magnetic field for meeting the ECR condition (~ 875 G for 2.45 GHz). Depending on the number of CEPS used and the direction of polarity of the magnets on adjacent sources, different plasma-source and magnetic field geometries can be configured. Three different configurations were tried in the present work for realizing a suitable source configuration for uniform H^- ion production over large area.

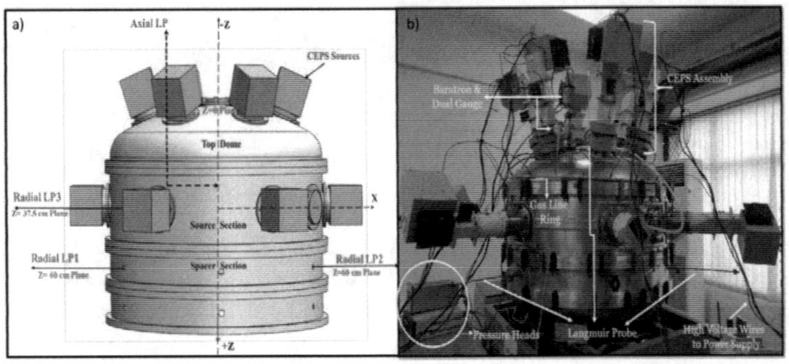

Figure 1(a) Schematic representation of the experimental system (b) Photograph of the enitre assembled system with sources (CEPS) mounted on all ports of top dome and sidewall.

2. Experimental System & Diagnostics:

The CEPS comprises of suitably designed permanent (NdFeB) ring magnets poled axially. The ECR surface lies within the CEPS and is quite complex. However, the magnets produce a simple, exponentially decaying (diverging) field within the expansion chamber to which it is attached. A detailed description of CEPS can be found elsewhere **[10].** Figure 1(a) shows the schematic of the cylindrical stainless steel (SS-304) experimental chamber (height $\approx$ diameter $\approx$1 m). The top dome is equipped with seven ports (ID $\approx$ 91mm), with six being centred symmetrically on a circle of dia ~ 60 cm and one located at the centre of the dome. (Although there is provision for attaching more CEPS on the vessel sidewall (Figure 1(b)), these were not used in the present experiments.) The system has a bottom dome to which is attached a diffusion pump (Varian, VHS-6) backed by a mechanical pump (Varian, DS 1002). Hydrogen gas flow into the chamber is controlled via a precision needle valve. For diagnostics, three radial Langmuir probes (LP1, LP2, LP3) with tungsten tip of diameter ~ 0.5 mm - 0.25mm and length ~ 2.2 mm – 4 mm) were introduced from ports on the chamber sidewall. The location of the probes is shown in Figure 1(a). LP1 and LP3 are located in the same vertical plane at z = 60 cm and z = 37.5 cm, respectively, while LP2 is located at z = 60 cm, diametrically opposite to LP1. Here z = 0 is measured from the center of the top dome inner surface. The tips of all three probes were oriented radially (i.e., perpendicular to the chamber or z-axis). The LP data was analysed to determine the bulk electron density (n_e) and its temperature (T_e), the plasma potential (V_p) and the warm electron density and its temperature (n_w, T_w). The base vacuum was ~ 4.0 × 10^{-6} Torr while the working pressure was typically, 0.13 -0.4 Pa. For monitoring the pressure, an MKS Baratron 626A was mounted. The input microwave power to each source (monitored from the power supplies) was kept in the range of 400 W-1000 W.

3. Experimental results:

As discussed above experiments with hydrogen were conducted using three CEPS configurations: (i) Six CEPS with opposite polarity on adjacent magnets (figure 2a), (ii) seven CEPS with identical polarity on all the magnets (figure 2b) and (iii) a single CEPS at the centre (figure 2c). As will be seen below, the magnetic field patterns produced in each configuration inside the chamber are very different and strongly influence the flow and filling of plasma in the chamber.

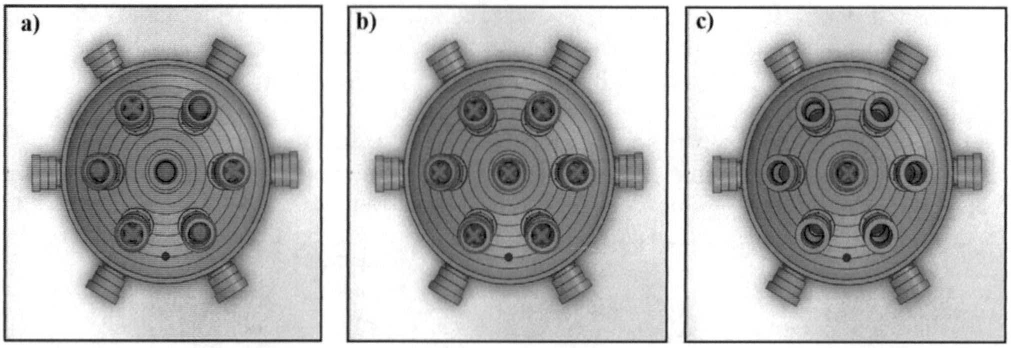

Figure 2: Top view of the top dome for the three configurations. The red circle indicates CEPS ring magnets. The cross (dot) indicate the field pointing into (out of) the chamber.

3.1 Six CEPS with opposite polarity on adjacent magnets [Cusp]:
In accordance with figure 2 a), six CEPS with alternating polarity of their magnets were attached to the top dome as shown in the figure (on a circle of dia ~ 60 cm). The resulting multi-cusp field pattern is shown in figure 3(a). Figure 3 (b) shows a typical image of the plasma inside the chamber in this configuration. Comparing the plasma profile with the magnetic field configuration clearly shows the plasma concentrates to the field lines. As a result, plasma from one source penetrates the adjacent sources on either side causing strong feedback and cross-talk between the sources that yield a highly unstable plasma and strongly fluctuating reflected power. Consequently, this configuration was not investigated further.

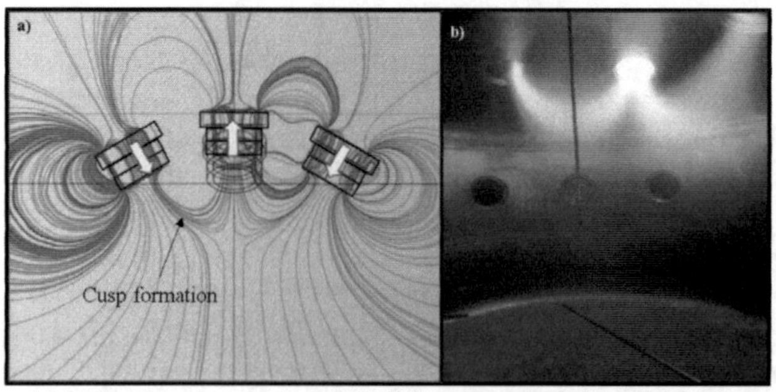

Figure 3: (a) Comsol Mutipysics simulation of CUSP magnetic field (Fig. 2(a)) with arrow showing polarity of the magnets, b) plasma inside the chamber under same configuration.

3.2 Seven CEPS with identical polarity on all the magnets [Uni-Polar]:
In this configuration, the CEPS were attached to all seven ports on the top dome, with like polarity on all magnets as in figure 2 b). Figure 4(a) shows the simulated magnetic field lines inside the chamber, which shows how the field lines repel and spread out into the chamber. Once again, the plasma shows a strong tendency to follow the field lines, yielding images of plasma formation and flow as in figure 4(b). It is seen that plasma from the sources flow into the chamber as separate columns without merging mutually. This tendency of hydrogen plasma to obey the field lines without mixing is completely different than that of argon plasma, whose cross-field diffusion is strong and allows for highly uniform plasma formation **[9]**. The reason for this is that the cross-field diffusion coefficient, $D_{\perp} \alpha M^{1/2}$, allowing heavier ions to diffuse more easily across magnetic fields.

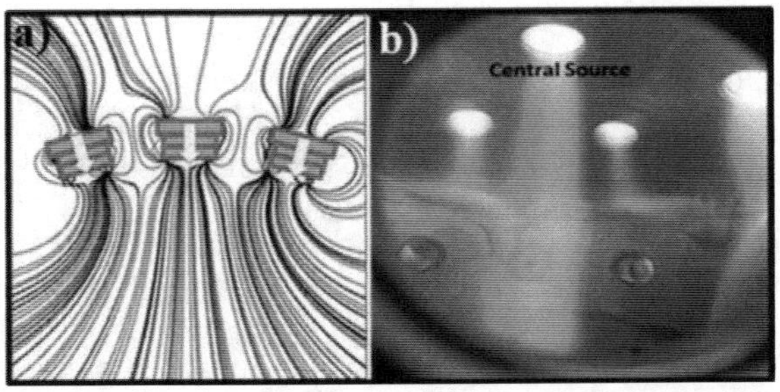

Figure 4: (a) COMSOL Multiphysics simulation of unipolar magnetic field (figure 2 b)), (b) plasma inside the chamber under same configuration.

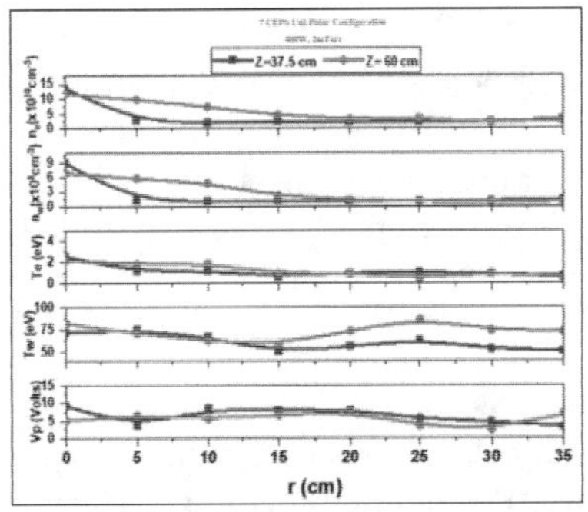

Figure 5: Radial plasma profiles at a pressure of ≈ 2 mTorr and microwave power ≈ 400W (at each CEPS) at z= 37.5 cm (Black) & Z = 60 cm (Red) using 7 CEPS ignited with uni-polar configuration.

Radial LP profiles were taken at two axial locations using LP3 (at z = 37.5 cm) and LP1 (at z = 60 cm) at a pressure of 0.27 Pa and microwave power of 400W for each source. The results are displayed in Figure 5, where n_e, T_e, and V_p are displayed for both probes. It is seen that n_e at z = 37.5 cm falls sharply from 1.5×10^{11} cm^{-3} (at r = 0) to 3×10^{10} cm^{-3} within a radius of 5 cm, following which it saturates at an average value of 1×10^{10} cm^{-3} up to r = 35 cm. The downstream density (at z = 60 cm) is a little lower at r = 0 (~1.2×10^{11} cm^{-3}) and falls linearly up to r = 15 cm, beyond which it falls more gradually up to r = 30 cm, saturating at ≤ 5×10^{10} cm^{-3}. The slower fall in the latter case may be due to a *higher* cross field diffusion towards the boundary as the magnetic field strength weakens [Figure 4(a)]. Nonetheless, the radial non-uniformity in n_e is still high in the downstream region since the magnetic field lines from each source are pushed out relative to the central source field because of their like polarities. The plasma electrons (as well as the ions through ambipolar fields) are strongly bound to the magnetic field of each source giving rise to discrete plasma beams. This prevents significant contribution from the peripheral plasma column towards filling the chamber uniformly. It may be noted, however, that the density at r = 0 is still quite high at z = 60 cm, because magnetic field lines of the peripheral sources prevent the field lines of the central source from fanning out. T_e, and V_p vary typically in the range 1.0 – 2.5 eV and 5 – 10 V, respectively. One also observes the presence of a warm electron population throughout the chamber with n_w ~ 10^8 cm^{-3} and T_w ~ 40-100 eV. The warm electron population as well as the tail of the bulk electron population could be responsible for the production of plasma within the expansion chamber. However, it is interesting to note that the profile for n_e follows closely the T_e profile (figure 5), indicating that ionization within the expansion chamber is due to bulk electrons, with T_e controlling the profiles. Overall, one may conclude that this configuration is also not suitable for uniform, large area H^- production even though the electron temperature is quite favourable at both planes.

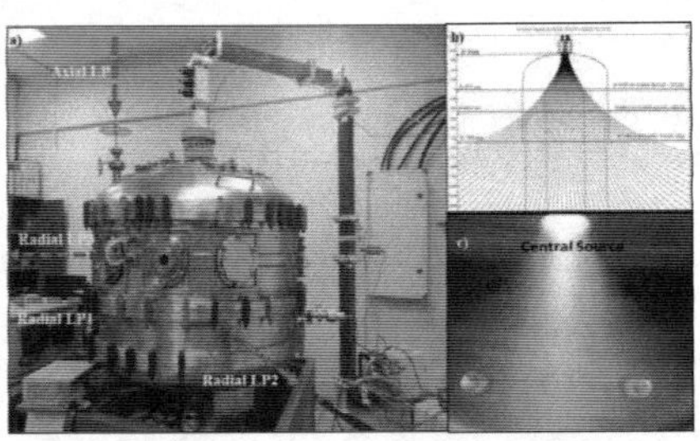

Figure 6: (a) Photograph of chamber with single CEPS Configuration of Fig. 2(c). (b) Simulated magnetic field structure inside chamber using single CEPS. (c) Fanning out of plasma observed inside the chamber with single CEPS mounted at top dome.

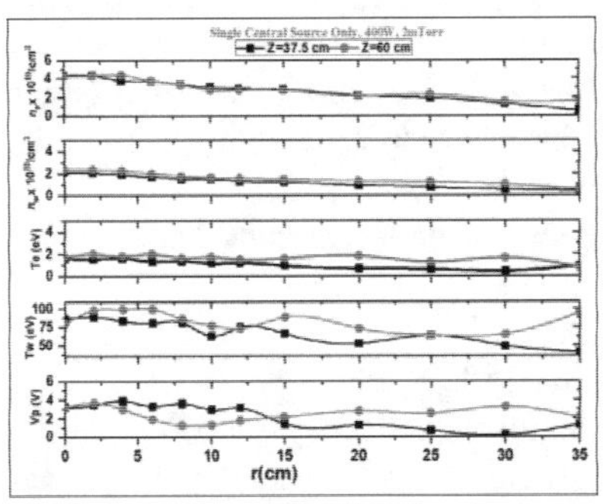

Figure 7: Radial plasma profiles at a pressure of ≈ 2 mTorr and microwave power ≈ 400W @ Z= 37.5 cm (Black) & Z = 60 cm (Red) using single CEPS ignited at central port of top dome.

Journal of Physics: Conference Series **2244** (2022) 012055 doi:10.1088/1742-6596/2244/1/012055

3.3 *Single CEPS at the center of top dome:*
Figure 6 (a) shows a photograph of the system with a single source placed centrally on the top dome, while Figure 6 (b) shows the simulated field lines in the chamber for this case. The field lines are seen to fan out radially with distance from the source, that yield a radially expanding, uniform plasma with z (Figure 6(c)) since the Hydrogen plasma follows the field lines. The radial density profiles (Figure 7) at both z planes (37.5 cm and 60 cm) are found to be fairly good, falling from ~ 4.5×10^{10} cm^{-3} (at $r = 0$) to about ~1.5×10^{10} cm^{-3} (at $r = 30$) at both the z-planes with T_e, typically in the range 0.5 – 2 eV and V_p ~ 4 V.

The results obtained from the single CEPS indicate that the latter configuration could be optimized further for developing a large area, uniform H^- ion beam source. In order to raise the plasma density (and H^- density) one would need to increase the power (~ few kW), however. Plasma confinement to reduce plasma loss on the walls also need to be addressed.

4. Summary and Conclusions:
Hydrogen plasma formation was investigated in a large volume chamber that allows one to mount multiple compact ECR sources (CEPS) on its top dome. Each configuration of the sources and the polarity of their magnets yields a different magnetic field geometry in the expansion chamber. Three different magnetic field configurations were investigated by mounting (i) six CEPS with opposite polarity on adjacent magnets, (ii) seven CEPS with identical polarity on all magnets and (iii) a single CEPS at the center of the top dome. It was observed that unlike argon plasma, hydrogen plasma sticks closely to the magnetic field lines in the expansion chamber yielding very poor radial uniformity for the first two configurations and a fairly uniform, large area plasma in the last case. It is envisaged that top dome single CEPS configuration can be optimized further to develop a large area, uniform, high current H^- ion beam source.

Acknowledgments
The authors acknowledge Prof P. K. Kaw (Institute for Plasma Research, Gandhinagar) for motivating the group in initiating Hydrogen plasma experiments at IIT Delhi and for the numerous related discussions. Authors would also like to thank IPR Gandhinagar for the partial funding provided by them. The partial funding in developing the large volume plasma system through DST funds are also acknowledged by the authors. Authors would also like to thank Mr. Sumit Kumar, Mr. Hasmukh Kabariya and Mr. A.J. Josekutty for providing their technical help at various stages of the experiments.

References:
[1] Speth E 1989 *Rep. Prog. Phys.* **52** 57-121
[2] Hemsworth R S *et al* 2017 *New J. Phys.* **19** 025005
[3] Fantz U *et al* 2008 *Rev. Sci. Instrum.* **79** 02A511
[4] Fantz U *et al* 2020 *Fusion Eng. Des.* **156** 111609
[5] Mukhopadhyay D, Pandey A, Bandyopadhyay M, Tyagi H, Yadav R, and Chakraborty A 2019 *Rev. Sci. Instrum.* **90**, 083103
[6] Geller R 1996 *Electron Cyclotron Resonance Ion Sources and ECR Plasmas* 1st ed (Bristol: Institute of Physics Pub.)
[7] Tarey R D, Ganguli A, Sahu D, Narayanan R and Arora N 2017 *Plasma Sources Sci. Technol.* **26** 015009
[8] Ganguli A and Tarey R D 2006 Permanent magnet compact electron cyclotron resonance plasma generator *Indian Patent* #301583 Patentee: IIT Delhi
[9] Verma A, Singh P, Narayanan R, Sahu D, Kar S, Ganguli A, and Tarey R D 2019 *Plasma Res. Express* **1** 035012
[10] Ganguli A, Tarey R D, Arora N and Narayanan R 2016 *Plasma Sources Sci. Technol.* **25** 025026

Space charged based Residual Ion beam recovery for the Neutral beam injection

V. Variale

INFN-BA, via Orabona 4, 70126-Bari, Italy

M. Cavenago

INFN-LNL, viale dell'Università 2, 35020-Legnaro (Pd), Italy

V. Valentino

INFN-BA, via Orabona 4, 70126-Bari, Italy

Vincenzo.variale@ba.infn.it

Abstract. Energy recovery of residual ions may be needed to increase the energy efficiency of Neutral Beam (NB) injectors for fusion plants as DEMO while a deflection-based system has been proposed until now to dump residual ions. As an alternative, a compact beam energy recovery system, based on space charge effects due to the residual ion deceleration into 2 Farady Cups (FC) with holes for D^0 passage, can replace the Electrostatic Residual Ion Dump (ERID) designed for ITER to stop the residual D^- and D^+ before the NB injection in the tokamak plasma. All parameter tunings and preliminary simulations are here described, also providing the suppression of back streaming to the ion source. Ion energy spread and rectangular geometry are considered. Collection of ions at low energy (a few percent of the full neutral beam energy Ek_i) instead of Ek_i as in ERID gives advantages that will be mentioned.

1. Introduction

ITER project [1] will use an electrostatic method to remove the unwanted residual charged component from the Neutral Beam Injectors (NBI) in place of the usual magnet separation system [2]. In ITER, the residual ions will be deflected at full energy by two electrodes of opposite tension (+20 kV). The removal of the ions emerging from the neutralizer of the NBI by means of the Electrostatic Residual Ion Dump (ERID) device presents some issues concerning, for instance, the power load suffered by the ERID panels and the possible formation of plasma. The last one might be particularly critical, since the creation of a plasma would modify the deflection properties of the device, and therefore would affect its proper functioning [3]. The study of ref. [3] show that a peak power load of about 8 MW m^{-2} can be reached. On the other hand, in order to increase the energy production efficiency of the DEMO project [4] a beam energy recovery system in NBI could be foreseen instead of ERID. Few years ago, a compact beam energy recovery system based on space charge has been proposed [5]. Moreover, the possibility of the further simplification of that beam energy recovery system by using a collector without decelerating electrodes could allow of using it as an alternative to ERID in the ITER project. The simplified beam recovery system proposed in this paper is based on the ion beam Space Charge (SC) effect, which increases with the ion deceleration and becomes big enough to allow the enlargement of the ion trajectories and the collection of them at very low energy (few hundred of eV) on the walls of a collector [5]. The using of a such simple collector for the residual ions at very low

Journal of Physics: Conference Series **2244** (2022) 012056 doi:10.1088/1742-6596/2244/1/012056

energies instead of deflecting them at full energy as foreseen with the ERID system, gives also advantages that will be mentioned.

2. The ERID device in ITER and the new collector

The Electrostatic Residual Ion Dump (ERID) is part of the neutral beam heating system. It is obtained by electrostatic deflection of the charged ions at full energy onto water-cooled plates system designed to accept the high power-density (for ERID details see for example [2]). The beam of neutral atoms is either injected into the ITER plasma or intercepted inside the injector by a calorimeter with movable panels which can accept the full neutral beam power. From ref. [2], ERID structure consists of five vertical dump walls. The five walls produce four channels crossed by the particle beam. It uses an electrostatic field to deflect the ions at full kinetic energy that are dumped on the five ERID panels. The field is produced by applying zero potential to the odd panels (the two external and the central panels) and negative potential (- 20kV DC +/- 5kV @50Hz) to the two even panels. Each ERID wall is composed of 18 vertical beam stop elements which have a rectangular cross-section with four internal channels for cooling water. The complexity of the ERID system structure is due to the very high power that the system must dissipate. Furthermore, as reported in ref. [3] plasma formation on the ERID plates could reduce the deflecting field and making inefficient the foreseen residual ion dump. On the other hand, one can think that a device to recover their energy could be used to increase the fusion power production efficiency as proposed for DEMO project [4]. In principle, however, a beam energy recovery system could be complex and require some modification to the original design of the NBI of ITER. Few years ago, a very simple and compact beam energy recovery concept was proposed for the next fusion project DEMO; that system was based on the space charge effect in the ion beam at very low energy [5]. To prove that recovery system a test experiment on the NIO1 source beam installed in RFX (PD) was proposed [5,6]. In that proposal the NIO1 source beamlet had to be collected at very low kinetic energy through a cylindrical collector, at voltage Vc, with decelerating electrodes and a repel electrode, at voltage Vr. Furthermore, in ref. [6], the using of an additional electrode, the grid (Vg), to reduce the secondary electron problem was suggested. The Vg values used to trap the most part of the secondary electrons, however, could be, in general, always 50 V lower than about Vc.

The original collector designed for the NIO1 beamlet recovery test [4,5] presented two problems: 1) the size of the collector was too large for the experiment vacuum chamber; 2) the HV feedthroughs number on the CF 200 flange was not enough to include all the needed electrodes. In fact, in order to carry out the beam energy recovery test, an efficient diagnostic system should be foreseen to drive the NIO1 beamlet precisely in the collector entrance. That system is composed by 4 diagnostic electrodes placed at the entrance of the collector and a beam steering device placed at the exit of the source. The limited space on the CF flange surface did not allow to place the feedthroughs for all the 7 electrodes needed. Furthermore, the collector size had to be reduced to fit it in the vacuum chamber to be used for the test. Then a simplified collector with reduced size and without the decelerating electrodes has been designed as shown in figure 1 where a simulation of both collectors for comparison is shown.

In the simulation of fig. 1, the reference case of the beam energy recovery test experiment, on NIO1 or TRIPS [7] sources, is presented. From comparison between the results of the two collector versions, it can be noticed that the ions are recovered, in both cases, with an energy lower than 700 eV.

The simplified collector version with rectangular geometry has also been studied. The collector with rectangular geometry is interesting because that geometry is the ITER type geometry. In the figure simulations the same 'aspect ratio' and scaled current density (equal perveance) of the ITER beam after the MAMuG extraction have been used [2]. The ion collection efficiency of the new simplified collector could be different from the case of cylindrical geometry simulations; for that reason, several simulations with different values of Vc and Vr have been carried out to optimize the ion collection efficiency.

As first step, in fig. 2, the case of Ek_i=20 keV (the same energy to be used in the NIO1 beamlet recovery test) has been considered also for the simulations in rectangular geometry with reduced

Journal of Physics: Conference Series **2244** (2022) 012056 doi:10.1088/1742-6596/2244/1/012056

dimension for simulation speed and with scaled current I=0.023 A, which corresponds to ITER perveance. Furthermore, the 'aspect ratio' of the ITER beam, as done also in SIMION [8] simulations of ref. [5], has been used to demonstrate how the space charge effect worked also in the simplified collector version, with rectangular geometry.

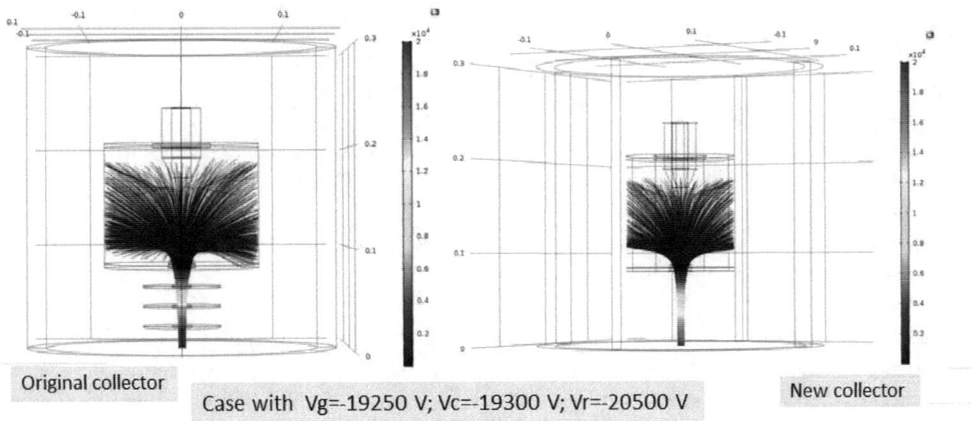

Figure 1. Ion trajectory simulations in the original and new collector for Ek_i=20 keV. Vg is the grid voltage, Vc is the decelerating voltage of the collector and Vr the repel voltage. The scale unit is in [m].

The simulation results of fig. 2 show that also for this simplified collector the space charge effect contribute to open efficiently the beam trajectories towards the walls without any back reflected ion. The SC effect acting on the ions, as already noticed, with the ion deceleration. On the other hand, the decelerating voltages (Vc) depend on the initial kinetic ion beam energy, Ek_i. The transverse electric field at the collector entrance hole depends also by Vc and then simulations for different initial kinetic energies which required different decelerating voltages (Vc) have been considered to verify the ion charge collection efficiency and their energy recovery for the related potential configurations at different ion energy.

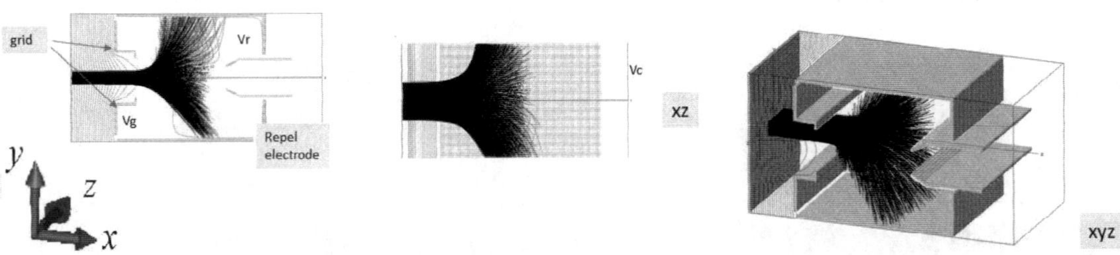

Figure 2. SIMION simulations for collector in rectangular geometry. For Ek_i=20 keV, with decelerating voltage, Vc=-19800 V and repel voltage Vr=-20500 V.

In figure 3, the ion trajectories for ion beams with Ek_i=5 keV and Ek_i=100 keV (with current, I=0.023 A) are shown as examples. At very low initial kinetic energy, the SC effect can become too strong before the collector entrance as shown in 5a). To avoid that problem a higher value of Vc could be used in such a way that the ions could be less decelerated and then enter the collector with higher Ek, but in this case, they could not have time to open towards to the walls before of hitting the repel electrode. For the case of 100 keV, however, the SC effect remains low before the collector entrance and the ions feel the hole lens effect before opening inside the collector.

Journal of Physics: Conference Series **2244** (2022) 012056 doi:10.1088/1742-6596/2244/1/012056

It must be noticed that SIMION simulations for very high initial kinetic energies (like 1 MeV in ITER) are not, practically, possible because they should require too much CPU time for large volume of electric field calculations.

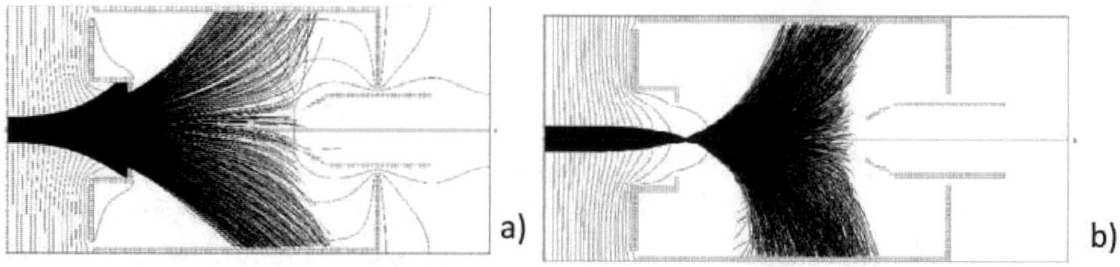

Figure 3. Ion trajectory simulations in rectangular geometry: a) Ek_i=5 keV with Vc=-4550V and Vr=-5400 V; b) Ek_i=100 keV with Vc=-99350 V and Vr=-101900 V.

However, since the simulation results depends only on the electric field and the doubling of the potential value can be compensated by doubling the distance (Dx) between the ground electrode and Vc, the simulation results for Ek_i=100 keV and Dx=33 mm (shown in fig.4b) should be like the results obtained for Ek_i=200 keV and Dx=66 mm, as confirmed by the results shown in 4a). Then, the simulation for 1 MeV and Dx=280 mm, should have similar results of 4b) where the case of Ek_i=500 keV and Dx=140 mm is shown. Notice that in fig. 4, the electric field values in the gap and near the collector hole are indicated to show that the fields used were still below the vacuum dielectric strength which is in the range of 15÷20 kV/mm (depending on the vacuum level and vacuum chamber wall shape). In the firsts tests we consider a vacuum pressure of about 10^{-6} mbar, which is typical of the ion source TRIPS to be used. It is still sufficient for good beam transport in real applications. Pressure may be larger for the neutralizer, but in energy recovery section it could be improved with differential pumping. It must be noticed that the compact simplified collector version shown above is characterized by only two voltage values (Vc and Vr) as the ERID device and if a double stage configuration is considered, as done in ref. [4, 5], it could recover both the residual ion polarities, D^- and D^+, in a simple way. For that reason, it could be thought as an alternative to ERID with some logical advantages. The ERID deflecting electrodes, in fact, should dump D^- and D^+ ions at full energy. The problems originated by this solution are studied and discussed in detail in ref. [3] as already mentioned. Essentially, they are due to the beam power to be dissipated on the ERID plates (which increases with the impact ion energy) and imposes strong water-cooling techniques. Furthermore, the plasma formation on the plates can perturb the deflecting electric field.

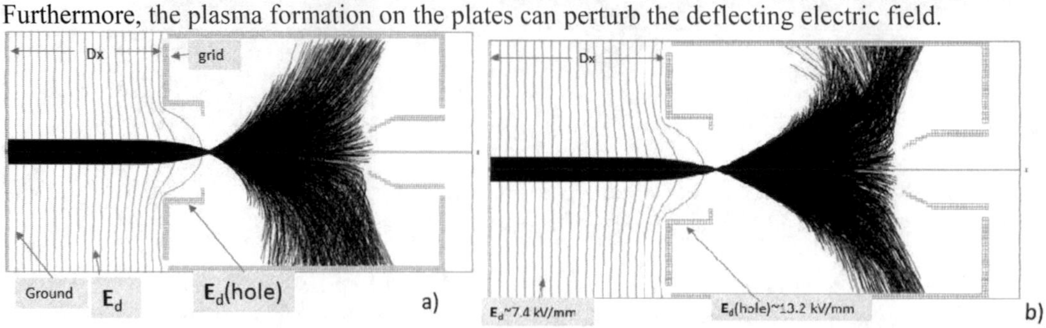

Figure 4 Ion trajectory simulations with a double sized Dx (66 mm): a) Ek_i=200 keV (very similar results of the case Ek_i=100 keV); b) Ek_i=500 keV which should be very similar to the case for Ek_i=1 MeV and Dx=280 mm.

On the other hand, the simulations carried out with SIMION on the simplified collectors have demonstrated that the ion charge collection efficiency could be maintained at 100% also for high Ek_i

(until to 1 MeV as in ITER) while the ion impact energy on the wall collector could remain lower of 1 keV. In fact, in the double stage simulations shown in fig. 5a), the potential value, Vc2, used to recover efficiently the D^+ ions, it is about 100 V lower than the |Vc1| value used for D^-. That means the D+ ions can be recovered with a final kinetic energy of about 100 eV higher than D^- ones.

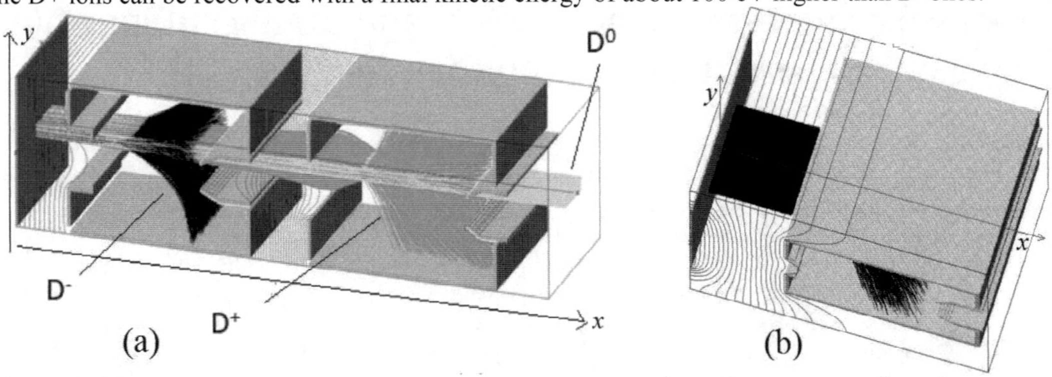

Figure 5 a) Simulation of the recovery for of D^- and D^+ ions with also neutral D^0, for Ek_i=100 keV with Vc1=-99250 V and Vc2=99400 V; b) ERID beam simulation for Ek_i=500 keV.

The double stage simulation of the compact simplified collector in fig. 5a) shows that this system could recover the beam residual ions in a very simple way and then it could be an alternative to ERID. A couple of electrode plates used in ERID, in fact, could be replaced by a double collector (for the double stage); so, the full ERID can be replaced by four system like fig 5a), stacked along y direction. Simulation results for the same beam 'aspect ratio' accepted by ERID for Ek_i=500 keV are shown in 5b). Those results show that all the ions can been collected on the walls at very low kinetic energy. These collectors, then, could be used instead of ERID without big changes in the NBI of ITER. This would allow to avoid the ERID plate complexity mentioned above. Furthermore, the energy recovery electrodes could also avoid the plasma formation since they would collect the ions at very low energy. However, further simulations with a more realistic geometry and with codes capable of more accurate SC calculations (like COMSOL and/or OPERA already used for comparison in ref. [9]) are needed to confirm the SIMION results. Moreover, the mentioned beam energy recovery experiment proposed on the NIO1, or TRIPS source beams should confirm the obtained simulation results.

Conclusion

The simplification of an ion beam energy recovery collector based on SC effect, capable to collect the residual ions at very low energies, proposed recently for DEMO, suggested that it could be used as an alternative system to ERID in ITER. That solution would avoid the power dissipation and plasma formation problems of ERID presented in [3]. Preliminary ion beam trajectory simulations carried out with the same ERID beam 'aspect ratio' have shown the full ion recovery at very low energies could be possible with the suggested beam recovery system. However, new simulations with more realistic geometry and accurate space charge calculations will need to confirm that results.

References

[1] ITER Design Description Document 5.3, N53 DDD 29 01-07-03 R0.1 2003 *IAEA*, Vienna.
[2] Technical Note - Mitica Beam Line Component, F4E_D_25FLDW, v1.0.
[3] Surrey E, Holmes A, McAdams R et King D, 2010 *Journal of Fusion Energy* **29**, 486–498.
[4] R. Mc Adams, 2014 *Rev. Sci. Instrum.* **85**, 02B319
[5] Variale V, et al., 2016 *Rev. Sci. Instrum.* **87** 02B305
[6] Variale V, et al., 2018 *AIP Conference Proceedings* **2052**, 070006
[7] Fagotti E, Palmieri A and Ren X, 2008 *Rev. Sci. Instrum.* **79**, 02B306
[8] Dahal D A, *SIMION manual 3D*, INEEL-95/0403
[9] Variale V, et al. 2017 *AIP Conference Proceedings*, **1869** 030033

Towards the optimization of the Cs evaporation configuration for long pulse operation of negative ion sources

A. Mimo, C. Wimmer, D. Wünderlich, M. Fröschle, U. Fantz
Max Planck Institut f. Plasmaphysik (IPP), Boltzmannstr. 2, 85748 Garching, Germany
E-mail: alessandro.mimo@ipp.mpg.de

Abstract. Negative ion sources for neutral beam injection rely on surface production of negative ions on a caesiated low work-function surface (plasma grid). To maintain the low work function in long pulses (one hour) and the desired source performance (extracted H^- /D^- ions and limited co-extracted electrons), Cs needs to be constantly delivered onto the plasma grid. The CsFlow3D code was applied to the RF driven negative ion source ELISE to simulate the evaporation and the plasma-assisted redistribution of Cs. The Cs flux stability is investigated for consecutive one hour plasma pulses and compared with experiments. The effect of different positions of the Cs oven for the case of ELISE equipped with two racetrack shaped RF drivers was investigated: a considerable increase of the Cs flux occurs when evaporating where the plasma is in contact with surfaces, as confirmed also by the back-tracking of the Cs particles reaching the PG. Additionally, an alternative evaporation method close to the plasma grid was designed with the help of CsFlow3D and tested in BATMAN Upgrade, with the aim of increasing the Cs flux onto the PG and achieve a better control of caesiation in long pulses.

1. Introduction

Neutral Beam Injection systems will be used in ITER to provide heating and current drive with a total power of 33 MW and 1 MeV energy in deuterium (0.87 MeV in hydrogen) [1]. These systems rely on large RF driven negative ion sources which need to deliver an extracted negative ion current density of 286 A/m^2 in deuterium for 1 hour and with co-extracted electron current below that of the negative ions. The RF prototype source at the BATMAN Upgrade test facility (1/8 of the ITER-NBI source size) [2] and the RF source at the ELISE test facility (half size of the ITER-NBI source size) [3] are currently in operation at IPP Garching. The full-size prototype SPIDER is in operation at NBTF, Padova [4].

In these sources negative ions are mostly produced by surface conversion of H/D atoms onto a metallic converter surface: the conversion efficiency is increased by reducing the work function of the surface through Cs evaporation [5]. The main limiting factor for the achievement of the required performances is represented by the amount of co-extracted electrons and their instability during long pulses, especially in deuterium [6]. This increase of co-extracted electrons during a pulse is attributed to a degradation of the work function of the plasma grid, i.e. the first grid of the extraction and acceleration system, which is facing the plasma and acts as a converter surface. The degradation of the work function depends on the high reactivity of Cs with the impurities (background pressure around 10^{-4} Pa, far from ultra-high vacuum condition), the complex Cs dynamics as well as the plasma induced removal of the deposited Cs layers [7]. The transport of Cs depends on the source geometry, the location and orientation of the Cs oven and the temperature of the surfaces.

To investigate the Cs dynamics the Monte Carlo test particle code CsFlow3D is used [8]. The code calculates the Cs fluxes (neutral and ions) onto and from the inner source surfaces, the Cs coverage and the neutral Cs density, which can be compared with Tunable Diode Laser Absorption Spectroscopy (TDLAS) experimental data [9]. TDLAS is routinely in operation at both IPP test facilities BATMAN

Upgrade and ELISE. Comparisons between modelling and experiments were performed for both the test facilities [10, 11].

The target of this work is gaining more insight on Cs dynamics in long pulses to understand if an optimisation of the Cs evaporation configuration can improve the long pulse stability in terms of co-extracted electrons. This is particularly relevant if the source will be upgraded by exchanging the four cylindrical drivers where plasma is generated with two racetrack shaped drivers, a solution proposed for the DEMO NBI with the advantage of RF problems such as breakdowns and matching problems and thus improve source reliability **[3].**

2. Effect of Cs ovens position for ELISE with racetrack shape drivers

Figure 1 shows on the left the source at ELISE as currently in operation, i.e. with four cylindrical drivers where plasma is generated and with two Cs ovens located on the side of the source evaporating towards the back-plate (standard reference for the simulations, indicated as *configuration A*). On the right, the CAD view of ELISE with two racetrack shaped drivers is shown, where possible alternative positions for the Cs oven are higlighetd, i.e. *configuration B* (oven on the sides, like the standard configuration), *C* (oven located on the top and bottom), *D* (a single oven located in the center of the source back-plate, with the same design of the one used in SPIDER, i.e. a pipe with six orifices on the sides [4]). A comparison of the Cs dynamics for long pulse between experiment and simulations were performed, and this reference compared with the alternative positioning of the Cs oven for the racetrack driver case.

In order to reach high performances several consecutive cycles of plasma pulses (~ 20 s) and vacuum phases (from 3 minutes up to 10 minutes) with constant Cs evaporation are performed in the experiment (conditioning process). Previous simulations showed that the conditioning process performed in experiments redistributes the Cs on the inner walls and increases the amount of Cs that can be transported towards the PG pulse after pulse, however, during a plasma pulse a decrease of the Cs density in times occurs due to a depletion of the Cs reservoirs on the source walls [10, 11].

To investigate if using long plasma pulses (~ 1 hour) could improve the beneficial effect of plasma redistribution, three identical plasma pulses of 1 hour were performed in ELISE with a 15 minutes vacuum phase in-between (P_{RF} = 35 kW/driver, filling pressure 0.3 Pa, T_{source} = 40 °C, T_{PG} = 125 °C, deuterium). The same cycle was simulated with CsFlow3D with a total oven evap. rate of 10 mg/h. Fig. 2 shows the comparisons between the measured Cs density in front of the plasma grid, along a horizontal line of sight in experiment (a) and simulations (b). A similar reduction of the density was observed for

Figure 1. Left: View of the ELISE source with cylindrical drivers, the location of the Cs oven is highlighted (standard case A); Right: ELISE source equipped with two racetrack drivers, configuration with the oven on the side (B), on the top and bottom (C) or with the SPIDER-like oven in the middle (D).

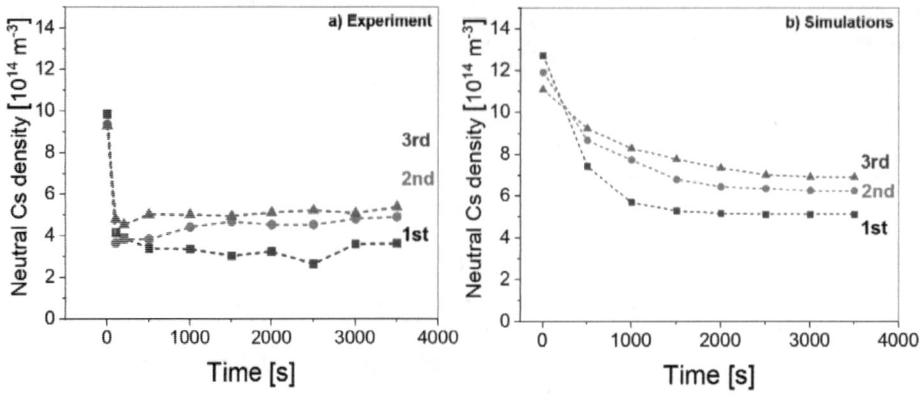

Figure 2. Neutral Cs density in front of the plasma grid at ELISE vs. time for three consecutive one hour pulses (with 15 min vacuum phase in between) for a) experiment b) simulations (total evap. rate 10 mg/h, T_{source} = 40° C, T_{PG} = 125 °C).

both experiment and simulations, however, the depletion occurs much faster in the experiment (within the first 100 s). It needs to be also specified that in the code the plasma parameters are assumed constant and the variation of them during the plasma turn on is not modelled: this can create and additional discrepancy between the code and the measurements before the steady state [10]. The absolute value of the Cs density is slightly lower in the experiment but still in good agreement with the simulations, considering that it strongly depends on the history of the source caesiation. An increase of the saturation value (after 500 s) of Cs density for the third pulse w.r.t. the first pulse is observable for both experiment and simulations (around 30 %). However, no significative reduction of co-extracted electrons was observed, indicating that the additional Cs is not enough to compensate for the work function degradation during the long pulse. Therefore alternative methods to increase the Cs flux onto the PG need to be investigated.

A change in the position of the Cs oven might reduce the Cs depletion and improve Cs management in long pulses. CsFlow3D was then used to simulate the effect on long pulses when changing the position and orientation of the Cs oven for the specific case of ELISE equipped with the racetrack driver. Figure 3 (a) shows the evolution of the total Cs flux (both neutrals and ions) averaged on the area of the inner beamlet group of the source during a 1 hour pulse. Fig. 3 (b) shows the relative contribution of the originating surfaces from which the Cs is transported towards the PG, obtained by backtracking the Cs particles. By comparing *A* and *B*, the effect of only changing from cylindrical to racetrack driver produced an increase of ~ 20% of the Cs flux delivered to the PG. This is associated with an increased contribution of the top and bottom side of the source: the horizontally elongated shape of the racetrack

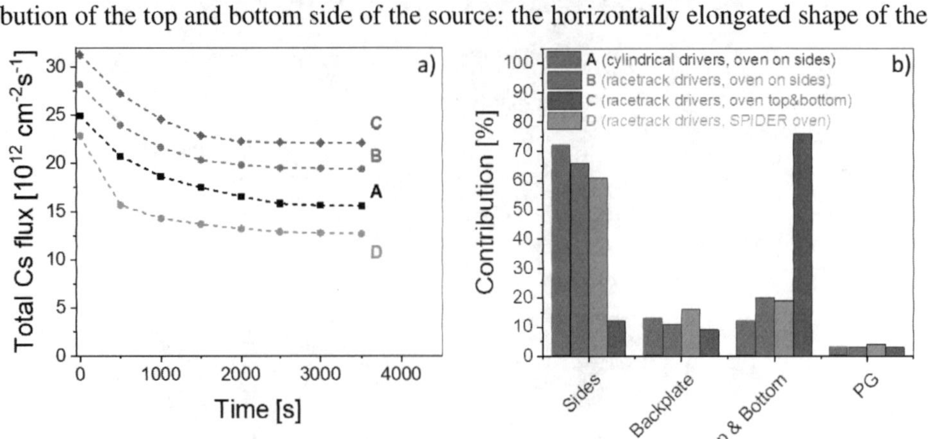

Figure 3. a) Cs flux for different oven configuration for three consecutive one hour pulses in the evaporation configuration shown in fig. 1; b) Contribution of the different surfaces on the Cs flux onto the PG for the different evaporation configuration shown in fig. 1 (total evap. rate 10 mg/h, T_{source} = 40° C, T_{PG} = 125 °C).

Journal of Physics: Conference Series **2244** (2022) 012057 doi:10.1088/1742-6596/2244/1/012057

driver provides an increased Cs removal by the plasma from the top and bottom sides. The Cs depletion is observable for all cases, being the worst for the case D of the SPIDER like oven (~45 % decrease in 1 hour pulse). The highest flux is achieved with the configuration C, i.e. when the evaporation towards the top and bottom side: in this case Cs is evaporated on the region where it is more likely to be removed and redistributed by the plasma exiting the racetrack driver (see the contribution of the top and bottom surfaces in fig. 3 (b)). The overall increase of the Cs flux is almost 50%, which considering the result discusses for fig. 1 might still not be sufficient for an efficient stabilisation of co-extracted electrons.

Additionally, CsFlow3D shows that in all cases the Cs flux consists for more than 70% of Cs ions, which are affected by the electric fields. In the experiment the PG is typically positively biased w.r.t. the plasma to limit the co-extracted electrons [13]: when this scenario is implemented in CsFlow3D a drastic reduction of the amount of Cs ions that can overcame the potential difference and reach the plasma grid can occur, as shown by previous calculations [11].

To avoid a 70% reduction of the Cs flux onto the PG and increase the effective flux towards the PG, an alternative Cs oven configuration relying on a distributed evaporation [14] of Cs close to the PG has been developed and tested at the smaller test facility BATMAN Upgrade, with the aim of extending the concept to larger sources as ELISE. By evaporating Cs close to the PG, the ionization probability of Cs is decreased, since the evaporation occurs in the region of the source with lower electron temperature.

3. Alternative Cs oven evaporation method and application at BATMAN Upgrade

Previous simulations showed that with evaporation close to the grid the Cs neutral flux is increased up to ~ 80% of the total Cs flux reaching the PG, compared with the 20 to 30% in the case of the standard evaporation configuration [11].

Due to the higher versatility and smaller size of BATMAN Upgrade, such an alternative Cs evaporation method (*Cs shower*) was tested at this test facility. The Cs shower, as shown in fig. 4, consists of a stainless steel pipe positioned around the plasma grid at an axial distance of 27 mm. The inner diameter of the pipe is 3 mm with a thickness of 0.5 mm. Cs is evaporated along each of the two vertical sections of the pipe through 2 × 27 equidistant orifices (0.4 mm diameter) oriented towards the PG. This geometry configuration is a result of numerical simulations oriented towards the achievement of an optimal coverage all over the PG and to ensure a uniform evaporation rate along the pipe (ratio between the diameter of the pipe and orifices to limit the pressure drop). The pipe is ohmically heated up to 300 °C to reduce as much as possible Cs sticking on the inner surfaces or orifices and it is connected to a temperature controlled Cs reservoir by means of an electrically controlled valve.

A dedicated campaign was performed with the TDLAS diagnostic positioned vertically, in-between the pipe and the PG (indicated by the red dashed line in the figure). Cs could be detected by TDLAS

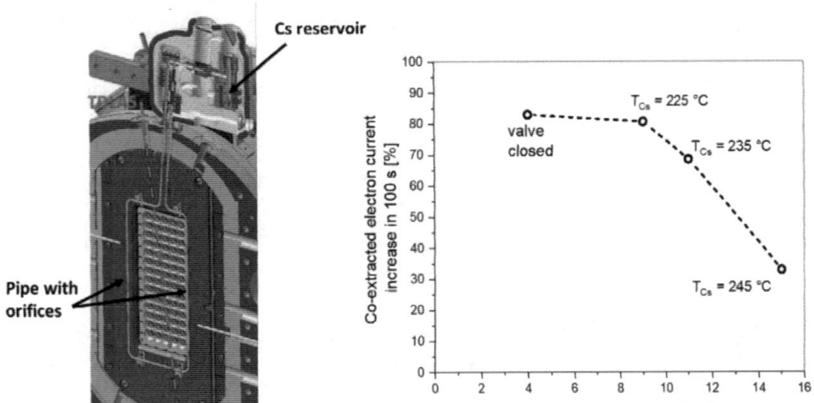

Figure 4. Left: CAD view of the Cs shower installed in BATMAN Upgrade; Right: percentage increase of co-extracted electron current during a 100 s pulse vs. neutral Cs density measured with TDLAS (leftmost point with Cs shower valve closed, other points with the Cs shower valve open and increasing Cs reservoir temperature)

between 10 to 20 s after the opening of the valve, proving a short reaction time of the system. A similar decay time is observed after the closing of the valve. The effect of the Cs shower on the source performance was tested by first conditioning the source with the standard oven and then performing a set of identical 100 s long pulses with different Cs reservoir temperature and by opening the valve at the beginning of the pulse [14]. The stability of the co-extracted electron current j_e for this set of pulses was then compared with a reference pulse with the Cs shower valve closed. Fig. 4 (b) shows the percentage increase of j_e as a function of the neutral Cs density measured by TDLAS during the plasma pulse. The value of j_e at the beginning of the pulse was similar for the four pulses. It can be observed that for densities higher than $8 \cdot 10^{14}$ m^{-3} (corresponding to a Cs reservoir temperature of 225 °C) j_e starts becoming considerably more stable, with the relative increase in 100 s dropping from 84% down to less than 30%. Relatively high reservoir temperature were needed to achieve the required flux and this was probably related to the presence of impurities in the source due to a water leak that was detected after the end of the campaign. Significantly lower temperature were achieved in the current campaign.
This proves that additional Cs flux can be provided by the Cs shower and it is effective in stabilising the co-extracted electrons. Ongoing tests are being performed to verify the vertical asymmetry of the evaporation and the high voltage holding in routine operation.

4. Summary and conclusions

The Cs dynamics simulations performed for long pulses at ELISE are in agreement with the experimental value obtained by TDLAS diagnostic and both show a strong depletion of Cs density in time. The effect of Cs oven location as a way to mitigate this depletion was investigated for the alternative racetrack driver concept in ELISE: a 50% increase of the flux and of the stability can be achieved by evaporating the oven in the top and bottom sides, where Cs redistribution is higher due to the elongation of the racetrack driver. However, the majority of Cs close to the plasma grid is ionised for all the explored cases, leading to an eventual reflection of the Cs ions away from the plasma grid when the plasma grid is positively biased. Therefore, an alternative evaporation method was designed and tested in the prototype source at BATMAN Upgrade and it showed that additional Cs flux is effective in stabilising the amount of co-extracted electrons. Further tests are needed before extending this concept to larger source such ELISE, to ensure the vertical uniformity of Cs evaporation.

Acknowledgments

This work has been carried out within the framework of the EUROfusion Consortium and has received funding from the Euratom research and training programme 2014-2018 and 2019-2020 under grant agreement No 633053. The views and opinions expressed herein do not necessarily reflect those of the European Commission.

References

[1] Hemsworth R *et al.*, Nucl. Fusion **49**, 045006 (2009)
[2] Kraus W *et al., AIP Conf. Proc.* **2052**, 040004 (2018)
[3] Fantz U, Hopf C, Wünderlich D *et al.*, Nucl. Fusion **57**, 116007 (2017).
[4] Toigo V *et al.*, *New Journal of Physics* **19,** 085004 (2017)
[5] Belchenko Yu I, Dimov G I, Dudnikov V G, Nucl. Fusion, **14**, 113 (1974)
[6] Wünderlich D *et al.*, *Nuclear Fusion* **61**, 096023 (2021)
[7] Gutser R, Wimmer C, Fantz U, *Rev. Sci. Instrum.*, **81** 02B102 (2010)
[8] Gutser R, Fantz U, Wünderlich D, *Rev. Sci. Instrum.*, **81** 02A706 (2010)
[9] Fantz U, Wimmer C, *J. Phys. D: Appl Phys.* **44**, 335202 (2011)
[10] Mimo A, Wimmer C, Wünderlich D, Fantz U, *AIP Conf. Proc.* **1869**, 030019 (2017)
[11] Mimo A, Wimmer C, Wünderlich D, Fantz U, *AIP Conf. Proc.* **2052**, 040009 (2018)
[12] Wünderlich D et al *Plasma Phys. Control. Fusion*, **60** 08500 (2018)
[13] Wimmer C, Mario I, Wünderlich D, Fantz U, *AIP Conf. Proc.* **2052**, 040003 (2018)
[14] Bansal G, *et al.*, *Rev. Sci. Inst.* **83**, 02B118 (2012)
[15] Wimmer C, et al. submitted to Journal of Physics: Conference Series, ICIS conference 2021

Journal of Physics: Conference Series **2244** (2022) 012058 doi:10.1088/1742-6596/2244/1/012058

A 2D semi-analytical electromagnetic model for cylindrical ICP sources with Faraday shield

P Chen, J Yang, L Zhou, C Zuo, D Li and D Chen

State Key Laboratory of Advanced Electromagnetic Engineering and Technology, Huazhong University of Science and Technology, Wuhan 430074, China

dzhchen@mail.hust.edu.cn

Abstract. Faraday shield(FS) is a metallic shield with slits through which electromagnetic energy is coupled into cylindrical ICP sources. Thus, the effect of FS on the RF field and power coupling is important. A 2D semi-analytical electromagnetic model considering both FS and plasma is proposed. The results compared to a 2D numerical model shows that the semi-analytical model is available in electromagnetic field calculations. It is found that both FS and plasma have no influence on the magnetic field outside them. FS strengthens the electric field around the slits and weakens the electric field around the middle of the stripes, and this effect is due to the gradient field generated by the electric charge accumulated on the FS surfaces. From the point of view of the pure 2D electromagnetic field, FS affects the electric field distribution in a small region, and it seems to not affect the total power coupling into plasma.

1. Introduction

The inductively coupled plasma (ICP) sources are used in RF ion sources for fusion and accelerator application[1, 2]. A sketch of a cylindrical ICP source is shown in figure 1. The dielectric tube which separates the solenoidal coil from the plasma is not shown since it has little effect on the field. The Faraday shield (FS), a thin-wall metal structure with slits, is placed between the coil and the plasma to reduce capacitive coupling[3]. Besides, in high-power RF ion sources, FS is placed in contact with the plasma to protect the dielectric tube from erosion[1]. The RF power from the coil is coupled to the plasma through the slits of FS, which makes FS an important component of ICP sources.

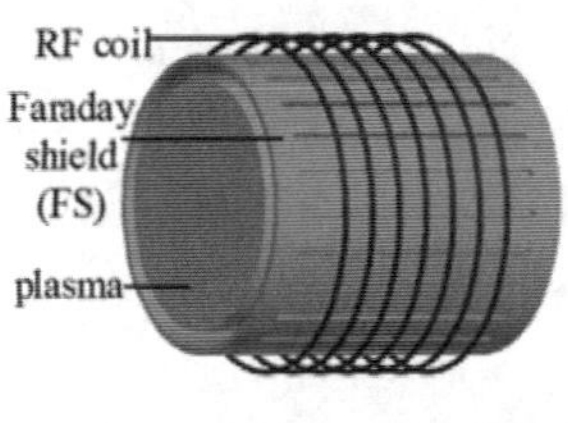

Figure 1. A sketch of a cylindrical ICP source. The dielectric tube is not shown.

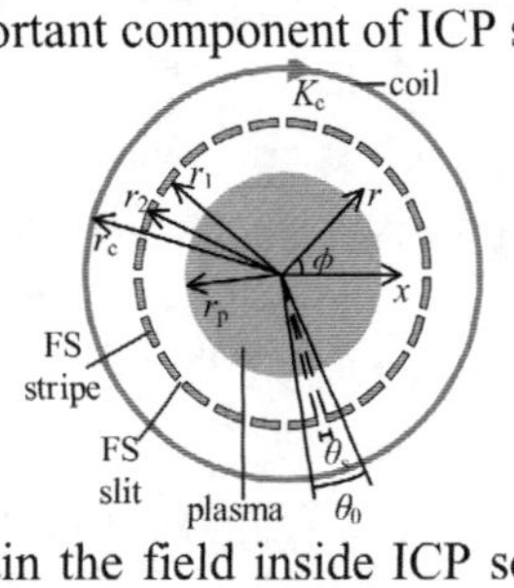

Figure 2. 2D model of a cylindrical ICP source. The dielectric tube is ignored.

There are some 2D electromagnetic models to obtain the field inside ICP sources with FS: semi-analytical one with FS and without plasma[4], numerical ones with FS and plasma approximated as a good conductor[1], and analytical one with infinitely thin FS and without plasma[5]. Based on the models with either plasma[3] or finitely thin FS[4], a 2D semi-analytical model considering both FS and plasma is proposed in this contribution. It won't give the field exactly but give qualitative results and physical insight, with less computing effort than numerical models. By this electromagnetic model considering both FS and plasma, the effect of FS on RF field and coupling are discussed.

2. Model

In 2D models, FS and/or plasma are assumed to be infinitely long straight, and the coil is assumed to be axially symmetrical and uniformly distributed. A sufficiently long straight ICP source can be simplified as a 2D model as shown in figure 2. The dielectric tube is ignored since it has little effect on the power coupling. The N-turns coil of radius r_c and length l, carries a current of rms value I_c. Thus, the surface current per unit length in figure 2 is $K_c = NI_c/l$. r_1, r_2, θ_0, and θ_s are the inner radius, outer radius, angle of a periodic sector, and angle of the silt of FS, respectively. The surface of FS stripes is denoted by S_{FS}. The plasma radius r_p is between 0 and r_1. Under RF excitation, the electromagnetic field in the ICP source is subjected to the Maxwell equations in phasor form

$$\nabla \times \boldsymbol{H} = (\sigma + \mathrm{j}\omega\varepsilon)\boldsymbol{E} \tag{1}$$

$$\nabla \times \boldsymbol{E} = -\mathrm{j}\omega\mu_0 \boldsymbol{H} \tag{2}$$

where $\boldsymbol{H}$ and $\boldsymbol{E}$ are the magnetic field strength and the electric field strength in the rms value phasor form. The initial phase of the coil current is set as 0. $\omega = 2\pi f$ is the driving angular frequency with f as the driving frequency. μ_0 is the vacuum permeability. σ and ε are conductivity and permittivity of different regions (FS stripe, plasma, vacuum containing FS slit), respectively.

The skin depth of copper FS is usually much smaller than the size of the FS stripe, so the FS stripe can be approximated as an ideal conductor. The plasma is treated as a lossy dielectric[3] characterized by the linear constitutive parameters $\sigma_{\text{plasma}} = \varepsilon_0 \nu_c \omega_{pe}^2/(\omega^2 + \nu_c^2)$ and $\varepsilon_{\text{plasma}} = \varepsilon_0\left(1 - \omega_{pe}^2/(\omega^2 + \nu_c^2)\right)$, where ε_0 is the vacuum permittivity. $\omega_{pe} = \sqrt{n_e e^2/\varepsilon_0 m_e}$ is the electron plasma frequency with n_e as the plasma electron density, m_e as the electron mass, and e as the elementary charge. ν_c is the collision frequency. When $\sigma_{\text{plasma}} \gg \omega\varepsilon_{\text{plasma}}$, viz. $\nu_c \gg \omega$, the plasma can be approximated as a good conductor, and then the quasi-stationary approach is applicable. The plasma material is assumed to be axisymmetric. For low-power sources, it is probably a safe assumption since FS is separated from the plasma by the tube. For high-power sources, the analysis presented in section 3 shows that FS affects the field at a very small depth for a typical discharge in a high-power source. Thus, this assumption may be available. Besides, the radial nonuniformity of plasma is ignored. The plasma is assumed to be unmagnetized. However, the RF magnetic field affects the power coupling[6, 7].

Now, the overall boundary value problem(BVP) of the 2D model in figure 2 is obtained. It is impossible to derive an analytical solution[5] and it is hard to obtain a numerical solution. To simplify the solution process, the overall BVP is solved semi-analytically as follows. The magnetic field is determined analytically first. From equations (1)-(2), the wave equation $\nabla^2 \boldsymbol{H} + k_c^2 \boldsymbol{H} = 0$ is obtained. $k_c = \omega\sqrt{\mu_0(\varepsilon - \mathrm{j}\sigma/\omega)}$ is the complex wavenumber. The magnetic field inside the ideal conducting FS stripe is fully screened by the eddy current on the surface of the FS stripe. Thus, the total magnetic field determined by the overall BVP is a superposition of the field of an ICP source without FS[3], and the fields of the surface currents on S_{FS}. Therefore, the wave equation is solved and the total magnetic field has only an axial component. $H_{z0} = K_c$.

$$H(r) = H_{z0} J_0(k_c r)/J_0(k_c r_p) \text{ inside plasma; } H_{z0} \text{ inside vacuum; 0 inside FS stripe.} \tag{3}$$

The total electric field is decomposed into two parts as shown in figure 3 in the way similar to [4]. The first part is a vortex field $\boldsymbol{E}_i$. It is defined as the electric field induced by the magnetic field as shown in figure 3(b) only. In figure 3(b), except $H = 0$ in the region $r_1 < r \le r_2$, the magnetic field is determined by equation (3). Therefore, $\boldsymbol{E}_i$ is axisymmetric and has only an azimuthal component. By the integral form of equations (2), $\boldsymbol{E}_{\mathrm{i}}$ is obtained analytically. The second part $\boldsymbol{E}_{\mathrm{c}}$ is defined by $\boldsymbol{E}_{\mathrm{c}}=\boldsymbol{E}-\boldsymbol{E}_{\mathrm{i}}$, and then approximated as a gradient field completely due to the electric charge accumulated on the surface of conductors. There is no free volume charge, so $\boldsymbol{E}_{\mathrm{c}}$ subjects to $\nabla\cdot\boldsymbol{E}_{\mathrm{c}}=0$. A scalar potential in the rms value phasor form can be introduced by $\boldsymbol{E}_{\mathrm{c}}=-\nabla\varphi$ and it satisfies $\nabla^2\varphi=0$. Due to the periodicity, this equation can be solved in a sector shown in figure 4, and it is set that $\varphi=0$, at OM, ON and MN. Then, the boundary conditions are obtained:

$$\varphi=\left[-\mathrm{j}\omega\mu_0 H_{z0}\left(r_{\mathrm{i}}^2-r_{\mathrm{p}}^2\right)/2+r_{\mathrm{p}}E_{\mathrm{i}}\left(r_{\mathrm{p}}\right)\right]\left(\phi\pm\theta_0/2\right),\ +\text{ at AB/BC/CD, }-\text{ at EF/FG/GH}. \quad (4)$$

All points of the gradient field are nearly in phase, so the quasi-static approach can be applied. After the gradient field is solved numerically, the total electric field is obtained. Then, The power deposited in plasma per unit length is $P_{\mathrm{plasma}}=\int_{\mathrm{plasma}}\sigma_{\mathrm{plasma}}\boldsymbol{E}\cdot\boldsymbol{E}^*\mathrm{d}S$. To characterize the ability of plasma to absorb the RF power[8], the plasma equivalent resistance (PER) is defined as $\mathrm{PER}=P_{\mathrm{plasma}}/I_{\mathrm{c}}^2$.

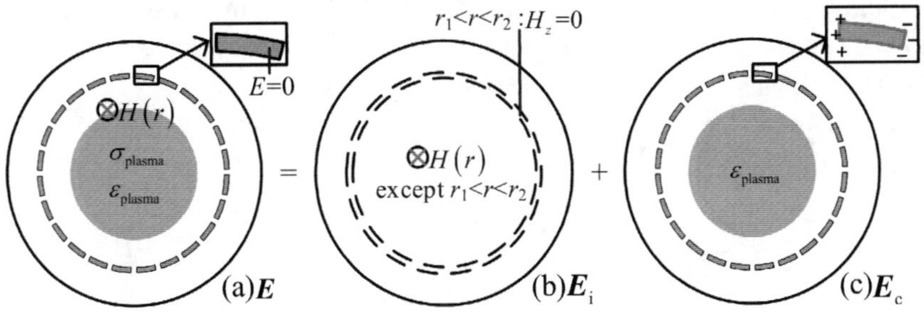

Figure 3. Decomposition of the electric field. (a) E subjects to the overall BVP. (b) The vortex field induced by H. (c) The rest field.

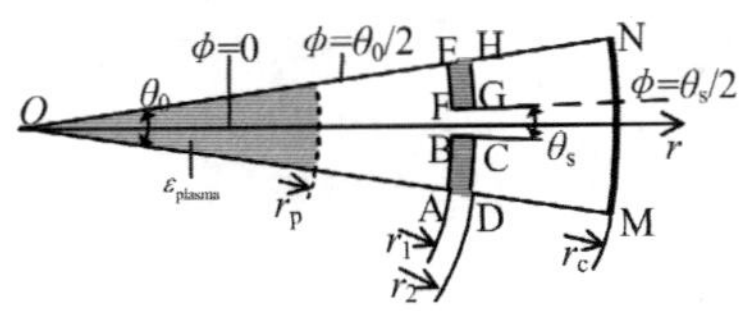

Figure 4. The solution domain of the gradient field. The capital letters denote points.

3. Results and discussion

The typical plasma in a high-power ICP source for fusion is a hydrogen discharge with $n_{\mathrm{e}}\approx10^{18}\ \mathrm{m}^{-3}$, electron temperature around 10 eV, f = 1MHz, and hydrogen gas density around $3.4\times10^{19}\ \mathrm{m}^{-3}$. Then, σ_{plasma} = 4.17e2 S/m and $\varepsilon_{\mathrm{plasma}}$ = −2.36e6 ε_0 according to the calculation process detailed in [9]. $N_{\mathrm{s}}=68$ and $w_{\mathrm{s}}=2$ mm are the number and width of the slit of FS, respectively. r_{c} / r_2 / r_1 / r_{p} = 133 / 119 / 116 / 115 mm. For the coil, N = 6, l = 72 mm, and I_{c} = 200 A.

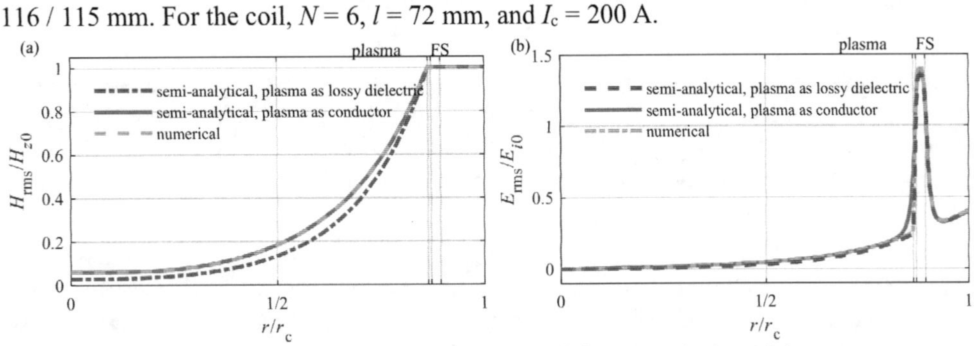

Figure 5. Field from the 2D semi-analytical model and the numerical model.
(a) $H(r)$ at Φ = 0. It has only z component. (b) $E(r)$ at Φ = 0. It has only Φ component.

First of all, the field from the 2D semi-analytical model and that from the 2D numerical model are compared in figure 5. The plasma is treated as a good conductor in this 2D numerical model, so two semi-analytical models treating plasma as a lossy dielectric and a conductor respectively are shown. As shown in figure 5(a), the magnetic field from the semi-analytical model (plasma as conductor) and that from the numerical model are in good agreement. As shown in figure 4(b), there are differences between the electric field from these two models treating plasma as a conductor around the plasma surface. The differences are due to the approximation in the decomposition of the electric field. The ratio of PER from the semi-analytical model (plasma as conductor) to PER from the numerical model is 1.10. Thus, the semi-analytical model is available in electromagnetic field calculations. Besides, the ratio of PER from the semi-analytical model (plasma as lossy dielectric) to PER from the numerical model is 0.81. This deviation is caused by the different treatments of the plasma mainly.

Figure 6 shows the field in four different situations, viz. with/without FS or plasma (treated as a lossy dielectric in the semi-analytical model). The plasma surface is denoted by black dash-dot lines in figure 6(a)-(b). E_{i0} in figure 6(b)-(c) is the electric field intensity at coil radius without plasma and FS. Figure 6(a) shows that neither FS nor plasma affects the magnetic field in the vacuum. The plasma screens the magnetic and electric field inside the plasma, and reduces the electric field in vacuum, as shown in figure 6(a)-(c). Figure 6(b)-(c) shows that FS strengthens the electric field around the slits and weakens the electric field around the middle of the stripes, and this effect is mainly due to the gradient field, viz. the charge accumulated on S_{FS}. Thus, the effect of FS on the field decays rapidly as the distance from S_{FS} increases. In the region slightly far from FS, the induction electric field is dominant and is the same as the electric field without FS.

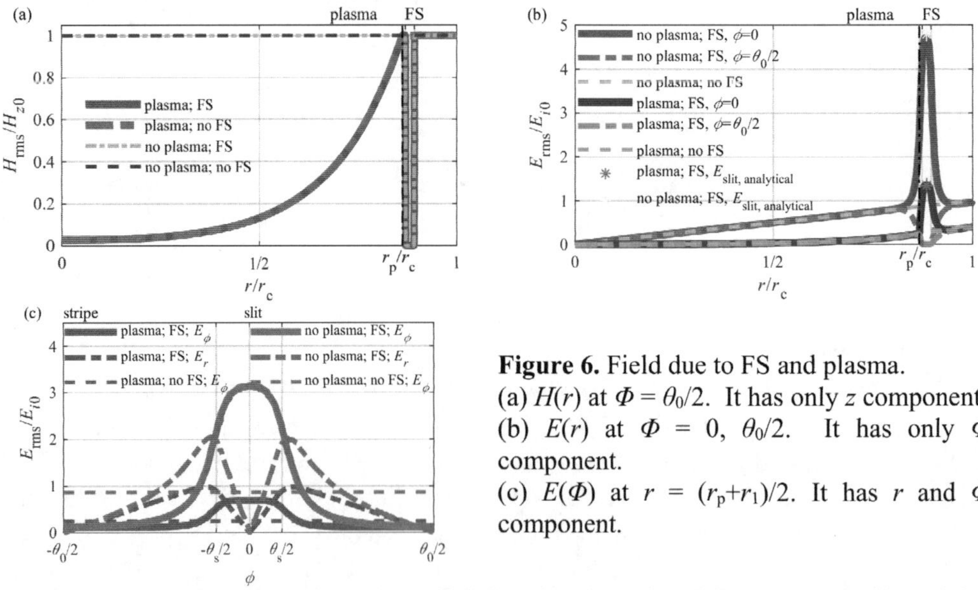

Figure 6. Field due to FS and plasma.
(a) $H(r)$ at $\Phi = \theta_0/2$. It has only z component.
(b) $E(r)$ at $\Phi = 0$, $\theta_0/2$. It has only Φ component.
(c) $E(\Phi)$ at $r = (r_{\mathrm{p}}+r_1)/2$. It has r and Φ component.

To further evaluate the effect of FS, the radial direction is analyzed first. The FS-affected depth is estimated in a way similar to [5]. Figure 6(b) shows the distance of the point where the gradient field exists from the FS inner surface, viz. the FS-affected depth. The FS-affected depth in plasma is much smaller than the skin depth of plasma which is around $0.17\,r_{\mathrm{c}}$. In the small FS-affected region, the electrons which gain energy are usually lost on S_{FS}. Thus, it seems that the effect of FS on RF coupling can be ignored. In [5], the FS-affected depth in vacuum (from a model without plasma) is used, which results in a larger value than the depth in plasma, as shown in figure 6(b). Besides, the point where the gradient field becomes weaker than the vortex field is used to estimate the FS-affected depth in [5], which results in a smaller value than the depth in vacuum here.

Then, the azimuthal direction is analyzed. The electric field inside the FS slit can be estimated analytically: the "electric voltage" $U_{\mathrm{slit}} = -E_i(r_1)2\pi r_1/N_{\mathrm{s}}$, and then the electric field $E_{\mathrm{slit,\ analytical}} = U_{\mathrm{slit}}/w_{\mathrm{s}}$ [4]. $E_{\mathrm{slit,\ analytical}}$ denoted by asterisks are in good agreement with the results from the semi-analytical model. Thus, it seems that the induction electric field at FS radius is squeezed into the slits by FS, while the integral of the electric field along the azimuthal direction remains unchanged.

Moreover, the ratio of PER with FS to PER without FS is 1.0. In summary, from the point of view of pure electromagnetic field, FS affects the electric field distribution in a small region, and it seems to not affect the total power coupling into plasma. Note that E_z in the region $r < r_1$ is screened by FS and out of the scope of 2D models. It is reported that the results from an axisymmetric fluid model without FS are in good agreement with those from experiments with FS[7], which also indicates that FS may have little effect on the total power coupling into plasma. The power deposited in FS due to eddy current and particle bombardment is out of the scope of the discussion here.

4. Conclusion

Based on the electromagnetic models considering only plasma or FS, a 2D semi-analytical electromagnetic model considering both FS and plasma is proposed. The model is applied to a high-power ICP source for fusion with a typical plasma. The results compared to a 2D numerical model shows that the semi-analytical model is available in electromagnetic field calculations. It is found that both FS and plasma have no influence on the magnetic field outside them. FS strengthens the electric field around the slits and weakens the electric field around the middle of the stripes, and this effect is mainly due to the gradient field, viz. the charge accumulated on the FS surface. The effect of FS on the field decays rapidly as the distance from the FS surface increases. The FS-affected depth in plasma is much smaller than the skin depth of plasma. From the point of view of a pure 2D electromagnetic field, FS affects the electric field distribution in a small region, and it seems to not affect the total power coupling into plasma.

Acknowledgments

This work was supported by the National Key Research and Development Program of China [grant numbers 2017YFE0300105]; the National Natural Science Foundation of China [grant numbers 11775088, 11705063].

References

[1] TOIGO V *et al* 2021 On the road to ITER NBIs: SPIDER improvement after first operation and MITICA construction progress, *Fusion Engineering and Design.* **168** 112622
[2] FAIRCLOTH D *et al* 2018 An overview of negative hydrogen ion sources for accelerators, *New Journal of Physics*. **20** 025007.
[3] CHABERT P *et al* 2011 Physics of Radio-Frequency Plasmas, Cambridge University Press
[4] ZHANG L *et al* 2016 A 2D semi-analytical model for Faraday shield in ICP source, *Fusion Engineering and Design.* **103** 74-80
[5] AKHMETOV T D *et al* 2021 On the Influence of a Faraday Shield on the Vortex Electric Field in a Radio-Frequency Plasma Generator, *Plasma Physics Reports.* **47** 732–741.
[6] ZIELKE D P *et al* 2021 Self-consistent fluid model for simulating power coupling in hydrogen ICPs at 1 MHz including the nonlinear RF Lorentz force, *Plasma Sources Science and Technology* **30** 065011
[7] ZIELKE D P *et al* 2021 Predictive fluid model for self-consistent description of inductive RF coupling in powerful negative hydrogen ion sources, in *19th International Conference on Ion Sources*
[8] Kralkina E *et al* 2016 Rf power absorption by plasma of a low-pressure inductive discharge, *Plasma Sources Science and Technology.* **25** 015016
[9] Cazzador M 2014 Analytical and numerical models and first operations on the negative ion source NIO1, *MSc Thesis* University of Padua, Padua

Study of the charged particle flow near the plasma grid in negative ion source

S. Masaki[1], H. Nakano[1,2], E. Rattanawongnara[1], K. Nagaoka[2,3], K. Ikeda[2], Y. Fujiwara[2], M. Osakabe[1,2], M. Wada and K. Tsumori[1,2]

[1]The Graduate University for Advanced Studies, SOKENDAI, Toki, Gifu 509-5292, Japan
[2]National Institute for Fusion Science, Toki, Gifu 509-5292, Japan
[3]Graduate School of Science, Nagoya University, Nagoya, Aichi 464-8603, Japan
[4]Graduate School of Science and Engineering, Doshisha University, Kyotanabe, Kyoto 610-0394, Japan

Corresponding author: masaki.shingo@nifs.ac.jp

Abstract. Flows of the charged particles in hydrogen (H) and deuterium (D) plasmas were measured with the single-tip directional Langmuir probe. The flow patterns of positive and negative ions have been constructed from the data by moving the probe near the plasma grid. The positive ion flow from driver to extraction regions showed the different velocity equivalent to the mass difference between H and D. The negative ion flow indicated the surface production of negative ions on the plasma grid, and changed the direction corresponding to the magnitude of the extraction electric field.

1. Introduction

High power beams of negative hydrogen (H^-) and deuterium (D^-) ions are indispensable to realize efficient neutral beam injector (NBI) systems for heating and current drive in fusion plasmas [1]. Negative ion beam extraction from ion sources accompanies co-extraction of electrons which causes heat load on the extraction grid. Comparing with H^- beam, D^- beam extraction shows higher co-extracted electron current with the same condition of the ion source operation. To avoid damaging the extraction grid with the high intensity of the electron current, the discharge power to generate the source plasma and the beam energy is limited in the D operation. This restriction results in the lower intensity of the negative ion beam current of D with respect to H [2], while the density of D^- ion near the plasma grid (PG) is higher than that of H^- ion [3].

The higher co-extracted electron current in the deuterium operation can be originated from the flow of charged particles toward the extraction aperture on the PG. Therefore, the flow measurement near the PG is motivated in H and D discharges to understand the difference in the plasma behavior between the hydrogen isotopes. The flow of positive ions has been measured with ion saturation current with a directional Langmuir probe [4]. For negative ions, laser photodetachment method is applied to measure the flow in the plasma [5,6]. To obtain the flow data of positive and negative ions simultaneously, the Single-tip Directional Langmuir Probe (SDLP) has been utilized in the National Institute for Fusion Science (NIFS) [7,8]. In this paper, we focused only positive and negative ion flows, and report the results of the flow measurement with SDLP in H and D plasmas.

2. Experimental setup

2.1. Negative ion source

The charged particle flows are measured in H and D plasmas generated in the NIFS R&D negative ion source (NIFS-RNIS). The volume of the ion source consists of 350 mm width, 700 mm length and 230 mm depth. A DC discharge is driven by the arc voltage (V_{arc}) of 80 V between hot tungsten filaments and the ion source wall. All data reported in this paper are obtained with the arc power of 50 kW in H_2 or D_2 gas pressure of 0.3 Pa. The generated plasma spreads to the extraction region across the filter magnetic field. The external voltage (V_{ex}) positively biased on the extraction grid with respect to the ion source wall takes out negative charged particles in the extraction region. In H operations, V_{ex} of 3 kV is applied as the potential realizes the optimum beam perveance of the NIFS-RNIS, while only up to 110 V can be applied on the extraction grid for preventing the generation of neutron radiation in deuterium operation. Electron deflection magnets (EDM) embedded in the extraction grid bent the trajectory of co-extracted electrons. Cesium (Cs) ovens are mounted on the backplate of the ion source and seed Cs vapor to the chamber wall to enhance the negative ion density and reduce the electron density. The flow measurement in this study was conducted during the Cs conditioning phase; H^- ion density is 1.4×10^{17} m^{-3}, D^- ion density was 1.7×10^{17} m^{-3}. Figure 1 shows a cross-sectional diagram of the NIFS-RNIS.

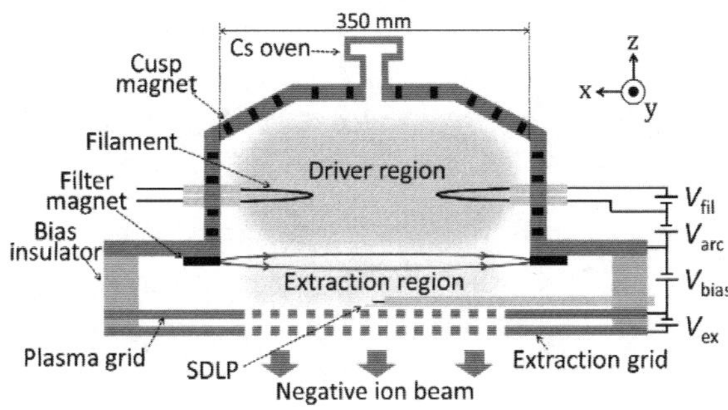

Figure 1. The cross-sectional diagram of the NIFS-RNIS.

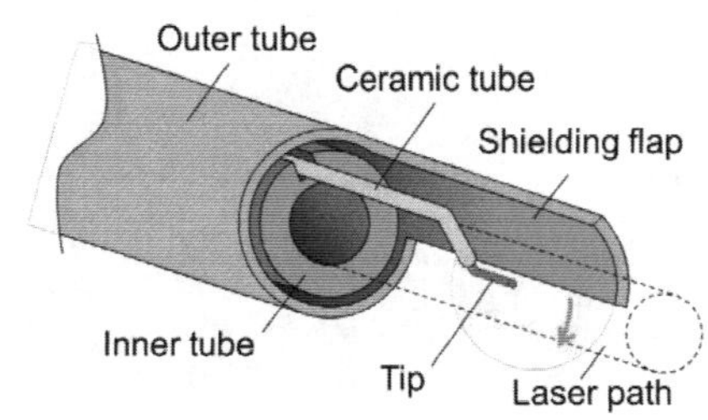

Figure 2. The structure of the SDLP.

2.2. Single-tip directional Langmuir probe (SDLP)

The flows of positive and negative ion are obtained with the SDLP mounted on the bias insulator. The structure of the SDLP is displayed in Fig. 2. The SDLP consists of 2 tubes: outer and inner tubes. The outer tube has 9 mm outer diameter and 8 mm inner diameter. A shielding flap equipped with the outer tube interrupts the particle flow to the probe tip. The shielding flap has 4 mm width and covered by ceramics to float the flap surface electrical potential. By rotating the outer tube, the flap can rotate around a tip co-axially. The rotating flap angle is detected with combination of a slit plate attached on the outer tube and a photocoupler installed in the other side of the flap. The molybdenum probe tip which has 0.3 mm diameter and 2 mm length is fixed with the inner tube. The inner tube has 7 mm outer diameter and 4 mm inner diameter. The probe detects photodetachment signals by irradiating Nd:YAG laser to the tip through the inner tube. The laser has 1064 nm wavelength and 50 mJ/cm^2 pulse energy density. The probe voltage is swept at 20 Hz to measure the *I-V* curve. The sweep timing of the probe voltage biased with respect to the ion source wall is synchronized with the laser trigger. The photodetached electrons are detected with 40 V which is higher than the plasma potential of ~3 V in this experiment. The position of the probe tip can sweep a square region with sides of 20 mm to obtain the flow pattern near the PG. Relative position of the measurement region with respect to the extraction apertures is shown in Fig. 3.

The flow speed of positive ions (v_{f+}) is estimated with Mach probe theory. The positive saturation current densities facing to the upstream and the downstream of the flow are used to calculate the Mach number [9]. In the measurement using the SDLP, the *I-V* curve shows the maximum density (j^+) and the minimum density (j^-) of the positive saturation current, when the flap faces to the downstream and the upstream of the flow, respectively. The flow speed is given as following equation.

$$v_{f+} = \frac{T_e}{2}\sqrt{\frac{k}{2m_i T_i}}\ln\left(\frac{j^+}{j^-}\right) \quad (1)$$

In this study, the electron temperature (T_e) is assumed as 0.8 eV, which is measured without Cs condition. The positive ion temperature is supposed to be same as the filament temperature: 0.3 eV. The flow direction of the positive ion is calibrated with the method of ref. [7] taking into account of the effect due to the magnetic field.

Negative ion flow is derived with recovery time of the photodetachment signals following the methodology in ref. [5]. In this study, the recovery time is associated with the shielding flap angle to derive the flow direction. A photodetachment signal is shown in Fig. 4. The recovery time (τ) indicates the time until negative ions reach the probe tip from the outside of the laser path. Laser beam radius (R_L) and the probe tip radius (R_p) give the negative ion recovery speed (v_f) with τ: $v_r = (R_L - R_p)/\tau$. In the same way as the positive ion flow evaluation, v_f takes the maximum and the minimum when the flap facing to the downstream and upstream of the negative ion flow, respectively. Thus, v_r depends upon the correlation between the flow direction (θ_{flow}) and flap angle (θ); $v_r = v_{th} + v_{f-}\cos(\theta - \theta_{flow})$, where v_{th} and v_{f-} the are thermal velocity and the flow speed of the negative ion, respectively. Figure 5 shows polar diagram of the positive saturation current and the recovery speed of negative ion.

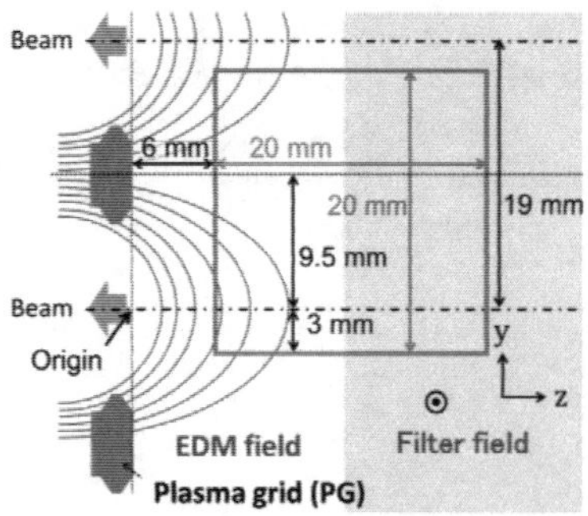

Figure 3. The measurement region.

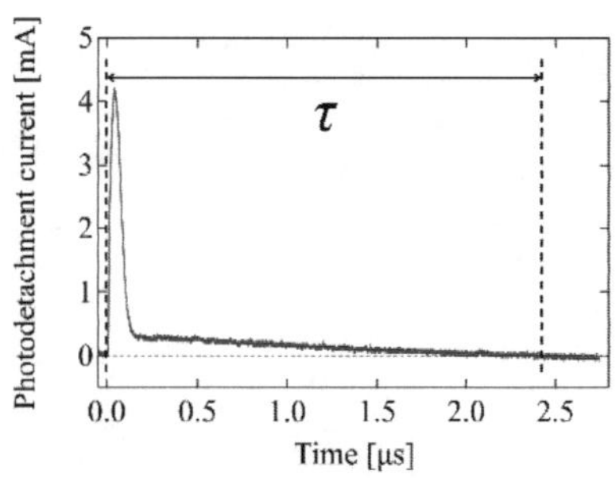

Figure 4. Photodetachment signal.

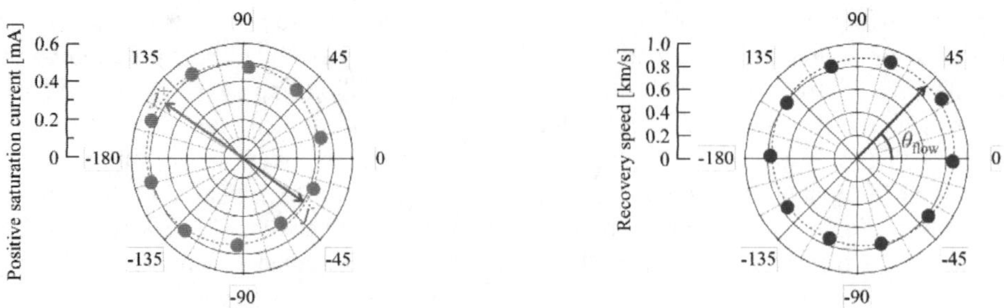

(a) (b)

Figure 5. The polar diagrams: (a) positive saturation current, (b) negative ion recovery speed.

3. Result

Positive ion flows measured in H and D plasmas with the output voltage of the power supply for PG bias (V_{bias}) of 2.4 V are shown in Figs. 6 and 7. The flows were detected locally at 6×6 lattice points in the measurement region. The flow patterns are drawn by linear scaling. The vertical position in the left axis of the figures indicates the distance from center of the extraction aperture on the PG. As the result, positive ions globally flow toward the PG and the top side of the ion source for both H and D plasmas. The averaged velocities decomposed into y and z components, $\overline{v}_y$ and $\overline{v}_z$, are listed in Table 1 when the vertical direction is defined as y and the direction perpendicular to the PG plane is defined as z.

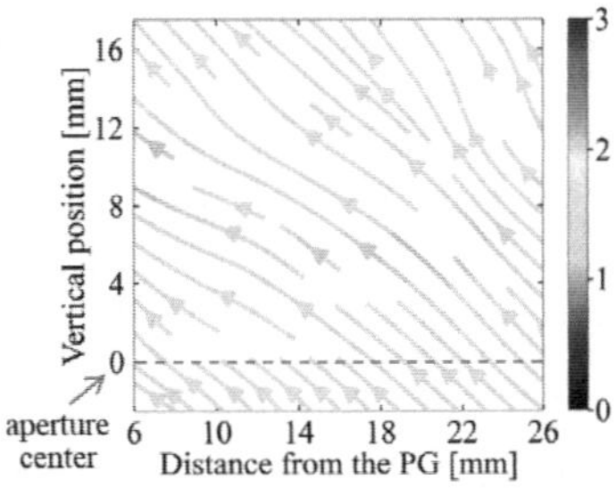

Figure 6. Flow patterns of the positive ion in H.

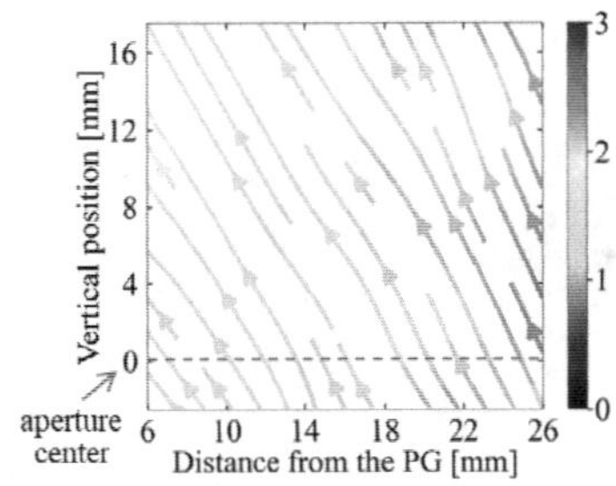

Figure 7. Flow patterns of the positive ion in D.

Table 1. The averaged velocity components of the positive ion flow.

Gas	$\overline{v}_y$ [m/s]	$\overline{v}_z$ [m/s]
H	854	-886
D	1046	-592

The $\overline{v}_z$ can be driven by the flow from the driver region because positive ions are generated in the driver region and flow into the extraction region. Assuming the positive H and D ions flow with the same energy, the flow velocity of the positive H ions is $\sqrt{2}$ times that of the positive D ions due to the mass difference, while the ratio of $\overline{v}_z$(H) to $\overline{v}_z$(D) is 1.5. Thus, the positive ions can flow with the same energy between H and D. The origin of $\overline{v}_y$ can be attributable to Larmor motion induced by the magnetic field in the direction perpendicular to the measurement plane. Positive ions rotate clockwise with ~10 mm radius in the measurement region while the mean free path of coulomb collision between ions (λ_{ii}) is ~8 mm (calculated with the measured positive ion density in H, 3.5×10^{17} m^{-3}). This indicates that positive ions coming from the driver region turn the direction to the top as Larmor rotation, then collide with the other ions before it completes the gyration. One of the possible reasons for the higher $\overline{v}_y$ in D compared to H is ambipolar diffusion. Measurement of I-V curve indicates that the D plasma has higher electron density compared with the H plasma. The flow of the positive ion can be affected by the flow of electron, which has higher mobility than ions. Thus, the positive ion in D plasma can flow faster than that in H plasma. The measurement of the electron flow is required to verify the difference in $\overline{v}_y$ between H and D.

Flow patterns of negative ions were obtained with the same condition of the positive ion flow measurements. The V_{ex} of 110 V was applied to investigate the effect of external voltage on the negative ion flow near the PG. Figures 8 and 9 show the flow patterns of the negative ion (a) before and (b) during the V_{ex} application in H and D plasmas. Note that each figure has different color scale of the flow speed. The averaged velocities $\overline{v}_y$ and $\overline{v}_z$ obtained from the flow before the V_{ex} application are shown in Table 2. According to the result without V_{ex}, H^- and D^- ions flow from the PG direction. This indicates that negative ions produced on the PG surface appeared in the measurement region. Assuming the flow direction is affected with the magnetic field, the global flow patterns of negative ions, except for some local flow structure, should have the same direction of the positive-ion flow. The flow direction of negative ions, however, are opposite to that of positive ones. The reason is not clear in this paper and is one of the objects in future. The subtracted flow from (b) to (a) is displayed in (c) which shows change of the flow by applying the V_{ex}. The difference in the flow patterns between the conditions without and applying the V_{ex} shows the flow trend toward the PG. Thus, D^-

ion distributed near the PG can respond V_{ex} of 110 V, which enables as to study optimization of the negative ion source without radiation problems.

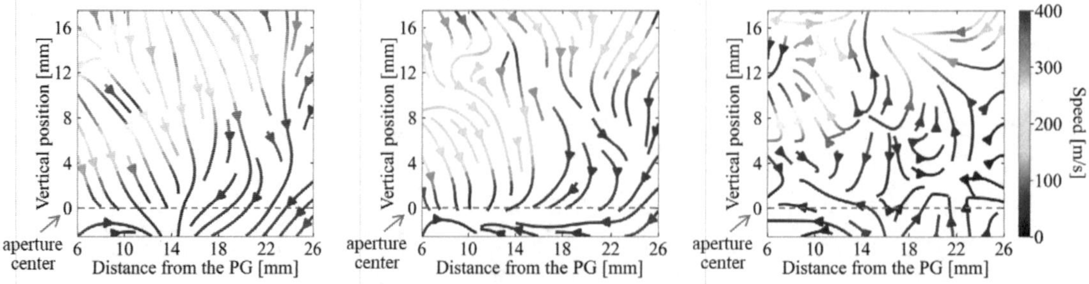

(a) (b) (c)

Figure 8. Flow patterns of H^- ion: (a) w/o V_{ex}, (b) w/ V_{ex}, (c) change from w/ V_{ex} to w/o V_{ex}.

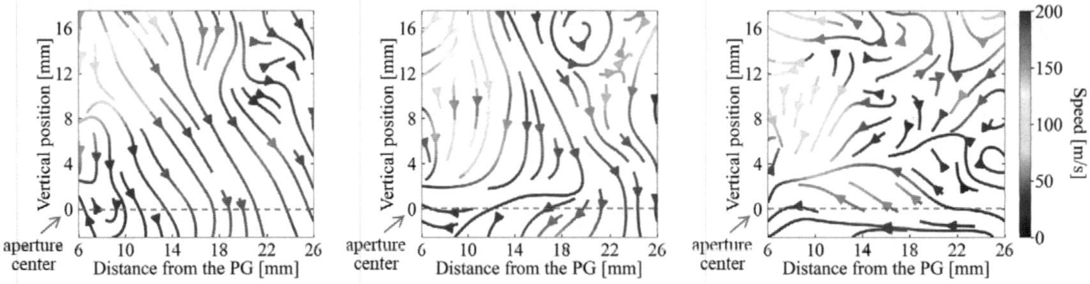

(a) (b) (c)

Figure 9. Flow patterns of D^- ion: (a) w/o V_{ex}, (b) w/ V_{ex}, (c) change from w/ V_{ex} to w/o V_{ex}.

Table 2. The averaged velocity components of the negative ion flow.

Gas	$\overline{v_y}$ [m/s]	$\overline{v_z}$ [m/s]
H	-106	31
D	-37	14

4. Conclusion

The charged particle flows were measured with the SDLP in the 20×20 mm square region near the PG. The flow data suggests that positive H and D ions flow from the driver region to the extraction region of the ion source with the same energy. The flow component of the positive ions toward the top side of the ion source can be explained with shorter collision mean free path of the positive ions than Larmor radius. The flow patterns of the negative ion indicates that both H^- and D^- ion flow from the PG where negative ions are produced on the surface. The effectiveness of V_{ex} of 110 V to observe the effect onto the negative ion flow was confirmed by subtracting the flow patterns. Investigations of the negative ion source for D^- ion beam extraction can be simulated with V_{ex} low enough to be free from nuclear radiation.

Acknowledgments

The authors would like to thank the technical staffs of the LHD-NBI group for supporting the experiment. This research is supported by the NIFS budget No. ULRR702 and JSPS KAKENHI Grant Number 19H01883.

References

[1] R. S. Hemsworth, D. Boilson, P. Blatchford, M. D. Palma, G. Chitarin, H. P. L. de Esch, F. Geli, M. Dremel, J. Graceffa, D. Marcuzzi *et al.* 2017 New J. Phys. **19**, 025005
[2] K. Ikeda, K. Tsumori, M. Kisaki, H. Nakano, K. Nagaoka, M. Osakabe, S. Kamio, Y. Fujiwara, Y. Haba, Y. Takeiri *et al.* 2018 AIP Conference Proceedings **2011**, 060002
[3] S. Masaki, H. Nakano, M. Kisaki, Y. Haba, K. Nagaoka, K. Ikeda, Y. Fujiwara, M. Osakabe and K. Tsumori 2020 Rev. Sci. Instrum. **91**, 013512
[4] M. Hudis and L. M. Lidsky 1970 J. Appl. Phys. **41**, 5011
[5] S. Kado, S. Kajita, T. Shikama, Y. Iida, D. Yamasaki, A. Okamoto and S. Tanaka 2006 Contrib. Plasma Phys. **46**, 367
[6] S. Geng, K. Tsumori, H. Nakano, M. Kisaki, K. Ikeda, M. Osakabe, K. Nagaoka, Y. Takeiri, M. Shibuya and O. Kaneko 2016 Rev. Sci. Instrum. **87**, 02B103
[7] S. Masaki, H. Nakano, M. Kisaki, E. Rattanawongnara, K. Nagaoka, K. Ikeda, Y. Fujiwara, M. Osakabe and K. Tsumori 2021 AIP Conference Proceedings **2373**, 090003
[8] K. Tsumori and M. Wada 2021 Appl. Phys. Rev. **8**, 021314
[9] K-S. Chung, I. H. Hutchinson, B. LaBombard and R. W. Conn 1989 Phys. Fluids B **1**, 2229

Journal of Physics: Conference Series **2244** (2022) 012060 doi:10.1088/1742-6596/2244/1/012060

Difference of co-extracted electron current and beam acceleration in a negative ion source with hydrogen-isotope ions

K. Ikeda[1], K. Tsumori[1,2] H. Nakano[1,3], K. Nagaoka[1,3], Y. Takeiri[1], S. Masaki[2] , E. Rattanawongnara[2] and M. Osakabe[1,2]

[1] National Institute for Fusion Science (NIFS), National Institutes of Natural Sciences, Toki, 509-5292, Japan
[2] The Graduate University for Advanced Studies, SOKENDAI, Toki 509-5292, Japan
[3] Graduate School of Science, Nagoya University, Nagoya 464-8603, Japan

E-mail: `ikeda.katsunori@nifs.ac.jp`

Abstract.
Improvement of the performance on a hydrogen/deuterium negative ion source for a nuclear fusion device is reported. In particular, the suppression of the co-extracted electron current, I_e, is an important issue to ensure the stable beam acceleration. Improvement of the I_e has been confirmed by optimizing the magnetic field of the electron deflection magnet in the extraction grid. Two other new methods for reduction of the I_e were validated. The first was an electron fence whose rods were set between the rows of apertures on a plasma grid. The electron and negative ion current ratio, approximately I_e/I_{acc}, was greatly improved from 0.7 to 0.25 in deuterium. The second was an outer iron yoke which enhanced the magnetic flux density 19% inside the arc discharge chamber. The I_e/I_{acc} using the outer yoke decreased by 0.1 compared with using a normal magnetic filter in a deuterium operation. These attempts have improved the total deuterium injection beam power of 8.4 MW by three negative ion based NBIs.

1. Introduction

High-energy and high-current neutral beam injection (NBI), using negative ion sources called N-NBI, is an important tool for efficient plasma heating and current drive, and for delivering fuel particles in the field of plasma-based fusion. These fusion devices require hydrogen isotopes such as deuterium (D) and tritium (T). N-NBI systems have been practically applied in the large helical device (LHD) [1–5], the JT-60U [6, 7] and the JT-60SA [8] where stable high-energy deuterium beams have been used for plasma experiments. The ITER tokamak, which is under construction in France [9], will use N-NBI to produce 1 MeV deuterium negative ion beams [10–12]. Tritium and deuterium as a fuel are essential for fusion plants, however the exsisting amount of tritium is limited on the earth [13]. On the other hand, tritium is produced by DD fusion reactions in the reactor, therefore it is desirable to reuse DT molecules in the evacuation gas, as a mixed isotope beam source [14]. For these reasons, the efficiency of negative ion production of hydrogen isotopes, the difference of isotope beam acceleration, and the suppression of co-extracted electron current are important issues in negative ion source engineering.

Three N-NBIs called BL1, BL2 and BL3 for plasma production and heating are operated in the LHD; their negative ion sources are optimized for hydrogen (H) [2]. Both hydrogen and deuterium beam injection have been carried out for deuterium experiments in the LHD [15,16]. It is possible to compare production of negative ions and their beam characteristics between hydrogen isotopes using a hydrogen optimized negative ion source. Therefore, the following four points of the isotope beam are interesting subjects. (i) Production of isotope negative ions and the extraction property of the negative ion current. (ii) Differences in stripping loss in the accelerator. These topics have a direct effect on the output beam power. (iii) Optical property of the isotope beam, which affects the heat load on the injector equipment and the port through efficiency. (iv) Co-extracted electron current [I_e], which is unique to negative ion sources and is very important in terms of preventing damage to the grid system by electrons.

This paper presents the history of the maximum neutral beam injection power of H and D beams on the NBIs in the LHD. Injection beam power in deuterium negative ions produced by a hydrogen optimized negative ion source decreased due to decreasing the negative ion current, increasing the stripping loss rate, and expansion of the beam width. The high power arc discharge required for high power beam generation is limited by the increase in the I_e. Two methods of effectively reducing the I_e will be reported.

2. Performance of hydrogen isotope beam injection

2.1. Injected beam power

The nominal acceleration drain current [I_{acc}] of hydrogen negative ions and the beam energy [E_{beam}] for the N-NBI in the LHD are 80 A and 180 keV, respectively, by two arc discharge type negative ion sources. The extracted beam power of negative ions is 14.4 MW, however the nominal beam injection power is 5 MW because the port through efficiency including neutralization and re-ionization is approximately 0.35. The standard beam pulse width is 2 seconds. Deuterium beam operation was initiated in 2017 using hydrogen optimized negative ion sources by switching the operation gas between H_2 and D_2. The hydrogen beam was used as an important heating source in high ion temperature plasma experiments, and the deuterium beam was used as a driver for neutron production in DD fusion experiments.

Figure 1 shows the history of the maximum port through power of both H (opened marks) and D (closed marks). The circle, square, and triangle marks indicate the values of BL1, BL2, and BL3, respectively. Maximum deuterium beam power improved from 7 MW to 8.4 MW in the latest experimental campaign. The negative ion source of TOSHIBA-type at BL1 uses a hexagonal arc chamber, which has good arc power efficiency on the hydrogen negative ion current [17]. The beam power of 6.2 MW in hydrogen was injected. However, the deuterium beam power was significantly reduced because of the increase in the I_e, which is defined by subtracting the I_{acc} from the extraction current [I_{ext}]. An electron fence performed well to suppress the I_e, and the injection beam power improved 3 MW in 2020. BL2 and BL3 are equipped with HITACHI-type ion source which uses a rectangle cross section arc chamber [2]. The two types of ion sources produce beams of the same size. The injection power of the hydrogen beam at BL2 reduced to 3.4MW, due to the decrease in the port through efficiency below 0.3 after the deuterium operation. The deuterium beam injection power was 2.4 MW in the latest campaign. On the other hand, the tangential injection direction of BL2 is opposite to the direction of BL1 and BL3. Therefore a stable supply of high power hydrogen beam is required on the BL2 for toroidal current control in the LHD. The BL2 will be operated on a conventional multi-aperture grounded grid [18] to focus on hydrogen beam operation in the next experimental campaign in 2021. The BL3 injected 4.4 MW hydrogen beam which slightly decreased before the deuterium operation, and the port through efficiency was maintained at 0.35. Deuterium beam injection of 3 MW has been achieved by modifying the accelerator gap setting [4]. Since this injector worked as a benchmark to verify suppression of the I_e on the

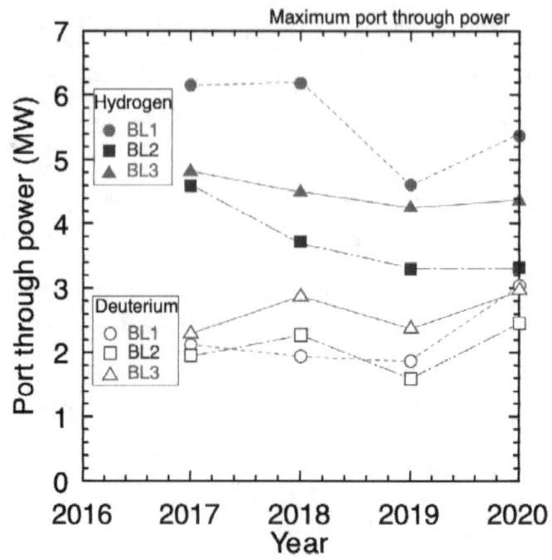

Figure 1. History of maximum port through power from N-NBI in LHD. Circle, square, and triangle marks indicate values of BL1, BL2, and BL3, respectively. Closed and opened marks are cases of hydrogen and deuterium beams, respectively.

deuterium beam operation, the effects of a bias voltage and a filter field strength on the I_e were also investigated.

2.2. Issues of deuterium beam acceleration

The negative-ion current linearly increased with respect to the arc discharge power and was not saturated. The averaged ratio of the deuterium current to the hydrogen current $[I_{acc(D)}/I_{acc(H)}]$ was approximately 0.66 for the same arc discharge power [4]. The reduction ratio in the injection beam power was below the reduction ratio in the negative ion current. Therefore, the discussion of the port through efficiency is important for the moderate beam injection power in deuterium. Figure 2 shows the shot history of the horizontal full width at half maximum (FWHM) of the negative ion beam, measured by the beam calorimeter, at the 8 m downstream from the ion source. The beam width in the horizontal direction consists of 14 beamlets, and in the vertical direction there are 55 beam rows. The FWHM of the hydrogen beam is 160 mm, however that of the deuterium beam expands to over 190 mm by switching operation gas. The divergence of the charged particle beamlet is caused by the self potential generated space-charge, perpendicular to the beam direction. Deuterium beams with larger mass numbers have larger divergence angles compared with hydrogen beams. Furthermore, the negative ion beam is bent by the magnetic field of an electron deflection magnet (EDM) installed in the extraction grids (EG) to prevent co-extracted electrons entering into the acceleration region. Deflection of the negative ion beams in hydrogen is corrected using the axial displacement of the steering grid apertures. However, this axial displacement is not corrected with the deuterium beam. For this reason, the deuterium beam width is expanded by shifting the beam direction alternately for each beamlet row group.

Another reason for reduction of the injection beam power is the stripping loss. Figure 3(a) shows the beam emission spectra both H (dashed line) and D (bold lime) observed at the 1 m downstream of the ion source. The optical line of sight is set 33° to the center of beam line. The E_{beam} is 171 keV including the extraction voltage of 9 kV in both cases. All peaked spectral lines are Balmer-alpha lines for hydrogen (H_α) and deuterium (D_α). The most intense line is the background light from the ion source, and the light from the full-energy beam appears at 646 nm and 649 nm for H and D, respectively, due to the Doppler effect. No clear peaks appear in the high-energy region near the full-energy peak; however asymmetric peaks are observed in the low-energy region near the background spectra. Figure 3(b) shows the beam emission

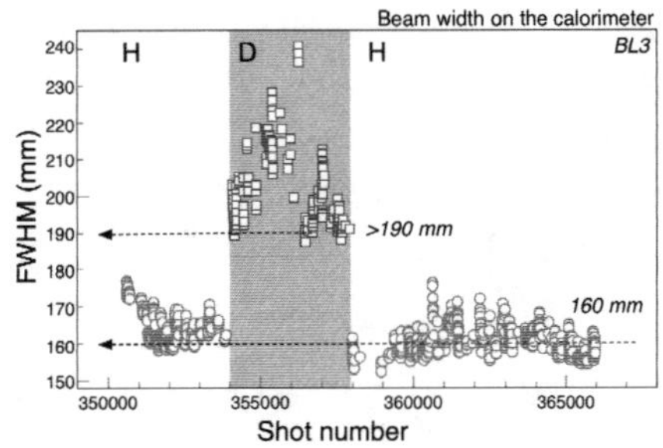

Figure 2. History of beam width (FWHM) on beam calorimeter for hydrogen (circle marks) and deuterium (square marks) beam operation.

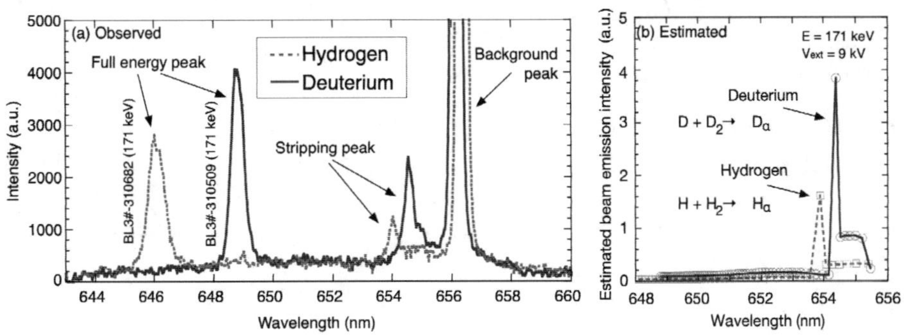

Figure 3. (a)Observed beam emission spectra for hydrogen (dashed line) and deuterium (bold line). (b) Estimated beam emission spectra by distributions of vacuum pressure and stripping loss neutrals in beam accelerator. Reproduced from [19], with permission of AIP Publishing.

spectrum estimated by calculating the number of neutralized particles from the vacuum pressure and energy distributions inside the accelerator from the plasma grid (PG) to the grounded grid (GG). The low-energy stripping peak corresponds to the energy of the extraction voltage (9 keV). This is a beam particle neutralized in the EG apertures by the high neutralization efficiency, due to 83% of the ion source pressure and low energy, and is a major component of stripping loss. In particular, the EG has a thickness of 14 mm due to the EDM and water cooling channels, which is a disadvantage in stripping loss. The neutralization reaction between the PG and EG also produces very low energy neutrals, therefore the stripping peak forms an asymmetric shape with a large tail on the low energy side. The stripping peak does not appear in the high energy acceleration region, which is consistent with the observed results. The amount of stripping loss in the accelerator is 16% and 24% for hydrogen and deuterium, respectively. Therefore, the stripping loss during beam acceleration of heavy isotopes is larger than that of light isotopes.

3. Improvement of the co-extracted electron current

The ratio of co-extracted electron current and the negative ion current is an important indicator for the safe operation of a negative ion source. Even in a large RF-discharge type negative ion source for ITER-NBI, which is currently under development, the electron current and ion ratio is required to be less than 1 for the deuterium operation. The low electron and ion current ratio of $0.2 < I_e/I_{acc(H)} < 0.25$ was achieved on hydrogen operation using the arc-discharge type negative ion source in LHD-NBI. On the other hand, the $I_e/I_{acc(D)}$ increased more than

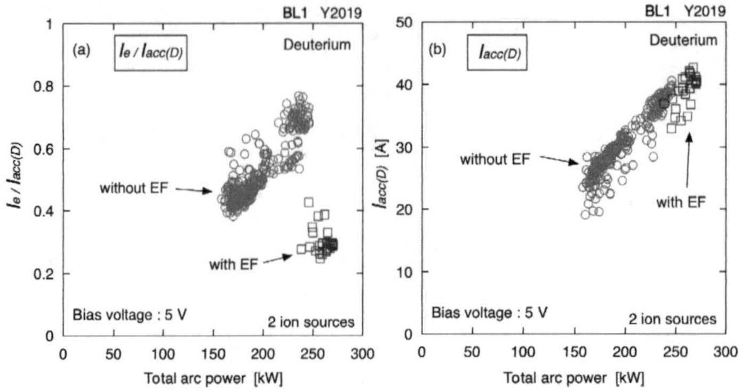

Figure 4. (a)Averaged $I_e/I_{acc(D)}$ using two ion sources in deuterium operation. Circle and square marks show original PG without EF and with EF, respectively. (b) Total $I_{acc(D)}$ using two ion sources in deuterium operation.

0.35 on deuterium operation, which is the critical issue for high power beam injection using high arc-discharge power. Increasing the bias voltage applied to the PG and changing the extraction gap length were effective in reducing the $I_e/I_{acc(D)}$ [5]. Recently, two other methods were found to reduce $I_e/I_{acc(D)}$ effectively.

3.1. Electron fence

An egg-box cell structure on the PG surface was tested in the small negative ion source at NIFS in the 1990s [20]. The egg-box cell was effective in reducing the electron current, however it did not increase the hydrogen negative ion current. The egg-box cell was a complex structure combination of 2 mm thin molybdenum plates. Therefore, thermal deformation was difficult to control in high-current ion sources for injectors that require high arc power discharge. An electron fence (EF) was developed to apply the electron suppression effect of the egg-box cell to deuterium discharges, and was installed into the negative ion source of BL1 [21]. The EF has a simple and robust structure made of 5 mm diameter molybdenum rods, which was placed between the rows of apertures on the PG surface in parallel to the EDM installed in the EG. The top of the EF was initially set at 5 mm from the PG, then this position was changed at 7 mm in 2020, which is as same height as the egg-box cell. The electrical potential of the EF was the same as the PG, and a positive bias voltage was also applied to the EF. The magnetic field lines produced by the EDM exit between the rows of PG apertures, pass through the EF, and cross the over the PG apertures; the EF is expected to absorb magnetized electrons trapped in the magnetic field lines produced by the EDM. Figure 4(a) and 4(b) are the averaged $I_e/I_{acc(D)}$ and the total $I_{acc(D)}$, respectively, in deuterium discharge with Cs vapor from two ion sources. In the case without an EF (circle marks), the $I_e/I_{acc(D)}$ exceeded 0.7 for a total arc discharge of 250 kW. This increase in the I_e limited the use of high arc power discharges. The TOSHIBA-type negative ion source for BL1 had good arc power efficiency of H^- ion current, however a large increase in electron current was observed compared with the HITACHI-type source in the deuterium operation. However, the $I_e/I_{acc(D)}$ decreased to 0.25 using the EF (square marks), which was low enough to operate the negative ion source in deuterium, as in the case of hydrogen discharge. The reduction of I_e enabled a high current negative ion beam by high arc power discharge, however the arc power efficiency was not changed. These properties of the EF were consistent with the egg-box cell.

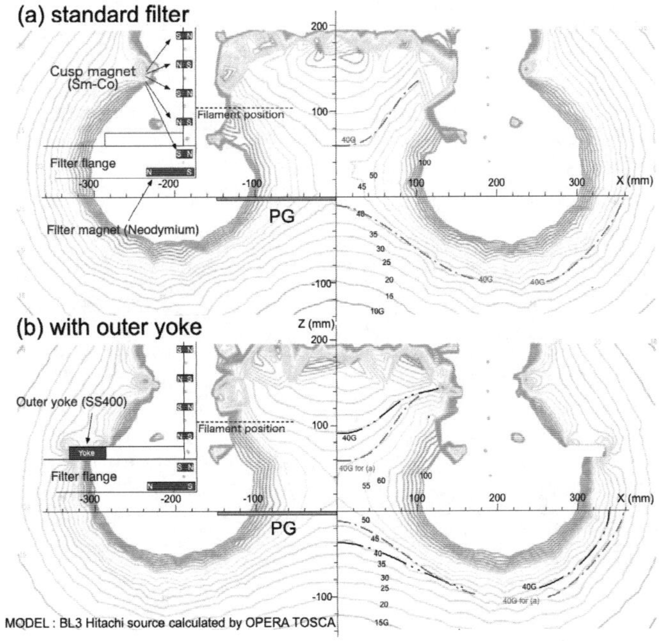

Figure 5. Distribution of magnetic flux density calculated by OPERA simulation software [24] using full magnetic model for arc discharge chamber of HITACHI-type source in BL3. (a) and (b) show cases of standard filter and extended filter using outer iron yoke, respectively. Left half of figure from center shows position of magnets and outer yoke. Magnets in right half are set symmetrical to center axis and its polarity is reversed.

3.2. Outer iron yoke

An external magnetic filter had been reported to be effective in enhancing the negative ion current and suppressing the I_e in an arc discharge type negative ion source [22, 23]. In the negative ion source for LHD-NBI, an insulating plate and a magnetic filter flange were inserted between the PG and the arc chamber to suppress the inflow of charged particles in the vicinity of the PG. Effects of the external magnetic filter were verified only for hydrogen discharge, therefore verification is needed whether the standard magnetic filter is matched for deuterium discharges. However, it is difficult to verify the filter effect by replacing the permanent magnets, because these magnets are installed inside the vacuum boundary of the magnetic filter flange. Therefore, an iron yoke is installed outside the vacuum vessel to increase the filter magnetic field strength inside the arc chamber. Fig. 5(a) shows a cross-sectional view of the arrangement of the magnets of the BL3 negative ion source in the short direction. The left half of figure from the center shows the position of the magnets. Samarium-cobalt magnets that create a cusp magnetic field are placed on the side wall and back plate. The external filter magnet is a neodymium magnet located at z = 30 mm from the PG surface. Both magnets in the right half are set symmetrical to the center axis and the polarity is reversed. Figure 5(a) also shows the contour plot of magnetic flux density [B] created by the cusp magnets and filter magnets, which is calculated by the OPERA simulation software [24]. The maximum B on the z-axis is 42 G, which corresponds to the position of the filter magnet at z = 30 mm. The magnetic field strength on the center axis, i.e. x = 0 mm, is the minimum value. The region above 40 G (single-dotted line) distributes to z = 60 mm in the standard filter case. Figure 5(b) shows the B distribution using the outer iron yoke. The yoke is made of structural steel (Japanese Industrial

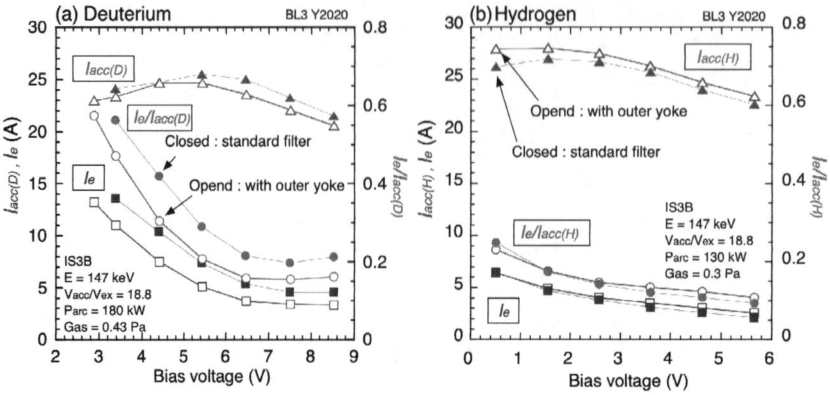

Figure 6. (a) Dependence of $I_{acc(D)}$ (triangle), I_e (square), and $I_e/I_{acc(D)}$ (circle) on bias voltage. (b) Dependence of $I_{acc(H)}$ (triangle), I_e (square), and $I_e/I_{acc(H)}$ (circle) on bias voltage. Closed marks and opened marks show cases of standard filter and extended filter, using outer iron yoke, respectively.

Standard SS400) with a cross-sectional size of 50 mm wide and 16 mm thickness. The side wall of the arc chamber has the terminal of the filament, which restricts the installation space for the outer yoke. Therefore, the outer yoke is placed on the filter flange around the four sides of the arc chamber flange. The outer yoke reduces the outer magnetic field strength and increases the magnetic field strength inside the vacuum vessel. The maximum B on the z-axis is 50 G, which corresponds to the position of the filter magnet at z = 30 mm. The outer yoke affects not only the filter magnet but also the cusp magnet, and the region above 40 G (double-dotted line) is extended to z = 90 mm. The weak magnetic field region, i.e. the confining plasma region, is considered to be shifted toward the back plate. This extended filter arrangement is expected to suppress the inflow of charged particles into the vicinity of the PG surface.

The effect of the outer iron yoke on hydrogen and deuterium discharges using a HITACHI-type negative ion source in BL3. Figures 6 show the bias voltage dependence of the negative ion and electron currents in the deuterium and hydrogen operation. Triangles, squares, and circles are I_{acc}, I_e, and I_e/I_{acc}, respectively. The closed and open marks are the standard filter and enhanced filter with outer iron yoke, respectively. Only one ion source was operated in this experiment which was different from the normal injector operation. The order of the NBI operation gas and filter condition was H_2 with the standard filter, D_2 with the standard filter, D_2 with the enhanced filter, and H_2 with the enhanced filter. In the deuterium discharge, a decrease in the I_e was clearly observed by the enhanced filter with the outer iron yoke in the entire bias voltage range, as shown in Fig. 6(a). The bias voltage of 5.6 V was applied in the normal deuterium operation with the standard filter, which produced the highest deuterium negative ion current and a reasonably low electron current. In low bias voltage operation, the operating region was limited by a large increase in electron current. However, a low bias voltage of 4.2 V provided good beam performance with low $I_e/I_{acc(D)} = 0.3$ using an enhanced magnetic filter by an outer iron yoke. At a bias voltage of 5.6 V, $I_e/I_{acc(D)}$ was 0.2 which was equally low as the $I_e/I_{acc(H)}$ in a high-arc power discharge in the hydrogen operation.

On the other hand, in the hydrogen discharge shown in Fig. 6(b), the I_e and the $I_e/I_{acc(H)}$ did not change when the filter field was enhanced by installing the outer iron yoke. The $I_{acc(H)}$ increased slightly on the enhanced filter. This result suggests that the use of an enhanced filter for deuterium does not have a negative effect on the hydrogen operation of the negative ion source. In the hydrogen operation of the surface-produced negative ion source, an ion-ion plasma was formed in the vicinity of the PG with the same amount of surface-produced negative hydrogen ions and positive ions supplied from the driver region. The amount of negative hydrogen ions was maintained in the enhanced filter condition, owing to the surface production by energetic hydrogen neutrals. The positive ion influx was reduced by the enhanced filter, however the electrons are not increased if there are enough negative ions. The mutual neutralization reaction between negative and positive ions should be decreased in the enhanced filter condition, that is the cause of increase in hydrogen negative ion current. In the deuterium discharge, the negative ion density at the center of the ion source was almost the same as that in the hydrogen discharge, although the deuterium positive ion density was observed to increase by a factor of 1.8 for hydrogen positive ions [25]. Therefore, ion-ion plasma was not formed in the deuterium discharge, and charge neutrality in the vicinity of the PG was maintained by increasing electrons from the driver region. When the positive ion influx is suppressed by enhancing the external magnetic filter, the amount of electrons can be decreased to maintain charge neutrality. For this reason, the enhanced external magnetic filter, using the outer iron yoke is effective in reducing the electron current in the deuterium operation. Similar suppression of the I_e has been observed in the enhanced PG filter in the RF negative ion source for ITER [26]. Therefore, suppression of the I_e in isotope negative ion beams can be efficiently produced by designing the ion source according to the conditions of isotopes with large mass numbers.

Acknowledgments
The authors would like to thank Dr. Fujiwara (NTT) and the technical staff of NBI in NIFS for their maintenance work and operational support. We also thank Dr. Kisaki (QST) for the significant contributions on the accelerator design which he has made it in the past at NIFS. This research is supported by the budget for the NIFS No. ULRR702.

References

[1] Iiyoshi A *et al* 1999 *Nucl. Fusion* **39** 1245
[2] Takeiri Y *et al* 2000 *Rev. Sci. Instrum.* **71** 1225
[3] Tsumori Y *et al* 2004 *Rev. Sci. Instrum.* **75** 1847
[4] Ikeda K *et al* 2019 *Nucl. Fusion* **59** 076009
[5] Ikeda K *et al* 2019 *Rev. Sci. Instrum.* **90** 113322
[6] Kuriyama M *et al* 1998 *J. of Nucl. Sci. and Technology* **35** 739 - 749
[7] Kojima A *et al* 2011 *Nucl. Fusion* **51** 083049
[8] Hiratsuka J *et al* 2020 *Review of Scientific Instruments* **91** 023506
[9] ITER project (*https://www.iter.org*)
[10] Hemsworth R *et al* 1996 *Rev. Sci. Instrum.* **67** 1120
[11] Hemsworth R *et al* 2008 *Rev. Sci. Instrum.* **79** 02C109
[12] Singh M J *et al* 2017 *New J. Phys.* **19** 055004
[13] Kovari M *et al* 2018 *Nucl. Fusion* **58** 026010
[14] Kojima A *et al* 2017 *Fusion Engineering and Design* **121** 145 - 151
[15] Osakabe M *et al* 2018 *IEEE Transactions on Plasma Science* **46** 2324 - 2331
[16] Ikeda K *et al* 2018 *AIP Conference Proceedings* **2011** 060002
[17] Tsumori K *et al* 2010 *Fusion Science and Technology* **58** 489 - 496
[18] Takeiri Y *et al* 2010 *Rev. Sci. Instrum.* **81** 02B114
[19] Ikeda K *et al* 2021 *AIP Conference Proceedings* **2373** 060001
[20] Tsumori K *et al* 1995 *Fusion Engineering and Design* **26** 473 -483
[21] Tsumori K *et al* 2021 *Proc. 28th IAEA Fusion Energy conference (virtual event)* TECH/3-4Rb, to be published Nucl. Fusion in press https://doi.org/10.1088/1741-4326/ac2d59
[22] Inoue T *et al* 1989 *Nuclear Instruments and Methods in Physics Research B* **37/38** 111 - 115
[23] Kuroda K *et al* 1996 *Review of Scientific Instruments* **67** 1114 - 1119
[24] Opera Simulation Software (*https://www.3ds.com/products-services/simulia/products/opera/*)
[25] Masaki S *et al* 2020 *Review of Scientific Instruments* **91** 013512
[26] Speth E *et al* 2006 *Nucl. Fusion* **46** S220 - S238

Reverse trajectory analysis of the hydrogen negative ion beam in a prototype accelerator for ITER

M Kisaki, A Kojima, G Q Saquilayan, J Hiratsuka, M Ichikawa, Y Shimabukuro, M Murayama, K Watanabe, H Tobari and M Kashiwagi

National Institutes for Quantum Science and Technology, 801-1 Mukoyama, Naka, Ibaraki 311-0193, Japan

kisaki.masashi@qst.go.jp

Abstract. Reverse calculation of negative ion trajectory based on the measured beam emittance has been performed for the first time in multi-stage accelerator in order to reconstruct negative ion profile near meniscus, which has been a long-term issue for negative ion sources. According to reverse calculation, negative ions are mainly extracted from the periphery of extraction area, and the particles extracted from the aperture edge is lost on acceleration grids. By taking into account the lost component with reconstructed negative ion profile, the negative ion trajectory became consistent with observed beam traces on the acceleration grids. This result can be applied directly to the design of ITER accelerator as well as the other Cs-seeded negative ion sources.

1. Introduction

In the neutral beam injector for ITER, acceleration of high-current and high-energy deuterium negative ion beam (40A, 1 MeV) is required during 3600 s [1]. To achieve these requirements, R&D activity is being carried out with a prototype multi-aperture multi-grid (MAMuG) accelerator for ITER in QST, so-called "MeV accelerator". So far, high-power-density hydrogen negative ion (H^-) beam of 184 MW/m^2 (0.97 MeV, 190 A/m^2) has been successfully produced for 60 s by tuning the geometry of MeV accelerator so as to meet an optimum beam optics [2]. Even though the grid heat load was kept lower than an allowable level, one of the crucial issues toward further extension of the pulse duration is to suppress the local grid heat load [3], which has been observed as unexpected beam traces around the aperture edge of the acceleration grids.

The beam traces were not predicted by the conventional beam analysis, in which negative ions are accelerated without direct interception by acceleration grids at optimum beam condition. In conventional beam trajectory codes, the negative ions are uniformly extracted through the meniscus with the physics model of the positive ion extraction [4,5]. However, it has been expected that negative ions (H^-/D^-) are not uniformly distributed in negative ion sources because the negative ions are mainly produced on the metal surface with low work function by Cs-seeding [6], and are influenced by magnetic field to suppress co-extracted electrons.

So far, to investigate the physics model of the negative ion extraction, two-dimensional negative ion trajectory had been reversely calculated from the measured beam profile on the beam target to the extraction region in a single-stage accelerator [7]. This method is so-called "reverse calculation of the negative ion trajectory". In this preliminary study, the divergence at the starting point was numerically assumed, and the magnetic field was not included in the two-dimensional beam trajectory code.

This time, the reverse calculation technique has been improved to three-dimensional code based on the measured emittance plot, in which the magnetic field and the variation of space charge along the beam path are also incorporated, to reconstruct the negative ion profile near the meniscus. In addition, the origin of local grid heat load has been investigated by the forward calculation based on the reconstructed H^- profile.

2. Calculation model

The cross-sectional view of the MeV accelerator is shown in Fig. 1(a). The MeV accelerator is composed of a plasma chamber, an extractor, and an accelerator, where each grid composing the extractor and the accelerator has 3 x 5 apertures. The extractor is composed of three grids, which are plasma grid (PG), extraction grid (EXG), and steering control grid (SCG). Aperture diameters of the PG, the EXG and the SCG are 14 mm, 13 mm, and 16 mm, respectively. Electron deflection magnets (EDMs) are embedded in the EXG in order to dump co-extracted electron. In addition, filter magnets are installed in the chamber in order to lower the electron temperature near extraction apertures. Figure 1(b) and (c) show profiles of filter field and EDM field along the beam path, respectively, where the origin of coordinates is set at the center of upstream side of the PG aperture. Apertures on the SCG are displaced with respect to those on other grids in order to compensate the beam deflection by the magnetic field. The accelerator has five acceleration stages. The lengths of the extraction and acceleration gaps are set to 6 mm and 88 mm, respectively, in order to demonstrate the ITER relevant beam acceleration [8].

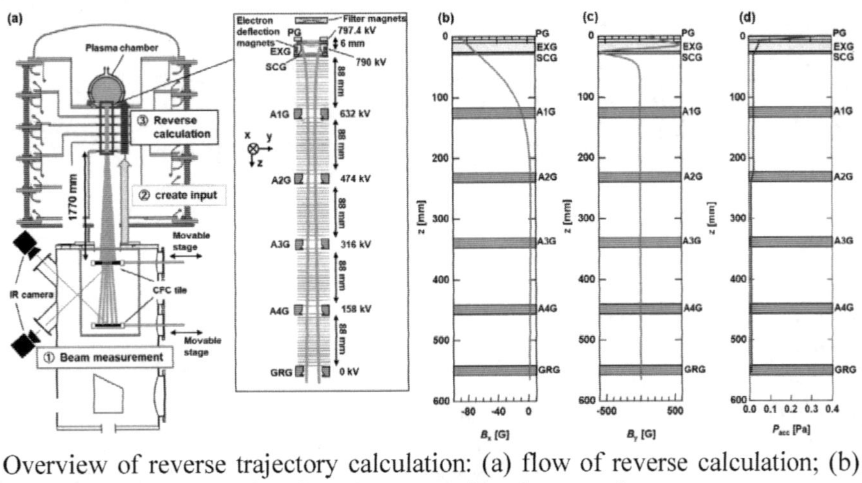

Figure 1. Overview of reverse trajectory calculation: (a) flow of reverse calculation; (b) filter field and (c) EDM field along beam path; (d) gas pressure along beam path.

The process flow of a reverse trajectory calculation developed in this study is divided into three parts as shown in Fig. 1(a): (1) to measure emittance on the one-dimensional carbon-fiber-composite (CFC) tiles, (2) to create input data of particle positions and velocities at the accelerator exit, which can be delivered from data (1), and (3) to calculate particle trajectories reversely from the accelerator exit to the beam extraction area. The detailed description of a beam diagnostic system can be found in Ref. [9].

In this study, the area covering 3 x 3 apertures was modeled to calculate the trajectory of the single beamlet self-consistently by Opera-3d [10], since the electric field near the aperture is affected by neighboring apertures. To simulate the beam deflection, the magnetic field and aperture displacement on the SCG were incorporated in the model. In addition, the stripping loss due to collision between the negative ions and background gas was taken into account to simulate the variation of the space charge along the beam path. Figure 1(d) shows a profile of an analytically estimated gas pressure along the beam path, where the gas inside the accelerator was assumed to be at room temperature.

3. Benchmark of reverse calculation

The reverse calculation was benchmarked by comparing between the forward calculation tracing the particle at the PG to the accelerator exit and the reverse calculation from the accelerator exit to the PG under the influences of the magnetic field and the space charge variation along the beam path. Note that the magnetic field has to be reversed in the reverse calculation to represent the gyromotion due to the magnetic field correctly.

The results of the forward calculation are summarized in Fig. 2(a-c), which are the input, the typical trajectories inside the accelerator, and the emittance plot at the GRG exit, respectively. The typical trajectories inside the accelerator in the backward calculation is shown in Fig. 2(d). As can be seen in Fig. 2(b) and (d), the particles in the reverse calculation follow the same trajectories as those in the forward calculation in the whole acceleration region, in which the particle energy changes over a wide range from several keV to 800 keV. As a consequence, the particle information at the PG was successfully reproduced from that at the GRG exit as shown in Fig. 2(e). Note that the reverse calculation must be done with the same calculation domain as that for the forward calculation otherwise the particle trajectories slightly deviate from those in the forward calculation.

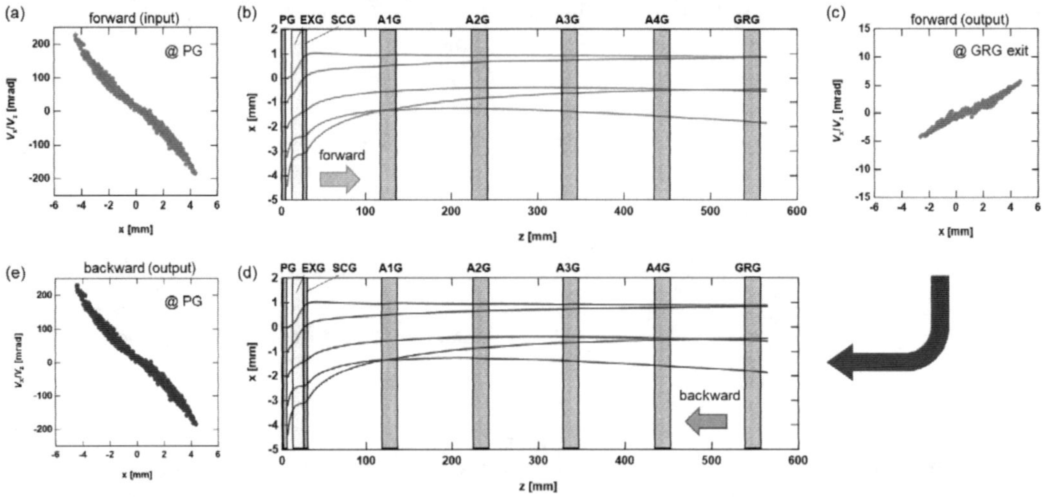

Figure 2. Results of benchmark: (a) emittance plot at PG exit, (b) particle trajectories, and (c) emittance plot at GRG exit in forward calculation; (d) particle trajectories and (e) emittance plot at PG exit in backward calculation.

4. Reconstructions of meniscus and negative ion trajectory

Figure 3(a) shows a measured emittance at current density of 200 A/m^{-2} and beam energy of 797.4 keV. The emittance plot at the GRG exit was reproduced from the measured emittance as an input data for reverse calculation as shown in Fig. 3(b). In order to create the input data, two assumptions were made: (1) the negative ion beam is fully charge compensated after the GRG. (2) phase space structure is the same in horizontal and vertical directions. Here, verification of these assumptions is planned after modification of the diagnostic system as one of the future works. On these assumptions, the input was created with random numbers using experimentally estimated e-folding width and velocity dispersion at GRG exit.

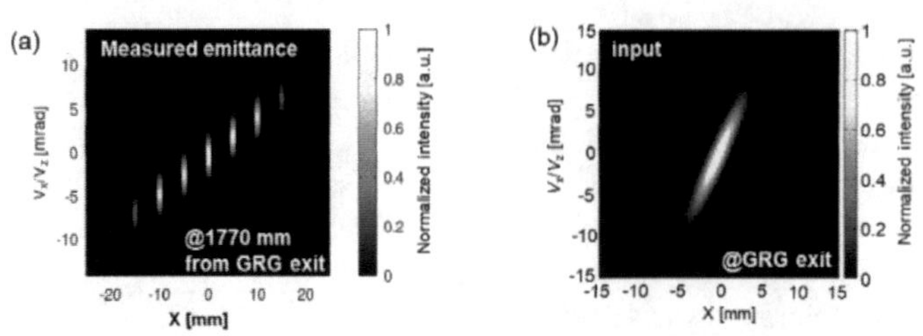

Figure 3. (a) measured emittance at a distance of 1770 mm from GRG exit and (b) emittance at GRG exit reproduced from experimental data.

Particle trajectory is shown in Fig. 4(a). 2000 macro-particles were tracked up to a plane intersecting the edge of conical part of the PG aperture in order to retrieve the information of negative ions near the meniscus. Hereinafter, the plane is referred as destination point. It was observed that more than 98% of macro-particles successfully reached the destination point. The current density at the destination point was estimated as shown in Fig.4(b). It was observed that the current density is higher in peripheral region than that for the center. This implies that the negative ion is distributed nonuniformly in the vicinity of the meniscus due to the surface production process. An asymmetric structure was also observed in higher current density region, which was not predicted by previous analysis [7]. This might be caused by the direct and indirect effects of the three-dimensional complex topology of the magnetic field near the extraction apertures [11], which may modify the transport of H^- ions and may also produce variation of the potential topology, producing an inhomogeneous potential barrier close to the aperture and thus determining an inhomogeneous amount of H^- ions escaping from the PG surface.

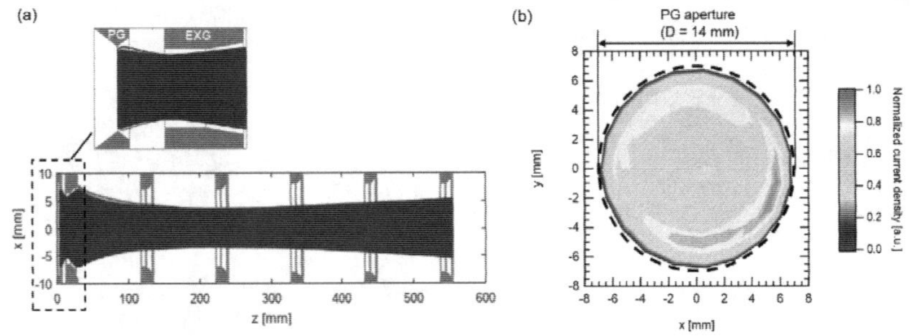

Figure 4. (a) Trajectories of negative ions tracked backward with experimental data; (b) Contour plot of normalized current density at PG.

As the next step, the result of reverse calculation was applied to the forward calculation in order to investigate the origin of local grid heat loads. We carefully examined the particle distribution at the destination point, and found the gap of about 0.6 mm between particles and inner wall of the PG aperture as shown in Fig. 5(a). Since the reverse calculation tracks only the particle passing through the accelerator, this gap might be one of the origins of the unknown local grid heat load. Therefore, the lost particles were assumed to be originated from this region as shown in Fig. 5(b). Considering the continuity of the emittance, the velocities of lost particles were estimated by linear extrapolation using the emittance plot from reverse calculation as shown in Fig. 5(c). Hereinafter, the particles from the reverse calculation and the additional particles are referred as core and edge particles, respectively.

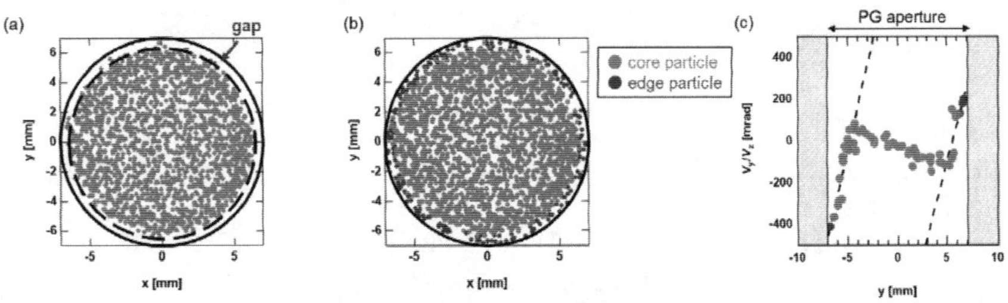

Figure 5. (a) particle distribution at PG; (b) particle distribution at PG with lost components; (c) emittance plot at PG with lost components.

The trajectories of core and edge particles are shown in Fig. 6(a) and (b), respectively, where the edge particle was assumed to deliver 10% of extracted negative ion current. The core particles traced on the similar trajectories as those in the reverse calculation. On the other hand, the edge particles were accelerated with large divergence angle due to the deformation of the meniscus caused by nonuniformity of the current density, and all of them were intercepted by the intermediate grids. This divergent component, so-called beam halo, is not reproduced with conventional trajectory codes as uniform current density is assumed for the meniscus formation. It was observed that the edge particles impinge on nearby the acceleration grid aperture as shown in Fig. 7(a). This implies that one of the possible candidates for the origins of beam traces observed after experiments is the negative ion extracted from the periphery of the extraction apertures. The grid heat loads estimated by the experiment and the forward calculation were compared as shown in Fig. 7(b), where the heat load was normalized by the input of the power supply for the beam acceleration. The measured heat load is larger than that by estimated from the calculation, especially on the A1G and the GRG. This is because the stripped and secondary electrons also contribute to the heat load in the experiment while not taken into account in the calculation. It was found that large part of heat loads on the A2G, the A3G and the A4G is originated from the negative ion while on the A1G and the GRG might be affected by the electron components.

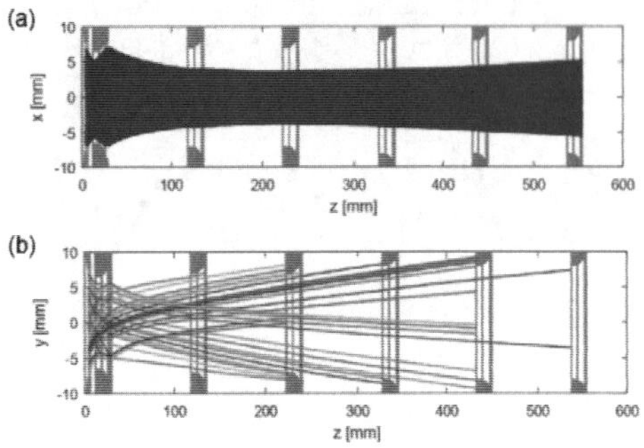

Figure 6. Trajectories of (a) core particle and (b) edge particle.

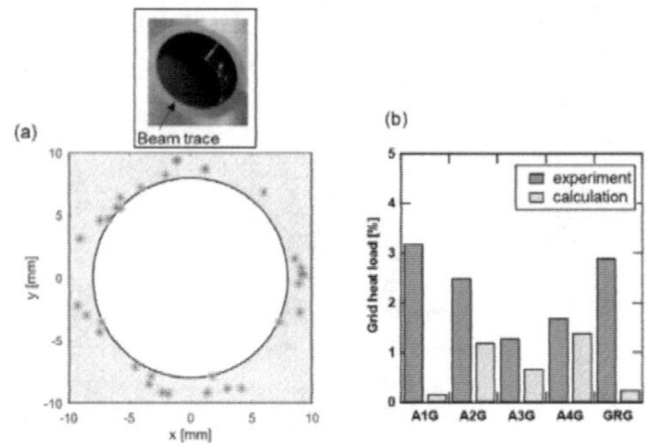

Figure 7. (a) distribution of negative ions impinging on upstream surface of A4G; (b) heat loads estimated by experiment and calculation.

5. Summary

The reverse trajectory calculation with the measured beam emittance has been performed for the multi-stage MeV-class accelerator for ITER. The applicability of the reverse calculation was carefully benchmarked taking into account the magnetic field and the variation of space charge during the beam acceleration. The hollow H^- density profile in the vicinity of the meniscus was identified for the first time in multi-stage accelerator by applying the reverse calculation based on the experiment. Beam trajectory was reconstructed by the forward calculation with nonuniform H^- profile including the one lost on the acceleration grids. As a result, it was found that the H^- near the inner wall of PG aperture is extracted with higher divergence angle, resulting in direct interception by acceleration grids. These findings contribute to the design optimization of the ITER accelerator.

References

[1] Hemsworth R S, Boilson P, Blatchford P, Palma M D, Chitarin G, de Esch H P L, Geil F, Dremel M, Graceffa J and Marcuzzi D 2017 New. J. Phys. 19 025005.
[2] Kojima A, Hiratsuka J, Umeda N, Hanada M, Kashiwagi M, Yoshida M, Ichikawa M, Nishikiori R, Watanabe K and Tobari H 2017 Fusion Eng. Des. 123 236
[3] Hiratsuka J *et al*, 2016 Rev. Sci. Instrum. 87 02B137
[4] Ohara Y, 1976 JAERI-M 6757
[5] Pamela J, 1991 Rev. Sci. Instr. 62 1163
[6] Bacal M and Wada M, 2015 App. Phys. Lett. 2 021305
[7] Ichikawa M *et al*, 2017 AIP Conf. Proc. 1869 030024
[8] Agostinetti P *et al*, 2016 Nucl. Fusion 56 016015
[9] Saguilayan G Q, Kisaki M, Kojima A, Shimabukuro Y, Murayama M, Hiratsuka J, Ichikawa M, Watanabe K, Tobari H and Kashiwagi M, this conference proceeding
[10] OPERA – Electromagnetic and Electromechanical simulation https://operafea.com/
[11] Taccogna F and Minelli P, 2017 New J. Phys. 19 015012

nuCARIBU: An upgrade for the CARIBU facility at the Argonne Tandem Linac Accelerator System

J. McLain[1], C. A. Dickerson[1], M. Gott[1], J. Greene[1], J. Nolen[1], G. Savard[1,2], J. Song[1], R. C. Vondrasek[1]

[1]Physics Division, Argonne National Laboratory, Lemont, IL 60439, USA

[2]Department of Physics, University of Chicago, Chicago, IL 60637, USA

Email: jmclain@anl.gov

Abstract. The Californium Rare Isotope Breeder Upgrade (CARIBU) facility is changing the mechanism for creating neutron-rich fission products. Spontaneous fission from a ^{252}Cf source has provided beams to support the low energy and accelerated-beams ATLAS programs. ^{252}Cf has a 2.65-year half-life, requiring the source to be replaced every three years to maintain high beam intensities. Fabricating an appropriately thin ^{252}Cf source to efficiently release the fission products has been challenging. The solution to these problems is nuCARIBU, a new system that provides neutron-induced fission on actinide foils. The Best Cyclotron B6P System (6-MeV proton beam at 0.5mA) is chosen, utilizing a multi-cusp negative ion source extracting into a cyclotron, which uses carbon foils to strip the H^- ions to protons. These protons are delivered to a ^{7}Li target to produce neutrons. The fast neutrons are moderated to thermal energies to induce fission in an actinide foil, providing neutron-rich fission products. These subsystems are described in detail along with comparisons between the current CARIBU facility and the presented nuCARIBU facility.

This work was supported by the U.S. Department of Energy, Office of Nuclear Physics, under Contract No. DE-AC02-06CH11357

1. Introduction

The Californium Rare Isotope Breeder Upgrade (CARIBU) was commissioned in 2011 at the Argonne Tandem Linac Accelerator System (ATLAS). CARIBU utilizes a Cf-252 source to provide a spectrum of neutron-rich fission fragments for use in both the low energy and reaccelerated physics programs. The fission recoils slow down through a degrader and the ultra-high purity helium gas inside of a state-of-the-art gas catcher. The gas catcher then holds 30-50% of the recoils in the 1+ or 2+ charge state because the helium has a higher binding energy for electrons than the other elements [1]. After using the complex RF guiding system in the gas catcher to transport the fission fragments to the gas catcher nozzle, the rare isotope beams are extracted into the beamline at up to 50kV [2]. The electron beam ion source (EBIS) is used as the charge breeder for the 1+ or 2+ neutron-rich isotopes to be accelerated in the ATLAS superconducting linac. Using the EBIS as a charge breeder allows for delivery of clean ion beams to state-of-the-art target areas because of the intrinsically low background level of an EBIS [3].

Although CARIBU successfully provided unique neutron-rich rare isotope beams to users for the past 10 years, replacing the Cf-252 source, which has a half-life of 2.65 years, is challenging. Every three years, a new source must be fabricated to replace the Californium source that has substantially decayed and is no longer capable of delivering acceptable beam intensities. Multiple attempts at manufacturing an appropriately thin source have not been successful and other novel approaches were considered to replace the passive Cf-252 source. A cyclotron-based neutron generator system that delivers thermal neutrons onto an actinide foil was chosen, and details are discussed.

2. nuCARIBU Motivation

The main obstacle with the passive Cf-252 source is that there is no longer a clear path forward to replace the depleted source, which limits the intensity capabilities of the low energy and reaccelerated CARIBU beams. Changing the mechanism of generating neutron-rich isotopes to a neutron induced fission on an actinide foil approach increases the isotope production intensities in some important regions, improves operation and maintenance simplicity, and provides a more reliable source. Another unique feature of nuCARIBU is the capability of changing target foils to allow for production of fission fragments that are only available from fission of specific isotopes.

The nuCARIBU system is a combination of multiple subsystems that work together to create an upgraded source of heavy neutron-rich fission products. The Best Cyclotron B6P System provides a 6-MeV proton beam at 0.5mA, utilizing a multi-cusp H^- ion source that is extracted into the cyclotron and then stripped using carbon foils to produce 6-MeV protons. These protons are delivered onto a ^7Li target where neutrons are produced by the p-^7Li reaction. The fast neutrons that are created are moderated to thermal energies and are delivered to an actinide foil to induce fission, creating the desired heavy neutron-rich fission products. These fission products are captured within the same gas catcher as is currently used by CARIBU and are extracted into the beamline at up to 50kV where an isobar separator selects the desired beam components with a designed mass resolution of 20,000 [1]. From here, the beam can be delivered to one of the low energy target areas [2], where charge breeding occurs to prepare the beam for reacceleration in the ATLAS linac.

CARIBU maintenance is currently a time-consuming process due to the radiological hazards present with the spontaneous fission of Cf-252 requiring many controls to be in place before carrying out work near the source. With nuCARIBU, the source can be switched off, greatly reducing the radiation present in the room. This decreases the complexity of the required controls, such as shielding, for a job to be completed. However, nuCARIBU will have some new scheduled maintenance tasks when compared with CARIBU. The multi-cusp ion source filament will need to be replaced every 500 beam-hours and the carbon foil cartridge will need to be replaced every 1260 beam-hours at 0.5mA. These tasks are designed to be simple and are expected to require less than 24 hours of maintenance down-time for every 20 beam-days of operation.

3. Best B6P Cyclotron System

The B6P Cyclotron System is a slightly modified version of a mature off-the-shelf product by Best Cyclotron, minimizing risk for this subsystem of the nuCARIBU source. The platform is a 15-MeV cyclotron that is extracted onto carbon stripping foils at 6 MeV. The cyclotron uses a four-hill isochronous, 14 ton, 1.2T magnet with two 75Mhz $\lambda/2$ resonators.

3.1. Ion Source and Injection

The B6P System begins with a multi-cusp negative ion source. Multi-cusp negative ion sources have been used for many years as a method to produce negative hydrogen and deuterium ions. The typical multi-cusp ion source uses a multipole magnetic arrangement around a cylindrical chamber for electron confinement, and the anode and cathode are simply the chamber wall and a negatively biased electrode within the vacuum chamber. Some multi-cusp sources are filament driven, while others are RF coil driven.

This multi-cusp ion source is driven by a tantalum filament that provides the necessary electrons to sustain the discharge. When the H^+ ions are accelerated through the sheath, they bombard the cathode surface and produce H^- ions that are accelerated through the sheath away from the cathode [4], and are extracted into the cyclotron with a 25kV extraction electrode. Once the beam is extracted out of the ion source, steering magnets are used to center the beam and the combination of a solenoid and quadrupole doublet are used to focus the beam into the cyclotron.

3.2. Extraction

At the 6 MeV orbit, two electrons are stripped from the H- ions using a carbon foil. The protons are then extracted from the cyclotron and are used as the incident particle for neutron production on a ^{7}Li target. The protons must be transported onto a high voltage platform which decelerates the beam a maximum of 250keV, but typically up to 150keV, and effects its ion optical properties. For this reason, a quadrupole triplet is installed upstream of the accelerator tubes to tailor the beam, giving it appropriate optical properties leaving the accelerator tubes.

Simulations of the extracted beam parameters were provided and used as an input for simulating the beam transport, determining that a ~3cm diameter spot size is achievable on the lithium target. Due to geometrical limitations, a dipole chicane is used to offset the beam from the cyclotron axis to the target and gas catcher axis. The beam optics and transport simulations from the cyclotron to the lithium target can be seen below in Figure 1. Additional diagnostics will be installed to further manipulate the beam and monitor the beam properties such as: 4-jaw slits, a collimator, a beam position monitor, a beam current monitor and radiation detectors.

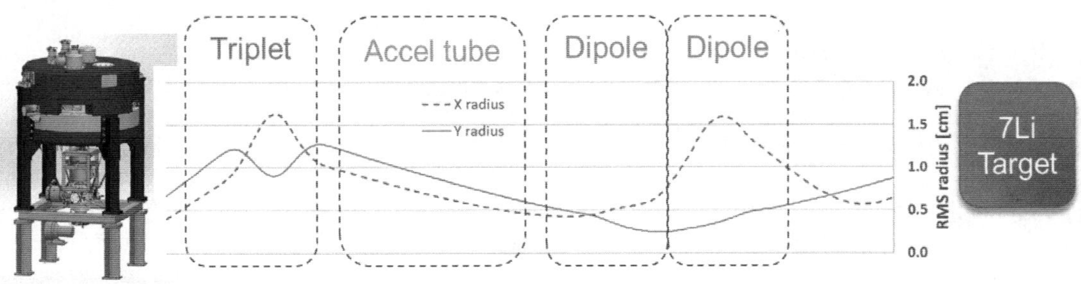

Figure 1. Beam optics from the B6P cyclotron to the lithium target.

4. Neutron Production Target and Moderation

Neutrons are produced by the p-^{7}Li reaction using a water-cooled lithium target with a 50-micron titanium window. Given 0.5mA of 5 MeV protons on ^{7}Li, the production is simulated to be at least 5 times the fission product yield of the Cf-252 passive source. The p-^{7}Li neutron production cross section is highest around 2.3MeV, with beam energies higher than 2.3-MeV still yielding more than 50% of the cross section at the 2.3-MeV resonance, shown in Figure 2. The target is designed to accept a 3cm diameter beam at 3kW, where the cooling design utilizes a micro-channel cooling approach that is currently being prototyped before field deployment.

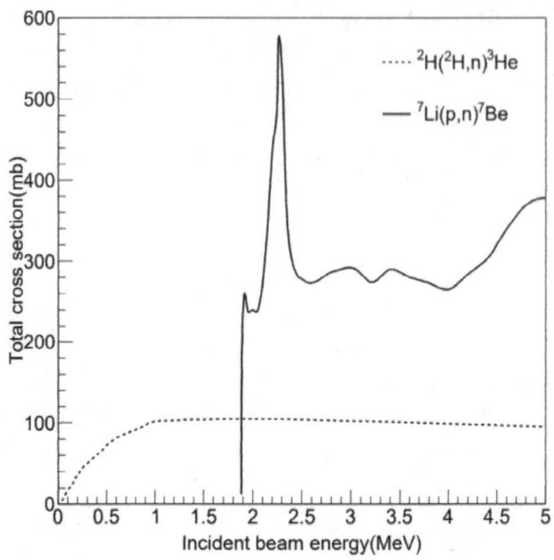

Figure 2. Total cross sections of the p-^{7}Li and D-D reactions.

A high-density polyethylene moderator is designed to thermalize the neutrons before they interact with the actinide foil to maximize the fission product yield, and therefore beam intensity. The fission yield was simulated while varying moderator thickness to determine the final design. The simulated fissions/second are as high as 7E8 with nuCARIBU, which is over an order of magnitude higher than CARIBU.

5. Neutron Induced Fission Target

An actinide foil is used as the production target for fission products in nuCARIBU. One of the compromises of nuCARIBU is that it has a different spectrum of available neutron rich fission products than the spontaneously fissioning Cf-252 source, but this compromise is only undesirable for the physics experiments that require the isotopes that are uniquely produced by Cf-252. Using target foils is also a strength in that it provides the flexibility of using different target foils, such as U-235, Pu-239, Th-229, and others. These different fissionable materials all have unique fission products that were not practically available with Cf-252.

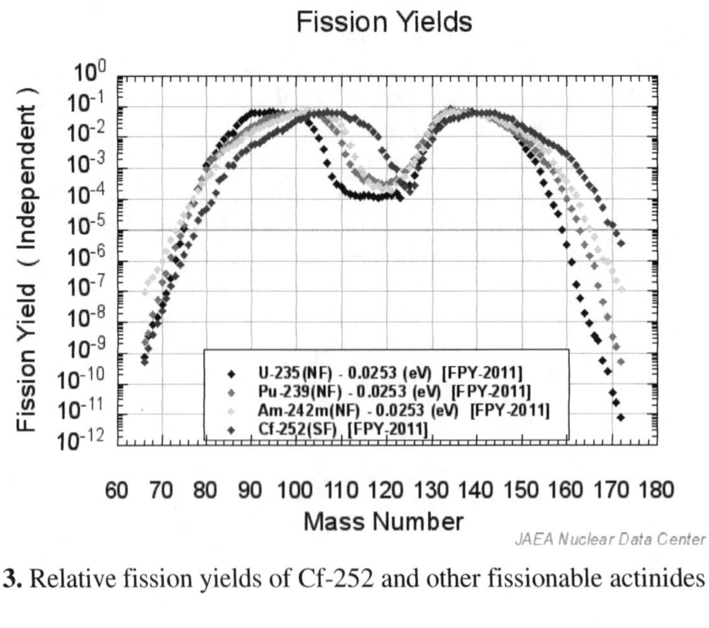

Figure 3. Relative fission yields of Cf-252 and other fissionable actinides [5].

The relative fission yields of Cf-252 and relevant foil materials can be seen above in Figure 3 [5]. Notice that there is fall off in the high masses, but for many products the intensity improvement of nuCARIBU can make up for the relative decrease in yield. A table of desired fission products for the physics community is also shared below in table 1 [5], highlighting the differences that nuCARIBU can provide for those isotopes.

Table 1. Desired fission products' relative yields for different fissioning materials [5].

		Fissioning Material			
		^{252}Cf	^{239}Pu	^{235}U	^{229}Th
Fission Product	^{82}Ge	5.4E-05	2.7E-04	1.3E-03	1.2E-02
	^{90}Se	4.1E-05	1.2E-05	1.3E-04	6.2E-05
	^{110}Ru	3.62%	0.57%	0.01%	1.4E-04
	^{132}Sn	0.14%	0.48%	0.59%	0.11%
	^{164}Gd	2.6E-04	1.7E-05	4.3E-08	7.9E-12

6. Conclusions

CARIBU currently provides 3000-4000 hours of low energy and reaccelerated beams per year, and many of the delivered beams are unique to the CARIBU facility. The demand is high and increases annually, where only about half of the requested beam hours are currently approved. More beam hours could be approved with the increased intensities, reliability, and maintenance simplicity that nuCARIBU will provide.

ATLAS is currently involved in upgrading the facility to a multi-user facility, where CARIBU and ECR beams can be delivered simultaneously, opening the door for even more CARIBU beam hours to be approved. The ability to change the target material to provide different CARIBU beams is a great opportunity, allowing for even more beams to be available to the physics community. With nuCARIBU, the increased available beams, increased reliability, increased beam intensities, and decreased maintenance complexity will all benefit the ATLAS and CARIBU facilities as well as the physics community.

Acknowledgment

This work was supported by the U.S. Department of Energy, Office of Nuclear Physics, under Contract No. DE-AC02-06CH11357. This research used resources of ANL's ATLAS facility, which is a DOE Office of Science User Facility.

References

[1] Savard G et al 2011 *Hyperfine Interact* **199** 301-309
[2] Savard G et al 2014 *JPS Conf. Proc.* **6** 010008
[3] Vondrasek R C et al 2017 *Review of Scientific Instruments* **89** 052402
[4] Leung K N 1994 *Review of Scientific Instruments* **65** 1165
[5] Shibata K et al 2011 *J Nucl. Sci. Technol.* **48**(1) 1-30

Recent developments and results of the LPSC PHOENIX type ECR charge breeder

J Angot[1], M Baylac[1], M Luntinen[2], M Migliore[1], O Tarvainen[3] and T Thuillier[1]

[1] Univ. Grenoble Alpes, CNRS, Grenoble INP, LPSC-IN2P3, 53 Avenue des Martyrs, 38000 Grenoble, France
[2] University of Jyväskylä, Department of Physics, Survontie 9D, 40500 Jyväskylä, Finland
[3] STFC ISIS Pulsed Spallation Neutron and Muon Facility, Rutherford Appleton Laboratory, Harwell, OX11 0QX, UK

E-mail: julien.angot@lpsc.in2p3.fr

Abstract. Four models of the PHOENIX ECR charge breeder have been manufactured for ISOL application. Two are currently under operation at TRIUMF (ISAC) and GANIL (SPIRAL 1) while the SPES one is being installed on the facility. The last model is set on the LPSC 1+N+ test bench where a R&D program is ongoing to improve its performances. The last modifications consisted in improving the beam line vacuum and the alignment. Commissioning experiments showed an improvement of the charge breeder performances for all the tested species. The global CB efficiency is close to 100% for Cs when correcting the measurements with the beam transmission. Na and K efficiencies have increased significantly to reach 18.7% for Na^{8+} and 22.7% for K^{9+}. In parallel, the charge breeder plasma was studied injecting short pulses of 1+ ions and using a zero-dimension model to estimate the plasma parameters. These experiments have provided a better understanding of the performance improvement. The last developments of the LPSC Charge Breeder together with the experimental results are presented.

1. Introduction

The PHOENIX type Electron Cyclotron Resonance Charge Breeder (ECRCB) has been developped at LPSC since the early 2000s [1]. Several versions have been manufactured for TRIUMF, ISOLDE, GANIL and SPES facilities. At LPSC, one model, installed on the 1+N+ test bench, is used to continue R&D and improve the charge breeding performances. One of the objectives is the reduction of the co-extracted contaminants, in the frame of the LNL-LPSC-GANIL collaboration [2] and on the basis of a previous agreement signed between LNL and LPSC in 2018. To perform the contaminants reduction experiments in the best possible conditions, the 1+N+ beam line was upgraded to improve the vacuum and the alignment of the optics and instruments.

The LPSC ECRCB is a 14.5GHz PHOENIX type ion source [2], with 3 coils to generate the axial magnetic field and a hexapole providing 0.8 T radial field on the poles at the plasma chamber wall surface, corresponding to a diameter of 72 mm. Two movable soft iron rings are set around the hexapole to act on the axial magnetic field gradients on the resonance surface. The objectives of the last major modifications were i) to improve the 1+ beam injection by making the injection soft iron plug symmetric and modifying the injection electrode shape [3, 4] and

ii) enhancing the plasma confinement by adding a soft iron plug under vacuum at injection to increase the axial magnetic field and tuning the position of the two soft iron rings surrounding the hexapole [5]. These modifications allowed to gradually improve the ECRCB single Charge State (CS) and the total efficiencies, while keeping the rise time short.

The LPSC PHOENIX ECRCB is qualified on the 1+N+ test bench [6], which is composed of a beam line that generates the 1+ beam to be injected into the ECRCB and another one that analyses the beams extracted from the charge breeder. Each beam line is equipped with a mass spectrometer, beam optics and diagnostics (Faraday cup (FC), horizontal and vertical Allison type emittance scanners) after the mass spectrometer. Two types of 1+ sources are used to produce the 1+ beams i) the COMIC source for gaseous elements and ii) the surface ionisation ion gun for alkali species. An electrostatic deflector is placed in front of the 1+ spectrometer to pulse the 1+ beam and measure the background for the efficiency calculations. On the N+ beam line, a removable FC is mounted between an Einzel lens and the mass spectrometer, to measure the total extracted beam intensity. The test bench configuration allows operation at 20kV maximum source potential. A high voltage supply is used to polarize the 1+ source. Another precision power supply sets the voltage difference (ΔV) between the ECRCB and the 1+ source.

In parallel to the charge breeder qualification experiments, the 1+N+ test bench was used for plasma studies [6, 7, 8]. Recently, a new method, based on short pulse 1+ injection of metal ions, was developed to estimate the ionisation, charge exchange and confinement times together with the plasma energy content and triple product [9]. This Consecutive Transient (CT) method, was applied to the LPSC ECRCB, injecting K^+ ions into an He plasma.

2. 1+N+ test bench upgrade

To prepare the contaminants reduction experiments, the 1+N+ test bench was upgraded in the period of 2018-2019, as described in detail elsewhere [2]. In summary, all the vacuum chambers were modified or replaced to be compatible with Ultra High Vacuum (UHV) techniques. The vacuum pumping configuration was also deeply modified and the turbo pumps were either refurbished by the manufacturer or replaced by new ones. The FCs were replaced together with the double Einzel lens at the ECRCB injection and the Einzel lense at the extraction. The emittance scanners were enhanced for UHV environment. Modifications were also done to the mass spectrometers to create view ports on each axis for the alignment of the beam line devices. Following the assembly, a vacuum leak detection was systematically performed on each section. After the upgrade, the vacuum level was measured to be between 3 to 6×10^{-8} mbar, which is one order of magnitude better than previously.

After this upgrade, qualification experiments were carried out to validate the assembly and check the effect on the charge breeder performances.

3. Commissioning experiments

Following the extraction Einzel lens replacement and the alignment, the N+ beam line transmission was checked. The charge breeder was operated with a H_2 plasma in a configuration optimized for potassium (K) charge breeding. The transport efficiency of the N+ beam line was measured to be 83% by comparing the total extracted N+ current and the sum of all mass analysed N+ ions.

Na, K, Rb and Cs were chosen for the commissioning of the PHOENIX ECR CB in the new 1+N+ test bench configuration. The source (microwave power, coils currents, support gas ...) and beam optics parameters were optimized to maximize the 1+N+ efficiency yield for each species. Table 1 compares the former (2017) and new (2020) results and tunings. The ECRCB efficiency Charge State Distributions (CSD) are plotted in Fig. 1.

Journal of Physics: Conference Series **2244** (2022) 012063 doi:10.1088/1742-6596/2244/1/012063

Table 1: The former (2017) and new (2020) performances (optimum charge state, efficiency, total efficiency and rise time) and tuning (support gas, injection - minimum - extraction axial magnetic field, microwave power and pressure at injection) of the LPSC ECRCB obtained for Na, K, Rb and Cs injected species. The 1+ (fly-through ions) efficiency is not included for the total efficiency calculation.

	Na 2017	Na 2020	K 2017	K 2020	Rb 2017	Rb 2020	Cs 2017	Cs 2020
Charge state	8+	8+	10+	9+	19+	18+	26+	26+
Eff. (%)	12.9	18.7	11.7	22.7	10.4	11.3	13.0	14.1
Total Eff. (%)	54.1	59.9	63.4	78.7	63.9	85.08	71.9	91.8
Rise time (ms/q)	12.9	26.8	8.2	13.5	29	16.7	44.2	12.8
Support gas	He	He	He	H_2	He	He	O_2	He
Binj (T)	1.57	1.56	1.57	1.58	1.57	1.58	1.57	1.56
Bmin (T)	0.47	0.46	0.44	0.45	0.45	0.46	0.46	0.47
Bext (T)	0.83	0.84	0.83	0.83	0.88	0.85	0.86	0.84
MW power (W)	486	770	290	520	480	700	500	670
Pinj ($\times 10^{-8}$ mbar)	46	5	44	14	40	5	52	4

4. Discussion

An increase of the total efficiency can be noticed for all the species and could be explained by a better capture (we see a decrease of the 1+ and 2+ charge state efficiencies in Fig. 1) and a better transport in the beam lines. The Cs total efficiency is 91.8%. Taking into account the N+ beam line transmission, the total capture efficiency is in fact very high and close to 100% for Cs. For Na and K, certain high charge states (Na^{10+}, $Na^{11+}$and K^{13+}) cannot be measured because the resolving power of the N+ spectrometer does not allow a good separation from contaminant beams. Therefore, the given total capture is underestimated for Na and K. For every species, an increase of the efficiency for the optimum charge state is noticed, the CSD efficiency being shifted toward lower charge states for K and Rb as clearly visible in Fig. 1. The rise times are longer for Na and K with respect to the former measurements. The new results were all obtained at lower pressure measured at injection and at higher microwave power.

For K, a great improvement is obtained with an efficiency increasing from 11.7% for K^{10+} to 22.7% for K^{9+}. After the upgrade, the first experiments were carried out with the same support gas (He) and the obtained performances were close to the former ones. Changing the support gas to H_2 seems to explain the increased efficiency in these measurements. To better understand this, short pulse measurements were done with H_2 and He plasmas, the tuning being optimized for K^{10+} and K^{9+}, respectively. The CT method was applied to the measured N+ responses to compare the plasma parameters. This study is described in detail elsewhere [10]. Table 2 summarizes the tunings and Fig. 2 compares the efficiency CSDs obtained during the short pulse experiments. The same trend is observed in Fig. 2 as in Fig. 1 b. The efficiency CSD modification and the high increase of K^{9+} efficiency can be explained by a change of the ionisation, charge exchange and confinement times and by an optimisation on K^{9+} which corresponds to a closed shell configuration [10]. The production of K^{9+} beam is favoured by a decrease of the charge exchange time for high charge states, a decrease of the ionisation time for the low charge states and a decrease of the K^{9+} confinement time. The ionisation time is inversely proportional to the electron density and to the ionisation rate coefficient [9]. The charge exchange time is inversely proportional to the neutral density and to the charge exchange rate coefficient.

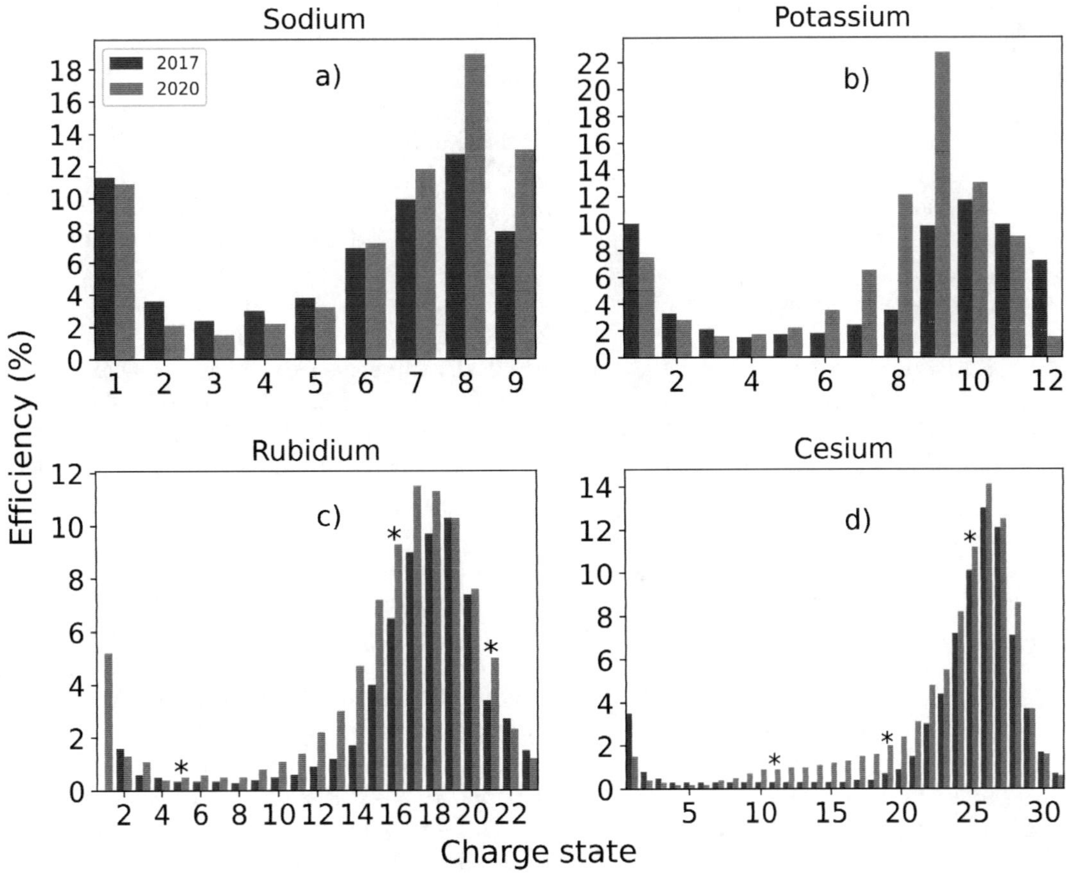

Figure 1: a) Na, b) K, c) Rb, d) Cs efficiency CSD comparison between former (blue) and new (red) results. CS efficiencies marked with * were extrapolated from adjacent charge states.

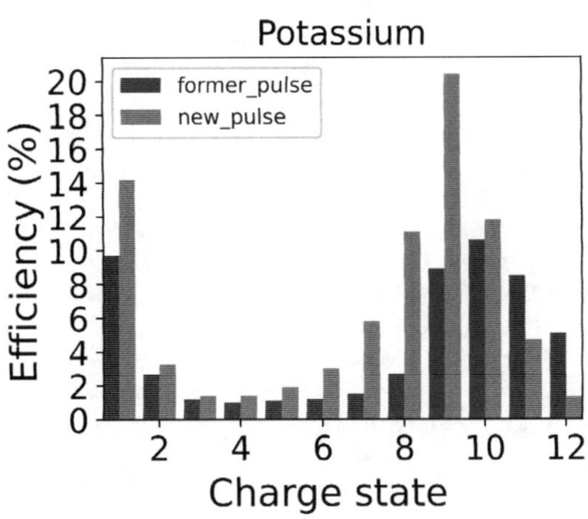

Figure 2: K efficiency CSD comparison for the pulse mode experiments with H_2 (red) and He (blue) support gas.

	K with He	K with H_2
Charge state	10+	9+
Eff. (%)	10.6	20.4
Total Eff. (%)	44.5	66.1
Rise time (ms/q)	13.2	14.6
Support gas	He	H_2
Binj (T)	1.58	1.57
Bmin (T)	0.45	0.44
Bext (T)	0.83	0.84
MW power (W)	500	530
Pinj ($\times10^{-8}$ mbar)	9	13.5

Table 2: The performances and tuning of the LPSC ECRCB for the pulse mode experiments with K obtained with He and H_2 support gas. The 1+ efficiency is not included for the total efficiency calculation.

In the new configuration, the electron density and the plasma energy content are found to be higher, which could in part be explained by a slightly higher microwave power. A higher neutral density and a higher relative speed between the ions and neutrals could explain the charge exchange time decrease. The higher relative speed between the ions and the neutrals is very likely due to the use of H_2 as support gas, instead of heavier He.

Regarding the charge breeding times, this value also depends on the plasma characteristic times (ionisation, charge exchange and confinement) but also on the different steps of ionisation and charge exchange experienced by the ions before their extraction; the confinement time represents a cumulative value from the charge state 1+ [10]. The rise time increase, from 9 to 13.5 ms/q, observed along with the decrease of most of the characteristic times would be due to a larger number of ionisation and charge exchange steps before extraction.

For Na, Rb and Cs, the modification of the ECRCB tuning led to different efficiency CSD and rise times. Na^{8+} efficiency also greatly increased from 12.9% to 18.7%. Short pulse measurements and CT method analysis for each species could help understanding the evolution of the performances, which is difficult based only on the efficiency CSD and rise time measurements.

5. Conclusion

The qualification experiments after the 1+N+ beam line upgrade showed an increase of the total efficiency and an increase of the efficiency per charge state for all the studied species. In addition to the alignment and vacuum improvements, some parameters of the ECRCB, such as the support gas species, microwave power and coil currents were optimized to obtain these results. A shift to lower charge states in the efficiency CSD was measured for K and Rb together with longer rise times for Na and K. A high efficiency increase was obtained for Na^{8+} and K^{9+}, this latter being now greater than 20%. For K, the short pulse measurements and the CT method allowed to better understand the influence of the support gas change on the plasma parameters and on the change of the efficiency CSD. The ECR plasma studies will continue, in particular based on the use of the CT method, to estimate the influence of the source parameters on the charge breeding performances. The next step of the LPSC PHOENIX ECRCB R&D program will consist in optimizing the magnetic structure and increasing the plasma chamber volume.

References

[1] Sortais P, Bruandet J, Bouly J, Chauvin N, Curdy J, Geller R, Lamy T, Sole P and Vieux-Rochaz J 2000 *Review of Scientific Instruments* **71** 617 (*Preprint* `https://doi.org/10.1063/1.1147232`) URL `https://doi.org/10.1063/1.1147232`

[2] Angot J, Galatà A, Maunoury L, Thuillier T, Baylac M, Migliore M and Sole P 2020 *ECRIS 2020 proceedings WEZZO02*

[3] Lamy T, Angot J, Thuillier T, Delahaye P, Maunoury L, Choinski J, Standylo L, Galata A, Patti G, Koivisto H and Tarvainen O 2014 *ECRIS 2014 Proceedings WEOBMH01*

[4] Lamy T, Angot J, Sortais P and Galatà T T A 2012 *Review of Scientific Instruments* **83**(02A909)

[5] Angot J, Thuillier T, Tarvainen O, Baylac M, Jacob J, Lamy T, Migliore M and Sole P *poster*

[6] Angot J, Tarvainen O, Thuillier T, Baylac M, Lamy T, Sole P and Jacob J 2018 *Phys. Rev. Accel. Beams* **21**(10) 104801 URL `https://link.aps.org/doi/10.1103/PhysRevAccelBeams.21.104801`

[7] Tarvainen O, Koivisto H, Galatà A, Angot J, Lamy T, Thuillier T, Delahaye P, Maunoury L, Mascali D and Neri L 2016 *Phys. Rev. Accel. Beams* **19**(5) 053402 URL `https://link.aps.org/doi/10.1103/PhysRevAccelBeams.19.053402`

[8] Tarvainen O, Lamy T, Angot J, Thuillier T, Delahaye P, Maunoury L, Choinski J, Standylo L, Galatà A, Patti G and Koivisto H 2015 *Plasma Sources Science and Technology* **24** 035014

[9] Angot J, Luntinen M, Kalvas T, Koivisto H, Kronholm R, Maunoury L, Tarvainen O, Thuillier T and Toivanen V 2020 *Plasma Sources Science and Technology* **30** 035018 URL `https://doi.org/10.1088/1361-6595/abe611`

[10] Luntinen M, Angot J, Tarvainen O, Toivanen V, Thuillier T and Koivisto H 2021 *ICIS 2021 proceedings*

Optimization of ion transport in a combined RFQ Cooler with axial magnetic field

M. Cavenago[1], M. Romé[2], G. Maero[2], L. Bellan[1], F. Cavaliere[2], M. Comunian[1], A. Galatà[1], M. Maggiore[1], N. Panzeri[2], A. Pisent[1], L. Pranovi[1], M. Sattin[1]

1 INFN-Laboratori Nazionali di Legnaro, Legnaro, IT
2 University of Milano, Dip Fisica, e INFN-MI, Milano, IT

E-mail: cavenago@lnl.infn.it

Abstract. The accurate mass spectrometry (with resolution goal 1:20000) of exotic ions requests beams with low energy spread (goal is about 0.5 eVrms or lower) and low transverse emittance, so it is necessary to cool ions produced by a fission source. In a radiofrequency (rf) quadrupole cooler (RFQC), collisions decrease ion kinetic energy, while rf and DC voltages confine and reaccelerate ions towards the extraction, where the cold ion beam is formed. Operation is based on carefully chosen tunings of voltages and of gas pressure p_g, which requires an adequate pumping system; efficient simulations of extraction systems, which critically depend on residual speed of ions, are also important. Progresses in the experimental setup are described. Indications from simple ray tracing and 'ray+collision' tracing are compared. Results are applied to a simple triode extraction system, and more flexible configurations are considered.

1. Introduction

Radio Frequency Quadrupole Coolers (RFQC) are a typical part of facilities for reacceleration of beams produced by fission targets, as in the SPES (Selective Production of Exotic Species[1]) project at LNL. Cooling is obtained by collisions of ions with a gas[2, 3] as Helium, while confinement is due to a radiofrequency (rf) field, which gives focusing, but also introduces an heating effect. Additional confinement may be provided with a solenoidal magnetic field $\mathbf{B}^s$, so that rf amplitude voltage V_{rf} can be reduced and ion transmission improved; an RFQC prototype built at LNL[4] has reduced dimensions (length $L_1 = 0.72$ m, outer diameter $\cong 0.15$ m), so it can be inserted in the Eltrap facility at Milan University[5], a $L_0 \cong 1.6$ m long solenoid, with a 0.25 m inner diameter vacuum chamber (see Fig. 1). Indeed, the required optimization of transmission, transverse cooling and longitudinal cooling is challenging[6, 7] and up to now calculations are based on many simplifying assumptions, here discussed in section III, so that experimental tests are necessary also to validate calculations. Set-up, whose status is reported in section II, also integrates an emittance meter EMI1, two Faraday cups FC1 and FC2, a commercial ion source IS1 for initial tests, and the gas differential pumping of RFQC.

Due the quadrupole pattern of electrodes and since wavelength $2\pi c/\omega$ is much larger than r_0, half distance between facing electrodes (see fig 2.a), inside RFQC the rf field is $\mathbf{E}_f = -\nabla\phi_f$ with an electric potential $\phi_f \cong 2V_{rf}(xy/r_0^2)\cos\omega t$ near z axis; in our case $\omega \cong 2.5 \times 10^7$ rad/s (about 4 MHz). The RFQC is divided in N sections by gaps (Fig. 2.b and 2.c); we have 4*N

Journal of Physics: Conference Series **2244** (2022) 012064 doi:10.1088/1742-6596/2244/1/012064

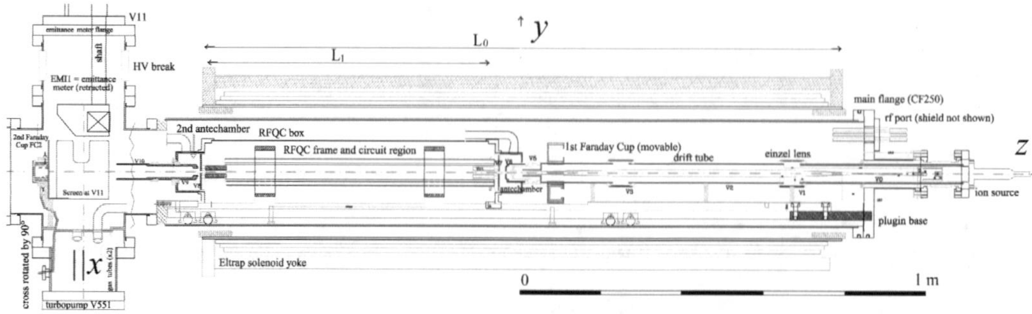

Figure 1. Eltrap+RFQC setup; see Figs 2,3 of Ref [7] and Fig 1 of Ref [11] for details.

electrodes, which can be biased to different static potentials $\phi_s = V_n^c$ (with $n = 1, ...10$), so to add a static field $\mathbf{E}_s = -\nabla\phi_s$.

Ions are singly charged with atomic number $A \leq 133$; kinetic energy K_i is typically 40 keV in our accelerator applications; our IS1 provides a maximum of $K_i = 5$ keV, with voltages quoted here adjusted for this case. The accurate mass spectrometry (with resolution goal 1:20000) of exotic ions requests beam with low energy spread (goal is $\leq$ 0.5 eVrms, while exotic ion source gives few eVrms spread) and low transverse geometrical emittance: we need $\epsilon_{geo} < 7$ mm-mrad for the beam exiting the RFQC with $K_i = 5$ keV, with input $\epsilon_{geo}^i \cong 24$ mm-mrad (or equivalently $\epsilon_{geo} < 2.4$ mm-mrad and $\epsilon_{geo}^i \cong 9$ mm-mrad with $K_i = 40$ keV[1, 8]). The so-necessary ion cooling requires several steps: 1) the injection, where the beam is focused to enter the RFQC, well between electrodes ($r_0 \cong 4.5$ mm in our case), and decelerated to $K_i \leq 0.1$ keV, so that collisions are more effective[9] and transit time longer; 2) transit in the RFQC, where collisions decrease ion kinetic energy, while rf and solenoid fields confine ions, and the bias potential ϕ_s (between RFQC sections) pushes ions towards the extraction; 3) extraction, where the cold ion beam is formed and accelerated again. This is critical since the beam size and velocity depends on previous steps, and reacceleration overfocusing must be controlled. We consider triode systems for extraction, while different solutions (microRFQ[10], tetrode) exist.

2. Setup

RFQC integration in the Eltrap test facility is shown in Fig. 1; the alkali metal ion source IS1, the injection line, the RFQC and its extraction line, are mounted on a rigid 2 m long plugin, which can be assembled, cabled, aligned, and tested outside, and then inserted inside the Eltrap vacuum chamber. Plugin support is a machined Al bar joined to a CF250/CF63 transition and a CF63 flanged isolator where IS1 is mounted. Let z be the solenoid axis, with origin at the

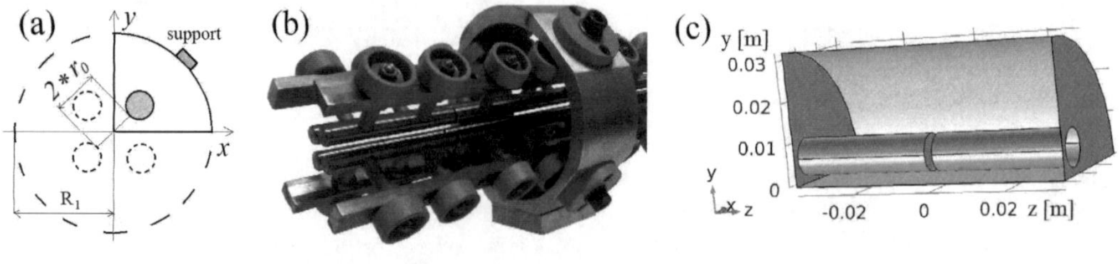

Figure 2. (a) xy section of RFQC; for symmetry, electrostatic simulation reduces to one-quarter geometry; (b) end of RFQC, showing the last two sections of electrodes and the support; (c) 3D geometry of a section of RFQC, showing the gap between electrodes.

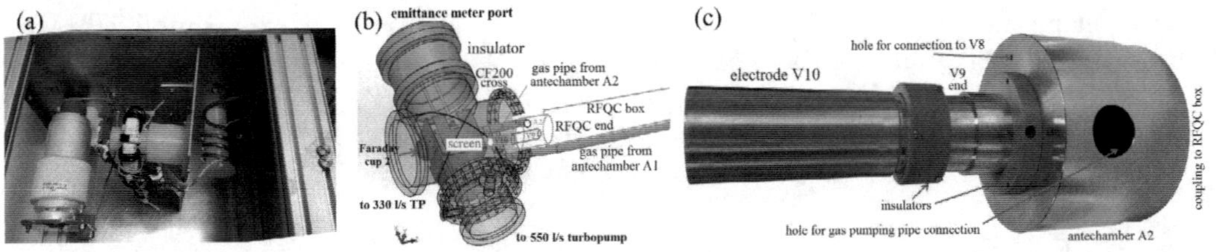

Figure 3. (a) matching box; (b) scheme of EMI1 and pumps; (c) view of triode extraction line

solenoid middle; ion source IS1 cathode is placed at $z_c = 1.021$ m, with its exit lens at $z_e = 0.996$ m; beam is directed towards negative z and exit lens produces a waist at $z_w = 0.85$ m (spot radius 0.5 mm, divergence 4 mrad). Ion beam propagates inside drift tubes V0 (joined to IS1 exit tube), V2, V4 held at say $V_0 = -4.8$ kV respect to ground, so that source cathode is at potential $V_s = K_i/e + V_0 = 200$ V, which can be precisely measured. Ion beam is refocused by einzel lens V1 and V3, and electrodes V5 and V6 decelerate it[11]. A Faraday cup FC1 can be inserted or removed inside drift tube V4.

The RFQC is enclosed in a gas tight box (-0.8 m $< z < -0.05$ m), at ground, except for the end plates V_7 and V_8 (within 100 V from ground). Extraction triode is formed by this plate V_8, an electrode V_9 (partly inside an antechamber A2) and a drift tube V_{10}; beam may finally reach the pepper pot emittance meter EMI1 or second Faraday cup FC2. The movable emittance meter EMI1 is mounted on a 40 kV ceramic break (with axis $z = -1.085$m plane) on a CF200 cross (with gas pumping). Beam image is formed on the EMI1 scintillating screen at $z = -1.055$ m and reflected to a camera. Current of Cs^+ ions is limited to $I_i \leq 200$ nA when the emittance meter is used. In principle FC2 can measure energy spread provided its gate voltage V_g is stable enough; stability $V_g - V_s$ within 0.1 Vpp is necessary, with electronics in development. Other critical equipment, IS1 and EMI1 were fully tested in a separated bench.

The matching box (Fig 3.a) splits the rf input power into two signals V_{rf1}, V_{rf2}, matched to capacitive loads about 200 pF, in phase opposition $V_{rf1} = -V_{rf2}$; RFQC electrodes are connected by capacitors to these signals and by resistors to the bias voltages V_n^c; this network is placed in the RFQC box, on a thermal sink (a copper bar with mica washers).

Differential gas pumping is needed[12]: inside RQFC a pressure $p_g = 2$ to 9 Pa must be provided, while we prefer $p < 0.001$ Pa in beamline and near EMI1; so RFQC box beam holes are restricted below 3 mm diameter and face antechambers A1 and A2, differentially pumped by 20 mm pipes (fig. 3.b); shaping these pipes to reach the V551 turbopump proved laborious. Two smaller pipes connects the RQFC box to the CF250 plugin flange: one connects to a capacitive p_g meter, the other is the gas input, operated with a small duty cycle (to save He). Most parts were acquired and tested, even if final assembly was delayed by emergency situation.

3. Ion transport and simulations

Let e and m_i be the ion charge and mass, and m_t the gas molecule mass. Due to rf field, each ion trajectory oscillates with angular frequency ω (micromotion) around an average trajectory $\mathbf{x}(t)$ (macromotion [13], driven by the so-called ponderomotive force $\mathbf{F}_p \equiv e\mathbf{E}_p = -e\nabla\phi_p$ with ϕ_p ponderomotive potential). The basic results for macromotion and friction force $\mathbf{F}_r$ due to gas collision was detailed elsewhere[2, 7, 11], with $\phi_p \cong (m_i/2e)\omega_M^2(x^2+y^2)$ and $F_r = -m_i\mathbf{v}\nu_m$ where ν_m is the collision frequency (depending on gas density n_g and on macromotion velocity $\mathbf{v} = \mathrm{d}_t\mathbf{x}$, see Fig. 4 in Ref[7]). Macromotion frequency ω_M and cyclotron frequency Ω_i are

$$\omega_M = \frac{e}{m_i\sqrt{2}}\frac{V_{rf}}{\omega r_0^2} \quad , \quad \Omega_i = \frac{e|B^s|}{m_i} \tag{1}$$

A convenient parameter of rf strength is $V_p = (m_i/2e)\omega_M^2 r_0^2$; simulations are simplified by axial symmetry of ϕ_p (2D3V) and of other beamline components (einzel lens, the triode, the ion source and the solenoid); $H_i = eV_s$ is value of energy $H = K_i + e\phi$ at ion source. Since $m_i/m_t \gg 1$, many gas collisions are needed to change $\mathbf{v}$ significantly; 1st order effect is the friction force given above, 2nd order effect is a straggling, which can be modeled by a Langevin-style equation:

$$\mathrm{d}_t\mathbf{v} = \frac{e}{m_i}\left(\mathbf{E}_s + \mathbf{E}_p + \mathbf{v}\times\mathbf{B}^s\right) - \nu_m\mathbf{v} + \boldsymbol{\eta} \quad , \quad \langle\boldsymbol{\eta}\rangle = 0 \quad , \quad \langle\eta^m(t)\eta^n(t')\rangle = D^{mn}\delta(t-t') \tag{2}$$

where $\langle\rangle$ indicates the statistical average, $\boldsymbol{\eta}$ is a random kick, with 2nd order moment given by diffusion tensor D, and $m, n = 1,2,3$ or x, y, z are coordinate indexes. A method[14] to simply account for D is summarized as follows. In paraxial approximation $v_x, v_y \ll |v_z|$, and the D non-zero components are $D^{xx} = D^{yy}$ and D^{zz}, calculated from interatomic potential, in the same way as the mobility and collision frequency[2, 7]. Let $\bar{x} = \langle x\rangle$; then $\tilde{x} = x - \bar{x}$ is the random fluctuation around the ray $\bar{x}$, computed with ray-tracing as usual. For the 2nd moments of the fluctuation, defined as

$$a_{mn} = \langle\tilde{x}^m\tilde{x}^n\rangle \quad , \quad b_{mn} = \langle\tilde{v}^m\tilde{v}^n\rangle \quad , \quad c_{mn} = \langle\tilde{x}^m\tilde{v}^n\rangle \quad , \tag{3}$$

time evolution follows from eq. 2 linearized around $\bar{x}$ and from symmetries. In our case (paraxial approximation) the non-zero independent components are not only a_{zz}, b_{zz}, c_{zz}, and $a_{xx} = a_{yy}$, $b_{xx} = b_{yy}$ and $c_{xx} = c_{yy}$ as trivially expected, but also $c_{xy} = -c_{yx}$ due to solenoid field (a total of 7 quantities out of 21 components). For example

$$d_t a_{xx} = 2c_{xx} \quad , \quad d_t c_{xy} = -\Omega_i c_{xx} - \nu_m c_{xy} \quad , \quad d_t b_{xx} = 2(D^{xx} - \nu_m b_{xx} - \psi_{,xx}c_{xx}) \tag{4}$$

with $\psi = e(\phi_s + \phi_p)/m_i$ and similarly for others. The pencil emittance, defined by

$$\epsilon_x^p = \sqrt{a_{xx}b_{xx} - c_{xx}^2 - c_{xy}^2} \qquad \rightarrow \qquad \mathrm{d}_t(\epsilon_x^p)^2 = 2a_{xx}D^{xx} - 2\nu_m(\epsilon_x^p)^2 \quad , \tag{5}$$

is constant where gas is absent (where D^{xx} and ν_m are zero). Evolution of reduced Hamiltonian $h = H/m_i = \frac{1}{2}\mathbf{v}^2 + (e\phi_s/m_i)$ and its fluctuation 2nd momentum $b_{hh} = \langle(\tilde{h})^2\rangle$ are also computed.

Simulation results shown in Figs 4 and 5 refer to standard tuning of RFQC; $B_z = 0.11$ T, $V_p = 1.95$ V, injection energy $H_i = 200$ eV or $H_i = 200 \pm 5$ eV, and n-th section bias voltage $V_n^c = (11-n)V_o$ with $V_o = 10$ V; both gas pressure p_g inside RFQC (from 2 to 9 Pa) and triode lens voltage V_9 were scanned. First note in Fig. 4.a (from ray-tracing with different H_i) that final average energy H_a differs at most by 1 eV for $p_g = 2$ Pa and is negligible for $p_g = 9$ Pa. On the other hand (Fig. 4.b), the rms contribution of fluctuation (or diffusion) roughly changes

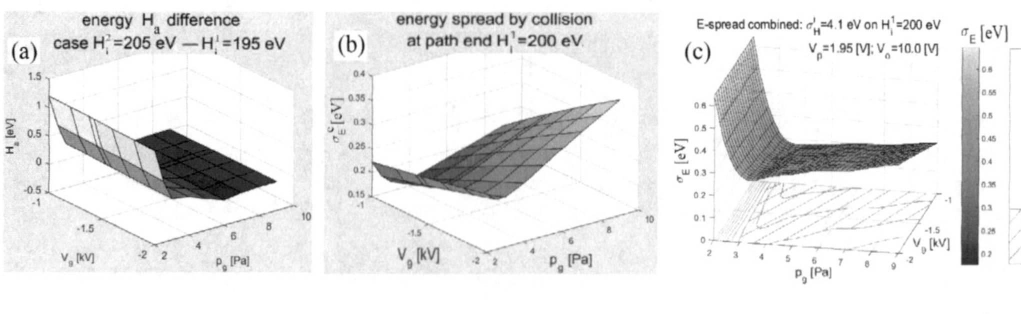

Figure 4. (a) difference of average final energies H_a vs gas pressure p_g and lens voltage V_9 for two H_i; (b) rms fluctuation induced by gas collisions; (c) combined rms energy spread.

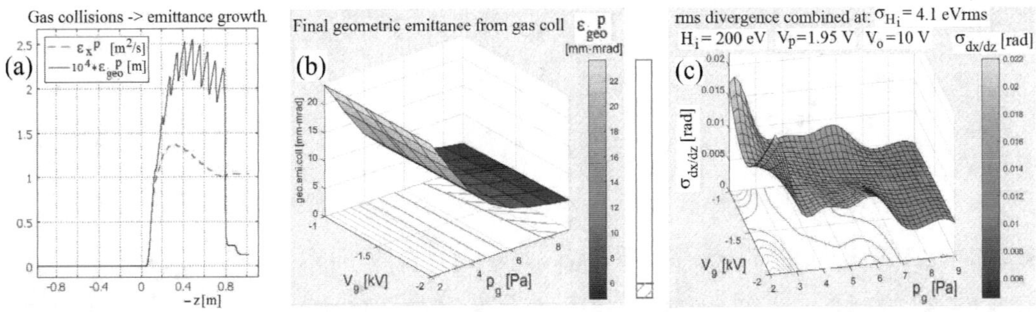

Figure 5. (a) pencil emittance ϵ_x^p vs $-z$; growth and damping due to gas collisions with $p_g = 4$ Pa; also plotted rms ϵ_{geo}^p; (b) output value of ϵ_{geo}^p; (c) output value of rms divergence.

as p_g. Relative weight of these effects depends from input energy spread σ_{H_i} (that is about 0.1 eV for IS1 and is expected ≥ 4 eV in exotic beam application[1, 8]). Assuming $\sigma_{H_i} \cong 4$ eV gives Fig. 4.c result, with a wide optimum plateau (from 3 to 9 Pa) for the output energy spread σ_E, which includes shown effects and also minor terms due to beam size; hatched region is $\sigma_E < 0.3$ eVrms. To mimic $\sigma_{H_i} \cong 4$ eV in Eltrap installation, we simply sum results at several H_i, for example at 195, 200 and 205 eV.

The pencil emittance is shown in fig 5.a for $H_i = 200$ eV (note $-z$ abscissa to follow beam direction); initial value is zero (no fluctuation) until gas region is met; emittance reaches roughly a value $(a_{xx}D^{xx}/\nu_m)^{1/2}$ slowly decreasing (since K_i is decreasing) until the RFQC exit is reached, after which ϵ_x^p is frozen. Evolution of pencil geometric emittance $\epsilon_{geo}^p \equiv \epsilon_x^p/|v_z|$ reveals more features: passing from one section to the other, each ion is accelerated (by bias voltage drop) in a $2r_0$ distance[7], so ϵ_{geo}^p has sharp repeated decreases; finally after RFQC exit we see a final decrease, due to the large reacceleration. A scan of output ϵ_{geo}^p in fig 5.b simply reveals a decrease with p_g (good region $\epsilon_{geo}^p < 6$ mm-mrad is hatched). A preliminary result for rms total divergence is shown in fig. 5.c; note that the $V_9 \cong -1.4$ kV optimum shows up, with a preferred trend for large $p_g \cong 9$ Pa.

Experimental tests of emittance ϵ_{geo} are expected in the beginning of 2022, while σ_E measurements request more time. Extraction tests with more electrodes (tetrode and so on) are envisioned through 2023, also to investigate a better balance of ϵ_{geo} and σ_E cooling and p_g effects.

Acknowledgments

Work set up in collaboration and financial support of INFN-Group 5.

[1] Comunian M et al. 2020 *J. Phys. Conf. Ser.* **1401** 012002.
[2] Schwarz S 2006 *Nucl. Instrum. Meth.* **A 566** 233.
[3] Herfurth F et al. 2001 *Nucl. Instrum. Meth. Phys. Research* **A469** 254.
[4] Maggiore M et al. 2014 *Rev. Sci. Instrum.* **85** 02B909; doi: 10.1063/1.4830357
[5] Amoretti M. et al. 2003 *Rev. Sci. Instrum.* **74** 3991.
[6] Guo-Zhong Li and Werth G 1992 *Phys. Scr.* **46** 587.
[7] Cavenago M et al. 2019 *Rev. Sci. Instrum* **91** 113324.
[8] Comunian M et al. 2018 *9th Int. Particle Acc. Conf., IPAC18* pp 3252-3254, online at www.jacow.org.
[9] Lieberman M A and Lichtenberg A J 1994 *Principles of Plasma Discharges and Material Processing*, John Wiley, New York, especially chapter 4.
[10] Boussaid R, Ban G, Quéméner G, Merrer Y and Lorry J 2017 *Phys. Rev, ST-AB* **20** 124701.
[11] Cavenago M et al. 2016 *Rev. Sci. Instrum.* **87** 02B504.
[12] Barquest B R et al. 2018 *9th Int. Particle Acc. Conf., IPAC18*, pp 3914-3916; online at www.jacow.org.
[13] Moore R B, Gianfrancesco O, Lumbo R and Schwarz S 2006 *Int. J. Mass Spectr.* **251** 190.
[14] Cavenago M, unpublished.

Design and thermal simulations towards a high intensity radioactive ion source for ISOL@MYRRHA

Sophie Hurier[1,2], Kim Rijpstra[1], Philip Creemers[1], João Pedro Ramos[1], Lucia Popescu[1], Thomas Elias Cocolios[2]

[1]Belgian Nuclear Research Centre, SCK CEN, Mol, Belgium,
[2]KU Leuven, Institute for Nuclear and Radiation Physics (IKS), Leuven, Belgium.

E-mail: `sophie.hurier@sckcen.be`

Abstract. MYRRHA will be the world's first large-scale Accelerator Driven System project at power levels scalable to industrial systems. ISOL@MYRRHA will produce Radioactive Ion Beams (RIBs) using the Isotope Separation On-Line (ISOL) technique, with increased isotope production by high intensity primary beams over a long period while maintaining a high-quality RIB. Higher atom flux produced prevalently affects the ISOL ion source. A surface ion source is chosen as a first source because of its reliability and simple design. To understand the hot cavity's behaviour, finite element thermal-electric simulations were performed. To start, a heating system study with experimental results from the SPES project was reproduced. This concept was then modified by: electrically insulating the source from its support, adding a feedthrough, transforming a passive thermal screen into an active part. With this heating system upgrade, the ion source temperature profile can be adjusted, especially at its exit part where high temperature is expected to play a crucial role in ion production and extraction.

1. Introduction

The MYRRHA project is the first prototype of a subcritical lead-bismuth cooled reactor driven by a particle accelerator [1]. ISOL@MYRRHA [2] will extract part of the proton beam (100-600 MeV proton, up to 0.5 mA) and have it impact on a target, installed in a vacuum vessel to produce Radioactive Ion Beams (RIBs) with the Isotope Separation On-Line (ISOL) technique [3]. This ISOL facility will increase its isotope production by using high intensity primary beams for a longer period of time than existing facilities while aiming at maintaining the RIB quality.

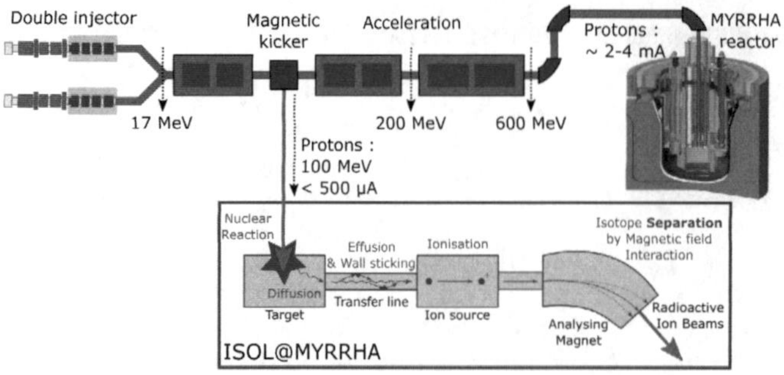

Figure 1: MYRRHA project and ISOL@MYRRHA diagram with the ISOL technique which combines: target bombardment with high-energy particles, release from the hot target, ionisation, extraction and acceleration by an electrical field towards analysing magnet for mass analysis and purification.

Journal of Physics: Conference Series **2244** (2022) 012065 doi:10.1088/1742-6596/2244/1/012065

An ion source for RIB production adapted to these new ISOL@MYRRHA conditions aims to have the same or higher total efficiency as already existing RIB ion sources at lower primary beam intensities. Because of the increased intensity and the longer operation time (several weeks), the source will face several challenges: saturation issues due to high atom influx, reliability issues due to thermal cycling and radiation damage. These issues have to be tackled while maintaining good RIB properties in terms of emittance, intensity and purity.

As a first ion source, the hot cavity is chosen. Used for surface ionisation [4] and resonant laser ionisation [5], they are simple and reliable sources for ISOL, but they are also known to get rather easily saturated. Cold spots at the cavity's end as well as a lack of drift field, are likely to contribute to this issue. In this work, we will study possible tools to improve upon these, through a thermal-electric design study using ANSYS® [6], a finite element solver.

2. Model and Validation

To validate our thermal-electric model, we compare it with experimentally validated simulations of an ion source heating system, presented in a SPES project study [7]. A 3D-model of the heating system was created and used in a simulation with boundary conditions, as shown in figure 2 and table 1. In the simulation settings, care was taken that only surfaces bordering the vacuum are allowed to emit heat radiation, which then also depended on the surface temperature.

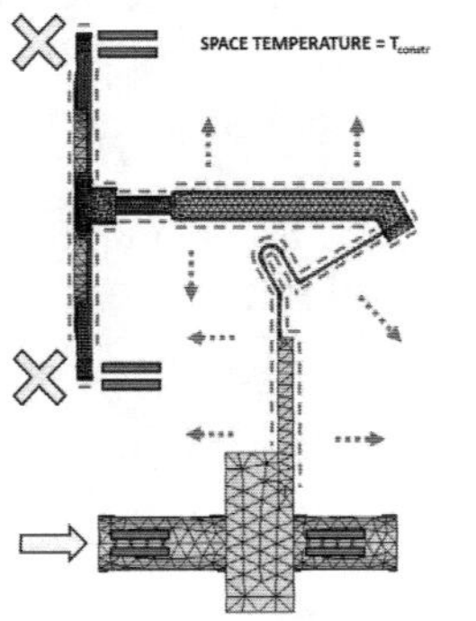

Figure 2: Model boundary conditions [7].

Temperature	=	T_{constr} = 25°C
Radiation emissivity	--→	ϵ(T°) of tantalum (ϵ=0.1 at 1°C to ϵ=0.3 at 2600 °C)
Voltage constraint	×	at 0 V
Current load	⇒	at 380, 350, 300 & 250 A
Material: Tantalum		Work function (ϕ=4.19 eV)
		Melting point (~ 3000 °C)

Table 1: ANSYS simulation boundary conditions following the representation given in figure 2.

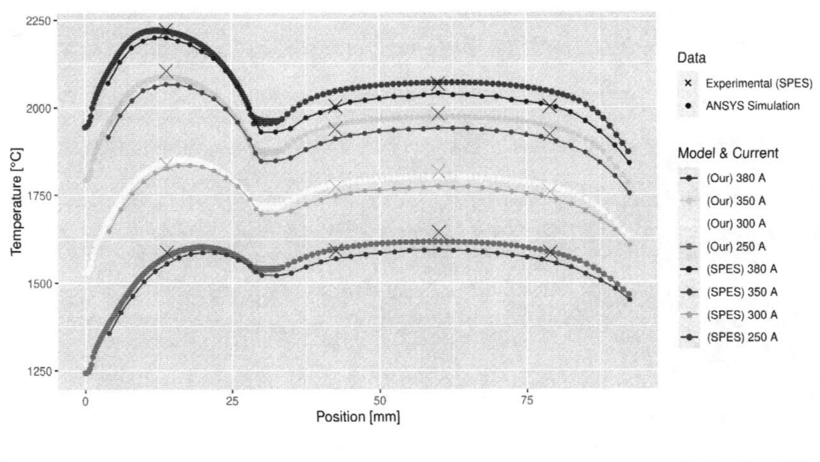

Figure 3: Temperature comparison between SPES and our model along the hot cavity (0-33 mm), the transfer line tube (30-95 mm). A linking pieces (29-35 mm) stand between those two tubes, but its temperature is not shown.

In figure 3, our model closely matches the experimental results obtained in Manzolaro et.al [7], especially the single measured point on the hot cavity. With this strong correspondence, we are confident that the correct physics and material data have been included in our model.

3. Results

3.1. Adding an output current feedthrough

Starting from the validated design above, our reference model, we adapted this ion source design. First the heating system was adapted to the feedthrough positions of the thermal test stand at SCK CEN, then the source heating system was insulated from its vacuum vessel. Instead of having an electric heating current from the transfer line to the base plate of the vacuum vessel, the current will run to an additional electrical feedthrough, see in figure 4). Three models, with different connections between the feedthrough and the ion source, were simulated and compared to our reference model (see part name in figure 4):

- A double connection to the ion source support ends,
- A single connection to the central part of the ion source support,
- A single connection to the thermal-screen end, see figure 4. This last one yielded the most promising result towards a higher temperature at the hot cavity exit.

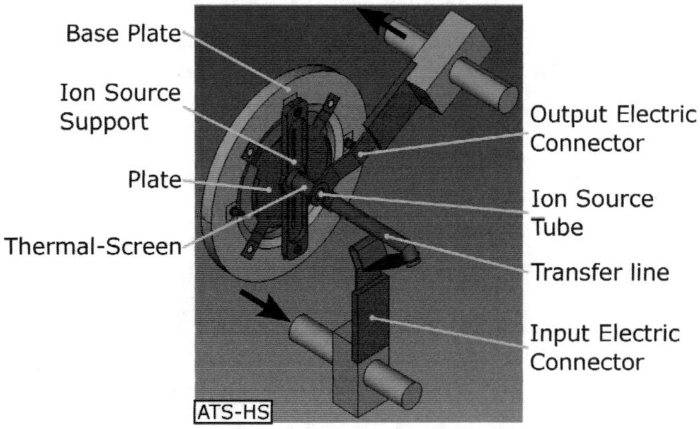

Figure 4: 3D-model design of the ATS-HS (Active Thermal-Screen Heating System), with the heating current connectors located at the bottom and top-right. A 0 V electrical constraint is imposed on the latter connector.

3.2. Transforming a passive thermal screen into an active part

With this thermal-screen connection of the output electric connector, see figure 4, a passive thermal-screen is transformed into an active part through which the electric current flows, see figure 5. This heated thermal-screen surrounds the ion source and keeps it hotter. Thus, we obtain an output connection with the added feedthrough and increase the thermal-screen body temperature in the process.

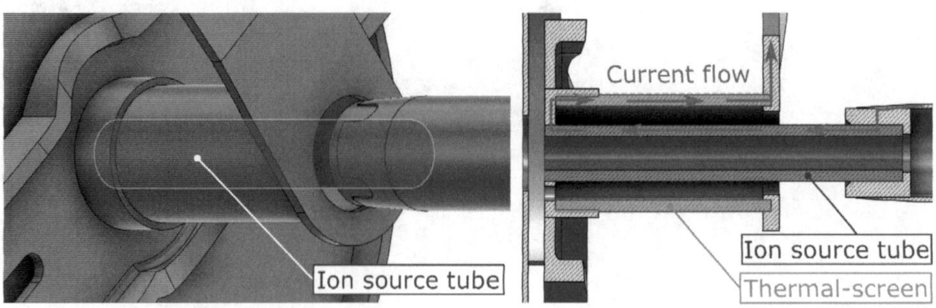

Figure 5: Ion source tube and thermal-screen. (Left) Ion source tube shape, in blue, inside the heating system; (Right) Model cross-section view along the ion tube centre plan. Ion source section is in blue and the thermal-screen section is in green.

The reference model with only one feedthrough, named "REF" in figure 6, has a thermal-screen cylinder shape with 10 mm length and 0.3 mm thickness. Given the mechanical stresses that may stem from the electric connector, two other thickness sizes (0.5 mm and 1 mm) were evaluated. The ion source temperature profile was tune with three other lengths (15 mm, 20 mm, 25 mm) of the thermal-screen. In figure 6, except for the "REF" model, all the other thermal-screen thicknesses were fixed at 0.5 mm.

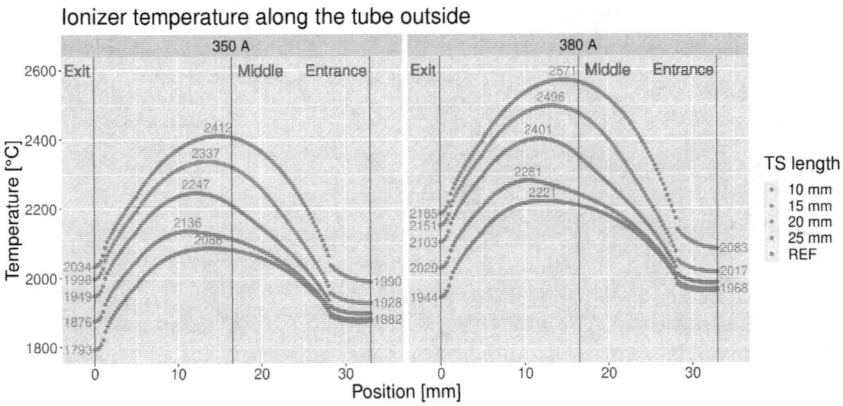

Figure 6: Ion source temperature with different thermal-screen (TS) lengths at two current load. Values represent temperatures at those positions, with in the middle the cavity maximum temperature.

In figure 6, the thermal-screen length influences the hot cavity temperature profile and its maximum temperature which increases as the the thermal-screen length increases. With ATS-HS design, the cavity exit will reach temperatures higher than the cavity entrance temperature. The model with a thermal-screen of 15 mm long and 0.5 mm thick was selected as a good compromise between the cavity temperature increase and the thermal-screen body strength. With this model, the current can also be lowered to 350 A while keeping a similar temperature profile as the reference model has at 380 A. This will help to reduce the thermal stress on the other heating system component.

3.3. Electrical and thermal insulation of the ion source heating system from its base plate

To insulate the heating system from its base plate, i.e. aluminium vacuum chamber surrounding the target and ion source, different washer layer configurations were tested, including one where the original plate is rotated by 45°. The configuration with the Base Plate/Plate/Washer/Support layer was selected, see figure 7. The base plate and the plate are connected electrically, but separated from the ion source support, using an aluminium oxide insulator. Then, the 45° plate rotation gives way to a thicker washer between the base plate and the support, allowing for a higher voltage difference there. It also allows to separate electrically the plate from the base plate with an additional washer in case we want to transform this plate into an other extraction electrode.

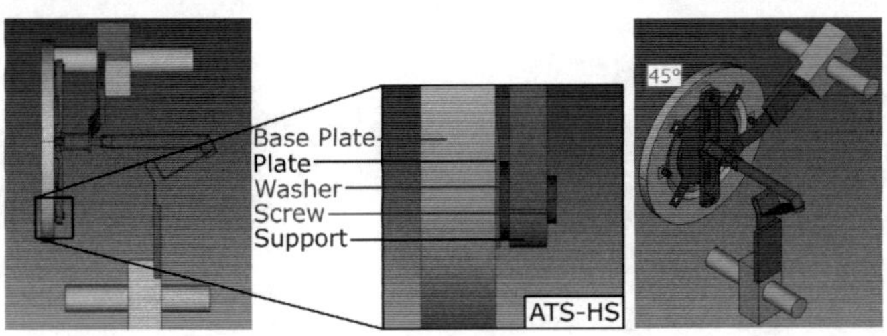

Figure 7: (Right) Washer layer configurations, (Centre) ATS-HS 3D-model side view without the 45° plate rotation, (Left) ATS-HS 3D-model with the 45° plate rotation.

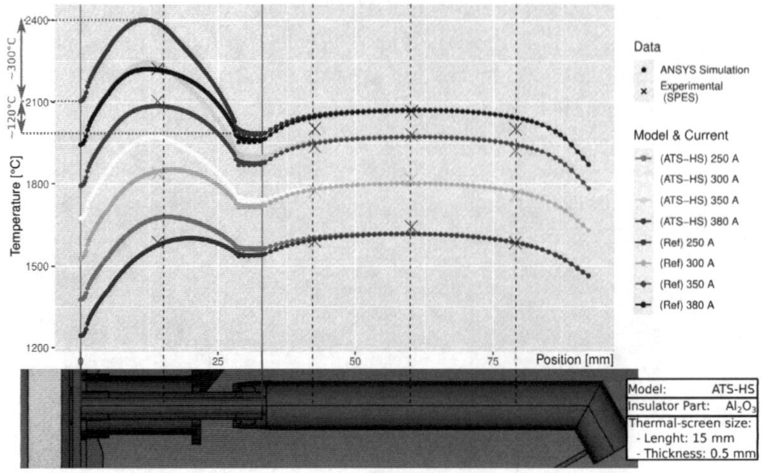

Figure 8: ANSYS simulation results: Temperature along the ion and transfer-line tube with a 15 mm thermal-screen length.

In figure 8, the temperature along the ion source tube is increased compared to the reference model without a second feedthrough and isolation system. The difference with the tube exit and the peak temperature remains 300°C, but this exit is now 120°C warmer than the ion source entrance, which is still at the same temperature as the reference model. This temperature difference might reduce the ion condensation in the cold spot at the ion source end and increase the ion extraction in this region with the help of the extraction electrode.

4. Conclusion

For the development of a high intensity surface ion source for ISOL@MYRRHA, a thermo-electric study is presented in this work. Modifications were made to an already existing source by adding an output current feedthrough, transforming a passive thermal screen into an active part in the process and insulating it electrically from the base plate. The cavity temperature profile can be changed by modifying this new thermal-screen size. The purpose of those changes is to get a higher temperature at the source exit which should lead to a better ionisation/extraction efficiency. Like the SPES design [7], to avoid misalignment of the cavity due to the material thermal expansion at high temperature, an alignment system will be added to our design. In a next phase, these thermal-electric results will be used as input for plasma simulations. Then finally, the hot cavity configuration will be prototyped and studied experimentally on a heating test stand recently constructed at SCK CEN.

Acknowledgements

This work has been carried out within the scope of a PhD supported by SCK CEN and KULeuven. We also wish to acknowledge M. Manzolaro from INFN-LNL for his help with details of the SPES Surface Ion Source.

References

[1] Popescu, Lucia 2014 *EPJ Web of Conferences* **66** 10011
[2] Ashford M, Popescu L, Houngbo D, Dierckx M and Aït Abderrahim H 2020 *Nucl. Instrum. Methods Phys. Res. B* **463** 244–247
[3] Van Duppen P 2006 *Isotope Separation On Line and Post Acceleration* (Berlin, Heidelberg: Springer Berlin Heidelberg) pp 37–77
[4] Kirchner R 1990 *Nuclear Instruments and Methods in Physics Research Section A* **292** 203–208
[5] Fedosseev V N, Kudryavtsev Y and Mishin V I 2012 *Physica Scripta* **85** 058104
[6] ANSYS®, Academic Research Mechanical, Release 20.1
[7] Manzolaro M, D'Agostini F, Monetti A and Andrighetto A 2017 *Review of Scientific Instruments* **88** 093302

Charge Breeder at GANIL: metal charge-bred elements

L. Maunoury[1], M. Dubois[1], S. Damoy[1], O. Bajeat[1], P. Chauveau[1], P. Delahaye[1], R. Frigot[1], S. Hormigos[1], P. Jardin[1], and J.C. Thomas[1]

[1]Grand Accélérateur National d'Ions Lourds, bd Henri Becquerel, BP 55027, F-14076 Caen cedex 05, France

O. Tarvainen[2]

[2] STFC ISIS Pulsed Spallation Neutron and Muon Facility, Rutherford Appleton Laboratory, Harwell OX11 0QX, United Kingdom

maunoury@ganil.fr

Abstract. The charge breeder of the SPIRAL1 (SP1CB) facility provided this year to physicists new Radioactive Ion Beams (RIB) for experiments and machine development, showing significantly improved performances compared to those obtained during the initial on-line commissioning of the SPIRAL 1 upgraded facility [1]. These improved performances were obtained thanks to thorough studies conducted off-line with 1+ ion beams produced by FEBIAD and ECR ion sources, successfully demonstrating the ability to couple SP1CB with them. In particular, the SP1CB established its capability to efficiently boost condensable elements such as $^{19}F^{n+}$, $^{32}S^{n+}$ and $^{54}Fe^{n+}$, in addition to more standard alkali elements. The charge breeding efficiencies have been investigated varying several parameters: buffer gas, beam transverse emittances. The ΔV curves of stable elements as well as radioactive ones were recorded for high charge states. Their trends are discussed in more detail. Finally, as molecular beams provide some advantages compared to atomic ion beams for selecting isobaric species and optimizing transport of the radioactive elements from the hot production target to the SP1CB plasma, we discuss if the charge-breeding efficiencies could also take advantage of beams in the molecular form. The performances of the 1+/N+ charge-breeding process were investigated using SF_6 molecules broken into SF_x^{1+} ions and compared to regular F^{1+} and S^{1+} ions. This contribution will deal with these topics and latest results will be showed.

1. Introduction

The GANIL charge breeder (ECRIS type charge breeder) of the SPIRAL1 facility is now regularly under operation [1-3]. After a commissioning phase at LPSC, the SP1CB has been implemented in the SPIRAL 1 beam lines, and has provided in the last 2 years charge-bred ions mostly from alkali elements (stable and radioactive), establishing its ability to deliver intense post-accelerated RIBs of high quality to nuclear physicists. The 1+ beams were delivered from a Target Ion Source System (TISS) based on a FEBIAD ion source [4] used in surface ionization (SI) mode [2]. Lately, the production of multicharged ions of non-alkali elements has also been investigated. For that purpose, the FEBIAD ion source was used in its more traditional plasma (P) mode to ionize a wide variety of condensable elements. In this mode, the injection of molecular beams was additionally investigated. Here, the production of charge-bred $^{56}Fe^{15+}$ ions involving the FEBIAD delivering the original $^{56}Fe^{1+}$ ions will be described and two examples of charge-bred $^{32}S^{q+}$ and $^{19}F^{q+}$ beams are presented, that are based on the monocharged molecule SF_x^{1+} produced with the NANOGANIII TISS [5]. The optimum ΔV values related to the various types of monocharged ion sources and beams used will be discussed in detail. ΔV value is defined later in the text as the voltage difference between the 1+ ion source and the SP1CB: $\Delta V = V_{1+\ \text{ion source}} - V_{SP1CB}$

2. SPIRAL1 charge breeder and iron charge-bred ions

The SPIRAL1 facility consists of a production cave where the monocharged ion sources are located [1]. Monocharged ions are transported and mass analyzed prior to their injection into the SP1CB. After the capture and stepwise ionization processes, charge-bred ions are extracted, transported and mass analyzed. M/Q spectra are recorded with a Faraday cup downstream from the N+ mass analyzer. Later in the text, outcomes will be showed for two different 1+ ion sources: FEBIAD and NANOGANIII.

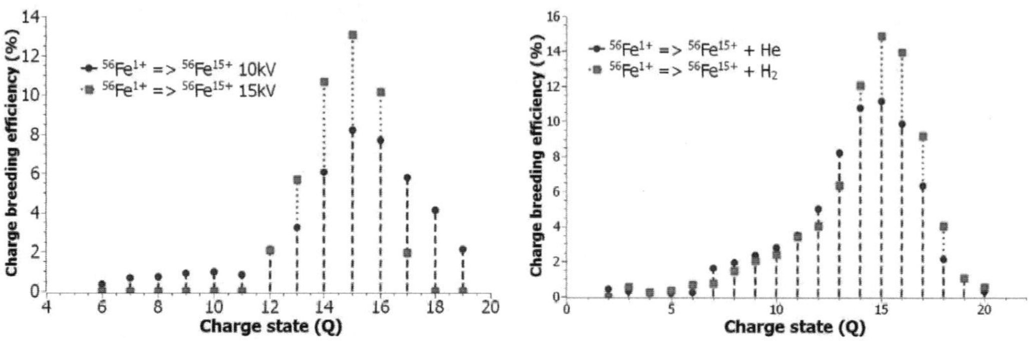

Figure 1. (left) $^{56}Fe^{q+}$ charge state distributions for two extraction voltages and He as the buffer gas; (right) $^{56}Fe^{q+}$ charge state distributions for two buffer gases (He and H_2) at 15kV of extraction voltage.

The FEBIAD ion source can be operated in two different modes: Surface Ionization (SI) or Plasma (P). The source is composed of an anode and a cathode. The cathode produces electrons, which are accelerated by the positive voltage between the cathode and the anode. Ions are produced within the anode volume by electronic impact. As the anode is surrounded by the electrical circuit of the cathode, the heating power also contributes to heat the anode. The SI mode has been found to correspond to 240A (1860W) heating for the cathode, and a difference of potential of ~ 0V between the cathode and anode. The P mode is ignited by using a higher heating current for the cathode, of 290A (2860W) and a difference of potential of 140V applied between the cathode and anode. For many years, the SP1CB performances have been evaluated based on noble gases and alkali elements. Thanks to the FEBIAD, it is now possible to produce other type of 1+ ions (P mode); a campaign of experiments has been achieved with the ^{56}Fe element: several hundreds of nA ($^{56}Fe^{1+}$) were extracted, pure from contaminants. For the tests, the SP1CB was tuned with helium as a buffer gas to sustain the plasma, and an optimum ΔV value of 15.3V was used. Figure 1 shows the $^{56}Fe^{q+}$ charge states distribution after the SP1CB as a function of the extraction voltage (Fig. 1 left) and buffer gas (Fig. 1 right). The beam losses with 10 kV are up to 40 %, and are reduced to 25 % with 15 kV extraction. By improving also the 1+ beam optics [7], the $^{56}Fe^{15+}$ charge breeding efficiency was increased by 58% while maintaining the total charge breeding efficiency (summed over all charge states). Finally, sustaining the SP1CB plasma with H_2 improves significantly the charge breeding efficiencies (Fig. 1 right) of high charge state ions, e.g. $^{56}Fe^{15+}$ from 11.2% to 14.9%. Furthermore, the total charge breeding efficiency increases to 78% with H_2 buffer gas. That behavior has already been established earlier [8, 9], and is confirmed here for a non-alkali element.

3. Charge-bred ions utilizing a monocharged molecular ion beam

The technique of using molecular 1+ beams (AlF, SeCO, SrF, BaF, BF_2) prior to their injection into an EBIS charge breeder has been formerly developed at the REX-EBIS facility (ISOLDE CERN) [10, 11]. Likewise, some preliminary attempts have been made with an ECRIS charge breeder with CO-CO_2 and LaO molecules to produce $C^{2,3+}$ or La^{23+} ion beams [12, 13]. The molecular form can enable chemical selectivity (depending on the reactant injected) leading to an improved isobaric selection while speeding up the release from the target and the transport of the radioactive elements to the charge breeder.

3.1. The experimental study

Here, the purpose is to produce S^{q+} and F^{q+} charge-bred ions and to study the behavior and performances of the SP1CB with the injection of molecular 1+ beams of S and F compounds. The NANOGANIII ECRIS provides $^{19}F^{1+}$, $^{32}S^{1+}$, $^{51}SF^{1+}$, $^{70}SF_2^{1+}$ and $^{89}SF_3^{1+}$ ion beams to the SP1CB. Table 1 sums up the NANOGANIII tuning to foster the production of such monocharged beams. The gas injected is sulfur hexafluoride without any buffer gas. 1+ beam transverse emittances (geometric) are limited with a series of three slits to 40 π.mm.mrad. Hence the 1+ beam is well defined by eliminating beam aberrations. The SP1CB extraction voltage, RF power, injection central and extraction solenoid current were set to 20kV, 270W, 1190A, 320A, 655A respectively; helium was injected as buffer gas.

Table 1. Settings of the NANOGANIII ion source to produce dedicated monocharged ion beams

	RF power input (W)	Bias disc (V)	Logarithm gas valve voltage (V) (0 = open / 10 = closed)	Intensity (nA)
$^{19}F^{1+}$	15	450	7.0	1100
$^{32}S^{1+}$	15	450	8.2	850
$^{51}SF^{1+}$, $^{70}SF_2^{1+}$, $^{89}SF_3^{1+}$	15	150	7.2	450 - 650

3.2. Sulphur and Fluorine charge-bred ions

Measurements have been done in the same way regarding the SP1CB; only the incoming 1+ ion beam was changed. Figure 2 displays the evolution of the ΔV curves for the injection inside SP1CB of S^{1+} and three molecular beams of SF_x^{1+}. For all 1+ beams, the optimum ΔV values are quite similar, around -20V. The FWHM of the peaks starts at 13V (S^{1+}) to 31.6V (SF_3^{1+}). The right part of the peaks have exactly the same rising slope (Fig. 2 left). It shows for the tails a second hump centered on around 35V and particularly pronounced for S^{1+} and SF^{1+}.

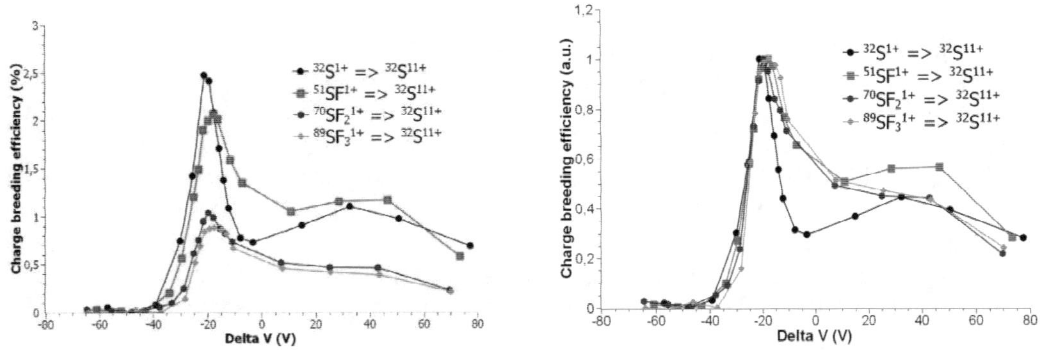

Figure 2. ΔV curves corresponding to S^{11+} charge-bred ions from elemental S or molecules $^{51}SF^{1+}$, $^{70}SF_2^{1+}$ and $^{89}SF_3^{1+}$. The right and left parts are the absolute and normalized data, respectively.

Measurements of F^{q+} (Fig. 3) charge breeding efficiencies were done in the same way as those with S^{q+} multicharged ions. The behavior is quite different as the optimum ΔV as well as the FWHM of the peaks increase nearly linearly from -19.3V to -9.5V and 9.5V to 29V, respectively. However, while the high ΔV part related to the peaks have the same slope as in the sulfur case (Fig. 3 left), no bump appears at high voltage.

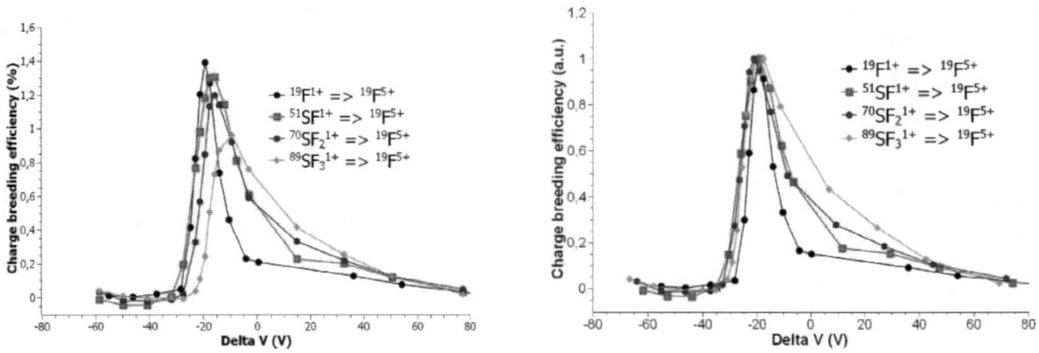

Figure 3. ΔV curves corresponding to the F^{5+} charge-bred ions from elemental F or molecules $^{51}SF^{1+}$, $^{70}SF_2{}^{1+}$ and $^{89}SF_3{}^{1+}$. The right and left parts are the absolute and normalized data, respectively.

3.3. Discussion

The charge breeding efficiencies are quite low with the maximum of ~2.5% and ~1.4% for S^{11+} and F^{5+}, respectively. A value of 5.5% was obtained for S^{1+}/S^{9+} but it was too close to a large contaminant peak to allow measuring methodically the ΔV curve. One explanation might be the large emittance combined with a large energy spread (4.8eV [7]) of the 1+ beam which are disadvantageous for achieving high charge breeding efficiencies [14]. The charge breeding efficiencies of single elements (S^{1+}, F^{1+}) are close to that of the simplest molecule SF^{1+}. With the increasing number of covalent bounds of the molecule to break (more F atoms), the charge breeding efficiency drops and even collapses for S^{11+}. Likewise (see the right part of Fig. 2 and 3), the slopes of the low energy part are really steep and equal for both cases. On the high ΔV side, the behavior are significantly different for S^{11+} and F^{5+}, which does not exhibit a hump. The subsequent open questions are: Does another process than coulomb collisions exists to slow down and capture 1+ ions? Why the ionization $F^{1+} \rightarrow F^{5+}$ appears in comparison so suppressed for these ΔV values? Are the molecules possibly more volatile than the atomic ions, and can be recycled from the plasma chamber wall? Finally one may wonder how the internal energy of the $SF_x{}^{1+}$ could possibly play a role in their capture by the ECR plasma.

4. ΔV curves of the SP1CB versus monocharged ion sources

Figure 4 displays the ΔV curves for SP1CB using He as buffer gas coupled with the NANOGANIII ion source and the FEBIAD ion source in P or SI mode. The black curve corresponds to the FEBIAD SI mode for radioactive $^{47}K^{10+}$ ($T_{1/2}$=17.5s). The optimum ΔV value is 4V for a FWHM of 11V. Reference 8 (Figure 7) provides ΔV curves for stable $^{39}K^{9+}$ using a thermal ion gun (which can be considered as a reference as it minimizes the energy spread), showed an optimal ΔV value and FWHM of -5.5V and 7V respectively. Thus if that case is considered as the reference, the optimum energy of the 1+ ion is equal to 5.5eV. The optimum energy measured with the FEBIAD in SI mode is equal to 7eV corrected of the additional 3V applied on the anode which is close to this nominal value. The blue ΔV curves were obtained using the NANOGANIII ion source providing S^{1+}/F^{1+} ions. The mean ΔV value is -20.2V for an average FWHM of 18.7V. NANOGANIII is an ECR type ion source with the plasma potential speeding up the 1+ ions. Therefore, the ΔV value is negative to slow 1+ ions down to 5.5eV (our reference) leading to a plasma potential value of 20.2V+5.5V=25.7V which is a typical value

[15] for ECR sources. Finally, the red ΔV curve was recorded with the charge distribution obtained for $^{56}Fe^{15+}$ ions using FEBIAD in P mode. The mean ΔV value is 15.3V for an average FWHM of 12V. With the same argument as developed for NANOGAN, the plasma potential of the FEBIAD should be of the order of -9.8V. In fact, the Figure 3 taken from the reference [16] reproduced in Fig. 4 (right) shows the electrostatic field distribution inside the anode of two different types of FEBIAD ion sources. It is clear from the picture that 1+ ions are extracted within an electrostatic field lower than the one applied. From our measurement, a value of -9.8V can be deduced, which is not so far to the one derived from the picture of the Fig. 4 (right) (VADIS "active" volume): 149V-[(125+135)/2]=19V.

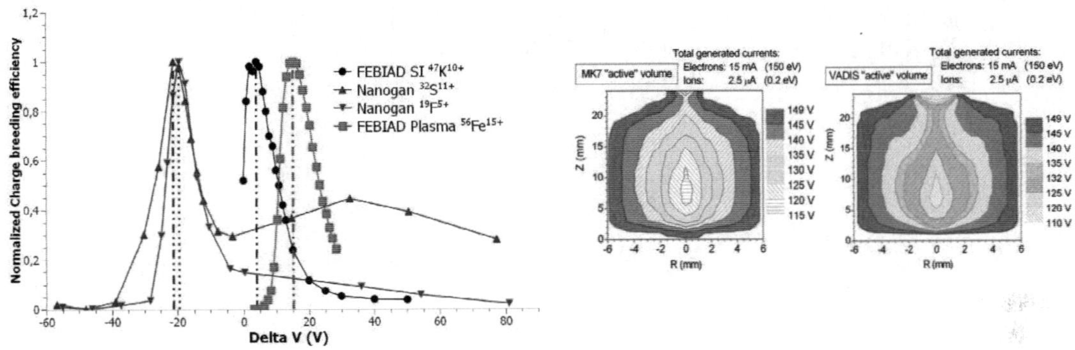

Figure 4. (left) ΔV curves for charge-bred ions for different monocharged ion sources and (right) electrostatic field distribution inside the anode of two FEBIAD types (MK7 and VADIS) [15].

5. Conclusion and outlooks

Each year the operation of SPIRAL1 leads to a better control and understanding of the 1+/N+ technique applied to the RIB production: FEBIAD TISS in the surface ionization mode, charge breeding performances, 1+ as well as N+ beam optics and optimum ΔV values have been investigated. In this article, the capability of SPIRAL1 facility to provide charge-bred ions of non-alkali ions, as well as the operation of the FEBIAD source in the plasma mode has been established with $^{56}Fe^{15+}$ reaching a charge breeding efficiency of 15%. The method of molecules as carriers for the radioactive atoms has been investigated with SF_x^{1+} molecular ions produced with NANOGANIII. The performances of the SP1CB appears to be very similar with atomic and molecular beams. The peculiarities observed (second hump, slopes in the high energy parts and slight shift of the optimum ΔV) in the ΔV scans might give new insights on the 1+ ion capture by an ECR plasma. For the future, hydride compounds would be the way to follow up. Last results obtained during a radioactive run in 2021 proved that the global TISS efficiency is similar to produce either $^{42}Cl^{1+}$ ($T_{1/2}$=6.8s) or $(H^{42}Cl)^{1+}$. Furthermore, with these two examples, on-line as well as off-line measurements showed evidence of the high potential of hydride molecules [17, 18]. Subsequently, next experiments to be performed with the SP1CB would be the charge breeding test of C^{q+} (physicists plan to perform experiment with ^{10}C $T_{1/2}$=19.29s) ions using CH_x^{1+} and CF_x^{1+} molecular ions from CH_4 and CF_4, respectively.

References

[1] L. Maunoury et al., Rev. Sci. Instrum. 91, 023315 (2020)
[2] M. Dubois et al., this proceedings
[3] L. Maunoury et al., Jour. of Instr. 13 C12022 (2018)
[4] P. Chauveau et al., Nucl. Instr. and Meth. In Phys. Res. B 376 (2016) 35
[5] L. Maunoury et al., Proc. of the 18th Intern. Work. on ECR Ion Source (College Station) (1997)
[6] V. Bosquet et al., xxx, this proceedings
[7] L. Maunoury, O. Kamalou, J.F. Cam, S. Damoy, P. Delahaye, M. Dubois, R. Frigot, S. Hormigos and C. Vandamme, Jour. of Instr. 16 P05013 (2021)

[8] L. Maunoury et al., Rev. of Scien. Instr. 87, 02B508 (2016)
[9] J. Angot, M Baylac, M. Luntinen, M. Migliore, O. Tarvainen and T. Thuillier, this proceedings
[10] F. Wenander, Nucl. Instr. and Meth. In Phys. Res. B 266 (2008) 4346
[11] J. Ballof et al., The Eur. Phys. Jour. A, (2019) 55
[12] L. Maunoury et al., Rev. Sci. Instrum. 85, 02A504 (2014)
[13] P. Delahaye and M. Marie-Jeanne, Nucl. Instr. and Meth. In Phys. Res. B 266 (2008) 4346
[14] A. Annaluru, P. Delahaye, M. Dubois and L. Maunoury, Phys. Rev. Acc. and Beams 24, 073403 (2021)
[15] O. Tarvainen, PhD report, University of Jyväskylä (2005)
[16] L. Penescu, R Catherall, J. Lettry and T. Stora, Rev.of Scien. Instr. 81, 02A906 (2010)
[17] P. Jardin et al., Rev. Sci. Instrum. 75, 1617 (2004)
[18] L. Maunoury, S. Kantas, R. Leroy and J.Y. Pacquet, Rev. Sci. Instrum. 77, 03A324 (2006)

Improvement of the Efficiency and Beam Quality of the TRIUMF Charge State Booster

J Adegun[1,2], F Ames[1,3] and O Kester[1,2]

[1] TRIUMF, 4004 Wesbrook Mall, Vancouver V6T 2A3 BC, Canada
[2] Department of Physics and Astronomy, University of Victoria, Victoria, BC, Canada
[3] Department of Astronomy and Physics, Saint Mary's University, Halifax, NS, Canada
E-mail: jadegun@triumf.ca

Abstract. An Electron Cyclotron Resonance Ion Source is used as the charge state booster (CSB) at the Isotope Separator and Accelerator facility (ISAC) of TRIUMF. Since its commissioning in 2010, the source has been used to charge breed radioactive ions ranging from potassium to erbium operating in single-frequency heating mode. Under this regime, the single charge state efficiency of the booster was measured up to 6 % for noble gases and the maximum charge state of Cesium that can be measured is 27+. The rf system of the source was recently upgraded to implement the two-frequency heating using a single waveguide. Preliminary operation of the booster in a two-frequency heating mode shifted the maximum charge state of Cesium that can be measured to 30+. Another point of improvement that is being addressed is the beam extraction system of the booster. A quadrupole scan technique using a thick lens approach has been developed to measure the emittance of the extracted beam and to analyze the quality of the extraction system. Overall, the system of the CSB is currently being optimized in preparation for the detailed determination of the effect of the two-frequency heating on the beam output.

1. Introduction

At the ISAC facility of TRIUMF, radioactive isotope beams (RIBs) with atomic mass $A > 30$ require charge breeding before being injected into the linear accelerator (LINAC) for post-acceleration. The first stage of post-acceleration, which is the radiofrequency quadrupole, was designed to accept RIBs at $A/Q \leq 30$, the second and third stages of post-acceleration, which are drift-tube linac and superconducting radiofrequency linac, were designed to accept RIBs at $A/Q \leq 7$. For charge breeding purposes at the ISAC facility, an ECRIS is used. Since it was commissioned in 2010 [1], only aluminum coating of the plasma chamber wall and the modification and replacement of the injection and extraction electrodes using aluminum have been done. This did improve the performance of the source by reducing the intensity of some background ions by about two orders of magnitude [2]. To further improve the overall performance of the TRIUMF charge state booster (CSB), two-frequency heating is being implemented using a single waveguide.

Furthermore, besides improving the performance of the ECRIS as a highly charged ions generator, optimization of the extraction system for better beam quality upon extraction is critical. Previous investigation of the extraction system of the ECR ion source [3] revealed that magnetic field greatly influences the beam emittance, and simulation codes are usually employed to systematically investigate and optimize the extraction system for good beam quality. Systematic investigation of the extraction system of the TRIUMF CSB using IGUN© code [4] has been embarked on. The plasma and the extraction parameters are being studied to understand their influence on the extracted beam. The insights gained from this systematic investigation will allow optimizing the extraction system of the CSB for better beam quality.

In this paper, the results of measurements of the extracted beam from the CSB during operation with the single frequency heating will be presented and compared with the preliminary result obtained after the implementation of the two-frequency heating. In addition, the results of the systematic investigation of the influence of extraction voltage and magnetic field on the beam emittance simulated in IGUN© compared with measurement are also reported.

2. The TRIUMF Electron Cyclotron Resonance Ion Source Charge State Booster

The TRIUMF CSB is a 14.5 GHz PHOENIX ECRIS from PANTECHNIK and was originally designed for single-frequency microwave plasma-heating to produce highly charged ions. The CSB is equipped with a three-electrode extraction system. The plasma chamber, the injection electrode and the extraction electrode are coated with pure aluminum. The TRIUMF CSB is routinely operated with helium as a support gas. Two cryo-pumps are connected to the booster at the injection and the extraction regions to achieve a low-pressure vacuum in the order of 10^{-8} *Torr* during operation. For axial plasma confinement, magnetic fields are generated axially by a set of three solenoid coils while radial confinement is achieved using the magnetic fields generated by hexapole permanent magnets. For the selection of the desired charge-bred ions, a Nier-type spectrometer [5] (combination of magnetic dipole and two electrostatic benders) is used. The typical operating magnetic field of the CSB are Binj = 1.14 T at I = 1050 A, Bmin = 0.35 T at 250 A, Bext = 0.87 T at I = 750 A.

3. Experimental Set-up and measurements

For all the results presented in this paper, the TRIUMF CSB high voltage bias was set to 10 kV. The puller electrode voltage was varied and the 2rms emittance of the CSB was determined via the Quadrupole Scan Technique (QST). For the single-frequency operation, the travelling wave tube amplifier power was set to 350 W. For the operation of the two-frequency heating, it was implemented using a single waveguide because the ECRIS CSB was not designed to accommodate two separate waveguides. A low-power, high-frequency combiner was used to add two microwave signals operating at two separate frequencies. The combined signal was fed into a broadband travelling wave tube amplifier (TWTA), and the combined RF power is transported into the plasma chamber via a WR-62 waveguide.

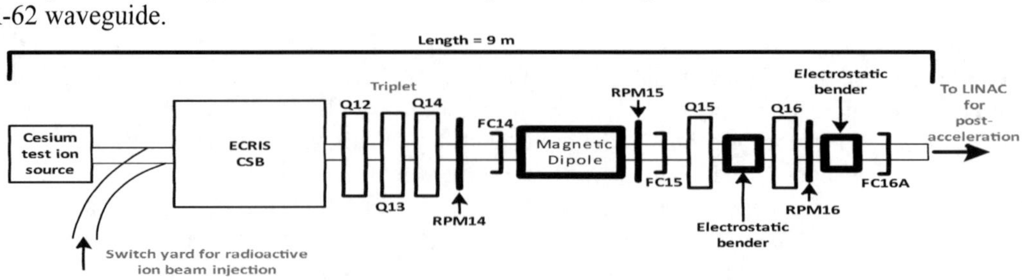

Figure 1: Schematic of the TRIUMF ECRIS CSB Beamline

Figure 1 shows the layout of the injection and extraction beamlines of the TRIUMF CSB. The Cesium test ion source, located upstream of the CSB, is used for injection, plasma and extraction system tuning. Extracted ions from the CSB are overall focused by the triplet Q12, Q13 and Q14, while the profile and the total current of the extracted beam from the CSB are monitored by the RPM-14 and Faraday cup FC14 respectively. The combination of magnetic dipole and two electrostatic benders act as a Nier-type spectrometer for the selection of the desired highly charged ions before injection into the linac for post-acceleration. The beam current measured on the Faraday cup FC16A is considered to have fewer background ions of similar A/q.

Singly charged Cesium ion beam of about 22 nA was injected from the test ion source, and the frequency of the microwave was tuned to determine the frequency that produces the maximum intensity of Cs^{24+} monitored on FC16A. The extraction system of the CSB, the adjacent beam transports up to the Faraday cup FC16A were optimized for the extraction and transport of the Cs^{24+} beam. The difference between the high voltage bias of the test ion source and the CSB, the delta-V value, was varied to determine the voltage at which the injected ions were efficiently captured by the plasma. During the variation of the delta-V, the intensities of Cs^{+}, Cs^{18+}, Cs^{21+} and Cs^{23+} were monitored on the Faraday cup FC16A. Figure 2 (left) shows the normalized current of Cs^{+}, Cs^{18+}, Cs^{21+} and Cs^{23+} plotted against delta-V. It can be seen from the plot that as the delta-V increases the intensities of Cs^{18+}, Cs^{21+} and Cs^{23+} increase until delta-V reaches 13.0 V where the intensities of the Cesium charge state are maximum. However, further increase of delta-V leads to a reduction in the intensities of the charge state. This behaviour indicates that below delta-V = 13.0 V, Cs^{+} ion beam injected into the CSB is strongly repelled by the positive potential of the plasma but efficiently captured when delta-V reaches 13.0 V and passed through the plasma without being captured at delta-V greater than 13.0 V. The plasma potential seems to be of the order of 13 V. On the other hand, the intensity of Cs^{+} extracted increases linearly between 10 V and 20 V which indicates that some or all of the Cs^{+} ions injected were captured by the plasma but were repelled below 10 V and not captured above 20 V, the behaviour that is evident in the shape of the distribution of the intensity of Cs^{18+}, Cs^{21+} and Cs^{23+}.

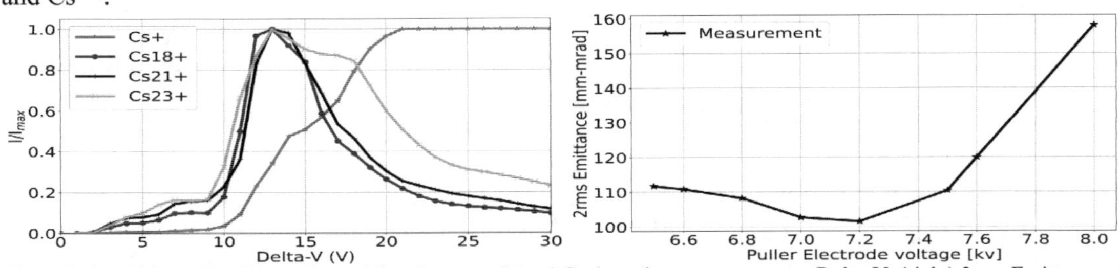

Figure 2: (left) Normalized Beam intensities of some selected Cesium charge states versus Delta-V. (right) 2rms Emittance of the total current extracted from the TRIUMF ECRIS CSB using Quadrupole Scan Technique

3.1. Emittance versus Puller electrode voltage

The Quadrupole scan technique (QST) was deployed via scanning through a range of focusing strength of a quadrupole and measuring the beam size at a given drift length downstream of the quadrupole, and by using the Twiss parameters transformation, the shape and the orientation of the emittance phase space can be determined at any location in the beam transport. The technique is well understood, and it has been used in several accelerator facilities [6]. Using the QST, the 2rms emittance of the CSB was measured as a function of the puller electrode voltage by scanning the strength of the quadrupole (Q14) of the triplet located after the extraction system and measuring the beam size on the profile monitor RPM-14 (see Figure 1). The puller electrode voltage was varied between 6.5 and 8.0 kV. Figure 2 (right) shows the plot of the 2rms emittance versus the puller electrode voltage. As seen, at 6.5 kV, the emittance was measured to be 111.7 mm-mrad, and as the puller electrode voltage is increased, the beam emittance decreases to 101.3 mm-mrad at 7.2 kV which is the minimum emittance. However, with a further increase in the voltage on the puller electrode, the beam emittance increased until it reached 156.9 mm-mrad at 8.0 kV. The behaviour of the emittance as the puller electrode voltage is changed suggests that the plasma boundary changed from the concave shape (low density, high extraction field) at 6.5 kV to a planar shape (moderate density, matched extraction field) at 7.2 kV and then convex shape (high density, low extraction field) at 8.0 kV, which has been observed in IGUN simulations of the extraction system.

3.2. Single Frequency Heating compared with the Preliminary result of Two-frequency heating

For the single frequency heating, the TWTA power was set to 350 W at a frequency of 14.5 GHz, while for the two-frequency heating, the CSB was operated at an additional rf frequency of 13.95

GHz. Meanwhile, regardless of the power settings on the signal generators, the same amplification gain was applied by the TWTA, thus the output power of the amplifier was set to 150 W for the two-frequency heating. Figure 3 (left) shows the mass spectrum of the CSB with and without Caesium between A/q of 3 and 10 with operation with single frequency heating. This A/q region is mostly dominated by residual ions from Carbon, Nitrogen and Oxygen. The peaks of the Caesium charge states are indicated by the arrows and the maximum charge state of Caesium that can be measured in the single frequency heating mode is 27+.

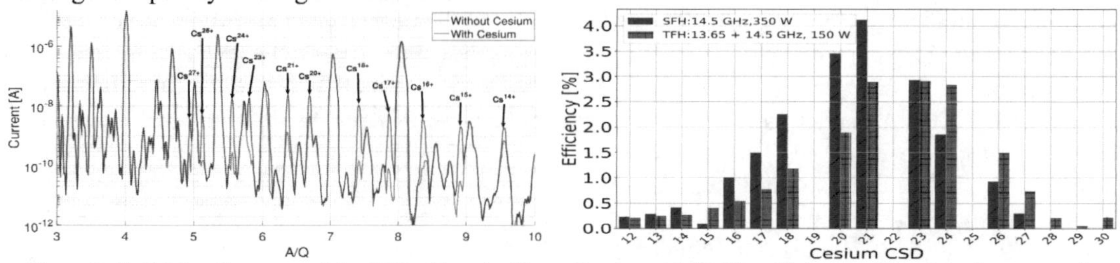

Figure 3: (left) Mass Spectrum of the CSB with and without Caesium. (right) The efficiency of Caesium charge state in single and two-frequency heating modes

In Figure 3 (right), the efficiency of Caesium in single frequency heating mode is compared with the efficiency of Caesium in a preliminary (not optimized) measurement of the two-frequency heating mode. Although, the charge state booster and the adjacent beam transport shown in Figure 1 were not properly optimized for the two-frequency heating, the maximum charge state of Caesium that can be measured shifted from 27+ for single-frequency heating to 30+ for two-frequency heating. In addition, while Cs^{21+} has the maximum efficiency of 4.1 % in the single-frequency heating mode, the maximum efficiency shifted to Cs^{23+} in the two-frequency heating regime with an efficiency of 2.9 %. Meanwhile, during the operation of the CSB in the two-frequency heating mode, the TWTA got damaged while performing frequency tuning. There was a high power reflection from the plasma and damaged the tube of the TWTA, and further measurements and optimization of the CSB using this approach of the two-frequency heating could not proceed further.

4. Extraction system simulation

To improve and optimize the emittance of the extracted beam from the TRIUMF CSB, its extraction system was simulated for the first time with IGUN© code. To gain the required input parameter, the magnetic field distribution of the CSB was modelled and simulated in OPERA and then benchmarked against the measured magnetic field distribution of the GANIL charge state booster [7], a twin source of the TRIUMF CSB.

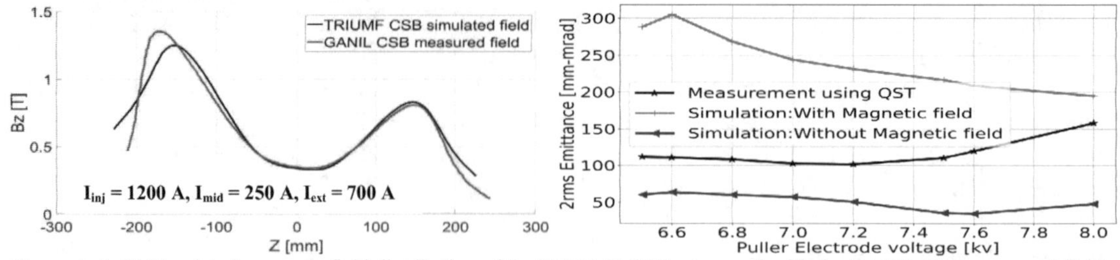

Figure 4: (left) Simulated magnetic field distribution of the TRIUMF CSB compared with the GANIL CSB measured field. (right) Comparison of the measured 2rms emittance with simulated 2rms emittance with and without the magnetic field

Figure 4 (left) compares the TRIUMF CSB simulated magnetic field distribution with the measured field of the GANIL CSB. The two fields are not matched at the injection region because the injection iron plug of the GANIL CSB is different from the TRIUMF CSB. The coils and other iron parts are similar. Since the simulated magnetic field of the TRIUMF CSB agreed with the measured magnetic field of the GANIL CSB at the extraction region, therefore the magnetic field values were imported into IGUN for the extraction system simulations. In the simulation, the difference between the biased voltage of the CSB and the voltage on the puller electrode which is defined as the extraction voltage was varied and beam emittance with and without magnetic field was calculated. The simulated emittances were compared with the measured 2rms emittance (black plot) of the CSB. As shown in Figure 4 (right), the measured emittance of the CSB does not agree with the simulated emittance assuming the magnetic field determined via OPERA (red plot). To gather information about the influence of the magnetic field on the beam formation, the magnetic field in the simulation was switched off and the extraction voltage was varied again, and the simulated emittance without magnetic field was calculated (blue plot). As it can be seen in Figure 4 (right), the emittance without the magnetic field is smaller compared with the case with the magnetic field. It demonstrates the strong influence of the magnetic field distribution in the extraction region on the beam emittance. The measurement reveals values somewhere in between these cases, which indicates that the magnetic field of the CSB is not yet correctly modelled with the geometry data available. Therefore it is planned to map the magnetic field distribution of the CSB in the near future.

5. Conclusion and Outlook

The research campaign and the deployment of techniques to improve the efficiency and beam quality of the TRIUMF charge state booster are in full swing. The implementation of the two-frequency heating of the plasma of the CSB indicated a significant increase in the electron energy and current density of the booster by measuring the charge state of Cesium up to 30+. A new approach to implement the two-frequency heating has been started. The new approach will involve using two separate TWTAs, a power combiner and a circulator to prevent reflected power from the plasma from damaging any of the RF components. It will be possible to individually adjust the power of each frequency of the microwave for better propagation into the plasma. Investigation of the effect of frequency gap and frequency range on the intensity of the maximum charge state will be conducted. Furthermore, the CSB will be opened during the next shutdown to directly measure its axial magnetic field distribution for benchmarking against the simulated magnetic field. Measuring the magnetic field will be beneficial for the optimization of the extraction system of the booster which is being modelled in IGUN©.

6. Acknowledgement

The project is funded by the Natural Sciences and Engineering Research Council of Canada (NSERC), TRIUMF, and the University of Victoria, BC.

References

[1] Ames F, Baartman R, Bricault P and Jayamanna K 2010 Commissioning of the ECRIS charge state breeder at TRIUMF *Proceedings of ECRIS* (Grenoble, France) pp 178–80

[2] Ames F, Baartman R, Bricault P, Jayamanna K and Mjøs A 2012 Operation of an ECRIS Charge State Breeder At TRIUMF *Proceedings of ECRIS2012, Sydney, Australia* pp 163–6

[3] Wutte D, Leitner M A, Lyneis C M, Taylor C E and Xie Z Q 2009 Design study of the extraction system of the 3rd Generation ECR ion source pp 384–95

[4] Becker R and Herrmannsfeldt W B 1992 IGUN - A program for the simulation of positive ion extraction including magnetic fields *Rev. Sci. Instrum.* **63** 2756–8

[5] Nier A O and Roberts T R 1951 The determination of atomic mass doublets by means of a mass spectrometer *Phys. Rev.* **81** 507–10

[6] Green A T 2015 Implementation of Quadrupole-Scan Emittance Measurement At Fermilab's Advanced Superconducting Test Accelerator (Asta) *Proceedings of IPAC2015* pp 669–71

[7] Maunoury L, Dubois M, Delahaye P, Annaluru A, Bajeat O, Frigot R, Hormigos S, Jacquot B, Jardin P, Osmond B, Predrag U, Retailleau B M, Savalle A, Toivanen V, Thomas J C, Angot J, Sole P, Lamy T, Koivisto H, Marttinen M and Tarvainen O 2020 Charge breeding at GANIL: Improvements, results, and comparison with the other facilities *Rev. Sci. Instrum.* **91**

Production of C-14 and stable ion beams at the Argonne Tandem Linac Accelerator System with the ECR3 ion source

R H Scott, J McLain, R C Vondrasek

Physics Division, Argonne National Laboratory, Lemont, Il 60439, USA

Abstract. The Argonne Tandem Linac Accelerator System (ATLAS) at Argonne National Laboratory (ANL) paused production of carbon-14 ion beams with the removal of the Tandem Van de Graaff in 2013. Installation of ECR3, an Electron Cyclotron Resonance Ion Source, returned that production capability to ATLAS, with the first C-14 beam delivered in December 2020. Information is presented on C-14 beam current, gas consumption and N-14 filtering techniques using stripping foils at different sections of ATLAS. ECR3 also fulfilled the operational goal of adding flexibility to ATLAS stable beam production capabilities. Beneficial impacts to ATLAS operation, beam development and experimental programs are discussed.

1. Introduction

For many years, carbon-14 and stable ion beams were produced at ATLAS using a SNICS II negative ion source coupled with a Tandem Van de Graaff Accelerator to inject positive ions into a linac for further energy gain. The Tandem was removed in 2013 to make room for a stopped beam experimental hall, halting C-14 production. A few years later the former ECRCB which provided stable beams and charge bred radioactive beams from CARIBU was replaced with an EBIS dedicated to charge breeding. With these changes, ATLAS stable beam production was entrusted to a single ion source ECR2. The ECR3 [1,2] ion source was installed to deliver beams of C-14 with the benefit of providing flexibility in stable beam production. High carbon-14 beam purity from an ECR source is intrinsically difficult to achieve because nitrogen-14 has an inseparable mass and is an unavoidable contaminant in ECR ion sources. Techniques developed to filter out N-14 in the ATLAS linac and the resulting target/ion source purity relationship are discussed. C-14 consumption and beam transmission under different accelerator configurations are also examined.

2. Carbon-14 beams from ECR3

ECR3 was commissioned in 2019 using stable beams of He and O. The C-14 would be produced using an ethylene gas with a specific activity of >100mCi/mmol. Experiments were performed with C-13 ethylene to develop baseline production information and to understand C:N ratios under different running conditions [2] before introducing C-14 into the system and complicating operations. First delivery of C-14 to target was achieved in December 2020 with some delay attributed to the COVID-19 outbreak. By conference submission (9/2021), over 30 days of C-14 beam time have been achieved at the 24 hour facility.

2.1. Experimental requirements

The latest approved ATLAS experiments requiring C-14 beams call for beam intensities as high as 300 pnA and energies reaching 210 MeV. These requirements can be satisfied with the flexibility

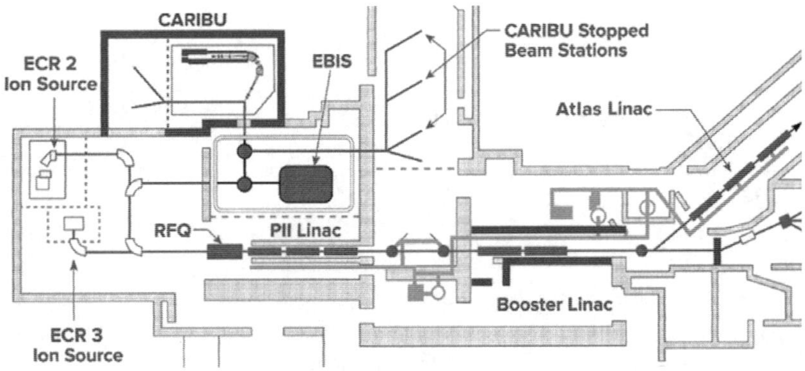

Figure 1. ATLAS front end layout including ion sources and accelerating sections

afforded by the accelerator configuration. Figure 1 shows ATLAS has three distinct accelerating sections: Positive Ion Injector (PII), Booster and Atlas with charge stripping capabilities after PII and Booster. There is a 40 degree bend after Booster allowing m/q selection. The C-13 tests determined that 3+ had the best ionization efficiency and the best LEBT transmission of the charge states that can be accepted by the ATLAS RFQ. Thus for material consumption reasons, unstripped 3+ is accelerated through the machine if the desired energy is achievable. For cases where the 3+ is not sufficient for the required energy, stripping to a higher charge state and achieving more energy gain per linac cavity is required. Table 1 shows the energies achievable with and without stripping under different linac

Table 1. Unstripped and stripped energy of C-14 beams under the following configurations: A) only active resonators turned on, B) active resonators with energy upgrade, C) all resonators with energy upgrade

Entrance charge state			Configuration energy Atlas exit (MeV)		
PII	Booster	Atlas	A	B	C
3	3	3	90	145	165
3	3	6	110	220	242
3	5	5	133	220	246
3	6	6	147	254	281

configurations. These numbers tend to be fluid as resonators are brought in and out of service, conditioned for better performance, or de-rated. The highest approved energy of 210 MeV cannot be achieved under the current configuration (A) having 5 resonator cavities awaiting repair and the last cryostat in the Atlas section removed for an energy upgrade scheduled for completion in early 2022. Up to 254 MeV is expected with current configuration combined with the energy upgrade. The energy range would be extended even further with continued resonator repairs approaching 281 MeV (configuration C). A stripping location that yields energy required, high C-14 purity and best possible transmission to target is to be considered for any given experiment.

2.2. Beam stripping

Calculations were performed using LISE^{++} ETACHA4 [3] and in-house semi-empirical based charge distribution codes to approximate stripping fractions using thin carbon foils at PII and Booster energies with results shown in Table 2. Measurements of N-14 stripping fractions were taken using a pure N beam from ECR2. The measured and ETACHA4 fractions for N-14 Booster stripping agreed.

Charge state intensities from an ECR3 mixed beam of C-14 and N-14 were measured after Booster. The N-14 charge state intensities (I_{Nq+}) were determined from the unique N7+ intensity (I_{N7+}) and the measured nitrogen stripping fractions (η_{Nq+}) from the pure beam:

$$I_{Nq+} = I_{N7+}\eta_{Nq+}/\eta_{N7+} \quad (1)$$

Journal of Physics: Conference Series **2244** (2022) 012068 doi:10.1088/1742-6596/2244/1/012068

Table 2. Calculated and measured stripping fractions of C-14 and N-14. PII exit energy 25 MeV and foil thickness 30 g/cm^2, Booster exit energy 65.7 MeV and foil thickness 75 ugcm^2.

		PII stripping fractions				Booster stripping fractions		
Source	Beam	4+	5+	6+	7+	5+	6+	7+
ETACHA4	C-14	0.03	0.26	0.71	-	0.04	0.96	-
In-house	C-14	0.17	0.47	0.35	-	0.36	0.58	-
ETACHA4	N-14	0.01	0.08	0.37	0.54	0.002	0.068	0.93
In-house	N-14	0.04	0.24	0.45	0.26	0.08	0.38	0.53
Measured	N-14	0.02	0.25	0.53	0.20	0.002	0.088	0.91

The C-14 charge state intensities were found by subtracting the nitrogen component from the mixed beam intensity of a given charge state. The stripping fraction of a given C-14 charge state could then be calculated by dividing its charge intensity by the sum of all C-14 charge intensities. The C6+ fraction was 0.955 and C5+ was .0455 which agree with the ETACHA4 values for carbon.

For stripping the pure N-14 beam at PII energy, the beam transport section was scaled for charge states which were individually identified and tuned in the bend after Booster. The ETACHA4 calculation for PII stripping disagreed considerably with the measured value, possibly due to the lower energy. While results more closely matched the in house code, they were still off by ~18% for N6+ and ~23% for N7+. The PII measurement had limitations. The distance between stripping and analysis is 35m. The stripped charge states were not accelerated in Booster, simplifying the measurement due to time constraints. Scheduling did not allow for a measurement of the mixed ECR3 charge states stripped after PII. A spur is being installed for a new target station 5 meters past the PII exit. It is planned to measure stripping distributions in this location of both pure N-14 from ECR2 and mixed beams from ECR3 upon completion.

2.3. Purity relationship

Stripping energy and the charge state being selected have an impact on beam purity at target. We define the beam purity as the percentage of carbon-14 in the mixed beam with the remaining being nitrogen-14, inseparable by m/q. A beam purity of 80% or greater (C:N $\geq$ 4:1) is the most stringent requirement of the currently approved experiments. The effect stripping has on beam purity by filtering out nitrogen can be derived mathematically with stripping fractions at a given energy:

$$P_{tgt} = \frac{P_{src}\eta_C}{P_{src}\eta_C + \eta_N(1 - P_{src})} \quad (2)$$

where P_{tgt} and P_{src} are the beam purities at target and ion source. η_C and η_N are stripping fractions of the same charge state for carbon and nitrogen respectively. Figure 2 plots this equation for mixed beams stripped at PII and Booster with energies and foil thicknesses used in Table 2. The PII error

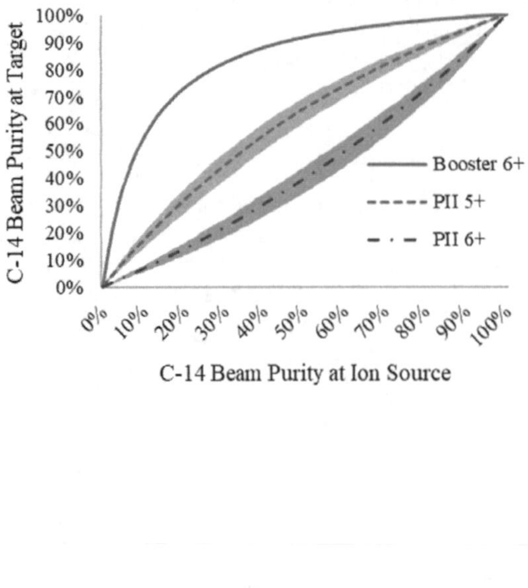

Figure 2. Target and ion source purity relationship for beams stripped to different charge states at Booster and PII.

bands come from +/- 20% C-14 stripping fraction uncertainty from using the in house code. Obtaining >80% target purity is most easily achieved by not stripping at PII and stripping to 6+ at booster, followed by stripping to 5+ at PII and not stripping at Booster. Stripping to 6+ at PII results in target purity that is worse than at the ion source.

2.4. C-14 experiment runs

Table 3 displays a series of nuclear physics experiments in order of operation that were run at ATLAS

Table 3. Carbon-14 experiments at ATLAS

		Beam parameters at target			Charge state		C-14	
Experiment number	Hours	Intensity (pnA)	Energy (MeV)	Purity requested	Booster	Atlas	Cons. rate (mg/hr)	Transmission
1728	88	1	140	~50%	6+	6+	0.025	8.5-12%
1739a	137	100	122	> 80%	3+	6+	0.075	36-45%
1732	165	16	100	> 80%	3+	6+	0	23-31%
1739b	216	95	122	> 80%	3+	6+	0	36-40%
1955	160	100	133	> 80%	5+	5+	0.16	6-14%

using different stripping schemes over the past year. Consumption rate was determined with pressure readings taken before and after a run of the fixed volume of gas using an Omega DPG2001B [4] pressure gauge. The C-14 transmission was determined by multiplying the C-14 purity and mixed beam intensity at the last common faraday cup before target stations and dividing by the same product at the source cup. Attenuation was also accounted for. Transmission varied over the course of a run due to progression of tuning, foil lifetime and changes in source purity.

The first experiment, 1728 required a mixed C-14/N-14 beam. Pure nitrogen beam from background gas was delivered to target for characterization, then C-14 was introduced until beam current at the source doubled. A purity of ~40% was verified from sampled ionization rate and total beam rate in an ionization chamber [5] and from elastic scattering (Rutherford regime) in a recoil E-dE silicon detector at the end station. Stripping was done at PII exit to reach energy and a factor of 5 attenuation was used to reach requested intensity.

The following experiment, 1739a was able to reach energy by stripping at Booster exit with a C-14 consumption rate of .075 mg/hr while delivering 100pnA. With a higher stripping fraction into 6+, transmission was improved from the previous run. For experiment 1732, the source was run at 7 watts while delivering 16 pnA on average to target without introducing C-14 gas. The drop in transmission is attributed to tuning required to reach a farther target station. Experiment 1739b immediately followed. We were able to increase output with an RF increase to 14W. Helium support was increased occasionally over the run to maintain ~95 pnA on target, also without C-14 gas feed. Remarkably, both experiments combined for 381 hours without gas feed. This can be attributed to wall recycling of the ethylene gas introduced previously. During the runs the target purity steadily dropped from 91% to 85% while the source purity fell from 48% to 34% indicating the usage of material from the walls. The effective consumption rate over 518 hours was 0.02 mg/hr.

Unfortunately the energy goal was not met with Booster stripping for experiment 1955. Poor C-14 transmission due to the PII stripping fraction into 5+ resulted in a return to active gas feed with higher consumption than previous to maintain intensity. It is possible that material from this run may be recycled during future runs depending on stable beam demands from this source.

3. ATLAS operational relief

As previously noted [1,2] the ECR3 installation has another benefit that cannot be overlooked. Aside from providing the C-14 beam which was lost to ATLAS in 2013, ECR3 has demonstrated that it can readily provide stable beams of z = 1 to 36 originating from gas. Between commissioning in October 2019 and September 2021, 71 operational and tune days were allotted to ECR3. Many of these days were used concurrently for ion source development and maintenance at ECR2. Without ECR3, these activities would have caused reduced programmatic hours or been deferred. Figure 3 shows a historical allotment of ion sources used for the production of stable beams. In the years leading up to 2019, ECR2 became the solitary source of stable beams at ATLAS. ECR3 produced 8% of the stable beam hours delivered this past year. Another 29% of those hours were readily producible from ECR3 but ran from ECR2 instead. With more operational knowledge and experience gained ECR3 can be used for more stable beams in the future.

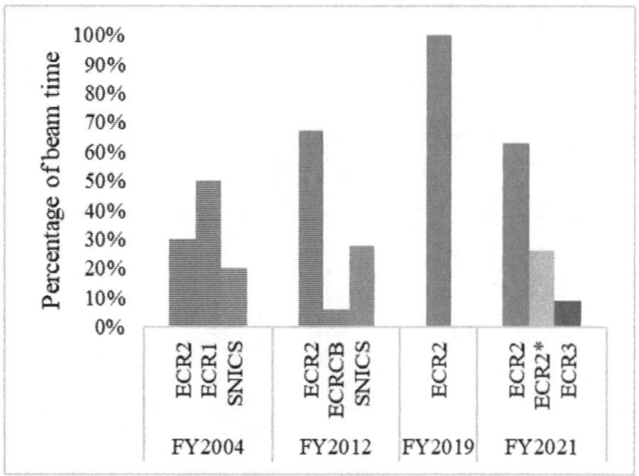

Figure 3. Sampling of ATLAS ion source usage for stable beams over time. *ECR2 beams that ECR3 is capable of producing.

4. Summary

The ECR3 ion source has provided C-14 beams while adding flexibility to the ATLAS experimental program as a backup and complementary source to ECR2 for stable beam production. Nitrogen filtering using a stripping foil has been demonstrated. From material consumption and beam purity standpoints, the ideal configuration has been found to be stripping to 6+ at Booster when energy cannot be achieved unstripped. After the pending energy upgrade of ATLAS, most experiments will use this configuration. Wall recycling of the ethylene material was also demonstrated without active gas feed, which can reduce the overall consumption rate. A wish to conserve material may conflict with the desire to use the source for stable beams since a part of the remaining C-14 would be consumed during operation. Strategic timing of experiments may provide a novel solution.

Acknowledgments

This material is based upon work supported by the U.S. Department of Energy, Office of Nuclear Physics, under contract number DE-AC02-06CH11357. This research used resources of ANL's ATLAS facility, which is a DOE Office of Science User Facility.

References

[1] Scott R H, Dickerson C, Pardo R and Vondrasek R 2016 *Proc. 22nd Int. Workshop on ECR Ion Sources (Busan)* pp 105-107

[2] Scott R H and Vondrasek *Proc. 24th Int Workshop on ECR Ion Sources (East Lansing)* "ECR3 commissioning and planning for C-14 ion beams at the Argonne Tandem Linac Accelerator System" (submitted, JACOW)

[3] Lamoure E, Fainstein P D, Galassi M, Prigent C, Ramirez C A, Rivarola R D, Rozet J P, Trassinelli M and Verhet D 2015 *Phys. Rev. A* **92**, 042703.

[4] www.omega.com

[5] J. Lai, L. Afanasieva, J. C. Blackmon, C. M. Deibel, H. E. Gardiner, A. Lauer, L. E. Linhardt, K. T. Macon, B. C. Rasco, C. Williams *et al.*, Nucl. Instrum. Methods Phys. Res., Sect. A **890**, 119 (2018).

First beams from the 1+ source of the ADIGE injector for the SPES Project

A Galatà, P Francescon, C Roncolato, G Bisoffi, M Ballan, L Bellan, J Bermudez, D Bortolato, M Comunian, A Conte, M De Lazzari, E Fagotti, F Gelain, M Manzolaro, D Marcato, V Martinelli, M Miglioranza, M F Moisio, A Monetti, E Munaron, A Pisent, M Rossignoli, M Roetta, G Savarese and D Scarpa

INFN - Laboratori Nazionali di Legnaro, Viale dell'Università 2, 35020 Legnaro (Padova), Italy.

E-mail: alessio.galata@lnl.infn.it

Abstract. The ADIGE (Acceleratore Di Ioni a Grande carica Esotici) injector consists of an electrostatic 1+ beam line, equipped with ion sources able to produce a wide variety of beams, coupled to a magnetic beam line, where charge multiplication is accomplished by implementing an Electron Cyclotron Resonance (ECR) based charge breeder. The injector is totally integrated in the SPES (Selective Production of Exotic Species) beam line, to allow the post-acceleration of radioactive ions and is now in an advanced phase of installation. The electrostatic 1+beam line has been put into operation and is now producing beams from alkali metals. This contribution concerns the first results of the beam commissioning of this part of the injector, with the description of the initial debug phase and the solutions adopted to ensure a reliable and continuous operation. Preliminary results of the 1+ beam line characterization will be shown, with a comparison between simulated and measured emittances.

1. Introduction

SPES [1] (Selective Production of Exotic Species) is an INFN project with the aim at developing a facility to produce and post-accelerate exotic beams, to perform forefront research in nuclear physics by studying nuclei far from stability. SPES is presently in an advanced phase of installation: its goal will be accomplshed by post-accelerating neutron-rich radioactive nuclei, with a mass range A = 80÷160, produced by inducing fissions in a multifoil UC_x target. Up to 10^{13} fissions/second will be induced by a proton beam accelerated by a commercial cyclotron [2] with a variable energy (30÷70 MeV) and a maximum current of 0.75 mA (upgradeable to 1.5 mA). The radioactive species produced will be delivered to ion sources coupled to the target by means of a tubular transfer line and extracted as a 1+ beam (RIB). Different ion sources can be used, in order to optimize the production of a specific ion beam: in the context of the SPES project a specific hot-cavity surface ion source (SIS) [3] and a plasma ion source (PIS) [4] have been developed and characterized in terms of ionization efficiency and transversal emittance. Tests of laser ionization has been also carried out, producing off-line ions from Al, Ge and Sn [5]. The online set-up will implement an all-solid-state laser system, tunable on transitions of all of the isotopes produced at SPES. To clean the RIB from the co-extracted isobars, it will be injected into a Radiofrequecy Quadrupole (RFQ) cooler [6] (under construction at LPC Caen)

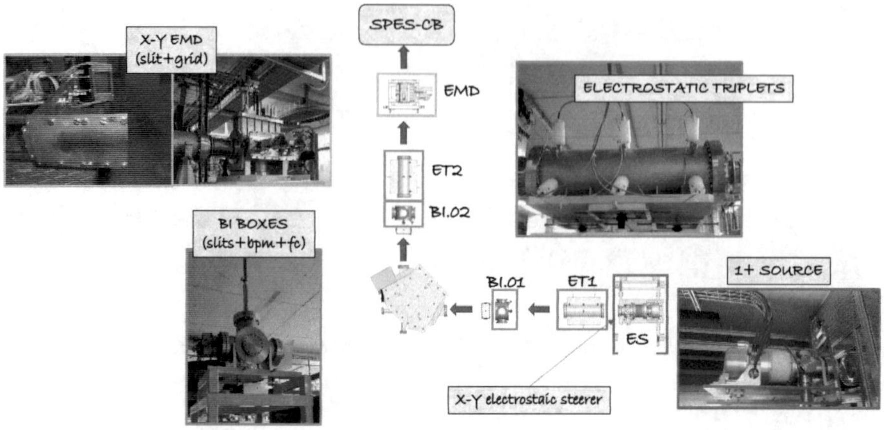

Figure 1. Schematic view of the 1+ beam line of the ADIGE injector with: a stale 1+ source, x-y electrostatic steerers, two electrostatic triplets (ET1 and ET2), two beam instrumentation boxes (BI.01 and BI.02) and an emittance measurement device (EMD). The beam will then be injected inside the SPES charge breeder.

and then into a High Resolution Mass Spectrometer (HRMS) [7], with an expected resolution of R=1/20000. Purified 1+ RIBs will then be delivered to the ADIGE (Acceleratore Di Ioni a Grande carica Esotici) injector [8], where an Electron Cyclotron Resonance based charge breeder [9] coupled to a Medium Resolution Mass Separator [10] will provide purified highly charged RIBs to the post-acceleration, accomplished coupling a new normal-conducting RFQ [11] to the superconducting linac ALPI. The ADIGE injector is equipped with stable 1+ sources and relative electrostatic beam line to test the charge breeder prior to the injection of radioactive ions: object of this contribution is the description of the first results of the commissiong of such part of the SPES project.

2. The ADIGE 1+ beam line

A schematic view of the 1+ beam line of the ADIGE injector is shown in figure 1: the ion sources available have been derived directly from the models will be coupled to the SPES target. The beam is extracted through a 3 mm hole, by applying a maximum voltage of 40 kV on a single gap extraction system, whose distance can be adjusted under vacuum. The beam passes through two couples of X-Y electrostatic steerers (±2 kV max), necessary to correct possible beam misalignments and is then be focused by an electrostatic quadrupoles triplet (ET1 in figure1) to the first beam instrumentation box (BI.01 in figure1), equipped with a faraday cup, two beam profile monitors and selection slits. Such box is mounted at the object point of the 1+ selection dipole: it is a 90°, 750 mm radius magnet, with entrance and exit angles of 26.6° and a pole gap of 110 mm. Another beam instrumentation box (BI.02 in figure1) is placed at the image point of the dipole and another electrostatic quadrupoles triplet (ET2 in figure1) further focuses the beam, that arrives at an emittance measurement device (EMD in figure1), consisting of two slit-grid systems with a spatial resolution of 0.2 mm and minimum and maximum measurable divergence of, respectively, 1.8 and 68 mrad.

The Control System of the whole apparatus is based on EPICS [12], with multiple Input Output Controllers (IOC) running on a remote server. These softwares are connected via network to all the field devices, with a common interface to operate them. The graphical user interface allows to monitor and control on a multi monitor PC the general layout of the injector, the beam transport interface, the vacuum and Machine Protection System (MPS) interface and the ion source. The control system is completed by some higher level services, such as automatic procedures for mass scan acquisition with the dipole magnet or the beam emittance measurement. Another service is provided by the Archiver, which stores all the parameters relevant for the operation of the injector, so that they can be retrieved later for offline analysis.

3. Results of the commissiong

The beam used for the tests described in this contribution has been Cs^{1+} at 25 keV, produced by the SIS. Basically a liquid solution containing cesium is dried up and inserted in tantalum capillar heated by a DC current (the oven). The vapour produced flow toward a tantalum ionizing cavity passing through a tubular tantalum transfer line, both kept at a temperature around 2000 °C. In the ionizing cavity, cesium atoms suffer surface ionization and are then extracted as a 1+ beam.

As a first step, the beam line transmission has been verified using a relatively low intensity beam (around 20 nA). First of all the intensity has been measured on both the farady cups available with all the optical elements off, obtaining 0.27 nA and 1 nA respectively. Then, the values of the first triplet has been optimized (1995 V/932 V/1303 V) in order to maximize the intensity on BI.02, obatining 23.6 nA. The values of ET1 has then been changed in order to optimize the intensity in BI.01 (1685 V/842 V/1293 V), obtaining 25.4 nA. The measured transmission turned out to be 93%, a value reproducible even a slightly higher intensty. It worth noticing that, in this condition, the impact of the electrostatic steerers on the optimization process hs been very limited.

The second step was a check of the alignment of the first part of the beam line, using a higher intensity beam (around 100 nA), values for ET1 very close to those corresponding to the optimim transmission to BI.02 (1975 V/ 932 V/1413 V) and checking the intenisty on the faraday cup installed in BI.01. The procedure applied is the following: (i) check of the intensity using only ET1; (ii) optimization of the intensity using the steerers; (iii) check of the intensity with all the optical elements off; (iv) optimization of the intensity with only the steerers; (v) check of the intenisty adding ET1. The maximum recorded intensity has been 136 nA, by switching on ET1 first and then optimizing the steerers. Anyway, the results revealed that the optimum values of the steerers changed depending on wheter ET1 was on or off, with a difference in the measured current of a factor of two. This was a clear indication of a misalignment between the ion source, the triplet and the faraday cup.

The tests prosecuted with an analysis of the beam stability with time, a fundamental parameter in view of the injection and tuning of the charge breeder. Basically, the evolution of the intenisty in BI.01 has been observed for a fixed value of the oven current: the results are shown in figure 2. It can be seen how the intensity increases, reaches a maximum, then decreases and this oscillation is the faster the higher is the oven current. Clearly, in this condition the source was not stable enough to test the charge breeder, so it has been decided to apply a feedback on the intenisty acting on the oven current. Considering that, during charge breeder injection, all the farady cups would be extracted, it has been decided to apply the feedback on the current acquired by the beam profile monitor. The first test of the feedback revealed a very strange effect: while the current acquired by the beam profiler was efficiently kept stable, the current measured at the same time on the farady cup showed a steady increase. This anomaly was investing more in detail, by acquiring the two intensities and other beam characteristics (position and width) for sixteen hours (applying the feedback on the faraday cup this time): the results showed us that the intenisty on the beam profiler decreased of 20%, the beam position drifted a little bit (around 0.1 mm in opposite directions for the two vertical directions) and the beam width increased (up to 1 mm in horizontal direction). In general, the results described up to now revealed several problems concerning the beam line alignment and the realiability of the beam instrumentation boxes. Once fixed, the situation improved considerably, as can be

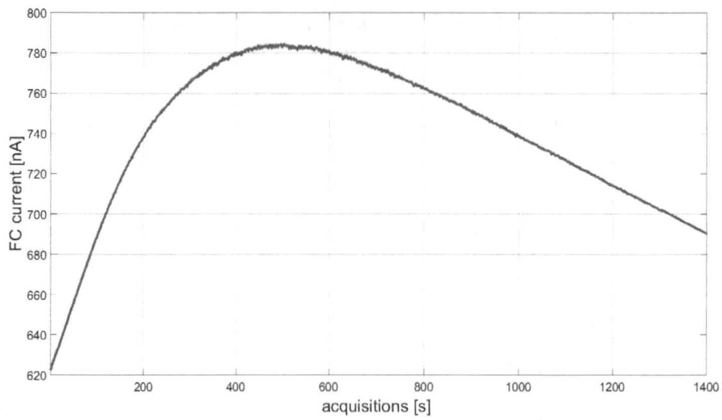

Figure 2. Evolution of beam intensity on the first farady cup (FC) for an oven current of 37 A.

seen in figure 3 showing the acquisition of the intensity of the beam profiler and the faraday cup acquired for more than sixteen hours (feedback applied on the beam profiler). The stability is now excellent, with a variation of less than 1% during the entire acquisition time.

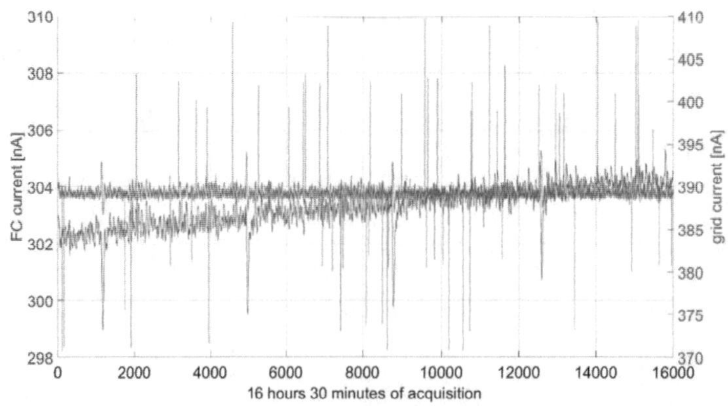

Figure 3. Intenisty acquired from the farady cup (FC) and the beam profile monitor (grid) during a stability test of more than sixteen hours. The intensity on the farady cup varies of less than 1%.

The final test regarded the validation of the beam optics calculations, by measuring the beam emittance. Simulations of the beam line optics have been carried out using TraceWin [13], starting some tens of centimeter out from the ion source. The ion beam envelopes, as well as the initial beam Twiss parameters, have been obtained by adjusting the simulations input to reproduce experimental rms profiles. The validation regarded two intensity levels: 60 nA and 300 nA. The beam exiting from the source has been simulated with a divergent transverse gaussian distribution at an energy of 25 keV and an energy spread of 1 eV. The source emittance has been evaluated at the exit of ET1, by applying the quadrupole scan technique embedded in the BOLINA high level software tools [14], looking at the beam in BI.01. The measurements gave a beam normalized emittance ϵ_{nx}=(1.30±0.20)·10^{-3} π mm mmrad and ϵ_{ny}=(0.84±0.20)·10^{-3} π mm mmrad at 60 nA and ϵ_{nx}=(1.25±0.11)·10^{-3} π mm mmrad and ϵ_{ny}=(1.10±0.20)·10^{-3} π mm mmrad at 300 nA, used as input for the simulations. Different envelopes have been measured to adjust the simulation input and reproduce experimental profiles: by acquiring a set of 17 data, a mean accuracy of 30% on the rms profiles has been obtained, probably due to residual problems on the beam profile monitors. As an example, figure 4 shows the TraceWin simulations of the beam envelope, with indicated the beam size in mm, calculated and measured, at the two focal points, respectively in red and black. Finally, the simulated emittance at the emittance measurement device (ϵ_{nx}=1.10·10^{-3} π mm mmrad and ϵ_{ny}=0.90·10^{-3} π mm mmrad) was compared with an experimental measurement: the results showed a very good agreement, giving values of ϵ_{nx}=(1.20±0.10)·10^{-3} π mm mmrad and ϵ_{ny}=(0.90±0.10)·10^{-3} π mm mmrad.

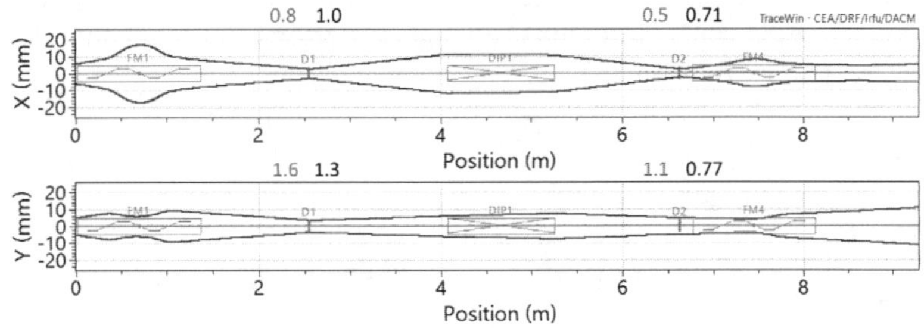

Figure 4. Simulated beam envelopes with indicated the calculated (in red) and measured (in black) dimensions at the two focal points. Dimensions are in mm.

4. Conclusions

The first results of the commissiong have been very promising. The beam line transmission was found to be close to 100%, and the beam stability adequate to test the SPES charge breeder. Emittance measurements, carried out with two different techniques, showed a very good agreement with the simulations, so benchmarking the beam optics calculations. A deep charaterization of the beam line is expected for the next future, with an upgrade of the beam instrumentation boxes with models developed especially for SPES. All the measurements will be repeated extracting rubidium ions, then the SIS will be replaced with the PIS to test the production of the stable counterparts of other elements of interest for SPES.

References

[1] T Marchi *et al* 2020 *J. Phys.: Conf. Ser.* **1643** 012036
[2] M Maggiore *et al* 2017 *Mod. Phys. Lett.* **A 32** 1740010
[3] M Manzolaro, F D'Agostini, A Monetti and A Andrighetto 2017 *Rev. Sci. Instrum.* **88** 093302
[4] M Manzolaro, G Meneghetti, A Andrighetto and G Vivian 2016 *Rev. Sci. Instrum.* **87** 033303
[5] D Scarpa, A Barzakh, D Fedorov, A Andrighetto, E Mariotti, P Nicolosi and A Tomaselli 2016 *Rev. Sci. Instrum.* **87** 02B708
[6] M Maggiore *et al* 2014 *Rev. Sci. Instrum.* **85** 02B909
[7] C Baltador, M Comunian, L Bellan, M Cavenago, A De Lorenzi, A Galatà, L Ferrari, F Mosio and A Pisent 2020 *J. Phys.: Conf. Ser.* **1401** 012014
[8] A Galatà *et al* 2019 *J. Phys.: Conf. Ser.* **1350** 012090
[9] A Galatà *et al* 2016 *Rev. Sci. Instrum* **87** 02B503
[10] A Galatà *et al* 2018 *JINST* **13** C12009
[11] A Pisent, L Bellan, M Comunian, B Chalykh, A Russo and L Calabretta *Spes beam dynamics*, *Proc. of HB2014*, East-Lansing, MI, U.S.A., 2014
[12] L R Dalesio, M R Kraimer, and A J Kozubal *EPICS architecture*, *ICALEPCS* **91**, 1991
[13] DACM softwares, http://irfu.cea.fr/dacm/logitiels
[14] V Martinelli, D Alesini, M Ferrario, A Giribono, S Pioli, C Vaccarezza, A Variola and A Bacci *High Level Software for Beam 6D Phase Space Characterization*, *Proc. 10th International Particle Accelerator Conference (IPAC'19)*, 19-24 May 2019 Melbourne, Australia

Radioactive and Stable Ion Beam production at GANIL

M. Dubois[1], O. Bajeat[1]; C. Barué[1], V. Bosquet[1], P. Chauveau[1], S. Damoy[1], R. Frigot[1], S. Hormigos[1], P. Jardin[1], N. Lechartier[1], F. Lemagnen[1], L. Maunoury[1], V. Métayer[1], B. Osmond[1]

[1]*Grand Accélérateur National d'Ions Lourds, bd Henri Becquerel, BP 55027, F-14076 Caen cedex 05, France*

Fax: +33 2314544538 E-mail address: dubois@ganil.fr

Abstract. GANIL Facilities deliver ion beams from Proton to Uranium up to 95MeV/A. The Cyclotron facility combined with the SPIRAL1 facility provides Stable Ion Beams (SIB) since 1983 and Radioactive Ion Beams (RIB) since 2001 for Physics experiments. R&D of new stable and radioactive beams are continuously under progress matching physicist requirements. During the last ten years, the SPIRAL1 facility was upgraded to extend its capabilities to deliver RIBs of condensable elements. After a new one delivered in 2019, off-line R&D's have been done on the Target Ions Source System and also on the Charge Breeder to improve efficiencies and achieve an operational and reliable facility.
In parallel, the commissioning of the LINAC-SPIRAL2 facility started in 2019 with a proton beam production. In 2020, a 16 kW proton beam was delivered for the first time on the Neutron For Science physics area, and a 4 mA beam of $^4He^{2+}$ was produced with the PhoenixV3 ECR ion source to mimic a deuteron beam. In this article, an overview of GANIL ion sources, beam properties and future developments are described.

1. Introduction

GANIL is a heavy ion beam accelerators facility dedicated to Nuclear, Atomic, Material Physics experiments and industrial applications. The former facility is based on cyclotron technology and delivers ion beams since 1983 to the experimental areas. The ion beam production ranges from carbon to uranium with an energy up to 95 MeV/A. These SIBs could be used either directly for physics experiments, or used as a driver for RIB production on LISE [1] and on SPIRAL1 [2], which are In-Flight and ISOL facilities respectively. SPIRAL 2 is a new facility based on a superconducting LINAC, which started in 2019 [3]. It is designed to produce high power deuteron beam up to 200 kW at 20 MeV/u or intense 1/3 mass to charge ratio ions up to mass ~60 with a power up to 44 kW at 14.5MeV/u. Proton and deuteron beams will be dedicated to Neutron For Science (NFS) experimental area or could be used in the future as a driver for a new RIB facility. Heavy ions will be used for S3 (Super Separator Spectrometer) in-flight RIB facility with the purpose to create and study properties of Super Heavy Elements (SHE).
For producing those beams, eight ion sources of different type operate all over GANIL (Fig. 1).

- To feed cyclotron injectors, two ECR4 (14.5GHz) [4] ion sources of ECR type operate alternatively to deliver beams of medium charge state with mass ranging from C to U.

- To feed the SPIRAL2 LINAC, a 2.45GHz ECR ion source (SILHI like [5]) and a 18GHz ECR ion source (Phoenix V3) can be used, The 2.45GHz ECRIS, dedicated for intense proton or deuteron beams (up to 10mA in CW or pulse mode) was developed by CEA lab. The 18GHz ECRIS was developed for heavy ions [6] and designed by LPSC lab.
- SPIRAL1 Target Ion Source Systems (TISSs) use two types of ion sources for the production of RIBs. For gaseous elements, a 10 GHz ECRIS labelled NanoganIII is preferred, and is able to produce multicharged ions, which can be directly post-accelerated by the CIME cyclotron. For metallic element, a FEBIAD TISS has been developed [7] and produces monocharged ions. To fulfil the injected beam characteristics of the CIME cyclotron, it is mandatory to charge bred the 1+ beam with an ECR charge breeder labelled SP1CB, based on the Phoenix ECRIS breeder [8]. Performances and last results of these ion sources are described in the following.

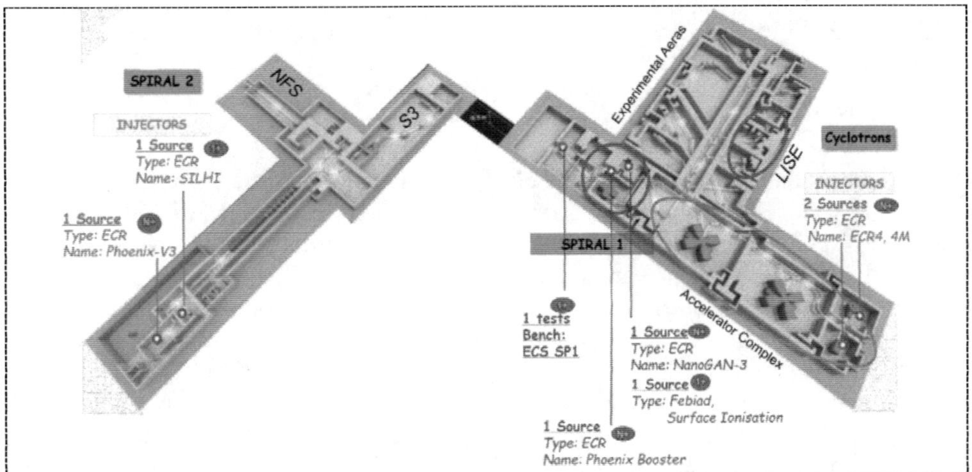

Figure 1. Layout of the GANIL facilities with the locations of the various ion sources

2. GANIL Facilities and Ions sources

2.1. Beam time statistics

The total beam time available for physics experiments with cyclotron accelerators was about 4000 hours per year from 2010 to 2016 (Fig. 2). That corresponds to the time when ECR4 injectors deliver SIBs. In 2016, there was a reduction of cyclotron beam time in order to initiate the commissioning of SPIRAL2 machine. The first step of the commissioning involved the RFQ (2016) and the characterization of the beam at the injection of the LINAC. 2018 was dedicated to the preparation of the LINAC, commissioned in 2019. The first proton and deuteron beam were accelerated in 2019 and 2021, respectively. Now, both facilities are under operation 4 to 5 months per year for physic experiments (twenty four hours a day) or commissioning (ten hours a day except for physics experiments).
Regarding SPIRAL1, the facility was off-duty for 4 years in order to upgrade it by setting up the new 1+/N+ method with the aim to produce metallic radioactive beams [9]. The facility was commissioned in 2018 and has delivered two new RIBs ($^{38m}K^{9+}$ and $^{47}K^{10+}$) to physicists.

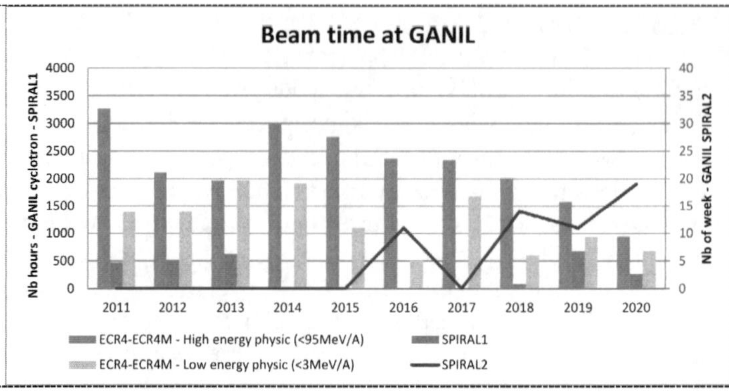

Figure 2. Total beam time provided by the GANIL complex over 10 years

2.2. GANIL – Cyclotrons

The two ECR4 produce a large variety of elements. Metallic ones are mainly produced using two techniques: the MIVOC method and the Large Capacity Oven (LCO) [10]. A rotation is done between the two ECRISs one delivering its beam to the ongoing experiment, while the other delivering beams to a low energy beam line or is in preparation for the next experiment.
Nowadays, the main developments are focused to produce new on-demand elements requested by physicists. For example, in 2019, a Tellurium beam was effectively developed. It was produced with the GANIL LCO heating up a metal sample [11] up to vapour state. In 2021, a thorium beam must be elaborated to foresee a future physics experiment requested by physicists. A sample of ThF_4 was supplied in June 2021. It will be tested in the second semester of 2021. Intensity requested by physicist is 1.2 10^{10} pps on target at an energy of 6 MeV/A, which involves a beam intensity in the range of 0.5 to 1µAe of $^{232}Th^{30+}$at the exit of the source.
The main limitation of the MIVOC technique is the conductance between the location of the chemical compound and the ion source chamber. The room is very limited in the injection region partly due to the presence of the coaxial RF injection. Tests are scheduled by the end of 2021 to increase the coaxial tube diameter and hence increase the conductance between the MIVOC sample and the plasma chamber. When an acceptable configuration will be determined, trials based on ^{58}Ni will be accomplished to evaluate the improvement in terms of intensity and stability.

2.3. GANIL – SPIRAL1

SPIRAL1 uses the stable primary beam coming from the GANIL cyclotron complex to produce RIBs by nuclear reaction of beam fragmentation on C target. Two types of ion sources are routinely employed. The 10GHz ECRIS(NanoganIII) is dedicated for radioactive gaseous elements as Helium, Neon, Argon and Krypton isotopes with half-lives longer than 100ms (^{8}He). This TISS was upgraded by the installation of a direct RF injection and by a target supporting up to 3kW of beam power. Since 2012, a new TISS was developed for the production of radioactive metallic beams. This ions source is a combination of the GANIL graphite target and a FEBIAD VADIS type ions source [12]. As the FEBIAD ion source produces single charge state, an ECR charge breeder [ref] has been implemented in the middle of the LEBT. Since 2018, an important effort was done on FEBIAD TISS as well as on SP1CB to increase efficiencies and get a high performant facility.

In 2019, FEBIAD TISS was able to produce new RIB [**7**] with intensities (10^{6}pps) large enough to be post-accelerated and delivered on physicist target (minimum of 10^{4}pps required). Unfortunately, since 2018, the stability as the reliability has not been sufficient for an operational machine. In 2020, the origin was identified [13]. After modifications and several tests, the TISS was able to produce 5 to 8% ionization efficiency of Ar^{1+} beam on the test bench without fluctuation. This new version was carried out during the first run in 2021 and provided a steady state beam of radioactive alkali ($^{47}K^{1+}$) for

seven days with an ionization efficiency close to 70%. After this experiment, the ion source worked on plasma mode for 7 days with an Ar^+ ionisation efficiency of 15%. At the end of the test, an ionization efficiency of 25% was obtained for 30 min (cathode current up to 350 A – 130 mA anode current). Finally, taking into account all the period of operation, the new version of FEBIAD TISS worked 25 days without failure.
Concerning the SPICB, there were many optimizations performed to increase the global efficiency since 2015 and the qualification done at LPSC (Fig 3.).

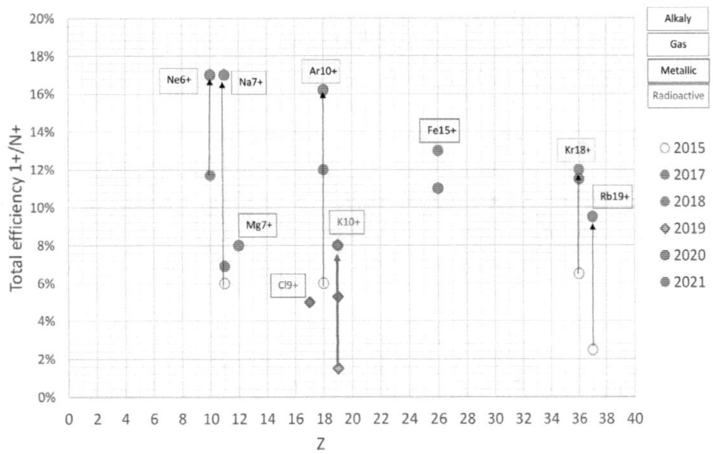

Figure 3. Evolution of the charge breeding efficiencies over the last 6 years

The main optimizations accomplished on SP1CB were the adjustment of the plasma electrode close to the highest magnetic field at extraction, the modification of the magnetic field profile (iron plug at extraction and location of soft iron rings), the use of hydrogen as a buffer gas and finally the studies and definition of a dedicated optic beam line for the injection of 1+ beam into the SP1CB [14]. Moreover, a procedure was completed to tune the 1+/N+ system in stable and radioactive modes. All these improvements enable a significant increase of SP1CB, performances either on gas, alkali and metallic ions. In addition, all process to tune the RIB (optic 1+ and injection on SPICB, DeltaV, SPICB parameter) are now totally under control had lead in 2021 to a charge breeding efficiency of 8% for the $^{47}K^{1+ \text{ to } 10+}$ ions factor 4 higher than the efficiency obtained for $^{38m}K^{1+ \text{ to } 9+}$ produced in 2019.
New developments are under progress. The first one consists on implementing 1+ molecular beam to limit the isobaric contamination and speed up the release – diffusion of radioactive elements towards SP1CB. Hence, SP1CB has demonstrated quite similar performances for singly charged atomic and molecular ion beams [15]. The second one figures out how is it possible to shift a charge state distribution (CSD) by employing a variable RF frequency amplifier while keeping constant the ionization efficiency on the most abundant charge state. That method will open up a larger range of operation for the CIME cyclotron. Some preliminary trials have been performed. In July 2021, tests were carried out with a variable 8-18GHz TWTA radio frequency at a power of 200W injected into SP1CB, which operated in source mode and only with gas. By adjusting the RF frequency and the magnetic field profile, the most abundant ion charge state in the CSD of Ne ions varies from 3+ to 8+while conserving a constant ionisation efficiency of 20% on the charge state optimized. The total efficiency taking into account all charge states was of the order of 80%. In pure double frequency mode (14.5GHz + 11.6GHz), it shows an increase of the high charge states for $^{84}Kr^{22+}$ by 60%. Next meaningful step is to reproduce those results in 1+/N+ mode with alkali elements.

2.4. GANIL – SPIRAL2

Commissioning of the LINAC started in 2019 after getting the green light of national safety authorities.

The first proton beam was accelerated by RFQ + LINAC in November 2019. In 2020, a 16kW proton beam was produced and in September this year, a 2kW of Deuteron beam was delivered for physics experiments on NFS experimental area. Ion source used for proton and deuteron beam production is a 2,45GHz SILHI type [ref] made of permanent magnet. That type of source is dedicated to the production of intense (tens of mA) light ion beams (p, d) with a high stability, which matches performances required by the linear accelerator. Unfortunately, its life time is presently too short regarding the machine operation objective, i.e. 3 months without maintenance. This limitation is due to the Boron Nitride (BN) plate located into the plasma chamber, which degrades steadily over the time and declining the source performances. Up to know, a maintenance was required after 1000 hours and 500 hours of proton and deuteron beam production respectively. Magnetic field profile as well as BN thickness have been modified but the result is not effective so far. A collaboration with INFN-LNS is under progress to study the suspicious RF coupling with the plasma and ion sputtering process. The final goal is to adjust the design of the source to increase its durability and total ionization efficiency (lowering gas consumption). The second injector, dedicated to heavy ions, is a PhoenixV3 18 GHz ECRIS. First beam of $^{40}Ar^{14+}$ was produced in January 2020, with an intensity of 100 μAe, twice the intensity delivered by the previous version Phoenix V2. The final objective of that ECRIS is to deliver beams of ^{58}Ni, ^{48}Ni, $^{48,50}Ti$, $^{50,54}Cr$,… to S3 project, with intensities up to 2pμA for A/Q around 3. For this purpose, oven technique will be used. In 2020, a first beam of $^{58}Ni^{19+}$@40kV was produced with the GANIL LCO, displaying an intensity of 40 eμA, which is in accordance with the specifications of the S3 project. For 2022, an experimental program is already scheduled to develop other beams, especially the Ti beam for which a High Temperature oven in under development [16]. A liner technique will also be implemented to limit the ion losses on plasma chamber walls.

2.5. GANIL – Future program

Due to the limitation of the current delivered by Phoenix V3 ECRIS, the study of a new project has been funded to develop a new LINAC injector (ion source + RFQ). The project, called NEWGAIN, consists, for the source part, of a superconducting ECRIS (28GHz +18GHz) able to produce up to 5 pμA of $^{235}U^{34+}$ or 15 pμA of $^{48}Ca^{11+}$. It will be combined with a new A/Q = 7 RFQ. Beams will be reinjected in the actual LINAC of SPIRAL2. To optimise the availability of the machine, the possibility to inject the current Phoenix V3 beams will be conserved, allowing the SC-ECRIS to be in preparation or maintenance.

3. Outlooks

Several improvements or developments are under progress on each ion source at GANIL. On one hand, there is a significant effort to carry on as to reach performances fitting with the objectives of the new facility (SPIRAL2) in terms of intensity, stability and reliability. Moreover, a new project is under study to design a new injector for SPIRAL2 to use all the powerful of this accelerator On the other hand, there are always new requests from physicists to develop new beams for the GANIL cyclotrons complex. In addition, some effort will continue for reliability of ion sources and optimize current beams in term of intensities (MIVOC method). About the main devices of SPIRAL1, numerous improvements were achieved and current performances now fulfil the initial specifications of the upgraded project. However, R&Ds on SP1CB are always under progress with LPSC, TRIUMF, LPSC, INFN-LNL, JYFL to improve that equipment (reduction of contamination, double frequency heating…). Moreover, New development is under progress to produce new element, like TULIP ANR project where the goal is to produce new neutrons deficient RIBs by fusion-evaporation process [17].

4. References

[1] Muller A.C 1991 *Nucl.. Phys. A*, Vol56-57 ; https://doi.org/10.1016/0168-583X(91)96095-3
[2] Villari A 2006 *Nucl. Phys. A*, Vol. 787, p126-133 ; https://doi.org/10.1016/j.nuclphysa.2006.12.023

[3] Dolegeviez P et al. 2019 JACoW 10th IPAC Proceeding ; https://doi.org/10.18429/JACoW-IPAC2019-MOPTS005
[4] Sortais P et al. 1990 *RSI.* 61, p. 228 ; https://doi.org/10.1063/1.1141324
[5] Tuske O et al. 2012 *RSI* 83, 02A316 ; https://doi.org/10.1063/1.3665963
[6] Thuillier T et al. 2016 Proceeding of ECRIS2016; https://doi.org/10.18429/JACoW-ECRIS2016-TUAO05
[7] Chauveau P et al. *NIMB* volume 376, page 35-38; https://doi.org/10.1016/j.nimb.2016.01.039
[8] Maunoury L and al. 2016 *RSI* 87, 02B508 ; https://doi.org/10.1063/1.4935215
[9] Dubois M 2015 oral communication *EMIS 2015*, SPIRAL1 Upgrade at GANIL: Status
[10] F. Lemagnen et al., Proceedings of ECRIS-2014, http://hal.in2p3.fr/in2p3-01057669/document
[11] Lemagnen F et al. 2020 Proceeding of *24th ECRIS*, New metallic stable beams for GANIL – MSU
[12] Bajeat O et al. *NIMB* volume 317, page 411-416; https://doi.org/10.1016/j.nimb.2013.07.049
[13] Bosquet V et al. 2021 *ICIS proceeding* #178 ; Triumf Vancouver Canada
[14] Maunoury L et al. 2021 *JINST* 16 P05013 ; Triumf Vancoucer Canada
[15] Maunoury L et al. 2021 *ICIS proceeding* #11 ; Triumf Vancouver Canada
[16] Bajeat O et al. 2021 *ICIS proceeding* #12 ; Triumf Vancouver Canada[17]
[17] Venkateswarlu K. Development of an innovative ISOL system for the production of short-lived neutron-deficient ions; HAL Id : tel-01994668, version 2

Improvement of the FEBIAD ion source at SPIRAL1

V. Bosquet, P. Jardin, P. Chauveau, S. Damoy, P. Delahaye, M. Dubois, S. Hormigos, L. Maunoury, C. Michel, J.C. Thomas
GANIL, Grand Accélérateur National d'Ions Lourds, Bvd H. Becquerel, 14000 Caen

bosquet@ganil.fr

Abstract. The behaviour of the Forced Electron Beam Induced Arc Discharge (FEBIAD) ion source used at GANIL within the Target Ion Source System (TISS) of the "Système de Production d'Ions Radioactifs Accélérés en Ligne" (SPIRAL1) has been recently investigated to improve its performances and render them reproducible. Off line systematic studies of the evolution of working parameters (cathode shape, cathode heating, anode current) with time and temperature configuration have clearly shown the origin of issues was mainly related to an excessive temperature of the anode insulators. Modifications of the TISS have allowed the reduction of the anode current drift and an ionisation efficiency of ^{40}Ar of 15%, stable over more than eight days. The system is still under improvement to stabilize its Ar ionization efficiency to 25%, which is the maximum observed during testing.

1. Introduction

Isotope Separator On Line (ISOL) installations are based on a combination of primary beams, targets, ion sources and a mass separator. Primary ion beam characteristics (Z, intensity and energy) and energies depend on stable ion sources and locally available accelerators. Coupled to the targets, they define the regions of the nuclide chart where the facility can efficiently produce radioactive nuclei. Once released by the target as atoms, the radioactive isotopes are ionized by a source, which is chosen regarding its ability to ionize chemical elements of interest.

Among existing source principles, the Forced Electron Beam Induced Arc Discharge (FEBIAD) ion source offers a large ionizing spectra and has therefore been implemented in several ISOL leading facilities as at ISOLDE CERN [1], ISAC TRIUMF [2], JAEA ISOL [3], and in 2012 at GANIL/SPIRAL1 to extend its beam offer. All these sources are based on the design initially developed by R. Kirchner [4], which uses the electron emission principle of hot metals firstly observed by Edison at the end of the 19th century and later studied by Richardson [5]. If the electrons emitted are sufficiently accelerated within the source chamber, they can ionize all elements, which makes the FEBIAD a powerful solution for ionizing radioactive atoms.

In principle, a FEBIAD ion source is a dynode installed in a vacuum chamber where atoms are injected and then ionized by the electrons going from the cathode to the anode. A magnetic field generated by a coil controls the electrons' trajectories and increases the electronic density. Performances and stability strongly depend on the temperature of the different parts and then on services, *i.e.* heating current of the cathode, cooling of the chamber, power dissipated within the dynode, and on electrode materials whose characteristics can change over time due to the high working temperature (of the order of 2000°C). Source temperature can also be influenced by the temperature of the target, and thus by the power of the

Journal of Physics: Conference Series **2244** (2022) 012071 doi:10.1088/1742-6596/2244/1/012071

primary beam through the mechanical, thermal and possibly electrical coupling of the target to the FEBIAD source.

A precise control of temperature would provide precious information on the evolution of the source, for diagnosis and operation. However, a direct measurement of temperature is difficult due to the small size and geometrical configuration of the source. The source temperature is generally deduced from indirect parameters as heating power/current/voltage settings, current exchanged between cathode and anode and previous simulations and calibrations.

Owing to the variety of primary beams and targets available in ISOL installations, the target-source couples are adapted case by case. The geometry, material and temperature are optimized for maximum atom-to-ion transformation in the Target Ion Source System (TISS). In the case of SPIRAL1, the mechanical configuration of the FEBIAD is close to the VADIS version of ISOLDE CERN, whose plans predate 2005. As carbon target material is well adapted to the production of radioactive elements by fragmentation of primary beams available at GANIL, the FEBIAD was coupled in 2012 to a graphite target dedicated to the production of isotopes with masses ranging from ~20 to ~80. Several coupling configurations have been tested up to 2019, still suffering from drifts of current exchanged between cathode and anode, and from hardly repeatable performances. In 2020, the different technical aspects of the TISS were reconsidered. Diagnosis, modifications and off-line tests led to a version, which has finally been used on-line after having demonstrated reproducible and satisfying off-line performances. Main modifications, improvements and performances are described below.

2. SPIRAL1 FEBIAD TISS configuration

The SPIRAL1 TISS (Figure 1) was initially based on the coupling of the Penescu [1] FEBIAD version coupled to a graphite target designed for fragmenting low masses (mainly from C to Kr) primary ions delivered by GANIL cyclotrons at a maximum energy of 95 MeV/A. Radioactive atoms effuse from target container to the FEBIAD through a 8 cm long Ta transfer tube. The tube is connected on one side to the target container and on the other side to the Ta disc of the cathode. Thermal expansion (~1.6 mm) of the transfer tube at 2000°C induces an effort on both extremities. As the target container is firmly fixed on the vacuum chamber, the source has to be slightly mobile to balance the thermal expansion of the transfer tube, and so avoid deformation of the cathode disc. This goal has been reached by using a system of coaxial sliders [6], the inner part of the slider being fixed on the vacuum chamber and the outer part being mobile and supporting the FEBIAD. Four flexible legs of carbon fibers connecting the outer graphite slider and the vacuum chamber (common electrical mass) allow circulation of the cathode heating current (~300 A). Former Ta thermal screens were conserved around the FEBIAD. Screens were also installed around the transfer tube to limit thermal losses.

The target is housed in a graphite container, closed by a 7 μm thick Ta window on the primary beam side. The container is surrounded by a graphite resistor (supplied by a ~120 A current), itself surrounded by Ta screens. To increase the mechanical strength of the target container to the effort due to the thermal expansion (~1.6 mm) of the transfer tube, the target container and the resistor were mechanically linked, what simultaneously induced an electrical coupling between the target and the cathode heating circuits.

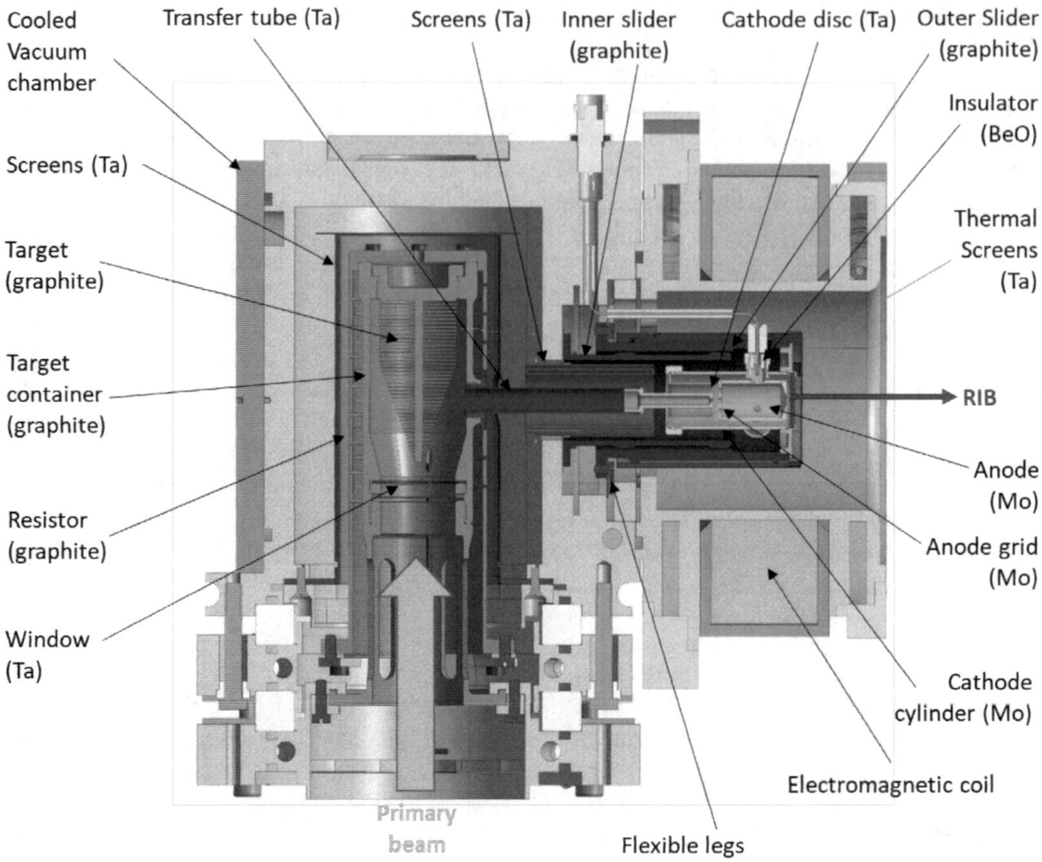

Figure 1: Drawing of the SPIRAL1 FEBIAD TISS at the beginning of this study (2019). Only main parts are identified.

The thermal configuration is thus different from the ones of ISOLDE, TRIUMF and BRISOL [4]. Simulations have been performed to determine if the modifications operated on the SPIRAL1 TISS could have a significant effect on its working performances. Such simulations give indications on relative temperature changes but need a stringent iterative confrontation with thermal measurements to be reliable and to finally allow reliable prediction of absolute temperatures. This process is time consuming and the final uncertainty on the results depends on several parameters such as the quality of the thermal model of the TISS, the reliability of the thermal code, and the precision on experimental temperature measurements, which are particularly difficult to obtain in a system as small and confined as the FEBIAD. As the source temperature is roughly known, its working regime can hardly be predicted, as it is directly related to the current of electrons exchanged between the cathode and the anode, which varies exponentially with the temperature. The diagnosis and modifications of the configuration presented below are thus mainly based on working parameters (currents and voltages) observed from outside, as they are typically known with a better precision than temperature.

3. Measurements

Off-line tests were performed on the SPIRAL1 test bench where all the TISSs are qualified before being used in the on-line production cave. The FEBIAD TISS was tested off-line during six periods totalling 19 days with a cathode current over 250 A and a target heating power over 1000 W. The different FEBIAD versions were compared using the ionization efficiency of Ar. During all the tests, the coil current was maintained between 9 A and 10 A, without any significant change of the Ar+ current.

This study was motivated by the observation of drifts in electronic current exchanged between the cathode and the anode (ICA), which can be up to several tens of percent per hour for Icathode ~ 300 A. Simultaneously, important drifts in time of the Ar+ current were observed but not clearly related to ICA. This behavior, requiring a constant monitoring by operators to be contained was unsuited to regular operation.

The drift of Ar+ current was observed simultaneously with the drift of the cathode heating current (Icath), and thus with a drift of the temperature or electronic emissivity of the cathode. The cathode heating power supply was power regulated. An accurate observation of Icath and of the cathode voltage revealed a drift in time of Icath, between 70 and 260 mA/h, for Icath between 280 and 315 A. This drift has been attributed to the evolution of the resistance of parts of the supply circuit over time. As the temperature of the cathode disc is mainly determined by Icath, the Icath supply has been modified to be current regulated.
Regarding the electronic emissivity, the surface of the six cathode discs used for the six tests have been visually examined after test, revealing no difference in terms of colour and surface roughness. The electronic emissivity was then assumed to be unchanged.

Despite Icath stabilisation, ICA drifts were still observed. After each test period, the FEBIAD was fully dismantled and the observations made on each parts were reported. Two main material degradations were observed in the FEBIAD: the deformation of the cathode disc and a black deposit on BeO insulators and on their immediate environment. Degradation of BeO was suspected to play a role in ICA drift if BeO resistivity changed. BeO degradation hypothesis was reinforced by the presence of Mg+ (which is a BeO contaminant with a relative atomic abundance of 0.1%) and by the presence of Be+ in the mass spectrum, which increased by a factor of 30 for Mg+ and by 100 for Be+ when Icath increases from 280 A to 308 A (in 5 hours). BeO resistivity can also vary by several orders of magnitude when its temperature changes [8]. Leaving aside temperature changes, it is thus difficult to determine if the ICA drift is due to the change of resistivity or to the BeO matter transformation.
To clearly identify the reasons for resistivity changes, iterative modifications described in paragraphs 3.1. 3.2. 3.3. where made with the aim of reducing the BeO insulator temperature. Figure 2 shows the current I_{Ar+} versus ICA for three different thermal configurations.

3.1. Initial thermal configuration

The first measurement (orange dots) was made with the Ta screens present on the VADIS FEBIAD source. The cathode heating current Icath was maintained at 281 A during 16h. The anode was biased at 150 V. During this period, a monotonic relationship relates IAr+ and ICA. The day after (gray dots), the heating current of the cathode was slightly increased, up to 286.5 A per step of 1.5 A within a period of 9 h. The relation between IAr+ and ICA was close to the one obtained with Icath = 281 A. For Icath = 288 A (red dots), ICA increased continuously with time and reached 256 mA after 6 h, which is the maximum current delivered by the anode HV supply. The relation between IAr+ and ICA was roughly linear but not reversible: setting back Icath at 281 A did not lead to the same IAr+ current as before applying Icath = 288 A. A higher ICA current was necessary to find the same IAr+ current. This behavior suggested a transformation of the system, namely a damage of the BeO insulators leading to a reduction of its resistance, possibly due to an excessive temperature. Compared to the VADIS version, the SPIRAL1 version comprises a graphite slider, which inevitably acts as an additional screen around the source and leads to a higher temperature of the insulators.

3.2. First modification of thermal configuration

In order to reduce this temperature, the two external Ta screens have been removed and a new measurement of IAr+ as a function of ICA has been done. For Icath heating current ranging from 280 A

and 304 A (light blue dots in Figure 2), the relationship between IAr+ and ICA was again monotonic. The behavior was reproducible as a modification of the temperature by playing on Icath led to the same (IAr+; ICA) couple. One deduces that these working conditions do not damage the system in an irreversible way. What must also be noticed is that the current of Ar+ obtained for a given current of electrons was greater than with Ta screens: only 40 mA of electronic current were required to obtain 50 nA of Ar+, whereas 140 mA were required with the screens. This shows that 100 mA of the ICA current were useless for Ar ionization, and escaped through other regions rather than between cathode and anode. As the source behavior was reproducible without screens, one can assume that this useless current obtained with screens was emitted between the cathode cylinder and the anode, and had disappeared due to the lower temperature of the cathode cylinder caused by the removal of the screens. Part of this current could also go through the insulators, not damaged for a temperature corresponding to Icath = 304 A.
For Icath = 308 A, IAr+ increased up to 150 nA (6% of efficiency) in 6 minutes with the increase of ICA. Then ICA globally increased continuously during 5 hours up to the limit of 256 mA without any clear relationship with IAr+ (dark blue dots). Moreover, the behavior was no longer reproducible. It is then assumed that the insulators were damaged again.

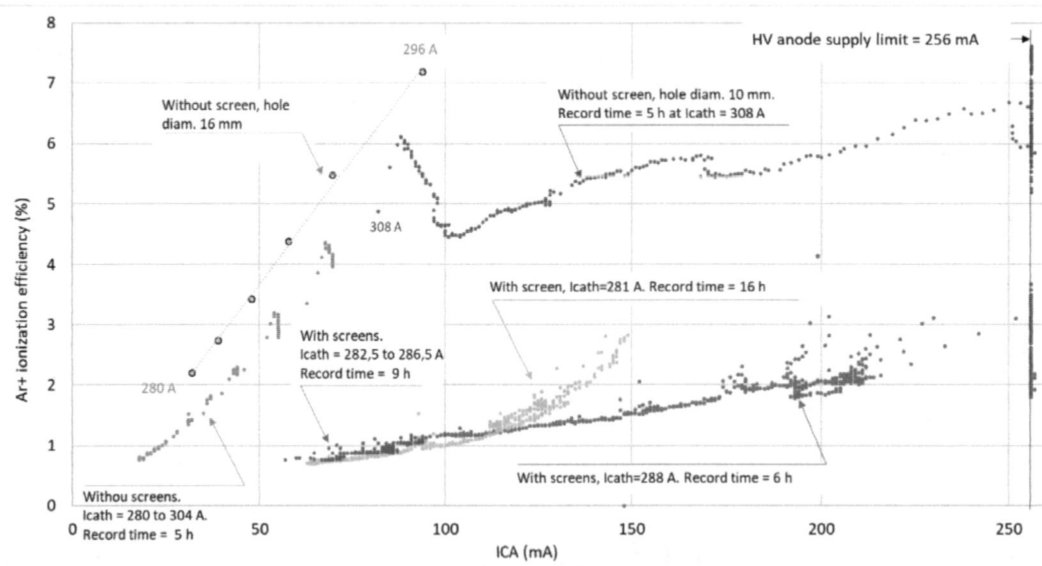

Figure 2: Ar+ ionization efficiency of the TISS versus current exchanged between cathode and anode (ICA), for different screening configurations and cathode heating currents. Grey, orange and red : with thermal Ta screens around the FEBIAD source. Light and dark blue: Without Ta screens, and for insulators hole of 10 mm in diameter. Green: Without Ta screens and for insulators holes of 16 mm in diameter. The green doted line is a linear fit. (see text for details)

3.3. Second modification of thermal configuration

To pursue the aim of identify working conditions suited to operations, thus avoiding damages of BeO insulators and non-reproducible behaviors induced by excessive temperature, the holes in the outer graphite slider facing the insulators and their nuts have been enlarged in order to improve radiative cooling. A large aperture certainly helps to reduce the insulator temperature but a too large aperture would also reduce the anode temperature, which would increase the sticking time of metallic elements in the anode and reduce the atom-to-ion transformation performance for sticky radioactive isotopes. The enlargement of this diameter was finally fixed at 16 mm, instead of the initial 10 mm. In these conditions, still without Ta screens, IAr+ versus ICA has been measured (green dots) by varying the cathode heating current from 280 A to 296 A. An ICA current of only 32 mA was needed to obtain a current of Ar+ egal to 50 nA (2.2% of efficiency) whereas a current of 42 mA was previously needed, showing that with the 10 mm hole, 10 mA were still useless for the Ar ionization.

4. Conclusion on behavior of the source

The behavior of the FEBIAD source as understood from the above measurements is schematically described in Figure 3. The current delivered by the anode supply can be divided in 3 parts: i) - the current ICAi exchanged between the cathode disc and the anode grid, which is the one necessary for the ionization; ii) – the current ICAu emitted by the cathode cylinder towards the anode cylinder, which is useless for ionization; iii) – and the leak current ICAl passing through the insulators, which is also a useless current for ionization.

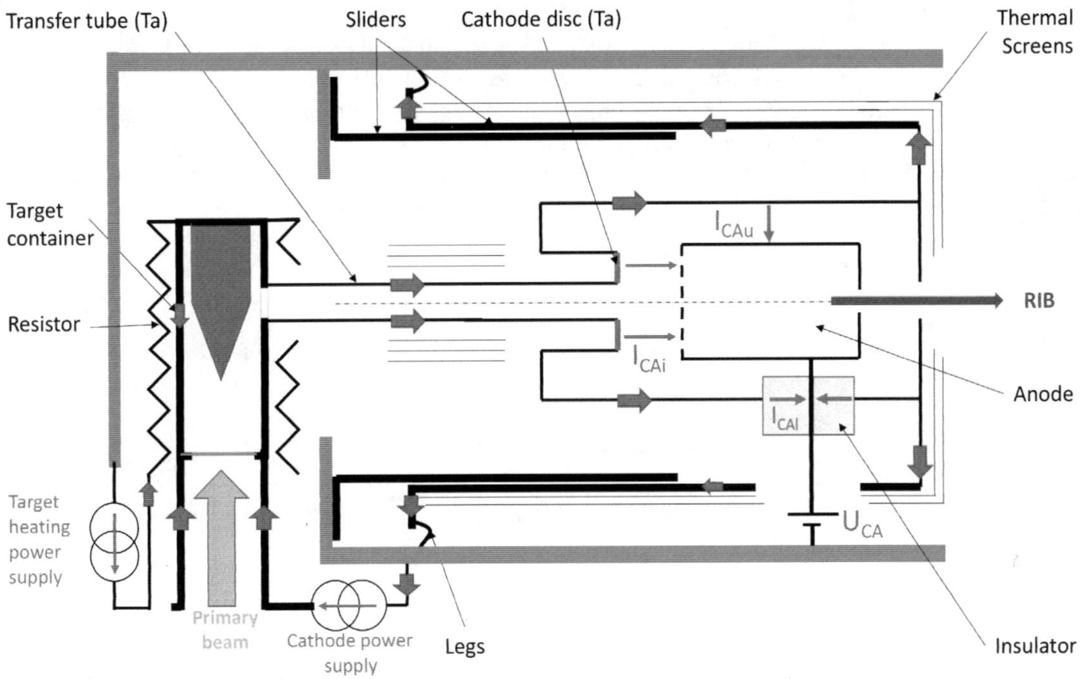

Figure 3: Schematic of the current circulation routes in the FEBIAD TISS

All the currents ICAi, ICAu and ICAl increase with the temperature but only ICAi and ICAu are reversible with the temperature. ICAl is not. It is directly related to the resistance of the insulator material, which decreases irreversibly when exposed too long to high temperature. A limited temperature of the insulators is thus required to ensure a life span of ~10 days, and simultaneously, a maximum temperature of the FEBIAD is desirable to maximize its atom-to-ion transformation efficiency. A compromise has been found for the present configuration using an empirical approach. It can thus certainly be improved.
An ongoing study is addressing this possibility.

5. Ar ionization efficiency

The following results have been obtained using the last version of the FEBIAD, *i.e.* the one without Ta screens and with holes of 16 mm in diameter performed in the graphite slider.
Limiting the cathode heating current at 296 A, the ^{40}Ar ionization efficiency slowly drifted from 6% to 9.6% in 48 h with a ripple lower than 1%, which is acceptable regarding the operation requirements. As 6% satisfied the requirement for an on-line experiment scheduled in March 2021, no attempt to reach a higher efficiency has been made before the experiment.

After eight days of $^{47}K^+$ on-line production in surface ionization mode (performed at a cathode heating current reduced to 240 A), the TISS has been tested in FEBIAD mode, maintaining the cathode heating current at 315 A during 8 days. Within this period, the ICA current and the Ar ionization efficiency, starting from 40 mA and 13%, increased continuously up to 72 mA and 19%. During the same period, the intensities of all the contaminants initially present in the mass spectrum, CO^+(571 nA), Mg^+(97 nA), C^+(81 nA), Fe^+(74 nA, introduced sample), O^+(56 nA), CO_2^+(18 nA), N^+(18 nA) and Cr^+(2 nA, introduced sample) decreased by a rough factor of 2, depending on the element. The intensity of Be^+ remained lower than 0.1 nA, showing that the insulators were not affected by the temperature and suggesting that the source could work at higher temperature.
In order to determine the maximum reachable Ar+ ionization efficiency, the cathode heating current was scanned (Figure 4) from 290 A to 350 A. Blue dots correspond to efficiency values measured the 14th and the 15th of April 2021, before the 8 days test at Icath = 315 A. Red dots correspond to Ar+ ionization efficiencies measured when Icath increases from 315 A to 350 A, at the end of the 8 days test. The red curve corresponds to Ar+ ionization efficiencies measured when Icath was reduced from 350 A to 300 A. The ionization difference between blue and red dots is due to the drift of the source regime mentioned above. The main result is that a maximum efficiency of 25% can be reached for a cathode heating current of ~335 A. As this last measurement has been realized in 3 hours, it is not possible to conclude on the possibility to conserve this efficiency over several days.

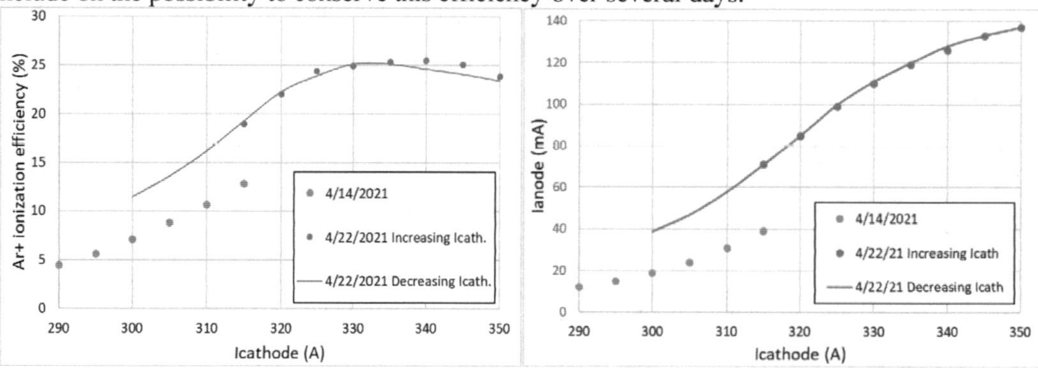

Figure 4: Ar^+ ionization efficiency and Ianode versus heating current of the cathode. UCA = 140 V.

6. Next steps

Excessive temperature of the anode insulators has clearly been identified as responsible for the drift of ICA in the SPIRAL1 FEBIAD TISS. The present technical solution allows an Ar ionization efficiency higher than 13% over a minimum duration of 8 days. 25% have also been obtained for 3 hours. The next goal and challenge of the version already under test are to maximize the anode temperature while maintaining an Ar+ ionization efficiency of 25% for ten days.

References

[1] L. Penescu, R. Catherall, J. Lettry, and T. Stora, Rev. Scient. Instrum. 81, 02A906 (2010)
[2] B.E. Schultz, J. Sandor, P. Kunz, A. Mjos, O. Kester, F. Ames, 9th Int. Part. Accel. Conference
[3] Y. Otokawa, A. Osa, T.K. Sato, M. Matsuda, S. Ichikawa, and S.C. Jeong, Rev. Scient. Instrum. 81, 02A902 (2010)
[4] B. Tang et al., Nuc. Inst. Meth. Phys. Res. B 463 (2020) 154-157
[5] R. Kirchner and E. Roeckl, Nuc. Inst. Meth. Phys. Res.133 (1976)187-204
[6] P. Chauveau et al., "A new FEBIAD-type ion source for the upgrade of SPIRAL1 at GANIL", Nucl Instrum Methods Phys Res B
[7] G. Reboul, Phénomènes thermo-électriques et électro-capillaires dans les gaz. J. Phys. Theor. Appl., 1908, 7(1), pp. 840-856. https://hal.archives-ouvertes.fr/jpa-00241415.
[8] A.W. Pryor, "The electrical conductivity of beryllium oxide", Journal of Nuclear Materials 14 (1964) 258-264

A cold electron-impact ion source driven by a photo-cathode – New opportunities for the delivery of radioactive molecular beams?

Jochen Ballof[1,2], Mia Au[1,2], Ermanno Barbero[1], Katerina Chrysalidis[1], Christoph E Düllmann[2,3,4], Valentin Fedosseev[1], Eduardo Granados[1], Reinhard Heinke[1], Bruce A Marsh[1], Michael Owen[1], Sebastian Rothe[1], Thierry Stora[1] and Alexander Yakushev[3,4]

[1] CERN, Accelerator Systems Department, 1211 Geneva 23, Switzerland
[2] Johannes Gutenberg - Universität Mainz, Department Chemie, Standort TRIGA, Fritz-Strassmann-Weg 2, 55128 Mainz, Germany
[3] GSI Helmholtzzentrum für Schwerionenforschung, 64291, Darmstadt, Germany
[4] Helmholtz-Institut Mainz, 55099 Mainz, Germany

E-mail: `jochen.ballof@cern.ch`, `thierry.stora@cern.ch`

Abstract. The thick-target ISOL (Isotope mass Separation OnLine) method provides beams of more than 1000 radionuclides of 74 elements. The method is well established for elements with sufficiently high volatility at ca. 2000 °C. To extract non-volatile elements the formation of a volatile molecule is required. While successful in some cases (*e.g.* carbon or boron), most of these elements are not yet available as ISOL beam. A variety of volatile carrier molecules has been proposed for all elements produced in the target material, but their probability of survival during the extraction and ionization process is often limited by the high temperatures required for isotope diffusion in the thick targets and for ion source operation. While cold target concepts have already been proposed, the normal mode of operation of the typically used Versatile Arc Discharge Ion Source (VADIS) with a hot cathode is not well suited. Here, we report about first measurements with an electron-impact ion source operated at ambient temperature using electrons that were liberated via the photo-electric effect from a copper cathode.

1. Introduction

Since the advent of the ISOL (Isotope Separation On-Line) method in the 1960s, facilities like CERN-ISOLDE [1, 2] have been erected, which provide beams of a variety of radioisotopes. To keep up with the demand for ever more exotic and intense radioactive ion beams, extensive upgrade programs are on the way, *e.g.*, at GANIL (Caen, France) or TRIUMF (Vancouver, Canada) [3]. The ISOL technique exploits thick targets (typically at least tens of grams per cm^2) in which nuclides are produced in reactions induced by an energetic driver beam. The nuclides are stopped within the target material and have to diffuse out of the target matrix and effuse into the ion source from where they are electrostatically extracted and separated by mass-to-charge ratio. The target is typically kept at high temperatures of up to ca. 2000 °C to promote diffusion and effusion processes. Even at these elevated temperatures, many elements cannot be extracted due to their refractory properties, *i.e.*, high melting and boiling points, which prohibit their transport.

A powerful method to volatilize elements is the *in-situ* formation of a volatile carrier molecule [4]. While volatile carrier molecules have been proposed for almost all elements [5, 6], the *in-situ* volatilization for radioactive ion beam formation was so far only successful for few refractory elements like carbon [7] or boron [8], which form compounds that are stable in a high-temperature environment [9, 10]. Suitable compound classes are mostly restricted to binary halides or chalcogenides. To increase the variety of volatile carrier compounds, the temperatures of the target and ion source need to be significantly reduced. Cold target concepts have been proposed, which exploit the fast diffusion in nano-materials [11] or fully avoid diffusion and exploit the recoil momentum of nuclear reaction products [12]. However, the ion sources typically used at ISOLDE are not well suited for the ionization of more delicate volatile compounds [13].

A review about ion sources for radioactive ion beams can be found, *e.g.*, in ref. [14]. For compounds with low ionization potential ($\lesssim 6\,\mathrm{eV}$), hot cavities facilitating surface ionization are typically used. While most elements with higher ionization potential can be efficiently ionized by element-selective resonant laser ionization [15], the ionization of compounds is addressed with electron impact ionization. Typically, the Versatile Arc Discharge Ion Source (VADIS) [16, 17] is used for such compounds at ISOLDE. For surface ionization, the efficiency directly depends on the temperature according to the Saha-Langmuir law [18]. In the VADIS, electrons are released thermionically according to the Richardson–Dushman equation [19]. This equally requires high temperatures. Radio-frequency heated plasma ion sources have been proposed [20–22] which, however, suffer from instabilities in pulsed primary beam operation [23] and an electron energy distribution that often favors breakup of the molecule over ionization [13, 24].

We investigated the operation of the VADIS at ambient temperature by exploiting laser-induced electron release. We have previously reported on a VADIS operated at ambient temperature in which electrons were released from the tantalum cathode by a femto-second laser [12]. We were able to show that the fragile carrier molecule $Mo(CO)_6$ can be ionized with efficiencies in the same order of magnitude as the noble gas krypton in the same setup. A parameter study was presented that indicates that an ion source based on electron production via the photo-electric effect could reach an efficiency of ca. 1 %. However, the mechanism of electron production was not identified in our previous experiment.

In this work, we have tailored our proof-of-concept experiment to the release of electrons via the photo-electric effect. A copper cathode was used which exhibits a relatively high quantum efficiency and does not require extreme vacuum conditions [25]. The wavelength, pulse energy and pulse length of the laser were chosen such that ablation and thermionic electron generation could be excluded. Non-volatile fragments can be produced in electron-molecule collisions and condense on the surface of a photo-cathode. This work provides a first insight into the influence of these deposited fragments on the quantum efficiency of a photo-cathode.

2. Experimental setup

The experiments were conducted at the ISOLDE Offline 2 separator [26]. The ion source used within this work is based on the VADIS version VD7 equipped with a water-cooled transfer-line [16, 17]. The VADIS VD7 is often referred to, quite incorrectly, as cold Forced Electron Beam Induced Arc Discharge (FEBIAD) ion source. It is schematically shown in fig. 1. The ion source is part of a Target and Ion Source (TIS) unit that is equipped with a coil that generates a longitudinal magnetic field (in the following referred to as target magnet). The coil can be supplied with currents of up to ca. 6 A which are expected to produce a magnetic field of up to ca. 30 mT [27]. The anode was biased up to 500 V.

The photo-cathode was built by pressing a copper cap on the tantalum cathode of a standard VADIS. The cap was manufactured from a sheet of polished oxygen-free copper. It was degreased in pentane and oxide layers were removed by an aqueous citric acid solution. The cap was rinsed with demineralized water and finally subjected to ultra-sound cleaning in ethanol. The cathode

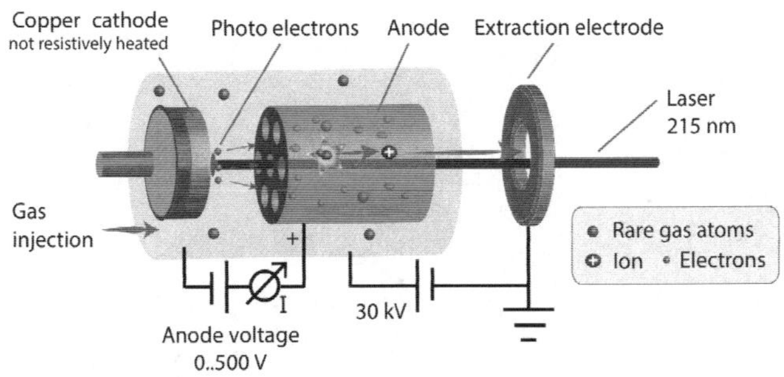

Figure 1. Schematic of the electron-impact ion source used in our experiment. The electrons were liberated from a copper cathode by a laser beam (photo-electric effect) in a cold environment.

Table 1. Parameters of the photo-cathode experiment.

parameter	value	unit
ion source properties		
anode-cathode gap	2.8(2)	mm
cathode material	Cu	
laser properties (on the cathode)		
center wavelength	215	nm
repetition rate	10	kHz
beam diameter	1.5	mm
average power	~ 10	mW
pulse length	~ 30	ns
pulse energy	~ 1	µJ
pulse energy fluence	~ 56	$\mu J\,cm^{-2}$

was electrically insulated from the target assembly to enable measurements of the photo-electron current. Additionally, the electron current was obtained as drain current of the anode power-supply. The anode drain current measurement does not differentiate between electrons impinging on the anode grid and electrons entering the anode volume. Differences between cathode and anode drain current might arise *e.g.*, due to the fraction of electrons emerging from the cathode but not arriving at the anode due to space-charge repulsion, or by electron generation by the laser beam on the ion extraction side where metal disks (typically serving as heat shields) are located that are electrically insulated from the anode. Electrons produced in this region are repelled by the extraction electrode and accelerated towards the anode.

The cathode was not baked prior to, or during operation. A deep-ultraviolet laser beam (215 nm, corresponding to 5.77 eV photon energy) was guided through the ion source outlet orifice, through the central hole of the anode grid (both 1.5 mm in diameter), and onto the cathode surface. The UltraViolet (UV) beam was generated by frequency quadrupling of a Ti:sapphire laser, identical to the ISOLDE resonance ionization laser ion source setup [28]. It provided 30 ns-long pulses at a repetition rate of 10 kHz. Prior to installation of the ion source on the beam line, the laser beam transmission through the mass separator window and ion source aperture was measured with a laser alignment device that resembles the geometry of a VADIS. The laser beam was focused such that ca. 73 % of the power arriving at the ion source reached the cathode surface. The remaining fraction was stopped by the aforementioned heat shields. During the experiment, the laser power was measured in front of the vacuum window. The transmission between the latter position and the cathode was measured to be 35(2) %. The power arriving at the cathode was ca. 10 mW and was varied during the experiment. Krypton (Carbagas, 99.998%) and carbon dioxide (Carbagas, 99.998%) at a pressure of 1.2 bar were supplied through a calibrated leak of 1.7×10^{-5} mbar L s^{-1} (1 bar abs., air). In addition to Faraday cups, the ion beam was assessed with a MagneToF detector (ETP DM291). The latter was used to determine the time-structure of ion release. The ion beam transmission through the mass separator was estimated to be 87 %.

3. Results and discussion

The previously described setup was operated under varying conditions for about 6 days. The laser was continuously irradiating the cathode during this period. During the first days, krypton was supplied though the calibrated leak. For the last 66 hours, carbon dioxide was injected. The ion and electron currents were measured as a function of anode voltage, magnet current

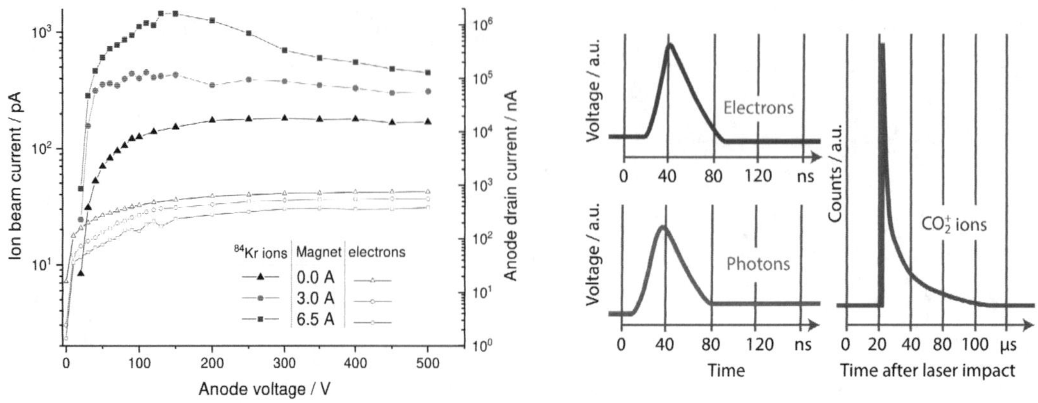

Figure 2. (left) Dependence of anode drain current and ^{84}Kr ion current on the anode voltage for different settings of the target magnet at a laser power and center wavelength of 8(1) mW and 215 nm, respectively. Connecting lines were added to guide the eye. (right) Time structure of electron, laser and ion pulses. A photo-diode was used to assess the laser time structure. Laser and electron pulse shape were estimated by measurement of the voltage drop over a 50 Ω resistor. The curve shapes were redrawn from an oscilloscope photograph. The offset between electron and laser pulses was not measured. The ion time structure was obtained with a MagneToF detector and is given relative to the laser pulse.

and laser power.

The photo-electron currents were measured as cathode current and anode drain current. Both were in agreement within ca. 10 % to 20 %, while the anode drain current was typically higher. The maximum observed ^{84}Kr current at a relatively strong magnetic field (5.5 A) was ca. 1 nA, which corresponds to an ionization efficiency of ca. 0.004 %. At the same time, the anode drain current was measured to be ca. 100 nA. At the same anode voltage of 100 V and without magnet field, the electron current increased to ca. 270 nA which is in agreement with the expected space-charge limited current of 263^{+42}_{-37} nA as given by the Child-Langmuir equation [29, 30] for a one dimensional emitter, and an anode-cathode distance of 2.8(2) mm. The highest drain current of ca. 1.5 µA was obtained at an anode voltage of 500 V and without magnetic field. The space-charge limit at 500 V is expected to be ca. 2.5 µA. The quantum efficiency computes to $8(2) \times 10^{-4}$. The value exceeds efficiencies reported by other authors by a factor of ca. 2 [31], and could indicate that the laser beam transmission during the experiment exceeded the estimate of 35(2) % that was obtained during alignment.

The evolution of ion and anode drain current while varying the anode voltage (anode scan) is given for three settings of the target magnet in fig. 2. The target magnet significantly increases the ion currents and decreases the electron currents. This effect is typically explained by an increase of electron density in the ionization volume, which in turn increases the ionization efficiency [17]. At the same time, the higher charge density might reduce the total electron emission due to space-charge repulsion. Without magnetic field, the ion and electron currents increased steadily with anode voltage, likely due to a combination of lower space-charge repulsion at higher voltages and a focusing effect of the electrostatic field. The plateau region (250 V to 500 V) could indicate that the electron supply is rather limited by quantum efficiency than space-charge repulsion. At elevated magnet currents (6.5 A), a pronounced maximum in the ion current emerges at ca. 150 V anode voltage, which can be explained by the electron-energy dependence of electron impact ionization cross sections [32].

The dependence of ion beam current and anode drain current on laser power is shown in fig. 3.

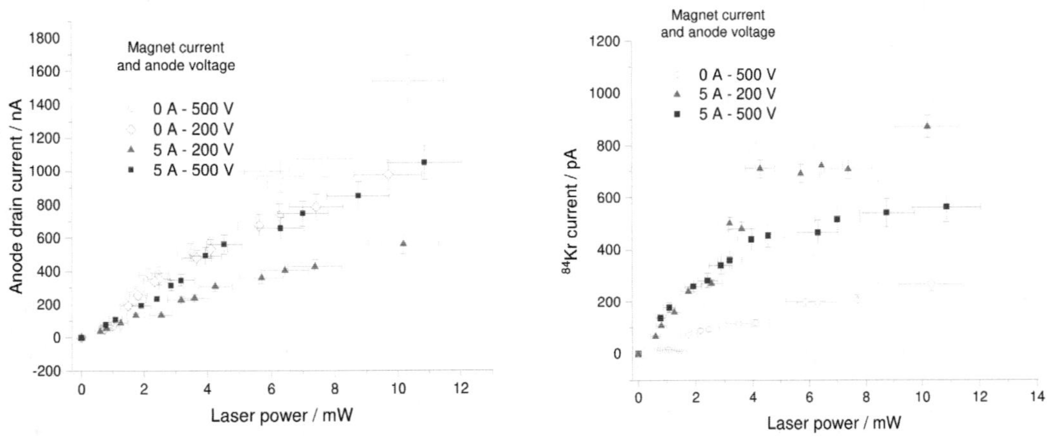

Figure 3. Dependence of anode drain current and ^{84}Kr ion current on laser power for different anode voltages and magnet currents. The electron current measurement does not differentiate between electrons impinging on the grid and electrons entering the anode volume. The ion currents for 500 V and 200 V at 0 A magnet current were indistinguishable, and only the data for 500 V are shown.

The measurements were suffering from relatively high errors due to instabilities of ion beams, data acquisition and laser power. Ion and electron currents steadily increased with laser power. The ion currents, in particular the data for a magnet current of 5 A and 500 V anode voltage, show saturation, *i.e.*, an increase in laser power and electron production does not translate into a proportional increase in the ion beam intensity. It seems likely that despite increased electron production, electrons either do not reach the anode volume, their energy is not sufficient for ionization or the probability of ion extraction is negatively impacted.

A mass spectrum obtained with the photo-cathode driven ion source in comparison to a regular VADIS is shown in fig. 4. Due to operation at ambient temperature the level of impurity is significantly lower for the photo-cathode ion source.

The quantum efficiency of a photo-cathode driven ion source could be impacted by poisoning of the cathode surface due to condensation of molecule fragments produced in electron collisions and the unusually high rest gas pressure for photo-cathode operation [31]. To evaluate the photo-cathode lifetime in these conditions, we have injected carbon dioxide into the ion source for ca. 66 hours. The ionization efficiency for the extraction as CO_2^+ was measured to be ca. 2×10^{-5}. An immediate effect on the anode drain current could not be observed. However, after 66 hours, the anode drain current reduced to ca. 35 % of its original value. The gas pressure inside the ion source was estimated to be ca. 1×10^{-4} mbar, following the calculation proposed in [17], and under consideration of the gas flow rate through the calibrated leak and the conductance of the outlet orifice. The typical pressure inside the ion source of a hot VADIS is far below this value and in the order of 1×10^{-6} mbar.

Measurements of the electron, photon and ion beam time structure are provided in fig. 2. The electron pulse shape closely resembles the laser pulse shape as measured by a photo-diode. The CO_2^+ ions are, in contrast to what has been observed for resonantly ionized atomic species inside a hot VADIS source [33], contained in a single bunch and arrive ca. 20 µs after the laser impact on the cathode at the MagneToF detector. This difference supports the assumption that, for this study, the VADIS is acting predominantly as an electron impact ion source without ion and electron dynamics that govern ion survival and extraction in standard operating conditions of the VADIS.

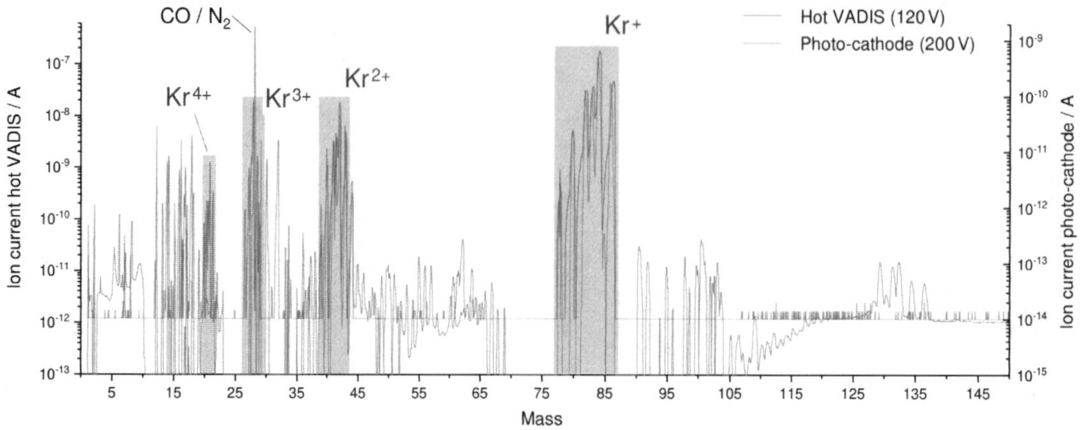

Figure 4. Mass scan obtained with the photo-cathode ion source obtained at an anode voltage of 200 V in comparison to a mass scan obtained with a standard VADIS (VD7, thermionic electron emission) operated at 120 V. Both sources were supplied with krypton. In addition, the standard VADIS was supplied with carbon monoxide gas and residual xenon was present from previous tests. The scales have been adjusted to show a similar level of sensitivity. Negative currents (possibly arising from secondary electrons created near the detector) for the photo-cathode source have been trimmed to 1×10^{-14} A.

To confirm that the electron production was governed by the photo-electric effect, blue light (430 nm, corresponding to 2.88 eV photon energy) at a power of 154 mW (second harmonic generation from the same laser system) was guided on the cathode. The work function of copper (4.5 eV to 4.9 eV [34]) is above the photon energy at that wavelength, so that a single-photon induced electron emission is not possible. In agreement with expectations, no electron production could be observed.

4. Conclusions and outlook

In this study we have demonstrated production of a molecular CO_2^+ ion beam using an electron impact ion source driven by a photo-cathode at ambient temperature. In combination with high in-target production rates of radioisotopes, the existing ion source design would in principle allow first experiments with fragile radioactive molecules. We have previously proposed a parameter study to increase the ion source efficiency to ca. 1 %, which would allow a wide range of experiments with exotic molecules [12]. A major efficiency increase requires the modification of the ion source geometry to resolve the space-charge limitation implied by the 1.5 mm laser beam spot on the cathode surface. A further efficiency increase, especially for molecules disintegrating to non-volatile species, could likely be obtained by embedding the anode volume in an ion guiding structure, *e.g.*, an electric quadrupole as applied in the Laser Ion Source and Trap (LIST) ion source [35]. A decrease of ionization efficiency with operation time was observed while carbon dioxide was admitted to the ion source. Molecule fragments of carbon dioxide are expected to deposit on the surface of the cathode and impact the quantum efficiency of the material. The extend of the observed ionization efficiency decrease (65 % after 66 hours) could possibly be reduced by lower gas pressures and application of laser cleaning procedures [36]. The new ion source will enable extraction and ionization of molecules at ISOLDE which are not stable at ultra-high temperatures. Development and implementation of this method at ISOLDE will expand its radioactive ion beam capabilities to refractory elements [12] and help to serve the increasing demand in radioactive molecules for spectroscopy applications [37–39].

Acknowledgments
We would like to thank Mathieu Bovigny and Julien Thiboud for the assembly of the target unit used in our experiments, James Cruikshank for the support regarding the mass separator and Eric Chevallay for sharing his expertise about copper photo-cathodes. This project has received funding from the European Union's Horizon 2020 research and innovation program under grant agreement No 654002.

References
[1] Borge M J G and Jonson B 2017 *J. Phys. G Nucl. Partic.* **44** 044011
[2] Catherall R *et al* 2017 *J. Phys. G Nucl. Partic.* **44** 094002
[3] Blumenfeld Y *et al* 2013 *Phys. Scr.* **T152** 014023
[4] Sidenius G and Skilbreid O 1961 E.M. separations with high efficiency of microgramme quantities *Electromagnetic Separation of Radioactive Isotopes* (Springer Vienna) pp 243–249
[5] Freeman J and Sidenius G 1973 *Nucl. Instrum. Meth.* **107** 477–492
[6] Herrmann G 1967 *Ark. Fys.* **36** 111
[7] Franberg H 2008 *Production of exotic, short lived carbon isotopes in ISOL-type facilities* Ph.D. thesis Universität Bern URL https://cds.cern.ch/record/1136904
[8] Ballof J and Seiffert Ch *et al* 2019 *Eur. Phys. J. A* **55**
[9] Köster U *et al* 2007 *Eur. Phys. J. Spec. Top.* **150** 285–291
[10] Kronenberg A *et al* 2008 *Nucl. Instrum. Meth. B* **266** 4252–4256
[11] Ramos J P *et al* 2016 *Nucl. Instrum. Meth. B* **376** 81 – 85
[12] Ballof J *et al* 2021 (*Preprint* arXiv:2108.01745) URL https://arxiv.org/abs/2108.01745
[13] Ballof J 2021 *Radioactive Molecular Beams at CERN-ISOLDE* Ph.D. thesis Johannes Gutenberg - Universität Mainz URL https://cds.cern.ch/record/2797475
[14] Stora T 2014 Radioactive Ion Sources CAS-CERN Accelerator School: Ion Sources Senec Slovakia 29 May - 8 June 2012 ed Bailey R CERN-2013-007
[15] Marsh B 2013 Resonance Ionization Laser Ion Sources CAS-CERN Accelerator School: Ion Sources Senec Slovakia 29 May - 8 June 2012 ed Bailey R CERN-2013-007
[16] Penescu L *et al* 2010 *Rev. Sci. Instr.* **81** 02A906
[17] Penescu L C 2009 *Techniques to produce and accelerate radioactive ion beams* Ph.D. thesis Universitatea „Politehnica" Din Bucuresti URL https://cds.cern.ch/record/2259078
[18] Langmuir I *et al* 1925 *P. Roy. Soc. A-Math. Phy.* **107** 61–79
[19] Wolf B 2017 *Handbook of Ion Sources* (CRC Press)
[20] Gaubert G *et al* 2003 *Rev. Sci. Instr.* **74** 956–960
[21] Kronberger M *et al* 2013 *Nucl. Instrum. Meth. B* **317** 438–441
[22] Suominen P *et al* 2010 Ionization efficiency of a COMIC ion source equipped with a quartz plasma chamber *Proceedings of the XIXth International Workshop on ECR Ion Sources (ECRIS 2010)* p MOPOT006 Aug 2010 Grenoble France
[23] Jardin P *et al* 2003 *Nucl. Instrum. Meth. B* **204** 377–381
[24] Suominen P 2011 Plasma ion sources for 1+ molecular radioactive ion beams ARIS 2011 Leuven Belgium URL https://indico.fys.kuleuven.be/event/0/contributions/22/attachments/82/87/3_Pekka_Suominen.pdf

[25] Pimpec F *et al* 2013 *Appl. Phys. A* **112** 647–661
[26] Warren S *et al* 2020 *Nucl. Instr. Meth. B* **463** 115–118
[27] Martinez Palenzuela Y 2019 *Characterization and optimization of a versatile laser and electron-impact ion source for radioactive ion beam production at ISOLDE and MEDICIS* Ph.D. thesis KU Leuven URL `https://cds.cern.ch/record/2672954`
[28] Fedosseev V *et al* 2017 *J. Phy. G* **44** 084006
[29] Child C D 1911 *Phys. Rev.* **32** 492–511
[30] Langmuir I 1913 *Phys. Rev.* **2** 450–486
[31] Palmer D 2005 A review of metallic photocathode-research *SLAC Publications and Report No. SLAC-TN-05-080* URL `https://www.slac.stanford.edu/pubs/slactns/tn04/slac-tn-05-080.pdf`
[32] Higgins M J *et al* 1989 *Recommended cross sections and rates for electron impact ionization of atoms and ions: copper to uranium* (Abingdon: Culham Lab.)
[33] Day Goodacre T *et al* 2016 *Nucl. Instrum. Meth. B* **376** 39–45
[34] Rumble J (ed) 2020 *CRC Handbook of Chemistry and Physics* (Taylor & Francis Ltd.)
[35] Fink D A *et al* 2013 *Nucl. Instrum. Meth. B* **317** 417–421
[36] Srinivasan-Rao T *et al* 1991 *J. App. Phys.* **69** 3291–3296
[37] Garcia R F *et al* 2020 *Nature* **581** 396–400
[38] Safronova M S *et al* 2018 *Rev. Mod. Phys.* **90**
[39] Athanasakis M *et al* 2021 Radioactive molecules at ISOLDE CERN-INTC-2021-017 URL `http://cds.cern.ch/record/2748712`

High-voltage breakdown in the injection-extraction region of the CANREB EBIS: A possible mechanism

Chris R.J. Charles[1], **Mathieu Cavenaile**[1,2], **Paige Harford**[1,3], **Friedhelm Ames**[1], **Brad Schultz**[1], **Oliver Kester**[1]

(1) TRIUMF. 4004 Wesbrook Mall, Vancouver, BC, V6T 2A3, Canada.
(2) Dept. of Astronomy and Physics, Saint Mary's University, Halifax, NS, B3H 3C3, Canada.
(3) Dept. of Physics and Astronomy, University of Waterloo, 200 University Avenue West, Waterloo, ON, N2L 3G1, Canada.

E-mail: `ccharles@triumf.ca`

Abstract. OmniTrak simulations were performed in the 40°K injection-extraction region near the collector trumpet of the CANREB EBIS to understand the nature of chronic electrical breakdowns of PEEK feedthroughs. These feedthroughs are designed to route high-voltage wires (up to +15 kV) from room temperature to each drift-tube at 4°K, while operating in the strong superconducting semi-Helmholtz field. The electric field strength |E| at the PEEK feedthroughs are $\sim 7 \times 10^5$ V/m within about 2-3 mm of the surroundings at ground potential. When simulating only bare wires (with no PEEK feedthroughs), |E| increases by about 1 order-of-magnitude to $> 10^6$ V/m in the same region. Assuming a 3 Tesla magnetic field, electrons starting with 0 eV near the PEEK feedthroughs undergo helical motions due to $\vec{E} \times \vec{B}$-drift and demonstrate reflections from magnetic mirror effects. A combination of this helical motion, reflecting mirror trajectories, and attraction of electrons to exposed HV wires may be responsible for the ongoing HV breakdown of insulating feedthroughs in this region of the EBIS. Solving this problem requires complete ground-shielding of all HV wiring to eliminate |E| and should encapsulate all wires to prevent breakdown paths.

1. Introduction

The EBIS (Electron Beam Ion Source) shown in Fig 1 is TRIUMF's newest system for charge-breeding rare and exotic radioactive ion beams (RIB) to beams of highly-charged ions (HCI). The EBIS is the heart of CANREB (the CANadian Rare isotope facility with Electron Beam ion source) dedicated to preparing and delivering bunched HCI beams for experiments. CANREB is part of the larger ARIEL (Advanced Rare IsotopE Laboratory) that will triple TRIUMF's RIB output by 2026 [1, 2]. Presently, RIB for charge-breeding are produced at the ISAC (Isotope Separator and ACcelerator) facility. RIB bunches spanning the mass range up to U are formed by a radiofrequency quadrupole cooler-buncher in CANREB upstream of the EBIS at $\leq$100 Hz repetition rates containing $< 10^7$ ions/bunch. Bunch energies are then adjusted by a pulsed drift tube to $6< E_{bunch} <14$ keV immediately before acceptance into the EBIS at the 'collector side', where the ionizing electron beam from the EBIS dissipates [3].

Fig 1(a) shows the system layout. The EBIS was designed and built at the Max Planck Institute for Nuclear Physics (MPIK) in Heidelberg [4]. A semi-Helmholtz style superconducting magnet with maximum 6 T field strength compresses an e^- beam (up to 20 keV at $\leq$500 mA)

Journal of Physics: Conference Series **2244** (2022) 012073 doi:10.1088/1742-6596/2244/1/012073

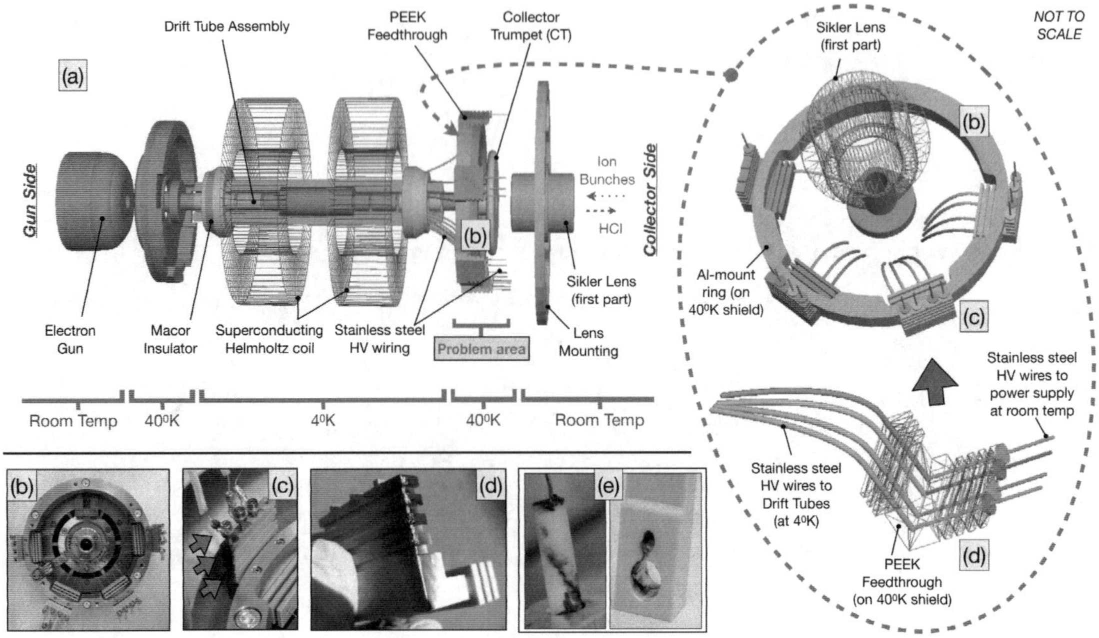

Figure 1. OmniTrak 3D rendering of the CANREB EBIS with photos of the chronic HV breakdown. Insets: (a) overall EBIS layout omitting the vacuum envelope, thermal shields and the full Sikler collector assemblies for clarity; (b-d) location of the breakdown at the PEEK feedthroughs mounted to the 40°K thermal shield (not shown), and (e) photos of the first breakdown discovered in 2019 with the original PEEK-alumina (Al_2O_3) feedthrough design.

emitted from a $\frac{1}{4}$-inch barium cathode at the 'electron gun' side. This e^- beam passes through the drift tubes and experiences an intense compression due to the $\vec{B}$-field, increasing the electron beam density to a few 10^3 A/cm^2 depending on settings.

CANREB commissioning began from summer 2020. However, the EBIS has continued to enjoy persistent high-voltage (HV) instability and breakdowns since January 2019 to the present. Fig 1 shows this HV breakdown, in the 40°K region on the collector side, where four PEEK feedthroughs route 12 high-voltage wires from room temperature to the drift tubes at 4°K (see insets (b) to (d)).

These feedthroughs are intended to (1) route 12 thin stainless steel wires (0.8 mm diameter) from ~300°K through a 40°K thermal shield (not shown) to each drift tube at 4°K; each wire carries up to 15 kV, (2) thermally isolate the drift tubes at 4°K from room temperature, by acting as a temperature 'stop' for the stainless steel wires, and (3) operate within ~50 mm of the strong field of the superconducting magnet. The thin 0.8 mm wire diameter was chosen due to the poor heat conduction of stainless steel to reduce thermal load on the 4°K stage, however, this diameter presents a challenge for higher $\vec{E}$-fields around the wiring. The combination of magnetic field and high-voltage is responsible for the chronic instabilities and breakdowns, possibly due to Penning discharge, however the mechanism remains poorly understood.

Deterioration of the PEEK feedthroughs by routine use of the EBIS has reached the point where voltage cannot be applied to the drift tubes, even in low magnetic fields. Here we present simulations of the EBIS in the area of Fig 1(b) to better understand the electric and magnetic fields around the feedthroughs and HV wiring. We aim to discover possible mechanisms involved in the breakdowns, and consider solutions to overcome this chronic debilitating issue.

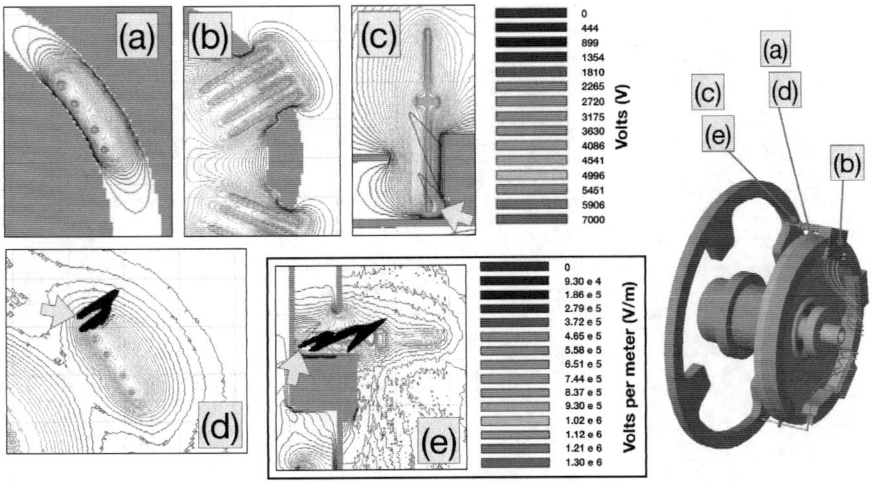

Figure 2. $\vec{E}$-field equipotentials and electron orbits in the collector-trumpet region of the EBIS with 7 kV applied to all wires. PEEK feedthroughs are included. Arrows show the staring position of the electrons. Insets: (a) perpendicular to one set of 4 HV wires and one PEEK feedthrough between the thermal shields; (b) same as in (a) but at the base of the feedthrough where the wires route to the drift tubes; (c) view parallel to one wire inside the feedthrough with one electron orbit shown (outside the feedthrough); (d) same view as in (a) but with 2500 electrons, and (e) same view as (d) with 2500 electrons but plotting |E|.

2. Background

2.1. High Voltage Breakdowns

Fig 1(b) to (d) show the location and aftermath of the most recent breakdown, in the 40°K region of the collector trumpet (CT) at the four PEEK feedthroughs. Similar failure occurred in 2019 with the original PEEK-alumina HV feedthroughs designed by MPIK in Fig 1(e), thought to be along a Stycast epoxy that glued the PEEK, Macor and wire assembly as one unit. Inspection revealed other discharge paths on the alumina and PEEK not directly glued by Stycast. Breakdown of that original MPIK design motivated our redesigned PEEK feedthroughs in Fig 1(d), that again failed similarly.

Tests in 2020 showed the PEEK insulators readily handled 15 kV to the drift tubes in no magnetic field. However, with magnetic field, the situation became progressively worse where the feedthroughs could only tolerate ~7 kV in a 1 T magnetic field. Higher magnetic fields and higher voltages worsened the electrical breakdown; voltages >7 kV were not possible to achieve at 1 T. This operation is significantly below the EBIS design specs (15 kV and 6 T) and are limiting the achievable extracted ion energy [5].

2.2. Simulations Setup

OmniTrak Professional (v. A3D0450 software suite) was used for all simulations. Fig 1(a) shows the 3D model accurately describing the full CANREB EBIS in its present build. The vacuum enclosure, thermal shields, and Sikler lens electrodes beyond the first lens element are included in the model but not shown for clarity.

Components were rendered from the MPIK SolidWorks drawings of the EBIS as symmetric turnings where possible, or imported as STL files and ro-translated to their correct orientations in the model (i.e. the PEEK insulators, wires, and parts of the 40°K shield). All stainless steel wires were numerically defined by a separate parabolic-linear model to render the actual wires and their routing in both the 40°K region in Fig 1(b) and inside the PEEK feedthroughs.

Variable resolution magnetic and electric meshes were defined separately but overlain in

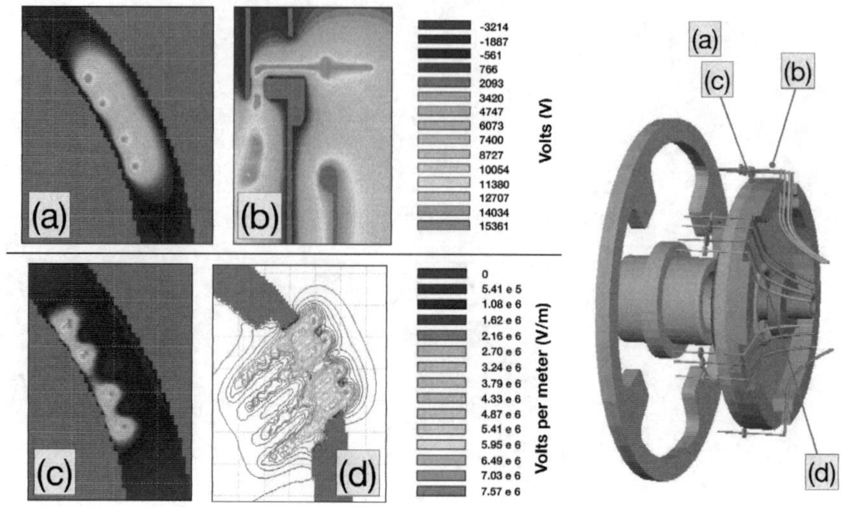

Figure 3. $\vec{E}$-field in the collector-trumpet region of the EBIS with the PEEK insulators removed from the simulation and 15 kV applied to all bare wires. Insets (a) perpendicular to 4 HV wires between the thermal shields; (b) parallel to one wire; (c) between the thermal shields as in (a) but plotting |E|; (d) HV wires at the base of the 40°K shield plotting |E|.

the final OmniTrak simulations. Mesh density was highest around the PEEK insulators and wires of Fig 1(b) and smoothly transitioned to coarser resolutions away from that region. A dielectric constant of 3.3 was assumed for the four PEEK insulators in the electric field solution. Solenoid dimensions and their position in the model were determined from the MPIK SolidWorks drawings. The free-air space was not enlarged to fully relax Neumann boundary conditions because we were only interested in the form and magnitude of the magnetic field in the 40°K region around the PEEK insulators.

3. Results

Simulations were done with 7 or 15 kV applied to all wires using a 3 T semi-Helmholtz magnetic field distribution. Electrons were defined with 0 eV starting energy at a given (x,y,z) position in the model, and allowed to follow the $\vec{E}$ and $\vec{B}$ fields around the wires and PEEK insulators. Electrons were emitted either as (1) single particle isotropic point sources with zero radius, or (2) 1×1 mm^2 squares of 2500 electrons.

Fig 2 shows electric potentials (ϕ), electric field magnitudes |E| and electron orbits around the PEEK insulators in different orientations. Fig 3 and Fig 4 show the same ϕ and |E| but with the insulators removed, to understand the fields and orbits around bare wires in this area.

With 7 kV applied to all wires in Fig 2, |E| near each of the four PEEK feedthroughs approaches $\sim 7\times10^5$ V/m in ~2-3 mm of nearby flanges at ground potential that are part of the surrounding 40°K shield. This is true for the feedthroughs in the horizontal and vertical directions (insets (a) and (b)). Electrons emitted near the base of the feedthroughs just above the 40°K shield follow helical trajectories, staying close to the PEEK feedthroughs and attracted to the positive 7 kV of the exposed wire (inset (c)). Electrons eventually terminate on the surrounding thermal shields at ground potential. This helical motion is the same for single electrons and 2500 electrons (insets (c) to (e) respectively). Simulations with 15 kV applied to the wires showed the same helical electron trajectories outside the PEEK feedthroughs.

When the PEEK feedthroughs are removed from the simulation in Fig 3 and the maximum 15 kV is applied to all exposed wires, electric field magnitudes increase by about one order of

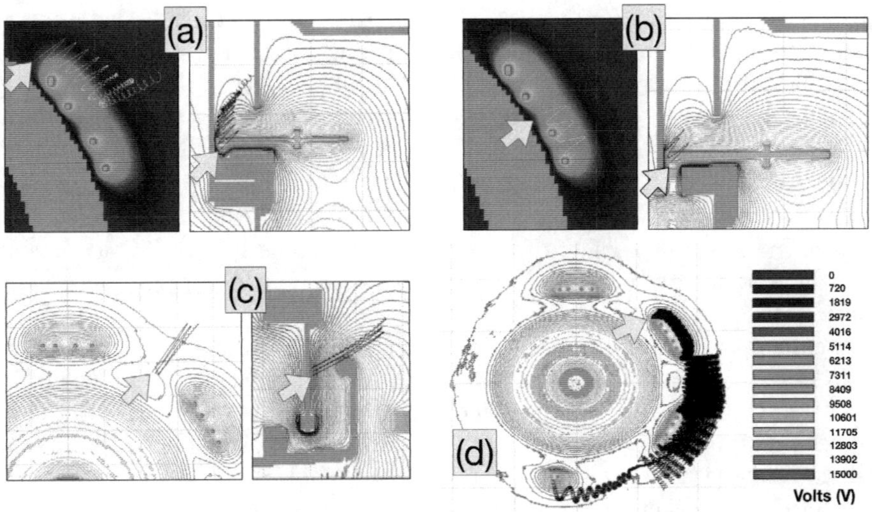

Figure 4. Electron orbits in the collector trumpet region with 15 kV applied to all wires. Arrows show the staring position of the orbits. Insets: starting (a) at the base of the 40°K shield on the left-side of 4 HV wires with the PEEK feedthroughs removed from the simulation; (b) same as in (a) except in-between the wires; (c) in-between two PEEK feedthroughs (this time included) just above the 40°K shield; (d) starting next to one PEEK feedthrough, just above the 40°K shield, showing 2500 electron orbits.

magnitude to $\sim 4.5 \times 10^6$ V/m in ~2 mm of the grounded shields. Fig 3(c) in particular shows a very high |E| developed adjacent to the flange on the 40°K shield where the wires are routed along. However, note that electric field magnitudes produced from bare wires throughout this area are consistently in the 10^5 to 10^6 V/m range.

Fig 4 shows electron trajectories starting near the base of the HV wires just above the 40°K shield with the PEEK feedthroughs removed. The applied potential on all wires is the maximum 15 kV. Electrons again follow helical trajectories around the wires, terminating on the nearby shields at ground potential, similar to the orbits in Fig 2. The helical trajectories are the same regardless of electron starting position, whether next to the wires, or between the PEEK feedthroughs (i.e. inset (c)). Electrons may also become partially trapped in the fields of the collector trumpet region, tending to spiral around the HV wiring and PEEK feedthroughs, terminating on different wires far from their starting position (inset (d)).

4. Discussion

All simulations show that electrons starting with 0 eV in the collector trumpet region of the 40°K shield in Fig 1(b) undergo helical motions outside and along the PEEK feedthroughs, and around bare wires at positive high-voltage.

The helical motions are due to the expected $\vec{E} \times \vec{B}$-drift of electrons in the magnetic field. Fig 4(c) in particular shows an intriguing possible magnetic mirror in the collector trumpet region for electrons spiraling along the field lines. If the magnetic field strength increases in the direction of motion, electrons will experience a force that reduces the component of velocity parallel to the field. This force slows the motion along the field line and reverses it, forming a magnetic mirror. Such a mechanism is plausible in the collector trumpet area, because the magnetic field will initially be oriented axially closer to the solenoid, but will gradually orient to a radial direction with distance from the solenoid where the field-lines become more compressed.

Magnetic mirrors together with the spiraling electron trajectories may be a mechanism causing electrons to oscillate back-and-fourth along the surfaces of the PEEK insulators, leading to ongoing damage of the PEEK feedthroughs. In particular, bare HV wiring in the collector trumpet region clearly influences electron orbits, causing attraction of electrons to the positive wires, and having the ability to cause very complex orbits and partial trapping seen in Fig 4(d).

These simulations clearly demonstrate that bare wires carrying positive potentials should never be used in the collector trumpet region of the EBIS. Instead, steps must be taken to reduce the electric potential in this area as much as possible, for example, by covering each individual HV wire with a grounded sheath to eliminate |E|. While shielding all HV wires will not stop helical motions or magnetic mirrors of free electrons in the $\vec{B}$-field, it will eliminate electrons attracted to positive potentials on the wires, that may be causing a sustained discharge along insulating surfaces.

Effects of temperature are also worth considering. The PEEK insulators rest on the cold 40°K shield, but are in direct line-of-sight to room temperature in-vacuum components. Gettering residual N_2, O_2 and other gases on the surface of the PEEK feedthroughs, over time, may also play a role in forming breakdown paths along insulating feedthroughs. While this idea remains speculative, if such a situation did occur, breakdown would be exasperated by the interplay of the helical electron motion, attraction of electrons to bare high-voltage wires and potential reflecting of electrons due to magnetic mirrors.

Eliminating cryogenic temperatures on parts of the HV wiring is impossible with this EBIS design. However, fully encased wires, for example in Teflon, Kapton or ceramics leading from room-temperature to the drift-tubes at 4°K, where each wire is also surrounded by a grounded sheath, would eliminate |E| in the collector trumpet region and remove discharge paths.

5. Acknowledgments

TRIUMF is funded by the National Research Council of Canada (NRC). The CANREB project is funded by grants from the Canadian Foundation for Innovation (CFI), the Natural Sciences and Engineering Research Council of Canada (NSERC), and collaboratively with Saint Mary's University and the University of Manitoba.

6. References

[1] L. Merminga, F. Ames, R. Baartman, P. Bricault, Y. Bylinski, Y-C. Chao, R. Dawson, D. Kaltchev, S. Koscielniak, R. Laxdal, F. Mammarella, M. Marchetto, G. Minor, A. Mitra, Y.-N. Rao, M. Trinczek, A. Trudel, V. Verzilov, V. Zvyagintsev (2011). *ARIEL: TRIUMF'S Advanced Rare Isotope Laboratory.* Proceedings of IPAC2011, San Sebastian, Spain, 1917-1919.

[2] J. Dilling, R. Krucken, L. Merminga. (2013) ARIEL overview. In: Dilling J., Krucken R., Merminga L. (eds) *ISAC and ARIEL: The TRIUMF Radioactive Beam Facilities and the Scientific Program.* Springer, Dordrecht.

[3] F. Ames, R. Baartman, B. Barquest, C. Barquest, M. Blessenohl, J. R. Crespo López-Urrutia, J. Dilling, S. Dobrodey, L. Graham, R. Kanungo, M. Marchetto, M. R. Pearson, and S. Saminathan (2018). *The CANREB project for charge state breeding at TRIUMF.* AIP Conference Proceedings 2011, 070010-1-3.

[4] M.A. Blessenohl, S. Dobrodey, C. Warnecke, M. K. Rosner, L. Graham, S. Paul, T. M. Baumann, Z. Hockenbery, R. Hubele, T. Pfeifer, F. Ames, J. Dilling and J. R. Crespo López-Urrutia (2018). *An electron beam ion trap and source for re-acceleration of rare-isotope ion beams at TRIUMF.* Reviews of Scientific Instruments 89, 052401-1-16.

[5] B.E. Schultz, C.R.J. Charles, M. Cavenaile, R. Kanungo, F. Ames and O. Kester (2021). *CANREB EBIS commissioning at TRIUMF.* Proceedings of ICIS21, Vancouver, Canada.

Journal of Physics: Conference Series **2244** (2022) 012074 doi:10.1088/1742-6596/2244/1/012074

Anomalous Ionization Regime in a Forced Electron Beam Induced Arc Discharge Ion Source for Singly Charged Radioactive Ion Beam Production

F Maldonado Millán[1,2], C Babcock[2], T Day Goodacre[2], and A Gottberg[1,2]

[1] Department of Physics and Astronomy, University of Victoria, 3800 Finnerty Road, Victoria, BC V8P 5C2, Canada

[2] Department of Target and Ion Sources, TRIUMF, 4004 Wesbrook Mall, Vancouver, BC V6T 2A3, Canada

E-mail: gottberg@triumf.ca

Abstract. At TRIUMF's ISAC facility, 1+ Radioactive Ion Beams (RIBs) of noble gases, halogens, and molecules are created using a Forced Electron Beam Induced Arc Discharge (FEBIAD) ion source. Reported ionization efficiencies for FEBIADs range from 10% to 25% for $^{40}Ar^{+}$, while TRIUMF-FEBIAD ionization efficiency seems to be < 1% with a 90 % emittance <15 µm. As RIB ion sources aim for a high ionization efficiency, an experimental and numerical campaign was conducted to investigate the comparably low efficiencies observed. The experimental results for $^{40}Ar^{+}$ indicate that up to a 10% ionization efficiency is possible by operating the source at different parameters. The measurements agree with theoretical estimations; however, certain combinations of parameters produce an anomalously high electron current which enhances the ionization efficiency threefold. Present investigations aim to characterize and model the anomalous electron current because, if proven reliable, the argon ionization efficiency could reach 30% with no significant impact on beam emittance.

1. Introduction

TRIUMF utilizes the Isotope Separation On-Line (ISOL) [1] method to produce Radioactive Ion Beams (RIBs) at the Isotope Separation and ACcelerator (ISAC) facility [2] and the new Advanced Rare IsotopE Laboratory (ARIEL) facility [3]. In the ISOL method [4], a driver beam impinges on a thick target to produce nuclear reaction products that effuse into an ion source to be ionized. The ions are extracted by an electric field and sent to a dipole magnet for separation based on the mass-to-charge ratio.

The ion source typically used for the production of singly charged RIBs of noble gases, halogens, or molecules is the Forced Electron Beam Induced Arc Discharge (FEBIAD) [5]. In a FEBIAD source, a hollow cathode [6] is resistively heated to temperatures of ≈2000 °C to generate electrons that are accelerated via a grid into an anode volume to induce electron impact ionization. The electrons are focused with a tunable electromagnetic coil that provides confinement in the axial direction.

The principle figure of merit for radioactive ion sources is the ionization efficiency and while the ISAC-FEBIAD (Fig. 1) shows offline $^{40}Ar^{+}$ efficiencies <1 %, values from 10 % to 25 % [7, 8]

are within reach. With the goal of addressing this, a combined numerical and experimental campaign has been undertaken to identify the underlying processes that improve the FEBIAD's performance.

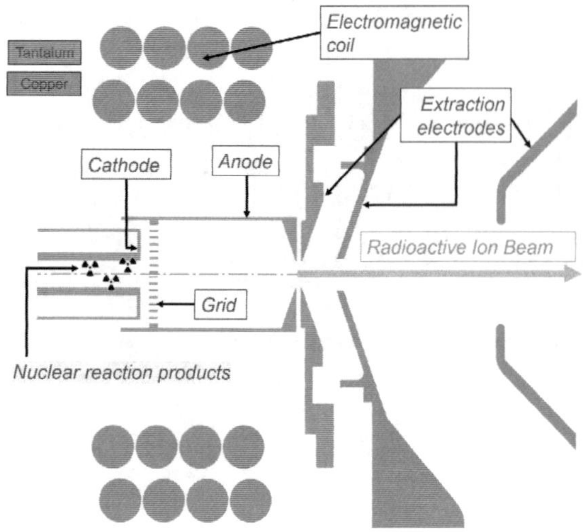

Figure 1. Cross-section of the ISAC-FEBIAD ion source to produce RIBs with the ISOL method. The cathode and anode/grid material is tantalum and the nominal cathode-grid distance is ≈1 mm.

2. Materials and methods

In order to explore the FEBIAD parameter space, a scanning algorithm was developed to automatically vary the cathode heating current, coil current, and anode voltage. The nominal operational parameters and their limits are driven by reliability, however, these are conservative values. Therefore, the scanning range of the algorithm was set to explore values beyond the operational combinations to gain insights into what could be limiting the performance of the source. The algorithm recorded simultaneously the ion current on a Faraday cup and the electron current sunk by the anode power supply. There is a ten-fold increase in measured data points compared to the manual paradigm within the same available experimental time.

To determine the ionization efficiency, a calibrated leak was installed on the ion source unit and the efficiency was computed by comparing the rate of ions extracted to the neutrals entering the system. The calibrated leak was characterized with helium and the flow was adjusted to argon and was determined to be in the range of 4.7×10^{-6} mbar L s^{-1}.

The electron current emission is described either by the Thermal Limited (TL) regime (Eq. 1) [9] or by the Space Charge Limited (SCL) regime (Eq. 2) [10]. The TL regime is described by the Richardson-Dushman equation

$$J_{cath} = A_0 T^2 e^{\frac{-W}{k_B T}} \tag{1}$$

where A_0 is the effective Richardson constant (material dependent), T the cathode temperature, W the work function and k_B the Boltzmann constant. The SCL regime is described by the Child-Langmuir equation

$$J_{cath} = \frac{4\epsilon_0}{9}\sqrt{\frac{2e}{m_e}}\frac{V^{3/2}}{d^2} \tag{2}$$

where ϵ_0 is the vacuum permittivity, e the elementary charge, m_e the electron mass, V the voltage applied between anode and cathode, and d the anode-cathode distance. The TL regime is valid when a voltage difference effectively extracts the electrons generated on the cathode at a set temperature. Conversely, the SCL regime is valid when, despite a temperature increase, the electron current decreases compared to the TL regime estimation and starts reaching a plateau. Equations 1 and 2 were used in a fitting routine to model the electron emission from the FEBIAD cathode.

Complementary to the experiments, a comprehensive simulation package was developed [11] with COMSOL Multiphysics [12] to gain understanding into the FEBIAD's fundamental mechanisms and correlate observables from the experimental studies.

3. Results and discussion

For a constant anode voltage of 300 V, the $^{40}Ar^{+}$ ionization efficiency increased from 0.3 % to 10% (see Fig. 2) when increasing the cathode heating current from the nominal 290 A to 335 A. According to an existing calibration between the cathode heating current and pyrometric measurements, this current difference corresponds to a cathode temperature increase from ≈1690 °C to ≈1910 °C. The efficiency likely improves as a result of an increase in the electrons available for ionization (Eq. 1).

The measured electron current follows the TL regime for cathode temperatures below 1900 °C at all anode voltages (Fig. 3). For temperatures above 1900 °C and voltages ≤100 V, the measured electron current appears to transition from the TL regime to the SCL regime. However, above 1900 °C and 200 V none of the regimes describe the measured electron current even if the material properties are unconstrained in the fitting routine. The anomalously high electron current causes the ionization efficiency to almost triple at anode voltages >350 V (Fig. 2).

The anomalous regime onset matches the grid geometry changes caused by intense heating. Given the identical geometry deformation observed in two different sources, sharp points and a reduced cathode-anode distance likely cause locally increased electric fields that translate into a higher electron current. The anomalous regime is transient and vanishes after keeping the grid under intense heating. With the deformed grid geometry, a new set of measurements were taken starting from nominal values; however, there are no signs of the anomalous regime even at the highest setting of anode voltage (Fig. 4, bottom). Moreover, the ionization efficiency contour plots at a fixed temperature of ≈1910 °C, differ between the grid geometries and show that different operational parameters provide the optimum efficiency (Fig. 4). The results suggest that a thermally robust grid with features that tailors and enhances the electric field at the center could improve the ionization efficiency by extracting more electrons from the central part of the cathode.

Assuming a homogeneous electron release from the planar cathode and the first 2 mm of the hollow cathode (see Fig. 5, left), thermal simulations indicate that at least half of the electron current must come from latter. Preliminary electron tracking simulation showed no significant extraction from particles further inside the hollow cathode even when computed with a magnetic field. The two part contribution arises because the more focused electron trajectories from the hollow cathode translate into sufficient power deposition to exceed the melting temperature of tantalum (3017 °C [13]) (Fig. 5, right). Using the measured electron current as an input, particle tracking, power deposition, and thermal simulations show that the temperature of the grid center rises above 3000 °C (Fig. 5).

Additionally, the beam quality was investigated by measuring the vertical emittance for selected cases with an Allison-type meter [14]. In this case, the 90 % emittance computed corresponds to the area of the phase space that contains 90 % of the ion beam current. For parameter combinations of interest, the 90% intensity emittance ranges between 8 µm to 12 µm, this range is compatible with the ISAC beam transport system, and shows that the increased efficiency does not negatively affect the ion beam extraction/formation.

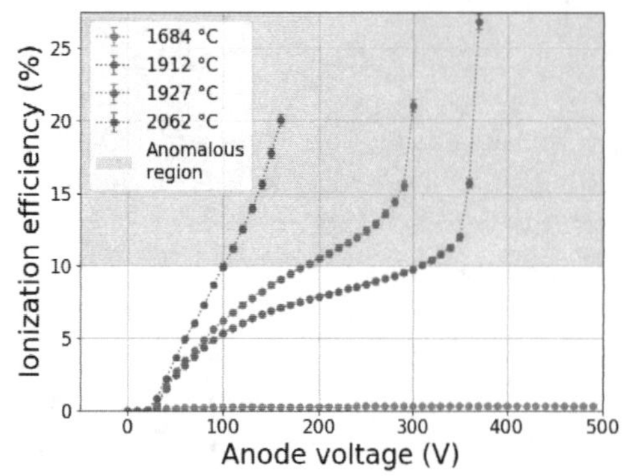

Figure 2. Efficiency as a function of anode voltage. The sharp increase in efficiency matches the region of increased electron current.

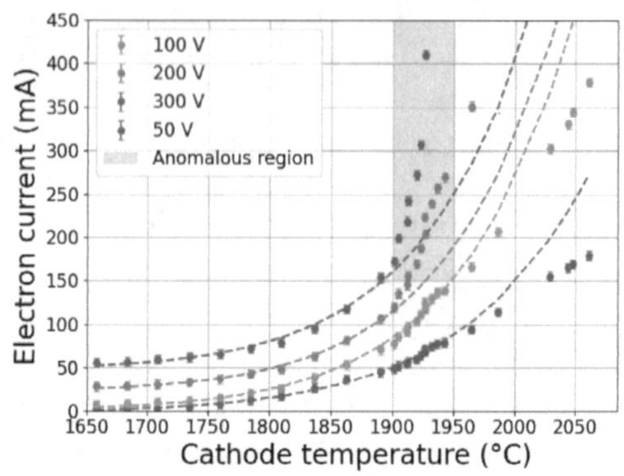

Figure 3. Measured electron current as a function of cathode temperature for various anode voltages. The electron current (dots) below 1900 °C is described well with the thermal limited regime fit (dotted lines) (see text).

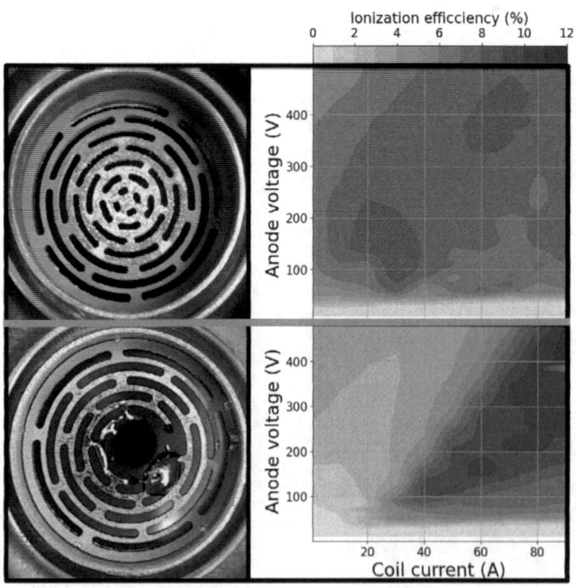

Figure 4. Top: Nominal grid geometry and ionization efficiency plot at ≈1910 °C, as a function of two main FEBIAD parameters, the voltage applied to the anode and the magnetic field used to confine the electrons, as measured by the current in the electromagnetic coil. Bottom: Deformed grid and efficiency plot for the same temperature. The maximum efficiency is obtained at different parameter combinations on each geometry.

4. Conclusions

The established methodology successfully matches the electron current to the two-regime electron emission model. With this model, incompatible trends in both electron current and ionization efficiency can be used to determine geometrical changes that ultimately provide insight into developing novel and more efficient designs. Increased localized electric fields in front of the hottest part of the cathode trigger the anomalous ionization regime. This means that the FEBIAD design can benefit from an increased electron current along the axis. Thermal simulations of the cathode and grid explain the observed deformation during testing and can be used to propose new designs with superior robustness and efficiency.

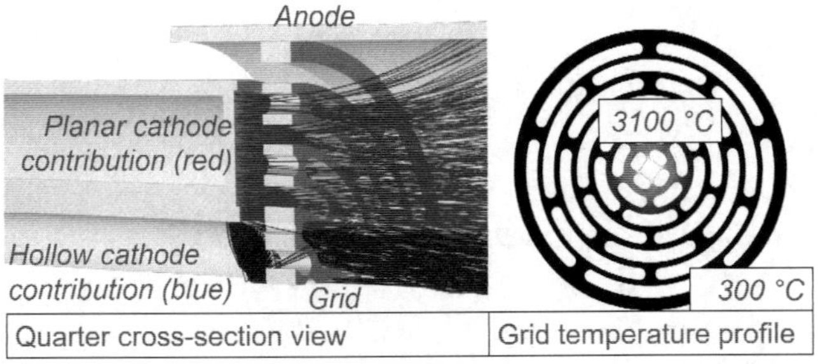

Figure 5. Left: Two part electron current contribution for grid heating. Electrons from further inside the hollow cathode and the magnetic field were not included in the simulation (see text). Right: Temperature obtained from thermal simulation using the electron power deposited.

5. Outlook

Electron particle tracking simulations predict the grid reliability and the ionization rate inside the anode volume. On-going ion particle tracking simulations account for the predominant hollow cathode electron current, and preliminary results agree well with the efficiency contour plots presented here. With the experimentally validated model, simulation-based optimization is foreseen to improve not only the FEBIAD grid reliability but also the FEBIAD ion source design and performance.

Acknowledgments

TRIUMF receives federal funding via a contribution agreement with the National Research Council of Canada. FM acknowledges additional support from their NSERC funding. FM also acknowledges A. Schmidt for the ion source construction and experimental support, and the High Level Application group which deployed the server in which the scanning algorithm was developed.

References

[1] Kofoed-Hansen O and Nielsen K O 1951 *Phys. Rev.* **82** 96–7
[2] Dilling J, Krücken R and Ball G 2014 *ISAC overview* (Dordrecht: Springer Netherlands) pp 1–8
[3] Dilling J, Krücken R and Merminga L 2014 *ARIEL overview* (Dordrecht: Springer Netherlands) pp 253–62
[4] Blumenfeld Y, Nilsson T and Duppen P V 2013 *Physica Scripta* (T152)
[5] Kirchner R and Roeckl E 1976 *Nucl. Instruments Methods* **133** 187–204
[6] Sundell S and Ravn H 1992 *Nucl. Inst. Methods Phys. Res.* B **70** 160–4
[7] Penescu L, Catherall R, Lettry J and Stora T 2010 *Rev. Sci. Instrum.* **81** 2–6
[8] Delahaye P *et al.* 2020 *Nucl. Inst. Methods Phys. Res.* B **463** 339–44
[9] Dushman S 1923 *Phys. Rev.* **21** 623–36
[10] Child C D 1902 *Science* **15**(379) 553
[11] Maldonado Millan F, Day Goodacre T and Gottberg A 2020 *Nucl. Inst. Methods Phys. Res.* B **463** 302–4
[12] Comsol multiphysics URL `www.comsol.com`
[13] Haynes W 2016 *CRC Handbook of Chemistry and Physics* (CRC Press)
[14] Allison P W, Sherman J D and Holtkamp D B 1983 *IEEE Trans Nucl Sci* **30** 2204–06

Simulations of a new electron gun for the TITAN EBIT

J D Cardona[1,2], K Dietrich[1], I Mukul[1], J Dilling[1], G Gwinner[2], O Kester[1,3] and A A Kwiatkowski[1,3]

[1]TRIUMF 4004 Westbrook Mall, Vancouver, BC V6T 2A3 Canada,
[2]Department of Physics and Astronomy, University of Manitoba, Winnipeg, Manitoba R3T 2N2, Canada,
[3]UDepartment of Physics and Astronomy, University of Victoria, Victoria, British Columbia V8P 5C2, Canada

Abstract. Penning trap mass spectrometry is the tool of choice for mass measurements to test the Standard Model or lay the nuclear-physics foundation of neutrino physics due to the high precisions achievable. This precision can be further boosted by higher charge states (Ettenauer et al 2011). For this purpose, an electron beam ion trap (EBIT) provides radioactive HCIs at the TITAN facility at TRIUMF. To improve the electron beam properties and its control, a new electron gun is under development. The electron gun within its TITAN EBIT environment was simulated using Field Precision's TRAK software. A new electrode geometry was chosen and optimized to extract up to 5A, 66 keV electron beams. Due to the strong fringe field of the unshielded 6T magnet, options for the passive and active shielding of the gun were explored to compress the electron beam. During the design process, careful attention was paid to safety and mechanical considerations. Simulations and the status of the new electron gun will be presented.

1. TITAN Setup

TRIUMF's Ion Trap for Atomic and Nuclear Science (TITAN) [2] is a series of traps at TRIUMF whose primary goal is high-precision mass measurements of short-lived nuclides. The most precise measurements are performed in a precision Penning trap (MPET), wherein the precision scales with the charge state for all measurement techniques [3, 4]. For a time-of-flight-ion-cyclotron-resonance (TOF-ICR) measurement performed in MPET, the uncertainty in the mass measurement is defined as [3]:

$$\frac{\delta m}{m} \propto \frac{m}{qBT\sqrt{N}}, \qquad (1)$$

where q is charge state of the trapped ion, T is the excitation time, and N is the number of ions. From this equation it follows that the higher the charge state of the trapped ion, the greater the precision of the mass measurement performed. The TITAN EBIT was installed to charge breed the radioactive ions. TITAN alone has charge-breeding capabilities combined with a Penning trap mass spectrometer, leading to other beneficial developments. The EBIT has also been used for purification [4] and RIB production [5].

In addition, radial ports around the EBIT allow optical access for in-trap spectroscopy [6]. Several advantages present themselves to in-trap decay spectroscopy compared to a typical tape-type decay station: The ions do not interact with the tape being stored in vacuum. The vacuum enables measurements of highly charged ions, which are of intrinsic interest themselves [7]. The strong magnetic field guides the electrons away from the ports and detectors, eliminating the 511 keV background.

1.1 TITAN-EBIT
TITAN-EBIT traps ions axially through electric potentials established by segmented drift tubes and radially through the space charge of an electron beam. The electron beam is compressed by the inhomogeneous magnetic field it crosses from gun to ion trap, reaching its maximal compression at the trap center. While the ion bunch is trapped, the electron beam will strip off bound electrons from the ion through electron impact ionization. Figure 1 depicts the cross-sectional view of the TITAN-EBIT.

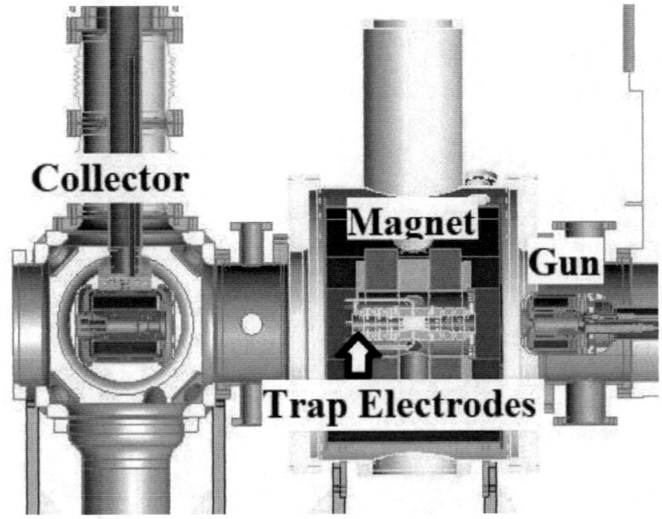

Figure. 1 TITAN-EBIT. 6T Hemholtz-style magnet from Cryogenics, current electron gun, 9 segmented drift tubes, and collector. Designed and modified to run with an upper limit of 66 keV energy and 5 A electron beam.

As TITAN is dedicated to studies of short-lived nuclides, the charge breeding time is of immense importance and contributes to nuclear decay losses. The charge breeding time of the EBIT is defined as [8]:

$$t_i = \frac{e}{J_e}\sum_{j=1}^{i}\frac{1}{\sigma_j^{El}}. \qquad (2)$$

Here e is equal to the elementary charge, σ_j^{El} is the cross section of the respective electron impact ionization, and J_e is the electron current density. From this equation it is shown that a large electron beam current density will lead to shorter breeding time. The TITAN EBIT has been successfully deployed for Penning trap mass spectrometry [1] and decay spectroscopy [6]. After ten years in operation, new avenues for research are being explored, particularly for the in-trap spectroscopy program. These experiments require better control of the electron beam, for example an easy on-off of the electron beam itself. These research interests motivated the development of a new gun.

2. Electron Gun Development
A decade in use has shown that the current electron gun is a fully functional design, but it has also revealed its faults. The current electrode structure is not as dynamic in its control and tuning capabilities as the use of a Wehnelt electrode has been shown to be capable of [9], and its reliance on active magnetic shielding in the form of two coils, bucking and trim, limits the amount of fringe field minimization possible without a massive gain in heat produced by the coils. The goals of the new design were to increase control of the electron beam, to improve electron beam properties, and to benefit from lessons learned over a decade of

operation. In order to achieve these goals, it was necessary to show that the new design could shape both the electric field and magnetic field inside the gun, as well as fit into the existing TITAN EBIT structure.

2.1 The New Design

In figure 2 the important electrodes and magnetic shielding components of the gun are shown. It is necessary that the Wehnelt electrode shapes the electric field around the cathode efficiently (shown in figure 3). This allows for fine control over the electron beam emission as well as the compression. The Wehnelt electrode potential can be used to switch the emission current on and off fast. Although the amount of current produced can be controlled, it will also create an inhomogeneous emission distribution on the cathode surface, that needs to be minimized.

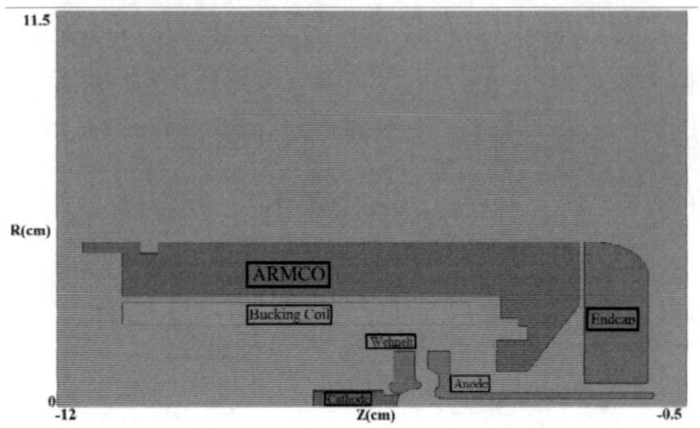

Figure 2. The proposed design makes use of a Wehnelt electrode to focus the electron beam.

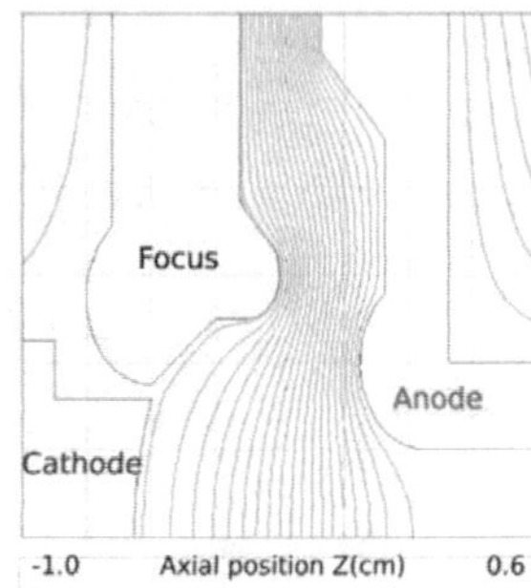

Figure 3. The equipotential lines (blue) shaped by the Wehnelt indicate that the beam will be focused toward the center of the anode aperture [10].

For maximum compression of the electron beam between the cathode and the trap, the fringe magnetic field from the 6 T superconducting coil must be as low as possible at the cathode surface. This will lead to compression of the electron beam dominated by the electrostatic potential. Passive and active shielding are used as a layer of pure ARMCO iron, and either as an 80-turn or 40-turn bucking coil respectively. The 40-turn model is pictured in figure 2. To meet the space constraints the 80-turn design had ~30% less iron.

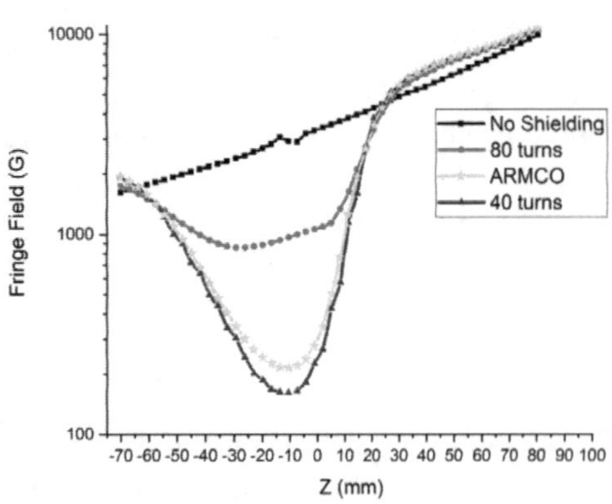

Figure 4. Simulated magnetic field on axis for both designs, as well as the 40-turn model without the bucking coil turned on (ARMCO), when the magnet is at 6 T.

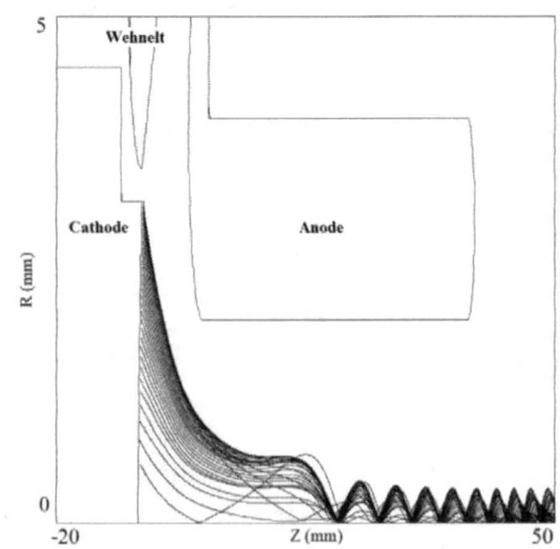

Figure 5. A 250mA simulated electron beam shows that most beam compression occurs in the gun.

While both versions reduced the EBIT magnetic field at the cathode surface, the 40-turn design with more iron shielding achieved the lowest magnetic field. As can be seen in figure 4, the increase in iron content induces the biggest change in shielding done, and the bucking coil fine tunes the field around the cathode. The thicker layer of ARMCO iron and 8 A through the 40-turn bucking coil results in a 165 G magnetic field at the cathode surface (figure 4). At a 62 keV cathode float (Wehnelt: -11 V, Anode: +5keV w.r.t cathode) using this magnetic field, the electron beam went from an initial J_e = 0.792 A/cm^2 at the cathode surface to a maximum of $J_e = 3.2$ kA/cm^2 at the trap center (figure 5).

The combined electric and magnetic fields were considered for three cathodes with diameters : 3mm, 6mm, and 12mm (Heatwave standard series barium tungsten dispenser cathode 10101, 10104, 101111 smallest to largest). An example of the electron trajectories, from a 10104 model with 6 mm diameter, is shown in figure 5. Current simulated electric and magnetic fields, show that the electron beam suffered no magnetic reflections and was compressed by a factor of 34, 21, and 31at the trap center for the three cathode sizes, respectively. Values for the compression factor are double to three times smaller than the calculated Hermann radius would lead to of 65, 88, and 124. Further investigation is underway to limit scalloping (figure 5) to avoid loss currents on the anode.

2.2 *Machine safety and gun material Considerations*

The ARMCO shielding will exert a force on the EBIT magnet which might break the magnet's struts. A limit of 800 N was provided by Cryogenics. To counteract this, simulations were carried out to find how close the electron gun can be placed to the Helmholtz coils while still achieving good electron beam properties. figure 6 shows the force on the struts as the gun is pulled away from the Helmholtz coils; at 3 cm from the initial anode starting position, the force on the struts is below the 800 N limit. Also shown in figure 6 is that the residual field on the cathode drops along with the shift in the gun. We verified that retracting the gun along its linear translator made no difference to the electron beam's transmission and total compression (figure 7). We even found that the displacement slightly improved the compression of the beam.

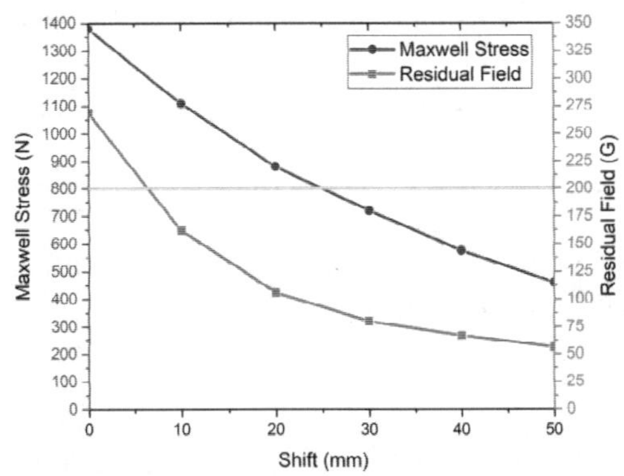

Figure 6. The residual field magnetic field at the cathode surface and resultant stress on the EBIT magnet as the gun assembly is retracted. 0 mm is the initial anode position. The horizontal line is the 800 N limit.

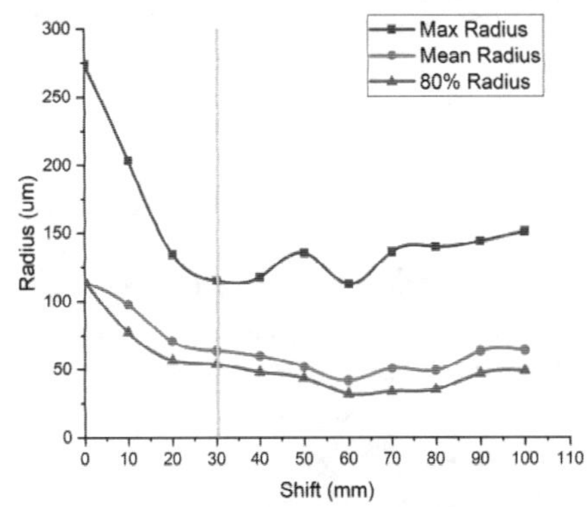

Figure 7. The electron beam properties are largely unaffected as the electron gun is moved along its translator between -20 and -100 mm relative to initial anode position. Vertical stripe denotes same 800 N limit.

Another significant consideration was heat. A substantial amount is generated by the bucking coil and the cathode. When the cathodes are heated to 1200 °C and the bucking coil is set to compensate the fringe field with 8 A of current, the emitted power has been calculated to be 4 W, 24 W, and 237 W [10] for the three different sizes. To counteract this, cold water will flow through the hollow bucking coil wire and a larger cooling pipe. The outer chamber of the gun will be filled with transformer oil to further increase the heat exchange. Aluminum nitride was chosen as the material for the insulators between electrodes, as its high thermal conductivity 200 W/mK [11] allows for proper heat transfer to the cooling pipes. Inside the vacuum, the cathode holder will be cased in molybdenum in order to withstand the heat from the cathode at 1200 °C.

3. Summary and Outlook

In order to meet the broadening research scope of the TITAN facility, a new electron gun for the EBIT has been designed to provide superior control and electron beam properties while preserving the EBIT's excellent charge breeding capabilities. Magnetic shielding was found to best through a combination of a thick ARMCO iron layer and a slim bucking coil. This combination and a Wehnelt geometry will accomplish all the new operating requirements. The heat load of the cathode and bucking coil as well as the longitudinal stress on the EBIT magnet struts will be mitigated through careful material choice, cooling circuits of oil and water, and additional separation, respectively. After the design is finalized, parts will be manufactured in house. Assembly, installation, and testing are planned for spring 2022 in order to be available for on-line experiments by the summer.

Acknowledgements

We thank M. Good for his technical support, especially in design, and M. Marchetto for fruitful discussion on machine safety. This work has been supported by the Natural Sciences and Engineering Research Council (NSERC) of Canada and through TRIUMF by the National Research Council (NRC) of Canada.

References

[1] Ettenauer, S. et al (2011). First Use of High Charge States for Mass Measurements of Short-Lived Nuclides in a Penning Trap. *Physical Review Letters*, *107*(27).

[2] Dilling, J., Bricault, P., Smith, M., & Kluge, H. J. (2003). The proposed TITAN facility at ISAC for very precise mass measurements on highly charged short-lived isotopes. *Nuclear Instruments and Methods in Physics Research Section B: Beam Interactions with Materials and Atoms*, *204*, 492–496.

[3] Simon, M. C. et. al. (2013). Charge breeding rare isotopes for high precision mass measurements: challenges and opportunities. *Physica Scripta*, *T156*, 014098.

[4] Frekers, D. et al. (2013). Penning-trap Q-value determination of the Ga71(ν,e−)Ge71 reaction using threshold charge breeding of on-line produced isotopes. *Physics Letters B*, *722*(4–5), 233–237.

[5] Leistenschneider, E. et al. (2020). Diversifying beam species through decay and recapture ion trapping: a demonstrative experiment at TITAN-EBIT. *Journal of Physics G: Nuclear and Particle Physics*, *47*(4), 045113.

[6] Lennarz, A. et al. (2014). In-Trap Spectroscopy of Charge-Bred Radioactive Ions. *Physical Review Letters*, *113*(8).

[7] Leach, K. et al. (2017). Electroweak Decay Studies of Highly Charged Radioactive Ions with TITAN at TRIUMF. *Atoms*, *5*(4), 14.

[8] Currell, F., & Fussmann, G. (2005). Physics of electron beam ion traps and sources. *IEEE Transactions on Plasma Science*, *33*(6), 1763–1777.

[9] Becker, R. (1981). Two electron guns with adjustable beam compression and perveance. *Nuclear Instruments and Methods in Physics Research*, *187*(1), 255–258.

[10] K. Dietrich, MSc. Thesis, University of Heidelberg, 2019

[11] Slack, G. A., Tanzilli, R., Pohl, R., & Vandersande, J. (1987). The intrinsic thermal conductivity of AIN. *Journal of Physics and Chemistry of Solids*, *48*(7), 641–647.

Journal of Physics: Conference Series **2244** (2022) 012076 doi:10.1088/1742-6596/2244/1/012076

Investigation of Ho ion production with a laser ablation ion source

R. Bhandari[1], M. Horana Gamage[1], N. D. Gamage[1], M. Redshaw[1,2]

[1]Department of Physics, Central Michigan University, Mt. Pleasant, MI-48859, USA

[2]National Superconducting Cyclotron Laboratory, East Lansing, MI 48824, USA

Abstract. At Central Michigan University (CMU), we are developing the CMU High Precision Penning Trap (CHIP-TRAP) with the aim of performing high-precision mass measurements on stable and long-lived radioactive isotopes. A major goal of CHIP-TRAP is to measure the ^{163}Ho EC Q value to ~1 eV precision to aid direct neutrino mass determination experiments with ^{163}Ho. CHIP-TRAP utilizes a laser ablation source (LAS) and a Penning ion trap source to produce ions from solid and gaseous samples, respectively. Ions from these sources are identified by measuring their times-of-flight from the source to their detection on a micro channel plate detector (MCP). With the goal of producing $^{163}Ho^+$ ions from a solution of ^{163}Ho dissolved in nitric acid and dried out on a backing target to be inserted into the LAS, we investigated the production of naturally occurring, stable $^{165}Ho^+$ following this method. We aimed to determine the minimum number of Ho atoms required on the target to produce a detectable number of Ho^+ ions, and to investigate different backing materials to determine if a particular material minimized contaminant ions from the backing material and maximized the number of Ho^+ ions compared to, for example, HoO^+ ions.

1. Introduction

The Standard Model (SM) does not account for the fact that neutrinos oscillate between flavor eigenstates and, as such, do have mass, see e.g. Ref. [1] for a recent review. This result has been demonstrated by a number of oscillation experiments with atmospheric, solar, accelerator and reactor neutrinos [2, 3, 4, 5, 6]. However, these experiments are not able to determine that absolute mass scale of the neutrino, which is still unknown. The discovery that neutrinos have mass has implications in a wide range of fields in physics, from theoretical particle physics to cosmology. A neutrino mass of only a few tenths of an eV indicates that neutrinos could have had significant impact on large-scale structure formation and the evolution of the universe [7]. The determination of the absolute mass of the neutrino is a high priority in the field and the focus of several large scale experimental efforts [8].

Journal of Physics: Conference Series **2244** (2022) 012076 doi:10.1088/1742-6596/2244/1/012076

Currently, the most stringent limits on the electron neutrino mass come from experiments that study the shape of β-spectra near the end-point using ^{3}H [9, 10], and also ^{187}Re [11]. A similar method can be performed via an electron capture (EC) decay, and a focus in recent years has grown around ^{163}Ho. The very low Q value and relatively long half-life for ^{163}Ho EC, along with recent developments in quantum sensor technology, make ^{163}Ho a promising candidate [12, 13, 14]. The ^{163}Ho EC Q value was recently measured at the SHIP-TRAP Penning trap facility with the result 2833(30)(15) eV/c^2 [15], with statistical and systematic uncertainties, respectively, shown in the parentheses. However, ultimately an uncertainty in the Q value of ~1 eV will be required for a meaningful test of systematics in experiments that will employ detectors with an anticipated resolution at this level. At Central Michigan University, we are developing the Central Michigan University High Precision Penning Trap (CHIP-TRAP) [16], for precise measurements on stable and long-lived radioactive isotopes. At CHIP-TRAP, we plan to measure the ^{163}Ho Q value via the mass difference of ^{163}Ho and daughter ^{163}Dy to a precision of ~1 eV.

2. Experimental Apparatus and Methods

CHIP-TRAP utilizes two ion sources—a laser ablation source (LAS) and a Penning ion trap (PIT) source [17]. The LAS can produce ions from solid target materials, and the PIT source from gaseous samples. The LAS utilizes a frequency doubled pulsed Nd:YAG laser that is focused to a spot size of $\lesssim$0.5 mm, providing energy densities of at least 10^8 W/cm^2, sufficient to ablate material from the target. The target is attached to a rotatable circular target holder that is connected to a motorized vacuum feedthrough, as shown in Fig. 1(b). Ablated ions are accelerated to 1 keV via a series of extraction electrodes, focused into a 90 degree electrostatic quadrupole bender and steered into the main beamline. LabView is used for experimental control and data acquisition, see e.g. Ref [18]. A schematic of the LAS and section of beamline used in this work is shown in Fig. 1 (a).

2.1. Commissioning and Characterization of the LAS

We initially installed a Sigradur® glassy carbon target in the LAS to enable the production of $^{12}C_n^+$ clusters to provide a wide m/q range to characterize ion production and transport. The ions are created at the surface of the target, which is biased with a potential U_o (typically 300 – 500 V) and are then accelerated and transported through the beamline. The potential along the beamline, $U(s)$, is typically –1000 V, but varies depending on local fields from electrostatic elements such as the extraction electrodes, einzel lenses, and the quadrupole bender. Hence, the time-of-flight (TOF) to the microchannel plate (MCP) detector a distance S away is

$$t = \sqrt{\frac{m}{q}} \int_0^S \left(\frac{1}{2(U_o - U(s))} \right)^{1/2} ds, \tag{1}$$

where q and m are the charge and mass of the ions, respectively. Since the different $^{12}C_n^+$ clusters are transported under the same beam conditions, the TOF is proportional to $(m/q)^{1/2}$. A fit to the TOF vs $(m/q)^{1/2}$ data for carbon clusters with

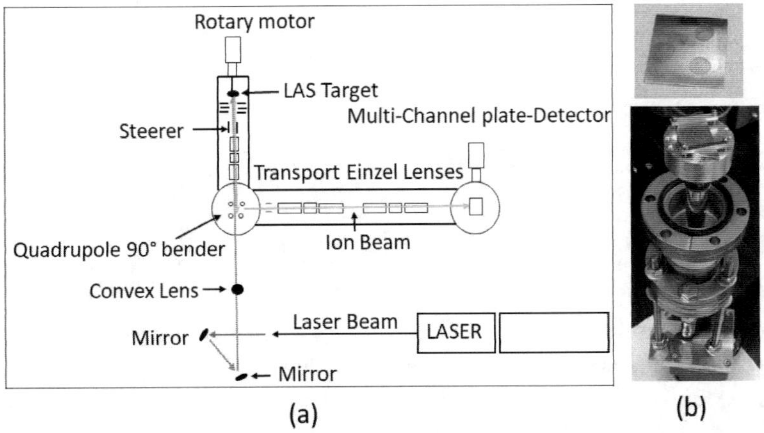

Figure 1. 1(a) Schematic of the Laser Ablation Source of the CHIP-TRAP. 1(b) Target foil where there are three dried holmium nitrate solution drops and Metal target on target holder before installing the rotary motor in the LAS.

$n = 2 - 14$ was performed and showed good agreement with the expected relationship from Eqn. (1) [17].

After performing initial tests with carbon clusters, we investigated ion production and measured the TOF for different metal ions, including aluminum, iron, nickel, zinc, silver, indium, holmium, tungsten, and gold. We found that for light elements (A $\leq$ 100) we were able to resolve different isotopes, whereas for heavier elements (A $>$ 100) we were not. In that case, e.g. as for ^{107,109}Ag, we took the average atomic mass weighted by isotopic abundance for the determination of m/q. Future improvements to our system include the installation of a Multi-reflection Time-of-flight Mass Separator (MR-TOF-MS) to improve resolution and enable the identification and separation of isotopes and isobars [19]. The resulting TOF data for all singly-charged atomic ions of these metals are plotted vs $(m/q)^{1/2}$ in Fig. 2(a). The calibration curve for the metal ions was in good agreement with that for the carbon cluster ions indicating that the initial conditions for the different ions, e.g. energy, energy spread, spatial position, and spatial spread, are similar.

2.2. Investigation of Ho$^+$ ion production from holmium powder with the LAS
^{163}Ho does not occur naturally, so must be synthesized and is only available in small quantities. Therefore, we investigated the production of ^{165}Ho$^+$, the only stable and naturally occurring holmium isotope. After ^{163}Ho is synthesized, it can be extracted by dissolving it in nitric acid, forming holmium nitrate [20, 21]. We made solutions of holmium nitrate by dissolving holmium powder in 6M nitric acid. We then dried out the holmium nitrate solution on a metal backing target to be installed in the LAS. We investigated holmium ion production for different holmium nitrate concentrations i.e. to vary the number of holmium atoms on the target, and for different backing materials. We used $\sim 10^{15} - 10^{21}$ holmium atoms deposited on a series of different

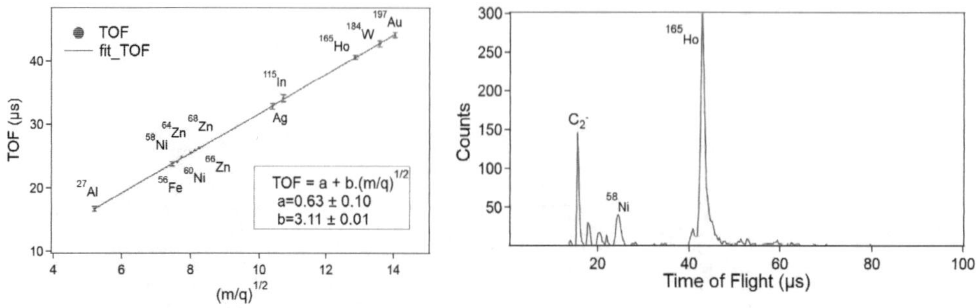

Figure 2. (a)Time-of-flight of different metals measured source to MCP detector. (b) Time-of-flight spectrum for ions produced with $\sim 10^{17}$ Ho atoms from a dried-out holmium nitrate solution on a nickel backing target. $^{165}Ho^{+}$ ions, as well as $^{58}Ni^{+}$ and other ions, e.g. $^{12}C_{+}$ were identified in the spectrum.

metal backing materials that included tungsten, titanium, nickel, zinc, and silver. An example TOF spectrum that shows Ho^{+} ions produced from a holmium nitrate solution dried out on a nickel backing target is shown in Figure 2(b).

3. Results and Discussion

For the Ti and Ni backing targets we could observe Ho^{+} ions on the MCP with only $\sim 10^{17}$ atoms on the target, whereas Zn and Ag required $\sim 10^{19}$ and W required $\sim 10^{21}$. For the Ni and Zn backing targets we observed only Ho^{+} ions, whereas for Ag we observed only HoO^{+} ions. For Ti we observed Ho^{+} and HoO^{+} in equal amounts, and for W we observed HoO_2^{+} at about half the rate of Ho^{+} (we expect that HoO^{+} ions were also made when using the W backing target, but they have a similar m/q as the W^{+} ions and it was not possible to resolve the different species). In the case of the Ti and Ag backing targets, the amount of Ho^{+} and HoO^{+} ions produced was small compared to the amount of Ti^{+} and Ag^{+} ions produced from the backing material ($\sim$10 %). As mentioned above, it was not clear whether we produced W^{+} or HoO^{+} ions, but the peak in the TOF spectrum that could include both of these species contained about twice as many ions as the Ho^{+} peak. Since we observed HoO_2^{+} and also observed W_2^{+}, the peak probably contained both W^{+} and HoO^{+} ions. Finally, for the Ni and Zn backing targets, we observed relatively small amounts of Ni^{+} and Zn^{+} ions ($\sim$10 – 20 %) compared to the number of Ho^{+} ions.

Our goal was to investigate whether the backing material on which the holmium nitrate solution was dried out on and that was used in the LAS had any effect on (1) the minimum number of holmium atoms that are required on the target, (2) the relative rates at which Ho^{+} ions were produced compared to HoO^{+} and HoO_2^{+} ions, and (3) the relative rates that any ions were produced via ablation of the backing material itself. We found that Ni and Zn enabled the production of Ho^{+} ions with the lowest number of ions produced from the backing material. These two backing targets also resulted in the production of Ho^{+} only and no HoO^{+} or HoO_2^{+}. However, for the Ni backing material only $\sim 10^{17}$ Ho atoms were required on the target, whereas

$\sim10^{19}$ were required for Zn. The Ti target also required only $\sim10^{17}$ Ho atoms on the target, but significant quantities of HoO^+ and Ti^+ were produced in this case. The Ag and W targets performed least well, requiring larger numbers of Ho atoms on the target and producing relatively large amounts of HoO^+, HoO_2^+, and/or ions from the backing material. Hence, Ni stands out as the most favorable backing material in all three of the criteria listed above.

4. Conclusions

A laser ablation ion source that is part of the CHIP-TRAP Penning trap apparatus at Central Michigan University has been used to create Ho^+ ions from a holmium nitrate solution dried out on various different metal backing targets. It was found that a nickel backing material was the most favorable material tested in terms of the least number of Ho atoms required on the target, and the relative abundance of Ho^+ compared to HoO^+, HoO_2^+, and any ions produced from ablation of the backing material itself.

We found that we could observe Ho^+ ions on the MCP detector with $\sim10^{17}$ Ho atoms on the target. However, we anticipate that the number of ^{163}Ho atoms that would be required for a measurement of the ^{163}Ho Q value using the CHIP-TRAP apparatus could be reduced by at least 2 orders of magnitude. This reduction can come firstly from the fact that not all Ho atoms on the target are ablated by the laser. We estimate the number of atoms on the target that the laser actually accesses is $\sim$10 % due to the size of the holmium nitrate solution drop that was placed on the target and the path and spot size of the laser. This ratio can be increased by more precise deposition of the holmium nitrate solution and an optimization of the laser path and spot size. Secondly, a measurement with ^{163}Ho will not require continuous ablation of the target and detection on the MCP as was the case in the work presented here. Single laser shots will be used to produce an ion bunch that will then be transported to the CHIP-TRAP Penning trap. Ions will then be detected in the Penning trap with significantly higher sensitivity and efficiency than on the MCP.

References

[1] King S F and Luhn C 2013 *Reports on Progress in Physics* **76** 056201 URL `https://doi.org/10.1088/0034-4885/76/5/056201`
[2] Fukuda Y, Hayakawa T, Ichihara E and Inoue K (Super-Kamiokande Collaboration) 1998 *Phys. Rev. Lett.* **81**(8) 1562–1567 URL `https://link.aps.org/doi/10.1103/PhysRevLett.81.1562`
[3] Ahmad Q R, Allen R C, Andersen T C, DAnglin J and Barton (SNO Collaboration) 2002 *Phys. Rev. Lett.* **89**(1) 011301 URL `https://link.aps.org/doi/10.1103/PhysRevLett.89.011301`
[4] Eguchi K, Enomoto S, Furuno K, Goldman J, Hanada H and Ikeda (KamLAND Collaboration) 2003 *Phys. Rev. Lett.* **90**(2) 021802 URL `https://link.aps.org/doi/10.1103/PhysRevLett.90.021802`
[5] Ahn M, Aliu E, Andringa S, Aoki S, Aoyama Y, Argyriades J, Asakura K, Ashie R, Berghaus F, Berns H *et al.* 2006 *Physical Review D* **74** 072003 URL `https://link.aps.org/doi/10.1103/PhysRevD.74.072003`
[6] Michael D (MINOS Collaboration) 2006 *Phys. Rev. Lett.* **97**(19) 191801 URL `https://link.aps.org/doi/10.1103/PhysRevLett.97.191801`
[7] Loureiro A, Cuceu A, Abdalla F B, Moraes B, Whiteway L, McLeod M, Balan S T,

Lahav O, Benoit-Lévy A, Manera M *et al.* 2019 *Phys. Rev. Lett.* **123**(8) 081301 URL https://link.aps.org/doi/10.1103/PhysRevLett.123.081301

[8] Drexlin G, Hannen V, Mertens S and Weinheimer C 2013 *Advances in High Energy Physics* **2013** URL https://www.hindawi.com/journals/ahep/2013/293986/

[9] Aker M, Altenmüller K, Arenz M, Babutzka M, Barrett J, Bauer S, Beck M, Beglarian A, Behrens J, Bergmann T *et al.* (KATRIN Collaboration) 2019 *Phys. Rev. Lett.* **123**(22) 221802 URL https://link.aps.org/doi/10.1103/PhysRevLett.123.221802

[10] Otten E W and Weinheimer C 2008 *Reports on Progress in Physics* **71** 086201 URL https://doi.org/10.1088/0034-4885/71/8/086201

[11] Arnaboldi C, Brofferio C, Cremonesi O, Fiorini E, Lo Bianco C, Martensson L, Nucciotti A, Pavan M, Pessina G, Pirro S, Previtali E, Sisti M, Giuliani A, Margesin B and Zen M 2003 *Phys. Rev. Lett.* **91**(16) 161802 URL https://link.aps.org/doi/10.1103/PhysRevLett.91.161802

[12] Gastaldo L, Blaum K, Chrysalidis K, Goodacre T D, Domula A, Door M, Dorrer H, Düllmann C E, Eberhardt K, Eliseev S *et al.* 2017 *The European Physical Journal Special Topics* **226** 1623–1694 URL https://link.springer.com/content/pdf/10.1140/epjst/e2017-70071-y.pdf

[13] Nucciotti A 2016 *Advances in High Energy Physics* **2016** URL https://www.hindawi.com/journals/ahep/2016/9153024/

[14] Gamage N D, Bhandari R, Horana Gamage M, Sandler R and Redshaw M 2019 *Hyperfine Interactions* **240** ISSN 0304-3843 URL https://www.osti.gov/biblio/1674968

[15] Eliseev S, Blaum K, Block M, Chenmarev S, Dorrer H, Düllmann C E, Enss C, Filianin P E, Gastaldo L, Goncharov M, Köster U, Lautenschläger F, Novikov Y N, Rischka A, Schüssler R X, Schweikhard L and Türler A 2015 *Phys. Rev. Lett.* **115**(6) 062501 URL https://link.aps.org/doi/10.1103/PhysRevLett.115.062501

[16] Redshaw M, Bryce R, Hawks P, Gamage N, Hunt C, Kandegedara R, Ratnayake I and Sharp L 2016 *Nuclear Instruments and Methods in Physics Research Section B: Beam Interactions with Materials and Atoms* **376** 302–306 ISSN 0168-583X URL https://www.sciencedirect.com/science/article/pii/S0168583X1600135X

[17] Horana Gamage M, Arnold A L, Bhandari R, Gamage N D, Purcell Z, Sandler R and Redshaw M 2019 *Hyperfine Interactions* **240** ISSN 0304-3843 URL https://www.osti.gov/biblio/1674970

[18] Kalkman C 1995 *Journal of clinical monitoring* **11** 51—58 ISSN 0748-1977 URL https://doi.org/10.1007/bf01627421

[19] Schultz B, Kelly J, Nicoloff C, Long J, Ryan S and Brodeur M 2016 *Nuclear Instruments and Methods in Physics Research Section B: Beam Interactions with Materials and Atoms* **376** 251–255 ISSN 0168-583X URL https://www.sciencedirect.com/science/article/pii/S0168583X16001737

[20] Eliseev S, Blaum K, Block M, Chenmarev S, Dorrer H, Düllmann C E, Enss C, Filianin P E, Gastaldo L, Goncharov M, Köster U, Lautenschläger F, Novikov Y N, Rischka A, Schüssler R X, Schweikhard L and Türler A 2015 *Phys. Rev. Lett.* **115**(6) 062501 URL https://link.aps.org/doi/10.1103/PhysRevLett.115.062501

[21] Ketelaer J, Krämer J, Beck D, Blaum K, Block M, Eberhardt K, Eitel G, Ferrer R, Geppert C, George S, Herfurth F, Ketter J, Nagy S, Neidherr D, Neugart R, Nörtershäuser W, Repp J, Smorra C, Trautmann N and Weber C 2008 *Nuclear Instruments and Methods in Physics Research Section A: Accelerators, Spectrometers, Detectors and Associated Equipment* **594** 162–177 ISSN 0168-9002 URL https://www.sciencedirect.com/science/article/pii/S0168900208008693

Non-destructive beam spatial profile measurement using a gas sheet monitor for a high-intensity ion beam

I Yamada[1,2], M Wada[1], J Kamiya[2], M Kinsho[2]

[1] Graduate School of Science and Engineering, Doshisha University, 1-3, Miyako-tataradani, Kyotanabe, Kyoto, 610-0321, Japan

[2] J-PARC center, Japan Atomic Energy Agency, 2-4, Shirakata, Tokai-mura, Naka-gun, Ibaraki, 319-1195, Japan

E-mail: ip_yamada@icloud.com

Abstract. A non-destructive beam spatial profile monitor system using a gas sheet has been developed. The 4 mm thick gas sheet of over 40 mm width enables increase of the local gas pressure high enough to record a two-dimensional beam profile. A CCD camera coupled to an image intensifier detects photons produced by beam-gas interaction with the signal intensity that makes the time evolution study of the beam possible. We demonstrated that the J-PARC 3 MeV, 60 mA H^- beam profile which is measured with intensity error of 8.4-12% and spatial resolution of 1 mm changed depending on the controls of the acceleration cavity in the beam pules of 50 μs.

1. Introduction

In a high-intensity ion accelerator, non-destructive or non-invasive beam diagnostic systems play an important role in minimizing beam losses and radio-activations of the accelerator systems. A wire-scanning-type invasive beam spatial profile monitor is utilized in the J-PARC linac which accelerates a negative hydrogen ion beam of 50 mA to 400 MeV [1]. It limits the beam-pulse length in profile measurement because the heat load of the energy deposition breaks the wire. To solve the problem, non-destructive beam profile monitors based on beam-gas interaction have been developed. An ionization beam profile monitor (IPM) using a residual gas is utilized in the operation of J-PARC rapid-cycling synchrotron which accelerates a proton beam from 400 MeV to 3 GeV after charge stripping [2]. While the IPM realizes non-destructive measurement, the obtained profiles are one-dimensional as same as the one obtained with the wire-scanner monitor. To improve the profile-measurement quality for investigation of high-intensity-beam dynamics, we have developed the gas sheet beam profile monitor shown in Fig. 1 and demonstrated that the monitor can measure a two-dimensional beam profile [3]. The gas sheet monitor injects a sheet-shaped gas into a beam line, induces photons, and detects the spatial distribution of the photons as a 2-D image. In principle, control of the sheet-gas density enables various 2-D profile measurements, such as pulse-by-pulse measurements and time-development measurements. To demonstrate the possibility of time-development measurement, we applied the gas sheet monitor to investigate the effect of parameters controlling a radio-frequency cavity for acceleration. When a beam is accelerated in a cavity, the beam induces an electromagnetic field on the cavity wall.

The wakefield weakens the electric field of the cavity for acceleration, and the beam becomes unstable: the beam loading effect. The feed-back and feed-forward controls are employed to cancel the wakefield [4] but are not optimized in terms of the beam profile. In this paper, we report the details of the gas sheet monitor and the result of time-development measurement of the beam profile to optimize the control system of the cavity.

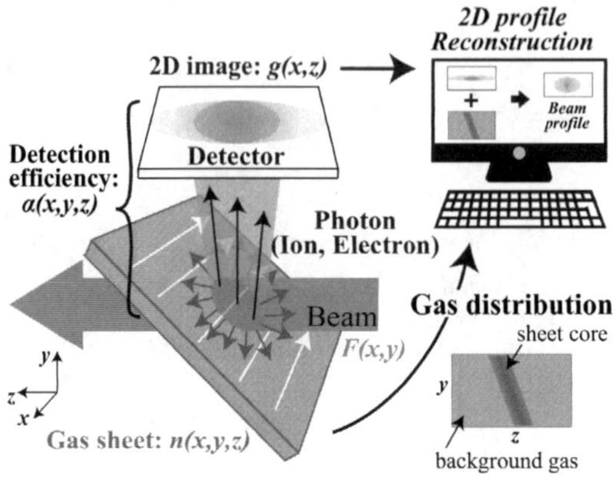

Figure 1. The beam profile diagnostic process using the gas sheet monitor.

2. Principle of gas sheet monitor

2.1. Gas sheet formation

The gas sheet formation is based on rarefied gas dynamics [5]. When the mean free path of gas molecules is longer than the typical vacuum chamber dimension, intermolecular collisions are negligible. For reflections on the chamber wall, the probability distribution function of the reflection angle is proportional to the cosine of the reflection angle: the cosine law. Thus, the motion of gas molecules can be regarded as the individual thermal motion with reflections being subject to the cosine law. On the basis of the principle, a long gas conduit with a thin cross section can form a gas sheet as shown in Fig. 2. The conduit increases the number of reflections in the thickness direction. Only molecules passing the conduit through large reflection angles can reach the exit to form a thin gas sheet.

2.2. Beam profile measurement with gas sheet monitor

Figure 3 shows the operational principle of the gas sheet monitor to measure a beam profile. A gas sheet flows along x axis, a beam passes through the gas sheet along z axis, and a CCD camera detects the photons along y axis. Considering the finite gas-sheet thickness and the background gas, the photons produced by beam-gas interaction are integrated on the photon detecter along y axis. The luminous intensity spatial distribution $g(x, z)$ of the image obtained with the CCD camera, the gas density spatial distribution $n(x, y, z)$, and the beam spatial profile $F(x, y)$ are correlated by an integral equation:

$$g(x, z) = \int_D \alpha(x, y, z)\, n(x, y, z)\, F(x, y)\, \mathrm{d}y = \int_D k(x, y, z)\, F(x, y)\, \mathrm{d}y \qquad (1)$$

where the integral region D is the diameter of the beam pipe. The distribution $\alpha(x, y, z)$ is the detection efficiency spatial distribution of the photon detector system consisting of the detection efficiency of the CCD image sensors and non-uniformity of the solid angle. The distributions $\alpha(x, y, z)$ and $n(x, y, z)$ are combined as the relative-sensitivity spatial distribution $k(x, y, z)$.

Thus, the two-dimensional beam profile can be reconstructed by solving the integral equation Eq. (1) with the relative-sensitivity spatial distribution measured at an off-line setup [6].

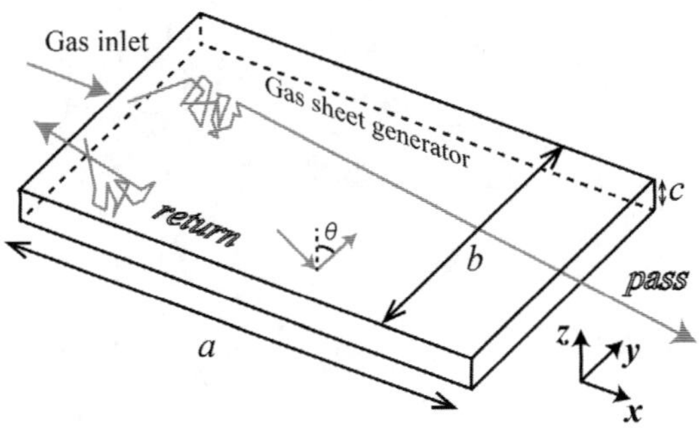

Figure 2. The gas sheet generator is a thin and long gas conduit.

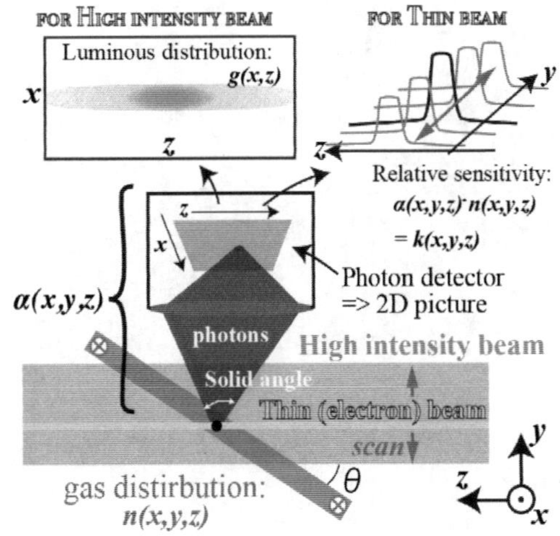

Figure 3. The principle of beam profile measurement using the gas sheet monitor. The relative-sensitivity spatial distribution is calibrated by injection of a thin beam at an off-line system.

2.3. Relative-sensitivity spatial distribution measurement

The relative-sensitivity spatial distribution should be measured by the same principle and the same system as the gas sheet monitor because the sensitivity spatial distribution strongly depends on the gas sheet monitor system. According to Eq. (1), a photon signal produced with a thin beam can be regarded as a one-dimensional sensitivity distribution with spatial resolution of the thin beam diameter. Thus, as shown in Fig. 3, injecting a thin beam instead of a high-intensity beam into the gas sheet monitor at an off-line setup and scanning the thin beam position give the 3-D relative-sensitivity spatial distribution.

3. Developed gas sheet monitor and beam line

3.1. Gas sheet monitor system

The developed gas sheet monitor consists of a gas sheet generator surrounded by a cover chamber, a main chamber attached to a beam line, a photon detector system, and three vacuum pumps. The gas sheet generator is a conduit of a 100 mm × b 50 mm × c 0.1 mm where a, b and c are described in Fig. 2. The cover chamber having a rectangular slit of 50 mm × 0.5 mm cuts the tail part of the sheet-gas distribution and evacuates the reflected molecules with a turbo-molecular pump. The gas sheet core passing through the slit interacts with the beam and is evacuated with a cryo pump and a turbo-molecular pump. Photons produced by the beam-gas interaction are detected with a photon detector system. The system consists of a set of lenses, an image intensifier whose gain is up to 10^4 and gate width is longer than 25 ns, and a CCD camera having 16-bit sensors of 1920 × 1080 pixels whose minimum shutter speed is 1 ms. The main wavelength of the photons is expected as 391 nm [7]. Figure 5 shows the measured relative-sensitivity spatial distribution of the developed gas sheet monitor system at $x = 0$ mm: $k(0, y, z)$. The distribution indicates the gas sheet tilt angle is 36.3 degrees which is consistent with the design angle of 36 degrees.

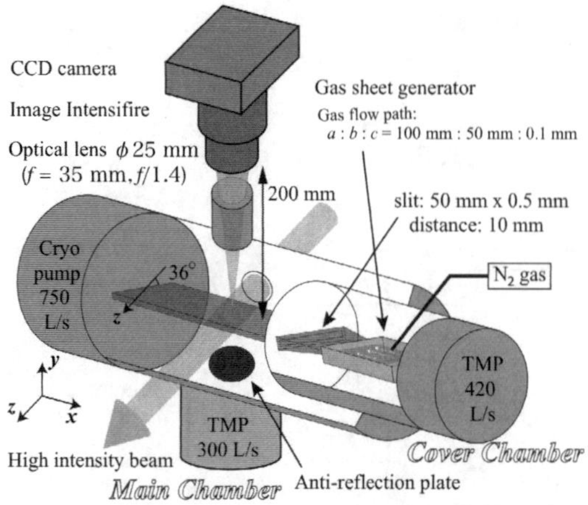

Figure 4. The developed gas sheet monitor system.

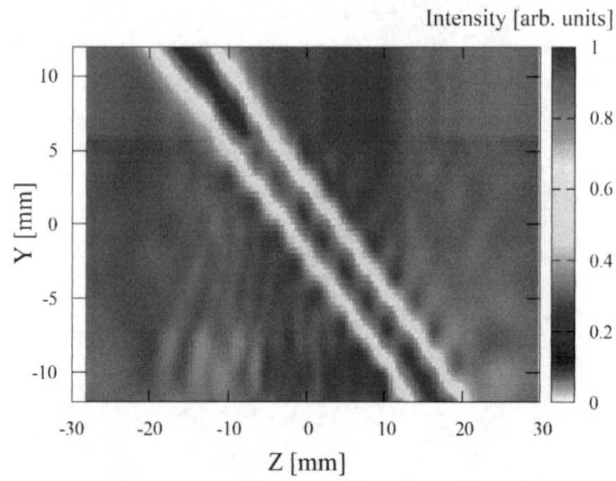

Figure 5. The relative-sensitivity spatial distribution at $x = 0$ mm: $k(0, y, z)$.

3.2. J-PARC MEBT test stand

The gas sheet monitor was installed into the J-PARC medium-energy-beam-transport (MEBT) line test stand [8]. The negative hydrogen ion source produces a 50 keV, 60 mA, 50 μs beam by up to 25 Hz. The radio-frequency quadrupole (RFQ) linac bunches the beam into 9 degrees (RMS) of 324 MHz and accelerates the beam to 3 MeV. The beam is transported with three quadrupole magnets and is dumped with a Faraday cup. A feed-back/feed-forward field compensation system mitigates the beam loading effect in the RFQ. The feed-back control system makes the intensity of the electric field to be a reference intensity. The feed-forward control system adds an electric field of table data prepared in advance and needs to be optimized for the beam.

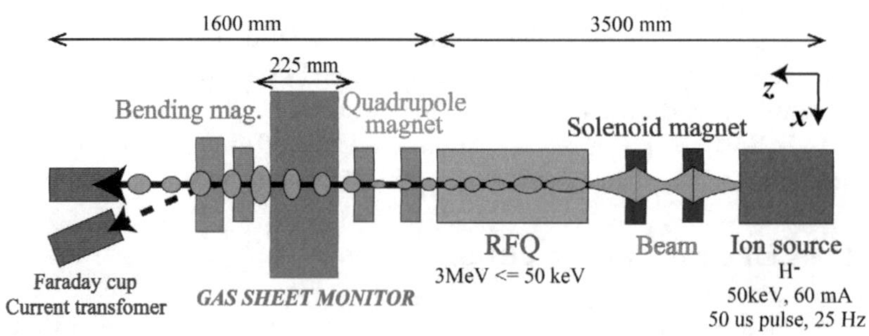

Figure 6. The J-PARC medium-energy-beam-transport (MEBT) line test stand.

4. Time-development measurement of J-PARC beam profile

Figure 7 shows the time evolution of the two-dimensional beam profile in the x-y plane for 50 μs of the beam-pules length with the feed-back and the feed-forward controls. Each profile was measured by the image-intensifier gate width of 1 μs, the CCD-camera exposure time of 200 s, the beam-repetition frequency of 25 Hz, and 15-frame averaging which correspond to a 1.5 μA DC beam of 3,000 s. The gas pressures at the inlet of the gas sheet generator, the cover chamber, and the main chamber were 100 Pa, 1.0×10^{-3} Pa, and 5.6×10^{-5} Pa, respectively. The beam profiles are reconstructed by fitting the parameters of a Gaussian basis function to satisfy the integral equation Eq. (1) with the captured images $g(x, z)$ and the measured sensitivity distribution $k(x, y, z)$. The residual error of the intensity caused by the profile fitting analysis and the standard deviations on the measurements of both the beam profile and the relative-sensitivity spatial distribution are ranged from 8.4% to 12% in total. The spatial resolutions of the beam profile which are determined by the thin beam widths for the relative-sensitivity spatial distribution measurement are 1 mm in x and y axes. Figure 8 shows the variations of the root mean square (RMS) values of the beam profile and the beam current against time with and without the feed-forward control. Both of the x and y profiles with the feed-forward control change in the first 10 μs and is almost constant after the 10 μs while the beam profiles without the feed-forward control change in the first 30 μs. On the other hand, the feed-forward control has little effect on the beam current. These results indicate that it is necessary to monitor not only the current-transmission rate of the cavity but also the beam profile for optimizing the feed-forward control system. The beam profile variation in the first 10 μs may indicate that the feed-forward control is not optimized or the beam profile depends on the space-charge force. As shown in Fig. 8b, the beam current almost doubles in the 10 μs.

In conclusion, we have demonstrated that the gas sheet monitor enables to measure time evolution of the two-dimensional beam profile. We have also clarified that the time evolution of the beam profile depends on the feed-forward control of the acceleration cavity. These facts indicate that the gas sheet monitor can provide a new perspective for optimization of the accelerator components and for study of beam dynamics by depicting a detail beam profile in two dimensions.

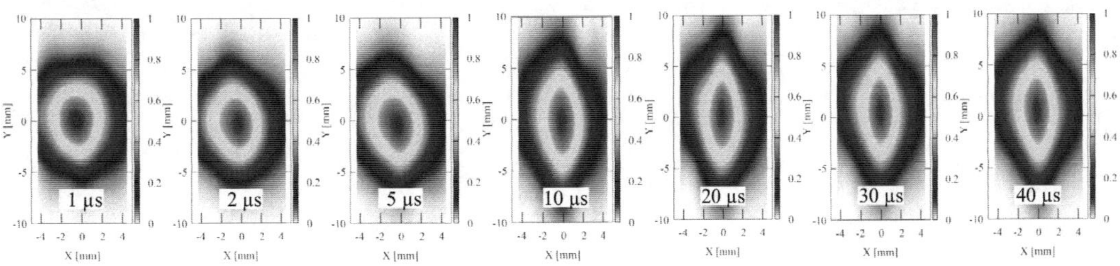

Figure 7. Time evolution of the two-dimensional beam spatial profile of 1 μs with the feed-forward control of the RFQ. Each panel shows the profile of each timing from the beam head.

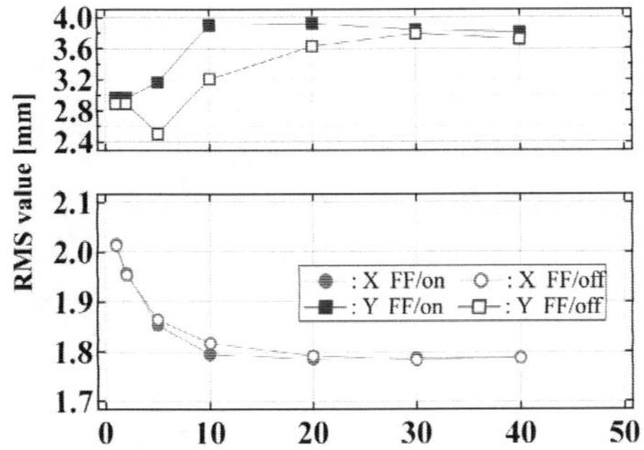

(a)

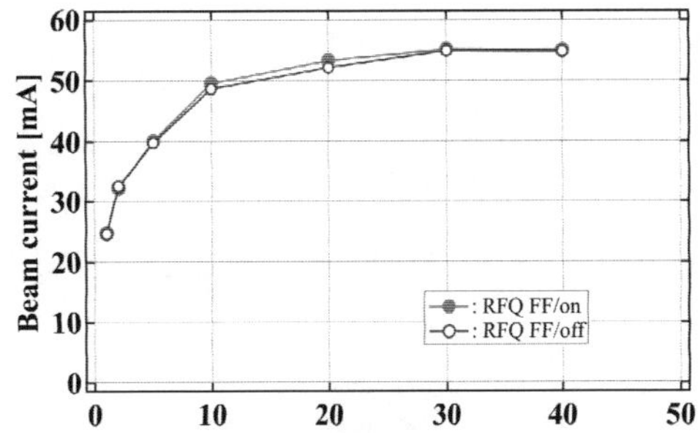

(b)

Figure 8. Variation of (a) the RMS values of the beam profile and (b) the beam current against time from the beam head with and without the feed-forward control of the RFQ.

References

[1] Miura A *et al.* 2016 *Journal of the Korean Physical Society* **69** 1005-1013
[2] Harada H, Yamamoto K, and Kato S 2013 *Proc IPAC2013, Shanghai, China* MOPME021
[3] Yamada I, Wada M, Moriya K, Kamiya J, Saha P K, and Kinsho M 2021 *Phys. Rev. Accel. Beams* **24** 042801
[4] Li S *et al.* 2019 *Nucl. Instrum. Meth. Phys. Res. A* **945** 162612
[5] Ogiwara N 2016 *Proc. IPAC2016, Busan, Korea* WEOBB03
[6] Yamada I, Wada M, Moriya K, Kamiya J, and Kinsho M 2021 *Proc. IBIC2021, Pohang, Rep., Korea* TUOA04
[7] Becker F *et al.* 2010 *Proc. BIW10, Santa Fe, New Mexico, US* TUPSM020
[8] Kondo Y *et al.* 2019 *Journal of Physics: Conference Series* **1350** 012077

Extraction and low energy beam transport models used for the IFMIF/EVEDA RFQ commissioning

L. Bellan[1], T. Akagi[2], B. Bolzon[3], P. Cara[4], Y. Carin[5], M. Cavenago[1], N. Chauvin[3], M. Comunian[1], H. Dzitko[5], E. Fagotti[1], F. Grespan[1], K. Kondo[2], I. Moya[5], A. Pisent[1], I. Podadera[6], F. Scantamburlo[1,5], M. Sugimoto[2]

[1] INFN-LNL, Legnaro, IT
[2] QST, Rokkasho Fusion Institute, Rokkasho, JP
[3] CEA/IRFU, Paris-Saclay University, Gif-sur-Yvette, FR
[4] IFMIF/EVEDA Project Team, Rokkasho, JP
[5] F4E, Garching, DE
[6] CIEMAT, Madrid, ES

Abstract. The Linear IFMIF Prototype Accelerator (LIPAc) is a high intensity D^+ linear accelerator; demonstrator of the International Fusion Material Irradiation Facility (IFMIF). In summer 2019 the IFMIF/EVEDA Radio Frequency Quadrupole (RFQ) accelerated its nominal 125 mA deuteron (D^+) beam current up to 5 MeV, with 90% transmission for pulses of 1 ms at 1 Hz. This success was possible thanks to an intense previous campaign of modelization and measurements in order to characterize the RFQ input beam, which is affected by the ECR ion source extraction and the low energy beam transport. The simulation models used with the measurement benchmarks are here presented.

1. Introduction

During the commissioning of the Linear IFMIF Prototype Accelerator (LIPAc)[1], an intense study campaign focused on the injector, composed by the ECR ion source and the LEBT line [2] has been perfomed, in order to characterize the beam at the RFQ input and to reach its nominal transmission (achieved in 2019 [3]). The experiments related to the injector are described in [4, 5]. The ECR (Electron Cyclotron Resonance) consists of a 2.45 GHz RF power source with two coils magnetic structure. The nominal CW beam extracted consists of 140 mA D^+ at 100 kV. For commissioning purposes, the source can extract also tens of mA of H^+ at 50 kV and can work in pulse mode. From the beam dynamics point of view, these beams are characterized by a high perveance beam transport: the general perveance (un-compensated) ranges from 5×10^{-4} for the probe up to 5×10^{-3} nominal beam. In order to preserve the beam quality from the ion source up to the RFQ injection point (the LEBT), two magnetic solenoids and two repellers, one in the extraction and one at the RFQ entrance, are used in order to allow the space-charge compensation to take place. The solenoids need to be tuned in order to match the output beam characteristics from the source exit to the RFQ input, while the steerers set to 0 the first order moment of the beam at injection point. Diagnostic boxes are installed in order to allow commissioning and control. In particular, the LEBT is equipped with two diagnostic boxes (the first one fixed between the two solenoids and the second one removable, used for injector commissioning). The diagnositcs considered for this paper are: current monitors a

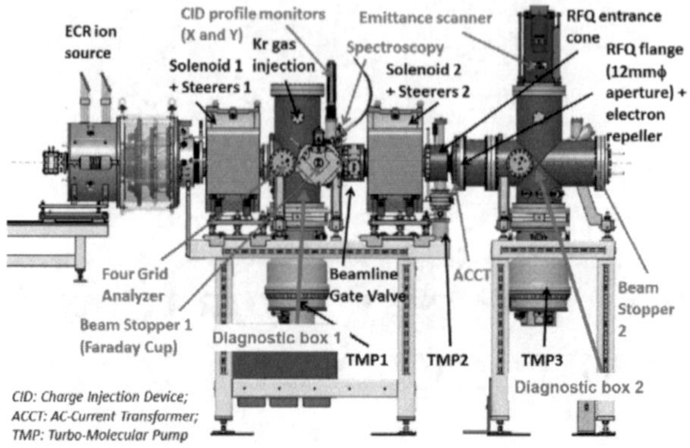

Figure 1: Sketch of the IFMIF/EVEDA injector during the commissioning. The main elements are indicated.

couple of beam stoppers, one in the Diagnostics Box 1 between the solenoids, and a second one in the removable Diagnostics Box 2, and an ACCT placed around the entrance cone at the RFQ injection point see Fig. 1, an Allison scanner type emittance meter, vacuum gauges and a residual gas analyzer, a Doppler Shift spectrometer for species fraction measurement.

2. Description and outcame of the model used during injector commissioning for RFQ injection

2.1. RMS trace forward method

We not only study deeply the injector behaviour, but also to test several beam dynamics models under different space-charge regimes. In this framework, we test the RMS trace forward method via the usage of TraceWin [6]. The method is based on matching the RMS emittance measurements (coming from the Allison scanner in the diagnostic box 2) changing the Courant-Snyder parameters at beam input. The input beam was located outside the extraction, at 19.6 cm from the plasma electrode, with a parabolic 4D distribution: it was of some interest to study the possibility to be independent from the source output. The iterations of the method are repeated with different measurements of the emittance for several solenoid configurations. Once a solution is found, it is possible to retrieve any information in any point of the line from the simulation model. In particular, we were interested at the Courant-Snyder parameters of the beam at the RFQ input, that are otherwise difficult to be measured directly. The space-charge compensation influences heavily on the method [7], by increasing the number of degrees of freedom to the iteraitve loop. In the first trial, we considered the effect of the space-charge compensation as a parameter η, which multiplies $(1-\eta)$ the current of the beam; the condition on the spatial derivatives were $\frac{\partial\eta}{\partial r}=0$ and $\frac{\partial\eta}{\partial z}\neq 0$. Three zones were individuated with fixed value of η: from the input point up to the RFQ repeller, the zone close to the repeller, where the space charge compensation is expected to drop to zero and the diagnostic box number 2. In particular, in this last section (diagnostic box 2) the vacuum level was always one order of magnitude less with respect to the LEBT line (which ranges from 10^{-5} mbar down to 10^{-6} mbar). We used the four grid analyzer, placed between the two solenoids, in order to measure the space charge compensation in that point [8].

Results (comparing simulations and measured values) for a total beam of 85 mA proton at 50 keV with 70±5% H^+ fraction are given in Fig. 2. It resulted, in average, a value of the parameter $\eta = 0.95$. The measurements of the emittance and the profile were taken at the Allison scanner position, in the second diagnostic box.

As it is possible to see from Fig. 2, (c), there is an overestimation of the tales of the distribution, but the overall behaviour is respected (a superposition of a core parabolic

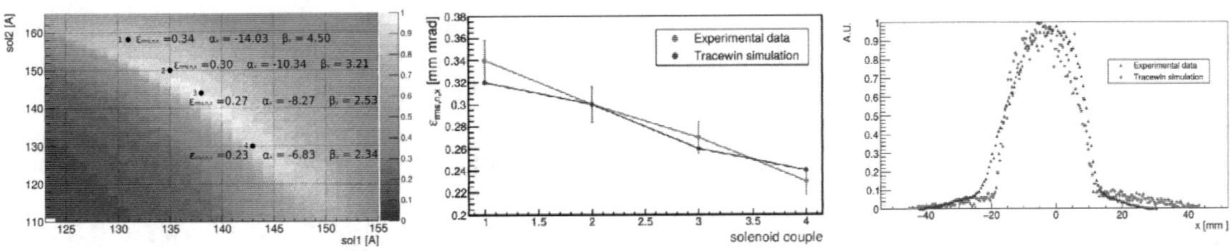

(a) (b) (c)

Figure 2: (a) solenoid scan with the z-coordinate showing the transmission at second diagnostic box. Emittances are reported. (b) Simulated (red) and measured (blue) emittances for the various solenoids values shown in (a). (c) x beam projections (simulated and measured) at the Allison scanner emittancemeter in the second diagnostic box, point 3. The beam considered is an 85 mA total proton beam at 50 keV.

distribution and gaussian distribution for the tales). In plot (b) the solenoid couple values ranges from 1 (see (a)), which is close to the matching for RFQ injection values ($B_{z,sol2} > B_{z,sol1}$), up to 4 , which refers to $B_{z,sol2} < B_{z,sol1}$ condition. The results of this method indicates that the emittance depends on the solenoid values. This is due to both the combined effect of the solenoid aberrations [9] and, in first approximation,the dependance of the space-charge compensation with the beam radius , as shown in [10]. This last point limits also the validity of a certain value of η, found in the loop, within 15% of fixed solenoid values. In order to remove this degree of freedom, we tested the RMS traceforward model trying to calculate each step of the loop, self-consistenly, the space-charge compensation. The calculation of the space-charge compensation was carried out with the software WARP [11]. The main difference with respect to the previous trial, is the fact that the compensation is determined by the dynamic balance between the electrons/ions generated by the ionization, the charge exchange with the residual gas and the collision of the beam with the metallic wall of the LEBT. There is no

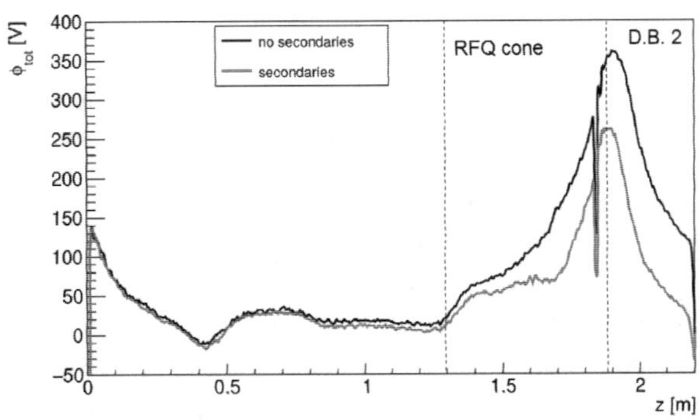

Figure 3: total potential (applied and self) at steady state along z, in case the effect of the secondary electrons from beam collisions are included or not. The RFQ cone region and the second dianostic box region (D.B.2) are shown. The beam considered is an 155 mA total deuteron beam at 100 keV.

any more a factor η in the loop that needs to be set. However, this came with a cost of time-consuming simulations and convergence issues [12]. Details of this simulation can be found in [8]. Unlike from previous simulation of this kind [13], we include [7, 14] the contribution given by the secondary electron emission from the metal. In particular we found a non-negligible effect on the space charge compensation given by these secondary electrons. Fig. 3 shows the total potential (applied and self) at steady state along the beam axis, in case the effect of the secondary electrons from collisions is included or not. The beam considered is composed by 155 mA total deuteron beam, with a ratio of 93% of D^+. It is worth to mention that the IFMIF/EVEDA LEBT does not implement any collimator before the entrance cone of the RFQ, where the majority of the contaminants collides with the copper wall of the cone. Moreover, in Diagnostic box 2, the deuterons impact on the tungsten shield of the emittance meter, during the measurement. Both collisions are driven by a secondary electron yield $\gamma_{e,Cu,W} > 1$. The net effect is that the compensation between these two regions (i.e. the RFQ cone and the Diagnostic Box 2) differs from the case of that one give only by the residual gas.

2.2. Semi-analyitic approach
During 2016 - 2017 the extraction column of the source was re-qualified, from the mechanical point of view [4], and the major inconsistency between the simulations and the measurement in that zone were solved. This opened the opportunity to include the extraction in the simulation. In order to build a faster to simulate model (order of half and hour with respect to months) but

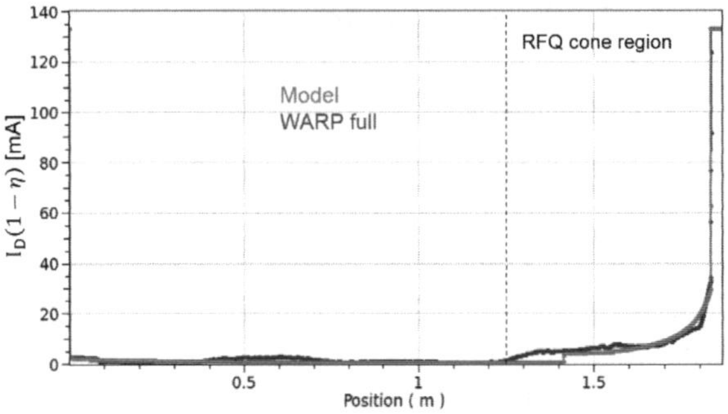

Figure 4: equivalent current $I(1 - \eta)$ calculated with the semi-analyitic approach and the full WARP model, which takes into account the generation of the secondary plasma through Monte-Carlo based processes.

keeping as much as possible the self-consistency in the space-charge compensation calculation, we decide to implement in our simulations the model described in [15], based on [10]. In few words, the model calculates η based on beam radius, current, vacuum level, residual gas composition and other minor factors at a certain z along the axis. The extraction simulation were perfomed with IBSimu code [16]. The beam evolution along the LEBT up to the input of the RFQ was preformed with WARP and with TraceWin. We decided to apply two modifications to the model [15] in order to treat the zone close to the repellers and the injection cones, where the hypothesis of the model down-estimates the residual potential, due to the presence of external applied electric field and secondary electron emissions. This can be seen looking at Fig. 4: the plot shows the equivalent current $I(1-\eta)$, and the estimate η factor along the LEBT from WARP simulation with realistic input and the other one calculated with the semi-analyitic approach. It is possible to see the correction applied, close to the interface with the RFQ injection cone as a discontinuity in the model curve, that it is otherwise continuous and follows quite well that one calculated by the WARP. The result of the models are shown in Fig. 5, with the measurement of the emittance taken from the Allison scanner positioned this time in the middle between the two solenoids. Table 1 summarizes the results.

3. Conclusions
The models described contributed to study the high intensity low energy beam transfer line phenomenology and to improve the new method based on semi-analytical approach, which was very useful for real beam commissioning and characterization. This work was undertaken under the Broader Approach Agreement between the European Atomic Energy Community and the

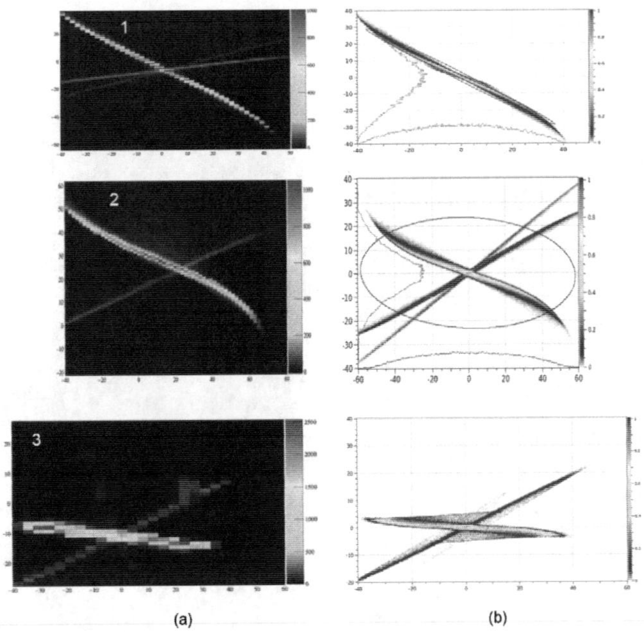

Figure 5: (a) measured emittances between the two solenoids for proton and deuteron species. (b) simulated emittances with the semi-analyitic approach. The numerical values are reported in Table 1.

Government of Japan. The views and opinions expressed herein do not necessarily state or reflect those of the Parties to this Agreement.

Table 1: Comparison between the simulated (semi-analytical approach) and measured phase spaces between the two solenoids. The I_{ext} refers to the total current extracted by the ion source. The index N refers to Fig. 5.

N	Sp.	Energy [keV]	I_{ext} [mA]	$\epsilon_{exp}/\epsilon_{sim}$	$\alpha_{exp}/\alpha_{sim}$	β_{exp}/β_{sim}
1	H^+	50	70	0.25/0.25	14.8/10.7	14.7/13.6
2	H^+	50	87	0.38/0.34	7.3/5.8	18.3/17.4
3	D^+	100	163	0.19/0.19	2.2/1.4	20.7/18.0

References

[1] Cara P *et al.* 2020 *Proc. FEC'20* TECH/4
[2] Gobin R *et al.* 2013 *Proc. IPAC'13* THPWO003
[3] Grespan F *et al.* 2020 *Proc. IPAC'20* TUVIR11
[4] Chauvin N *et al.* 2019 *Nuc. Fusion* **59** 106001
[5] Akagi T *et al.* 2020 *Rev. Sci. Instrum.* **91** 023321
[6] Uriot D and Pichoff N 2015 *Proc. of IPAC'15* MOPWA008
[7] Bellan L 2017 Ph.D. thesis Padova University
[8] Bellan L *et al.* 2018 *AIP. Conf. Proc.* **2011** 080013
[9] Comunian M *et al.* 2016 *Proc. HB'16* TUPM4Y01
[10] Gabovich M D *et al.* 1975 *Fizika Plazmy* **1** 304–308
[11] Grote D *et al.* 2005 *AIP. Conf. Proc.* **749** 55
[12] Noll D *et al.* 2016 *Proc. HB'16* WEPM8Y01
[13] Chauvin N *et al.* 2012 *Rev. Sci. Instrum.* **83** 02B320
[14] Gérardin F 2018 Ph.D. thesis Paris-Saclay University
[15] Winklehner D and Leitner D 2015 *JINST* **10** T10006
[16] Kalvas T *et al.* 2010 *Rev. Sci. Instrum.* **81** 02B703

Empirical study of multidimensional Child-Langmuir law with plasma ion source extraction using round apertures

S T Kosonen[1], T Kalvas[1], O Tarvainen[2] and V Toivanen[1]

[1]Department of Physics, University of Jyväskylä, P.O. Box 35, FI-40014 Jyväskylä, Finland
[2]STFC ISIS Pulsed Spallation Neutron and Muon Facility, Rutherford Appleton Laboratory, Harwell OX110QX, United Kingdom

E-mail: `sami.t.kosonen@jyu.fi`

Abstract. One dimensional Child-Langmuir (CL) law is commonly used in ion source physics to describe space charge limited ion extraction from the plasma. Recently 2D and 3D CL laws have been derived, but plasma ion extraction does not strictly meet the assumptions in these CL laws. Investigations on the applicability of the CL laws to ion source extraction were conducted using filament-driven ion source, measuring the beam currents as a function of extraction voltage. The experiments are complemented with simulations where IBSimu is used for studying the behaviour of plasma meniscus. Three different voltage regions were identified with low, moderate and high voltage. With moderate voltages the extracted beam current density shows signs of 3D CL law. The results indicate that for the total beam current CL laws are not valid but instead require collimation for the CL effects emerge.

1. Introduction

The classical Child-Langmuir (CL) law describes the maximum current density J for electron beam in a vacuum diode formed by inifinitely large planar anode and cathode with a given distance D and voltage difference V between them [1, 2]. The generalized 1D CL equation for the maximum current density is

$$J = \frac{4}{9}\epsilon_0\sqrt{\frac{2q}{m}}\frac{V^{3/2}}{D^2}, \tag{1}$$

where q and m are the charge and the mass of the particles and ϵ_0 is the vacuum permittivity. The physical interpretation of the CL law is that excessive space charge present in the diode gap forms a virtual cathode, which restricts the maximum current that can be extracted from the electrode. The 1D CL law assumes: (i) Indefinitely many particles to emit, (ii) particles do not have initial velocity, (iii) electrodes are infinite in the plane normal to the beam, and (iv) particle velocities are non-relativistic. The 1D CL law has recently been extented to 2D and 3D, i.e. the geometry of the emitter is taken into account [3, 4]. The 3D CL law is derived as an enhancement factor for the classical 1D case, i.e.

$$J_{3D} = J_{1D}(1 + F \times G), \tag{2}$$

where F is the mean-position factor (for non-relativistic beam $F = 1/4$) and G is the geometrical factor. For round emitter $G = \frac{D}{R}$, where R is the radius of the emitter. The 3D enhancement for the 1D CL equation is derived assuming that the particle flow is parallel and the emitter geometry is larger than the gap length, which is not always the case in plasma ion source extraction systems.

The classical 1D CL law is commonly used with plasma ion sources to describe the space charge limited ion emission at extraction voltages below the plasma density limited regime. However, ion source extraction is clearly not a 1D phenomena and even with the 3D expansion the dynamically changing plasma meniscus doesn't meet the criteria of fixed planar electrode in 3D CL law. Furthermore, the extracted ions do not have zero initial velocity as assumed in both 1D and 3D CL laws, but instead they are accelerated by the plasma sheath [5]. Here the meniscus is defined to be the equipotential surface where V is the plasma chamber voltage. The shape of the meniscus is determined by the extraction geometry, electric fields in the extraction region and plasma density and temperature. Nevertheless, it has been argued that the deviations from the assumptions can be considered small enough to apply the CL law for plasma extraction [6]. 3D CL effects on the beam current density extracted from a plasma ion source have been previously reported for a Penning type ion source with H^- beam [7].

In this study, the applicability of the 3D Child-Langmuir law to positive ion extraction is examined through experiments with different size round apertures, comparing the results to simulations. The beam currents were measured as a function of the extraction voltage in space charge and plasma limited regions.

2. Experimental setup and measurements

The experiments were carried out with a filament-driven DC-discharge multicusp ion source using argon plasma at the University of Jyväskylä Accelerator Laboratory [8]. Filament driven ion source was used as it produces highly repeateble and stable beams.

Two different extraction systems were built to study the 3D CL law effects on Ar^+ beam current density. Figure 1 **A** shows the first one, which is a grid based extraction consisting of the plasma electrode with a 3 mm diameter aperture, a fine mesh puller and a Faraday cup (FC) made of several concentric rings with 9 mm width and 1 mm spacing between them. The center of the cup is a round plate with 9 mm diameter. The gap length between the plasma electrode and the puller is 21 mm. This extraction system was designed to study the radial spread and distribution of the extracted beam, and to measure the total beam current, i.e. the sum of the currents of each ring. Figure 1 **B** represents the second system consisting of plasma and puller

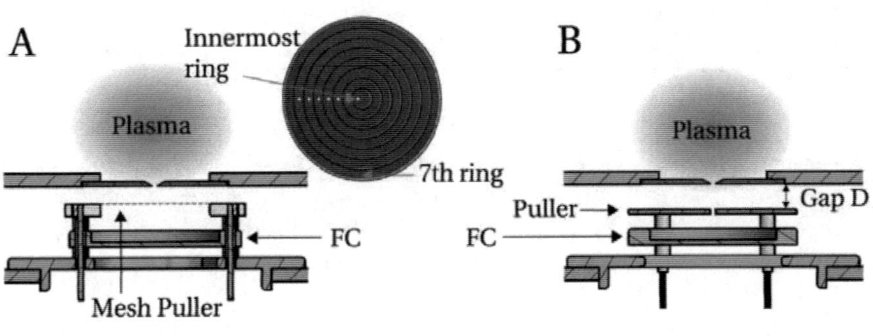

Figure 1: Extraction systems used in the study. **A** 1st extraction system with fine mesh puller and FC of concentric rings. **B** 2nd extraction system with equal diameter extraction and puller aperture.

electrodes with the same size round apertures ($R = 0.9, 1.8$ and 2.5 mm were used), and a FC measuring the beam current passing through the puller. The electrode apertures were chosen to be the same following the 3D CL theory which assumes that the emitter and collector have the same size and shape (beam has no transverse spread). The gap length was 16 mm which was chosen to achieve a flat plasma meniscus with reasonable extraction voltage (< 8 kV). This was inferred from extraction simulations. The beam current passing through the puller was measured as a function of the extraction gap voltage with several discharge currents. Discharge voltage was kept constant of 72 V, hence the plasma density was adjusted with discharge current.

The ion optical simulation code IBSimu [9] was used to study the behaviour of the extracted beam and the shape of the meniscus in simulations using the second extraction system geometry. The simulation parameters were chosen to match the known plasma parameters [10] and only the saturation current was matched to the measured value. After matching the saturation currents the extraction voltage was varied and the resulting current recorded. The voltage sweep was repeated with the same simulation parameters for all the different geometries (R of electrodes). The simulations provide insight especially to the changes in the meniscus geometry.

3. Results and discussion

Figure 2 **A** shows an example of the total beam current measurements with different discharge currents using the first extraction system (mesh puller) and ring FC. Higher extracted currents of Ar^+ were achieved with higher discharge current (plasma density). The beam current was always non-zero even without the extraction voltage, increasing rapidly when a low voltage was applied. After the initial jump the beam currents rise slowly, almost linearly, with the extraction voltage. It is clear that the measured total currents show no signs of the $V^{3/2}$ behaviour implied by the CL law in Eqs. 1 and 2. The beam currents recorded from different rings of the FC, shown in figure 2 **B** , reveal that with low extraction voltages the beam spreads in transverse direction. For example around 300 V the current from every ring is around 40 μA except for the innermost plate, which is about 20 μA. Increasing the extraction voltage focuses the beam and the currents near the center increase while the currents of the outermost rings decrease. With

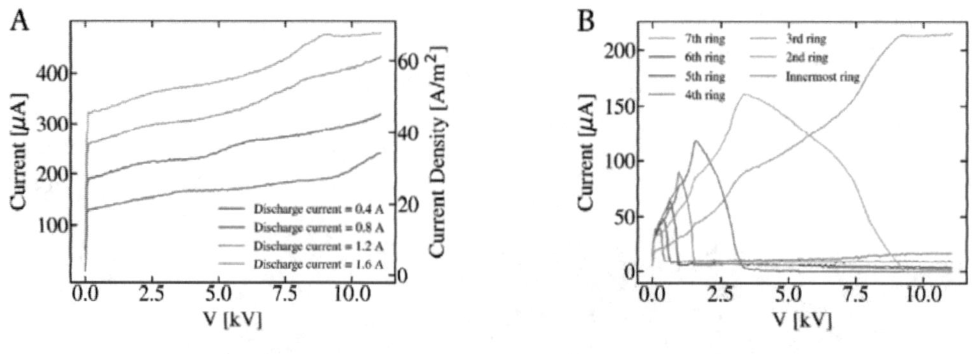

Figure 2: Measurements with first extraction system. Gap length is 21 mm and plasma aperture diameter 3 mm. **A** Total Ar^+ currents measured by the ring FC as a function of extraction voltage with different discharge currents. **B** Currents measured by the different ring sections of the Faraday cup as a function of the extraction voltage.

zero extraction voltage only a very small number of ions are autoextracted to the FC by the positive plasma potential. When the extraction voltage is applied a plasma sheath is formed in the proximity of the plasma electrode, which explains the abrupt increase of the beam current. Increasing the voltage further only moderately increases the total beam current due to the sheath accelerating more particles towards the plasma aperture. On the whole this experiment shows that CL law doesn't hold for the total current extracted from plasma ion source.

Figure 3 **A** shows the measured current densities with the second extraction system where the electrodes aperture sizes are matched and plasma densities are equal. All the current densities are calculated as the measured beam current divided by the extraction aperture area. These data reveal three extraction voltage regions: The emission (or plasma density) limited saturation currents are reached with high extraction voltage, the beam is space charge limited at moderate voltages and the current densities are almost equal at low extraction voltage. The main results are: (1) in space charge limited region bigger R leads to lower current density, (2) bigger R leads to higher saturation current density, not only current and (3) with bigger R the emission limited region is reached with higher voltage.

The first observation follows directly from the 3D CL law as for bigger R the geometrical factor G is smaller. Intuitively, in the space charge limited region, if the transverse charge distribution of the beam is constant, then bigger R means more charge and stronger space charge forces which leads to smaller beam current density.

The second result is explained by the behaviour of plasma meniscus: The location and, therefore, the area of the meniscus is more sensitive to voltage with larger aperture size (this point is elaborated on when discussing the simulations). So when comparing different extraction aperatures for example with high voltages the area of the concave meniscus has increased more with bigger R than with smaller R. This bigger meniscus harvests more ions from a larger effective area and explains the second result.

The third result follows from the previous two: with smaller R the beam current density is higher in the space charge limited region but lower in the emission limited region, so saturation current densities must be reached wih lower voltages (when compared to bigger R). After the saturation current densities are reached, increasing the extraction voltage results in larger area of the meniscus harvesting more ions from the plasma. At the same time the beam is overfocused and collimated, which reduces the measured current densities at high gap voltages, as seen in the measurement (see fig. 3 with $R = 0.9$ mm).

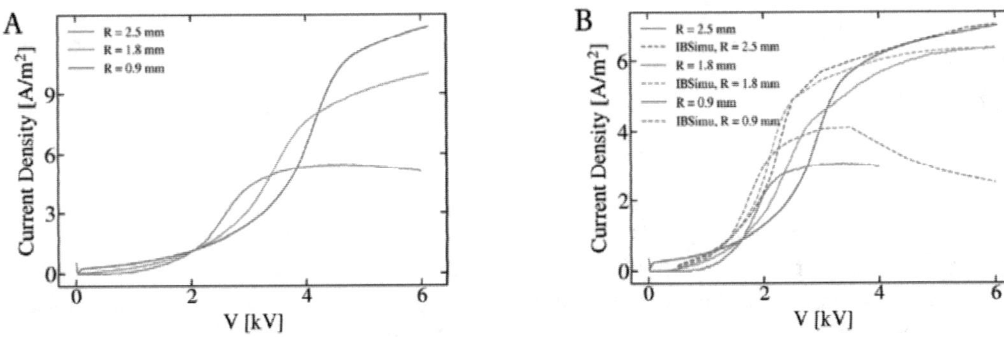

Figure 3: Measurements using second extraction system with gap length of 16 mm. **A** Measured currents densities as a function of extraction voltage. **B** Current densities compared with IBSimu simulations. Note that **A** and **B** have different discharge currents and plasma densities.

Figure 3 **B** compares IBSimu simulations to the measurements. Here the simulated beam current density was matched with measured emission limited current with 2.5 mm radius. For the $R = 1.8$ mm case the simulated saturation beam current densities are in good agreement with the measured ones. In the $R = 0.9$ mm case IBSimu overestimates the saturation currents slightly. This happens because the beam current density was matched with larger radius hence the effective areas of the menisci are not equal. In the space charge limited region the simulated current densities follow the CL law similar to the experiment, i.e. bigger R leads to slightly smaller current density. However, the space charge limited current densities are overestimated for all the radii. Regardless, these comparisons show that, overall, IBSimu reproduces qualitatively the trends observed in the experiment.

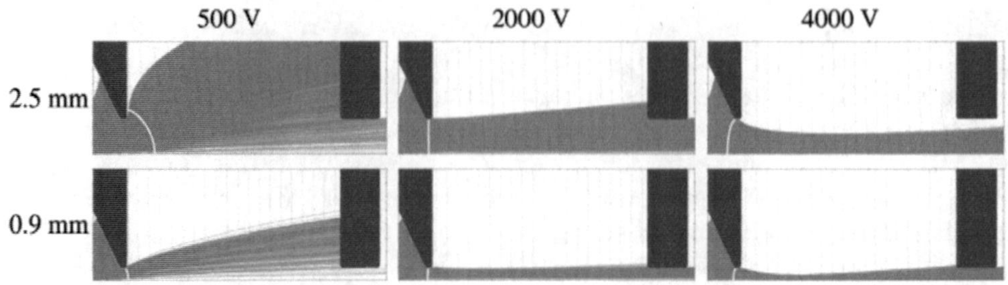

Figure 4: Simulation results, where the upper row is with $R = 2.5$ mm and the lower one with $R = 0.9$ mm and the gap length was 16 mm. Meniscus is highlighted with yellow arc.

Figure 4 shows examples of the simulations with the extraction voltages of 500, 2000 and 4000 V. With 500 V the beam is dispersing heavily with both apertures, though the meniscus is more convex with the $R = 2.5$ mm case than with $R = 0.9$ mm. At 2000 V almost all of the beam goes through the puller aperture in the $R = 0.9$ case while with larger radius the beam is collimated substantially. With 4000 V the beam is collimated only in the $R = 0.9$ mm aperture case. The aforementioned observation that with bigger R the meniscus is more sensitive to extraction voltages is clearly visible in the simulations. Furthermore, the area of the meniscus can be calculated with spherical surface approximation and then compared to the extraction aperture area. For example in the 4 kV case the area of the meniscus is only 4% bigger with $R = 0.9$ mm but 21% bigger with $R = 2.5$ mm i.e. the effective area of the meniscus changes more with bigger R.

4. Conclusions

The measurements in this paper show that the CL law is not applicable to the total beam current extracted from plasma ion sources, which is mainly due to the role of the plasma meniscus. Instead, a puller electrode with a beam limiting aperture is required for the CL law to emerge. In that case three voltage regions are found, the beam current being space charge limited only in the intermediate region.

The dependence of the beam current density on the aperture radius R in the space charge limited region is predicted by the 3D CL law, where bigger R leads to smaller current density. Direct comparison between theory and measurements are not done here, as the deviation of the plasma extraction from the CL law assumptions (meniscus vs. planar electrode) makes the comparison inaccurate.

Similar experiments are planned with rectangular apertures investigating the 3D CL effect on plasma ion sources even further. The experiments will be complemented by particle-in-cell simulations to study the plasma meniscus behaviour in the three voltage regimes identified here.

References

[1] Child C D 1911 Discharge From Hot CaO *Phys. Rev. (Series 1)* **32** 492 – 511
[2] Langmuir I 1913 *Phys. Rev.* **2** 450 –86
[3] Lay Y Y 2001 Simple Theory for the Two Dimensional Child-Langmuir Law *Phys. Rev. Letters* **87**
[4] Koh W S, Ang L K and Kwan T J T 2005 *Physics of plasmas* **12**
[5] Riemann K U 1991 The Bohm criterion and sheath formation *J. Phys. D: Appl. Phys.* **24** 493
[6] Rautenbach W L 1961 *Nuclear Instr. & Methods* **12** 169 – 79
[7] Faircloth D *et al.* 2020 *Rev. Sci. Instrum.* **91**
[8] Leung K N *et al.* 1991 *Rev. Sci. Instrum.* **62** 100 – 4
[9] Kalvas T, Tarvainen O, Ropponen T, Steczkiewicz O, Ärje J, and Clark H 2010 *Rev. Sci. Instrum.* **82**
[10] Hwang Y S 2006 *Rev. Sci. Instrum.* **77**

A Low Energy H^- Beamline for the ALPHA Antihydrogen Experiment

W. A. Bertsche[1], D. Faircloth[2], M. A. Johnson[1], T. Kalvas[3], S. Lawrie[2] and O. Tarvainen[2]

[1] School of Physics and Astronomy, The University of Manchester, Manchester M13 9PL, UK
[2] STFC ISIS Pulsed Spallation Neutron and Muon Facility, Rutherford Appleton Laboratory, Harwell OX11 0QX, UK
[3] University of Jyväskylä, Department of Physics, Survontie 9D, 40500 Jyväskylä, Finland

E-mail: mark.andrew.johnson@cern.ch

Abstract. The CERN ALPHA experiment makes precision measurements of antihydrogen atoms, confined in a superconducting magnetic minimum trap. Recent measurements of the antihydrogen spectrum have already provided high-resolution tests of fundamental symmetries, and ALPHA has now embarked on an ambitious upgrade programme aimed at directly comparing hydrogen and antihydrogen within their existing atom trap. One aspect of this upgrade will be the development of a low-energy (50 eV) hydrogen ion source that is compatible with ALPHA's existing magnetic charged particle beamlines. PELLIS, previously developed at JYFL, is a 5 keV filament-driven source that generates H^- beams with low emittances and currents of up to 50 μA. Here, we explore the feasibility of a proposed electrostatic beamline design to transport H^- ions from a PELLIS-type ion source into ALPHA's various particle traps. We present SIMION simulations that were used to develop the beamline, focusing on components such as a quadrupole switchyard and drift tube deceleration stage.

1. Introduction

The ALPHA experiment at the CERN Antiproton Decelerator makes precision measurements of antihydrogen ($\overline{\mathrm{H}}$) atoms confined in a superconducting magnetic minimum trap. Comparing the spectrum of the familiar hydrogen atom to that of its antimatter counterpart provides unique, high-resolution tests of fundamental symmetries, and may help to explain the dominance of matter over antimatter in our universe. In recent years, the ALPHA collaboration has made landmark measurements of several features in the antihydrogen spectrum, including its narrow 1S-2S transition [1] and 1S-2P Lyman-alpha transitions [2]. To improve on these measurements, ALPHA is pursuing an ambitious upgrade programme to carry out high-precision measurements of both hydrogen and antihydrogen atoms in the environment of its existing atom trap.

In a typical ALPHA experimental cycle, antihydrogen is produced by mixing around 10^5 antiprotons ($\bar{\mathrm{p}}$) with a cold plasma of 3×10^6 positrons (e^+) inside a nested Penning-Malmberg trap. The $\bar{\mathrm{p}}$ and e^+ are initially confined in separate Penning traps several metres apart, and must be transferred into one of ALPHA's two $\overline{\mathrm{H}}$ mixing traps before antihydrogen synthesis can take place. At ALPHA, charged particles are transferred between Penning traps as low-energy (50 eV) pulsed beams, produced by rapidly (1 μs) manipulating the electric potential along the axis of one trap so that particles are ejected in one direction. These $\bar{\mathrm{p}}$ and e^+ beams are born in strong axial magnetic fields, and have a number of properties that make them difficult to transport using conventional magnetic or electrostatic beamlines. Instead, positrons and antiprotons are guided between ALPHA's various particle traps through a series of solenoid magnets, which provide continual steering and focusing in the transverse plane. At low energies, the $\bar{p}$ and e^+ will adiabatically follow the direction of the magnetic field, and so steering can be achieved using a region of curved magnetic field lines [3].

In order to produce hydrogen atoms using techniques similar to those described above, a source of low-energy protons or H^- ions would need to be added to the ALPHA apparatus. This ion source should be compatible with ALPHA's existing magnetic beamlines, so that low-energy ions can be transported into either of the $\overline{H}$ mixing traps and and used to form atomic hydrogen. The PELLIS ion source [4], previously developed at the University of Jyväskylä (JYFL), is a 5 keV filament-driven source that generates H^- beams with low emittances and currents of up to 50 μA. PELLIS could potentially be used as an H^- source for ALPHA due to its modest beam current and low extraction energy. However, this ion source cannot be connected directly to the existing ALPHA experiment, and its integration presents a number of challenges.

In this work, we explore the feasibility of using an electrostatic beamline to transport H^- ions from a PELLIS-type ion source into ALPHA's various particle traps. We present SIMION simulations that were used to develop and characterise the proposed beamline design, focusing on key elements including an electrostatic quadrupole deflector and decelerating drift tube section.

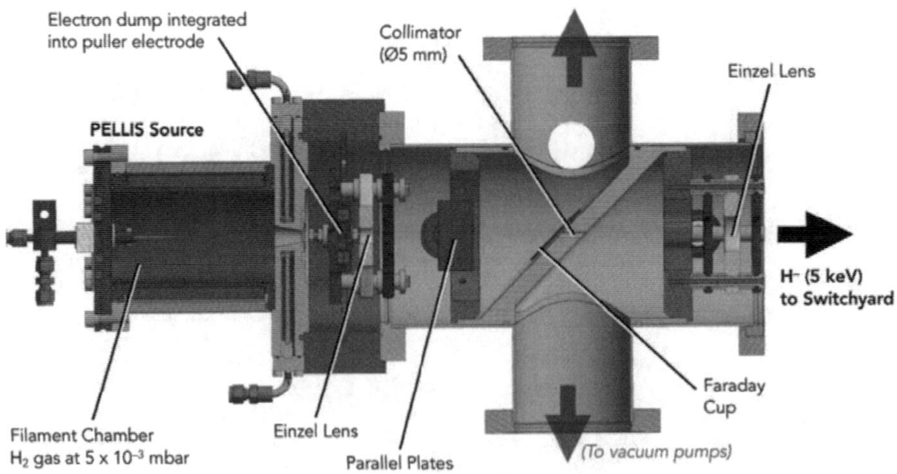

Figure 1. Diagram showing a cross-section of the PELLIS ion source, together with its extraction optics installed inside the differential pumping chamber.

2. PELLIS Source

The PELLIS ion source (shown towards the left of Figure 1) was originally developed by Kalvas et al. [4] to provide low-emittance H^- beams for the Pelletron accelerator at JYFL. A detailed description and characterisation of the PELLIS source itself is beyond the scope of this work, and has already been given elsewhere. However, we will give a brief overview of the H^- source parameters and extraction optics as pertains to this study.

PELLIS is a filament-driven multicusp ion source that produces H^- beams with energies in the range of 5 – 15 keV. Typically, the filament chamber is filled with H_2 gas at a pressure of 5 $\times 10^{-3}$ mbar, such that a filament (arc) current of 5.6 A will produce 50 μA of H^- beam current. In this study, the source and its extraction optics have been configured to operate at a nominal energy of 5 keV. The H^- beam emerges from the 2 mm plasma electrode aperture with a 95% normalised rms emittance of 0.015 mm mrad.

Figure 1 shows the PELLIS source and its associated extraction optics, which have been modified slightly for this application. Beam extraction simulations were carried out using IBSIMU [5]. Several electrostatic elements are integrated into a custom vacuum chamber, which acts as the first stage of a differential pumping system between PELLIS and ALPHA's existing beamline. A 20 mm long, 5 mm aperture at the centre of the chamber acts as both a collimator for the H^- beam and a pumping restriction between the ion source and the rest of the beamline.

Electrons that are co-extracted with the H^- beam are deflected into an electron dump using a pair of permanent magnets, which produce opposing dipole-antidipole fields. The ion beam is largely unaffected by these deflector magnets, but emerges from the puller electrode at a small (~5 mrad) angle to its original velocity. This offset will be corrected using a set of parallel plates approximately 128.5 mm from the plasma electrode. Pulsing the voltage on the parallel plates will chop the H^- beam into short pulses, so that ions are only delivered to ALPHA when required. A typical pulse will be around 1 μs in duration, corresponding to a bunch charge of 50 pC or $\sim 3.1 \times 10^8$ H^- ions. A Faraday cup mounted around the collimator will be used to monitor the beam current between shots.

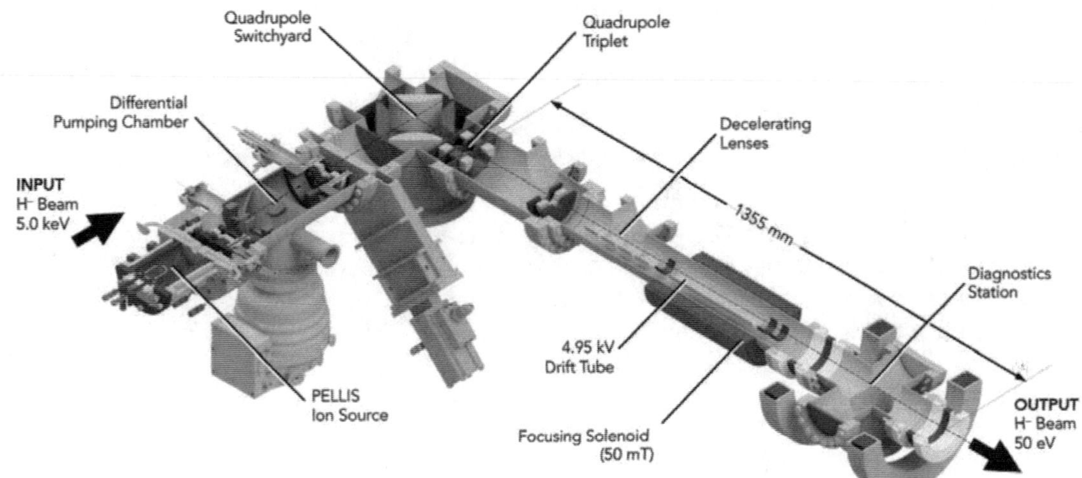

Figure 2. Diagram showing a cross-section of the proposed beamline design to transport H^- ions into the ALPHA experiment from a PELLIS-type ion source.

3. Beamline Design

Figure 2 shows a cross-section of our proposed beamline design to decelerate H^- bunches and transport them into the ALPHA experiment. The PELLIS source is connected to the main beamline through an electrostatic quadrupole deflector (the 'switchyard'). The switchyard is intended to maximise the flexibility of the beamline, so that additional ion sources and particle traps can be installed in future without further modifications to the apparatus.

The four electrodes that form the quadrupole switchyard each have a radius of 80 mm, and an inscribed radius of 53.3 mm. Based on an analytical model for an ideal quadrupole deflector [6], the required field strength for a 5 keV H^- beam is 0.864 V/mm^2, corresponding to electrode voltages of $\pm$2.46 kV. Simulations of the transport beamline, including the switchyard, were carried out using SIMION [7]. Figure 3 shows the simulated trajectories of 5 keV H^- ions through the quadrupole deflector. Within the switchyard, space charge effects were modelled

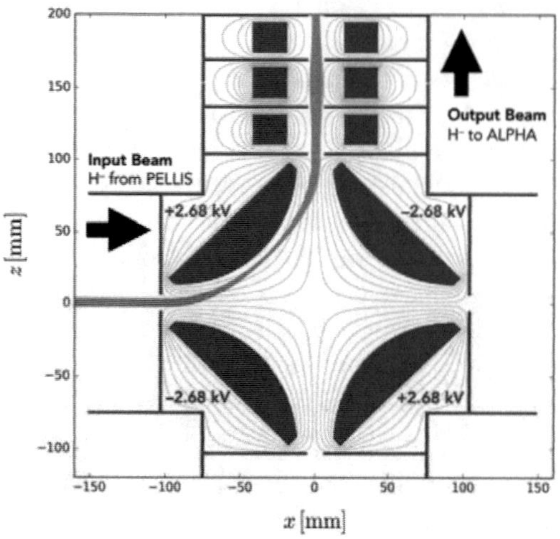

Figure 3. SIMION simulation showing the trajectories of 5 keV H$^-$ ions through the quadrupole switchyard. The contour lines show the electric potential in 400 V intervals.

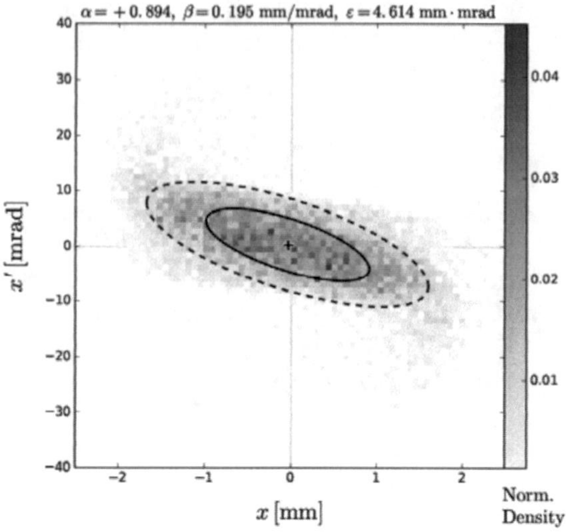

Figure 4. Simulated transverse (horizontal) phase space of the H$^-$ beam after the final focusing quadrupole.

using SIMION's built-in beam repulsion method. In this simulation, the electrode voltages were increased to ±2.68 kV to compensate for field errors around the input and output of the deflector, so that the beam is well-aligned as it leaves the switchyard.

Upon leaving the switchyard, the H$^-$ beam is no longer circular, with an rms radius of 0.86 mm along its horizontal axis, and 1.40 mm along its vertical axis. An electrostatic quadrupole triplet was added after the switchyard to correct this, and to match the two transverse phase spaces of the beam. The beam emerges from the quadrupole triplet with a slightly converging envelope and an rms beam radius of 0.96 mm. Figure 4 shows the transverse (horizontal) phase space immediately after the final quadrupole.

Before entering ALPHA's existing beamline, H$^-$ pulses are decelerated to ~50 eV using a 250 mm long drift tube, as shown in Figure 2. The drift tube voltage is switched from 4.95 kV to ground within 100 ns as the beam transits along its length. Since the ions lose 99% of their initial kinetic energy upon entering the drift tube, care must be taken to avoid blow-up of the beam profile inside the narrow (20 mm radius) electrode. A series of decelerating lenses were added before the drift tube to prevent large transverse electric fields from rapidly defocusing the beam. Figure 5 shows the configuration of these lenses and their effect on the H$^-$ trajectories calculated using SIMION. Within the drift tube, space charge effects were represented using SIMION's built-in particle-in-cell solver.

To produce atomic hydrogen using H$^-$, ions must be introduced into one of ALPHA's two $\overline{\text{H}}$ mixing traps, where there is a strong (1 T) axial magnetic field. To avoid losses due to magnetic mirroring, H$^-$ pulses must therefore transit from the electrostatic portion of the beamline into ALPHA's existing magnetic beamlines with very small transverse velocities. As shown in Figure 2, the drift tube is enclosed by a solenoid magnet that generates an axial field of ~50 mT. This solenoid provides transverse focusing for H$^-$ bunches in transit along the drift tube, and also helps to guide ions into the magnetic field of ALPHA's existing beamline. The voltages applied to the decelerating lenses were optimised to limit the transverse energies of H$^-$ within the drift tube, thereby minimising the number of ions lost due to magnetic mirroring.

Upon leaving the drift tube, each 50 eV H$^-$ pulse has a normalised emittance of

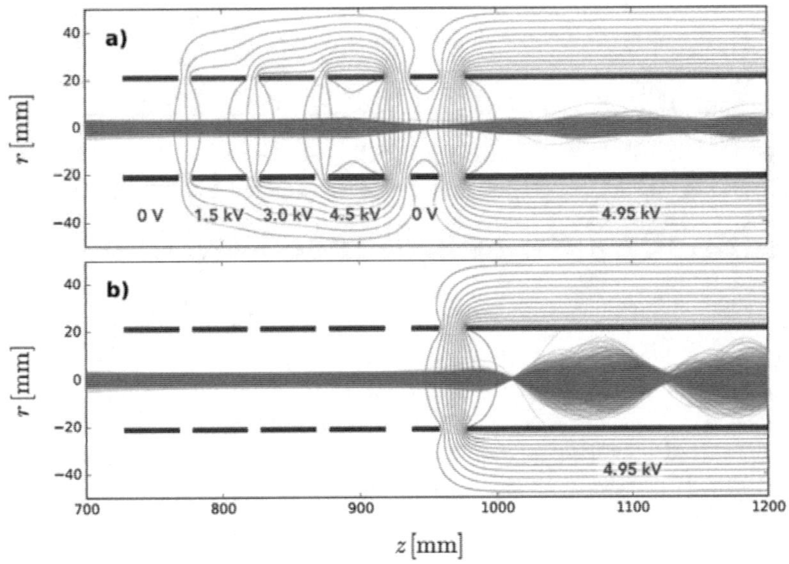

Figure 5. SIMION simulation showing H^- ion trajectories through the decelerating lenses and drift tube a) with the lenses turned on and b) with the lenses grounded. The contour lines show the electric potential in 400 V intervals.

0.019 mm mrad, indicating some emittance growth during deceleration. The final beam radius is on the order of 1 mm and can be adjusted by changing the magnetic field within the drift tube section. Extending the simulation to track ions into either of ALPHA's $\overline{\mathrm{H}}$ mixing traps suggests that only $< 1\%$ of ions are lost in transit, primarily due to magnetic mirroring effects and ballistic losses within the drift tube section.

4. Summary and Outlook

In these proceedings, we have briefly outlined the motivations for adding an H^- source to the ALPHA antihydrogen experiment. We have explored the feasibility of an electrostatic beamline design to transport 5 keV H^- ions from a PELLIS-type ion source into ALPHA, and to decelerate the beam to a final energy of just 50 eV. Using extensive SIMION simulations, we have shown that the beamline meets its design criteria and maintains the low emittance of the PELLIS source. Initial assembly of the ion source and differential pumping chamber are scheduled for November 2021, with commissioning to take place on the VESPA test stand at STFC ISIS [8].

References

[1] Ahmadi M *et al.* (ALPHA Collaboration) 2018 *Nature* **557** 71–75
[2] Ahmadi M *et al.* (ALPHA Collaboration) 2018 *Nature* **561** 211–215
[3] Johnson M 2019 *Design and Commissioning of Beamlines for the ALPHA Antihydrogen Experiment* Ph.D. thesis Department of Physics and Astronomy The University of Manchester, UK
[4] Kalvas T, Tarvainen O and Komppula J e a 2013 *AIP Conference Proceedings* **1515** 349–358
[5] Kalvas T, Tarvainen O, Ropponen T, Steczkiewicz O, Ärje J and Clark H 2010 *Rev. Sci. Instrum.* **81** 02B703
[6] Zeman H D 1977 *Rev. Sci. Instrum.* **48** 1079–1085
[7] Manura D J and Dahl D A 2006 *SIMION 8.0 User Manual* Scientific Instrument Services Inc. Ringoes, NJ 08551, USA
[8] Lawrie S, Faircloth D, Letchford A, Perkins M, Whitehead M and Wood T 2014 Installing the VESPA H^- Ion Source Test Stand at RAL *5th International Particle Accelerator Conference*

Mass distributions of ions extracted from target cavity confined plasmas produced by nanosecond laser pulses

J E Hernandez II and M Wada

Graduate School of Science and Engineering, Doshisha University, Kyotanabe, Kyoto 610-0321, Japan

E-mail: cyjd3302@mail4.doshisha.ac.jp

Abstract. Ion and electron current signals produced from a solid graphite target under 1064 nm nanosecond laser pulse irradiation were investigated. A cavity is formed by laser incidence at the center of the target surface at 10 Hz repetition rate while being rotated along its axis. Ion mass distributions were recorded via a time-of-flight spectrum coupled with a detector assembly attached along the propagation axis. Peak shifts towards slower times were observed for increasing number of pulses. The effect of magnetic detection on the ion current signals is investigated, and showed significant peak shifts for 20 GW/cm^2 laser power density.

1. Introduction

Pulsed laser ablation has been utilized as a source of positive and negative ions which can be singly or multiply charged [1-3]. One application of nanosecond laser ablation as an ion source is the extraction of highly charged ions such as C^{6+} produced by direct laser injection on the target [4]. Laser irradiation above the target ablation threshold involves the plasma plume production consisting of a broad species distribution of ions and electrons whose motions are directed perpendicular to the target surface. Fast electrons first escape the target which induces an electric field to extract ions, accelerating them further away from the target. Ion production via laser ablation mainly depends on the laser parameters such as intensity and wavelength, as well as background environment such as ambient pressure [5,6]. Ions and electrons accumulate near the target surface at higher ambient pressure and collide with gas particles in which their propagation is slowed down, while in high vacuum conditions the expansion is unconstrained. Aside from the gas phase ambient, the propagation pathway of the laser plasma plume can be hindered by the target itself. When the laser drills a hole on a solid target, the target cavity walls can act as obstacles of plume expansion. As the plasma is initiated deep within the cavity, the wall promotes collisions between ions and effectively cools down the plasma. However, the mechanisms of ion production under the condition of cavity confinement have yet to be fully investigated. In this work, constituent ions within the plasma produced via nanosecond pulsed laser ablation ion source under expansion on a fresh target are compared with that under target cavity confinement. Ion distributions are characterized by time-of-flight spectroscopy. The influence of magnetically steering the ion beam trajectory on the collected ion signals is also investigated via a deflection assembly placed along the ion flight path.

2. Experimental details

The laser ablation ion source consists of a Q-switched Nd:YAG laser ($\lambda = 1064$ nm, 5 ns pulse width) incident on a cylindrical graphite target at 12 degrees from the target surface normal axis. The source geometry with the beam diagnostics unit is schematically shown in figure 1(a). The laser light is focused at the center of the target surface, forming a spot size of approximately 1 mm diameter. An electrostatic lens assembly that can form as an energy selector consisting of an aluminum bias plate between two grounded plates is shown in figure 1(b).

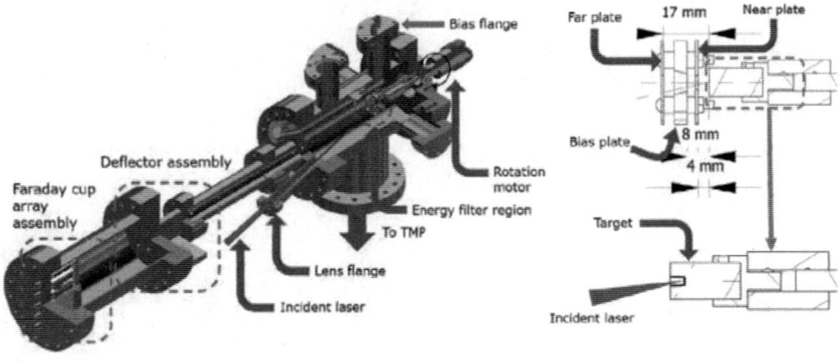

(a) (b)

Figure 1. (a) Laser ion source with beam diagnostics (b) Electrostatic lens assembly schematic

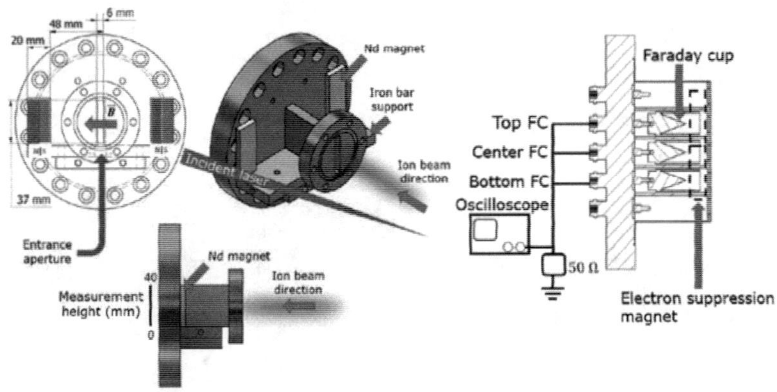

(a) (b)

Figure 2. (a) Deflector assembly schematic (b) Structure of the Faraday cup array

The ion beam trajectory is steered via a magnetic deflector attached in front of the flight tube, whose schematic is shown in figure 2(a). The magnetic flux density at the exit slit with 37 mm aperture is measured at approximately 80 Gauss. Current signals due to ions from the laser plasma are recorded using a Faraday cup array placed 650 mm from the target surface whose schematic is shown in figure 2(b). Permanent magnets of 300 Gauss magnetic flux density are attached at a position 5 mm from the entrance of the Faraday cup array to suppress incoming electrons. During continuous laser ablation at 5 GW/cm^2 laser intensity, the target is rotated at a rate of 18 revolutions per minute using a DC motor. The signals collected were averaged over 16 shots. Experiments were performed inside a vacuum chamber at 6×10^{-6} Pa base pressure.

3. Results

Figure 3 shows the time-of-flight (TOF) signals of the Faraday cup (FC) array for irradiation for the fresh target without the cavity. For increasing laser intensities shown in figure 3(a), increase in peak current signal at $t = 4.5$ μs was approximately 400 μA for the top and bottom Faraday cups, while 600 μA increase was observed for the center FC. Shifts in the signal rise to earlier times were observed, with approximately 1 μs difference shift to earlier times comparing the ion signals due to 5 to 20 GW/cm^2 laser intensity ablation. With the applied deflection field, current increase also results to peak shifts to earlier times with a time difference of 1.7 μs from the top FC to the bottom FC peak, as shown in figure 3(b). Figure 3(c) shows the signal overlap for cases with and without the magnetic deflection field. Around 600 μA current decrease was observed for both top and bottom signal peaks, while approximately 700 μA reduction was shown for the center FC. Peak shifts were also observed for the signals, where the peak with the highest current amplitudes were from 6, 8, and 4 μs from top to bottom FCs, respectively.

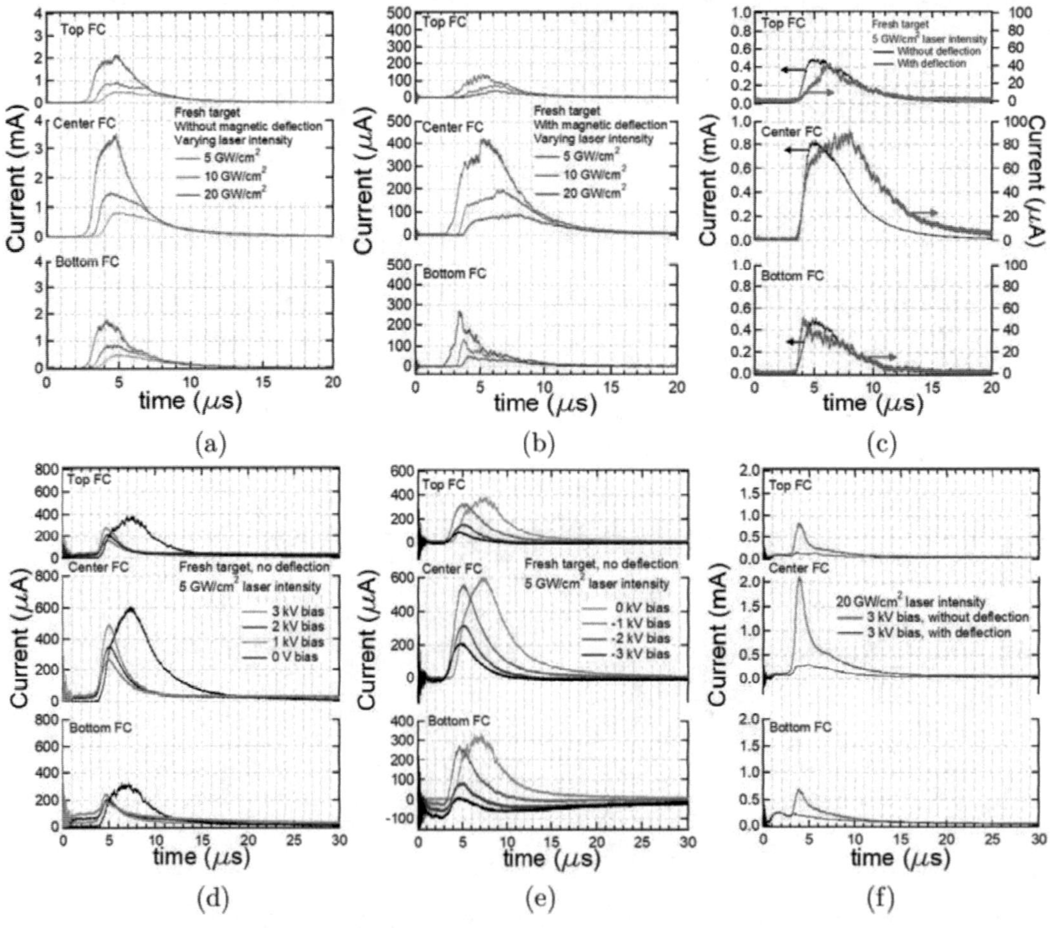

Figure 3. Ion time-of-flight signals for fresh target condition (a) Faraday cup array signals for varying laser intensity (b) Array signals with and without magnetic deflection assembly for varying intensities (c) Overlap of current signals for 5 GW/cm^2 laser intensity (d) Ion signals for increasing electrostatic filter bias from 0 V to 3 kV (e) Ion signals from 0 to -3 kV filter bias (f) Ion signal overlap for 20 GW/cm^2 laser intensity and 3 kV lens bias with and without magnetic deflection field

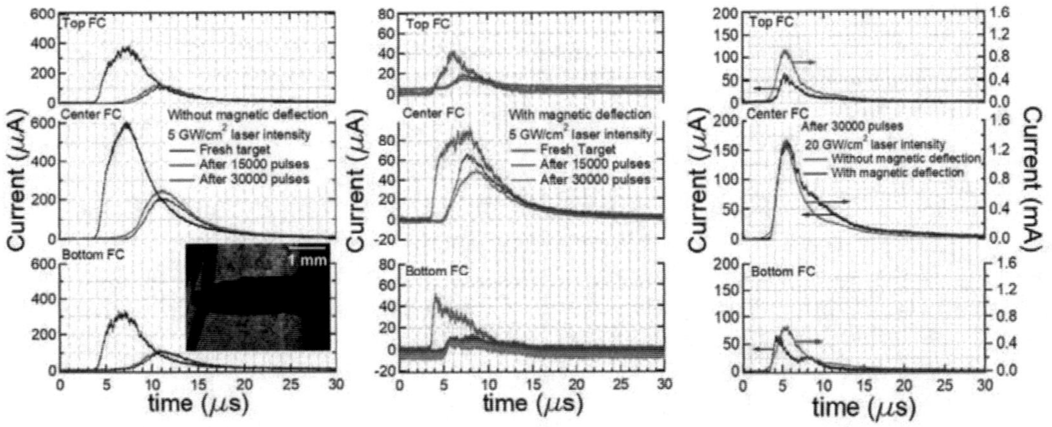

(a) (b) (c)

Figure 4. Ion signals due to fresh target ablation and after 5 GW/cm^2 drilling intensity ablation (a) Ion current signals for varying number of pulses, inset shows the cavity shape after 30000 pulses (b) Array signals under magnetic deflection for varying number of pulses (c) Comparison of ion signals with and without magnetic deflection at 20 GW/cm^2 laser intensity after 30000 pulses

Figures 3(d) and (e) show the effect of increasing and decreasing electrostatic filter bias, respectively. For both increasing and decreasing bias, the peak shift from 7 to 5 μs is shown. Further increasing or decreasing the filter bias higher than 1 kV or lower than -1 kV is not observed to significantly increase the peak shift. The shifted peak to 5 μs decreases, while signals earlier than 4 μs increased or decreased by approximately 100 μA for increasing and decreasing bias voltage, respectively. Figure 3(f) shows the effect of deflection field on the ion signals at 20 GW/cm^2 intensity. Decrease of about 600 μA, 1 mA, and 400 mA are observed on the top, center, and bottom FC signals are observed, respectively, where the decrease is shown after approximately 3 μs. Figures 4(a) and (b) show the effect of number of pulses on the TOF signals with and without magnetic deflection. After 15000 pulses, peak current amplitudes for all FCs reduce to approximately half of that the fresh target condition. Furthermore, 7 to 11 μs peak shift occurred from the first 15000 pulses, where the shift is not significant after 30000 pulses. Peak current decrease of 200 μA for the top and center FC, and 400 μA decrease are shown for the bottom FC. For 20 GW/cm^2 intensity, the effect of deflection field on the TOF signals is shown in figure 4(c). Peak current decrease of 1 mA is shown for the center FC, while 500 and 800 mA decrease is observed for the top and bottom FC, respectively. Peak shifts were not observed for the top and center FC, while two peaks with 4 μs and 8 μs amplitude were shown for the bottom FC signal.

4. Discussion

Plasma initiated from laser vaporization at the fresh target results to uninhibited motion of fast electrons and ions away from the surface. Therefore, ions from the plasma traverse freely towards the Faraday cup array. Increasing the laser power density also increases the ionization probability, thus increasing the positive current peak. Peak shifts towards earlier times are due to increased laser energy input which increases the ion velocity. From figure 3(a), a 1 μs shift to earlier times due to an increase from 5 to 20 GW/cm^2 laser intensity, with rise time of 4.6 μs would result to 30% increase in ion velocity, assuming an energy-dependent mass distribution.

Under the influence of magnetic field for the fresh target as shown in figure 3(c), ion signals in all the FC arrays represent broad energy and mass distributions entering the collectors. Ion signals collected at the top FC which peak at 6 μs TOF correspond to singly charged molecular carbon ions, or cluster ions. On the other hand, ion distribution with a 4 μs peak is detected at the bottom FC, where these suggest singly and multiply charged atomic carbon due to earlier TOF. The center FC represents the molecular carbon ions whose motions were undeflected by the magnetic field. Increasing or decreasing the electrostatic lens bias filters the ions which can pass through the lens aperture. Since the laser plasma involves the production of fast electrons followed by slower ions, the initial electrons form an electric field with the ions causing a charge separation near the target surface [7]. The increase in positive bias on the lens plate serves as a collector of the fast electrons while decelerating the positive ions. Then, the positive ions exceeding the bias potential are accelerated via the field induced between the bias plate and the front plate. The opposite case occurs with the negative bias, where the positive ions with energy below the bias potential are collected by the negatively charged plate. Acceleration of these ions is shown on the signal peak shift of 2 μs towards earlier TOF in figures 3(d) and (e).

The current decrease after subsequent pulses is due to the constricted ion propagation when the plasma is initiated within the cavity. Increase in collisions within the plasma constituents and the cavity walls result to energy losses shown in the shift towards later times, as shown in figures 4(a) and (b). Further increase in the number of pulses reduces the rate of current decrease due to the stagnation in the rate of drilling with the formation of the deep hole [8]. With the application of magnetic field, current amplitude decrease for the top and bottom FCs after 15000 and 30000 pulses is less than 10 μA, whose peak at 7 μs is approximately half of that measured from the center FC which indicates fewer deflected ions from the target axis. As the laser power is raised to 20 GW/cm^2 after 30000 pulses irradiation, the current waveforms are shifted towards later and earlier times for the top and bottom FCs, respectively. Shift towards later times may occur due to the influence of the collision-induced energy losses, while the shift to earlier times, as well as the presence of a broad distribution after $t = 7$ μs suggests the presence of negative ions as they propagate opposite the Lorentz force direction. The ion peak at $t = 5.5$ μs at the center FC shifts to approximately $t = 0.5$ μs, suggesting these ions along the axis are not significantly affected by the deflection field.

5. Summary

The effect of cavity formation on the current signals of target-induced confined cavity laser ablation plasmas is investigated. A deflection magnet assembly is installed to characterize propagating ions from the target surface. Current signals increase with laser intensity and decrease with increasing the number of pulses. Current decrease, as well as shifts in later times were observed for ion signals after a cavity is formed after repeated pulse ablation. Further characterization of the TOF distribution is underway to resolve the masses from the broad ion distributions in the laser plasma.

6. References

[1] Munemoto N, Takayama K, Takano S, Okamura M and Kumaki M 2014 *Rev. Sci. Instrum* **85** 02B922
[2] Balki O, Rahman Md. M, and Elsayed-Ali H 2018 *Opt. Comm.* **412** 134-140
[3] Okamura M, Stifler C, Palm K, Steski D, Ikeda S, Kumaki M and Kanesue T 2016 *Rev. Sci. Instrum* **89** 02A906
[4] Yamaguchi A, Sako K, Sato K, Hayashizaki N and Hattori T 2014 *Rev. Sci. Instrum* **32** 02B921
[5] Nemes P and Vertes A 2007 *Anal. Chem.* **79** 8098-8106
[6] Verhoff B, Harilal S S, Freeman J R, Diwakar P K and Hassanein A 2012 *J. Appl. Phys.* **112** 093303
[7] Krasá J Delle Side D Giuffreda E and Nassisi V 2015 *Laser Part. Beams* **33** 601-605
[8] Kononenko T V, Klimentov S M, Konov V I, Pivovarov P A, Garnov S V, Dausinger F and Breitling 2001 *Phys. Rev.* B **39** 3317

Effects of axial magnetic field in a magnetic multipole line cusp ion source

Bharat Singh Rawat[1,2], S K Sharma[1,2], B Choksi[1], P Bharathi [1], B Sridhar [1], L N Gupta[1], D Thakkar[1], S L Parmar[1], V Prahlad [1] and U K Baruah[1,3]

1. Institute for Plasma Research, Bhat, Gandhinagar, Gujarat - 382428, India
2. Homi Bhabha National Institute, Anushaktinagar, Mumbai, Maharashtra - 400094, India
3. ITER-India, Koteshwar, Ahmedabad, Gujarat -380005, India

Email:bharat.rawat@ipr.res.in

Abstract. Experiments have been performed to study the effects of axial magnetic field on plasma and ion beam parameters in a magnetic multipole line cusp ion source. Studies performed on ring cusp and Kaufmann type ion sources suggest that the axial component of magnetic field may help in improving the ion extraction current at a low discharge power. The line cusps ion sources have been used since long to produced uniform beams, however, they lack axial component of the magnetic field. These ion sources generally suffer from low efficiency possibly due the absence of axial magnetic field. In this work, an additional magnetic coil is added at various axial positions between the back-plate and the plasma grid of the multipole line cusp ion source. We have investigated the effects of axial magnetic field on the discharge efficiency and source parameters like beam current and uniformity in the multipole line cusp ion source. The beam profiles are obtained using an eleven-channel faraday cup array to estimate the effects of the axial magnetic field on beam uniformity and divergence. Initial studies suggest a reduction of beam divergence with increasing axial magnetic field. A significant rise in the beam current and the discharge current is also observed when the axial magnetic field is increased. Particle trajectory simulation using the CST-Studio suite is utilised in understanding the role of confinement of primary electrons behind the improved performance of the ion source.

1. Introduction

Multipole line cusp ion sources have been utilised in various applications such as neutral beams injectors, high energy accelerators and ion beams for surface modification and ion thrusters[1,2]. One of the key features of the line-cusp ion sources is its ability to generate uniform beam plasma profile due to its field free region at the centre. The line cusp configuration is not utilised in ion thrusters mainly due to low arc efficiency, which is the ratio of required arc power and beam current (0.5 – 1 A/kW). Although ring-cusp ion sources have poor plasma uniformity as compared to other configurations, they are commonly used as ion thrusters due to their higher arc efficiency (4 A/kW) [3]. Ring cusp and Kaufmann type ion sources have an axial component of the magnetic field present, which helps in improving the arc efficiencies. On the other hand, the line cusps ion source lacks the axial component of the magnetic field. In this work, we have investigated the effect of axial magnetic field produced by the coil on the arc efficiency in the line-cusp ion source.

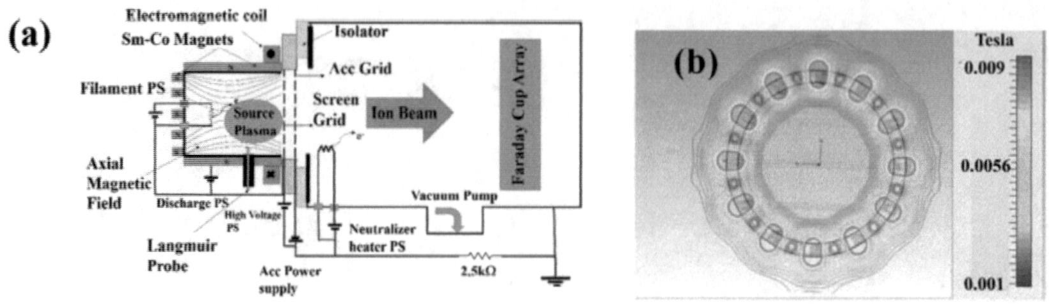

Figure 1: (a) Schematic of the experimental set up of ion source (b) Contours of magnetic field in line cusp configuration.

2. Experimental setup

The ion source system is comprised of arc chamber, extraction grids, hot filament neutralizer, vacuum vessel for beam transport (Figure 1(a)).The arc chamber of the ion source is a water-cooled cylindrical chamber of both diameter and depth of 200 mm made-up of stainless steel. A hot tungsten filament cathode of diameter 1.0 mm and length 100 mm is used to emit energetic electrons for ionizing the background gas (Argon). Magnetic multipole line cusp configuration is utilised for the confinement of the plasma and the primary electrons which ensures a magnetic field free region (< 10 gauss) over a circular area of diameter about 100 mm near the extraction grids(figure 1 (b)). A magnetic coil of 40 turns is utilised to produce axial magnetic field of 50-100 G near the grid region. The ion source has been operated in both modes i.e. with and without the axial field. The ion extraction system comprises of screen grid and the acceleration grid made of SS304 with a thickness and inter-grid spacing of 1mm each having 1050 apertures. The diameter of the screen grid and acceleration grid apertures is chosen to be 2mm and 1.2mm respectively. The beam extraction is carried out by fixing the acceleration grid at around -350V and varying the screen grid potential at 800 and 1650V. A Pfeiffer make turbo molecular pump is used for producing a base vacuum of $\leq 5\times10^{-6}$ torr.

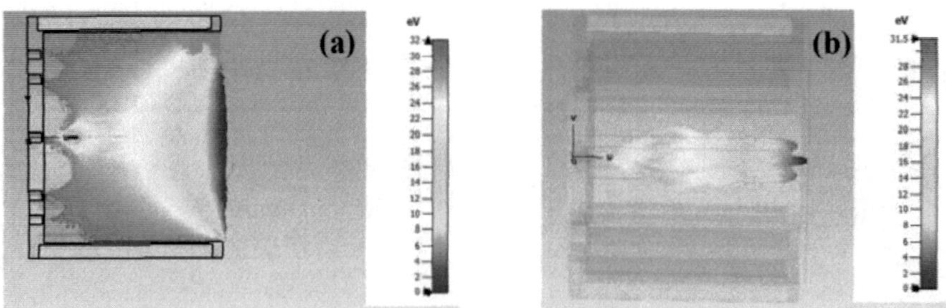

Figure 2: Trajectories of primary electrons in a line cusp magnetic configuration with (a) axial magnetic field off, and (b) axial magnetic field on

3. Theoretical studies on improvement of plasma parameters with axial magnetic field

Magnetic coils of 40 turns are mounted near the screen grids and back plate of an existing ion source with line cusp configuration. The currents in the coils can be varied from 0 – 20 A to produce axial magnetic field of different strength. The direction (configuration) of the axial magnetic field can also be changed by reversing the direction of coil current. The experiments have suggested that the axial magnetic field directed towards the screen grid produced by the coil fixed near the screen grid has most noticeable effect on the ion source parameters. Therefore, the studies only with this configuration are included in this work. The simulations of primary electron trajectories in CST studio may give an idea about the role of axial magnetic field in the confinement of primary electron[4]. Figures 2(a) and (b) shows the trajectories of primary electron produced by the simulation using CST without and with axial magnetic field, respectively. The energy (in eV) mentioned in the plots is basically the energy of electron at a particular position while moving along the trajectory. In this simulation, the electron source is considered at the centre of ion source. In case of no coil current, the magnetic field at the position of filament is < 10 G and with coil current it is about 50-100G. It can be observed that the electrons are lost in all direction in the absence of axial magnetic field and are confined in a narrow region along the axis when the axial field is present. The loss area of the primary electrons is greatly reduced hence; the ionisation can take place only in a central narrow region of the plasma chamber. The plasma ions cannot be confined by the axial magnetic field but they diffuse along the perpendicular direction at a slower rate. The available CST module does not support the modelling of plasma and hence diffusion of the ions is estimated analytically. The confinement of the plasma ions can be understood by the ion diffusion coefficient in a plasma under axial magnetic field. The diffusion coefficient of ions in presence of magnetic field is given by equation (1) [5].

$$D_+ = \frac{D}{(1+(\omega_c\,\tau)^2)} \qquad (1)$$

Where D is the ion diffusion coefficient in absence of magnetic field
ω_c is the ion cyclotron frequency
τ is the mean collision time between ion and neutrals

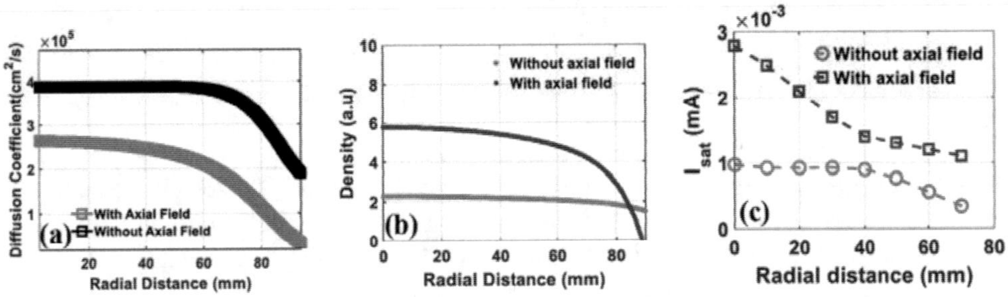

Figure 3: (a) Estimated radial profiles of diffusion coefficients of ions, (b) plasma density, and (c) experimental ion saturation current profiles at source pressure of ~ 2mTorr.

The diffusion coefficient of ions in line cusp configuration with and without axial magnetic field using eq. 1 are plotted in figure 3(a). The values of diffusion coefficient as a function of radial position are calculated using the magnetic field produced by the coils and permanent magnets in both cases. The magnetic field values are estimated using CST. The profiles of diffusion coefficient of ions for both cases are flat in the central region (Fig. 3(a)). The ion diffusion coefficient is reduced by a factor ~1.4 by applying the axial magnetic field. The decreased diffusion coefficient suggests that the ions can be confined in a better way by applying axial magnetic field. If we consider the generation of plasma ions in the central region, the plasma profiles in the ion source can been estimated from the solution of the diffusion equation given by a zero order Bessel function as shown in eq. 2 [5].

$$n = J_0\left(\frac{r}{\sqrt{D\tau}}\right) \qquad (2)$$

where, D is the diffusion coefficient, r is Radial distance , τ is the decay constant. The plasma density profiles estimated using eq. 2 are depicted in figure 3(b). It is seen that the application of axial magnetic field of about 50 - 100 G produced by the 10 A coil current can improve the plasma density by a factor of ~ 2.5 times. In these calculation, the ions are considered to be produced along the central axis and diffused radially. In case of axial magnetic field, the confinement of primary electrons in the central region of the plasma chamber produces the ions in a local region in contrast to the case without the axial magnetic field. The observed density profiles are nearly flat in both cases. These calculation does not consider the axial diffusion of ion i.e. along axial magnetic field. The ions produced in the central region, which diffuse along the axial direction, can however give a density peak at the centre.

4. Experimental studies on improvement of plasma parameters with axial magnetic field

The improvement of plasma parameters from the axial magnetic field are confirmed by the measurement of ion saturation current using a Langmuir probe biased at -90V. The ion saturation current has been measured with/without the axial magnetic field (Coil current 10 A) at fixed discharge parameters. The radial profiles of ion saturation current are plotted in figure 3(c).The ion saturation current at the centre of the plasma chamber is increased by a factor of 3 times with the axial magnetic field. The experimental profiles of the ion saturation current with axial magnetic field is different from the theoretically profiles, which is possibly due to axial diffusion of ions. Detailed calculations using a suitable model may be required to calculate the exact profile of density. In an ion source, a rise of screen grid current by a factor of 1.5 times has also suggested an improvement of plasma density with axial magnetic field (not shown here).

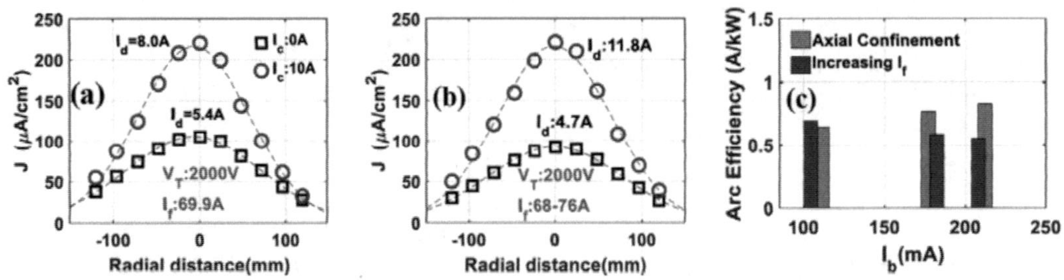

Figure 4: (a) Radial profiles of ion current density with and without axial magnetic field. Other input parameters are kept constant. (b) Beam profiles estimated as a function of discharge current ($\propto I_{fil}$) for extraction voltage (V_T) of 2000V. (c) Arc Efficiency as a function of beam current with and without axial confinement (All measurements taken at a source pressure of 2mTorr).

5. Improvement in beam current with axial magnetic field

To see the effect of the axial field variations on ion beam current (I_b), the ion beam extraction is carried out with/without axial magnetic field at constant discharge conditions. The ion beam current density profiles are obtained using an eleven-channel Faraday cup array, which is located at 465mm from the acceleration grid. The ion current density is estimated by measuring potential drop across a shunt resistor of 100Ω. The radial profiles of ion beam density are plotted with (coil current I_c=10 A) and without axial magnetic field (I_c=0 A).The ion current density all along the radial profile increases significantly by applying axial magnetic field as shown in figure 4(a).The peak ion current density is increased by a factor of more than 2. The beam width is reduced with applying axial magnetic field either due to asymmetry in the plasma density or reduced beam divergence at better perveance. It is observed that the discharge current also has been increased with applying axial magnetic field. This suggests that the plasma density and beam current are partially improved by the additional emission of primary electrons. So the emission of primary electrons is also enhanced by the additional axial magnetic field and may require more studies to understand the phenomenon. As the beam width is reduced with increasing beam current by applying axial magnetic field, the same effect is checked with increasing the beam current by increasing filament current beyond 70A. The rise of ion beam current density (at various radial positions) with filament temperature can be seen in figure 4(b). Here two different mode of operation have been studied. In the first mode, the improvement of plasma density and beam current is increased by the axial magnetic field by keeping other parameters constant. In the second mode, the plasma and beam parameters are improved by raising the hot cathode filament current without any axial magnetic field. It is obvious that both modes can be utilised to achieve similar beam parameters as shown by radial profiles of ion beam current density. The use of axial magnetic field at low filament current improves the filament life as it is operated at a lower temperature.

With axial magnetic field, an ion current density of 200 μA/cm^2 is obtained at a lower discharge current (I_d) of 8.1A as compared to the case where the same ion current density is achieved without axial magnetic field at a higher filament current and higher discharge current of 11.8A. The discharge current is lower in case of axial magnetic field corresponding to a similar beam current. This suggests that an additional axial magnetic field can improve the arc efficiency of the ion source. A higher beam current ($\propto$ plasma density) is desirable at low discharge current for a higher arc efficiency.

In figure (4(c)), the arc efficiency is plotted as a function of beam current with axial magnetic field and without axial magnetic field where the density is improved by increasing the filament current. It shows that arc efficiency is slightly improved by adding the axial magnetic field due to the better confinement of primary electrons and plasma ions. The arc efficiency could not be improved much by adding axial magnetic field due to a rise in discharge current while adding axial magnetic field. The rise in discharge current can be explained by the increase in plasma parameters such as plasma density and plasma potential by adding axial magnetic field. A rise in plasma density decreases the sheath thickness, which can increase the emission of primary electrons or discharge current. Similarly, the rise in plasma potential can increase the emission of electrons and therefore the discharge current. The rise in plasma density is already verified by the increase in the ion saturation current with axial magnetic field. The plasma potential however could not be measured by the Langmuir probe due to arcing at the probe tip while biasing it at the positive potential near the plasma potential. The use of emissive probe for measuring plasma potential may be required which is out of the scope of this work. It is observed that the beam divergence (from width) is reduced by increasing coil current. The reduction in the divergence is partially due to the fact that the plasma profile in the source becomes narrow by the axial magnetic field. In a ring-cusp and Kaufman type ion sources, the axial magnetic field is responsible for better confinement as well as higher non-uniformity.

6. Conclusion

The experimental and simulation results show that ion source parameters like beam current, arc efficiency can be improved by adding the axial magnetic field in the range of about 50-100G. The improvement in the arc efficiency effect is slightly low due to the rise in discharge current with increasing axial magnetic field. In the presence of axial magnetic field, improved beam parameters are obtained at a lower filament temperature, which ensures the better filament life for the given beam current. A strong non-uniformity in plasma density profile is observed by increasing the axial magnetic field. The non-uniformity observe in experimental measurements is possibly due to the axial diffusion of plasma ions in the core region. The non-uniformity in beam is partly due to beam divergence as verified by increasing the beam current by increasing filament current and temperature. The simulations of trajectories of primary electrons in the composite magnetic field suggests that the enhanced beam parameters are due to the improved confinement of primary electrons in the axial magnetic field.

7. References

[1] Ehlers K W and Leung K N 1980 Multicusp negative ion source *Rev. Sci. Instrum.* **51** 721–7
[2] Pachchigar V, Ranjan M and Mukherjee S 2019 Role of Hierarchical Protrusions in Water Repellent Superhydrophobic PTFE Surface Produced by Low Energy Ion Beam Irradiation *Sci. Rep.* **9** 1–10
[3] Wirz R E 2005 *Discharge Plasma Processes of Ring-Cusp Ion Thrusters(Dissertation (Ph.D.))* (California Institute of Technology)
[4] Systems D 2019 CST studio suite
[5] Francis F.Chen 2016 *Introduction to Plasma Physics and Controlled Fusion* (Springer, Cham) p 490

Studies of bremsstrahlung and characteristic X-Ray lines using the VENUS ECR

Janilee Benitez, Damon Todd and Daniel Xie

One Cyclotron Road, Berkeley, CA 94720

jybenitez@lbl.gov

Abstract. Though known for their production of high currents of highly-charged ion beams, advanced electron cyclotron resonance (ECR) ion sources like VENUS at Lawrence Berkeley National Laboratory (LBNL) also generate significant numbers of x-rays. The LBNL ECR ion source group has spent many years studying the x-rays emitted from VENUS in order to gain a better understanding of the ECR plasma. Based on the emitted bremsstrahlung continuum, a spectral temperature T_s can be calculated which is a relative indication of the temperature of the plasma electrons. We will present correlations between T_s with respect to parameters such as magnetic fields. In addition to the bremsstrahlung continuum, the plasma ions emit characteristic x-rays. We will show that additional information can be gained by investigating spectral line shifts. A summary of this recent research using VENUS will be presented and discussed.

1. Introduction

Third generation electron cyclotron resonance (ECR) ion sources, such as the fully superconducting source VENUS at Lawrence Berkeley National Laboratory [1], continue to be pushed to their limits in order to better understand their capabilities. Utilizing its high confining magnetic fields and heating frequency, VENUS has produced many medium and high charge state ion beams with record intensities [2]. The increased performance of third generation ECR ion sources such as VENUS have brought about great improvements in accelerator capabilities due to their ability to produce higher energies with the higher charge states. As a result, there are now several groups working towards a 4th generation source [3-4]. Knowledge and experience with the 3rd generation ECR ion sources will be used to help predict the performance and operation parameters of 4th generation sources. Because 4th generation ion sources will be superconducting due to their high magnetic confining field requirements they will be sensitive to the hot electrons and the radiation produced by them and deposited in the cold mass [5]. For this reason, the ECR group at Lawrence Berkeley National Laboratory's 88 Inch Cyclotron has been investigating the temperatures of the hot electrons of the VENUS ECR Ion Source.

The VENUS ECR Ion Source is the world's first 3rd generation ECR ion source optimized for 28 GHz operation. It is built with a combination of three solenoid coils and a sextupole structure, all NbTi, that produce a high-strength, minimum-B-field configuration with a magnetic field magnitude that increases in all directions from the source center. This coil configuration allows considerable flexibility in the source magnetic field configurations, and previous studies have investigated the effects of the magnetic field gradient at the axial resonance zone and how it affects the spectral temperatures [5,6]. This effort to understand how the magnetic field configuration affects the temperature of the plasma electrons continues and has expanded to studies of both the emitted bremsstrahlung spectra and the characteristic x-rays.

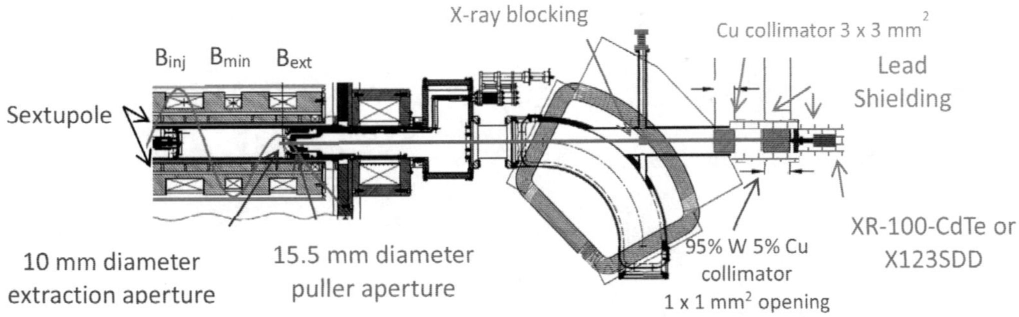

Figure 1. The VENUS ECR ion source, solenoid lens, and analysing dipole are shown, along with the setup for the bremsstrahlung and characteristic x-ray measurements. A typical axial magnetic field shape is overlaid on the source. Indicated distances are in mm.

2. Studies

2.1. Bremsstrahlung Studies

Bremsstrahlung x-rays are produced when electrons are decelerated as they interact with another charged particle or nuclei. For ECR ion sources this either occurs when electrons are decelerated by plasma ions or when they strike a solid surface such as the chamber wall or extraction aperture. The bremsstrahlung x-rays coming from a source form a continuum spectrum, and from the inverse of the slope of this spectrum a spectral temperature T_s is extracted, as described in [6]. This spectral temperature is related to the hot electron temperature, and we can use the spectral temperature to gauge how changing a source's confining magnetic field affects the hot electron temperature.

Electrons gain energy when they cross the resonance zone and it is expected that the shallower the gradient at resonance, the higher the average energy gained by electrons when crossing this resonance. The gradient at the resonance zone is often characterized by the ratio B_{min}/B_{ECR} as for more simple sources, with only two confining solenoid coils, changing this ratio changes the gradient at the resonance zone. Previous bremsstrahlung studies showed that T_s increases as the resonance zone gradient decreases, or, alternatively, as B_{min}/B_{ECR} increased [5]. As it is often difficult to separate varying B_{min} and varying the gradient at the resonance zone, the T_s dependence on only B_{min}/B_{ECR} was only previously investigated.

VENUS' confining magnetic fields, on the other hand, are produced by the superposition of three solenoids and a set of sextupole coils, all superconducting, which can produce a wide range of field configurations and allows for things like fixing the minimum field and varying the gradient at resonance. In our previous study using the more variable fields of VENUS, we found that T_s depends on the magnetic field parameter B_{min} rather than B_{min}/B_{ECR}, as shown in Figure 2 [6,8]. This result goes against expectations and the results of simple simulations that show electron energy gain dependence on the gradient at the resonance zone. The sole dependence of T_s on B_{min} also removes the dependence on heating frequency as B_{ECR} is proportional to the heating frequency. Building on this result, recent measurements from other laboratory's sources [9,10] are also included in Figure 2, as well as results from the other two ECR ion sources at the 88-Inch Cyclotron: the ECR and AECR-U. This data makes clear that the T_s dependence solely upon B_{min} is not only specific to VENUS, but is also true for all other ECR ion sources built that operate near typical, optimal performing fields. We continue this study here to further probe how the magnetic field configuration affects T_s. To further understand these effects, we have added the capability of measuring characteristic x-ray lines and present results here.

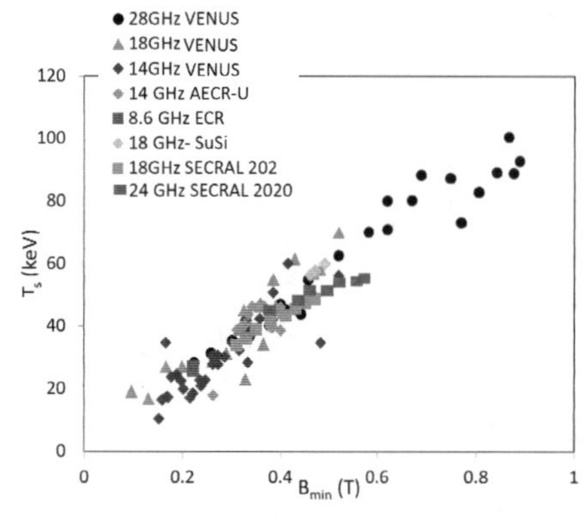

Figure 2. Results show that the spectral temperature T_s depends on the source B_{min}. The plot includes results from several frequency sources [6,8].

2.2. Characteristic X-Ray Studies

Like the study of bremsstrahlung x-rays, characteristic x-rays serve as an indicator of the electron distribution within the source and give information on how it is affected by varied source parameters. These characteristic x-rays are produced when an atom or ion is bombarded by high energy particles moving a bound electron from one shell to another or completely ejecting it. When an electron falls back into this vacated inner shell, it emits x-rays of energy equal to the difference of orbital states that are characteristic to the given atomic species. When electrons create these vacancies, the incident electrons must have energies greater than this energy difference, and therefore the x-rays serve as an indicator of electron temperature. Additionally, as inner shell electrons are lost the remaining electrons are attracted more strongly to the nucleus and their binding energy increases [11]. As a result, the energy of the characteristic x-ray lines increases as more electrons are removed and the plasma becomes more highly ionized [12]. In addition to a shift in energy, the lines have also been shown to broaden as more highly charged ions are extracted [13]. This is due to each peak containing several individual lines representing different charge states, which cannot be resolved with our detection system. By measuring the shifts and broadening of the characteristic x-ray lines when various parameters of the ECR ion source are changed, we gain information on the electron distribution within the source.

The characteristic x-ray spectra presented below were taken from a xenon-124 plasma that used oxygen-16 as a support gas. Centroid energies for the peaks were determined by Gaussian fitting the peaks using OriginPro Software [14]. Before each peak was fitted, a linear estimate of the background was made and subtracted from each peak.

3. Setup

The bremsstrahlung and characteristic x-ray lines investigated here were detected by viewing the plasma through the extraction aperture along the source's long axis with the same setup, shown in Figure 1, used for each set of measurements. Details of the setup can be found in the following reference [6]. The bremsstrahlung spectra were recorded using Amptek's XR-100-CdTe detector and the characteristic x-ray spectra were recorded using Amptek's X123SDD detector, with only one detector being used at a time [15]. For characteristic x-ray lines below 20-25 keV the X123SDD offers better resolution [16]. However, after 10 keV its detection efficiency falls off quickly and the CdTe detector is preferable. Both detectors were placed in vacuum and the only thing between the detector and the extraction electrode of the ECR ion source was collimators. These collimators narrowed the detectors' viewing area to a 20-mm square at the extraction plane that was centered on the 10-mm diameter extraction aperture, meaning that detected x-rays had come from either the extraction aperture or the source.

4. Bremsstrahlung X-Ray Measurement Results

As discussed in section 2.1 and shown in Figure 2, the spectral temperature T_s determined from the bremsstrahlung spectra, depends only on B_{min} and not the resonance zone gradient. Previous studies held the radial field B_r constant while B_{min} and the gradient at the resonance zone were investigated independent of one another [6,8]. Below we present results when varying the radial field B_r over a wider range than previously covered. In Figure 3, B_r was varied from 0.5 T to 1.9 T for an 18 GHz-heated plasma (B_{ECR}=0.64 T) while the injection, extraction, and middle field B_{min} were held constant at 2.23 T, 1.47 T, and 0.33 T, respectively. When going from very low B_r to high B_r the spectral temperature T_s decreases significantly. As seen in Figure 3, when B_r is below 0.5T (below the resonant field) there is no plasma. Figure 4 shows how the spectral temperature varies with with B_r for different magnetic field configurations. When B_r approaches B_{ext} the T_s appears to level off, although this requires further investigation to see if this really the case.

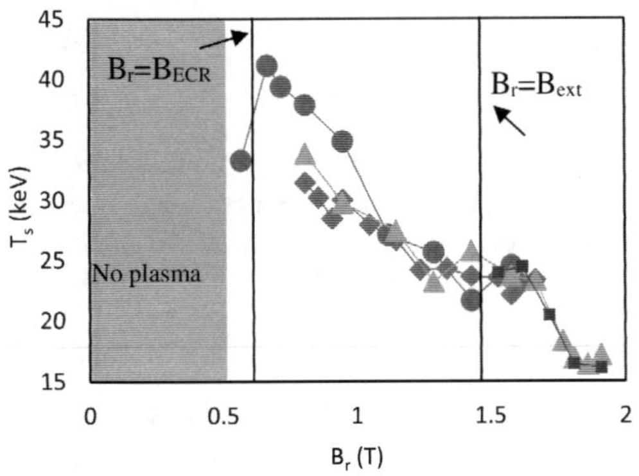

Figure 3. The radial field is varied while the injection, extraction, and middle field are held constant at 2.23T, 1.47T, and 0.33T, respectively.

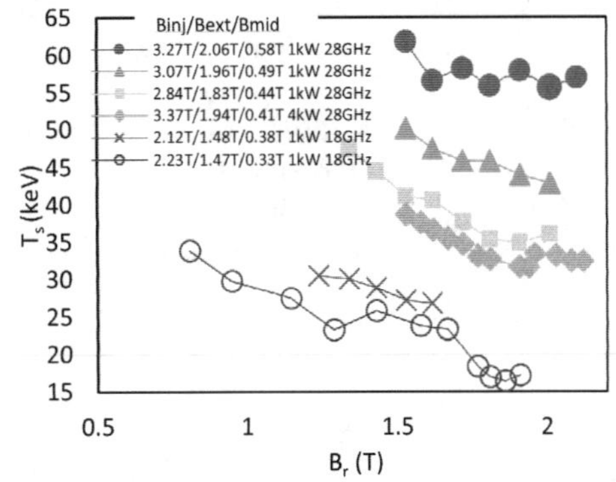

Figure 4. The radial field is varied while the injection, extraction, and middle fields are held constant for each curve.

Table 1. Semi-empirical scaling laws and decades of experiment have yielded the following magnetic field ratios for optimal operation [17,18].

Magnetic Field Parameter	Optimal Value
B_{inj}	$3\text{-}4 \cdot B_{ECR}$
B_{min}	$0.4\text{-}0.8 \cdot B_{ECR}$
B_{ext}	$\sim B_r$
B_r	$2 \cdot B_{ECR}$

ECR ion sources are designed around a set of scaling laws, given in Table 1, that have been shown to produce optimal performance over a wide range of field magnitudes [17,18]. As the delivery of ion beams to accelerators usually requires optimum performance, typical VENUS operation utilizes fields meeting these criteria and therefore uses $B_r \sim B_{ext}$. In Figure 5, the normalized $^{16}O^{6+}$ production is plotted against the ratio B_r/B_{ext} for various magnetic field configurations. The results agree with the scaling laws as they show optimal production around $B_r \sim 0.8\text{-}1.0 \cdot B_{ext}$. As the spectral temperature T_s typically levels off around near $B_r \sim B_{ext}$, Figure 5 shows that in the region of optimal operation T_s does not vary significantly with B_r. Regions where B_r are far from B_{ext} are of little interest when operating the ECR ion source for maximum performance, although it is worth acknowledging that there is significant variation when the range is extended beyond normal operation.

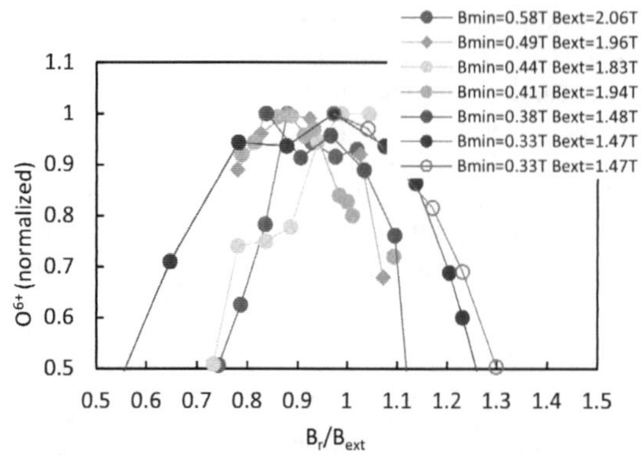

Figure 5. $^{16}O^{6+}$ production normalized to the maximum $^{16}O^{6+}$ production for a given magnetic field is plotted against the ratio B_r/B_{ext}.

5. Characteristic X-Ray Measurement Results

When operated near the well-established optimal confining field values, the results presented in Section 4 continue to point towards a linear dependence of T_s on B_{min}. However, it is still not clear whether the linear spectral temperature T_s on B_{min} is a result of a change in electron confinement, a change in heating efficiency of the plasma electrons, or indicative of something not previously considered. The study of characteristic x-ray lines is intended to help answer these questions.

5.1. Varying Bmin

As our primary interest is seeing how varying B_{min} affects the temperature of the plasma electrons, we first varied B_{min} and observed the peak centroid energy of the characteristic x-ray lines. For this, and subsequent sections the $L_{\alpha1}$ and $L_{\beta1}$ characteristic x-ray lines for xenon, at 4.110 keV and 4.415 keV, are observed [19].

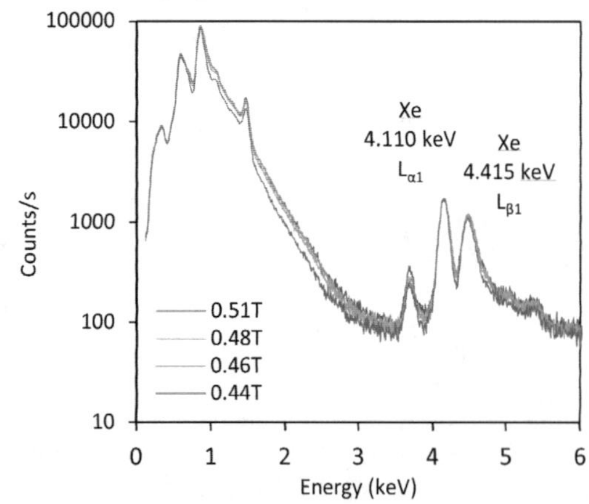

Figure 6. Bremsstrahlung spectrum taken of an oxygen-16 plasma with xenon-124 added and its characteristic lines visible as B_{min} is varied.

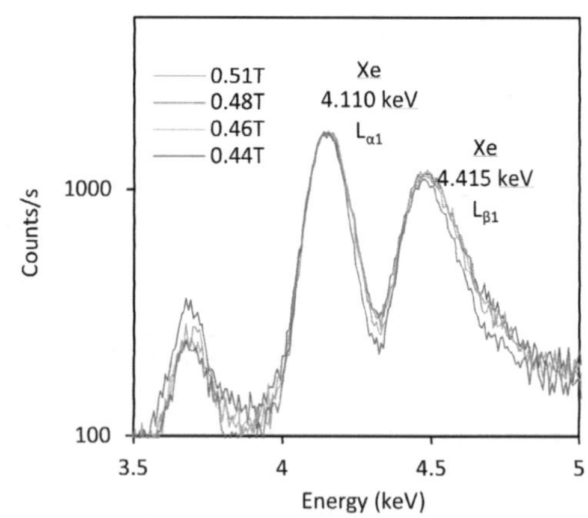

Figure 7. An expanded view of the xenon characteristic x-ray lines shows a slight shift towards higher energy as B_{min} is increased.

In Figure 6 the xenon $L_{\alpha1}$ and $L_{\beta1}$ characteristic x-ray lines are shown on the bremsstrahlung continuum as B_{min} was varied from 0.44 T to 0.51T. Figure 7 shows an expanded closer view of the peaks of interest. Table 2 shows the centroid energy of the $L_{\beta1}$ peak as B_{min} is varied. When increasing B_{min} from 0.44 T to 0.51 T, the centroid shifts up in energy by 0.15%. This small energy shift should be compared to the linearly-fit change in T_s from Figure 1 of just over 14% for the same increase in B_{min} [20]. The $L_{\alpha1}$ line reflects an M shell electron filling an L shell vacancy, while an $L_{\beta1}$ line reflects a higher energy N shell electron filling an L shell vacancy. For this section and subsequent characteristic x-ray sections the $L_{\beta 1}$ always has a larger energy shift. In addition to understanding why the $L_{\beta 1}$ presents a larger shift, the authors hope to expand this B_{min} range in the future measurements. A larger B_{min} range may help better understand whether varying B_{min} does contribute to changes in the electron heating and temperatures. Table 2 also lists the full width at half maximum (FWHM) for each B_{min} and shows that as B_{min} increases the line does broaden, as shown in [13].

Table 2. Peak centroids for the xenon $L_{\beta 1}$ obtained from the Gaussian fits when the B_{min} is varied.

B_{min}	Gauss Fit Peak Center	FWHM
0.44	4.48401	0.17492
0.46	4.49078	0.18456
0.48	4.49406	0.18936
0.51	4.49083	0.19234

5.2. Varying Power

In this section, the characteristic x-ray lines of xenon are observed as the power is varied. Figure 8 shows the characteristic x-ray lines on top of the bremsstrahlung continuum and Figure 9 is an expanded view. While the 18 GHz heating power is held at 1 kW, the 28 GHz power is varied from 1 kW to 3 kW. Table 3 shows that in this range of power the centroid of the xenon $L_{\beta 1}$ shifts by 0.27 %. This indicates that as the power is increased the plasma becomes more highly ionized.

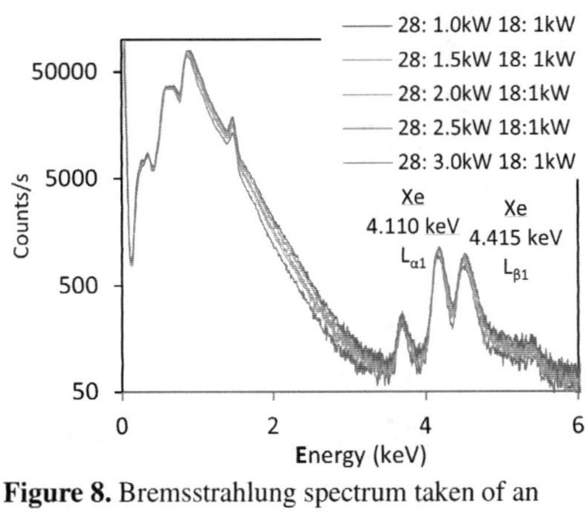

Figure 8. Bremsstrahlung spectrum taken of an oxygen-16 plasma with xenon-124 added and its characteristic lines visible as the total 18 + 28 GHz power is varied.

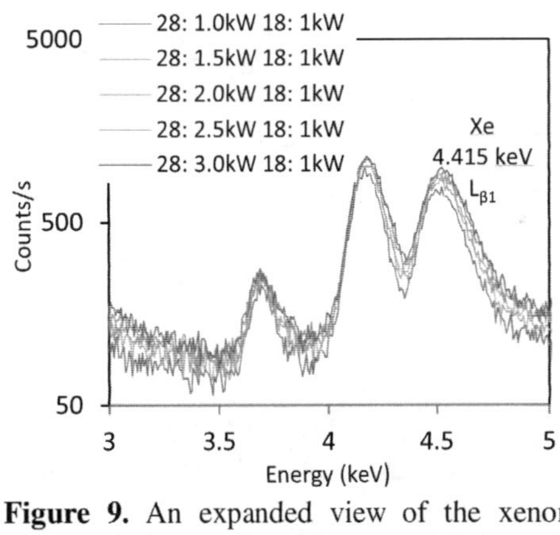

Figure 9. An expanded view of the xenon characteristic x-ray lines shows a s shift towards higher energy as the total heating power is increased.

Table 3. Peak centroids for the xenon $L_{\beta 1}$ obtained from the Gaussian fits when the total heating power is varied.

28 GHz Power (+1kW 18GHz)	Gauss Fit Peak Center	FWHM
1.0	4.5217	0.1980
1.5	4.5269	0.2022
2.0	4.5249	0.2020
2.5	4.5324	0.2133
3.0	4.5338	0.2116

5.3. Varying Pressure

In this section the xenon-124 gas pressure is varied for a plasma composed of oxygen and xenon. Figure 10 shows the characteristic x-ray lines and the bremsstrahlung continuum and Figure 11 shows an expanded view of the Xe $L_{\alpha1}$ and $L_{\beta1}$ peaks. In the range that the pressure is varied there is a more pronounced shift in the peak centroids than when B_{min} and total heating power were varied. Table 4 shows that as the pressure is increased the peak centroids shift towards lower energy by 1.25 %. This is indicative of a less ionized plasma. In general, in order to produce higher charge state plasmas lower pressure is needed and this agrees with this result.

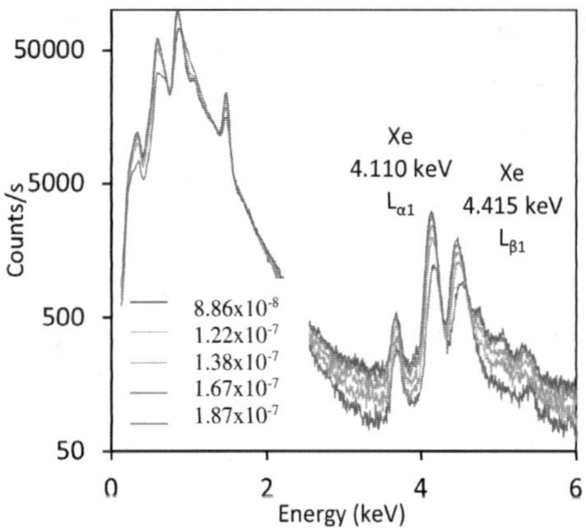

Figure 10. Bremsstrahlung spectrum taken of an oxygen-16 plasma with xenon-124 added and its characteristic lines visible as the gas pressure (in Torr) is varied.

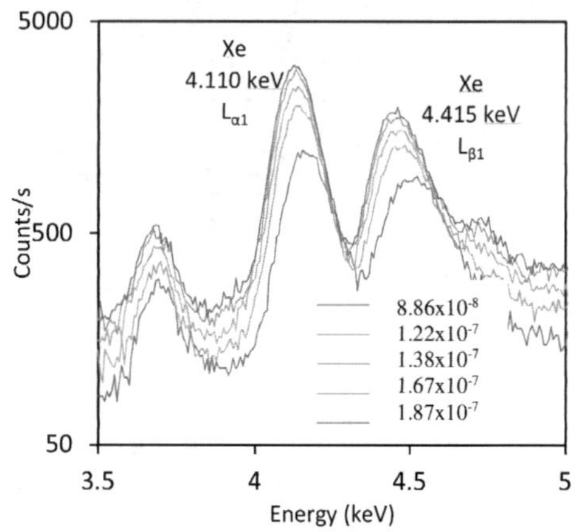

Figure 11. An expanded view of the xenon characteristic x-ray lines shows a shift towards lower energy as the gas pressure (in Torr) is increased.

Table 4. Peak centroids of the xenon $L_{\beta1}$ obtained from the Gaussian fits when the pressure is varied.

Pressure (Torr)	Gauss Fit Peak Center	FWHM
8.86×10^{-8}	4.5170	0.2151
1.22×10^{-7}	4.4832	0.1835
1.38×10^{-7}	4.4706	0.1741
1.67×10^{-7}	4.4645	0.1698
1.87×10^{-7}	4.4608	0.1639

6. Future Experiments

Though the spectral temperature measured via bremsstrahlung shows a strong, linear dependence on B_{min}, this is not reflected in the characteristic x-rays where there is very little indication of ionization level change from increased electron energies, while increasing power and density give characteristic x-ray shifts that seem consistent with current source understanding. For completeness, characteristic x-rays studies will be extended over a wider range of B_{min}, but these initial results hint that there will be little change for these lines. The limited set of results presented here suggests that the increase in T_s might be more due to decreased confinement of high energy electrons than increased average energy with B_{min}. Further studies will aim to tease this out.

References

[1] D Leitner, C M Lyneis, S R Abbott, D Collins, R D Dwinell, M L Galloway, M Leitner and D S Todd 2005 *Nuclear Instruments and Methods in Physics Research B* **235** 486–493
[2] D Z Xie, W Lu, J Y Benitez, C M Lyneis, D S Todd 2016 *Proc. of 22nd International Workshop on ECR Ion Sources (Busan)* THAO01
[3] D Z Xie, J Y Benitez, A Hodgkinson, T Loew, C M Lyneis, L Phair, P Pipersky, B Reynolds and D S Todd 2016 *Rev. Sci. Instrum.* **87** 02A702
[4] H Zhao, L Sun, J Guo, W Lu, Y Yang and X Zhang 2015 *A 45 GHz Superconducting ECR Ion Source FECRAL and Its Technical Challenges* (*Book of Abstracts: 16th Int. Conf. on Ion Sources*) ThuPS26
[5] D Leitner, J Y Benitez, C M Lyneis, D S Todd, T Ropponen, J H Koivisto and S Gammino (2008) *Rev. Sci. Instrum.* **79** 3
[6] J Benitez, C Lyneis, L Phair, D Todd and D Xie 2017 *IEEE Transactions on Plasma Science* **45** 7
[7] D Leitner, J Y Benitez, C M Lyneis, D S Todd, T Ropponen J Ropponen, H Koivisto, S Gammino (2008) *Rev. Sci. Instrum.* **79** 033302
[8] J Benitez, C Lyneis, L Phair, D Todd, D Xie 2016 *Proc. of 22nd International Workshop on ECR Ion Sources (Busan)* MOCO04
[9] T. Ropponen 2010 Presented at ECRIS2010, Grenoble, France
[10] J B Li *et al.* 2020 Plasma Phys. Control Fusion **62** 095015
[11] K Slabkowska *et al.* 2014 Physics of Plasmas **21**, 031216
[12] H F Beyer, H J Kluge, and V P Shevelko 1997 *X-ray Radiation of Highly Charged Ions* (Berlin: Springer)
[13] G Douysset, H Khodja, A Girard and J P Briand 2000 *Phys. Rev. E* **61**, 3015
[14] OriginLab, viewed 27 September 2021 < https://www.originlab.com/>
[15] Amptek viewed 27 September 2021 <www.amptek.com>
[16] R Redus, A Huber, T Pantazis, J Pantazis and B Cross 2012 *X-Ray Spectrometry* **41**, 393–400
[17] D Hitz, A Girard, G Melin, S Gammino, G Ciavola and L Celona 2002 *Rev. Sci. Instrum.* **73** 509
[18] C M Lyneis 2016 *Proc. of 22nd International Workshop on ECR Ion Sources (Busan)* MOAO01
[19] X-ray absorption edges, characteristic X-ray lines…viewed 27 September 2021 <https://edisciplinas.usp.br/pluginfile.php/2171854/mod_resource/content/1/X-ray%20absorption%20edges%2C%20characteristic%20X-ray%20lines.pdf >
[20] J Benitez 2016 Presented at ECRIS2016, Busan, Korea

Journal of Physics: Conference Series **2244** (2022) 012084 doi:10.1088/1742-6596/2244/1/012084

Analysis of electron behaviour around a spring-shape filament inside a low-energy electron gun

Matsumoto Y[1], Nakano H[2,3], Kisaki M[4], Shinto K[5], Sasao M[6] and Wada M[6]

[1] Tokushima Bunri Univerisity, Nishihama, Tokushima 770-8514, Japan

[2] National Institute for Fusion Science, 322-6 Oroshi-cho, Toki, Gifu 509-5292, Japan

[3] SOKENDAI (The Graduate University for Advanced Studies), 322-6 Oroshi-cho, Toki, 509-5292 Japan

[4] National Institutes for Quantum and Radiological Science and Technology, 801-1 Mukoyama, Naka-shi, Ibaraki 311-0193 Japan

[5] Japan Atomic Energy Agency, J-PARC center, Tokai-mura, Naka-gun, Ibaraki 319-1195, Japan

[6] Graduate School of Science and Engineering Doshisha University, Kyoto, 610-0394, Japan

E-mail address: ymatsumoto@tokushima.bunri-u.ac.jp

Abstract. Electron behaviour inside a low-energy electron gun based on a spring-shape filament were studied with experiment and Particle-In-Cell simulation. The energy range of the electron beam which we expected was from 1 to 20 eV. The analysis told that smaller size of filament is more useful to improve electron density in front of an extraction hole in the gun to enhance beam current. Relation between a space potential distribution in the gun and electron transport was also studied. A heater voltage to drive a filament for thermionic electron emissions has another role to form a spatial potential distribution in the gun. The potential guides electrons, which are at a distant area from the beam extraction hole, toward the hole. It can be a significant help to realize efficient electron extraction by a beam extraction electric field induced near the hole.

1. Introduction

Negative hydrogen ion (H^-) sources have been utilized as core devices in neutral beam injection (NBI) systems to heat plasmas in thermo nuclear experimental devices [1], including the international thermo-nuclear experimental reactor (ITER) [2]. In order to achieve requirements for the purpose, numerous studies to understand physics in the ion source plasmas leading to improvements to higher efficient sources have been conducted for past decades [3]. According to the requirements, the sources should realize large H^- ion current outputs while keeping small electron drain currents [1]. To achieve the requirement optimizing ion source plasmas, low-energy electron transport in H^- source plasmas is one of fundamental knowledge. Low-energy electrons, whose energy is considered around 1 eV, have an essential role in H^- production through the dissociative attachment process in a plasma [3]. As a result, the spatial distribution of low-energy electrons affects the spatial distribution of H^- ions in a plasma which determinates H^- source performance. In addition, the spatial distribution of electrons also affects an electron drain current. The spatial distribution of electrons is the consequence of electron transport in a plasma. Engineering design and operation of H^- source based on knowledge of the electron transport are expected to improve both H^- beam and electron drain currents.

To understand the electron transport in an H^- source, we initiated a simulation experiment using an electron source [4]. We install another electron source inside an H^- source chamber, in addition to the main electron source to generate and maintain the ion source plasma. Using the additional source, we inject additional electrons into an ion source plasma from a position close to or inside of the plasma. The energy of the injected electrons from the source is adjusted to the energy similar to the bulk electrons in the plasma. Then, the injected electrons are expected to be transported with the same manner to the bulk electrons in the plasma. After the injected electrons locally elevate the electron density around the injection point, they diffuse from the point with the transport process similar to the original plasma. As a result, the local density increment gradually decreases against the increasing distance from the injection point. The spatial profile of the density slope is expected to clarify the transport process of the injected electrons in the plasma (See figure 1 of reference 4). Comparison of the experimental results on the injected electrons with a Particle-In-Cell (PIC) simulation [5] should give us better understanding of the low-energy electron transport.

We considered that a low-energy, small electron gun is one of possibilities for realizing the additional electron source [6] in this research. However, the gun is required to achieve some features for our purpose, including extraction of 1 mA/cm^2 beam current density in low beam energy range from 1 to 20 eV. We have been developing the device to enlarge the beam current density. In this paper, we analyse experimental data of our prototype electron gun using PIC simulation in order to understand electron behaviour, especially transport, inside the gun linking to their extractions. It involves a plenty of physics related to improvement of the beam current density.

2. Experimental setup and simulation settings

2.1. Experimental setup

Experimental apparatus of the prototype electron gun is shown in figure 1, which was based on our previous work [6]. The prototype electron gun was cube-shape. The width, depth and height of the gun were 2.2 cm, 2.7 cm and 5.1 cm, respectively. It was attached on the back-end plate of a cylindrical-shape ion source chamber. The electron gun was assembled with a W-filament electron emission cathode, Wehnelt cylinder, extraction electrodes and chassis [7]. The cathode was a spring-shape filament made of a 0.2 mm diameter tungsten wire. The winding diameter (width of the spring) and height of the spring filament were 3 mm and 10 mm, respectively. A cylinder-shape Wehnelt electrode was used to confine thermal electrons emitted from the filament. The disk-shape extraction electrode was located at downstream from the Wehnelt cylinder to extract electrons from the electron dense region around the filament. The chassis worked as deceleration electrode to adjust beam energy and focusing. All beam extraction apertures in these electrodes had 3.8 mm diameter.

The current density extracted from the electron gun was measured with a Faraday cup. The Faraday cup position was 0.3 cm away from the extraction hole of the electron gun in the z direction. Now, since we focus to improve performance of the electron gun, beam experiments in this paper were conducted in a vacuum, without hydrogen gas loadings and discharges in the ion source chamber.

2.2. Settings of the PIC simulation

Features of the 2D3V(2-dimensional position, 3-dimensional velocity) PIC code in this paper were based on our previous work except for the filament [6]. The two-dimensional simulation domain is shown in figure 1 with the red broken line. The area was divided into 600 x 400 grids in the x and z direction, respectively, to have 0.1 mm mesh size. The electron gun in the simulation had the same size and the same electrode configuration to the real one reducing the dimension into the planar geometry.

In this paper, we carefully treated the filament spatial structure to discuss electron transport around the filament. To simulate the filament material, we utilized a special electrode, which we called "effective electron source box", in this simulation. The box can store electrons in the inside like the real filament. The electrons inside the box can move without receiving any forces by electric fields, magnetic fields and collisions among particles. When the electrons reach the boundary of the box after their transports, they can get out and leave the box like real emission events. For electrons outside the box, the box works as a normal electrode. The motions of the electrons are affected by the electric potential of the box. When electrons reach the boundary of the box, they can collide and vanish from a vacuum region like the real collision events on a metal wall. The box was used to simulate the cross-sectional view of the real spring-shape filament in the *x-z* plane shown in figure 1. In the view, some small circle-shape electrodes, whose sizes are determined by the tungsten wire diameter, are arranged in the two rows along and both sides of the *z* axis. To simulate the structure, we set the size of the box same as the diameter of the real filament wire. In addition, we put the boxes on same positions of the wire footprints on the *x-z* plane in figure 1.

The real filament was driven by the heater voltage, V_h, to generate thermal electrons. It was applied between both edges of the filament. The voltage gave a spatial potential distribution along the filament. The potential distribution was determined depending on a distance from the one of edges to an observation point on the filament. To simulate the potential distribution, we gave a unique potential to each box as a function of the distances from the edge. The distances were calculated with a procedure that adjacent two boxes in the order shown in figure 1 were connected with a straight line. In this treatment, the potential of the series of boxes gradually changes from one edge to the other.

The number of electrons inside the effective electron source box was kept by introductions of new electrons in every time steps. Then, energies of these introduced electrons were decided by Maxwell distribution with 0.1 eV mean energy. Extracted electron beam current from the gun was estimated

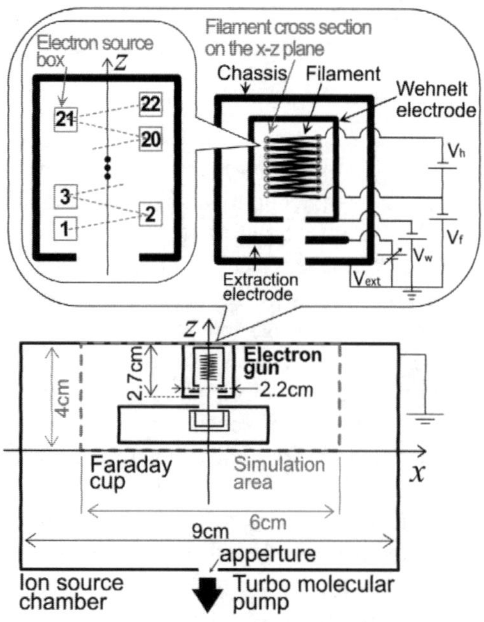

Figure 1. The lower figure is an experimental apparatus. The upper right figure shows the configuration of the electron gun. The upper left figure is a rough sketch of electrodes to simulate the filament in the model. The numbers in box-shape electrodes show the order which is used to decide the electric potentials of the boxes.

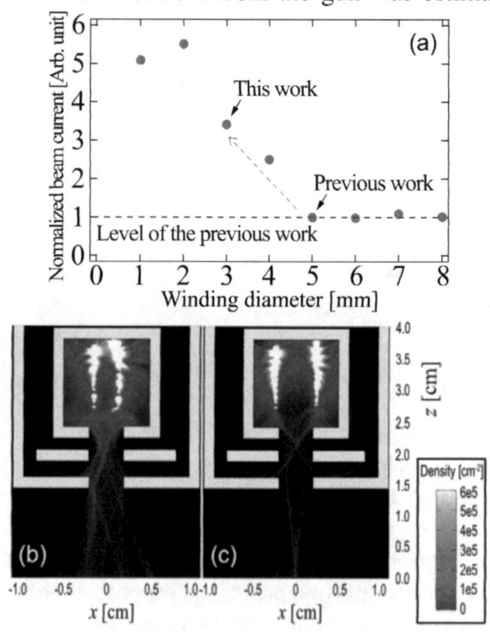

Figure 2. (a) Dependence of the beam current on winding diameter of the filament in simulation. The beam currents were normalized based on the current in the previous work whose winding diameter was 5 mm.
(b) and (c) Contour plots of electron density in the electron gun by the simulation. The winding diameters of the filaments are (a) 3 mm, (b) 5 mm. Electrodes were hand-drawn to guide positions roughly.

0.3 cm away from the extraction aperture of the gun on the z axis, which was the same position to the experiment.

3. Results and discussion

3.1 Influence of filament size on electron beam current

We set the winding diameter of the spring-shape filament to 3 mm, which was 5 mm in our previous work. In a beam extraction experiment with the smaller filament, we confirmed an enhancement of maximum electron beam current density available with the prototype gun. It was 0.04 mA/cm^2 in case of the 5 mm winding diameter. After the replacement of the filament, it was enhanced, at least, to 0.1 ~ 0.2 mA/cm^2, which was around 3 times larger than before the filament shrinkage.

We conducted the PIC simulation to understand the reason of the filament size effect. Our aim in this paper is to understand how electron transport changes depending on the filament size. Therefore, the number of emitted electrons stored in filaments is assumed independent on the filament size in the simulation. The figure 2(a) shows normalized beam currents received by the Faraday cup in the simulation. A smaller winding diameter filament basically produces a larger beam current. The enhancement ratio by the shrinkage from 5 mm to 3 mm is about 3 in figure 2(a), which is in good agreement with our experimental result. The figure 2(b) and (c) show contour plots of electron densities in the simulation. It is clearly seen that the smaller filament is more suitable to create an electron dense area in front of the extraction hole. It successfully connected to the enhancement of the electron beam current. According to the result in figure 2(a), we should continue to shrink the filament. It may increase the beam current about 2 times larger than that of this work. Meanwhile, the beam current is almost constant in the range of the filament size from 5 mm to 8 mm in figure 2(a). The result indicates that the size of the electron cloud around the filament has an upper limitation defined by the potential wall induced by the Wehnelt cylinder. The electron confinement by the potential wall seemed to keep not only the cloud size, but also electron density in front of the hole with a certain level, if we use larger size filaments over 5 mm winding diameter.

3.2 Roles of the heater and the extraction voltages for electron extraction

We discuss roles of the extraction voltage V_{ext} and the heater voltage V_h for electron transports and extractions inside the electron gun. In figure 3, we show an experimentally measured beam current dependence on V_{ext}, using a constant V_h values which were +10 V. Here, all beam currents are normalized by the highest beam current obtained for V_{ext} = 30 V, V_h = +10 V case, for convenience in comparison with simulation results to be discussed later. The beam current shows an increasing trend with increment of V_{ext}. The reason is found in figure 4, which are spatial profiles of electron density and space potential in the Wehnelt cylinder on the z axis by PIC simulation. The result in figure 4(b) shows that the increment of V_{ext} elevates the potential in the vicinity of the extraction hole. It gradually enhances the number of electrons attracted toward the extraction hole. The growing influence on these electrons links to an enhancement of the simulated beam current with V_h = +10 V in figure 3. However, we may find one question in the experimental results in figure 3. The beam current is not zero, even in case of no extraction electric field condition promised by V_{ext} = 0 V [6]. The result indicates an existence of electron extractions which is independent of the extraction field induced by V_{ext}. In addition, the electron extraction is negatively affected by a potential barrier which is induced by the Wehnelt cylinder and is higher than the thermal electron energy. It means that the extraction is realized overcoming the potential barrier without a support of V_{ext}. This extraction mechanism is provided by V_h.

The reason will be found in figure 4. In figure 4(b) with V_h = +10 V and V_{ext} = 15 V case (shown with a red line), V_h forms a potential well to accelerate electrons from z = 2.5 to 3.5 cm along the z axis toward the hole. These accelerated electrons are expected to reach the extraction hole overcoming the potential barrier, even without a support by the extraction electric field induced by V_{ext}. We check the contribution of the potential well in the simulation. The depth of the potential well can be controlled by changing V_h. The beam currents for several V_h values are shown in figure 3. The beam current at V_{ext} =

0 V has positive dependence on increment of V_h. Then, we also examine the case of a negative V_h, -10 V. In this case, the potential distribution shown in figure 4(b) is formed to confine electrons inside the Wehnelt cylinder rather than extractions, whose density profile is shown in figure 4(a). As a result, in case of a reversed polarity of V_h, a high beam current is not obtained even applying high V_{ext} voltages. We already confirmed the tendency in an experiment. A very low beam current was observed in a reversed polarity of V_h with a similar voltage setting to figure 3 and 4, whose example were V_h = -10 V, V_{ext} = 35 V, V_w = 3.6 V and V_f = 1 V.

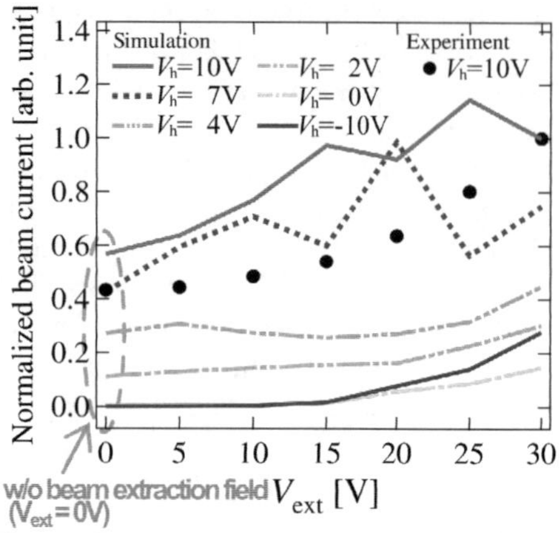

Figure 3. Dependence of the electron beam current on V_{ext} in experiment and simulation. Several V_h conditions were used. V_f = 1 V, V_w = 2.4 V, respectively.

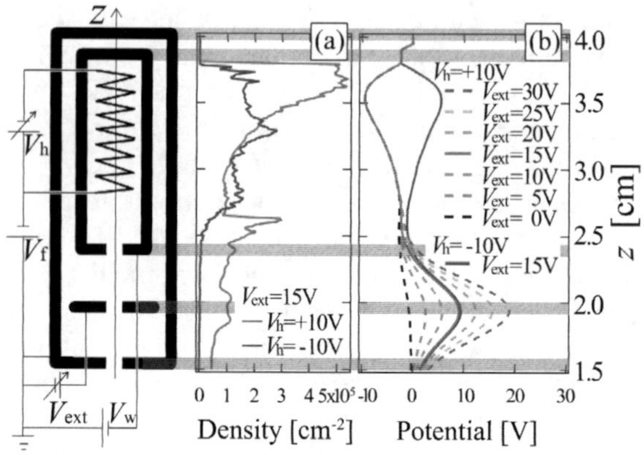

Figure 4. Spatial distributions on the z axis of (a) electron density, (b) space potential in PIC simulation. The results were calculated changing V_{ext} and V_h. V_f = 1 V, V_w = 2.4 V. The gray transparent bars roughly show positions of electrodes.

4. Conclusion

Electron behaviour in a low-energy electron gun based on a spring-shape filament was discussed with experiment and PIC simulation focusing on filament settings. A smaller size filament produces a denser electron region in front of an extraction hole to enhance an electron beam current. Electron extraction is affected by two important components for electron transports: the filament heater voltage V_h and the extraction voltage V_{ext}. Spatial potential structure induced by V_h conveys electrons from an inner region of the Wehnelt cylinder toward the hole to form the base part of the extraction beam. The electric field by V_{ext} boosts the beam current generated by V_h extracting more electrons in the vicinity of the extraction hole.

Acknowledgements
Both experimental and numerical research activities in this work were performed under the collaboration research program of National Institute for Fusion Science (NIFS) in Japan (Grant No. NIFS20KLER102, NIFS20KNSR013 and NIFS21KNSR015).

References

[1] Ikeda K, Tsumori K, Nakano H, Kisaki M, Nagaoka K, Kamio S, Fujiwara Y, Haba Y, Osakabe M 2019 Nucl. Fusion **59**, 7
[2] Bigot B 2021 Fusion Engineering and Design **164**, 112207
[3] Bacal M, Sasao M and Wada M 2021 J. Appl. Phys. **129**, 221101
[4] Matsumoto Y, Kisaki M, Sasao M, and Wada M 2018 AIP Conf. Proc. 2011, 020011
[5] Birdsall C K and Langdon A B 1991 Plasma Physics via Computer Simulation (Institute of Physics Publishing, Bristol, UK)
[6] Matsumoto Y, Kisaki M, Shinto K, Nakano H, Sasao M and Wada M 2020 Rev. Sci. Instrum. **91**, 013333
[7] Fan J, Peng Y, Xu J, Xu H, Yang D, Li X and Zhou Q 2019 Vacuum **164**, 278

Double-sided detector for electron beam alignment and measurement of back-streaming electrons in ExtendedEBIS at BNL

Shunsuke Ikeda, Edward Beebe, Takeshi Kanesue, Sergey Kondrashev, and Masahiro Okamura
Collider-Accelerator Department, Brookhaven National Laboratory, New York 11973, USA

sikeda@bnl.gov

Abstract. We developed an electron beam detector installed between the superconducting solenoids in ExtentedEBIS. The detector is two-sided. Each side has 4 quadrant plates with an aperture slightly larger than the electron beam radius. The gun-side is for alignment of primary electron beam. The collector side detector is to measure the electrons back-streaming from the collector or the downstream electrodes to the cathode. It is important to understand how the back-streaming electrons behave and how to control it. The detector was designed with considerations of heat load. The influence on the electric potential distribution was investigated. A detector test showed that the back-streaming electrons were affected by the external electromagnetic fields and the space charge of the primary electron beam. It was proved that the behaviour of the electrons can be observed by this detector.

1. Introduction

An electron beam ion source (RhicEBIS) at Brookhaven National Laboratory has been operated to provide several types of ions to RHIC and NSRL [1]. A high current electron beam (up to 10 A) is compressed in 5T-superconducting solenoid to have 500 A/cm2. In the test laboratory, an updated EBIS (ExtendedEBIS) is being developed [2]. Figure 1 is a schematic drawing. There are two identical 5T-superconducting solenoids in series. They are separated by 200 mm. The development is to increase the ion trap capacity and to have a pulsed gas injection system in the superconducting solenoid. The gas system is for polarized helium-3 [3] and other gases especially for hydrogen and helium-4.

As a part of ExtendedEBIS development, a double-sided detector for electrons was developed to install between the superconducting magnets. One of the purposes is helping to align the electron beam from the cathode with the gun-side detector. In ExtendedBEIS, transverse correction coils are needed to adjust the magnetic field from the cathode to the collector [4]. If there is a detector for electron beam position, it will be helpful for this correction.

The other purpose is to observe the back-streaming electrons with the collector-side detector. The primary electrons from the gun enter the collector and are deflected by a reflector toward the collector wall. Some electrons are backscattered in the wall without large energy loss and go back to the gun [5, 6]. In addition, some electrons moving near the central axis are elastically reflected directly to the gun [6]. These back-streaming electrons are reflected by the gun and then enter the collector again. Some of them may repeat the reflection several times. The back-streaming electrons may cause the beam to stop by affecting the primary beam or hitting an electrode as an overloading current to trip the power supply. To avoid this, the transverse magnetic fields and drift tube potentials are adjusted. It is expected that the back-side detector can be used for the field adjustment and for more basic research to understand the behavior of the back-streaming electrons.

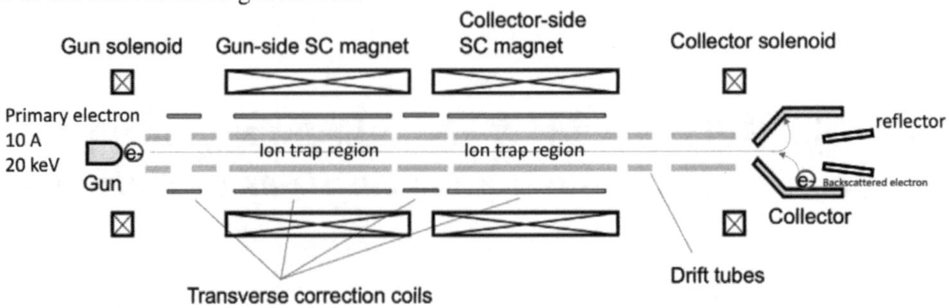

Figure 1. Schematic drawing of ExtendedEBIS.

2. Detector design

Figure 2 (a) shows a drawing of the detector in the chamber between the superconducting solenoids. The primary electron beam emitted by the gun travels from left gun-side to right collector-side. The detector is seen like Fig.2 (b) and (c) from the gun-side and the collector-side, respectively. The detector is composed of Molybdenum quadrants with apertures. The aperture diameter is 6 mm. This is larger than the beam diameter (4 mm). Each quadrant was connected to an oscilloscope with a 50-ohm resistor termination. The detector was actuated by a pneumatic linear feedthrough. It was inserted between drift tubes. The gap between the drift tube and the detector was 7 mm. The vertical position was able to be adjusted by changing the position of a stopper of the linear actuator. The detector could be pushed horizontally by screws. To do this, there was a bellow between the chamber and the flange of the linear feedthrough. The entire system is isolated from the vacuum chamber by a ceramic break.

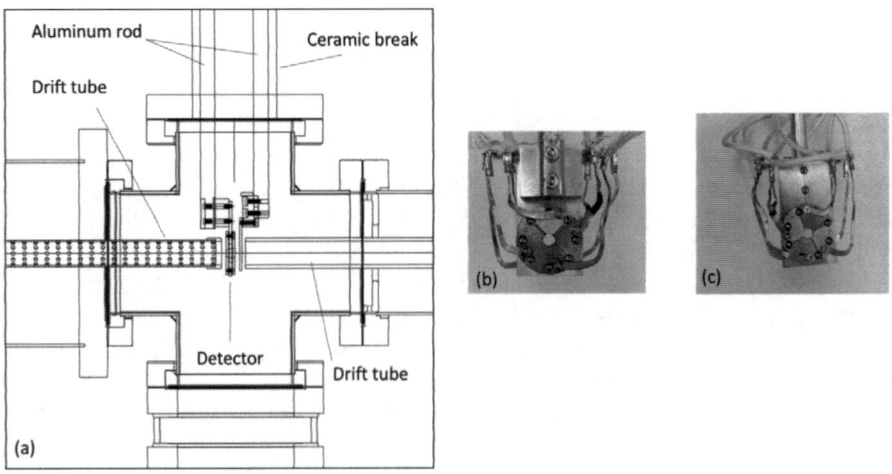

Figure 2. (a) Drawing of detector in mid-chamber between superconducting solenoids. The electron beam comes from the gun-side (left) to the collector-side (right). The detector is connected through an aluminum rod to a linear feedthrough to retract when it is not needed. A ceramic insulator is placed between the detector and the aluminum rod. (b) Gun-side detector viewed from the gun-side (c) Collector side detector from the collector-side.

Since the primary electron beam from the gun has a high power, the temperature rise was considered. The part of the primary beam and the back-streaming electrons hit detector. The current is expected to be up to 100 mA. The beam pulse is around 100 ms for ionization. Repetition rate is about 1/4 Hz. The electron energy is about 30 keV. The beam power averaged as continuous heat is 75 W. If the steady heat is conducted from the detector to the feedthrough at the room temperature, the temperature rise at the detector was estimated to be 150 degrees Celsius. The melting point of Molybdenum is 2617 degrees. Therefore, this temperature rise is acceptable. Meanwhile, the temperature should be higher right after a single beam pulse irradiation. The deposited energy by a single pulse is 300 J. The heat capacity of the detector is 2.4 J/degree Celsius. Therefore, the instant temperature rise is estimated to be 130 degrees Celsius. This is also acceptable.

The influence of the detector insertion on the potential distribution was investigated. A 2D beam simulation was carried out with TRAK [7]. The model is shown in Fig.3 (a). The electron beam simulation started from the inside of the solenoid where the field is 5 T on the left in the figure. The beam propagated in a drift tube with the radius of 10 mm. Then the beam passed through the detector aperture with the 3-mm radius. There is a gap between the drift tube and the detector. The simulation was done with the gaps of 7 mm and 20 mm. The tube and the detector were surrounded by the mid-chamber wall. The potentials are 30 kV on the drift tube and the detector, and 12 kV on the camber wall. The electrons were assumed to be emitted from the cathode at 0 V and 0.18 T. The total energy of the electrons was 30 keV. The beam radius in the solenoid was determined from the cathode radius and the ratio of the magnetic fields. The beam current was 10 A. The black curve in Fig.3 (b) represents the trajectory of the outermost electron, or the beam radius. The horizontal axis, z, is zero at the detector. The beam radius becomes larger as moving toward the detector. This is because the magnetic field is smaller in the mid-chamber region than in the solenoid. Figure 3 (b) represents the potential distribution on the axis with 7 mm (red) and 20 mm (blue) gaps. The potential is lower for $z < -400$ mm than that for $z > -200$ mm. This corresponds to the beam radius change because of the reduction of the magnetic field. The 2 kV potential rise around $z = 0$ mm is due to the small radius of the detector aperture. So, when ions are extracted, the detector potential should be lowered with a consideration of the 2 kV potential rise, otherwise it blocks the ions. On the other hand, with 20 mm gap, there is a potential drop around -5 mm while almost no drop with 7 mm. This means that 7 mm gap, which is used in the detector design, is small enough to avoid large deceleration of the electron beam.

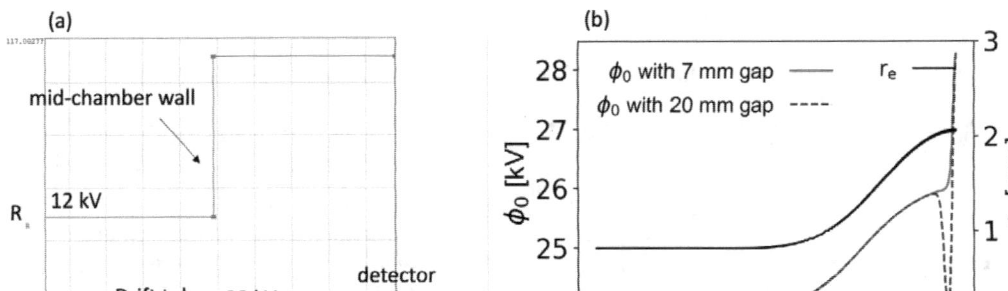

Figure 3. (a) Axisymmetric model in mid-chamber to calculate electric potential. The left border is in 5 T field in the gun-side superconducting solenoid. The right border is at the place where the detector is inserted in the mid-chamber. The gap between the drift tube and the detector was 7 mm or 20 mm in the simulation. Electron beam propagates from the left to right. (b) Electric potential on the central axis as a function of the distance from the detector. The potential with 7 mm gap (red) shows almost no drop while the potential with 20 mm (blue) has a drop to one in the 5 T field.

3. Detector test results

The detector was tested with a 2.7-A primary beam from the cathode. The pulse shape was trapezoidal (200 us rise - 500 us flat - 200 us fall). The cathode bias was -9.7 kV. The detector and most of the drift tubes were at 0 V. Firstly, we adjusted the transverse correction field to see no signal on the detector. This means that the primary beam and the back-streaming electrons did not hit the detector. The signals on the quadrants of the collector-side detector are shown as black lines. Then we added more transverse fields after the detector, namely in the collector-side superconducting magnet and in the chamber in front of the collector solenoid. With this configuration, beam current was observed on the collector-side detector (blue line). Since the transverse field after the detector does not affect the position of the primary beam at the detector, the primary beam still did not hit the detector. Therefore, the observed beam current was of the back-streaming electron. This result proves that the back-streaming electrons can be observed with the detector. In addition to the transverse field, 2 kV voltage was applied to the drift tube in the collector-side superconducting magnet. The shapes of the signals were changed (green line). This indicates that the electric field also affect the amount of the back-streaming electrons.

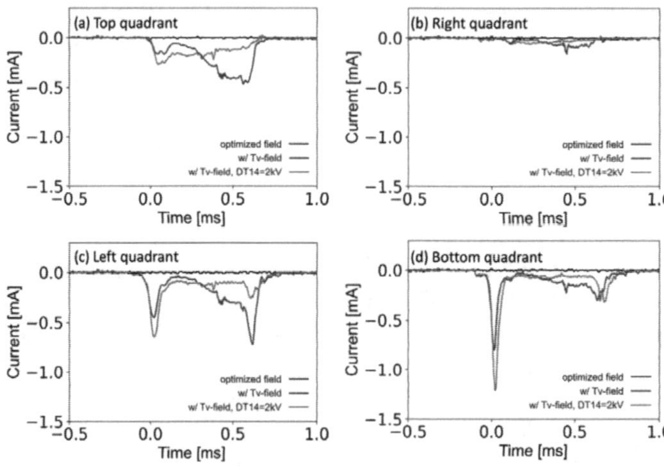

Figure 4. Signals on (a) top quadrant, (b) right quadrant, (c) left quadrant, and (d) bottom quadrant of the collector-side detector. Black curves are signals with an optimized transverse correction magnetic field. Blue curves are signals with more transverse field. Green curves are with more transverse field and 2-kV voltage on a drift tube behind the detector.

For different primary beam currents, the back-streaming electrons were measured. Figure 5 (a) shows the beam current waveforms on the top quadrant of the collector-side. The red curve shows the current waveform when the primary beam was 1.3 A. The trapezoidal beam shape was similar to the primary beam. As the primary beam became larger to 2.1 A (blue) and 2.8 A (green), the beam shape of the back-streaming electrons was distorted. This indicates that the space charge of the primary beam affects the trajectories of the back-streaming electrons, and the effect changes during the beam pulse. We integrated the current waveform to obtain beam charges. The integrated charge for each quadrant was plotted in Fig.5 (b). The sum of the integrated charges was plotted as black points. The sum was increased as the primary beam was increased from 1.3 A to 1.7 A. The charges at top (red), left (green), right (blue) were also increased while the bottom (magenta) was decreased. The interpretation is that the larger number of the primary electrons hit the wall of the collector, and thus more electrons were backscattered. At the same time, the beam shape of the back-streaming electrons was distorted to have smaller integrated charge at the bottom quadrant. On the other hand, when the primary current was increased from 1.7 A, the sum of the charge was decreased, the top and the right were increased, and the bottom and the left were decreased. This means that the back-streaming electrons were pushed toward upper-right by the space charge of the primary beam. The decrease of the sum may mean that the back-streaming electrons were skimmed somewhere like the entrance of the collector. The result shows that the space charge effect is not just repelling outside but distortion of the shape of the beam of the back-streaming electrons. It was shown from the test that the detector can be used to investigate this kind of complicated space charge effect on the back-streaming electrons.

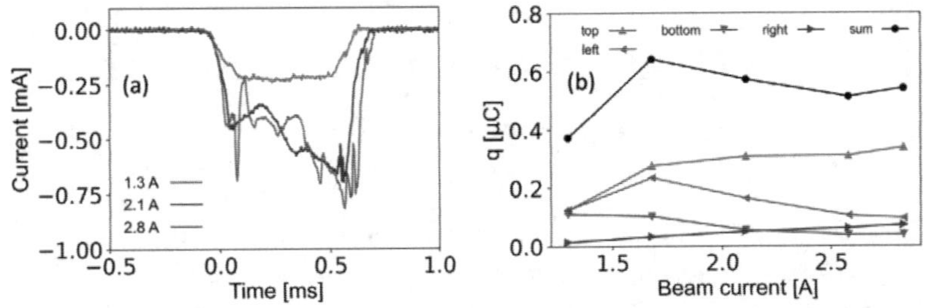

Figure 5. (a) Beam current of back-streaming electrons detected by top quadrant on the collector-side. The primary beam current were 1.3 A (red), 2.1 A (blue), and 2.8 A (green). (b) Integrated charge of the back-streaming electrons on each quadrant for different primary beam current. Red, green, magenta, and blue represents top, left, bottom, and right quadrant, respectively. Black points show the sum of the integrated charges of the quadrants.

4. Conclusion

A double-sided detector was developed for alignment of the primary electron beam and investigation of the back-streaming electrons in ExtendedEBIS. Calculation shows that the detector withstands the heat load on the primary beam, and it does not significantly affect the electrical potential. Experiments were conducted to test the detector. It was shown that the back-streaming electrons can be observed and the reaction to the space charge of the primary can be investigated with the detector.

References

[1] E. Beebe, et al., AIP Conf. Proc. **1640**(1), 5 (2015).
[2] S. Ikeda, E. Beebe, M. Okamura, and A. Pikin, AIP Conf. Proc. **2011**(1). 070008-1 (2018).
[3] M. Musgrage, et al., "Development of a Polarized 3He++ Ion Source for the EIC", The 18th International Workshop on Polarized Sources, Targets, and Polarimetry (PSTP2019), Knoxville, Tennessee, 23-27 September, 2019, PoS, 043
[4] S. Ikeda, E. Beebe, T. Kanesue, S. Kondrashev, and M. Okamura, Rev. Sci. Inst., 91(1), p.013327, (2020).
[5] A. I. Hershcovitch and A. E. Kponou, NIM A, 330, 1-2, 144-149, 1993.
[6] R.Mertzig, E. Beebe, A. Pikin, and F. Wenander, AIP Cof. Proc. 1640, 28 (2015)
[7] http://www.fieldp.com/trak.html

A Low Energy Beam Transport to Match a Multicusp Ion Source to an RFQ

Loyd Waites, Janet Conrad, Daniel Winklehner
Massachusetts Institute of Technology, Cambridge, MA, USA
E-mail: lwaites@mit.edu

Abstract. The RadioFrequency Quadrupole Direct Injection Project (RFQ-DIP) uses a filament-driven multicusp ion source to produce H_2^+. A Low Energy Beam Transport (LEBT) injects the beam from the ion source into the RFQ. The LEBT shapes the beam to the optimum input Twiss parameters for transmission through the RFQ while minimizing the emittance. The LEBT system has been designed using the accelerator codes IBSimu and Warp. To ensure the accuracy of the simulations, a series of diagnostic tools to assess the beam quality have been integrated into the LEBT design. The LEBT also integrates a beam chopper for machine protection and duty factor control, as well as steering capability to compensate for small misalignments. Presented here are the beam dynamics simulations and the design of the LEBT and matching to the RFQ.

1. Introduction

The IsoDAR experiment requires a high current proton driver capable of delivering 10!mA of 60 MeV protons on target in order to produce a high enough flux of neutrinos to reach its physics goals within five years [1–4]. To overcome space-charge limitations, we have designed the IsoDAR cyclotron to accelerate H_2^+.

To produce H_2^+ we have built a new filament-driven multicusp ion source [5–7]. We chose this source for its low emittance and high H_2^+ ratio. Both of these aspects contribute to a higher quality beam, allowing higher transmission through the machine, while limiting space charge and emittance growth in high intensity, low energy regions.

Another important technology we are leveraging in the IsoDAR system is direct injection using a Radio-Frequency Quadrupole (RFQ) embedded in the cyclotron yoke (see Figure 1). The RFQ accelerates and bunches the beam. By bunching the beam, the RFQ allows more of the beam to be in the limited phase acceptance window of the cyclotron, reducing losses in the central region. We use this to limit the current requirement from the ion source, which decreases emittance growth from space charge in low energy regions as well as strain on the ion source filament. This also allows the system to be far more compact than a traditional LEBT injection system, that often uses several magnets.

We connect the ion source and the RFQ with a short LEBT. This system serves to steer and shape the beam to match it to the RFQ acceptance while also providing beam diagnostics and machine protection. We optimized the LEBT parameters to maximize the transmission through the system.

Based on simulations completed by the IsoDAR collaboration, transmission through the full RFQ system from ion source to RFQ end is expected to be over 95 % [2, 8, 9]. However, we expect additional loses in the central region of the cyclotron on the order of 30 % [10, 11].

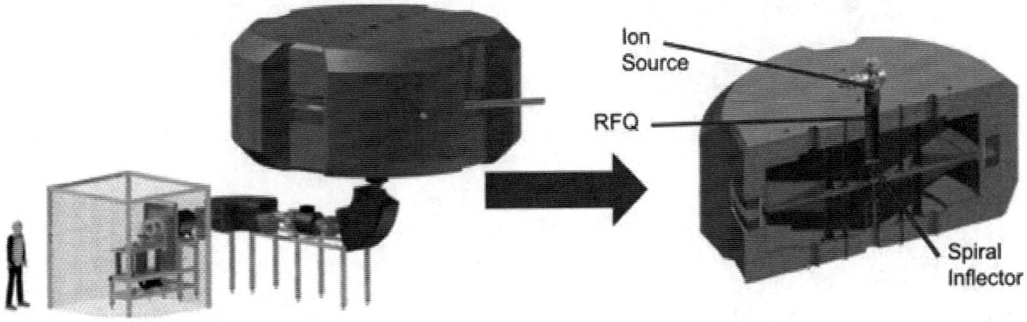

Figure 1. (Left) Traditional LEBT injection system, using a series of magnets to axially inject beam into the cyclotron. (Right) half cut view of the IsoDAR cyclotron using an RFQ direct injection system. Labeled is the ion source, RFQ, and spiral inflector in the central region of the cyclotron. This shows a much more compact system than the traditional LEBT system.

2. LEBT requirements

While matching the beam from the ion source to the RFQ, the utility of the LEBT goes beyond beam dynamics alone. We must design the LEBT to fit several design constraints.

These requirements include:

(i) The output energy of the LEBT must be 15 KeV (RFQ input energy).

(ii) The beam Twiss parameters out of the LEBT must be matched to the beam dynamics of the RFQ. These parameters are shown in Figure 3.

(iii) The electrode array should be short and accessible for underground installation.

(iv) A diagnostics section to ensure proper beam parameters into the RFQ from the LEBT. A CAD rendering of the diagnostics section can be seen in Figure 4. The diagnostic systems include:

- (a) Four button pickups to measure the position of the beam at the entrance to the RFQ.
- (b) An ACCT to measure beam current in pulsed mode.
- (c) A faraday cup to measure the beam current in DC mode.

(v) Space for vacuum pumps.

(vi) Separation from the RFQ by a gate valve.

(vii) Small angle steering for alignment using magnetic steerers.

(viii) The LEBT must include a chopper to facilitate low power commissioning and to act as a machine protection system.

We addressed these requirements in the latest LEBT design, which we describe in more detail in the next section. Then we present beam dynamics studies demonstrating the feasibility of our design choices.

Journal of Physics: Conference Series **2244** (2022) 012086 doi:10.1088/1742-6596/2244/1/012086

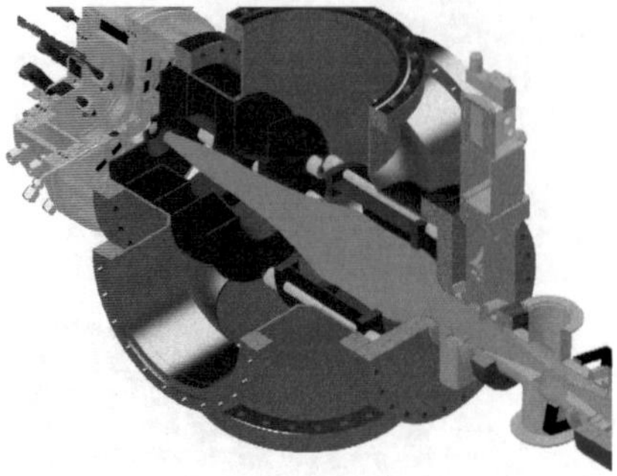

Figure 2. A beam dynamics simulation imported into a CAD model of the LEBT to show the shaping of the beam through the electrodes. The six-way cross is 450 mm in length.

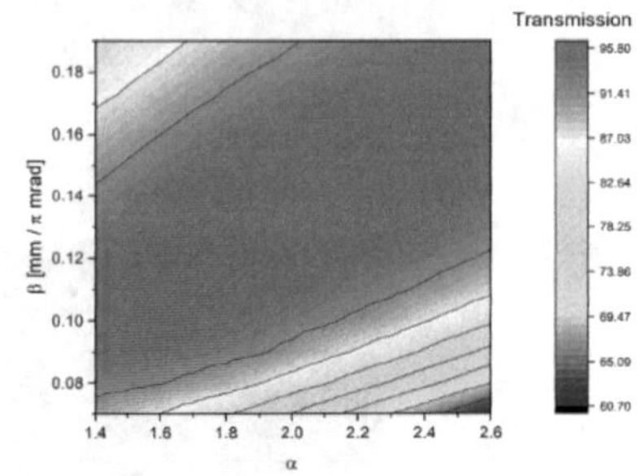

Figure 3. A temperature plot indicating the Twiss parameters needed for maximum transmission through the RFQ.

3. Mechanical Design

We keep the voltage on the source plate of the ion source at 15 kV to maintain the injected beam energy at 15 KeV. Following this, we have designed a series of electrodes which are based on the optimum orientations from our beam dynamics simulations. We use a series of electrodes which are axially aligned within the six-way cross to manipulate the beam by their shape, position, and voltage to best match the beam parameters for the RFQ at the lowest possible emittance. We use a six way cross for the electrode LEBT to maintain a system that is compact and easily accessible. The layout of this can be seen along with a beam dynamics simulation in Figure 2. The six-way cross can then be connected to a gate valve for machine protection and followed the diagnostics section, see Figure 4. The diagnostic section ensures that the beam is properly aligned and matched going into the RFQ. In the diagnostic section we added a cross which can be connected to an additional vacuum pump (which will allow for additional pumping over the diagnostic region with the vacuum pumps present in the 6-way cross), as well as a solenoid in order to allow for an additional degree of freedom for tuning the beam. We must ensure the input beam to the RFQ allows for high transmission through the RFQ. A plot showing transmission corresponding to various Twiss parameters is shown in Figure 3.

We use three rods connected at the ion source to confine each electrode to a plane. The final electrode is split, requiring a total of 6 rods from the opposite end of the six-way cross. We ensure each electrode is properly spaced by a series of insulating ceramic spacers; an inner spacer insulates the electrode from the aligning rod, and the outer spacers provide distance to match the beam dynamics simulations.

3.1. Beam Steering and Chopping

Mechanical misalignments of the ion source and lens system can cause problems in the final injection into the RFQ. To compensate for sub-millimeter misalignments, we use additional fields to manipulate the beam. These fields can be generated in two ways. First, we can include a small magnetic dipole field generated by a magnetic steerers made up of four coils placed around the beamline. We control the magnetic steerer with an external power supply and increase the dipole field to properly align the beam into the RFQ. Another way would be splitting an electrode and applying a potential difference across the two shells. A similar design

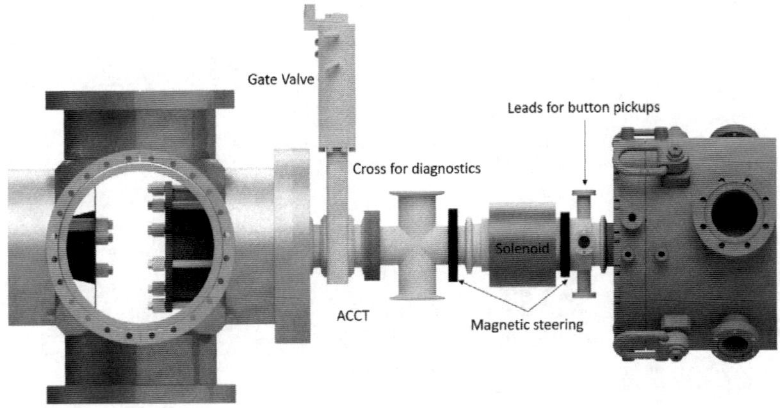

Figure 4. CAD rendering and labels of the diagnostic section of the LEBT. At the left is the six-way cross with tetrode array, seen in Figure 2. This is separated by a gate valve, followed by a miniature four-way cross and extension line to the RFQ on the right (dark gray). Outside the beam pipe is an AC current transformer (ACCT) and two magnetic steerers. Four leads for button pickups extend out from the end of the beamline to give a final position measurement. The six-way cross is 450 mm in length for scale.

was used in the LEBT at the SNS [12].

It is important to note that in order to properly align the beam into the RFQ, the beam must enter not only at the proper position, but at the proper angle. Due to this, two sets of steerers must be used for each coordinate, one to adjust the position of the beam, and one to adjust the relative angle. This would require 2 sets of magnetic steerers, or two sets of electrodes cut into 4 independent shells.

Once we make the potential difference between two electrode shells sufficiently high, it is possible to change the angle of the beam so much that the beam terminates before reaching the end of the beamline. This is known as chopping the beam. We do this to create a duty factor in an otherwise DC beam, lowering the total power from the ion source without directly affecting the plasma. This is also an important tool for machine protection, as it is possible to terminate the beam early on before it is injected into the rest of the system.

The chopper in this LEBT is the final electrode, which is cut into two shells. When we ground a single shell, the beam terminates upon the grounded shell. We have the shell water-cooled to prevent thermal damage. This has been simulated using the beam dynamics codes described in the following section.

The magnetic fields used for the magnetic steerers as well as the solenoid were calculated using COMSOL [13] and were imported into Warp. Warp then used the fields to model the impacts on the particle paths and distributions. The effects of different fields on the centroid on the beam can be seen in Figure 5.

4. Beam Dynamics Simulations

Beam dynamics for the LEBT were investigated using IBSimu [14] and Warp [15]. IBSimu has the advantage of an accurate plasma model, therefore this code was used in regions of high plasma density (ion source extraction). However, IBSimu is computationally expensive in large simulation spaces, and so the particle distibution was handed off to Warp to complete the

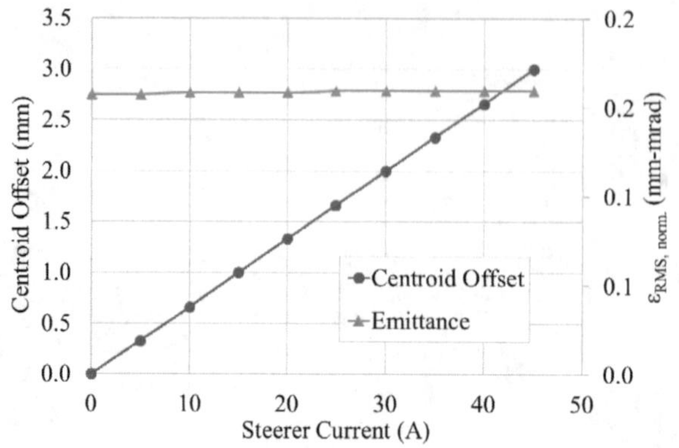

Figure 5. Plot showing the centroid position of the beam when being influenced by the magnetic steerers with different amounts of current, showing a well controlled linear relationship between current through the magnetic steerer and position.

transport through the LEBT. In our IBSimu simulations we used a total of 12 mA of beam with a ratio of 80% H_2^+ and 20% protons (as to be consistent with [6]). This provides adequate beam current for our IsoDAR system while assuming a 50% full system transmission efficiency with a current density that has been observed in our ion source experimentally in [5]. The particle distribution is then analyzed at the entrance point of the RFQ in order to compare to the desired beam input parameters. This beam dynamics simulation was optimized to demonstrate a well-matched beam for the RFQ with emittance that surpassed that used for the baseline RFQ simulations. This can be seen in Table 1.

Table 1. Comparison of LEBT Twiss Parameters

Parameter	LEBT Output	Baseline
Norm. RMS Emittance	0.157 π mm mrad	0.3 π mm mrad
Alpha	2.1	2.1
Beta	0.13 mm/mrad	0.17 mm/mrad

5. Conclusion

We designed the LEBT to properly match the beam into the RFQ, as well as provide several other important functions to the beamline including machine protection, and running in pulsed and DC mode. We have reached these goals, as demonstrated using the simulation tools IBSimu and Warp. Using these simulations, we estimate transmission of over 95 % through the RFQ direct injection system, with additional losses of approximately 30 % in the central region. This high transmission is crucial to the goal of the parent project, IsoDAR of delivering 10 mA of 60 MeV protons on target, a factor 10 higher current than commercial cyclotrons can deliver at the moment.

References

[1] Abs M, Adelmann A, Alonso J, Axani S, Barletta W, Barlow R, Bartoszek L, Bungau A, Calabretta L, Calanna A *et al.* 2015 *arXiv preprint arXiv:1511.05130*

[2] Winklehner D, Bahng J, Calabretta L, Calanna A, Chakrabarti A, Conrad J, D'Agostino G, Dechoudhury S, Naik V, Waites L *et al.* 2018 *Nuclear Instruments and Methods in Physics Research Section A: Accelerators, Spectrometers, Detectors and Associated Equipment* **907** 231–243

[3] Conrad J, Shaevitz M, Shimizu I, Spitz J, Toups M and Winslow L 2014 *Physical Review D* **89** 072010

[4] Bungau A, Adelmann A, Alonso J R, Barletta W, Barlow R, Bartoszek L, Calabretta L, Calanna A, Campo D, Conrad J M, Djurcic Z, Kamyshkov Y, Shaevitz M H, Shimizu I, Smidt T, Spitz J, Wascko M, Winslow L A and Yang J J 2012 *Physical Review Letters* **109** 141802 publisher: American Physical Society URL `https://link.aps.org/doi/10.1103/PhysRevLett.109.141802`

[5] Winklehner D, Conrad J M, Smolsky J and Waites L 2020 *arXiv preprint arXiv:2008.12292*

[6] Ehlers K and Leung K 1983 *Review of Scientific Instruments* **54** 677–680

[7] Winklehner D, Axani S, Bedard P, Conrad J, Corona J, Hartwell F, Smolsky J, Tripathee A, Waites L, Weigel P, Wester T and Yampolskaya M 2017 *Proceedings of the 17th International Conference on Ion Sources*

[8] Winklehner D, Hamm R, Alonso J and Conrad J 2015 *6th International Particle Accelerator Conference (IPAC2015)*

[9] Winklehner D, Hamm R, Alonso J, Conrad J M and Axani S 2016 *Review of Scientific Instruments* **87** 02B929 (*Preprint* `http://dx.doi.org/10.1063/1.4935753`) URL `http://dx.doi.org/10.1063/1.4935753`

[10] Winklehner D, Adelmann A, Conrad J M, Mayani S, Muralikrishnan S, Schoen D and Yampolskaya M 2021 *arXiv preprint arXiv:2103.09352*

[11] Alonso J, Axani S, Calabretta L, Campo D, Celona L, Conrad J M, Day A, Castro G, Labrecque F and Winklehner D 2015 *Journal of Instrumentation* **10** T10003 (2015), publisher: IOP Publishing

[12] Nath S, Billen J, Stovall J, Takeda H, Young L, Jeon D and Kim S 2002 *Proc. of the XXI Intl. Linac Conf*

[13] Multiphysics C 1998 *COMSOL Multiphysics, Burlington, MA, accessed Feb* **9** 2018

[14] Kalvas T, Tarvainen O, Ropponen T, Steczkiewicz O, Ärje J and Clark H 2010 *Review of Scientific Instruments* **81** 02B703

[15] Friedman A, Cohen R H, Grote D P, Lund S M, Sharp W M, Vay J L, Haber I and Kishek R A 2014 *IEEE Transactions on Plasma Science* **42** 1321–1334

Design and simulations for a Penning Ion Trap Source for the CHIP-TRAP Mass Spectrometer

M. Horana Gamage[1], R. Bhandari[1], N. D. Gamage[1], D. Keblbeck[1] and M. Redshaw[1,2]

[1]Department of Physics, Central Michigan University, Mount Pleasant, MI 48859, USA
[2]National Superconducting Cyclotron Laboratory, East Lansing, MI 48824, USA

E-mail: horan1mv@cmich.edu, redsh1m@cmich.edu

Abstract. At Central Michigan University (CMU) we have been developing the CMU High Precision Penning Trap (CHIP-TRAP) mass spectrometer for precise mass measurements on stable and long-lived radioactive isotopes. As part of this apparatus, we have designed a Penning Ion Trap (PIT) source to produce singly-charged, low intensity (∼100s to 1000s of ions in ∼1 μs duration) ion pulses from gaseous samples. The PIT source is similar to a PIG type source, composed of an electric field superimposed on a magnetic field. The electric field is produced by a cylindrical Penning trap structure consisting of two end caps and a center ring with a trap volume of about 0.8 cm^3. The trap structure is housed inside a permanent neodymium ring magnet bore. Gas is inserted into the trapping region and ionized by an ∼1 μA electron beam from a thermal emitter. Ions are extracted from the PIT source by lowering the voltage on one of the end caps. They are then accelerated into the beam line. In this paper, we report on the design of the PIT source and simulations to investigate and characterize the expected beam properties.

1. Introduction

Ion sources are an important component in apparatus and equipment in a wide range of experiments and applications across the physical sciences and in industry and medicine [1]. A wide range of ion sources exist to meet the needs of these applications by providing ions from elements across the periodic table with various chemical properties, and with intensities, energies, spatial properties and charge state optimized to the particular application.

In the field of Penning trap mass spectrometry (PTMS), the need for ion sources that can access a wide range of chemical elements, and the means to deliver low emmittance short duration pulsed beams of ions to the apparatus is paramount. Sophisticated techniques to trap and cool continuous ion beams using a combined He gas-filled RFQ ion guide and Paul trap have been developed to prepare ion bunches for injection and capture in the Penning trap e.g. [2, 3]. These techniques were developed mostly to enable measurements with beams of rare isotopes produced at various accelerator facilities. However, they also apply to ions delivered from offline ion sources. In recent years, the laser ablation ion source (LAS) has become a workhorse of PTMS experiments for the delivery of stable and long-lived isotopes for use as reference ions [4], and for mass measurements with applications in nuclear and neutrino physics e.g. [5]. Stable and long-lived isotopes are also regularly obtained from surface ionization sources, and plasma ion sources e.g. [6], or are made *in situ* in the Penning trap via electron impact ionization of gas and vapors e.g. [7].

In this paper we describe the design and simulations for a Penning Ion Trap (PIT) source. This ion source uses electron impact ionization to create ions from gaseous samples. Ions are confined via the combined magnetic and electric field of a simple cylindrical geometry Penning trap. The PIT source is designed to produce singly-charged, low intensity, small bunches ($\sim$ 100s to 1000s) of ions in short ($\sim$1 μs) duration [8].

2. Principle and design of the Penning Ion Trap source

2.1. Principle of the Penning Ion Trap source

The Penning Ion Trap (PIT) source is based on ion production via electron impact ionization of gas samples that are leaked into the ion source. After they are produced, ions are confined to a small region ($\sim$0.8 cm^3) via a combined magnetic and electric field in a cylindrical Penning trap geometry. Trapping enables the accumulation of ions to form a bunch that can then be released as a short duration pulse.

The magnetic field is provided by a permanent NdFeB ring magnet with 1.27 cm height, 4.45 cm outer diameter, and 1.27 cm bore diameter (K&J Magnetics part #RXC88 [9]). The magnetic field through the bore is 0.55 T. The ring electrode is located inside the magnet bore, and the end caps are at the bore openings. The electrodes produce a weak quadrupole electric field using nominal trap potentials of $\sim$5 V. Ions confined in the trapping region undergo the usual motions for an ion in a Penning trap: the three normal trap-cyclotron, magnetron and axial modes of motion. The end-caps have 3 mm diameter holes in the centers to allow the electron beam to enter from one end, and to admit gas into the trap and to allow ions to be extracted from the other end.

2.2. Design of the Penning Ion Trap Source

For confinement of ions in a Penning trap, the following condition must be satisfied [10]

$$\omega_c^2 > 2\omega_z^2, \tag{1}$$

where $\omega_c = qB/m$ is the free-space cyclotron frequency for an ion with charge-to-mass ratio m/q in magnetic field B, and $\omega_z = \sqrt{qV/md^2}$ is the axial frequency of an ion in a trap with characteristic trap size d and potential difference V applied between the end-cap and ring electrodes. Hence, for stable trapping, the magnetic field strength must satisfy

$$B > \sqrt{\frac{2Vm}{qd^2}}. \tag{2}$$

This condition indicates that to trap higher m/q ions a stronger B-field is required. This can be alleviated to some extent by using a lower trapping potential, but reducing V will reduce the trap depth and the number of ions that can be confined. The trap size, d, also influences the B-field requirements, with a larger trap size reducing the B-field required to trap a particular m/q.

With these considerations in mind, and the goal to be able to create, and hence trap, ions with m/q up to $\sim$40 – 50 in the PIT source, we found that a 0.55 T magnetic field with an approximately 10 mm length by 10 mm diameter trapping region, and a 5 V trapping potential would serve our needs.

For simplicity, and because a precise control over the frequencies of the normal modes of motion of ions in the trap was not required for this application, we used a cylindrical trap geometry with no correction electrodes. Nevertheless, the axial trap length, $2z_0$, and trap

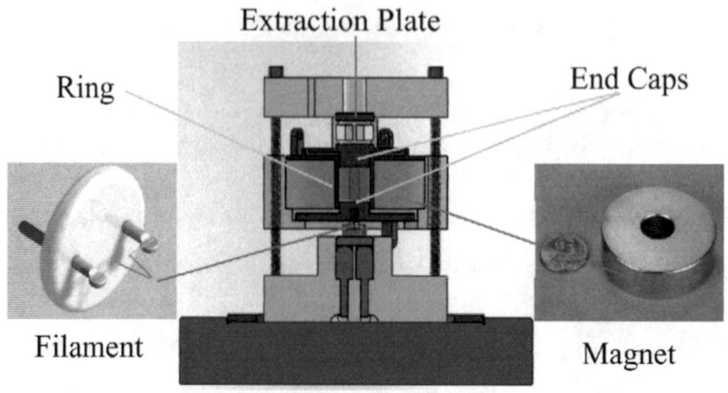

Figure 1: Cross sectional CAD drawing view of the PIT source (middle), the thermal emitter (bottom left), and NdFeB ring magnet (bottom right).

diameter, $2\rho_0$, were chosen to follow the relationship $\rho_0 = 1.203z_0$ to minimize the lowest order electric field imperfection, C_4 [11].

To produce a narrow diameter electron beam for electron impact ionization of gas in the trap, we used a tungsten thermal emitter (Ted Pella AEI filament #1403 [12]), capable of producing electron beam currents up to $\sim$50 μA with an emission area of 50 $\times$ 100 μm. The trap electrodes were machined from OFHC copper, and the magnet, trap electrodes, filament and extraction electrodes were housed in a PEEK structure, attached to a 4.5" ConFlat® flange, and installed inside a 6-way cross. Gas can enter the ion source through a precision leak valve connected to a 1/8" copper tube that opens over the hole in the end-cap at the opposite end to the end-cap hole that the electron beam enters the trap. A picture of the magnet and filament, and cross-section schematic view of the PIT source is shown in Fig. 1.

3. Simulation of the Penning Ion Trap Source

3.1. Model of PIT source and simulated experimental runs

SIMION, see e.g. Ref. [13, 14], was used for modelling the PIT source and beamline, and for simulating the motion of ions in the trap, their extraction, and their transport along the beamline. The electric field in the trap was calculated in SIMION from the model of the trap electrodes, as described above. The magnetic field inside the bore of the NdFeB magnet was mapped out using a Hall probe. These data were then used to model the magnetic field in SIMION.

In the PIT source, electrons pass through the trapping region along the central axis. Any atoms and molecules that are ionized by the electron beam along this path and between the two end caps will be automatically confined in the trap. Ions were modeled to have an initial energy of 30 meV, corresponding to room temperature, and a randomized initial velocity. The ions were created with a random position along the axis of the PIT in a cylinder with radius 0.25 mm. Initial simulations were performed to verify ion confinement in the trap, then ion extraction was simulated by lowering the potential on the ejection end-cap of the PIT, reducing it linearly as a function of opening time. The voltages of the extraction electrodes, quadrupole bender, and einzel lenses in the beamline were optimized to transport the beam and focus it on to a surface in the simulation that represented the location of the MCP. An image showing ion transportation from the PIT to the MCP is given in Fig. 2(a).

Besides testing the ability of the PIT source to trap and release bunches of ions, we used our simulation to benchmark experimentally measurable quantities, such as time-of-flight (TOF) and, in particular, the FWHM of the TOF peak for ion bunches arriving at the MCP. We used our simulation to investigate the effect of trap opening time, axial distribution of ions in the

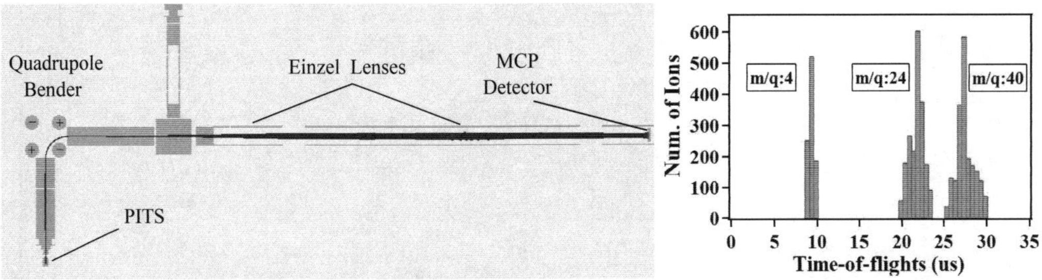

(a) (b)

Figure 2: (a) Simulation of ion beam transportation from the PIT source to the MCP detector. (b) Separation of ions with different m/q ratios at the MCP via their time-of-flight.

trap, and damping due to background gas in the trap on the TOF peak width. We varied the axial distribution length from 0 mm up to 8 mm (close to $2z_0$), and varied the trap opening time from 0.5 – 50 μs. For the damping studies, the pressure was varied from 10^{-3} – 10^{-7} torr. Collisions were modelled using the hard-sphere collision model in which ions and neutral background gas undergo elastic collisions and the velocity of the gas particles follow a Maxwell-Boltzmann distribution [15, 16]. This model was based on on existing SIMION Lua codes [17]. In the simulations reported here, ions of $m/q = 24$ u/e were used as a representative middle ground for the range of m/q ions we expect to use in the PIT source (e.g. up to $m/q \sim 40 - 50$).

3.2. Simulation results

The results from simulations of ion transport from the PIT source to the MCP in which the trap opening time and axial distribution of ions in the trap were systematically varied are shown Fig. 3(a). Each data point included 1000 trapped ions that were then released from the PIT source and transported to the MCP. The ions' TOFs to the MCP were recorded, as in Fig. 2(b), and the FWHM of the TOF distribution was obtained. Fig. 3(a) shows that the FWHM of the TOF peak increases with the length of the axial distribution of ions in the trap. The FWHM remains constant until the trap opening time increases above 3 μs, after which it starts to increase. The increase in the FWHM is more significant for the longer initial axial distributions of ions in the trap.

Results from simulations in which the collisions with background gas in the trap were modelled are shown in Fig. 3(b). Again, for each data point 1000 ions were confined in the PIT source before being released and their TOF to the MCP recorded. In this case, the time that the ions spent in the trap before being released i.e. the damping time in the trap, was varied from 0 – 3000 μs, and this was done for four different background gas pressures in the trap. The simulations show that no decrease in the FWHM of the TOF peak of the ion bunch was observed during the 3000 μs trapping time for pressures of 10^{-5} and 10^{-7} torr. In other words, no reduction in the energy spread of ions occurred due to damping effects. For pressures of 10^{-4} and 10^{-3} torr there was a reduction in the FWHM of the TOF peak.

3.3. Comparison with preliminary experimental data

A preliminary commissioning and study of our PIT source has been completed. We were able to make ions in the trap by ionizing background gas and by leaking in helium or argon gas. These ions could then be released from the trap and their TOF to the MCP recorded. An example of the preliminary background gas ionization is shown in figure 4. The trap opening time for the switch used in our experimental set-up is 500 ns. A calibration of the TOF of ions to the MCP (see e.g. Ref. [8]) determined that the background gas ions had an $m/q \approx 19$ u/e,

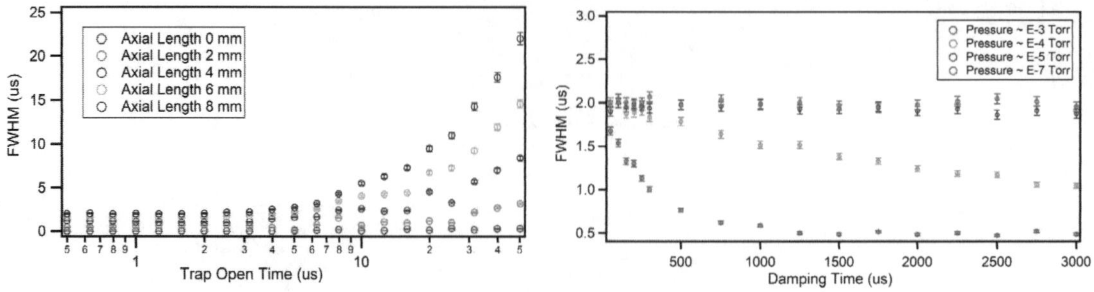

(a) (b)

Figure 3: Results from simulations showing the FWHM of the ion distribution on the MCP for (a) different trap opening times and initial ion distribution lengths, and (b) different trapping times to investigate the damping effect for different background pressures in the trap.

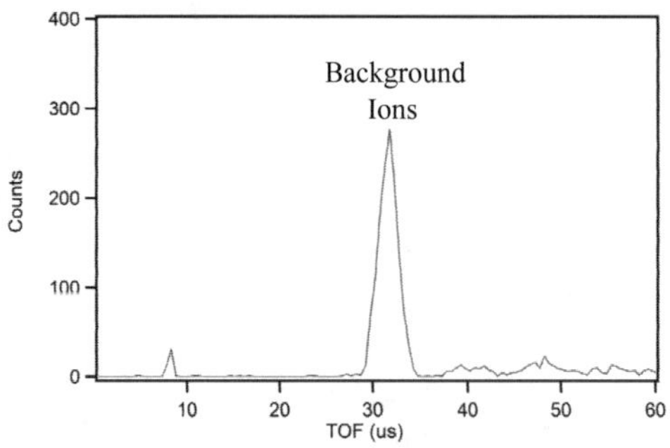

Figure 4: Histogram of counts on the MCP detector from ions made via ionization of background gas in the PIT source. The FWHM of the peak is about 2.5 μs.

indicating that they could be H_3O^+. This ion peak had a FWHM of about 2.5 μs, consistent with our simulations for an axial distribution length of 8 mm, trap opening time of 500 ns, and pressure $>10^{-5}$ torr (a nearby Penning gauge recorded the pressure as $\sim 10^{-8}$ torr during these measurements). By increasing the leak rate of He or Ar, the pressure in the vacuum chamber could be increased to $\gtrsim 10^{-6}$ torr. No change in the FWHM of the ions' TOF peak was observed when varying the pressure from $\sim 10^{-6} - 10^{-8}$ torr, consistent with the simulation results shown in Fig. 3(b).

4. Conclusions

In this paper, we have discussed the design and simulations for a novel Penning ion trap (PIT) source to provide low intensity pulsed ion beams produced from gaseous samples. The PIT is intended to be used as an ion source for the CHIP-TRAP mass spectrometer. Our simulations show that the initial axial distribution of ions in the trap has a significant effect on the temporal spread of the pulse, as measured via the FHWM of the TOF peak. The opening time of the trap has no effect on the FWHM of the TOF peak for opening times ≤ 3 μs. Our simulations indicate that damping via collisions with background gas will not affect the ion motion for the expected operating pressures of $\leq 10^{-5}$ torr.

Comparison of simulations and preliminary experimental results that gave a measured TOF FWHM of 2.5 μs lead us to conclude that the ions are initially distributed over the entire

Journal of Physics: Conference Series **2244** (2022) 012087 doi:10.1088/1742-6596/2244/1/012087

axial length of the trap. This is consistent with the ions being produced via electron impact ionization all along the trap axis. We also conclude that the ions' axial motion is not damped via interaction with background gas at pressures of $\sim 10^{-8} - 10^{-6}$ torr. i.e. they retain their initial axial amplitude while they are trapped, and that our trap opening time of 500 ns does not contribute to the temporal spread of the ion pulse.

References

[1] Wolf B 1996 *Handbook of ion sources* (CRC Press, Boca Raton)
[2] Herfurth F, Dilling J, Kellerbauer A, Bollen G, Henry S, Kluge H J, Lamour E, Lunney D, Moore R, Scheidenberger C, Schwarz S, Sikler G and Szerypo J 2001 *Nuclear Instruments and Methods in Physics Research Section A: Accelerators, Spectrometers, Detectors and Associated Equipment* **469** 254–275 ISSN 0168-9002 URL https://www.sciencedirect.com/science/article/pii/S0168900201001681
[3] Schwarz S, Bollen G, Ringle R, Savory J and Schury P 2016 *Nuclear Instruments and Methods in Physics Research Section A: Accelerators, Spectrometers, Detectors and Associated Equipment* **816** 131–141 ISSN 0168-9002 URL https://www.sciencedirect.com/science/article/pii/S0168900216001194
[4] Blaum K, Bollen G, Herfurth F, Kellerbauer A, Kluge H J, Kuckein M, Sauvan E, Scheidenberger C and Schweikhard L 2002 *The European Physical Journal A* **15** 245–248 URL https://doi.org/10.1140/epja/i2001-10262-4
[5] Kandegedara R M E B, Bollen G, Eibach M, Gamage N D, Gulyuz K, Izzo C, Redshaw M, Ringle R, Sandler R and Valverde A A 2017 *Phys. Rev. C* **96** 044321 URL https://link.aps.org/doi/10.1103/PhysRevC.96.044321
[6] Lincoln D L, Holt J D, Bollen G, Brodeur M, Bustabad S, Engel J, Novario S J, Redshaw M, Ringle R and Schwarz S 2013 *Phys. Rev. Lett.* **110** 012501 URL https://link.aps.org/doi/10.1103/PhysRevLett.110.012501
[7] Cornell E A, Weisskoff R M, Boyce K R, Flanagan R W, Lafyatis G P and Pritchard D E 1989 *Phys. Rev. Lett.* **63**(16) 1674–1677 URL https://link.aps.org/doi/10.1103/PhysRevLett.63.1674
[8] Horana Gamage M, Arnold A, Bhandari R, Gamage N, Purcell Z, Sandler R and Redshaw M 2019 *Hyperfine Interactions* **240** 1–9 URL https://doi.org/10.1007/s10751-019-1617-4
[9] K&J Magnetics, Inc 2019 Strong neodymium magnets URL https://www.kjmagnetics.com
[10] Brown L S and Gabrielse G 1986 *Rev. Mod. Phys.* **58** 233–311 URL https://link.aps.org/doi/10.1103/RevModPhys.58.233
[11] Gabrielse G and Mackintosh F 1984 *Int. J. Mass Spectrom. Ion Process.* **57** 1–17
[12] Ted Pella, Inc 2019 Microscopy products for science and industry URL http://www.tedpella.com
[13] Dahl D 2000 *International Journal of Mass Spectrometry* **200** 3–25
[14] Appelhans A and Dahl D 2002 *International Journal of Mass Spectrometry* **216** 269–284
[15] Ding L, Sudakov M and Kumashiro S 2002 *International Journal of Mass Spectrometry* **221** 117–138
[16] Gillig K J, Ruotolo B T, Stone E G and Russell D H 2004 *International Journal of Mass Spectrometry* **239** 43–49
[17] Adaptas Solutions, LLC 2003-2021 Scientific instrument services (sis) URL https://simion.com/

Generation of proton and carbon ion beam from laser ion source using hydrocarbon oil target

K Takahashi[1], H Katane[1], K Miyazaki[1], N Harukawa[1], K Ishikuro[1], T Sasaki[1] and T Kikuchi[1,2]

[1] Department of Electrical Engineering, Nagaoka University of Technology, Nagaoka, Niigata 940-2188, Japan
[2] Department of Nuclear System Safety Engineering, Nagaoka University of Technology, Nagaoka, Niigata 940-2188, Japan

E-mail: kazumasa@vos.nagaokaut.ac.jp

Abstract. The supply of high-flux proton and carbon ions are required for applications such as accelerator-driven neutron sources and heavy ion radiotherapy. So far, it is shown that a laser ion sources using the target of plastics made of hydrocarbons can provide protons and carbon ions. On the other hand, due to the damage of the target by laser irradiation, it is necessary to change the laser irradiation position every time, and the number of ion supply is limited to the area of the laser target. To solve the problem of target damage, we suggest a laser target in which the surface is continuously renewed by flowing hydrocarbon oil instead of solid hydrocarbon. In this study, we demonstrated the generation of protons and carbon ions using the target made of a hydrocarbon oil for a diffusion pump.

1. Introduction

A laser ion source provides an ion beam from the plasma generated by irradiating the target with a laser, and is expected as an ion source that can supply a large current ion beam. It has been used to supply the ions of metal species required for high energy physics, simulating biological effects of space radiation [1] and studied as a carbon ion source to use for particle beam therapy [2] by directly injecting a laser generated plasma into an RFQ linear accelerator [3-7], and a lithium driver for an accelerator-driven neutron source with a lithium beam [8,9]. In a laser ion source, the pulse width of beam is proportional to the transport distance of the plasma, so if the transport distance of the laser generated plasma is shortened and ions are extracted from the plasma near the laser target, short pulse ions can be extracted. For applying to an accelerator driven neutron source for neutron imaging with good resolution, short pulse protons are required to make a short pulsed neutron [10]. Since protons are gaseous elements at room temperature, solid compounds must be used to supply them with a laser ion source, but metal hydride [11,12] and hydrocarbon plastic targets [13,14] can be used.

To generate multi-charged ions or high current ion beam, a laser irradiation with high power density is required. In addition, since the amount of desired ion species decreases using a compound target containing multiple elements such as alloys, ceramics, and plastics is used, compared with the target made of a single element, it is necessary to increase the power density of the laser in order to compensate for this decrease. Increasing the laser power density increases the damage to the target due to laser irradiation. In such use, the number of laser irradiations is limited to the target area, thereby the life-time of the laser target becomes an issue for the steady operation of the laser ion source.

Therefore, in order to solve this problem, we suggest a laser ion source with a target of a circulating liquid that can continuously replace the target surface. In this study, a liquid consisting of hydrocarbons is used for supplying protons and carbon ions. Among such candidates, we adopted oil for oil diffusion vacuum pumps with low vapor pressure as a liquid that can be handled in vacuum. A target whose surface is continuously replaced by circulating this oil. We constructed a laser ion source using the oil, and demonstrated the generation of plasma and investigated the reproducibility of the ion current waveform of the plasma.

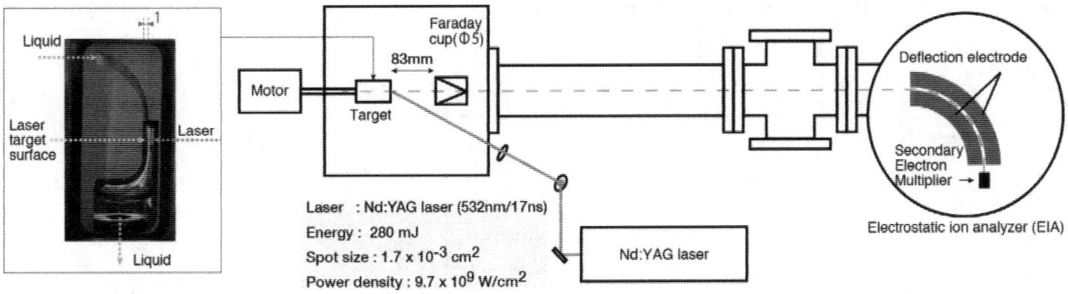

Figure 1. Schematic of experimental setup

2. Experimental setup

As shown in Fig. 1, the experimental setup was constructed to circulate the liquid in a vacuum as a liquid target, and the liquid was circulated by driving a magnet pump from outside the vacuum chamber using a rotary motion feedthrough. The liquid target was vacuum pump oil (ULVOIL D-11) for an oil diffusion pump consisting of hydrocarbons (mainly alkylnaphthalene), and the vapor pressure at 20 °C in the experimental environment was 7.3×10^{-5} Pa. The liquid pump was driven using a motor with 900 RPM. As a result of measuring the flow velocity by observing the movement of small bubbles in the liquid with a moving image, the flow velocity was 0.9 cm/s.

In this experiment, plasmas were generated at a pressure of 5×10^{-4} Pa in a vacuum chamber. The laser used for plasma generation was a Nd: YAG laser with a wavelength of 532 nm, the pulse width of 17 ns, and the energy of 280 mJ. The laser was focused on the target using a lens with a focal distance of 300 mm and irradiated with a laser power density of 9.7×10^9 W/cm^2 on the target.

The ion current of the plasma was measured using a Faraday cup. The Faraday cup (FC) with an aperture of 5 mm in diameter was installed at 83 mm from the laser target, and a retarding voltage of -300 V was applied to the cup. It should be noted that this FC does not have an electron-suppressor to prevent emitting secondary electrons induced by hitting ions on the FC. Therefore, the FC signals indicate the actual ion signals with induced current by emitted secondary electrons, which means the FC signals may indicate a larger ion current than the actual ion current extracted from the laser produced plasmas. In addition, an electrostatic ion analyzer whose detector was a secondary electron multiplier (PHOTONIS, MAGNUM5900) was used to investigate the generated ion species and charge state distribution. In the electrostatic ion analyzer, only ions having specific velocity and charge state can pass between the deflection electrodes to which positive and negative voltages are applied. The distribution of the ion charge state was obtained by changing the applied voltage to the electrodes.

Results and discussion

Figure 2 indicates a typical signal waveform obtained using the electrostatic ion analyzer and the generation of plasma composed of proton and carbon ions was confirmed based on the time-of-flight as shown in this result. The ion charge-state distribution can be obtained by scanning the voltage applied to deflection electrodes in the electrostatic ion analyzer.

Figure 3 shows the ion charge state distribution obtained by averaging signals of three laser shots. The proton was relatively faster than carbon ions and the carbon ions with higher charge state tend to be faster as well as other laser ion sources.

Figure 4 indicates FC signals with 20 laser shots under the same condition. The pulse width was a few μs (FWHM) and the FC signals with on order of 0.1 mA were obtained. Sometimes two peaks in the waveform were observed corresponding to proton and carbon ions.

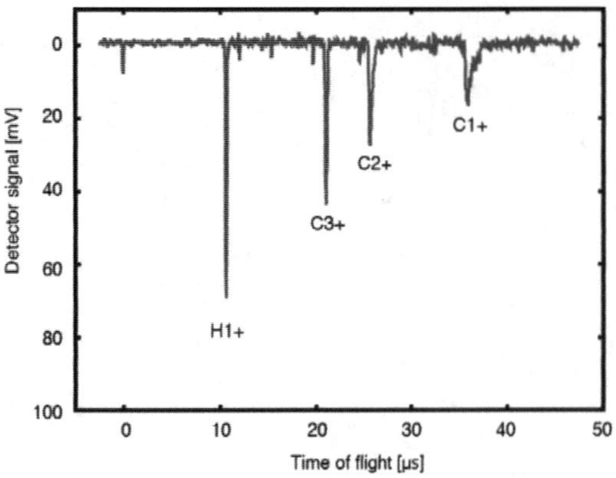

Figure 2. The typical proton and carbon ion signals obtained using the electrostatic ion analyzer.

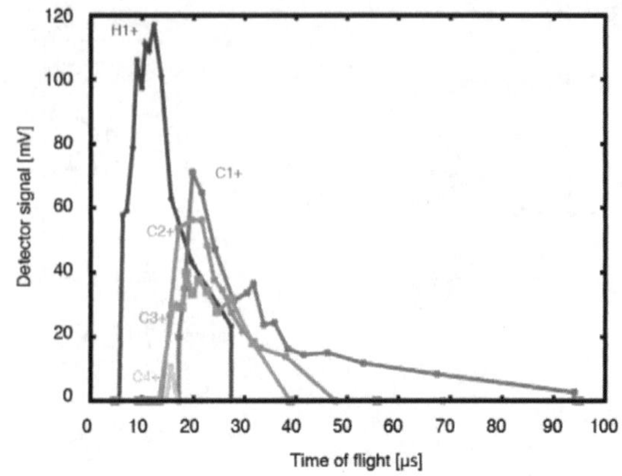

Figure 3. The ion charge-state distribution obtained by scanning the voltage applied to deflection electrodes in the electrostatic ion analyzer.

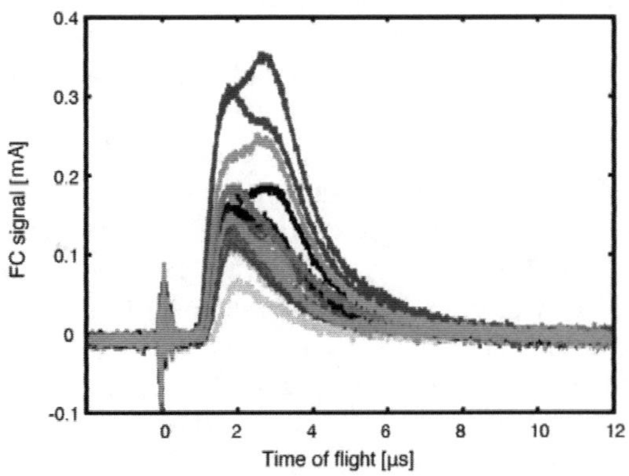

Figure 4. FC signals measured with the same condition. The signal waveforms with 20 shots are shown.

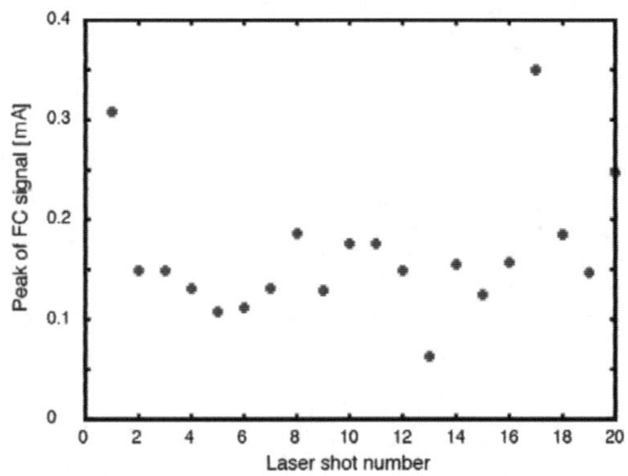

Figure 5. Variation of the peak value of FC signals as a function of the number of laser irradiation.

Furthermore, the reproducibility of the ablation plasma obtained with a liquid target was examined. Figure 5 shows the variation of the peak value of FC signals as a function of the number of laser irradiation. This result does not indicate clear tendency of increase or decrease in FC signals as increasing the laser shot number. The deviation in the FC signals may be caused by the fluctuation of the surface of the liquid target. Moreover, splashes of liquid target at laser irradiation were observed every laser shot. This phenomenon reduced the circulating target rapidly and would change the target condition.

Conclusion

In order to avoid target damage due to laser irradiation, we proposed a laser ion source using a target in which the laser irradiation surface is continuously replaced using a circulating liquid instead of a solid target. We developed a liquid target laser ion source using hydrocarbon oil for oil diffusion vacuum pumps with low vapor pressure for supplying protons and carbon ions. We demonstrated the generation of protons and carbon ions by the laser ion source with the liquid target.

References

[1] Okamura M *et al.* 2016 *Proc. 2nd North American Particle Accelerator Conference (NAPAC2016), (Chicago)* pp 49-53
[2] Sako T *et al.* 2016 *Rev. Sci. Instrum.* **87** 02C109
[3] Okamura M *et al.* 2002 *Rev. Sci. Instrum.* **73** 761
[4] Takeuchi T *et al.* 2002 *Rev. Sci. Instrum.* **73** 764
[5] Takeuchi T *et al.* 2002 *Rev. Sci. Instrum.* **73** 767
[6] Kashiwagi H *et al.* 2004 *Rev. Sci. Instrum.* **87** 1569
[7] Okamura M *et al.* 2008 *Rev. Sci. Instrum.* **79** 02B314
[8] Ikeda S *et al.* 2020 *Rev. Sci. Instrum.* **91** 023304
[9] Cannavò A *et al.* 2020 *Rev. Sci. Instrum.* **91** 033317
[10] Shinohara T *et al.* 2016 *Journal of Physics: Conference Series* **746** 012007
[11] Sekine M *et al.* 2012 *Rev. Sci. Instrum.* **83** 02B318
[12] Okamura M *et al.* 2016 *Rev. Sci. Instrum.* **87** 02A906
[13] Torrisi L *et al.* 2012 *Rev. Sci. Instrum.* **83** 02B310
[14] Takahashi K *et al.* 2019 *Rev. Sci. Instrum.* **90** 123311

Resolution improvement of electrostatic ion analyzer using additional slit for laser ion source

K Takahashi[1], **Y Matsumoto**[1], **T Sasaki**[1], **T Kikuchi**[1,2]

[1] Department of Electrical Engineering, Nagaoka University of Technology, Nagaoka, Niigata 940-2188, Japan
[2] Department of Nuclear System Safety Engineering, Nagaoka University of Technology, Nagaoka, Niigata 940-2188, Japan

E-mail: kazumasa@vos.nagaokaut.ac.jp

Abstract. A laser ion source provides an ion beam extracted from a drifting plasma produced by irradiating a high power laser on a solid target. Ion species contained in the plasma can be analyzed with an electrostatic ion analyzer (EIA). However, to distinguish ion species with close mass-to-charge ratio, high resolution is required due to the high drifting velocity. In order to improve the resolution of the EIA, we investigated the cause of spectral width broadening by simulation. The simulation result showed that the spectral broadening occurs because ions go through the paths in different angle rather than it is caused by difference of ion velocity in the conventional analyzer using two slits. To limit the angular distribution of the paths, we suggested to use an additional slit for the EIA. To demonstrate the effect of the additional slit on the spectral width broadening for the EIA, we investigated the variation in the spectral signals obtained with the suggested EIA. The experimental result indicated that the resolution of EIA can be improved by adding a slit to limit the trajectories in angular distribution of ion paths.

1. Introduction

A laser ion source is an ion source that supplies an ion beam from a plasma generated from a solid by irradiating with a high-power pulse laser [1-4]. It is easy to generate high flux multi-charged ions by increasing the laser power density. The laser ion source has been used to supply metal ion species for high energy physics experiments, galactic cosmic ray simulation [5] and expected to be applied to carbon ion sources for heavy ion radiotherapy [6], and lithium ion sources for accelerator neutron sources [7,8]. To identify the ion species obtained from the laser ion source, an electrostatic ion analyzer (EIA) is often used. The signal waveform is obtained by the EIA as spectrum and the plasma ion components can be derived based on the mass-to-charge ratio estimated by the time-of-flight (TOF) for the ion trajectory bended by the electric field generated between the two deflection electrodes.

The laser ablation plasma has a drift velocity perpendicular to the target surface. In particular, when a plasma is generated by laser irradiation with high power density, ions enter the analyzer with a high drift velocity. Then, the difference in flight time is small for ion species with a close mass-to-charge ratio such as $^{7}Li^{3+}$ and $^{16}O^{7+}$ and the separation of signals of these ions requires a narrow spectral width and high mass resolution [9]. In order to improve the resolution of the EIA by minor modification, we investigated the cause of the broadening in spectral width by simulation. Moreover, we suggested the method to reduce the spectral width and demonstrated the improvement of resolution of the EIA experimentally.

2. Simulation of ion trajectory

The calculation of particle trajectories in an EIA was performed to investigate a structure that reduces the difference of the TOF for ions. In this calculation, it is assumed that the EIA has the inner electrode of R_1 = 150 mm, and the outer electrode of R_2 = 160 mm with thickness of 20 mm as shown in Fig. 1. The region in x = −200 to 0 mm is defined as an analyzer and the distance from the edge of electrodes and the boundary of the analyzer was set to be 15 mm. The particle trajectories were calculated based on the velocity and position obtained by the Leap-Frog method. The ion trajectories with incident velocity v_i ranging 199.0 to 201.0 km/s in 0.1 km/s increments and incident angle θ ranging −2.0 to +2.0 ° in 0.1 ° increments were simulated for investigating the influence on the TOF. In the simulation, Li^{3+} ions were assumed as particles.

Figure 1 shows the particle trajectories with the incident velocity v_i = 200.0 km/s and incident angles θ = −2.0 to +2.0 °. The incident angle θ is negative on the inner electrode side and positive on the outer electrode side. The result indicates that the focal position of the ion trajectory is around 113 mm from the exit of the analyzer. Figure 2 shows the particle trajectory near the focal position of ion trajectories when focal slit with a 0.5 mm width is set at the focal position (x = 113 mm). Since the focal position depends on the velocity of the ions, the velocity of the ions passing through the focal slit is limited and ions with faster than 200.2 km/s and slower than 199.9 km/s cannot go through the slit in this case.

Figure 3 shows the TOF as a function of the incident angle as the transport distance from the laser target to the detector was set to 1 m on the central orbit. The result showed that the TOF increased as

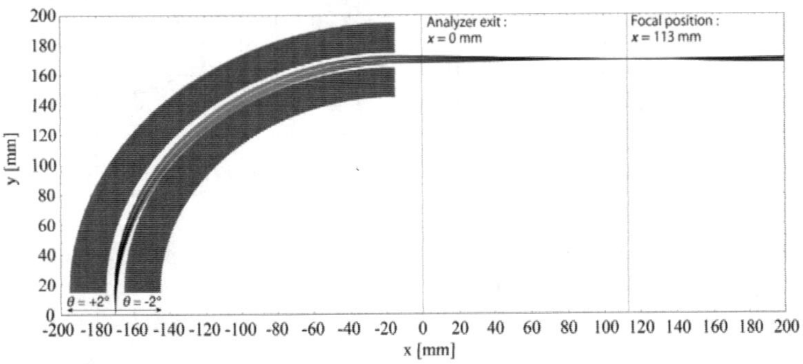

Figure 1. The trajectories of ions with velocity ranging 199.0 km/s to 201.0 km/s and angle ranging −2.0 to +2.0 °.

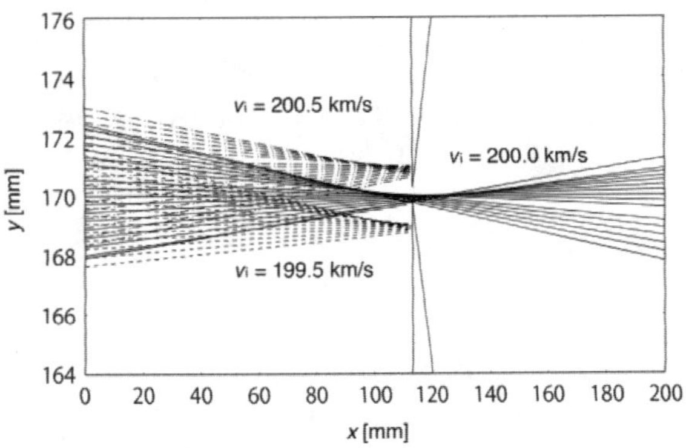

Figure 2. Ion trajectories near focal position. The slit with width of 0.5 mm is set at 113 mm from the analyzer exit.

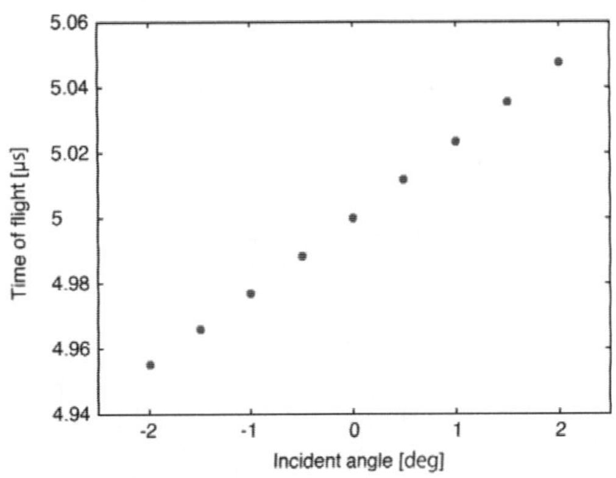

Figure 3. The time of flight for ions with the velocity of 200.0 km/s as a function of incident angle to the analyzer.

the incident angle of the ion increases because the transport distance is longer when passing through the outer electrode side (incident angle is positive), hence the ions arrive the focal position later than ions passing through the central orbit. Conversely, ions passing through the inner electrode side (incident angle is negative) arrive the focal position earlier because ions pass through shorter distance. The obtained difference of TOF ranging 4.95 to 5.05 μs, which corresponds to velocity difference of v_i $\pm 1\%$, is much larger than that due to velocity difference of v_i $\pm 0.1\%$. Therefore, the acceptable angle of ion trajectory needs to be limited to obtain spectrum with a narrow width.

3. Experiment for resolution improvement of EIA

3.1. Experimental setup

Since the ions having the same velocity are focused after passing through an analyzer as shown in the simulation, only ions with an identical velocity can go through the slit set at the focal position. On the other hand, ions with the same velocity cannot be restricted to pass the focal slit even if the slit is narrowed. Therefore, the spectral width cannot be reduced only by the focal slit. In order to make the smaller spectral width, we suggest a structure in which a slit that restricts ions passing through different orbitals at the same velocity was added between the analyzer exit and the focal point. In order to the effect of the additional slit on the reducing spectral broadening with the suggested EIA, we experimentally investigated the change in the spectral width as a function of the width of the additional slit.

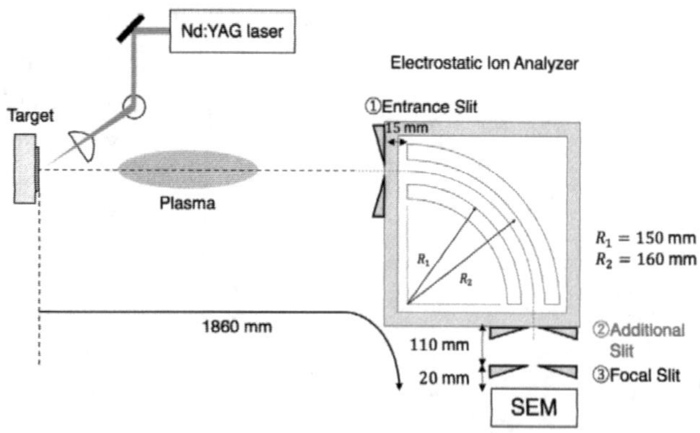

Figure 4. Schematic of experimental setup.

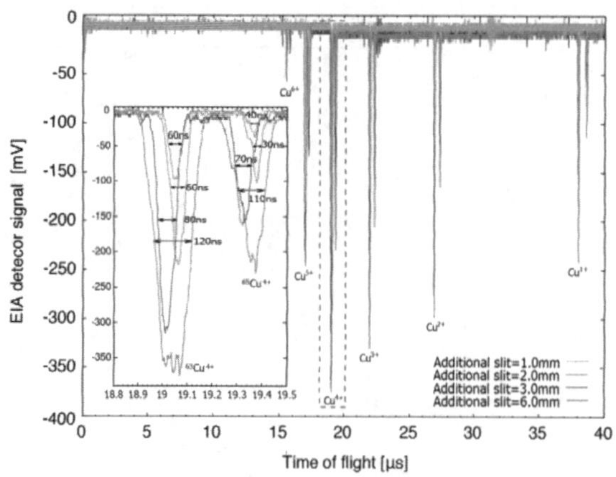

Figure 5. Typical spectral diagram obtained using the EIA with the additional slit. The width of entrance and focal slit were fixed 1.0 mm and the deflection electrodes was biased at ± 50 V.

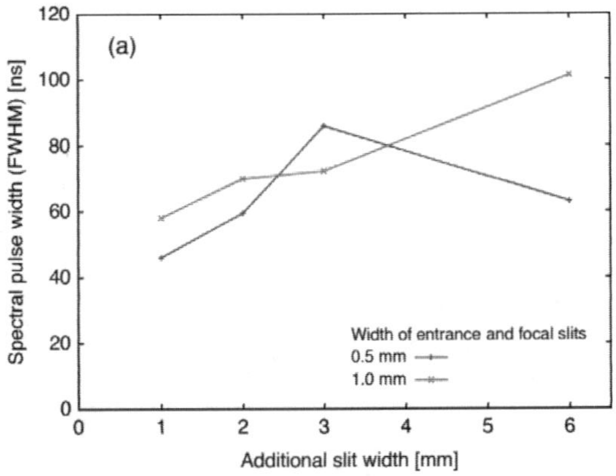

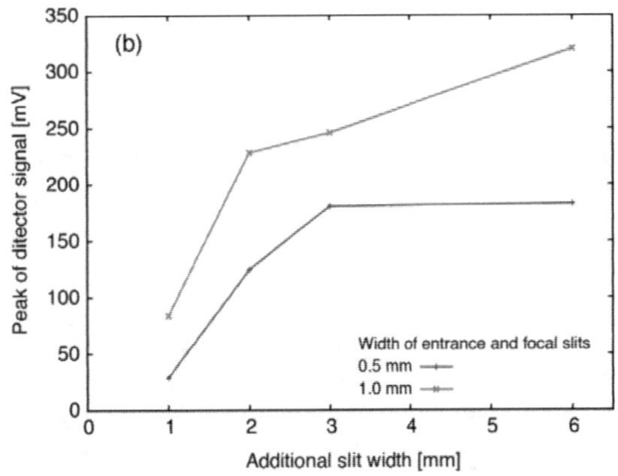

Figure 6. (a) Spectral width and (b) peak value of the signal of $^{63}Cu^{4+}$ ion as a function of the additional slit width. The points are the data of the averaged values for 3 shots.

Figure 4 shows an experimental setup for analyzing ion species using an EIA with three slits located at the entrance, the focal position, and analyzer exit added to limit the angular distribution of ion trajectory. The experimental apparatus composed of a Nd: YAG laser, a Cu target, and the EIA. The Nd: YAG laser has a wavelength of 532 nm, an energy of 250 mJ, and a pulse width of 17 ns. The laser was focused on the target to a spot diameter of about 0.1 mm and the laser power density was estimated to be 2.0×10^{11} W/cm^2. The EIA has two deflection electrodes arranged in an aluminum box with a size of 210 mm × 210 mm. The inner electrode has R_1 = 150 mm, and the outer electrode has R_2 = 160 mm. The entrance slit and the additional slit were located 15 mm in front of and after the electrodes, respectively. The focal slit was fixed at 110 mm from the additional slit. A secondary electron multiplier (R2362, Hamamatsu Photonics) was used as the detector and set at 1860 mm from the target. The experiment was conducted under the pressure of 6.0×10^{-3} Pa.

3.2. Results and discussion

As a result of analyzing the plasma using the EIA with both width for the entrance and focal slits with 1.0 mm, a spectrum having the maximum peak intensity was obtained for $^{63}Cu^{4+}$ when the voltage applied to the deflection electrodes were ±50 V as shown in Fig. 5. Therefore, we evaluated the magnitude and the pulse width of the spectrum of $^{63}Cu^{4+}$.

Figure 6 shows the spectral width (FWHM) and the peak intensity of $^{63}Cu^{4+}$ ions as a function of the additional slit width when both of the entrance and the focal slit widths are 0.5 mm and 1.0 mm. The values in the figure are the average of the signals obtained in the three experiments. The result indicates that the spectral width decreased by reducing the additional slit width for both cases using the entrance and the focal slits with the width of 0.5 and 1.0 mm. This implies the additional slit limits the ions passing through the trajectories in different angles at the same velocity. In addition, it is shown that the peak intensity also decreases as the additional slit width decreases as shown in Fig. 6(b). In particular, when the additional slit width was 0.5 mm, the signal could not be confirmed. This indicates that the additional slit limits the ions that reach the detector and although the signal strength is reduced by reducing the additional slit width, the resolution of the EIA can be improved by reducing the spectral width.

4. Conclusion

In this study, in order to improve the resolution of the EIA, we investigated the cause of spectral width broadening. The simulation result showed that the spectral broadening occurs because ions go through the paths in different angle rather than it caused by difference of ion velocity in the conventional analyzer which uses two slits. To limit the trajectories in angular distribution of the paths, we suggested to use an additional slit for the EIA. As the result of experiment, we demonstrated the additional slit can make the narrow spectrum. It was shown that the resolution of the EIA can be improved by adding a slit.

References

[1] Sharkov B Y *et al.* 2016 *Matter Radiat. Extremes* **1** 28-47
[2] Roudskoy I V 1996 *Laser Part. Beams* **14** 369-384
[3] Dubenkov V, B *et al.* 1996 *Laser Part. Beams* **14** 385-392
[4] Haseroth H *et al.* 1996 *Laser Part. Beams* **14** 393-438
[5] Okamura M *et al.* 2016 *Proc. 2nd North American Particle Accelerator Conference (NAPAC2016), (Chicago)* pp 49-53
[6] Sako T *et al.* 2016 *Rev. Sci. Instrum.* **87** 02C109
[7] Ikeda S *et al.* 2020 *Rev. Sci. Instrum.* **91** 023304
[8] Cannavò A *et al.* 2020 *Rev. Sci. Instrum.* **91** 033317
[9] Okamura M *et al.* 2016 *Rev. Sci. Instum.* **87** 02A901

Space Charge Compensation Study of High Intensity Ion Beams

Z Shen[1,2], L T Sun[1,2], Z H Jia[1], X Fang[1], C Qian[1], and H W Zhao[1,2]

[1] Institute of Modern Physics (IMP), Chinese Academy of Sciences (CAS), Lanzhou 730000, China
[2] School of Nuclear Science and Technology, University of Chinese Academy of Sciences, Beijing 100049, China

shenzhen@impcas.ac.cn

Abstract. The space charge effect is one of the key factors affecting ion beam transport and ion beam quality, especially for low-energy high-intensity ion beams. It can be partially compensated by the secondary electrons produced from the ionization of residual gas molecules in the beam pipe, and we are interested in the accurate value of this compensation degree. In this paper, we use a Three-grid Energy Analyzer (TEA) to measure the secondary ion energy distribution of the beam, which corresponds to the compensated beam potential distribution. To have accurate evaluation of the compensation degree, it is essential to adopt a proper transverse distribution of the ion beam, other than using the Gaussian theorem for uniform beams. The results of a space charge compensation degree study for singly charged ion beams with a 2.45GHz microwave ion source and for multiply charged ion beams with a high charge state ECR ion source are presented.

1. Introduction

The influences of space charge effect on ion beams, such as emittance growth and envelope enlargement, increase with beam intensity, and become most severe for low energy ion beams extracted from electron cyclotron resonance ion sources (ECRIS) [1]. The space charge effect can be partially compensated by the secondary electrons produced from the ionization of residual gas molecules in the beam pipe [2], while the secondary ions are expelled by the space charge fields of the beam. The degree of space charge compensation (SCC) η of an ion beam is defined as:

$$\eta = 1 - \frac{\Delta\phi_{\mathrm{c}}}{\Delta\phi_{\mathrm{un}}}, \tag{1}$$

where $\Delta\phi_{\mathrm{c}}$ and $\Delta\phi_{\mathrm{un}}$ are the electric potential differences between the center and the edge of compensated and uncompensated beams, respectively. Measurements of the SCC degree for ion beams of various charge states have been conducted in early works at IMP [3], CEA-Saclay [4], Peking University [5], MSU [6] and so on.

By measuring the energy distribution of secondary ions expelled from the beam, we can obtain the potential distribution of the compensated beam and therefore $\Delta\phi_{\mathrm{c}}$. For a uniform cylindrical beam of total current I and velocity v, $\Delta\phi_{\mathrm{un}}$ can be calculated according to the Gaussian theorem:

$$\Delta\phi_{\mathrm{un}} = \frac{I}{4\pi\varepsilon_0 v}. \tag{2}$$

Journal of Physics: Conference Series **2244** (2022) 012090 doi:10.1088/1742-6596/2244/1/012090

However, the distribution of beams extracted from an ECRIS is hardly uniform, especially for highly charged ion sources with hexapole magnets and beam optical elements on the beamline.

In this contribution, in order to have accurate evaluation of the compensation degree, we managed to calculate the uncompensated beam potential distribution by measuring the beam current distribution. Measurements with a 2.45 GHz ECRIS and an 18 GHz ECRIS are presented later.

2. Experimental setup

2.1. A 2.45 GHz ion source

An isotope enrichment system for ^{39}Ar has been developed at Institute of Modern Physics (IMP) [7], which includes a high current 2.45 GHz ECRIS capable of producing 1–10 mA Ar^+ beams. The layout of the ion source and beam diagnostic chamber right behind the extraction electrodes is shown in figure 1. The TEA was installed on the side of the diagnostic chamber, together with a faraday cup and a fluorescent screen to measure the beam profile. Up to 1.08 mA Ar^+ beams and 5.4 mA He^+ beams were used in the measurements.

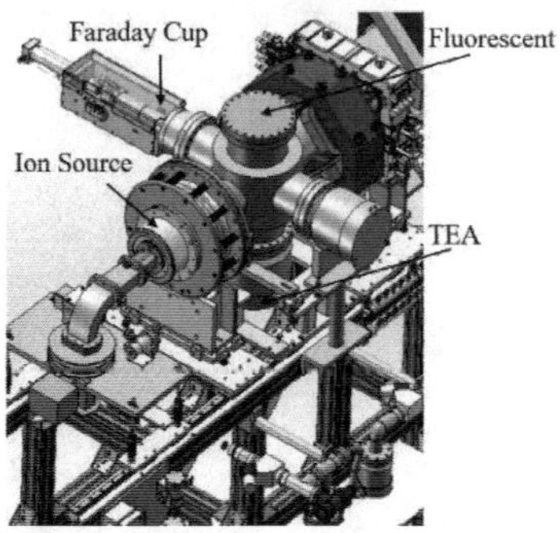

Figure 1. Layout of 2.45 GHz ion source.

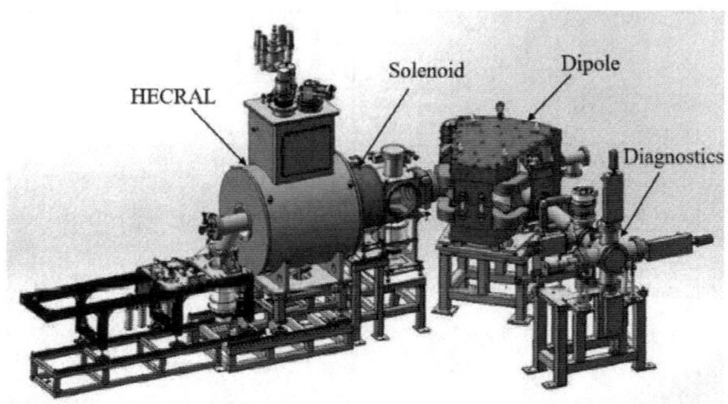

Figure 2. Layout of HECRAL ion source and LEBT.

2.2. HECRAL ion source

The Hybrid-magnet superconducting ECRIS with Advanced design in Lanzhou (HECRAL) has been designed and constructed at IMP, which is mainly operated with an 18 GHz microwave power supply. The magnetic fields of HECRAL are produced by a permanent hexapole magnet and a set of superconducting coils, similar to AISHa [8]. Oxygen beams with total extracted currents of up to 3.4 mA were used in these measurements. The layout of the HECRAL ion source and Low Energy Beam Transport (LEBT) can be seen in figure 2. The TEA and a Beam Profile Monitor (BPM) [9] were mounted about 45 cm downstream of the diagnostic chamber, which includes a faraday cup and two slits from both horizontal and vertical directions perpendicular to the beam direction. This position is after the analyzing magnet, where essentially only one ion species is present in the beam.

2.3. Three-grid Energy Analyzer (TEA)

A non-interceptive TEA has been designed and fabricated at IMP [3]. The mechanical design and picture of the TEA are shown in figure 3. It mainly consists of nine collimating apertures to restrain the incident angle of the secondary ions, three parallel girds (G_1, G_2, G_3) and a secondary ions collector C connected to a picoammeter (Keithley 6485) through a coaxial cable. G_1 is grounded and G_3 is biased with a negative high voltage (typically set to −300 V) to suppress the secondary electrons. By sweeping the voltage of G_2, we can obtain the energy spectrum of the secondary ions.

2.4. Beam Profile Monitor (BPM)

A BPM has been constructed and modified based on a 128-channel picoammeter system [10]. The mechanical structure of the BPM can be seen in figure 4. The range of each channel can be regulated in electronics to accommodate to the beam current to be measured. A total of 128 picoammeters are placed in a row behind a slit which is connected to a negative bias voltage (typically set to −500 V). By scanning the picoammeter system from top to bottom with a resolution of 0.5mm, the current distribution and the total current of the beam can be obtained.

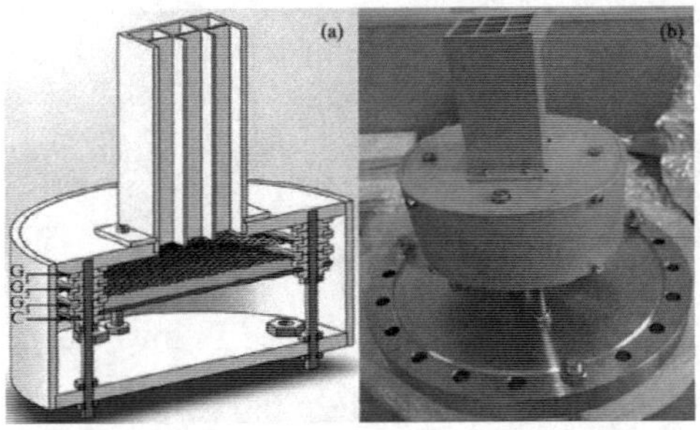

Figure 3. The cut isometric view (a) and picture (b) of the TEA.

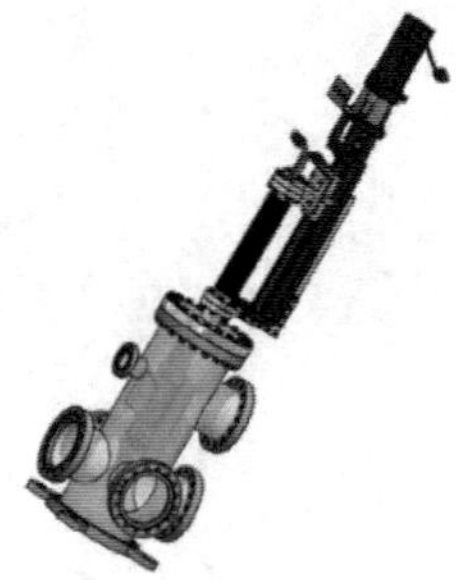

Figure 4. The mechanical structure of the BPM.

3. Measurements and Analyses

3.1. Measurements with 2.45GHz ion source

The secondary ion energy distributions of Ar^+ and He^+ beams were measured using TEA. An example of ion spectrums from 5.4 mA He^+ beams of 33 kV extraction voltage and different puller electrode voltages (U_P) under the vacuum pressure of 2.8×10^{-4} Pa is shown in figure 5. By fitting the curves with 3 straight lines and taking the x-values of 2 crossing points as the potentials of beam edge and beam center, we can obtain the potential difference $\Delta\phi_c$. Figure 5 indicates that for both -1 kV and -3 kV of U_P, $\Delta\phi_c$ of 5.4 mA He^+ beams have almost the same value of 12.5 V.

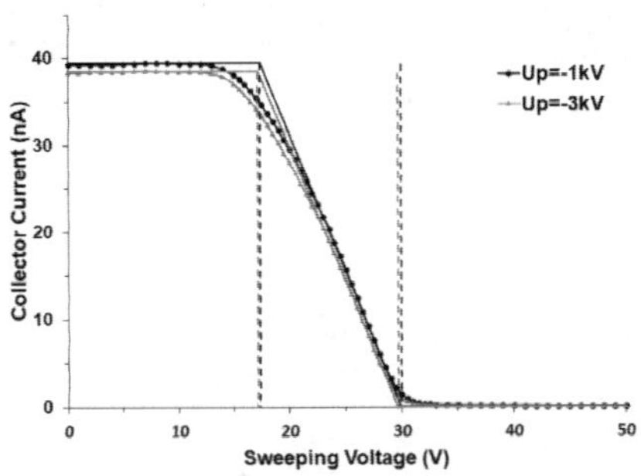

Figure 5. The secondary ion energy distributions of 5.4 mA He^+ beams of different puller electrode voltages U_P. The pressure in the beamline was 2.8×10^{-4} Pa.

Based on the image of the fluorescent screen at the same position of beamline with the TEA, we considered a gray level image segment as a two-dimensional current distribution function. According to the beam intensity measured with the faraday cup and the gray value of each pixel, the current I_j of any pixel j can be distributed. Therefore, the electric potential ϕ_i of any pixel i can be calculated by means of the superposition principle of electric potentials:

$$\phi_i = \sum_{j\neq i} \frac{I_j}{2\pi\varepsilon_0 v} \ln \frac{d_{j\leftrightarrow 0}}{d_{j\leftrightarrow i}}, \tag{3}$$

where v is the velocity of the beam, $d_{j\leftrightarrow 0}$ is the distance between pixel j and beam pipe where electric potential equals to 0, and $d_{j\leftrightarrow i}$ is the distance between pixel j and pixel i. In combination with the beam size information acquired from the fluorescent image, the potential difference of uncompensated beam $\Delta\phi_{un}$ can be obtained. Using equation (1), the degree of SCC can be calculated. The fluorescent image and the uncompensated beam potential distribution of He^+ beams can be seen in figure 6.

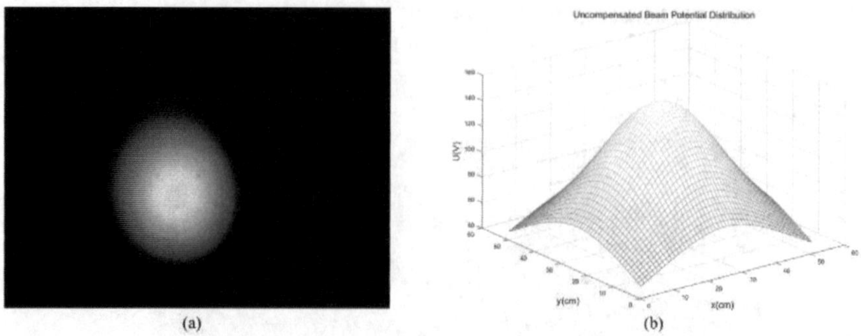

Figure 6. The fluorescent image (a) and the uncompensated beam potential distribution (b) of He^{+} beams.

3.2. Measurements with HECRAL

The TEA was used to measure the secondary ion energy distributions of O^{6+} beams same as *3.1.* As shown in figure 7, the focusing solenoid fields (see figure 3) were changed to study the influence of beam distribution on the measurement results. The $\Delta\phi_c$ of 1060 eμA O^{6+} beams of different solenoid currents vary from 5.8 V to 8.4 V.

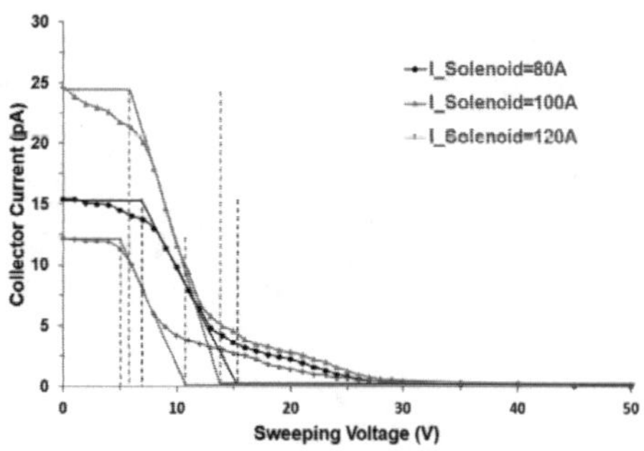

Figure 7. The secondary ion energy distributions of 1060 eμA O^{6+} beams of different solenoid currents. The extraction voltage was 15 kV and the pressure in the beamline was 9.6×10^{-6} Pa.

To measure the two-dimensional beam current distribution more directly and precisely, a BPM was used to replace the fluorescent screen. A two-dimensional current matrix of size 128×183 can be obtained after a single scanning process. The total current of the beam and each element of the matrix measured by the BPM were calibrated with a faraday cup. The electric potential of each element of the matrix can be calculated using equation (3). The current distribution and potential distribution of the uncompensated O^{6+} beams are shown in figure 8.

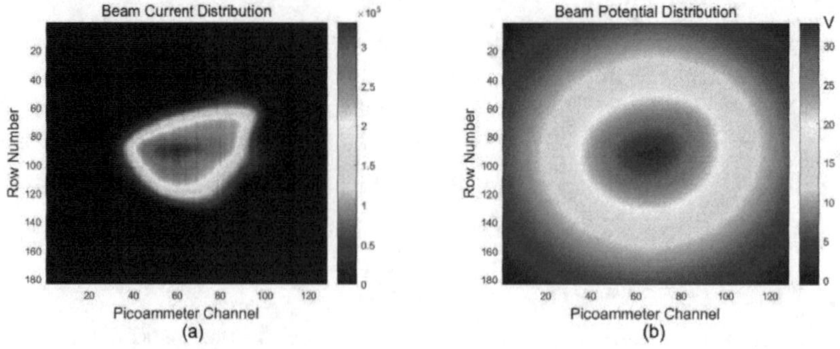

Figure 8. The current distribution (a) and potential distribution (b) of the uncompensated O^{6+} beams.

4. Results and Conclusions

The comparison of $\Delta\phi_{un}$ and SCC degree η between using uniform approximation and using current distribution calculation method measured with He^+, Ar^+ and O^{6+} beams are listed in table 1 and table 2, respectively.

Table 1. Comparison between uniform beams and gray level analysed beams.

	U_P (kV)	Beam Intensity (emA)	$\Delta\phi_c$ (V)	Uniform $\Delta\phi_{un}$ (V)	Gray $\Delta\phi_{un}$ (V)	Uniform η	Gray η
He^+	-1	5.40	12.5	38.5	64.2	0.675	0.805
He^+	-3	5.38	12.5	38.3	68.6	0.674	0.818
Ar^+	-1	1.08	10.0	24.3	39.6	0.588	0.747
Ar^+	-2	1.09	10.0	24.5	38.2	0.592	0.738
Ar^+	-3	1.07	10.0	24.1	38.9	0.585	0.743

Table 2. Comparison between uniform beams and current analysed beams.

	Solenoid Current (A)	Beam Intensity (eμA)	$\Delta\phi_c$ (V)	Uniform $\Delta\phi_{un}$ (V)	Current $\Delta\phi_{un}$ (V)	Uniform η	Current η
O^{6+}	80	520	6.6	4.5	12.2	-0.471	0.460
O^{6+}	100	535	5.4	4.6	10.5	-0.170	0.487
O^{6+}	120	525	5.0	4.5	9.0	-0.104	0.442
O^{6+}	80	1025	8.4	8.8	23.4	0.050	0.641
O^{6+}	100	1060	8.0	9.1	20.5	0.125	0.61
O^{6+}	120	955	5.8	8.2	14.6	0.296	0.604

In conclusion, measurements of the SCC degree for He^+, Ar^+ and O^{6+} beams at 2 different ECRIS LEBT systems have been conducted using a TEA, a fluorescent screen and a BPM. The results indicate that adopting proper transverse beam distributions influence greatly on $\Delta\phi_{un}$ and SCC degree η, especially for highly charged ion beams. In addition, η calculated by beam current distributions are almost unchanged with different puller electrode voltages or focusing solenoid fields.

Measurements with hybrid charge states ion beams at highly-charged ECRIS will be made in the near future.

Acknowledgments

This work was supported by NSFC with contract No. 11427904 and No. 12025506. The authors would like to thank J Liu and X Song from Institute of Modern Physics for their great help on the installation and utilization of the BPM.

References

[1] Peter Spädtke 2014 *Rev. Sci. Instrum.* **85** 02A744
[2] Chauvin N, Delferrière O, Duperrier R, Gobin R, Nghiem P A P and Uriot D 2012 *Rev. Sci. Instrum.* **83** 02B320
[3] Tan B, WU Q, Yang Y, Sun L, Zhang X and Zhao H 2016 *Nucl. Phys. Rev.* **33** 36-40
[4] Ferdinand R, Sherman J, Stevens R R and Zaugg T 1997 *Proc. of the 1997 Particle Accelerator Conf.* vol 3 pp 2723-2725
[5] Lu P *et al.* 2010 *Rev. Sci. Instrum.* **81** 02B711
[6] Winklehner D, Leitner D, Cole D, Machicoane G and Tobos L 2014 *Rev. Sci. Instrum.* **85** 02A739
[7] Jia Z *et al.* 2020 *Rev. Sci. Instrum.* **91** 033309
[8] Celona L *et al.* 2019 *Rev. Sci. Instrum.* **90** 113316
[9] Li X *et al.* 2019 *Rev. Sci. Instrum.* **90** 076107
[10] Yu D, Liu J, Xue Y, Zhang M, Cai X, Hu J, Dong J and Li X 2015 *Rev. Sci. Instrum.* **86** 115102

Journal of Physics: Conference Series **2244** (2022) 012091 doi:10.1088/1742-6596/2244/1/012091

High Power Density Beam Measurement of a Single Beamlet Multi-Grid Prototype H^- Negative Ion Accelerator

G Q Saquilayan, M Kisaki , A Kojima, Y Shimabukuro, M Murayama, J Hiratsuka, M Ichikawa, K Watanabe, H Tobari, M Kashiwagi

Naka Fusion Institute, National Institutes for Quantum Science and Technology, Naka, Ibaraki, Japan, 311-0193
mae.glynnis@qst.go.jp

Abstract. Beam optics of the ITER prototype accelerator with 88 mm acceleration gap length has been examined through H^- beam acceleration tests at the ITER target beam perveance. The beam optics has been examined for beam energies up to 790 keV under the ITER target perveance. The total grid heat load was successfully reduced to 10 %, which was lower than the allowable value of 15 %. The beam emittance of the ITER perveance has been measured for the first time by using newly developed emittance measurement system consisting of carbon-fiber-composite (CFC) plate with pin-holes. The emittance of 790 keV H^- beam shows the divergence angle of the beam core is satisfied with the ITER requirement of <7 mrad. Measurement at the beam periphery was made possible by modifying the pin-hole size of pepper-pot apertures, and the observed divergent components were 20 mrad. The ratio of the diverged beam became lower with the increase of the beam energy. This result contributes the design of the ITER accelerator and the beam line components.

1. Introduction

Neutral beam injector (NBI) for heating and current drive in the fusion experimental reactor ITER requires 1 MeV deuterium negative ion (D^-) beams of 40 A (200 A/m^2 current density) with 7 mrad maximum beam divergence and pulse lengths of up to 3600 s [1]. For hydrogen negative ion (H^-) beams, the perveance matched condition is 870 keV beam energies at 46 A (230 A/m^2). The H^- beam operation for 50 s is required in the first experimental stage of ITER using the NBI [2]. To achieve the required beam performance for ITER, acceleration of H^- beams up to 1 MeV is being examined using a prototype 5-stage negative ion accelerator [3]. It is necessary to finalize the ITER accelerator design and contribute to the upcoming full power test to be performed in the NB test facility (NBTF) in Consorzio RFX [4]. Presently, the H^- beam acceleration for 60 s at 0.97 MeV, 190 A/m^2, which mostly simulates the rated power density of 200 MW/m^2, have been achieved by using the prototype accelerator with 109 mm acceleration gap length [5]. In the next phase of experiments, stable H^- beam acceleration toward > 100 s with the ITER perveance matched beam conditions of 0.87 MeV energies at 230 A/m^2 is being planned by applying the designed accelerator gap length of 88 mm for the ITER NBI [6].

One of the issues in high power density beam acceleration is the unknown divergent components which have been observed as the molten marks at the aperture edges despite that the total heat load on each grid was lower than the allowable value. In the conventional measurement of the beam footprint visualized through an infrared (IR) camera, the peripheral of the beam footprint could not be clearly determined whether it is a divergent beam component or measurement background noise. To understand the divergent beam components, an emittance measurement system using one-dimensional carbon-fiber-composite (1D-CFC) plates have been newly developed to measure high power beams of 200 MW/m^2 [7]. In this paper, improvements were applied to the emittance measurement aiming to measure the peripheral region of the beam. Through the improved emittance measurement, the ITER perveance matched high power H$^-$ negative ion beam is investigated for the first time.

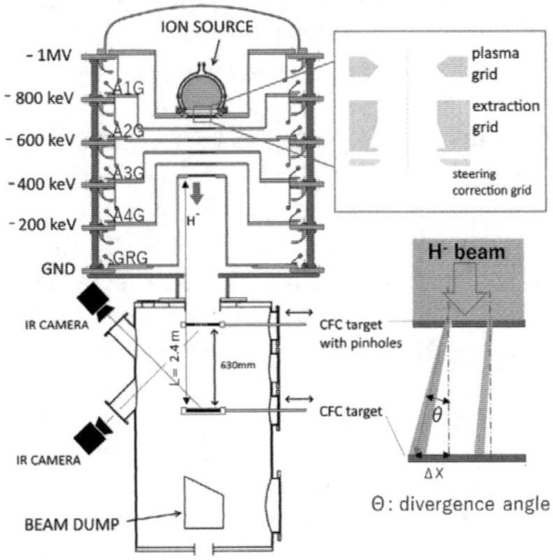

Figure 1. Cross-section of the five-stage accelerator with the emittance measurement setup located 2.4 m away from the GRG.

2. Experimental Setup and Beam Diagnostics for High Power Beams

The cross-section of the five-stage accelerator is shown in Figure 1 (a). The negative ions produced in the cesiated filament driven ion source are extracted from a plasma grid (PG) to an extraction grid (EXG), and accelerated through four acceleration grids (A1G, A2G, A3G, A4G) and a grounded grid (GRG). For this multi-grid multi-aperture accelerator, nine apertures are drilled in the lattice pattern of 3 × 3 for the beam extraction. In this experiment, only one aperture is used for beam acceleration and the emittance of a single beam was measured. The aperture diameters of the grids are 14 mm in the PG, 13 mm in the EXG, 14 mm in the A1G and A2G, and 16 mm in the A3G, A4G and the GRG. The gap length between each acceleration grids was modified from the 118 mm in the previous beam acceleration experiment to 88 mm which is the same as the ITER design.

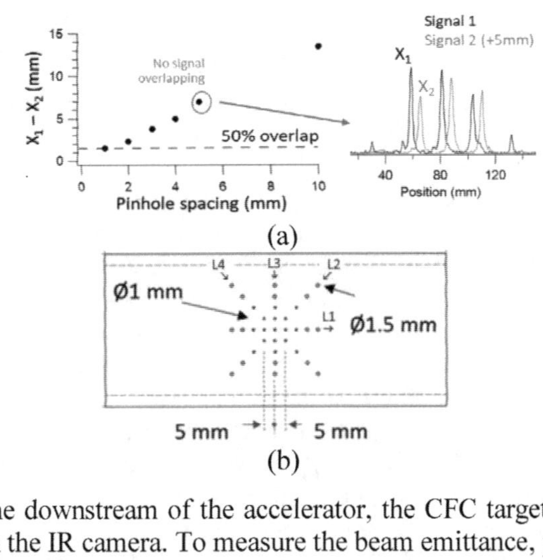

Figure 2. Design of the emittance measurement system: (a) beamlet signals with respect to pinhole spacing, (b) star-shaped pepper pot arrangement.

In the downstream of the accelerator, the CFC target was installed to visualize the beam footprint through the IR camera. To measure the beam emittance, pinholes for discrete measurement of the beam were drilled in the upper CFC. The divergence angle was calculated from the distance between the pinhole location on the first CFC and the discrete beamlet pattern on the second CFC. A beam dump is located 3 m away from the accelerator to measure calorimetrically the accelerated beam energy.

The design for the emittance measurement is reconsidered from the previous one [7] with the objective to investigate the peripheral region of the beam. First, to minimize the possible overlapping of signals in the emittance patterns of a pepper pot arrangement, 1 mm diameter pinholes and the distances of their spacing were examined. The corresponding beamlet signals having distances between peaks X_1 and X_2 for 5 mrad beams were examined as shown in Figure 2(a). To clearly distinguish between two peaks, 5 mm spacing was selected as the limit which permits well-defined fitting analysis with clearances suitable for up to <7 mrad deflections. Second, aiming to measure the weak signals along the peripheral region and beam edge, the pinhole diameters and pepper pot arrangement were adapted accordingly. Because the signals are measured thermally, 1 mm pinhole diameters allowed sufficient intensity for 5 mrad beam core signals to be detectable by the IR camera which includes the image spatial resolutions. In addition, to measure the beam edge region, the diameter of the outer perimeter pinholes were enlarged to 1.5 mm to strengthen the measureable signal intensity of the thermal footprint.

Lastly, a star shaped pattern was utilized to analyze emittance profiles across the beam as shown in Figure 2(b). As the special resolution is determined by the distance between the pinholes, for precise measurement of the emittance, the CFC target with the pinholes can be moved along lines of L1 or L3.

3. Beam Acceleration under perveance matched condition with ITER

For the beam acceleration test, the accelerator was modified by reducing the gap length from 109 mm to 88 mm, to optimize the beam optics for higher beam current acceleration at the target perveance for the ITER accelerator. The beam optics was analyzed using a two-dimensional (2D) beam trajectory analysis code BEAMORBT [7]. Figure 3 shows the beam trajectories at the minimum divergence angle for the H- beam with the currents and energies of 200 A/m^2 and 1 MeV for the gap of 109 mm and 230 A/m^2 870 keV for the gap of 88 mm. In both results, there were no direct interceptions of the H- beam on the accelerator grids. For the accelerator gap length of 109 mm, the H^- beam trajectories were partially diverged from the A4G.

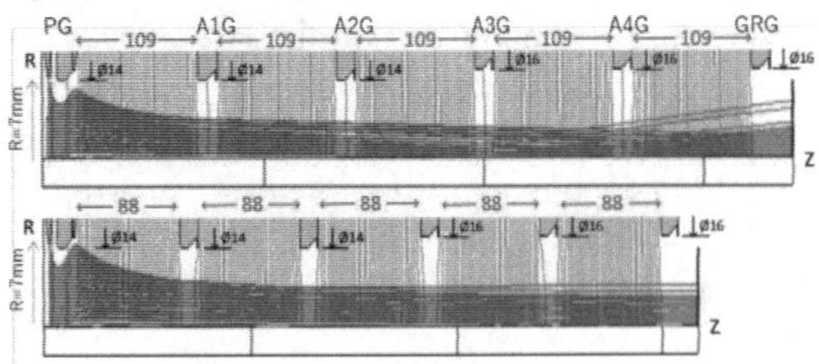

Figure 3. Calculated beam trajectory: (top) 109 mm gap for 1 MeV beam energy and, (bottom) 88 mm for 870 keV beam energy.

The acceleration gap length was determined from the analysis of the 2D beam trajectory by confirming the minimum beam divergence angle for each beam perveance. For the perveance of 1 MeV, 200 A/m^2 H^- beam, the acceleration gap was 109 mm. For the target in this test, which as the same as the ITER requirement of 870 keV, 230 A/m^2 H^- beam, the acceleration gap is 88 mm.

Figure 4 (a) shows the experimental data of the accelerated beam along the perveance curves of the accelerators with different gap lengths. The operation condition at the gap length of 88 mm for the optimum perveance of 0.2 10^{-6} A/$V_{acc}^{1.5}$ were maintained for the following beam energies, current densities, and extraction voltages respectively: 440 keV at 80 A/m^2 and 4.3 kV, 640 keV at 145 A/m^2 and 6.3 kV, 790 A/m^2 at 195A/m^2 and 7.3 kV. It was confirmed experimentally that the target current density for each acceleration voltage has been obtained as designed. Thus, the beam optics measurement can be performed at to the ITER perveance matched condition for the H- ion beam.

Figure 4 (b) shows the experimental result of the grid heat load. The allowable heat load is 3% in each grid and 15 % in the total. In the gap lengths of 109 mm and 88 mm, the grid heat loads were lower than the allowable value. However, the total grid heat load in the gap of 88 mm was slightly smaller than that of 109 mm. This result indicates that the diverged components in the beam for the acceleration gap of 88 mm is smaller than that of the case of 109 mm similar as the evaluated 2D beam trajectories.

Figure 5 shows the beam experimental result with 790 keV and 195 A/m^2 measured on the lower CFC target, i.e., (a) the footprint, (b) the beam profile and (c) the emittance along the L1 direction of the pinhole arrangement. This emittance diagram was measured by moving the CFC target with the pinholes to increase the special resolution. The observed beam footprint was not a coaxial circle and is asymmetric. The beam profile shown in Figure 5 (b) was observed to be a gaussian beam with a width of a standard deviation of 15 mm. The divergence angle obtained from the beam footprint was estimated to be 6 mrad. In the emittance diagram shown in Figure 4 (c), more detailed information was examined. The beam divergence was mostly within the ITER requirement (<7 mrad). Intensity of the peripheral beam was small, but was detectable and not part of the measurement background noise. It was confirmed that the peripheral beam is within 10 mrad. The divergence angle of the peripheral beam for the ITER-relevant intense beam has been obtained for the first time.

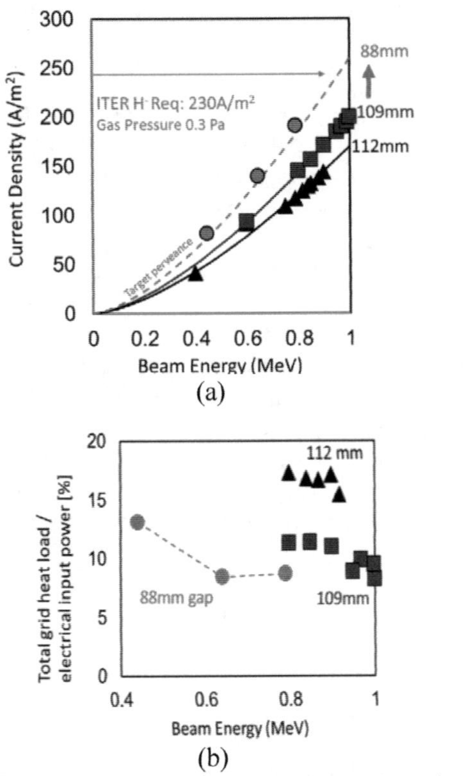

Figure 4. Beam measurements in the acceleration gap of 112, 109 and 88 mm (a) beam perveance, and (b) total grid heat load.

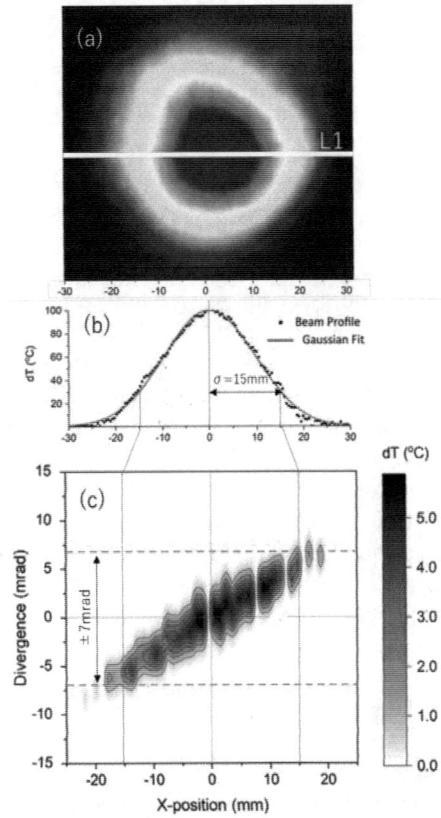

Figure 5. Measurements of the single beamlet at 790 keV. (a) Beam footprint (b) Beam profile, and (c) Emittance diagram.

Figure 6 shows the beam profile and the beam emittance in the case of 640 keV and 790 keV beams under the perveance matched condition. To cover a longer range for the beam emittance measurement, the diagonal line profile from the L2 pinholes were used to extend the emittance measurement up to 30 mm. Also, to clearly examine the signals of the peripheral region of the beam, the vertical axis is shown in the logarithmic scale. In both cases of beam energies of 640 keV and 790 keV beams, the beam components which are outside the gaussian distribution were observed. Emittances and the beam profiles at the beam edge reveal the beam divergences extend to 20 mrad. The diverged beam component is simply assumed by a difference between the measured profile and the gaussian profile. The divergent components were at 8 % in 640 keV and 3% in 790 keV of the beam cross-section. It was experimentally confirmed that the diverged component decreases with the increase in beam energy. This beam edge estimation only considers the beam after the ground grid which have exited the accelerator. The overall picture of the beam edge should also include portions of the beam that were directly intercepted on the accelerator grids, measured as part of the grid heat loads [9]. Investigation of the beam periphery is currently in ongoing to determine the susceptible parts of the accelerator and aim to suppress the beam edge which will contribute to extend the operational limit for long beam pulses.

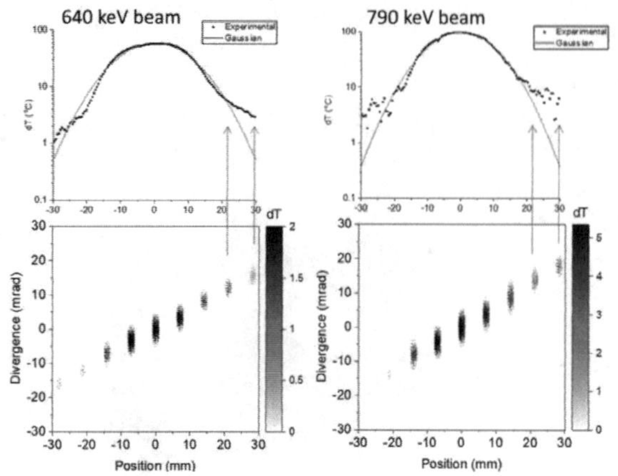

Figure 6. Measured beam profile and emittance diagram for 640 keV and 790 keV along the L2 profile.

4. Summary

In the high energy beam acceleration test up to 790 keV by using the prototype five-stage accelerator for the ITER accelerator, the acceleration gap was adjusted to 88 mm as the ITER accelerator's one to produce the beam under the required perveance matched condition for ITER, that is, 870 keV, 230 A/m^2 for the H^- beam. Simultaneously, to evaluate the H^- beam optics by the beam emittance for the first time, an IR imaging method for beam emittance measurement has been newly developed. This system has been reconsidered from the previous one aiming to measure the peripheral region of the beam. As the result, the calorimetric measurements of the beam current density show the accelerated H^- beam has reached the target perveance, with grid heat loads below 10% indicating optimal beam optics. The emittance measurement realized to detect the peripheral beam with small intensity and revealed the beam core according to a gaussian beam to have <7 mrad divergences which was within the ITER beam requirement. The measured divergent components had a maximum of 20 mrad. This emittance measurement is continued to get more information of the beam precisely under various condition, and then contributes to understand the mechanism of the negative ion extraction, the cause of the grid heat load, and the ITER accelerator design.

References

[1] Hemsworth R S, Boilson D, Blatchford P, Palma M D, Chitarin G, de Esch H P L, Geil F, Dremel M, Graceffa J and Marcuzzi D 2017 New J. Phys. 19 025005
[2] ITER Organization, ITER Technical Report ITR-18-003 2018
[3] Taniguchi M, de Esch H P L, Svensson L, Umeda N, Kashiwagi M, Watanabe K, Tobari H, Dairaku M, Sakamoto K and Inoue T 2009 AIP Conf. Proc. 1097, 335-43
[4] Toigo V et al 2017 New J. Phys. 19 085004
[5] Kashiwagi M, Umeda N, Tobari H, Kojima A, Yoshida M, Taniguchi M, Dairaku M, Maejima T, Yamanaka H, Watanabe K, Inoue T, and Hanada M 2014 Rev. Sci. Instrum. 85 02B320
[6] Kashiwagi M, Umeda N, Kojima A, Yoshida M, Tobari H, Dairaku M, Yamanaka H, Maejima T, Yamashita Y, ShibataN, K. Watanabe, and M. Hanada 2015 Fusion Eng. Des. 96-97, 107- 112
[7] Saquilayan G Q, Kashiwagi M, Tobari H, Hiratsuka J, Ichikawa M, Umeda N, Watanabe K, Wada M, and Sasao M 2018 AIP Conf. Proc. 2052 040014
[8] Ohara Y, JAERI-M-6757 1976
[9] Kisaki M, Kojima A, Saquilayan G. Q., Hiratsuka J, Ichikawa M, Shimabukuro Y, Murayama M, Watanabe K, Tobari H, and Kashiwagi M, Presented in this conference proceeding for ICIS 2021 (ID 151).

Design of the proton injector for compact neutron source DARIA

V.A. Skalyga[1], I.V. Izotov[1], S.S. Vybin[1], T.V. Kulevoy[2], G.N. Kropachev[2], A.L. Sitnikov[2], S.V. Grigoriev[3,4]

[1]Institute of Applied Physics of the Russian Academy of Sciences, 603950, 46 Ulyanova st., Nizhny Novgorod, Russia
[2]Alikhanov Institute for Theoretical and Experimental Physics of NRC ''Kurchatov Institute'', 117218, Moscow, Russia
[3]Petersburg Nuclear Physics Institute Named by B.P. Konstantinov of NRC ''Kurchatov Institute'', 188300 Gatchina, Russia
[4]Saint Petersburg State University, Ulyanovskaya 1, 198504 Saint Petersburg, Russia
E-mail: skalyga@ipfran.ru

Abstract. Project of the proton accelerator-driven compact neutron source DARIA (Dedicated for Academic Research and Industrial Application) is developed in order to replace small and middle flux neutron sources based on the nuclear reactors. DARIA has a uniquely high ratio of efficiency to cost due to deep optimization of each key element of the system (proton injector and accelerator, target, neutron moderator and neutron instruments. A unique ECR ion source, developed at the IAP RAS, would be used as a proton beam injector. In such device the plasma is heated by the powerful 28 GHz gyrotron radiation, providing a record level of volumetric energy input for such systems over 100 W/cm^3. The high plasma density and the optimal electron temperature provide proton beams formation with a current of up to several hundred mA and an emittance that meets the requirements of modern accelerators. The paper discusses the advantages of using such an ion source, its scheme and design performance.

1. Introduction

For the successful development of synchrotron and neutron research in the Russian Federation, in addition to the flagship mega-science facilities, such as the SKIF and SILA synchrotron radiation sources or the PIK research reactor, it is necessary to create an infrastructure supporting these facilities based on relatively small, compact sources that allow scientists training to carry out the research on the above mentioned installations, as well as to carry out preliminary work for the development of new equipment, techniques, and methods. Project of the proton accelerator-driven compact neutron source DARIA (Dedicated for Academic Research and Industrial Application) [1] is aimed at solving this problem, namely: creating serial compact sources of photons and neutrons based on the latest technologies of linear ion accelerators. DARIA will be based on a proton injector, 13 MeV linear accelerator consisting of a RFQ and a few DTLs, beryllium target, neutron moderator, and a set of neutron research stations. It is designed as a pulsed machine providing acceleration of 70-100 mA proton beams with 100 microsecond duration and 100 Hz repetition rate, and yielding about 10^{13} neutrons per second. A concept of a gasdynamic ECR source [2-4] was chosen as the ion injector to fulfill all proton beam requirements.

A distinctive feature of gasdynamic ECR ion sources being under development at IAP RAS is a confinement regime of a high density plasma. It was demonstrated [2–4] that the use of high-frequency gyrotrons is very beneficial for a dense plasma production and high-current ion beams formation. Previously, ion beams of both light (hydrogen) and heavy (nitrogen, argon) elements were extracted from the plasma of ECR discharge sustained by 37.5 [2] and 75 GHz [4] radiation, demonstrating ion beam currents on the level of 100–500 mA with normalized RMS emittance below 0.2 π·mm·mrad in a pulsed operation mode. This success stimulated the construction of the facility named GISMO (Gasdynamic Ion Source for Multipurpose Operation) [5] at IAP RAS. The scheme of the facility is shown in Fig. 1. 28 GHz/10 kW CW gyrotron is used for the plasma heating, manufactured by Gycom. This microwave generator is equipped with power supplies suitable for CW or pulsed operation. The plasma trap has an all-permanent magnet design. To ensure sufficient plasma confinement, the magnetic field configuration is designed to be similar to a simple mirror trap close to the system axis, and the magnetic field strength is 1.5 T at mirrors and 0.25 T at the trap center, yielding the mirror ratio of 6. The distance between magnetic mirrors is 12 cm. The magnet inner hole diameter is 50 mm, thus allowing the use of a water-cooled plasma chamber with a vacuum diameter of 32 mm. An ion beam extraction system has been designed to use the acceleration voltage of up to 100 kV. A high source potential requires a development of an appropriate high-voltage insulation of the plasma chamber from unbiased components of the facility. A quasi-optical system, shown in Fig. 1, is implemented instead of a conventional waveguide DC-break. The electromagnetic radiation of the gyrotron (TE02 mode of a circular waveguide) is converted to the Gaussian beam, which then propagates in open air through the gap of 15 cm, acting as a DC-break. The radiation is received by the horn antenna, converted into a TE11 mode of a circular waveguide, and then focused to the plasma with a microwave-to-plasma coupling system embedded into the vacuum chamber. Overall transmission efficiency of the beam line is 95%, (estimated when no reflection from the plasma is present).

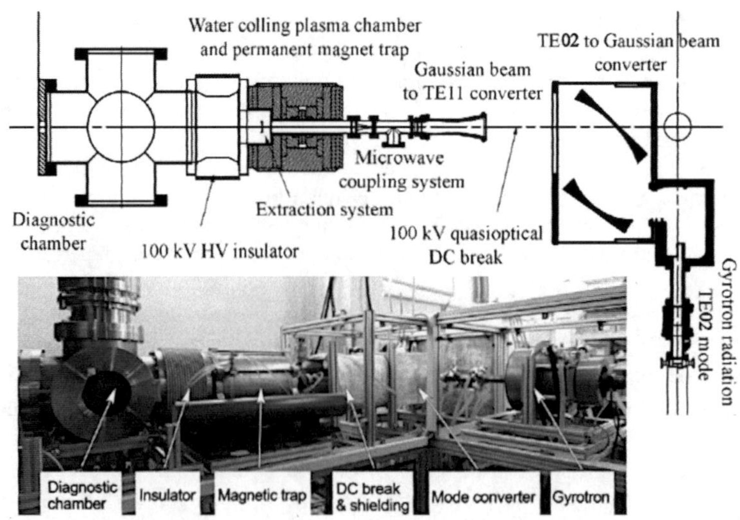

Figure 1. GISMO experimental facility layout.

It has been shown earlier that the plasma density can reach a level of 10^{13} cm^{-3} [6]. As a first step, we have used a simple 2-electrode extraction system for a beam formation. Plasma electrode aperture had a diameter of 3 mm. This system is obviously non-optimal for plasma parameters achieved at GISMO, as it has a very long (250 mm) and narrow (30 mm) puller tube, which leads to a more than 90% of beam losses, as was confirmed with IBSimu code [7].

Thus, instead of the extracted ion beam current measured with the conventional Faraday cup, the power supply drain current has been used for a beam current evaluation in the first experiments. The dependence of the drain current on the hydrogen pressure in the plasma chamber is shown in Figure 2, right axis shows the corresponding current density, assuming 3 mm extraction aperture, the microwave power was equal to 2 kW. For sure the flux of backstreaming secondary electrons is not known in the case, but the fig. 2 shows the general trend in plasma flux growth.

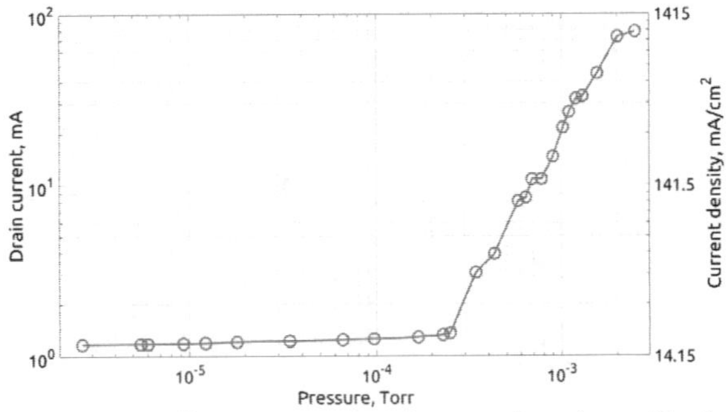

Figure 2. Drain current and corresponding current density as a function of hydrogen pressure in the plasma chamber.

2. Spherical extractor

In order to address the problem of high current density beam formation, the new type of extraction geometry is proposed [8]. The main idea is to change the shape of the electrodes, which exploits an inhomogeneous distribution of the electric field in the accelerating gap of the ion beam formation system. This method may be especially effective in systems with a high current density, mitigating the negative effect of the space charge of the beam on its quality. It allows for significant enhancement of the current and/or reducing the emittance at a certain accelerating voltage and characteristic size of the system, especially when the extraction system operates in the space charge limited regime, which is the case for GISMO facility. The problem of the dense ion beam formation is due to the harmful influence of the self space charge, which distorts the shape of the meniscus, making it convex. Accordingly, the accelerating field near the meniscus forms a diverging beam, which cannot be transported efficiently. The increase in the electric field near the meniscus provides higher beam acceleration gradient and reduces the region with high space charge level. It is proposed to consider the combination of electrodes shown in Fig. 3a. Hereinafter we denote such a system as a "spherical" because the puller and the tip of the plasma electrode form plates of a hollow hemispherical capacitor (this region is highlighted with a rectangle in Fig. 3a), whereas the conventional flat structure shown in Fig. 3b forms plates of a flat capacitor.

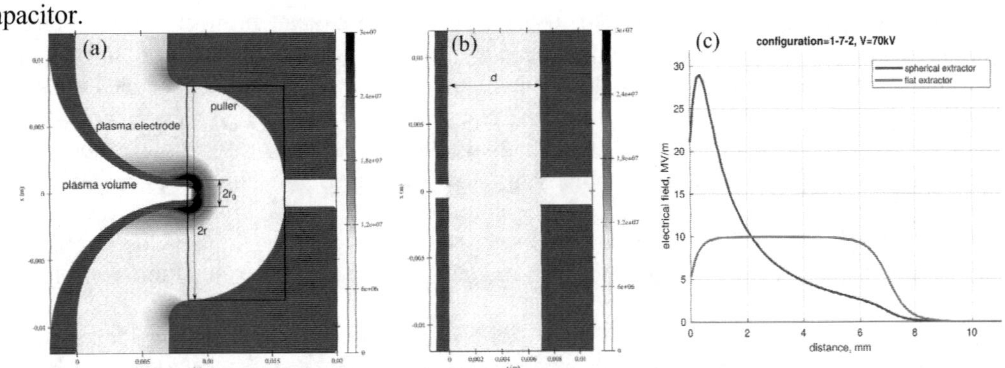

Figure 3. Distribution of the electric field modulus in the interelectrode space (a - spherical, b - flat). Distribution of the electric field along the axis of symmetry (c). The voltage between electrodes is the same and equal to 70 kV.

The use of inhomogeneous accelerating field allows an increase of the electric field magnitude near the plasma meniscus without raising the extraction voltage or reduction of interelectrode distance (see

Fig. 3c). In this work we consider a new electrode geometry which produces an inhomogeneous electric field providing higher gradients of ion beam acceleration and reducing the space charge influence on its quality.

It is of note that the new geometry can be used for a wide range of the ion beam sources and its applications. This approach in extraction system design allows for increase of the current of the beam at proton injectors [9, 10] and multicharged ion sources for accelerators [11]. The current density can exceed the level of 1 A/cm^2. On the other hand, the use of a new geometry allows for significant decrease of the extraction voltage and hence it is useful for low energy ion beam production. Such beams are required for ion implantation [12].

Another technical advantage of the proposed system is the possibility to use puller with much greater aperture than in conventional extraction systems, as the puller aperture barely affects the shape of the meniscus and the mode of extraction of the beam. The example is shown in Figure 4, where two significantly different puller apertures are simulated. The reason for that being the puller part highlighted in the Fig. 4a has a weak effect on the electric field distribution, while the highlighted part in the Fig. 4b makes the main contribution. Such concept has been already successfully test at pulsed SMIS 37 ion source [13]

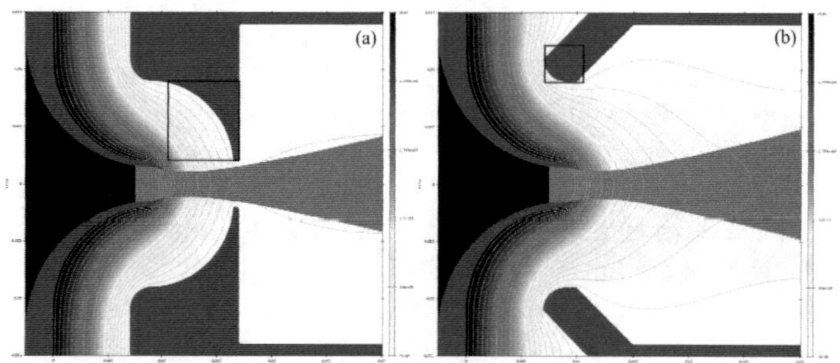

Figure 4. Formation of an ion beam with various puller designs. A significant increase in the aperture of the puller slightly affects the beam.

A set of tungsten electrodes of a new type (as shown in Fig. 4b) was designed and fabricated at IAP RAS for initial test on GISMO facility.

3. Ion beam properties

A use of the spherical extractor allowed us to greatly enhance the extracted beam current. Dependence of the Faraday cup current on the extraction voltage as a function of hydrogen pressure in the plasma chamber is shown in Fig. 5 on the left. The microwave power used in the experiment was 4.8 kW. It is seen from the Fig. 5 that the maximum current obtained at Faraday cup, which was located 400 mm from the plasma electrode, was equal to 65 mA. It is of note that despite drastic improvement in the beam formation efficiency, the extraction geometry is yet far from the optimal one, as puller current was in the range 50-150 mA (with unknown part of secondary electron current), what means the existence of some extra ion current which could be utilized. Indeed, further optimization of extraction system geometry is needed, and is ongoing at the moment, including the design and simulation of 4-electrode extraction system based on the new spherical design.

Yet another distinctive feature of GISMO facility is a very high power density deposed in the plasma. The maximum injected power density may reach the value of 400 W/cm^3. Such level of deposited power allows to run the ion source in quasi-gasdynamic mode (which is naturally featured with very high energy losses due to low plasma lifetime) while keeping the electrons hot enough to maintain 100% ionization degree and 100% H_2 dissociation level. As an example of that, hydrogen ion beam spectra are plotted in Fig. 5 on the right for different input microwave power. The spectra obtained at 20 kV of extraction voltage. It is seen from Fig. 5 that H^+/H_2^+ ratio increases with power.

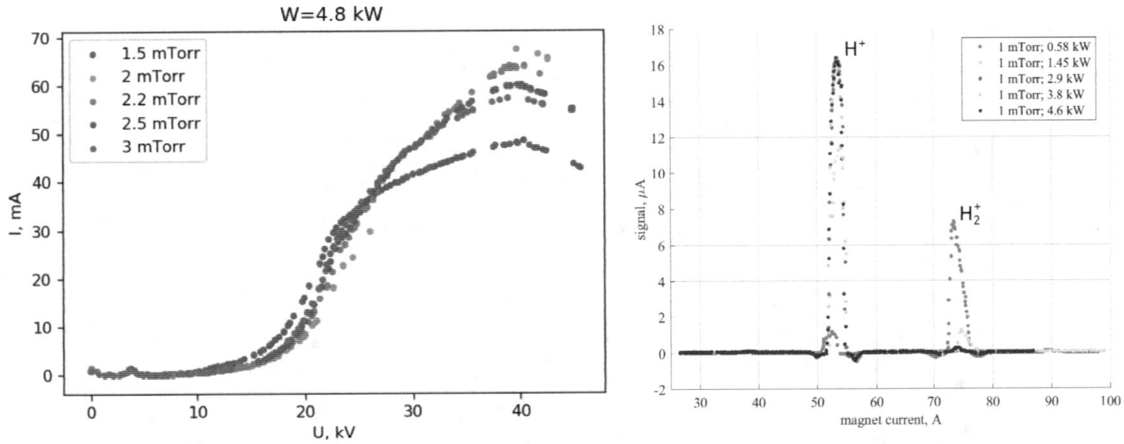

Figure 5. Dependence of Faraday cup current on the extraction voltage as a function of hydrogen pressure in the plasma chamber (on the left), and ion beam q/m spectra at different injected microwave power.

4. Conclusion

Hydrogen plasma emissivity of >1 A/cm^2 has been demonstrated. A new type of extraction system geometry able to handle beam current densities of >1 A/cm^2 is developed and experimentally tested. Hydrogen beams with current of 65 mA were successfully obtained at extraction voltage of 40 kV. Judging from conducted experiments and IBSimu numerical investigations, the extraction system consisting of 4 electrodes and new spherical geometry would be able to extract the beam current on the level of 250 mA (trough 5 mm plasma electrode aperture). Preliminary estimations of normalised RMS emittance suggested that the emittance of such a beam should be lower than 0.2 π·mm·mrad. Thus, GISMO facility is close to fulfill the requirements of the proton accelerator-driven compact neutron source DARIA.

5. Acknowledgments

The work was supported by the project of the Russian Science Foundation Grant No. 21-19-00844

References

[1] Kropachev G et al *Journal of Surface Investigation: X-Ray, Synchrotron and Neutron Techniques* **13(6)** 1126–1131
[2] Golubev S et al 2007 *Nuclear Instruments and Methods in Physics Research Section B: Beam Interactions with Materials and Atoms* **256(1)** 537–542
[3] Skalyga V et al 2016 *Review of Scientific Instruments* **87(2)** 02A716
[4] Skalyga V A et al 2017 *Review of Scientific Instruments* **88(3)** 033503
[5] Skalyga V A et al 2019 *Review of Scientific Instruments* **90(12)** 123308.
[6] Skalyga V et al 2016 *Review of Scientific Instruments* **87(2)** 02A715
[7] Kalvas T et al 2010 *Review of Scientific Instruments* **81(2)** 02B703.
[8] Vybin S S et al 2020 *Plasma Sources Sci. Technol.* **29** 11LT02
[9] Knaster J et al 2017 *Nuclear Fusion* **57** 102016
[10] Grigoriev S, Iashina E, Pavlov K 2019 *Journal of Surface Investigation: X-Ray, Synchrotron and Neutron Techniques* **13(6)** 1132–1134
[11] Yang J C et al 2013 *Nuclear Instruments and Methods in Physics Research Section B: Beam Interactions with Materials and Atoms* **317** 263–265
[12] Smith N S et al 2006 *Journal of Vacuum Science & Technology B: Microelectronics and Nanometer Structures* **24(6)** 2902
[13] arXiv:2107.10772 [physics.plasm-ph]

Preliminary results of positive ion mass spectrometry based on a 2.45 GHz ECR ion source and a non-metallic gas target

K. Li[1], S.X. Peng[1, *], T. H. Ma[1], W. B. Wu[1], Y. X. Jiang[1], B. J. Cui[1], A. L. Zhang[2], J. Sun[1], Z. Y. Guo[1], and J. E. Chen[1]

[1]State Key Laboratory of Nuclear Physics and Technology & Institute of Heavy Ion Physics, School of Physics, Peking University, No.201, Chengfu Road, Haidian District, Beijing, China
[2]State Key Laboratory of Particle Detection and Electronics, Department of Modern Physics, University of Science and Technology of China, Hefei 230026, China

*correspondence author, E-mail: sxpeng@pku.edu.cn

Abstract. Positive-ion mass spectrometry (PIMS) is a positive-to-negative ion conversion metrology that just operated on the opposite of traditional accelerator mass spectrometry (AMS) on high-precision radiocarbon dating. With a higher efficient simple structure plasma ion sources instead of AMS sputter ion sources, lower ion energy (tens keV) in place of MeV, and limited footprint facility, PIMS will be a powerful competitor to AMS in the future. To precisely measure ^{14}C, interference of molecular isobars (such as $^{12}CH_2$, ^{13}CH) will be eliminated with higher charged ions (charge state ≥ 3; molecular cations $^{12}CH_2^{2+}$ is metastable, can be eliminated by higher stripping pressure.), nitrogen will be eliminated inside the charge-exchange cell (CXC) through charge exchanging with target gas. To demonstrate the probability of PIMS with a 2.45 GHz ECR ion source, an experiment was launched at Peking University (PKU). A compact permanent magnet 2.45 GHz ECR ion source is designed to produce carbon ions from CO_2. A CXC unit was installed after the extraction system and ethylene is chosen as charge exchange gas. Preliminary experiments prove that more than hundreds of microamp of C^{2+} beam can be produced with this ECR source. By injecting ethylene into CXC, several microamp C^- beam was detected and the C^{2+} ions charge exchange efficiency to C^- ions can reach close to 5%.

1. Introduction

Accurate measurements of radiocarbon are widely used in atmospheric sciences, bioscience, archaeological science and pharmaceutical science [1], etc. AMS with high accuracy (with isotopic abundances as low as 10^{-12} to 10^{-16}), low sample consumption, and short measurement time has become the most important tool for radiocarbon measurement since 1977 [2-4]. However, AMS requires large-scale facilities with correspondingly high costs. Moreover, most of these devices are equipped with Cs sputter negative ion sources, which may lead to low sample availability, beam instability, fractionation effect, etc. Besides, the graphitization of samples also poses potential risks.

In 1978, Roy Middleton proposed the possibility of counting ^{14}C with low energy mass spectrometer based on charge exchange without accelerator [5]. According to his proposal, high charge state (≥ 3) ions will be generated and extracted from ion source (with CO_2 sample) to eliminate the interferences of molecular isobars (such as $^{12}CH_2$, ^{13}CH). Then, the positive ions will be converted to negative ions by passing through a CXC to eliminate ^{14}N interference, since nitrogen cannot form negative ions. Consequently, molecular and atomic isobars are suppressed as well as the AMS. This is the basic principle of PIMS. However, due to the limitation of the high-density multi-charged ion sources and charge exchange efficiency, PIMS was not practically realized for a long time.

In the 1990s, electron cyclotron resonance (ECR) ion sources are rapidly developed and applied for multiply charged ion beams, which makes it possible for the construction of PIMS systems [6, 7]. In the past ten years, to produce high-intensity C^{3+}/C^{2+} beams for PIMS, Australian Nuclear Science and Technology Organization (ANSTO), Oak Ridge National Laboratory (ORNL) and Scottish Universities Environmental Research Centre (SUERC) have carried out relevant researches on ECR ion sources with high microwave frequency of 7 GHz (C^{3+}), 14 GHz (C^{3+}) and 10 GHz (C^{2+}, molecular cation $^{12}CH_2^{2+}$ is metastable, can be eliminated by change exchange target.) respectively [8-10]. The feasibility of high-frequency ECR ion sources in PIMS has been proven by these works. Although these sources work very well, it will be a preferable choice if the 2.45 GHz ECR ion source is available for the PIMS.

Besides the ion source, a high-efficiency CXC is required for PIMS. Many studies have shown that the alkali metal vapor target has the best charge exchange efficiency. However, due to its low optimal charge exchange energy, high maintenance and operation cost, and short service life of the alkali metal CXC, it is not convenient for the stable operation of PIMS. ANSTO, ORNL, and SUERC use Rubidium vapor target, single-crystal alkali halide (LiF) target, and non-metallic gas isobutane target to form C^- ion by charge exchange, respectively [8-10]. At present, there is still no consistent recommendation on the selection of efficient, stable, and reliable charge exchange targets, which also limits the development of PIMS.

To overcome these two challenges and to establish a device for radiocarbon detection with high efficiency and low cost, a new design of the PIMS system is proposed by PKU. The PIMS system is based on a 2.45 GHz ECR ion source and a non-metallic gas CXC. The main contributions of this paper can be summarized as follows:

1) It is proven that more than hundreds of microamp of C^{2+} beam current can be produced with the specially designed 2.45 GHz ECR ion source.

2) By injecting ethylene into CXC, several microamp C^- ion was detected and C^{2+} ions charge exchange efficiency to C^- ions can reach close to 5%.

2. PIMS: Positive ion mass spectrometry

The PKU-PIMS system is presented schematically in Fig 1. It consists of

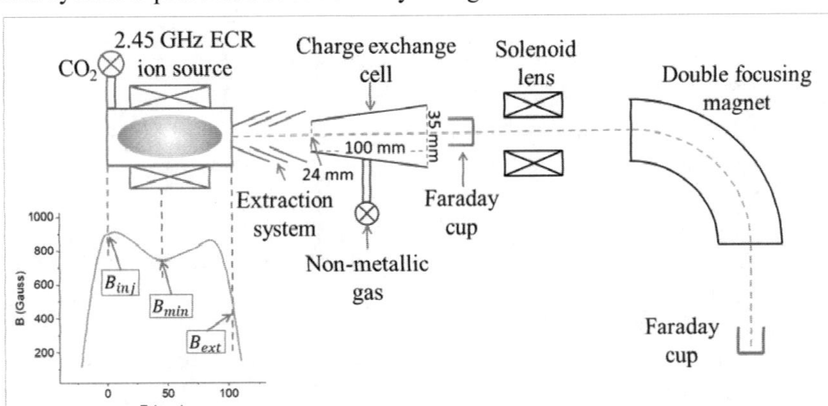

Figure 1. Schematic of the PKU-PIMS system.

- A 2.45 GHz ECR ion source body and extraction system.
- A CXC is installed after the extraction system to convert the positive C^{2+} ions to negative ones.
- A solenoid lens for beam focusing.
- A 90° double-focusing magnet (radius 500 mm) to analyze the negative ions from the CXC according to their mass-to-charge ratio (m/q).
- A faraday cup installed at the end of the beamline is used for beam current measurement.

In the ECR ion source, the CO_2 sample discharges to produce C^{2+} ions. Then the extracted beam will pass through the CXC to form negative ions. Finally, after focusing and analysis, the C^- beam intensity is measured with a Faraday cup at the end of the beamline.

3. Results and Discussion

3.1. *Ion source performance*

During the experiment, the carbon beam was generated by injecting CO_2 into an ECR ion source. The gas bled into the ion source through a low flow rate leakage valve. The background pressure of the vacuum chamber is about 2.0E-5 Pa, and the optimal pressure for producing C^{2+} ions is about 2.0E-4 Pa. The consumption rate of carbon is about 20 ng/s. Under average power of 100 W and extraction voltage of 35 kV, the ability to produce C^{2+} ions at different pressures have been tested, up to about hundreds of microamp of C^{2+} beam can be reached with the aluminum liner in the plasma chamber. The corresponding results are presented in Fig 2.

It can be seen that, on the one hand, with the increase of discharge pressure, the C^{2+} current intensity gradually increases to the maximum value. Then the current intensity of C^{2+} drops sharply as gas pressure increases. Compared with the previous results [11], the optimal discharge pressure is reduced to 2.0E-4 Pa, which is an order of magnitude lower. Low-pressure discharge is a necessary condition for the generation of high-charged ions. The alumina layer on the surface of the aluminum lining has a higher secondary electron emission coefficient and can provide more extra cold electrons, which is beneficial to increase the generation of high-charged ions.

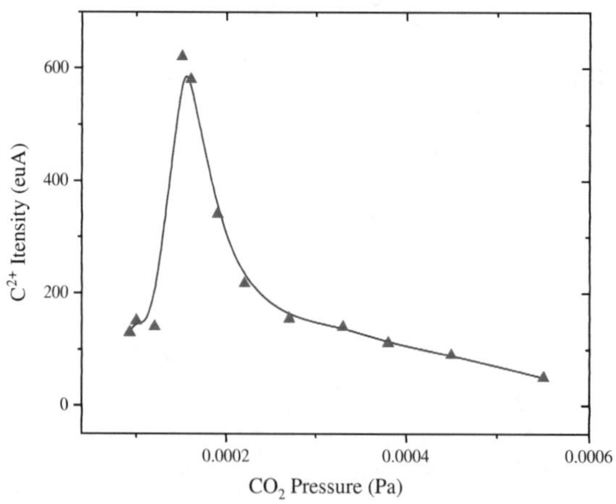

Figure 2. The current intensity of C^{2+} ions varies with the CO_2 pressure, the extraction voltage is 35 kV, and the average power is 100 W.

3.1.1. *Charge exchange result*

Charge-exchange anion formation is a very complicated process. Take the incident C^{2+} ion and the target Y charge exchange as an example, it is generally through a step-by-step charge exchange process (1), (2), and (3) to form C^- ion. ΔE_1 and ΔE_2 are equal to the difference between ionization energies E_i of the colliding partners. ΔE_3 is the difference between the electron affinity $E_A(C)$ of Carbon and the ionization energy $E_i(Y)$ of Y.

$$C^{2+} + Y \rightarrow C^{+} + Y^{+} + \Delta E_1 \quad (1)$$

$$C^{+} + Y \rightarrow C + Y^{+} + \Delta E_2 \quad (2)$$

$$C + Y \rightarrow C^{-} + Y^{+} + \Delta E_3 \quad (3)$$

Since the first ionization energy of the gas molecules around a dozen electron volts (metal vapor ionization energy is only a few electron volts). The first ionization energy of C is 11.26 eV and the second ionization energy is 24.38 eV [12]. Therefore, the charge exchange process (1) is generally an exothermic reaction with a large charge exchange cross-section. Process (2) the energy difference $|\Delta E_2|$ generally smaller within a few electron volts, when $|\Delta E_2| \approx 0$ can occur near-resonance charge exchange, thus process (2) also has a larger charge exchange cross-section. For process (3), due to the electron affinity of C is 1.26 eV [13], so $|\Delta E_3|$ is generally very large. It is an endothermic reaction. Therefore, the charge exchange cross-section of the process (3) is generally very small, which is the main factor limiting the formation of charge exchange anions. The probability of double and triple charge exchange is so small that will not be considered here.

To operate the CXC, a non-metallic mixed gas of 94.1% ethylene and 5.1% argon was fed into CXC through the Alicat MC-5sccm mass flow controller. It should be noted that there is no particular purpose in selecting a mixture of gases, except that they are readily available. However, we believe the experiments based on this target can give us some inspiration for the chosen of nonmetallic gases. In the case of incident C^{2+} ion at 80 keV, the charge exchange efficiencies P_{C^+} and P_{C^-} of incident C^{2+} ions charge exchange to form C^+ and C^- ion with the flow rate of this mixed gas target are studied, as shown in Fig 3. It can be seen that C^{2+} ions charge exchange efficiency to C^- ions can reach close to 5%, and the conversion efficiency of C^+ ions first reached the peak and then decreased. This further explains that the charge exchange of C^{2+} ions to form C^- ions is a process dominated by single electron capture process (1), (2), and (3) instead of multiple electron capture.

The charge exchange efficiency P_{C^-} and P_{C^+} are determined by comparing the negative and positive ions intensity I_{C^+} and I_{C^-} with the corresponding incident C^{2+} ion intensity $I_{C^{2+}}$:

$$P_{C^-} = \frac{-2I_{C^-}}{I_{C^{2+}}}$$

Where q is the charge state of the injected C^{2+} ions.

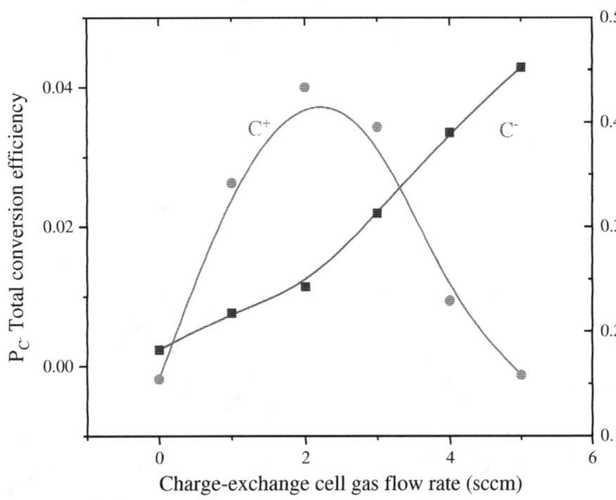

Figure 3. The efficiency of charge exchange to form C^+ and C^- ions at 80 keV of C^{2+} ion energy, dependence on charge-exchange cell gas flow rate with a mixed gas target of 94.1% ethylene and 5.1% argon. (It should be noted that there is no particular purpose in selecting a mixture of gases, except that they are readily available.)

4. Conclusion

Positive ion mass spectrometry is a new mass spectrometry technique that can measure radiocarbon with high precision. We have shown that the Peking University 2.45 GHz ECR ion source can efficiently generate more than hundreds of microamp of C^{2+} beam current. The consumption rate of C is about 20 ng/s. The potential of 2.45 GHz ECRIS to replace high-frequency ECRIS for PIMS is demonstrated by this work.

The charge exchange efficiency of C^{2+} ions to C^- ions using a non-metallic gas target is measured. The charge exchange efficiency of about 5% can be achieved using a mixed gas target of 94.1% ethylene and 5.1% argon. In addition, by measuring the efficiency of C^+ and C^- ions formation by C^{2+} ions charge exchange with the thickness of the target, it is confirmed that C^{2+} ions convert to C^- ions through a step-by-step charge exchange process.

At present, C^{2+} ions and C^- ions can be effectively generated, and the C consumption rate can be further reduced. In the future, the system will be further developed to test more efficient charge exchange targets and count ^{14}C.

Acknowledgments
This work is supported by the National Natural Science Foundation of China (Grant Nos. 11975036, 11775007, and 11575013). The support from the State Key Laboratory of Nuclear Physics and Technology, Peking University is appreciated.

References

[1] W. Kutschera, "Applications of accelerator mass spectrometry," *International Journal of Mass Spectrometry,* vol. 349-350, pp. 203-218, 2013, doi: 10.1016/j.ijms.2013.05.023.

[2] D. E. Nelson, R. G. Korteling, and W. R. Stott, "Carbon-14: direct detection at natural concentrations," *Science,* vol. 198, no. 4316, pp. 507-508, 1977.

[3] K. H. Purser *et al.*, "An attempt to detect stable N-ions from a sputter ion source and some implications of the results for the design of tandems for ultra-sensitive carbon analysis," *Revue de Physique appliquée,* vol. 12, no. 10, pp. 1487-1492, 1977.

[4] H.-A. Synal, "Developments in accelerator mass spectrometry," *International Journal of Mass Spectrometry,* vol. 349-350, pp. 192-202, 2013, doi: 10.1016/j.ijms.2013.05.008.

[5] R. Middleton, "On the possibility of counting 14C-ions without an accelerator," *Proceedings of the First Conference on Radiocarbon Dating with Accelerators held at The University of Rochester April 20 and 21,* pp. 157-164, 1978.

[6] A. G. Drentje, "Techniques and mechanisms applied in electron cyclotron resonance sources for highly charged ions," *Review of Scientific Instruments,* vol. 74, no. 5, pp. 2631-2645, 2003, doi: 10.1063/1.1569408.

[7] R. Schubank, "A low-energy table-top approach to AMS," *Nuclear Instruments and Methods in Physics Research Section B: Beam Interactions with Materials and Atoms,* vol. 172, no. 1-4, pp. 288-292, 2000, doi: https://doi.org/10.1016/S0168-583X(00)00358-X.

[8] M. Hotchkis and T. Wei, "Radiocarbon detection by ion charge exchange mass spectrometry," *Nuclear Instruments and Methods in Physics Research Section B: Beam Interactions with Materials and Atoms,* vol. 259, no. 1, pp. 158-164, 2007, doi: 10.1016/j.nimb.2007.01.153.

[9] F. W. Meyer, E. Galutschek, M. Hotchkis, F. D. McDaniel, and B. L. Doyle, "Low-Energy Grazing Ion-Scattering From Alkali-Halide Surfaces: A Novel Approach To C-14 Detection," presented at the AIP Conference Proceedings, 2009.

[10] S. P. H. T. Freeman, R. P. Shanks, X. Donzel, and G. Gaubert, "Radiocarbon positive-ion mass spectrometry," *Nuclear Instruments and Methods in Physics Research Section B: Beam Interactions with Materials and Atoms,* vol. 361, pp. 229-232, 2015, doi: 10.1016/j.nimb.2015.04.034.

[11] W. B. Wu *et al.*, "A 2.45 GHz microwave ion source for carbon positive-ion mass spectrometry," *Journal of Instrumentation,* vol. 15, no. 03, pp. P03028-P03028, 2020, doi: 10.1088/1748-0221/15/03/p03028.

[12] A. Kramida, Yu.~Ralchenko, J. Reader, and N. A. Team. NIST Atomic Spectra Database (ver. 5.8)

[13] R. D. J. III. NIST computational chemistry comparison and benchmark database, NIST standard reference database number 101

Concept design of new compact electron cyclotron resonance ion source with permanent magnets for multi-ion radiotherapy

Masayuki Muramatsu[1], Fumihisa Ouchi[2], Taku Suzuki[2], Katsuyuki Takahashi[2], Toshinobu Sasano[2], Tadahiro Shiraishi[2], Shinji Nomura[3], Nobuaki Takahashi[3], Yushi Kato[4], Ken Katagiri[1], Kota Mizushima[1], Yoshiyuki Iwata[1], Taku Inaniwa[1], and Atsushi Kitagawa[1]

1 National Institutes for Quantum and Radiological Science and Technology (QST), 4-9-1 Anagawa, Inage, Chiba 263-8555, Japan

2 Accelerator Engineering Corporation, 3-8-5 Konakadai, Inage, Chiba 263-0043, Japan

3 Technology Research Center, Sumitomo Heavy Industries Ltd, Yokosuka, Kanagawa, 237-8555, Japan

4 Division of Electrical, Electronic and Infocommunications Engineering, Graduate School of Engineering, Osaka University, 2-1 Yamadaoka, Suita-shi, Osaka 565-0871, Japan

muramatsu.masayuki@qst.go.jp

Abstract. The multi-ion radiotherapy with dose distribution and Liner Energy Transfer optimization is being studied at QST. Helium, carbon, oxygen and neon ions are considered as ion species for multi-ion therapy. A basic experiment using these ion beams has been carried out at the HIMAC. For heavy-ion radiotherapy facility, it is desirable to operate with only one ECR ion source with permanent magnets from the viewpoint of cost and maintenance. We considered the operating method with only one ion source for multi-ion radiotherapy. Ionization gases were helium, methane, oxygen, and neon to produce He^{2+}, C^{4+}, O^{6+} and Ne^{7+} ions. Requirement values of beam current were 940 μA correspond to He^{2+}, 290 μA to C^{4+}, 330 μA to O^{6+}, and 245 μA to Ne^{7+}, respectively. This value was obtained from the number of particles required for the cancer treatment and the transmission efficiency of the accelerator in the HIMAC. We performed some beam tests for design of a new compact ion source with the existing 18 GHz ECR ion source (NIRS-HEC). From the results of the beam tests, we estimate the mirror magnetic field of NIRS-HEC by using POISSON/SUPERFISH code. Then, the structure of the permanent magnets for new ECR ion source is determined so that they reproduce the values of the upstream mirror peak (Binj), B minimum (Bmin) and the downstream mirror peak (Bext) at the NIRS-HEC. The magnetic field of Binj, Bmin and Bext at NIRS-HEC were 1.14 T, 0.475 T and 0.9 T, respectively.

Journal of Physics: Conference Series **2244** (2022) 012094 doi:10.1088/1742-6596/2244/1/012094

1. Introduction

Carbon-ion radiotherapy using the Heavy Ion Medical Accelerator in Chiba (HIMAC) has been carried out at the National Institutes for Quantum and Radiological Science and Technology (QST) since 1994 [1]. Over 11000 cancer patients have been treated with carbon beams having energies of between 56-430 MeV/u.

A compact electron cyclotron resonance ion source (ECRIS) with all permanent magnets, named Kei2 [2], was developed for the high-energy carbon-ion radiotherapy facility at QST. The Kei2 source was designed for producing a sufficient current of C^{4+} ions for medical treatments. In 2010, the Gunma University Heavy Ion Medical Center (GHMC [3]) started operating as a carbon-dedicated demonstration facility. A compact ECR ion source for GHMC, the KeiGM, is also based on the development of compact ECR ion sources (Kei series) at NIRS [4]. Kei series were installed at the GHMC, the Saga Heavy Ion Medical Accelerator in Tosu, the Ion-beam Radiation Oncology Center in Kanagawa, the Osaka Heavy Ion Therapy Center and the East Japan Heavy Ion Center in the Yamagata University.

The multi-ion radiotherapy with dose distribution and Liner Energy Transfer optimization is being studied at QST [5,6]. Helium, carbon, oxygen and neon ions are considered as ion species for multi-ion therapy. However, in the case of compact accelerator for high-energy heavy-ion radiotherapy facility, it is desirable to use with one ECR ion source with all permanent magnets from the viewpoint of cost, operation and maintenance. We considered the switching method with only one ion source for multi-ion radiotherapy. Ionization gases were helium, methane, oxygen and neon to produce He^{2+}, C^{4+}, O^{6+} and Ne^{7+} ions. Requirement values of beam current were 940 eμA correspond to He^{2+}, 290 eμA to C^{4+}, 330 eμA to O^{6+}, and 245 eμA to Ne^{7+}, restrictively.

2. Beam test at the NIRS-HEC ion source

First, we searched the maximum microwave power for production of these ions at the NIRS-HEC [7] to determine the specification of the microwave amplifier. Second, optimal value of the mirror magnetic field for permanent magnets was checked.

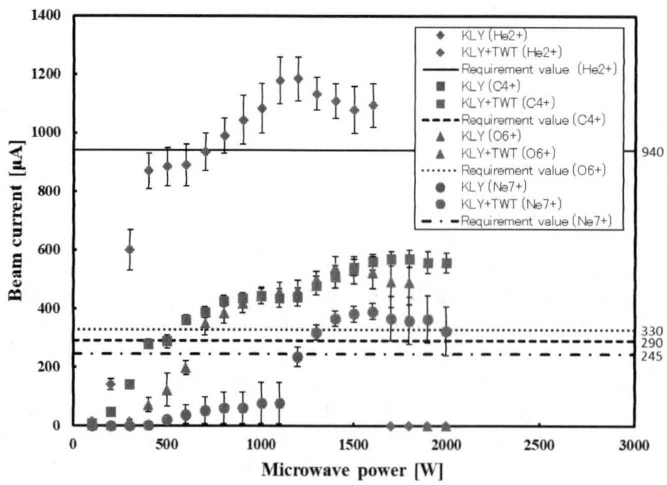

Figure 1. Dependence of the beam current on microwave power at He^{2+}, C^{4+}, O^{6+} and Ne^{7+}. Microwave frequency of TWT was 17.32 GHz. Extraction voltage were 26.5 kV for He^{2+}, 30 kV for C^{4+}, O^{6+} and Ne^{7+}. Extraction gap were 15 mm for He^{2+}, C^{4+}, O^{6+}, 16 mm for Ne^{7+}. The current values of the upstream and downstream mirror coils were 820A and 500A for He^{2+}, 860A and 490A for C^{4+}, 840A and 500A for O^{6+}, 860A and 540A for Ne^{7+}. The ionization gas of Helium, CH_4, O_2 and Neon were used in this investigation.

Figure 1 shows dependence of beam current on microwave power. Helium, CH_4, O_2 and neon gases were used for production of each ion. There are two microwave amplifiers for two-frequency heating method, a klystron amplifier (KLY) and a traveling wave tube amplifier (TWT) in the NIRS-HEC. The microwave frequency and maximum output power of amplifiers are 18 GHz, 1400 W for KLY and 17.10-18.55GHz, 1200 W for TWT, respectively. The KLY was used daring output power of 0 to 1200 W (blue dot). Two-frequency heating method [8,9,10] was used from 1300 to 2000 W (red dot). Dotted line shows requirement values. Beam instability is shown at each dot. Higher microwave power tends to make the beam unstable for all ion species. In the case of He^{2+}, the beam could not be extracted due to the effect of electric discharge when the total microwave power was over 1700 W. The beam currents of He^{2+}, C^{4+} and O^{6+} were reached to requirement value under only the KLY operation. However, two-frequency heating is necessary for production of Ne^{7+}. In the case of He^{2+}, C^{4+} and O^{6+}, maximum microwave power needs less than 1000 W with single microwave amplifier. However, production of Ne^{7+} ion needs about 1500 W and two-frequency heating method for stabilized beam current. From these results, we need two microwave amplifiers with total power of more than 1500 W.
From the results of microwave power dependence, production of Ne^{7+} ion is difficult compared to production of other ions. The beam current of Ne^{7+} was sufficient when the combination of mirror coils with the current value of the upstream/downstream mirror coil was 860 A/540A, respectively. However, it is difficult to reproduce the magnetic field of 860 A with permanent magnets because of its structural problems. Therefore, the current value of the downstream coil was investigated by setting the upstream mirror coil current value to 840 A.
Figure 2 shows dependence of the He^{2+} (square) and Ne^{7+} (circle) beam current on downstream coil current under the upstream coil current of 840 A. The microwave powers for production of He^{2+} and Ne^{7+} were 1200 W and 1200+400 W (KLY+TWT), respectively. The beam currents of He^{2+} and Ne^{7+} were higher than each requirement values when the downstream coil currents were 500 and 520 A. From the result of the coil current dependence, we were decided to design the permanent magnet for the magnetic field corresponding to the current values of the upstream and downstream mirror coils of 860 A and 500 A.

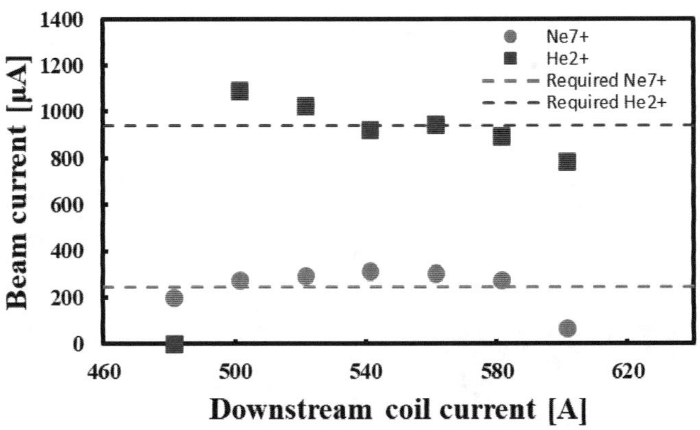

Figure 2. Comparison between He^{2+} and Ne^{7+} beam current at the dependence of downstream coil current under the upstream coil current of 840 A.

3. Calculation of magnetic field

From the results of coil dependence, we estimate the mirror magnetic field of the NIRS-HEC by using the POISSON/SUPERFISH code [11]. Then, the structure of the permanent magnet for new ECRIS is determined so as to reproduce the value of the upstream mirror peak (Binj), B minimum (Bmin) and the downstream mirror peak (Bext) at the NIRS-HEC. The magnetic fields of Binj, Bmin and Bext at the NIRS-HEC were 1.14 T, 0.475 T and 0.9 T, respectively.

3.1. Mirror confinement field
Input data for the POSISSON/SUPERFISH of ring magnets for mirror confinement field was used the Shin-Etsu Neodymium Magnet N Series (N50H-MF), which is composed mainly of neodymium (Nd), iron (Fe), and boron (B), is the high-performance rare earth magnets [12]. The permanent magnets for mirror field are consists of 5 ring magnets (RM1-5, see figure 3). The RM1 is the microwave injection side and the RM5 is the beam extraction side. The inner dimeter of the RM1, RM4 and RM5 are 47, 62 and 70 mm, respectively. A hexsapole magnet is installed inside the RM2 and RM3. An iron yoke is attached to the outside of the ring magnet. The outer diameter of the ring magnet including the iron yoke and the length are 400 mm and 357 mm, respectively. The magnetic fields of Binj, Bmin and Bext at the result of this calculation were 1.24, 0.488 and 0.97 T, respectively. Based on our experience, the actual value of the magnetic field has decreases by about 5-10% from calculated value, therefore, it is considered that the magnetic field reached to the target value.

3.2. Radial confinement field
A hexapole magnet is used to produce the radial confinement magnetic field. The inner diameter of the hexapole magnets was 56 mm so that the inner diameter of the plasma chamber was 50 mm as in the previous Kei series. This includes a plasma chamber wall thickness of 2 mm and an air layer of 1 mm for thermal insulation. Input data for the hexapole magnet was also used the Shin-Etsu Neodymium Magnet N Series (N43TS). The inner diameter on the pole is 56 mm. A groove with an inner diameter of 62 mm was made in the hexapole magnets to pass the cooling water of the plasma chamber. The distance between the plasma chamber wall and Becr must be sufficient. The distance is 9.4 mm which the microwave frequency is 14 GHz.

4. Design of new compact ECR ion source
We design the structure of new compact ECRIS based on the result of the calculation and our experience of medical use. Figure 4 shows schematic drawing of new compact ECRIS named Multi-Ion production at Kei5 (MIK).

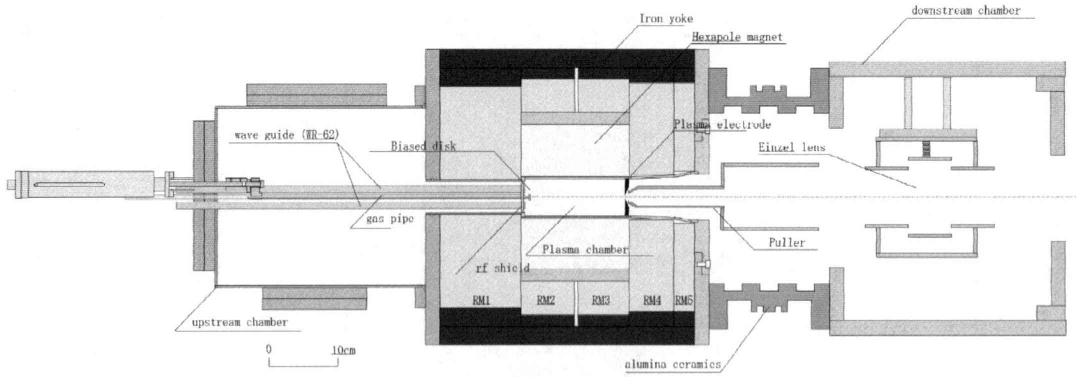

Figure 3. schematic drawing of new compact ECRIS.

The MIK source consists of an upstream vacuum chamber, a permanent magnet, and a downstream vacuum chamber. A rectangular waveguide, a gas pipe, and a biased disk are installed in the upstream vacuum chamber. An extraction electrode and an Einzel lens are installed in the downstream vacuum chamber. A high voltage for ion extraction is applied to the permanent magnet including the upstream vacuum chamber and the plasma chamber. The maximum extraction voltage is 30 kV.
The plasma chamber is made of copper with water cooling. The inner diameter of the plasma chamber is 50 mm and the outer diameter is 54 mm. In beam extraction regions, it is necessary to increase the inner diameter in order to suppress the discharge. The inner diameter of the flange of the plasma chamber is 62 mm. Cooling water is introduced into the plasma chamber from the flange of the beam extraction

side and flows in the chamber body in the upstream direction. It passes between the poles of the hexapole magnet and is returned to the flange and extracted. As a result of thermal calculation of the plasma chamber, this structure has sufficient cooling capacity. The calculation condition is that the heat input from the plasma is 100 W (microwave: pulsed operation, maximum power 2000 W, pulse width 50 msec, repetition frequency 1 Hz), and the discharge at the time of beam extraction is 30 W, total 130 W heat input. And the temperature rise of the cooling water was 3 ° C. Two rectangular waveguides, a gas pipe, a biased disk, and an RF shield drive are installed in the upstream vacuum chamber. The vacuum chamber is made of stainless steel. From the experimental results of the NIRS-HEC, two microwave frequency heating method is required to produce highly charged ions of neon. In addition, microwave of 13.7 GHz or higher is required from the result of magnetic field calculation. Therefore, microwaves are introduced into the plasma chamber using two WR-62 waveguides. The end of the waveguide is located at the peak of the upstream mirror magnetic field. The material of the waveguide is copper. As a result of the previous test, a sufficient effect was obtained even when the output of the second microwave was about 100 W. Therefore, the first microwave amplifier requires a maximum output of 2000 W, a frequency band of 13.7-14.6 GHz, a maximum pulse width of 100 msec, and a maximum repetition frequency of 1 Hz, and the second amplifier has a maximum output of 10.0-18.0 GHz and a maximum output of 0.25 kW, a maximum pulse width of 100 msec and a repetition rate of 1 Hz is sufficient.
He, CH_4, O_2, and Ne gases are used to produce four kinds of ions, respectively. The gas lines merge outside the ion source and the gas is introduced into the plasma chamber with a single gas pipe. These gases are switched on and off using a solenoid valve for fast ion species switching. The diameter of the biased disk is 8 mm. The maximum applied voltage to the biased disk is -100 V because there is damage due to ion sputtering. The biased disk and RF shield are movable in the axial direction. Because, it has been known that the position of the biased disk and the RF shield can be changed to effectively produce highly charged ions. The bias disk and RF shield are made of molybdenum and copper, respectively. An insulator made from alumina ceramics for extracting the ions is installed between the permanent magnet and the downstream vacuum chamber. An extraction electrode and an Einzel-lens is arranged in the downstream vacuum chamber. The material of the extraction electrode is made from molybdenum. The aperture of extraction electrode is 10 mm. A plasma electrode made from molybdenum having a beam extraction hole of 8 mm is arranged on the plasma chamber side. The extraction electrode is movable so that the distance of extraction electrode and plasma electrode can be optimized for the extraction of four ion species.

References

[1] Y. Hirao et al., Nucl. Phys. A 538, 541c (1992).
[2] M. Muramatsu et al., Rev. Sci. Instrum. 76, 113304 (2005).
[3] Satoru Yamada et al., Proceedings of NIRS-Etoile Joint Symposium 2009 on Carbon Ion Radiotherapy, Lyon, France, NIRS-M-218, 170 (2009).
[4] M. Muramatsu et al., Review of Scientific Instruments 87, 02C110 (2016).
[5] T. Inaniwa, N. Kanematsu, Phys. Med. Boil. 61 542-550 (2016).
[6] T. Inaniwa et al., Phys. Med. Boil. 62-5180 (2016).
[7] A. Kitagawa et al., Rev. Sci. Instrum. 69 674 (1998).
[8] Z. Q. Xie and C.M. Lyneis, "Improvements on the LBL AECR source", ECRIS1995, RIKEN, April 1995, INS-J-182 p. 24 (1995).
[9] S. Biri et al., Rev.Sci.Instrum. 85, 02A931 (2014).
[10] A. Kitagawa et al., Rev.Sci.Instrum. 87, 02C107 (2016).
[11] Reference Manual for the POISSON/SUPERFISH Group of Codes, LA-UR-87-115 (1987)
[12] https://www.shinetsu-rare-earth-magnet.jp/e/index.html

Duoplasmatron type molecular carbon ion source

M Wada, H Aki, N Miyamoto

Graduate School of Science and Engineering, Doshisha University
Kyotanabe, Kyoto 610-0321, Japan

Email: mwada@mail.doshisha.ac.jp

Abstract. An ion source capable of producing molecular ions delivers a beam of impurity dopant for shallow ion implantation of semiconductor device fabrication. Production and transport of low energy molecular ions is indispensable for tabulating fundamental data on plasma-wall interactions that should be used to design a future nuclear fusion reactor. An ion source plasma can produce molecular ions of solid materials by sputtering, while low temperature plasma efficiently transports produced molecular ions to the source extraction hole. A duoplasmatron ion source with the carbon plasma electrode has successfully produced poly-atomic carbon ions up to C_6^+. The intermediate electrode has six holes to guide cathode plasma to the carbon anode performing as the plasma electrode. A 1 mm diameter extraction hole is opened at the center of the carbon plasma electrode, where is no direct exposure to the plasma from the cathode region. The ion source maintains a stable discharge with both H_2 and Ar as discharge support gas, but works better at lower pressure with Ar. A hydrogen plasma produced molecular ions of hydrocarbons with the mass separated current of $C_3H_5^+$ with the intensity comparable to other ion species including atomic, diatomic and triatomic hydrogen ions.

1. Introduction

A duoplasmatron ion source produces a small volume of high-density plasma with a high degree of ionization. The cathode of the discharge is contained in the intermediate electrode that geometrically compresses the produced plasma toward the anode. The intermediate electrode is made of high-permeability magnetic material and is coupled to an electromagnet for further magnetic compression of the plasma. The plasma density between the intermediate electrode hole and the anode becomes large enough and the high-density plasma passes through an orifice opened at the center of the anode (anode orifice) to reach the other side of the anode facing the extraction electrode. A beam is formed by the extraction electric field, and a volume called the expansion cup downstream of the anode orifice reduces the plasma density to efficiently extract ion beams [1]. The traditional source design is still utilized for proton beam source for large accelerator facilities [2]. The first attempt to reduce the plasma source size using permanent magnet has been made by Samson and Liebl to make a VUV light source [3]. Tawara, Suganomata and Suemitsu, accomplished further miniaturization of a duoplasmatron source using 6 cm outer diameter permanent magnets [4]. The invention of high energy density permanent magnets enabled the further miniaturization of duoplasmatron plasma source. A stable operation of the source that can be attached to an ICF-34 flange with a hollow cathode was realized [5], while a thin hydrogen plasma column was suspended in a linear magnetic field to serve as the particle source to study plasma surface interaction using optical spectrometry [6].

An ion source that can produce molecular ions is necessary for shallow implantation processes of semiconductor industry [7]. Duoplasmatron sources produce high density plasma of the high degree of ionization and may not be suitable to produce beams rich in molecular ions with its original design. However, Collins and Gobbett demonstrated that a duoplasmatron ion source can produce negative ion of hydrogen by extracting the ions from the edge region of the plasma passing through the hole of the intermediate electrode [8]. Thus, we investigated the possibility to form molecular ion beam using a duoplasmatron ion source through modifying the intermediate electrode geometry.

2. Experimental apparatus

2.1. Ion source test stand

Figure 1 (a) schematically illustrates the configuration of the ion source test stand. A dc beam is extracted from a test ion source mounted in a high voltage cage. After travelling 1120 mm from the extractor, the beam is deflected by a 90-degree bending sector magnet (magnet is not shown in the figure) with the 285 mm Larmor radius. Two 22 mm high, variable width (20 mm maximum) slits are attached at both the entrance and the exit of the mass analyzer section. Both slits were set the width to 3.5 mm for realizing enough mass resolution for this series of experiment. The beam travels another 820 mm to reach Faraday cup to be detected. The pressure of the system was monitored with a vacuum ionization gauge down stream of the accelerator. (Ionization vacuum gauge in figure 1 (a).)

2.2. Duoplasmatron ion source

The structure of the designed duoplasmatron source is shown in figure 1 (b) with a conventional intermediate electrode opened at the center. A 0.5 mm diameter 90 mm long tungsten (W) filament works as a high temperature cathode with the 22 mm spacing between the tip of the filament and the hole of the intermediate electrode. The intermediate electrode is made of mild steel with the 56 mm inner diameter. An 80 mm inner diameter, 120 mm outer diameter and 20 mm long toroidal Nd-Fe magnet creates the magnetic field in the region between the inter mediate electrode and the anode. The anode is made of the same material with the intermediate electrode, and forms the magnetic circuit to concentrate the magnetic field intensity at the end of the intermediate electrode. The material of a 30 mm diameter 1 mm thick anode with the 1 mm diameter hole is graphite of 99.98% purity.

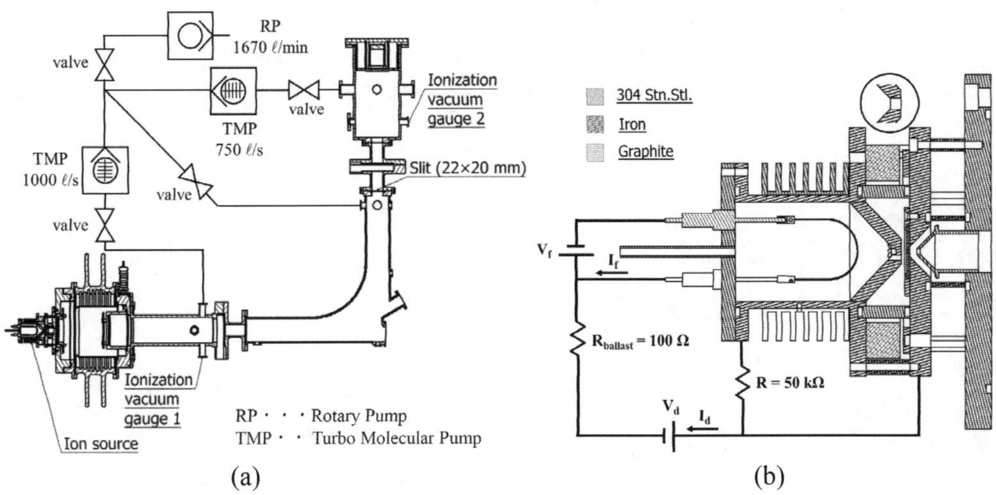

Figure 1. Experimental apparatus. (a) Schematic illustration of the ion source test system with a magnetic deflection mass analyzer. (b) Structure, materials of components and the discharge circuit of the duoplasmatron ion source.

2.3. Intermediate electrode

Three types of intermediate electrode were tested. The first structure is the ordinary single hole intermediate electrode shown in figure 1 (b) above the ion source. The diameter of the hole opened at the center of the electrode is 6 mm. Two others have six 3 mm diameter holes arranged on a circle of 7 mm diameter; the hole centers are located 3.5 mm from the center axis of the intermediate electrode as shown in figure 2. These electrodes with six holes are called multi-hole electrodes and a 3 mm diameter 3 mm deep pits are arranged on the axis of the electrodes. In one design, the pit faces toward the anode orifice to act as a volume to enhance collisions in the plasma. The pit was made at the opposite side of the intermediate electrode in the other design, to see the effect of the pit in front of the anode orifice. These structures of the electrodes are schematically shown in figure 3.

Figure 2. Picture showing the multi-hole intermediate electrode.

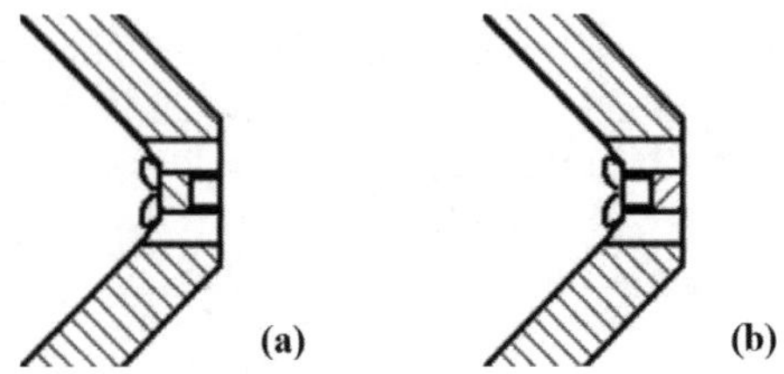

Figure 3. Pits arranged on the intermediate electrode. (a) the pit forms a reaction volume in front of the anode orifice. (b) the pit is opened on the opposite side without forming a volume in front of the anode orifice.

3. Experimental Results

3.1 Ar operation

The effect of multi-hole intermediate electrode on carbon cluster ion formation was studied using Ar gas. The plasma striking the surface of carbon anode orifice produces carbon atoms and molecules by sputtering processes. The released carbon atoms and molecules are ionized in the plasma, pass through the anode orifice and extracted. The results for single hole intermediate electrode, or ordinary duoplasmatron geometry is shown in figure 4. As shown in the figure, C^+ and C_2^+ ions are found in the mass spectrum. The peak position of C_3^+ was covered by low energy tail of Ar^+ ion spectrum. No peak was observed at the position of C_4^+. When the ion source was operated at the same discharge current with multi-hole intermediate electrode, the mass separated ion beams increased by an order of magnitude.

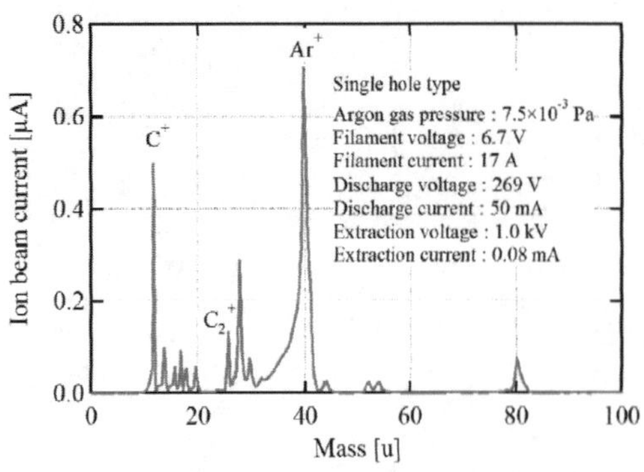

Figure 4. Mass spectrum of the ion beam extracted from a single-hole intermediate electrode ion source operated with Ar.

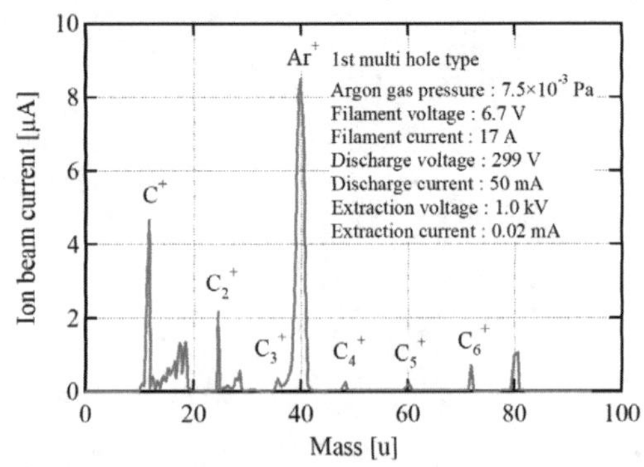

Figure 5. Mass spectrum of the ion beam extracted from a multi-hole intermediate electrode ion source operated with Ar.

This is because the single-hole structure produced a plasma too dense to extract ions without an expansion cup. The mass spectrum for 1 kV extraction for the source with multi-hole electrode shown in figure 5 clearly indicates the production of carbon cluster ions up to C_6^+. When we compare the ratio of C^+ ion intensity to Ar^+ ion intensity single-hole electrode shows a higher ratio, which possibly indicate that produced carbon molecules from the carbon anode surface are dissociated in an intense plasma channel directly touching down to the carbon anode from the cathode.

The multi-hole electrode used to take the mass spectrum in figure 5 had the structure without volume in front of the anode orifice: the geometry of figure 3 (b). The 3 mm diameter 3 mm deep volume of the electrode of the type illustrated in figure 3 (a) did not show any substantial change in mass spectrum from figure 5. Peaks of C^+, C_2^+ and C_3^+ were a little higher than larger cluster peaks, presumably due to dissociation in plasma volume when the source was operated with the figure 3 (a) electrode.

3.2 Hydrogen operation

When the source is operated with hydrogen gas, the source delivers beam rich in hydrocarbon ions. As shown in figure 6, hydrocarbon molecular ions show mass separated current peaks with the amplitudes nearly the same with those of H^+, H_2^+ and H_3^+. The multi-hole electrode did not change the mass spectrum largely except some better resolution enough to distinguish C_3^+ peak. This can be partly due to high intensities of the mass separated ion current. The intensity of $C_2H_2^+$, for example, is more than one order of magnitude larger for the multi-hole electrode source, while the extraction current of the source was nearly one order of magnitude smaller than the single hole electrode source. The expected smaller plasma density at the anode orifice made the ion extraction at low energy efficient.

The effect of volume in front of the anode aperture was investigated by replacing the intermediate electrode structure from the one shown in figure 3 (b) to the one shown in figure 3 (a). Firstly, we have observed the extraction current increased from 20 μA to 60 μA. Secondly, the discharge voltage became smaller for the same discharge current indicating higher plasma density in the intermediate electrode. All peak intensities of mass separated ion current increased about several percent, but the peak of C_3^+ and $C_3H_5^+$ showed the increase of several 10%. Thus, the volume may have advanced the ionization of molecular ions while the reaction did not contribute to dissociation of hydrocarbon molecules and carbon clusters.

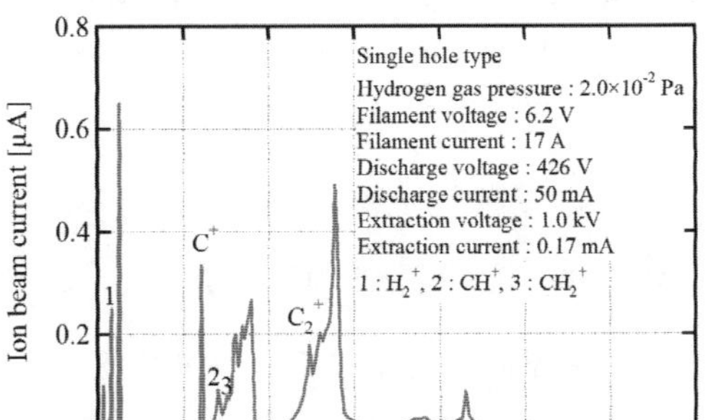

Figure 6. Mass spectrum of the ion beam extracted from a single-hole intermediate electrode ion source operated with H_2.

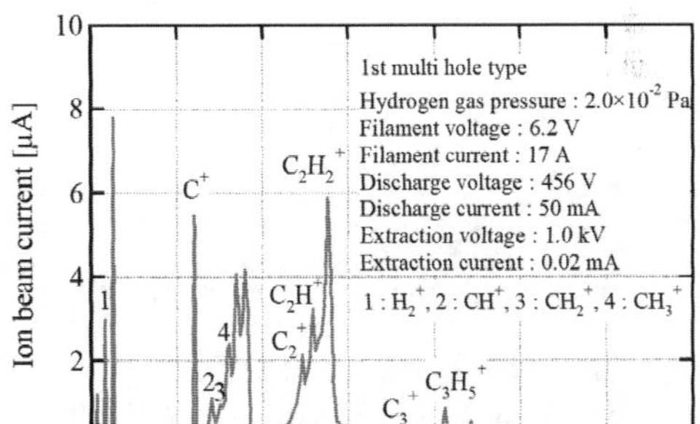

Figure 7. Mass spectrum of the ion beam extracted from a multi-hole intermediate electrode ion source operated with H_2.

4. Discussion

The geometry of transporting the source plasma through the anode orifice toward the ion extraction region is considered ideal to realize an ion source making use of plasma-surface interaction processes. The plasma striking the anode aperture should produce and ionize carbon clusters and carbon hydride molecules. Data on hydrocarbon emission from solid carbon can be found in literatures reporting plasma wall interaction in fusion experiment device of the previous generation [9]. Carbon clusters were known to be produced by Ar ion bombardment, but with energy higher than 75 eV [10]. As the particle retention largely changes the sputtering characteristics of solids at lower energy [11], the plasma bombardment may enhance molecular emissions from carbon surface in low energy range.

The clusters are ionized in a plasma volume after they leave the anode surface while energetic electrons can dissociate the formed cluster ions in the plasma. The electron energy necessary to ionize carbon clusters are 13 eV, 13 eV, 12 eV and 9.7 eV for C_3, C_4, C_5 and C_6, respectively [12]. Meanwhile, dissociation thresholds for C_2^+ to C_6^+ vary from 4.4 eV ($C_4^+ \rightarrow C^+ + C_3$) to 7.4 eV ($C_3^+ \rightarrow C_2^+ + C$) [13] and produced cluster ions can survive collisions before they are extracted as beam particles from a low temperature plasma. Only large molecular ions produced at the hot plasm channel can traverse the region of strong magnetic field to reach the region near the anode orifice. Thus, the magnitude of the extracted molecular ion current must be dependent upon the spacing between the hot plasma channels and the anode orifice. Optimization of the source should be possible by tuning the spacing between the intermediate electrode holes and the anode orifice as well as properly designing a plasma expansion structure (expansion cup) downstream of the anode orifice.

Regarding the transport of ions toward the anode orifice, the present source design does not take the plasma interaction with the surface of intermediate electrode into account. The electrode surface facing the anode should reflect ions back to the plasma, and the particle/energy reflection coefficients will change the plasma characteristics near the anode orifice. The effect due to surface roughness [14] suggests the possible improvement of particle transport to the anode orifice region with the smoother surface of the intermediate electrode tip.

5. Summary

Carbon cluster ions and carbon hydride molecular ions are produced and extracted form a duoplasmatron ion source by changing the anode orifice material of the source to carbon. The observed ion species in the extracted beam show these ions are formed from molecular emission at the anode carbon surface immersed in plasmas. The addition of a small volume around the anode orifice region increased the heavy hydrocarbon molecular ions. The cluster ions up to C_6^+ for Ar operation and $C_3H_5^+$ for H_2 operation are produced with a multi-hole intermediate electrode. The power efficiencies to produce these ions are 0.07 μA/W for C_6^+ and 0.04 μA/ W for $C_3H_5^+$. The present efficiency at 100 W discharge power will form 100 C monolayer on a 10-inch wafer with 1,000 s for C_6^+ and 3,500 s for $C_3H_5^+$. Further improvements of the source are possible as plasma conditions and ion extraction conditions can be controlled separately by adding a plasma expansion cup.

References

[1] Collins L E, Brooker R J 1962 Nucl. Instrum. Methods **15** 193
[2] Zhou B, Zhang H, Zang P, Li W An S 2021 Nucl. Instrum. Methods A **1010** 165550
[3] Samson J A R, Liebl H 1962 Rev. Sci. Instrum. **33** 1340
[4] Tawara H, Suganomata S, Suemitsu S 1964 Nucl. Instrum. Methods **31** 353
[5] Kuzumi T, Wada M 2020 Rev. Sci. Instrum. **91** 013505
[6] Guhit J R M, Gines A R B, Doi K, Wada M 2020 Plasma Fusion Res. **15** 1401048
[7] Hershcovitch A *et al* 2016 Rev. Sci. Instrum. **87** 02B702
[8] Collins L E, Gobbett R H 1965 Nucl. Instrum. Methods **35** 277
[9] Philipps V, Pospieszczyk A, Erdweg M, Schweer B, Vietzke E, Winter J 1996 Physica Scripta T **64** 71
[10] Oyarzabal E, Doerner R P, Shimada M, Tynan G R 2008 J. Appl. Phys. **104** 043305
[11] Kenmotsu T, Wada M, Hyakutake T, Muramoto T, Nishida M 2009 Nucl. Insrum. Methods B **267** 1717
[12] Ramanathan R, Zimmerman J A, Eyler J R 1993 J. Chem. Phys. **98** 7838
[13] Sowa-Resat M B, Hintz P A, Anderson S L 1995 J. Phys. Chem, **99** 10736
[14] Mayer M, Scherzer B M U, Eckstein W 1994 Nucl. Inst. Methods **B** 85 560

Laser ion source for semiconductor applications

T V Kulevoy[1], A A Losev[1], P N Alekseev[1], Yu A Satov[1], A V Shumshurov[1], P B Lagov[2,3], M E Letovaltseva[4], S A Zinoviev[2]

[1] NRC "Kurchatov Institute" - ITEP, Moscow, Russia
[2] National University of Science and Technology "MISiS", Moscow, Russia
[3] A.N. Frumkin Institute of Physical Chemistry and Electrochemistry, Moscow, Russia
[4] Russian Technological University MIREA, Moscow, Russia

E-mail: kulevoy@itep.ru

Abstract. Carbon implantation can be effectively used for axial minority charge carriers' lifetime control in various silicon bulk and epitaxial planar structures. When carbon is implanted, more stable recombination centres are formed and silicon is not doped with additional impurities, as for example, when irradiated with protons or helium ions. Economically, such a process competes with alternative methods, and is more efficient for obtaining small lifetimes (several nanoseconds). I-3 ion injector with laser-plasma ion source at Institute for theoretical and experimental physics (ITEP) is used as ion implanter in semiconductors. The ion source uses pulsed CO_2 laser setup with radiation-flux density of 10^{11} W/cm^2 at target surface. The ion source produces beams of various ions from solid targets. The generated ion beam is accelerated in the two gap RF resonator at voltage of up to 2 MV per gap. Resulting beam energy is up to 4 MV per charge. Parameters of carbon ion beam generated and used for semiconductor samples irradiation during experiments for axial minority charge carriers lifetime control in various silicon bulk and epitaxial planar structures are presented.

1. Introduction

Many silicon devices, including power MOSFET, IGBT, FRD etc. require applying technologies to control the lifetime of minor charge carriers [1, 2]. As a result, one utilizes Au- or Pt-thermodiffusion, radiation, and combined methods. In practice the most widely utilized method is treatment by accelerated electrons, protons or He^+ ions. The accelerated (usually 4–8 MeV) electrons provide uniform formation of defects in the volume of semiconductor wafers arranged one after the other (~10 pcs). However, achieving small switching times ($\leq$5 ns) requires high fluences and irradiation time. Proton irradiation at high fluences leads to doping by small donors and some instability of the achieved electrical parameters. Therefore, for these purposes, He^+ ion irradiation is often used, which produces displacements more efficiently, and the behavior of device structures after irradiation and subsequent annealing is more stable. Due to the fact that carbon ions also have a relatively large range at relatively low (no more than 20 MeV) accelerating voltages, the possibility to obtain low switch time for reference diode by carbon implantation was investigated. In addition, carbon in sufficiently high concentrations (10^{16}–10^{17} cm^{-3}) is present in single-crystal silicon, i.e., additional contamination with foreign impurities is minimal. The purpose of this study is to show in practice that carbon ion implantation with subsequent annealing can be effectively utilized to achieve optimal charge carriers lifetime in active silicon semiconductor structures and thus increasing their speed

Journal of Physics: Conference Series **2244** (2022) 012096 doi:10.1088/1742-6596/2244/1/012096

without degradation of other parameters. This technological operation can also be efficient in supressing thyristor effect in bulk silicon structures. To demonstrate the efficiency of carbon implantation planar diode was used as a test device.

2. Setup layout

The setup consists of a laser plasma ion source, a buncher, an accelerating RF resonator, a bending magnet and ion beam transfer system as shown in figure 1.

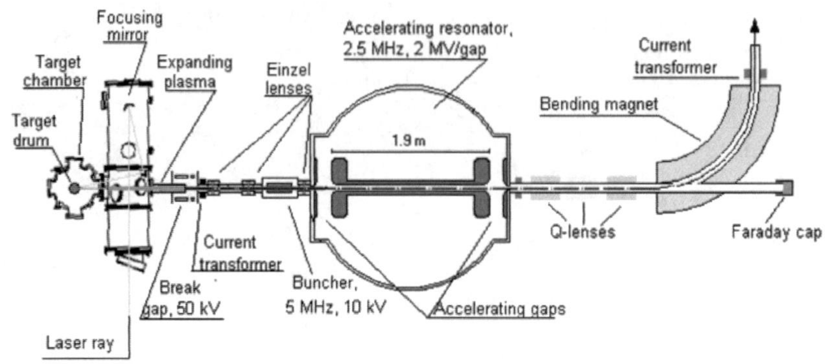

Figure 1. Setup layout.

2.1. Ion source

The laser-plasma ion source is based on the pulsed CO_2 laser which is described in [3]. It provides laser beam with high quality of temporal and spatial characteristics: pulse energy 6 J, peak power 60–70 MW, FWHM duration 30 ns, repetition rate 0.1 Hz, beam spatial profile is close to Gaussian. Its radiation is transfered by flat copper mirrors to the vacuum chamber and focused by combination of spherical ($f = 1600$ mm) and flat mirror. Carbon target surface is irradiated by laser pulses at a radiation flux density of 10^{11} W/cm^2. Generated plasma contains a set of charge states of carbon ions, of which C^{3+} and C^{4+} are most represented in terms of the number of particles as shown in figure 2.

Carbon ion beam is extracted from expanding plasma by high-voltage gap with grids, placed 1.68 m away from target surface. Total beam current from the source is measured by current transformer installed behind extraction gap and is shown in figure 3. 30 pulses were used to calculate avearge output current and its standard deviation. Extracted beam is accelerated by the I-3 ion injector, which is a single drift tube linac designed to accelerate ions in a wide range of charge to mass ratio.

2.2. I-3 ion injector

The I-3 ion injector is a single drift tube linac designed to accelerate ions in a wide range of charge to mass ratio. It can slightly vary accelerated beam energy. Its parameters are presented in table 1.

A 90° bending magnet is used to select charge state of ions in accelerated beam therefore this setup allows to change beam energy in wider range to meet irradiation requirements.

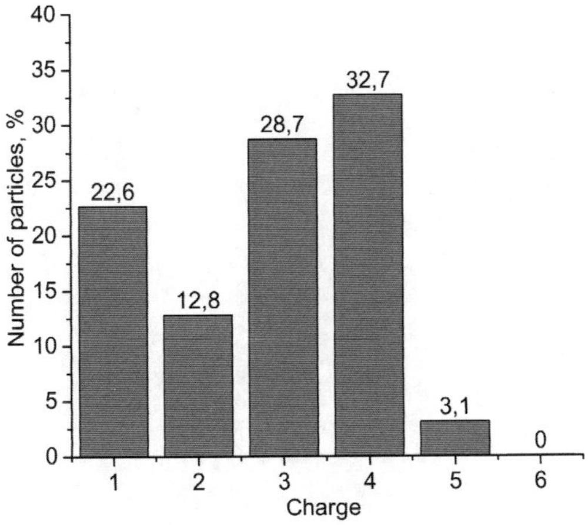

Figure 2. Charge state distribution.

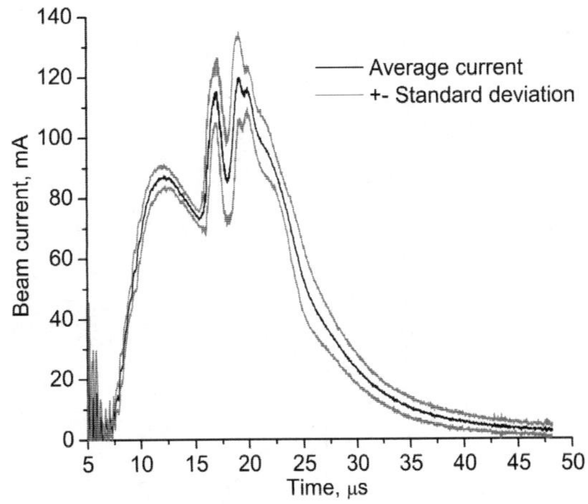

Figure 3. Laser plasma ion source output current (contains all generated charge states).

Table 1. I-3 linac parameters.

Parameter	Value
Frequency	2.504 MHz
Number of accelerating gaps	2
Voltage per gap	Up to 2 MV
Length of drift tube	1920 mm
Aperture	70–90 mm
Range of Z/A ratio	0.2–0.5
Input energy	50 kV per charge
Transverse acceptance	2000 π·mm·mrad

3. Irradiation conditions and results

Fluence was calculated by integration of accelerated ion beam current signal measured as voltage drop caused by 50 Ohm load connected to the target holder. This calculation was carried out for each ion current pulse. Single pulse is shown in figure 4. Irradiated area of the sample was determined using target holder photos and four marker holes as shown in figure 5. Irradiation uniformity was calculated as standard deviation of ion beam luminosity in irradiated area of the sample using scintillator images (figure 6).

C^{3+} ion was chosen for sample irradiation, accelerating voltage was 2.9 ± 0.3 MV resulting in 8.7 ± 0.9 MeV beam energy. Irradiation uniformity of the sample area was not worse than 18%.

Experimental reference diode structures were obtained by forming a 3 μm junction in 10 μm epitaxial n-type layers, followed by vacuum deposition of the 1 μm anode Al-contact. The optimal carbon implantation energy range of 8.4–8.7 MeV and fluences of $1 \cdot 10^{12} - 4 \cdot 10^{12}$ cm^{-2} were selected using the SRIM [4] software package and real experience of previous radiation experiments with different particles and device structures [1, 2, 5, 6, 7]. All samples were subjected to rapid thermal annealing at a temperature of 425°C during 15 seconds after irradiation. No other methods were used to reduce lifetime of minor charge carriers.

Three parameters were taken in consideration: t_{rr} – reverse recovery time, I_R – reverse (leakage) current, U_F – forward voltage drop at forward current of 10 mA. The results are

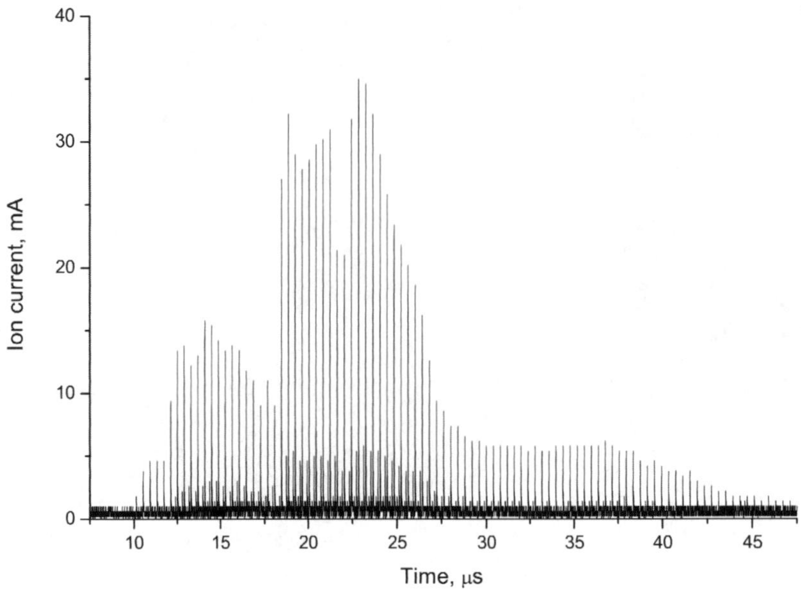

Figure 4. Single pulse of carbon ion beam current.

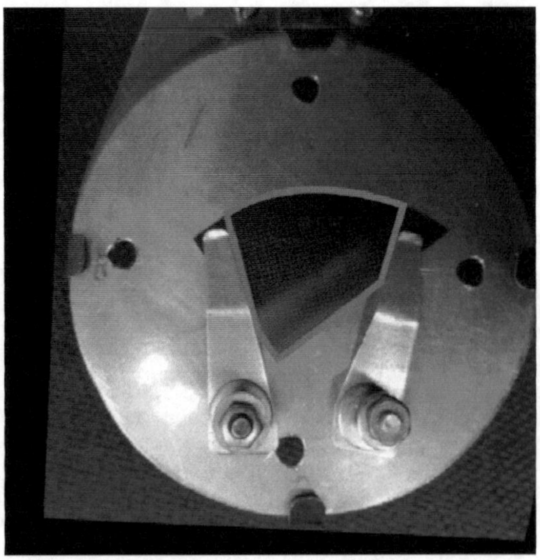

Figure 5. Sample in a holder.

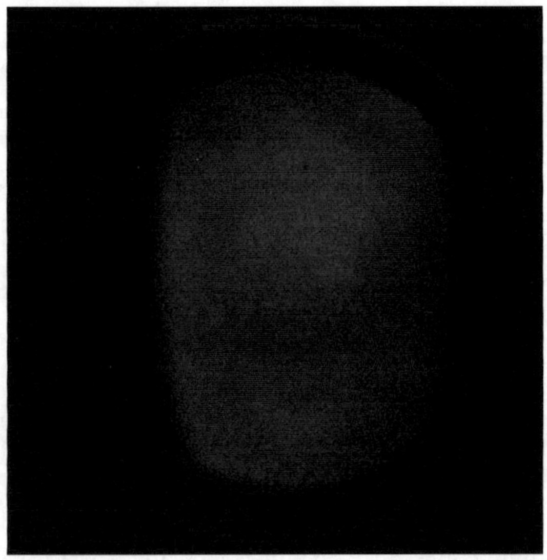

Figure 6. Beam profile on the sample.

shown in table 2.

The combination of these three parameters is considered to be the best when the lowest t_{rr} is reached with minimum I_R increase. Decrease in U_F due to growth of recombination component of forward current is also a positive effect. Such a decrease is typical for epitaxial diode structures with thick base width and low reverse voltage (less than 100 V), to which the investigated diode structure belongs. In our case the best combination of parameters is achieved at an energy of 8.7 MeV and a fluence of $2 \cdot 10^{12}$ cm^{-2}. This combination provides necessary frequency and pulse characteristics of the diode for operation in certain pulsed circuits.

Table 2. Parameters of reference diodes.

Irradiation by C^{3+}	t_{rr}, ns	I_R, nA	U_F, V (at $I_F = 10$ mA)
Initial	~ 100	< 1	0.81
8.4 MeV, F $= 1 \cdot 10^{12}$ cm^{-2}	18 ± 2	3.3 ± 0.3	0.78
8.4 MeV, F $= 4 \cdot 10^{12}$ cm^{-2}	2.7 ± 0.2	4.5 ± 0.4	0.75
8.7 MeV, F $= 2 \cdot 10^{12}$ cm^{-2}	4.9 ± 0.3	1.0 ± 0.1	0.76

4. Conclusion

The I-3 ion injector equipped with laser-plasma ion source is supposed to be used to determine optimal conditions for irradiation of various semiconductor devices. Wide range of ion charge states generated by ion source extends range of available beam energy for experiments. The diodes in consideration performed best after irradiation by C^{3+} ion beam at an energy of 8.7 MeV and a fluence of $2 \cdot 10^{12}$ cm^{-2}. The data obtained is valuable when designing industrial accelerators for semiconductor production.

References

[1] Lagov P B, Drenin A S and Zinoviev M A 2017 *J. Phys.: Conf. Series* **830** 012152
[2] Pavlov Y S, Surma A M, Lagov P B, Fomenko Y L and Geifman E M 2016 *J. Phys.: Conf. Series* **747** 012085
[3] Satov Yu A, Shumshurov A V, Vasilyev A A, Balabaev A N, Losev A A, Khrisanov I A and Rerikh V K 2016 *Instrum. Exp. Tech.* **59** 412–419
[4] Ziegler J F, Ziegler M D and Biersack J P 2010 *NIMB* **268** 1818–23
[5] In-Hwan Lee et al 2017 *ECS J. Solid State Sci. Technol.* **6** Q127
[6] Aslanov L A et al 2020 *Surf Interface Anal.* **52** 957–961
[7] Polyakov A Y et al 2020 *J. Phys. D: Appl. Phys.* **53** 274001

Low-voltage extraction characteristics of ion beams and space potential measurements on ECRIS

I Owada, W Kubo, S Harisaki, K Tsuda, K Sato, M Anan, T Maenaka and Y Kato

Division of Electrical, Electronic and Infocommunications Engineering, Graduate School of Engineering, Osaka University, 2-1 Yamada-oka, Suita-shi, Osaka 565-0871, Japan

owada@nf.eie.eng.osaka-u.ac.jp

Abstract. We have been using an electron cyclotron resonance ion source (ECRIS) for ion beam production, and it is desirable to construct a universal ECRIS that can produce ions with a wide range of mass/charge ratios, e.g. from several to several thousand. We investigated the characteristics of ions with high mass/charge ratios, e.g., iron endohedral fullerenes production, which requires low voltage beam extraction. We investigated the characteristics of the ion beam extraction at low extraction voltage. It is found that the beam current almost obeyed the Child-Langmuir law for various ion species. The space potential of the plasma in the ion source can be obtained from the relation between the extraction voltage and the square of the magnetic field strength of the dipole magnet. It is found that space potential values are larger with large charge state of ions, in low gas pressure condition and in high microwave incident power. At the same time, the plasma space potential was measured by using a Langmuir probe and compared with the ion beam method. It is found that same trend is confirmed with probe method.

1. Introduction

We have been using an electron cyclotron resonance ion source (ECRIS) for ion beam production. [1] The ECRIS has been applied to various fields such as heavy particle cancer therapy and space applications. In the bionanomaterials field, research has been conducted to encapsulate various types of atoms in fullerene molecules. Fullerenes have a cage-shaped construction, and their chemical and physical properties vary depending on the atoms encapsulated. [2] Iron endohedral fullerenes ($Fe@C_{60}$) are expected to have a variety of applications, including quantum computers and high-intensity MRI contrast agents. [3,4] Various past techniques such as laser vaporization, arc discharge, and ion implantation have been used to produce endohedral fullerenes. [5-8] We aim to synthesize $Fe@C_{60}$ by collisions in the plasma gas phase. We have achieved an ECR plasma in which iron and fullerene ions coexist. [9] The mass/charge distribution produced suggests the possibility of $Fe@C_{60}$ synthesis. [10]

In this study, we have been extracting ion beams at low extraction voltages, but it is unclear whether the extraction is being performed properly. Therefore, we investigated the characteristics of the ion beam extraction at low extraction voltage. It is found that the beam current almost obeyed the Child-Langmuir law. Because ECRIS is volume source of ion beams, the space potential is formed at the ECR plasma, the space potential affects the low-energy beam. We carried out space potential measurements using the extracted ion beam and showed dependence of the space potential on charge state and plasma parameters. It is found that space potential values are larger with large charge state of ions, in low gas pressure condition and in high incident microwave power. We also compared these space potentials with those using Langmuir probes. It is found that same trend is confirmed with the results from probe method.

2. Theoretical Background

In general speaking, the upper limit of the beam current of an ion source is proportional to the three halves power of the extraction voltage applied to the extraction electrode when the emittance of the beam is not a problem and there is sufficient source. [11] We conducted checking in our case of experimental conditions.

The plasma has a positive potential with respect to the grounded chamber wall, because the mobility of electrons is higher than that of ions. Using the plasma potential Φ_p with respect to the potential at the sheath and pre-sheath boundary and the potential Φ_w of the chamber wall with respect to Φ=0, the space potential V_s with respect to the wall can be expressed as follows:

$$V_s = \Phi_p - \Phi_w = \frac{kT_e}{2e}\left\{1 + \ln\left(\frac{m_i}{2\pi m_e}\right)\right\}, \quad (1)$$

where m_e is the mass of the electron, k is Boltzmann's constant, m_i is mass of ion, and T_e is the electron temperature.

When ion beam reflected perpendicularly through a dipole magnet following relationship is satisfied between magnetic field strength $B_\perp$ and space potential $V_s(q)$ [12,13]

$$B_\perp^2 = \frac{\kappa m_i}{q}(V_{PE} + V_s(q)), \quad (2)$$

where κ is a constant, q is the charge of the ion, and V_{PE} is the extraction voltage. The $V_s(q)$ is determined from the value of the magnetic field strength at the peak of each ion beam current in the m/q distribution. The value of $B_\perp$ corresponding to the spectrum is obtained while the V_{PE} has been reduced. The V_{PE}-$B_\perp^2$ relation is plotted and the offset of this line represents the value of $V_s(q)$.

The Langmuir probe has been also used to measure the space potential. After the electron current I_e derived with the probe current and probe voltage I_p-V_p characteristics, the space potential is also derived conventionally from the intersection of the two lines fittings in the deceleration field region and the electron saturation current region at I_e characteristics. The space potential derived by this method is defined as V_{s1}. Similarly, there is a method to derive V_s from the parameter of the electron temperature T_e obtained using a Langmuir probe. The space potential derived using the probe based on the theory described above as V_{s2}. These values are used for comparison with $V_s(q)$ obtained by beam method.

3. Experimental Apparatus

The top view of the ECRIS in our laboratory is shown in Fig. 1. The ECRIS consists of a vacuum vessel 160 mm in diameter and 1054 mm in length. Magnetic field is formed by four permanent magnets, two large mirror coils A and B, and an auxiliary coil C.

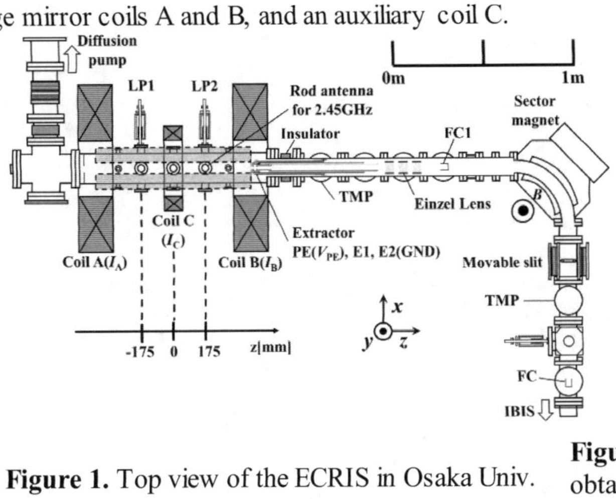

Figure 1. Top view of the ECRIS in Osaka Univ.

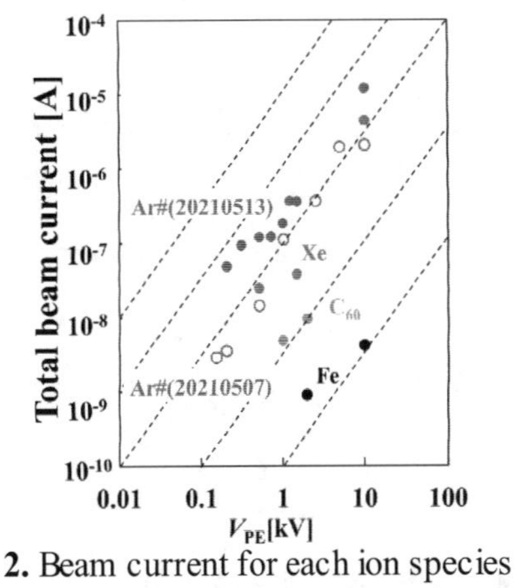

Figure 2. Beam current for each ion species obtained by changing the extraction voltage.

2.45 GHz microwave is introduced from the magnetron through a rod antenna. The incident power is adjusted to 20W~100W. Ar and Xe are used as support gases. In addition, Langmuir probes LP1 and LP2 are inserted at z = -175 mm and z = 175 mm, respectively, to obtain the plasma parameters in the x-axis direction upstream and downstream of the ion source. The plasma electrode (PE, z = 425 mm), the middle electrode (E1), and the extraction electrode (E2) are composed of the extractor, and the voltage can be applied independently. The voltage of PE electrode V_{PE} is applied normally 10kV with E2 as the ground potential. V_{PE}, while it is about 2kV for C_{60} beam extraction. For the space potential measurement experiment, the voltage was set at 1.5kV~0.15kV. The einzel lens is installed after the extraction electrode and the voltage of V_{el}=0~10 kV is applied to control the beam focusing. The ion beam is reflected perpendicularly by the $B_{\perp}$ of the sector magnet and separated by its mass/charge ratio. The charge state distribution (CSD) of the ion beams can be obtained. Beyond the faraday cup FC, the ion beam irradiation system (IBIS) can be used for applications.

4. Experimental Results

4.1. Beam extraction characteristics

Fig. 2 shows the total beam current for various ion species versus the extraction voltage V_{PE}. Summation value of the charge state is q = 1~4 for Ar and Xe, 1~3 for C_{60} and 1 for Fe. However, C_{60} beam current contains their fragmentation spectrum such as C^{q+}_{60-2n}. There are two data plots for Ar ion beam current. The blue closed circle plots show the gas pressure of 1.7×10^{-4}Pa and the incident microwave power of 100W. The blue open circle plots show the gas pressure of 1.2×10^{-3}Pa and the incident microwave power of 10W. The red plots show the Xe ion beam current, where the Xe gas pressure is 2.5×10^{-4}Pa and the incident microwave power is 20W. The green plots of C_{60} and black plots of Fe ion beam current, where the Ar gas pressure is 1.8×10^{-4}Pa and the incident microwave power is 10W. The C_{60} and Fe ion beam currents were obtained while attempting the coexistence of both ions and the Fe@C_{60} synthesis experiment, with the coil currents I_A=158A and I_B=110A. The dotted lines show the dependence of $V_{PE}^{3/2}$, and it is confirmed that the relationship between V_{PE} and the total beam current of each ion species is represented by the Child-Langmuir law.

4.2. Typical Ar CSD and gas pressure dependence of $V_s(q)$

Fig. 3(a) shows the typical CSD of Ar. The Ar gas pressure is 1.8×10^{-4}Pa, the incident power is 100W, and the extraction voltage is 10kV (upper figure), 1.5kV (middle figure), and 1.0kV (bottom figure). The applied voltage of the V_{el} was optimized with each extraction voltage, and it is confirmed that the ion peaks are shifted to the left. Figure. 3(b) and 3(c) show the V_{PE}-$B_{\perp}^2$ relationship for each spectrum of Ar^+~Ar^{4+} ions and show the results of $V_s(q)$ measurements performed at different Ar gas pressures. The incident microwave power was kept constant at 100W, and Ar gas pressures of 1.8×10^{-4}Pa (Fig. 3(b)) and 1.2×10^{-4}Pa (Fig. 3(c)), respectively. In Fig. 3(b) the space potentials are estimated to be

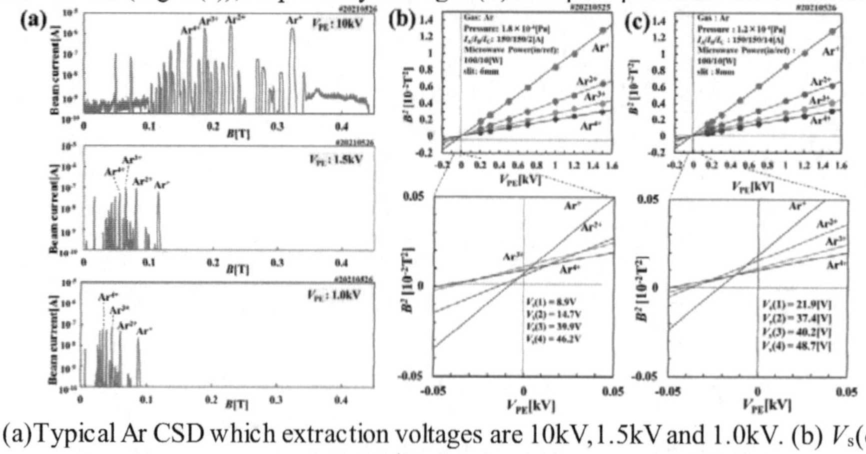

Figure 3. (a)Typical Ar CSD which extraction voltages are 10kV,1.5kV and 1.0kV. (b) V_s(q) obtained at Ar gas pressure 1.8×10^{-4}Pa, (c) Ar gas pressure 1.2×10^{-4}Pa

$V_s(1)$=8.9V, $V_s(2)$=14.7V, $V_s(3)$=39.9V, and $V_s(4)$=46.2V. Similarly in Fig. 3(c) the space potentials are estimated to be $V_s(1)$=21.9V, $V_s(2)$=37.4V, $V_s(3)$=40.2V, and $V_s(4)$= 48.7V. It is confirmed that the value of $V_s(q)$ becomes larger with higher charge state. The $V_s(q)$ values of Ar^+ and Ar^{2+} ions were larger at lower Ar gas pressures, while those of Ar^{3+} and Ar^{4+} ions were similar (see Fig. 5(a)).

4.3. Incident power dependence of $V_s(q)$

Fig. 4 shows the results of $V_s(q)$ measurements performed at different incident microwave powers, where the Ar gas pressure was kept constant at 1.1×10^{-4}Pa and the incident microwave power was 20 W (Fig. 4(a)) and 100 W (Fig. 4(b)). In Fig. 4(a) the space potential was estimated to be $V_s(1)$=30.6V and $V_s(2)$=29.3V. Similarly in Fig.4(b), the space potential was estimated to be $V_s(1)$=50.0V and $V_s(2)$=62.7V. Figure 5(a) shows the Ar gas pressure versus $V_s(q)$ plots and Fig. 5(b) shows the incident microwave powers versus $V_s(q)$ plots. In Fig. 5(b) it is confirmed that the value of $V_s(q)$ becomes larger under the condition of high incident microwave power.

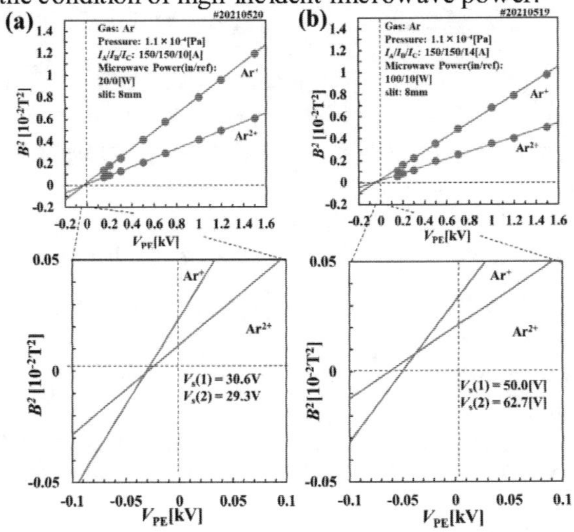

Figure 4. $V_s(q)$ measurement of (a)20W microwave incident power, (b)100W microwave incident power

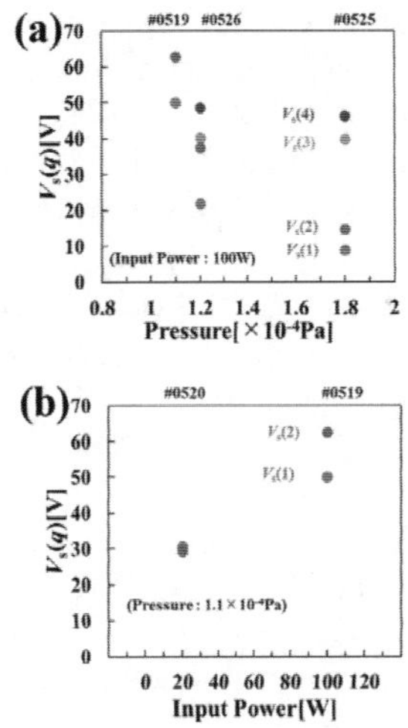

Figure 5. (a)Ar gas pressure-$V_s(q)$ plot, (b)Microwave incident power-$V_s(q)$

4.4. Parameter measurement using Langmuir probe

The plasma parameters acquired by the LP2 are shown in Fig. 6. The data were acquired three times at each point within x = 0~50 mm. The plots show the electron density n_e (Fig. 6(a)), the electron temperature T_e (Fig. 6(b)), and the space potential V_{s1} and V_{s2} (Fig. 6(c)), respectively. In addition, the red and black dots show the conditions for the incident microwave power of 100W and 20W, respectively. At the position x = 30 mm when the incident microwave power was 100W, the probe

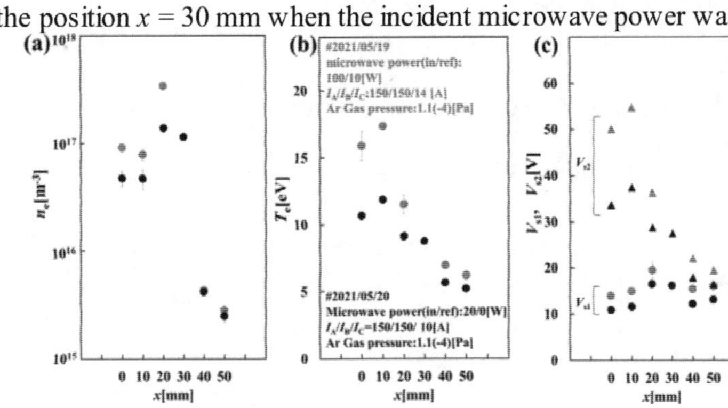

Figure 6. Results of x-axis probe measurements both incident powers are 100W and 20W. (a) electron density n_e, (b) electron temperature T_e, (c) space potential V_{s1}, V_{s2}

emitted light and the reliable I_p-V_p characteristics could not be obtained, so the data for that parameter was not obtained. It is measured that n_e = 2.80×10^{15}~3.43×10^{17} m^{-3}, T_e = 6.18~17.3 eV, V_{s1} = 14.0~19.5 V and V_{s2} = 19.5~54.7 V, respectively when the incident microwave power is 100W. It is also measured that n_e = 2.47×10^{15}~1.38×10^{17} m^{-3}, T_e = 5.20~11.9 eV, V_{s1} = 10.9~16.5 V and V_{s2} = 16.4~37.4 V, respectively when the incident microwave power is 20W. Both n_e and T_e are higher at the center of the plasma. The value of V_{s2} is similar to that of the T_e at the center of the plasma.

5. Discussions & summary

We investigated the extraction characteristics of the ion beam at low voltages for Ar, Xe, C_{60} and Fe ions. It is found that the beam current almost obeyed the Child-Langmuir law. Furthermore, we carried out space potential measurements using the extracted ion beam, and we showed dependence of the space potential on charge state and plasma parameters. It is found that space potential values are larger with large charge state, in low gas pressure conditions and in high incident microwave power conditions. We also compared these space potentials with those using Langmuir probes. It is found that same trend is confirmed with the results from probe method.

For both Ar^+ and Ar^{2+} ions, the values V_{s2} and $V_s(q)$ show incident microwave power dependence. Furthermore, we can also obtain space potential values using ion beam by ion beam deceleration characteristics in IBIS. [14] In the same way, it was confirmed that the high space potential values were obtained under the experimental conditions of high incident microwave power.

In the process of Fe@C_{60} synthesis experiments, it is also important to obtain the space potential values. In the future, we plan to evaluate the space potential of ion species with large mass/charge ratios, such as Fe and C_{60}.

References

[1] Kato Y, Kimura D, Yano K, Kumakura S, Imai Y, Nishiokada T, Nagaya T, Hagino S, Otsuka T, and Sato F 2016 *Rev. Sci. Instrum.* **87**, 02A711-1-3.
[2] Nagaya T, Nishioka T, Hagino S, Uchida T, Muramatsu M, Otsuka T, Sato F, Kitagawa A, Kato Y, and Y. Yoshida 2016 *Rev. Sci. Instrum.* **87**, 02A723.
[3] Harneit W 2002 *Phys. Rev.* A 65, 032322
[4] Bolskar D R, et.al. 2003 J. Am. Chem. Soc. 125, 5471
[5] Chai Y, Guo T, Jin C, Haufler E R, Chibante F P L, Fure J, Wang L, Alforz M J, and Smalley E R 1991, J. Phys. Chem. 95, 7564
[6] Lian Y, Shi Z, Zhou X, He X, and Gu Z 2000 Carbon 38, 2117
[7] Uchida T, Minezaki H, Ishihara S, Muramatsu M, Ra'cz R, Asaji T, Kitagawa A, Kato Y. Biri S, Drentje G A, and Yoshida Y 2014 *Rev. Sci. Instrum.* **85**, 02C317
[8] Cho C S, Takane T, Ishida H, and Hatakeyama R 2015 J. Appl. Phys. 117, 123301
[9] Kato Y, Tsuda Y, Watanabe T, Onishi K, Hamada K, Takeda T, Omori T, and Okumura K 2019 IIT2018, IEEE Conf. Publ, pp 172-175
[10] Kato Y, Omori T, Owada I, Kubo W, Harisaki S, Sato K, Tsuda K, Maenaka T, Anan M, Muramatsu M, Kitagawa A and Yoshida Y, to be submitted to Crystals.
[11] Lieberman A M 1984, Principle of Plasma Discharge and Material Processing, Plenum Pres, pp.154-166.
[12] Xie Q Z and Lyneis M C 1994, " Plasma potentials and performance of the advanced electron cyclotron resonance ion source ", *Rev. Sci. Instrum.* **65**, 2947.
[13] Kumakura S, Kimura D, Yano K, Imai Y, Nishiokada T, Sato F, Kato Y and Iida T 2014 IIT2014,IEEE Conf. Publ, pp.1-4.
[14] Sato K *et al.* contribution in this conf.

Optimized deceleration characteristics of multicharged ion beams from electron cyclotron resonance ion source

K Sato, S Harisaki, W Kubo, I Owada, K Tsuda, M Anan, T Maenaka and Y Kato

Division of Electrical, Electronic and Infocommunications Engineering, Graduate School of Engineering, Osaka University, 2-1 Yamada-oka, Suita-shi, Osaka 565-0871, Japan

k.sato@nf.eie.eng.osaka-u.ac.jp

Abstract. At present it is necessary the satellite lifetime 10-15 years for operating in space. Xenon is used as fuel for ion engines of satellites. There are problems of accumulated damages at irradiation and sputtering by low energy Xe ion from the engine. It is required to construct database of sputtering yield of ion beams in the low energy region from a hundred eV to 1keV. We are trying to investigate experimentally sputtering yield on satellite component by irradiating the low energy Xe^{q+} ion beams. We use the electron cyclotron resonance ion source (ECRIS) in the irradiation experiments. We decelerate the beam energy to several hundred eV after extraction at high voltage 10kV. It is found we cannot neglect the contribution of the space potential of the plasma in the ECRIS of several tens eV. We measured the plasma parameters and ion beam deceleration characteristics in operating conditions on ECRIS. As the results, it was found in operating ECRIS condition of low charge state ion being dominant that the space potential of the ECRIS has an effect of 10~20% at the beam energy at about 100eV in conducting irradiation experiments on satellite component materials.

1. Introduction

An electron cyclotron resonance ion source (ECRIS) are widely applied for plasma processing and ion beam applications.[1] In our ECRIS, we aim at producing and extracting various ion beams in the single device. It plays an important role in medical field of carbon cancer therapy and aerospace engineering field of artificial satellite engine.[2,3] The satellite engine uses an ion engine, and Xe is widely used as fuel. There are problems of accumulation damages at irradiating and sputtering by low energy Xe ion from the engine. It is required to construct database of sputtering yield since there are not enough data of satellite component materials by ion beams in the low energy region from a hundred eV to 1keV. In recent years, research on the sputtering yield of satellite materials has been conducted.[3] Therefore, we are trying to investigate experimentally sputtering yield on their materials by irradiating the low energy single species Xe^{q+} ion beams. However, there is a problem that if the extraction voltage is low and Xe^{q+} is extracted, the amount of beam currents is not enough for irradiation. Therefore, we attempted to decelerate the Xe^{q+} ion beam to the required energy after extracting at high voltage, normally 10kV.[4] In low energy ion beam irradiation experiments, it is necessary to control the beam energy of a hundred eV. We cannot neglect the contribution of the space potential of the plasma in the ion source of several tens of eV. In order to conduct more accurate irradiation experiments, it is necessary to understand the effect of ion source plasma parameters on the ion beam deceleration characteristics. As a result, it was found that the deceleration characteristics varied with the operating conditions. We considered the operating conditions of the ECRIS affected the ion beam deceleration characteristics. And we measured the plasma parameters under different ion source operating conditions. As a result, it was found that the space potential tends to be higher for all space potential measurement methods when multicharged ions are dominate when the experimental conditions, and tends to be lower when low charge state ion being dominant.

2. Experimental apparatus

2.1. ECRIS

Fig.1 shows the top view of the ECRIS and the beamline. The magnetic field of the ECRIS consists of octupole magnets and mirror field by two coils (Coil A, B) and additional coil (Coil C). The coil A and B currents $I_{A,B}$ are usually fixed 150A. The coil C current I_C is adjusted to optimize the ion beam current. We use the Xe and Ar gases. The Xe and Ar gases are introduced to the vacuum chamber by mass flow controllers. The typical operation pressure of Xe and Ar gases are $1\times10^{-3\sim-4}$Pa. The z-axis indicates the distance from the center of the ECRIS along to the geometrical axis. 2.45 GHz frequency microwaves are launched from the rod antenna through a coaxial window. The microwave incident powers are usually several ten to several hundred watts.

Plasma parameters such as the electron density n_e, the electron temperature T_e, and the space potential V_S are measured by Langmuir probe LP1(z=-175mm) and LP2(z=175mm) inserted from side wall (x=0~50mm, y=0mm). The space potential is measured by using the conventional space potential V_{S1} obtained from the probe current-voltage I_P-V_P characteristics. V_{S2} is estimated from the electron temperature T_e obtained by LP2. $V_S(q)$ measured by the ion beam extracted from the ion source.[5] V_{S3} obtained from the ion beam deceleration that describes the later.

The ion beam is extracted by three electrodes from ECR plasma. The extraction electrodes consist of PE, E1, and E2 electrodes. The electrode PE, E1 and E2 voltages are V_{PE}, V_{E1} and V_{E2} respectively. V_{PE} is usually 10kV and V_{E2} is grounded. V_{E1} varies from 0 kV to -6 kV. V_{E1} is optimized for beam current.[6] Einzel lens EL1 is set at the front of the sector magnet. Ion beams are transported into the beam line. Sector magnet is used to separate ion currents of mass/charge. We obtain charge state distribution (CSD) of the ion beam currents by the faraday cups (FC). The diameter of the FC is 50mm. We install the movable slit in front of the FC. It can control absolute value of currents and resolution of the CSD by changing the width with 0~40mm at the height is 40mm. The beam deflection can be adjusted by the two steering coils (S1 and S2). The beamline is connected to Ion Beam Irradiation System (IBIS).

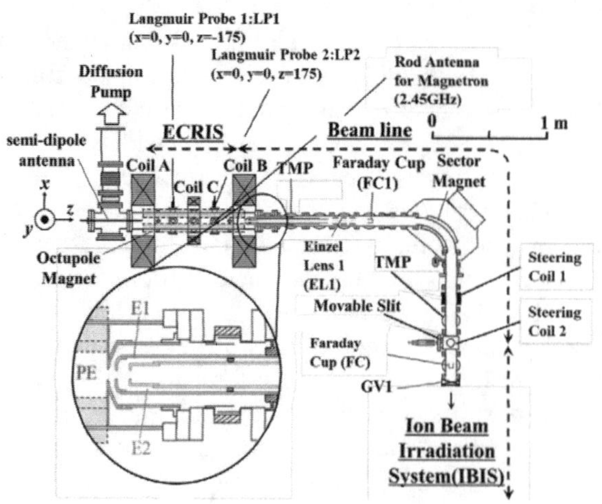

Figure 1. Top view of ECRIS (Osaka Univ.)

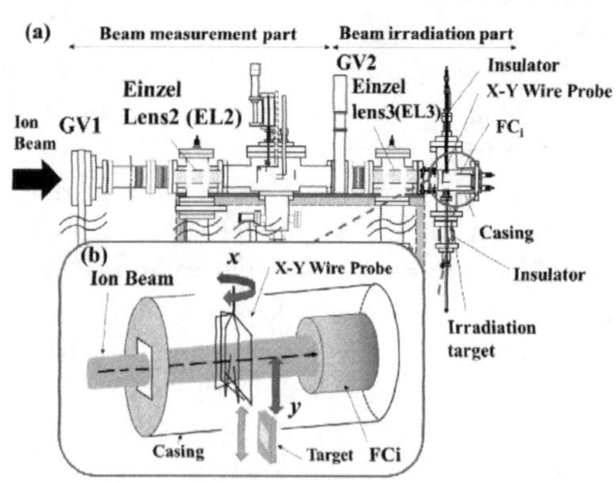

Figure 2. (a)Side view of Ion Beam Irradiation System (IBIS) and (b)details of the beam irradiation part

2.2. Ion Beam Irradiation System (IBIS)

Figure.2 (a) shows the side view of the IBIS. The IBIS and the ECRIS are separated by gate valve (GV1). The IBIS is divided two parts by the gate valve (GV2). IBIS consists of beam measurement part and beam irradiation part. Even during the irradiation experiment, we can exchange irradiation sample easily and quickly by load lock system of the beam irradiation part. Einzel lens 2 (EL2) is installed between the GV1 and GV2. And we use einzel lens 3 (EL3) in the irradiation part. The voltage of the EL2 and EL3 is V_{EL2} and V_{EL3}. Fig.2 (b) shows the details of the IBIS beam irradiation part. The casing covers the FC_i and the wire probe to reduce the effects of secondary electrons. We measure beam current at FC_i in deceleration experiment. The diameter of the FC_i is 50 mm. X-Y wire probe is used to measure the two-dimensional profile of the decelerated beam.[7] In the irradiation experiment, the target can be irradiated with decelerated ion beam currents by making the voltage of V_D the same as the voltage of the target. To obtain the ion beam deceleration characteristics, V_D is applied to decrease the beam energy $q(V_{PE} - V_D)$ and the beam current is measured. The deceleration voltage V_D is applied to the wire probe, casing, FC_i, and target. V_D is varied from 0 to 10.05kV. We evaluate the absolute values of the beam energy at which the beam current is zero as the space potential V_{S3} in these experiments.

3. Experimental results

3.1. Deceleration characteristics under multicharged ion dominated (low pressure & high-power microwave) and low charge state ion dominated (high pressure & low-power microwave) conditions

3.1.1 Deceleration experiment under conditions where multicharged ions are dominant (low pressure & high-power microwave) Figure.5 (a) shows the CSD obtained from FC. The ECRIS operating conditions are 100W of incident microwave power and 1.4×10^{-4}Pa of Ar operating pressure. The movable slit is 40mm and the extraction voltage V_{PE} is 10kV. The ion beam is optimized for Ar^{6+}. The Ar^{2+} beam current was the highest, confirming that the ion source operating conditions are dominated by multicharged ions.

Figure.5 (b) and (c) show the results of the Ar^{q+} ion beam deceleration characteristics obtained with FC_i. The horizontal axis represents the ion beam energy and the vertical axis represents the ion beam current. Figure.5 (c) shows the detailed behavior of the beam energy from -150eV to 250eV. experimental results for Ar^+ are shown in black, Ar^{2+} in red, and Ar^{3+} in green. V_{PE} was fixed at 10kV and V_D was varied from 0kV to 10kV. We confirmed that the beam current rapidly decreased when the beam was decelerated several tens eV. V_{S3} are estimated 27eV for Ar^+, 46 eV for Ar^{2+}, and 72eV for Ar^{3+} respectively from deceleration characteristics. The beam energy values which the beam current becomes zero are proportional to the charge state, and the beam deceleration characteristics show dependence of them on the charge state q.

3.1.2 Ar deceleration experiment under conditions low charge state ion dominate (high pressure low-power microwave) Figure.6 (a) shows the CSD obtained by FC. The incident microwave power is 10W and the Ar operating pressure is 1.1×10^{-3}Pa. The ion beam was optimized for Ar^+. Figure 6 (b) and (c) show the results of the Ar^+ ion beam deceleration characteristics obtained by FC_i. V_{S3} are estimated 8eV from deceleration characteristics. Comparing Fig.5 (c) and Fig.6 (c), V_{S3} estimated from Fig.5 (c) is larger than that estimated from Fig.6 (c). These results indicate that the different ion source operating conditions affect the ion beam deceleration characteristics.

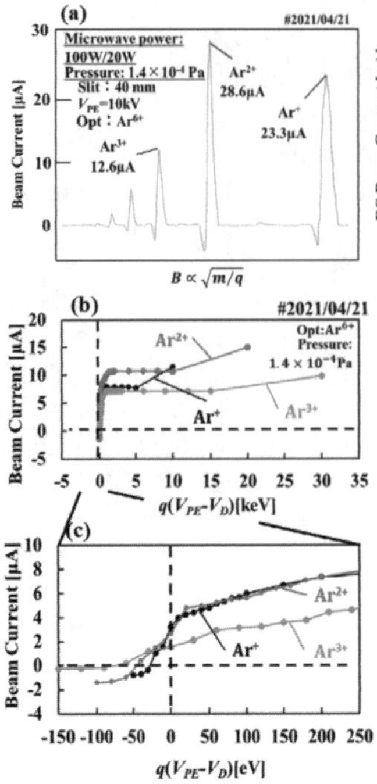

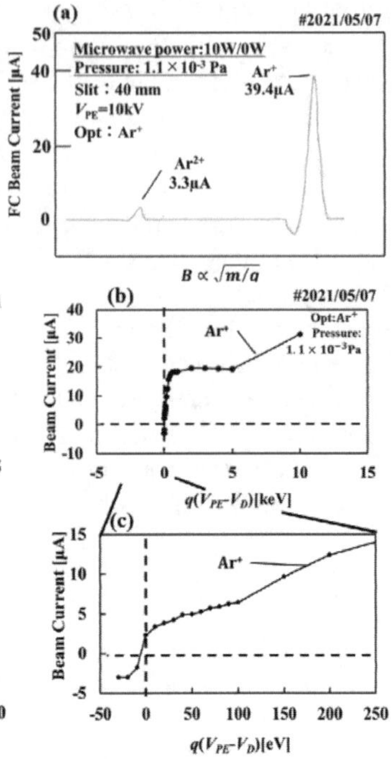

Figure 3. CSD obtained by FC and deceleration characteristics under multivalent ion dominated conditions obtained by FCi

Figure 4. CSD obtained by FC and deceleration characteristics under low charge state ions dominated conditions obtained by FCi

3.1.3. Results of plasma parameter measurements at low and high Ar pressures Figure.7(a) and (b) show the results of the measurement of plasma parameters by LP2 (a) and deceleration characteristics (b). The horizontal axis shows the probe position along the x-axis. The vertical axis shows space potential V_{S1} and V_{S2} by LP2 (a), and space potential V_{S3} obtained by the deceleration characteristics (b). Red shows the measurement results under the condition when multicharged ions are dominant. Black shows the measurement results under the condition when low charge state ions are dominant. In Fig.7 (a), the closed circle shows the space potential V_{S1}, and the triangular show the space potential V_{S2}. In Fig.7 (b), the horizontal axis shows the charge state q. The horizontal axis shows the Ar the space potential V_{S3}. Red shows the measurement results under the condition when multicharged ions dominate. Black shows the measurement results under the condition when low charge state ions are dominant. Comparing Fig.7 (a) and (b), the results of the measurement under the condition when multicharged ions are dominant tend to be higher, and the trend is coincidence the results of V_{S1}, V_{S2}, and V_{S3}.

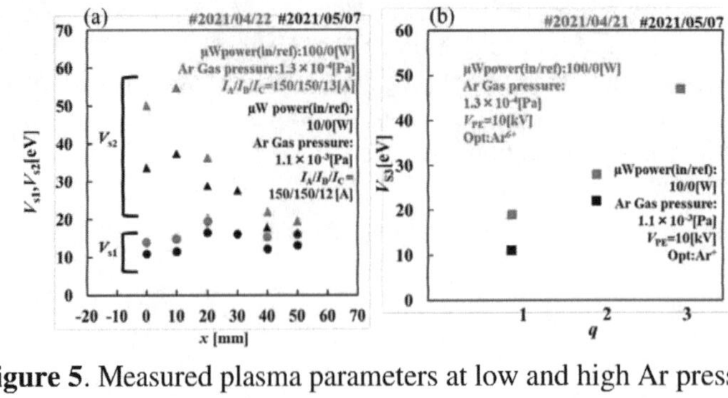

Figure 5. Measured plasma parameters at low and high Ar pressure

3.2 Deceleration characteristics under high ~~power~~ and low power conditions of microwave incident power under constant low pressure

In this section, we describe the measurement results of the space potential under high power microwave and low power microwave conditions at constant pressure. In the experiment under high power microwave condition, the incident microwave power is 100W and the Ar operating pressure is 1.4×10^{-4}Pa. The ion beam was optimized for Ar^{6+}. V_{S3} estimated from deceleration characteristics are 19eV for Ar^{+}, 27eV for Ar^{2+}, and 48eV for Ar^{3+} respectively from deceleration characteristics.

In the experiment under low power microwave condition, the incident microwave power is 20W and the Ar operating pressure is 1.4×10^{-4}Pa. The ion beam is optimized for Ar^{6+}. V_{S3} estimated from deceleration characteristics are estimated 11eV for Ar^{+}, and 22 eV for Ar^{2+} respectively from deceleration characteristics.

We measured the plasma parameters by LP2. The values of the measurement results tend to be higher under the condition of high microwave incident power. Comparing the results of the space potential measurement by LP2 (V_{S1}, V_{S2}) and the deceleration characteristics, It is found that the V_{S3} values obtained under the high incident microwave power condition tend to be higher than those obtained under the low incident microwave power condition, and the trend is coincidence the results of V_{S1}, V_{S2}, and V_{S3}.

4. Summary

We investigated the Ar ion beam deceleration characteristics for different ion source operating conditions at ECRIS. As a result, it was found that the deceleration characteristics are varied with the operating conditions of the ion source. The operating conditions affected the ion beam deceleration characteristics. We compared the space potential obtained from the deceleration characteristics that obtained from different measurement methods. As a result, it was found that the space potential tended to be higher for all space potential measurement methods when the ECRIS operating conditions were dominated by multicharged ions. On the other hands, space potentials of 10~20 eV were measured in the ion source operating conditions when low charge state ion is dominant. which were the experimental conditions for conducting ion beam irradiation experiments. In similar experiments using Xe, space potentials of about 10 eV were measured in the experimental conditions when low charge state ions are dominant. Therefore, it is necessary to evaluate the beam energy error when constructing database for sputtering yield. We will optimize beam deceleration system and improve the extractor to increase the low-energy beam current, to shorten the irradiation time. We are going to conduct irradiation experiments on satellite components using multicharged ions such as Xe^{q+} in the low energy region. We will also evaluate sputtering yield of satellite materials in low energy region, more over with incident angle dependence.

References

[1] Nagaya T, Nishiokada T, Hagino S, Uchida T, Muramatsu M, Otsuka T, Sato F, Kitagawa A, Kato Y and Yoshida Y 2016 *Rev. Sci. Instrum.*, vol. **87**, no. 2, p. 02A723
[2] Kitagawa A, Drentje G A, Fujita T, Muramatsu M, Fukushima K, Shiraishi N, Suzuki T, Takahashi K, Takasugi W, Biri S, Rácz R, Kato Y, Uchida T, and Yoshida Y 2016 *Rev. Sci. Instrum.*, vol. **87**, p. 02C107
[3] Takeda T, Otsuka T, Hagino S, Onishi K, Hamada K, Omori T, Okumura K, and Kato Y 2020 *Rev. Sci Instrum.*, vol. **91**, p. 013308
[4] Okumura K, Takeda T, Ishihara M, Onishi K, Hamada K, Omori T, Kubo W, Harisaki S, and Kato Y 2020 *Rev. Sci. Instrum.* vol. **91**, 023311
[5] Owada I, *et al* Contribution in this Conference
[6] Otsuka T, *et.al.*, 21st Int. Conf. on. IIT, IEEE Xplore Digital Library, IEEE Conf.Pub, 7882900, pp.1-4
[7] Sato K, *et.al*, 28-30 September 2020, Proceedings of the 24th International Workshop on ECRIS, Michigan, USA, http://www.jacow.org

Performance of a reactive magnetron sputtering ion source using water vapor plasma

Allen Vincent Catapang[1], **James Edward Hernandez II**[1], **Magdaleno R. Vasquez Jr.**[2] **and Motoi Wada**[1]

[1] Graduate School of Science and Engineering, Doshisha University, Kyotanabe, Kyoto, Japan 610-0394

[2] Department of Mining, Metallurgical and Materials Engineering, University of the Philippines, Diliman, Quezon City, Philippines

E-mail: cyjf3301@mail4.doshisha.ac.jp

Abstract. One of the methods used to produce ions of metals and metal oxides is through extracting them from a magnetron sputter-type ion source. Metal nanoparticles, such as zinc oxide (ZnO), can be deposited using this method. However, low-energy conditions are required to prevent damage to the substrate. The extraction of ions from a differentially pumped, reactive magnetron sputtering source with a pulsed, conduit-type extraction electrode is investigated, with Zn as the metal target and argon and water vapor as discharge support and reactive gases, respectively. The use of water vapor has been shown to improve the properties of ZnO, through hydrogen doping. Using a quadrupole mass analyzer, the species produced from the ion source were detected and identified. Ar, H_2O, H_3O, and Zn ions were extracted using the pulsed extraction electrode, and the intensity of the extracted ions were found to be frequency dependent. Extraction of ions were observed starting from -40 V extraction potential using a Faraday cup, and the ion beam current appeared to increase in proportion to the extraction voltage. A mechanism for realizing ion extraction using an alternating extraction potential is proposed.

1. Introduction

One of the methods used to produce ions and neutrals of metals and metal oxides is through the use of a magnetron sputter-type source. Particularly in nanoparticles, this method realizes a continuous, controlled supply of material and a homogeneous size distribution and composition of nanoparticles can be achieved by varying the discharge parameters (i.e. discharge current, gas admixture, etc) [1]. A typical method involving magnetron sputtering sources is a gas aggregation source, which involve cluster growth at relatively higher pressures and the aggregated particles are carried to the substrate by the flow of gas through an aperture [2, 3].

One of the metal oxide nanoparticles of interest is ZnO, a wide band-gap semiconductor with many potential applications, such in semiconductors, photovoltaics, or as a flexible, transparent conductive oxide. For flexible substrates, ZnO needs to be deposited in low temperature and low-energy conditions to minimize damage to the substrate [4]. An advantage of using ZnO is that the properties can be modified to suite the application, by metallic doping, or with a light ion such as hydrogen. In ZnO, hydrogen acts as a shallow donor dopant, which improves the film conductivity and reduces crystal lattice stress without excessive modification of the band

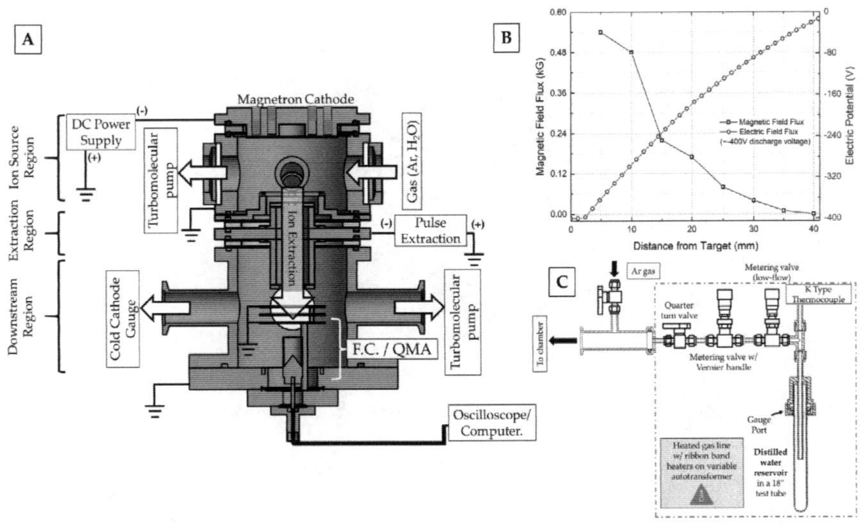

Figure 1. (A) Cross-section schematic of the reactive magnetron sputtering ion source experimental system, (B) Magnetic and Electric Field at the axis of the magnetron, and (C) the water vapor reservoir used in the experiment

structure [5]. Hydrogen doping of ZnO can be achieved by substituting O_2 gas with water vapor, however, control of the chamber pressure can be difficult due its high adsorption and reactive properties.

In this study, the use of a magnetron sputtering ion source for ZnO film deposition with water vapor and Ar plasma is investigated. Since a gas aggregation source utilizes gas flow based transport of the material to the substrate, the use of water vapor makes it a difficult process to control. A different configuration is explored, utilizing a differentially pumped ion source and a pulsed extraction system to realize ion transport from the source.

2. Experimental System

The experimental system (Figure 1.a) is composed of three regions, the ion source, the pulsed extraction region, and a downstream region. The ion source consists of a externally-cooled magnetron with two concentric, toroidal NdFeB magnets in an unbalanced configuration, and a 70-mm diameter metallic Zn (99.2%) as the target. The magnetic field strength at the central axis was measured using a hall effect gaussmeter (LakeShore 421 Gaussmeter) and is shown in figure 1.b. The plasma is sustained using a DC-regulated power supply (Takasago HV1.0-10), with ballast resistor (1 kΩ, 100W) and choke inductor (11.2 mH). Without the plasma, the electric potential at typical discharge voltages (≈-400V) is shown in figure 1.b, obtained through the finite element method (Field Precision AMaze software). The gases used in the study are argon (Ar, 99.999%) and water (H_2O, distilled water 99.999%). The constant flow of water vapor is achieved through a vaporization-driven water reservoir (Figure 1.c), with external heating (≈60°C) and flow control using two metering valves connected in series to set the intended pressure. The pressure is monitored at the two regions using two separate full-range, cold cathode gauges (Tokyo Electronics CC-10, Pffifer). The chamber walls of the ion source are grounded, and a 2.0 mm diameter aperture to the extraction region is positioned 40.0 mm away from the target. This also serves as the orifice to allow differential pumping of the ion source and the downstream region, as the ion source is operated at a working pressure of 1.0 Pa and 100 mA discharge current, while the downstream region is kept below $1.3\text{x}10^{-2}$ Pa

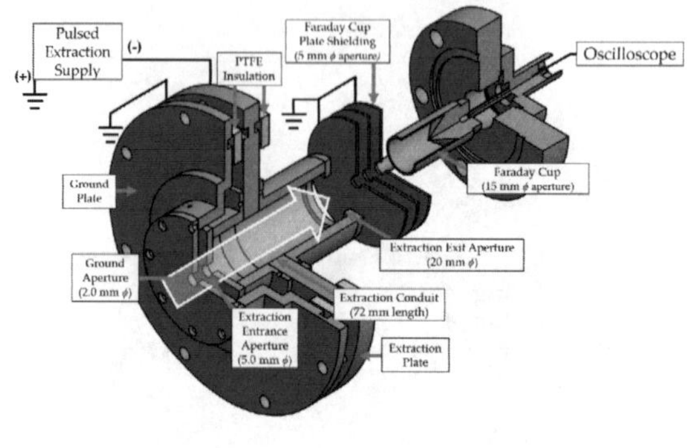

Figure 2. Pulsed extraction configuration and detection using a Faraday cup

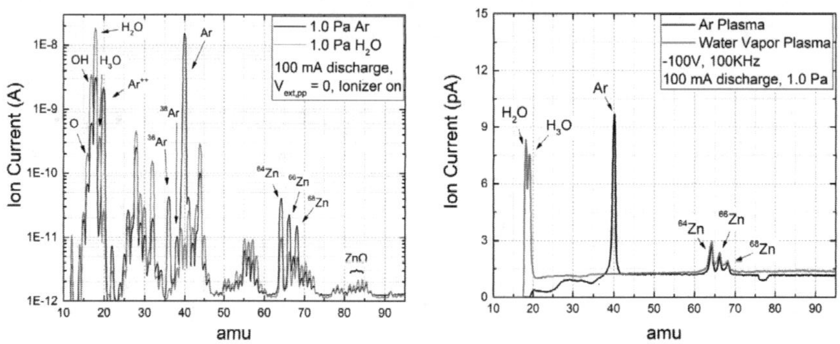

Figure 3. Mass spectra for 1.0 Pa Ar and 1.0 Pa H_2O plasma in the reactive magnetron sputtering ion source obtained (a., left) without pulsed extraction, and (b., right) with pulse extraction at -100V at 100 kHz

range for quadrupole mass analyzer (QMA, SRS RGA100) and Faraday cup measurements.

The pulsed extraction region, in figure 2, consists of the ground plate with a 2.0 mm diameter aperture, PTFE insulators, and a conduit-type electrode, with a extraction gap of 4.8 mm and a total length of 72.0 mm. The conduit-type electrode has an entrance aperture of 5.0 mm in diameter and an exit aperture of 20.0 mm in diameter. The voltage pulse is induced by a function generator (EZ Digital DFG6020) coupled with a high-speed amplifier/bipolar supply (NF Electronic Instruments, NF4020) to deliver up to a -100 V square wave pulse from 1 to 500 kHz with a 50% duty cycle. The ion species extracted in this setup is measured using the QMA and a shielded Faraday cup connected to an 1MΩ AC-coupled oscilloscope (Iwatsu DS-5424), positioned 28.0 mm and 36.0 mm away from the tip of the extraction conduit.

3. Results and Discussion

3.1. Species in the Reactive Magnetron Sputtering Ion Source

The ion species formed in the reactive magnetron sputtering source were confirmed by measurements using a QMA without pulsed extraction. In figure 3.a, neutral species produced from the ion source were detected by filament post-ionization in the QMA. The transport is caused by molecular flow from the ion source to the downstream region. The peaks are confirmed for the injected gas, Ar (40 amu) and H_2O (18 amu), and for the sputtering product, Zn (64,66,68 amu). When water vapor plasma is used, the reacted compound ZnO (80,82,84 amu) is observed.

To detect the ions transported by the pulsed extraction system, the QMA is operated without

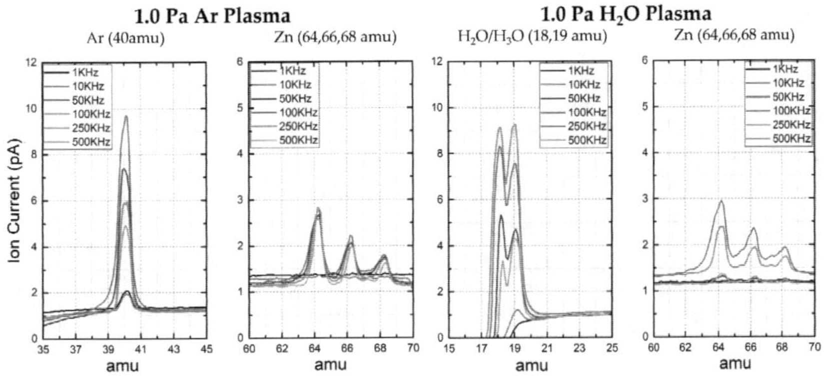

Figure 4. Effect of pulse frequency to the peak intensity for Ar, H_2O, H_3O, and Zn ions

the filament ionizer. In figure 3.b, the species extracted using -100V at 100 kHz from 1.0 Pa Ar plasma are the Ar and Zn ions. For the 1.0 Pa H_2O plasma, the species are the H_2O, H_3O (19 amu) and Zn ions. This confirms that ion extraction does occur, however, the presence of the ZnO peak was not detected.

3.2. Pulsed Ion Extraction

The characteristics of the pulsed extraction system was investigated by varying the pulse frequency and the extraction voltage. Shown in Figure 4, the QMA peak intensity of the detected ion changes as the frequency is increased. At low frequencies of 1 kHz and 10 kHz, only ions from the gas, Ar, H_2O and H_3O, are observed, and the Zn metal peak only appears at 50 kHz and greater. The maximum peak intensity for the detected species, in both the 1.0 Pa Ar plasma and 1.0 Pa H_2O plasma, was in the 100 to 250 kHz settings.

Shown in figure 5.a is the effect of varying the peak bias potential of the pulse in Ar plasma, at a fixed frequency of 100 kHz and measured using a shielded Faraday cup 28.0 mm away from the conduit electrode. The behavior of the measured current shifts as the potential is increased. At -20V, the measured current was in phase with the extraction pulse, however, at -40V to -100V, the current has a 180° phase shift. This indicates that at -20V and below, the transport of ions from the source to the conduit-type electrode is inefficient due to an insufficient potential difference between the extraction electrodes. At -40 V to -100 V, the peak is observed at the onset when the potential is reverted to ground, indicating the ion current contribution. To quantify the effect of the increasing potential, the peak height, measured from the onset of the ground potential, is plotted against the pulse extraction potential (figure 5.b). A linear dependence was observed and indicates that more ions are extracted from the source at higher potentials. Saturation due to the Child-Langmuir law is not observed, as the current is significantly lower in the order of magnitude compared to the theoretical values.

To explain the observed behavior, a mechanism for the pulsed extraction of ions is proposed(Figure 5.c). During the negative pulse, the positive ions arriving at the ground electrode aperture at ion acoustic speed is accelerated by the potential drop towards the downstream region. If the pulse duration is sufficiently near the ion transit time, which is defined as the time it takes an ion to pass through the conduit length at a certain velocity, then, as the conduit reverts to ground potential, the ions present in the conduit is accelerated to the substrate target by the potential difference between the local positive space potential and the substrate target. This would result to the maximum possible ion current extracted from the ion source, and is calculated using Equation 1. For an electron temperature of 5.5eV

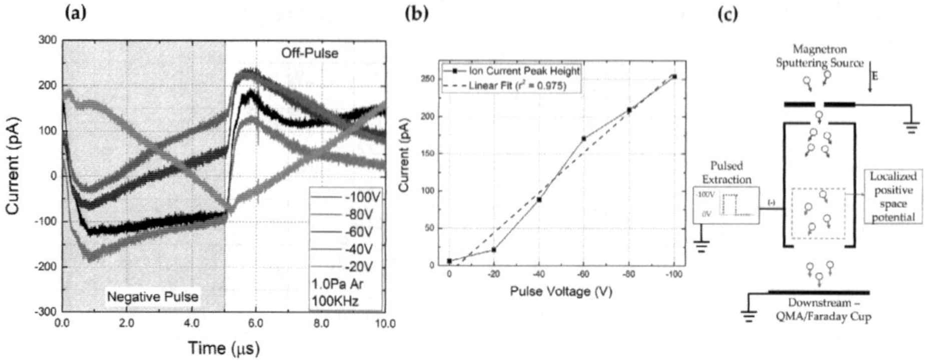

Figure 5. Effect of varying the pulse extraction potential to the current measured using a shielded Faraday cup, and the proposed mechanism for the conduit-type pulsed ion extraction

at the ion source, the value of t_{ion} is 2.81 μs for Ar, 4.23 μs for Zn, and 1.89 μs for H_2O, which is at 177.9 kHz, 118.2 kHz, and 265.1 kHz respectively, for a 50% duty cycle square wave extraction potential that was used in this study. These values are close to the frequencies where maximum values were observed from figure 4. If the pulse duration is significantly lower or greater than t_{ion}, the ions are either decelerated or deflected towards the conduit walls. This, and possible ion-ion or ion-electron interactions which could result to a decreased space charge in the conduit, can account for the ion current measured in the extraction frequencies. The possible space charge neutralization mechanism in the conduit when the ions are present is either through a self-neutralization mechanism [6], arising from the pulsed extraction, or by low energy electrons present in the conduit from the ion source [7], which realizes the low energy ion beam production.

$$t_{ion} = \frac{d_{conduit}}{\sqrt{\frac{k_b T_e}{m_i}} + \sqrt{\frac{2qE}{m_{ion}}}} \qquad (1)$$

4. Conclusion

A reactive magnetron sputtering ion source that uses a differentially pumped ion source and downstream region with a pulsed, conduit-type extraction electrode has been designed and tested. Using Ar and H_2O plasma in the ion source, the species that were transported to the downstream region with and without pulsed extraction were identified. Ar, H_2O, H_3O, and Zn ions were extracted using the pulsed extraction electrode, and the intensities of the extracted ions were found to be frequency dependent. The extraction of ions were observed starting from -40 V through Faraday cup measurements, and the ion current was linearly dependent on the extraction voltage. A mechanism for the pulsed extraction of ions using the conduit-type electrode resulting to low energy ion deposition is proposed.

5. References

[1] Gunnarson R, Helmersson U and Pilch I 2015 *J. Nanopart. Res.* B **17** 353
[2] Peter T, Polonskyi O, Gojdka B, Ahadi A, Strunskus T, Zaporojtchenko V, Biederman H and Faupel F 2012 *J. Appl. Phys.* **112** 114321
[3] Shelemin A, Kylian O, Hanus J, Choukourov A, Melnichuk I, Serov A, Slavinska D and Biederman H 2015 *Vacuum* **120** 162-169
[4] Kawamura Y, Hattori N, Miyatake N and Uraoka Y 2013 *J. Vac. Sci. Technol. A.* **32** 1105
[5] Van de Walle C 2000 *Physical Review Letters* **85** 1012-1015
[6] Chen Y, Sawadichai R, Tian S, Donnelly V, Ruchhoeft P and Economou D 2019 *J. Appl. Phys* D **52** 355205
[7] Vasquez M, Ilasin M, Pengson L, Ramos M and Cuevas A 2021 *Vacuum* **187** 110067

Production of Radioactive Molecular Ions in Radiofrequency Quadrupole Gas-Reaction Cells

Chris R.J. Charles[1], Friedhelm Ames[1], Oliver Kester[1], S. Malbrunot-Ettenauer[2], Carla Babcock[1], Brad E. Schultz[1], Cameron Peters[1,3], Erin L. Flannigan[4], Jean-Francois Alary[5], Aurelia Laxdal[1], Peter Kunz[1], Jens Lassen[1], Andrea Teigelhoefer[1], Phil J.A. McCausland[6], Roberta Flemming[6]

(1) TRIUMF. 4004 Wesbrook Mall, Vancouver, BC, V6T 2A3, Canada.
(2) CERN-ISOLDE. Esplanade des Particules 1, P.O. Box 1211, Geneva 23, Switzerland.
(3) Dept. of Physics and Astronomy, University of Waterloo, 200 University Avenue West, Waterloo, ON, N2L 3G1, Canada.
(4) Andre E. Lalonde Accelerator Mass Spectrometry Laboratory, University of Ottawa, 25 Templeton St., Ottawa, ON, K1N 6N5, Canada.
(5) Isobarex Corp., 60 Equator Crescent, Vaughan, ON, L6A 2Y9, Canada.
(6) University of Western Ontario, 1151 Richmond St., London, ON, N6A 3K7, Canada.

E-mail: `ccharles@triumf.ca`

Abstract. Limited types of radioactive molecules (RM) can be made inside hot-cavity targets at ISOL facilities like TRIUMF. However, extreme conditions in these targets present formidable unsolved challenges to efficient production and delivery of RM's. Here we propose using RFQ gas-reaction cells to produce RM from radioactive ion beams (RIB) by room temperature RIB-gas chemical reactions at eV energies. Two options are possible: (1) using an ion reaction cell (IRC) that is a linear RFQ ion guide and reaction cell used as an 'on-line ion source', and (2) using the ARIEL RFQ cooler-buncher (ARQB). RFQ gas-cells are a controllable and efficient method to produce RM from chemical reactants that cannot be used in ISOL targets. This 'on-line chemistry' offers a way to enable groundbreaking Beyond Standard Model (BSM) physics research, using a wide diversity of new rare and exotic RM beams that would be difficult or impossible to produce in hot-cavity targets.

1. Introduction

Radioactive molecules (RM) including RaF_x, $Ra(OH)_x$, RaH_x, ThO_x, BaH_x, YbO are exotic molecular species containing rare radioisotopes. These RM decay spontaneously, dissipating energy by the emission of radiation. Isotope separation on-line (ISOL) facilities like TRIUMF and CERN-ISOLDE synthesize a limited set of molecular species alongside radioactive ion beams (RIB) in their hot-cavity target ion sources [1, 2, 3, 4]. The RM form in cavity targets and extraction systems by a complex and poorly understood cascade of reaction kinematics following fission, spallation and fragmentation after the primary p^+ beam impinges on solid targets containing elements such as Si, Ta, Nb or U in the form of metal foils or composite ceramics.

Next-generation subatomic experiments wish to use RM in new Beyond Standard Model (BSM) physics searches, i.e. charge (C), parity (P) and time (T) symmetry violation effects in

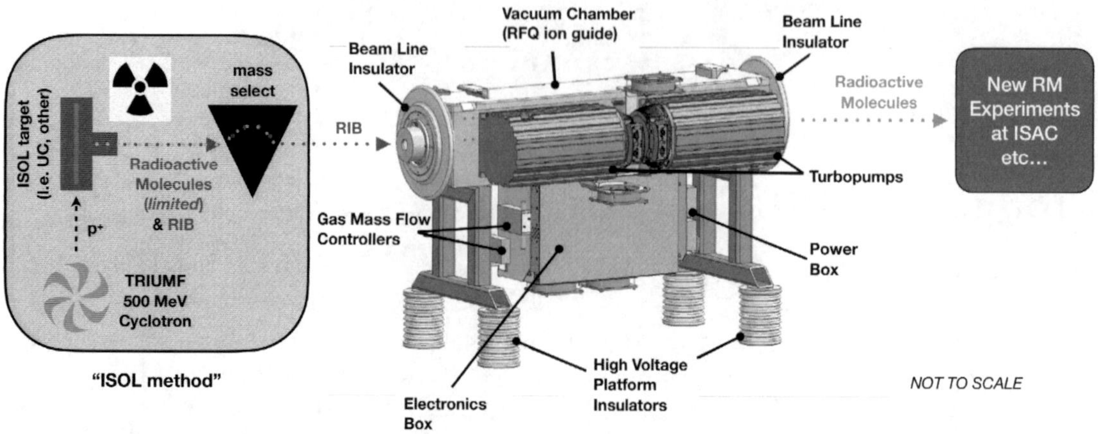

Figure 1. Sketch of the proposed RM production setup at TRIUMF, using an ion reaction cell (IRC). Mass-selected RIB from the ISOL target enter the IRC and undergo controlled ion-gas chemical reactions inside the RFQ reaction cell. Product molecules exit the IRC to downstream experiments. A model of the IRC is shown (*reproduced with permission from Isobarex Corporation*).

molecules, implications for dark matter, and fundamental topics in quantum chemistry [1]. These studies require efficient RM production of the required molecule at high purity and yield. However, the extreme conditions in hot-cavity on-line targets (e.g. high temperatures, poor vacuum, high radiation fields, isobaric interferences, unfavourable chemical conditions and reaction energies) present a major challenge to targeted molecule production for experiments.

RM creation from RIB in radiofrequency quadrupole (RFQ) ion guides by ion-gas chemical reactions at room temperatures is a more versatile, straightforward and efficient technique that allows controlling reaction conditions as well as reactants. Molecules containing radioactive isotopes are known to form unintentionally in cooler-bunchers [12]. However, in those cases, undesired charge exchange with the primary radioactive beam occurs. The rich array of ion-gas reaction chemistries in RFQ's to make and destroy molecules are very well established [6, 7] but remain unexploited for subatomic physics. Here we discuss the idea of using RFQ gas-reaction cells at ISOL facilities as an efficient and robust method for producing RM by 'on-line' RIB-gas chemistry.

2. Production of Molecules in RFQ Gas-Cells

Presently at TRIUMF, RIB are produced at the ISAC (Isotope Separator and ACcelerator) facility by the ISOL method where protons accelerated by the main cyclotron (up to 520 MeV at 100 μA) impinge on solid targets (see Fig 1). By 2026, ARIEL (the Advanced Rare IsotopE Laboratory) will add one additional p^+ target, and one novel 50 MeV electron photofission target to produce rare neutron-rich isotopes [8, 9].

To fully exploit ARIEL and ISAC capabilities, two RFQ systems may be considered for RM production. Fig 1 and Fig 2(a) shows the ion reaction cell (IRC), while Fig 2(b) shows an internal schematic of the ARIEL radiofrequency quadrupole cooler-buncher (ARQB).

2.1. Ion Reaction Cell (IRC)

The IRC in Fig 1 and Fig 2(a) is a room-temperature gas-filled RFQ ion guide and reaction cell that can accept positive or negative atomic ion beams at ISAC energies up to 40 keV (currently, only positive ions are extracted from ISAC targets). Molecules are produced (or destroyed) at low eV energies inside the RFQ gas cell from a wide variety of pure gaseous reactants under controlled conditions [5]. Product molecules are finally delivered from the instrument exit port to experiments. Fig 1 shows the overall IRC construction, including vacuum system components, the chemically resistant mass flow controllers, PLC and HV hardware, and a conceptual integration into TRIUMF radioactive beamlines. Fig 2(a) shows the IRC internals, including the optics at beam entry (deceleration), the RFQ ion guide and reaction cell where molecules are produced by ion-molecule interactions with gases or vapours, and optics for beam exit (reacceleration).

The IRC column in Fig 2(a) contains 13 RFQ segments made from chemically resistant Ultem, modular and reconfigurable. DC voltages are applied to individual rod segments and apertures. This flexibility means the gas reaction cell is alterable and can be tailored, e.g. cell volume and length, gas pressure and number of collisions, use of multiple reaction gases, collisional cooling and energy distribution of ions along the RFQ column. Fine control over the energy transferred to the ions is possible, for example, when forcing endothermic reactions to proceed. Efficient optimization of molecule production may then occur at eV energies by highly selective chemical reactions, where formation of other undesired molecules is minimized. Molecule production rather than destruction is achieved by providing favourable energy conditions for a suitable reactive gas with an electronegativity close to the ion of interest. This promotes chemical bond formation for a large diversity of molecules in high-yields and high-purities from many chemical reactants that could not be used inside hot-cavity on-line targets.

Typical operating parameters in negative ion mode used by [5] to induce ion-gas chemical reactions were 5 mTorr gas pressure at room temperatures and $V_{DC\ rods} = 14$ V, with $V_{ion\ source} = -20931.2$ V, $V_{IRC\ deck} = -20924.2$ for a 7 V difference. Analyte ions had the opportunity to experience at least several collisions. Simulations are now being performed for positive ions at ISAC, to determine the optimal time in the cell for a reaction to occur, required cell length, energy profiles of the ions, and gas mixtures to simultaneously collisionally cool and promote chemical reactions.

The system rests on four 40 kV high voltage insulators allowing the kinetic energy of the incoming atomic ions to be adjusted within <1 eV of the appropriate energy to react with gases introduced to the cell. The RFQ column is removable so that the empty vacuum box can be used telescopically to transfer beam between the entry and exit Einzel lenses without the column in place. The RFQ column has a current limit of ~ 1 μA. Electronics on the high voltage platform are connected to computer control at ground by a 5G wi-fi or fibre optic link. Reaction gases are introduced separately as required.

Initial setup and tests of the IRC will occur with stable-isotope beams at TRIUMF's ISAC Test Stand Facility (ISTF). Here, stable molecule production will be quantified with a diverse variety of gaseous inorganic and organic compounds reacting with positive or negative stable atomic ion beams (ISAC uses only positive beams, but negative beams can be implemented). This will (i) establish a clear understanding of stable molecular reactions that are relevant to future RM production from RIB, and (ii) offer research opportunities to both subatomic physics and broader scientific problems. Elements of interest include Al, Ca, Ba, Mn, Fe, Rb, I, W, Mo, Ti, Cr, Cs, Re, Pb, La, U. Lanthanides and actinides are of special interest. We aim to systematically study the kinetics of the formation of molecules for hydrides, oxides, fluorides, carbides, hydroxides, chlorides and other more complex molecules, i.e. polyatomic, heavy element and dia/paramagnetic molecules, and use these data to quantify IRC performance. These molecules are of high interest to RM subatomic experiments and theory. Following successful tests and further system refinement, the IRC will be moved to ISAC for long-term integration with the radioactive beam systems for RM production from RIB.

The IRC system is being developed by a collaborative TRIUMF-industry partnership. Developments are being explored to deliver bunched beams from the system. This method of rare molecule production and preparation offers a significant advancement for fundamental particle

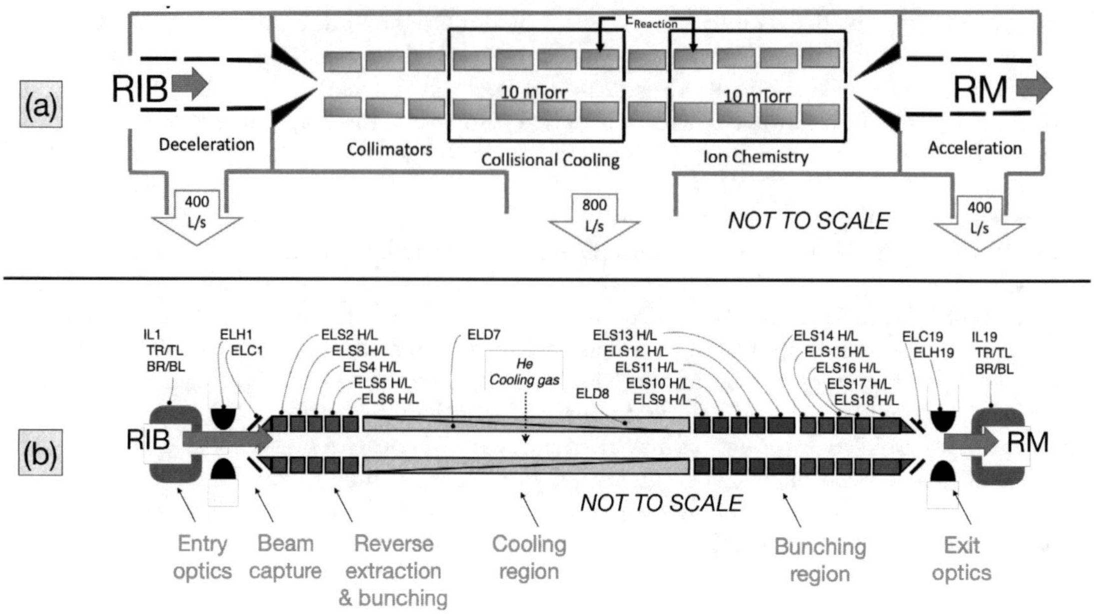

Figure 2. Internal sketches of (a) the ion reaction cell (IRC) and (b) the ARIEL RFQ cooler-buncher (ARQB).

and nuclear physics with molecules, and strengthens the science opportunities for upcoming low-energy precision experiments at TRIUMF.

2.2. ARIEL Radiofrequency Quadrupole Cooler-Buncher (ARQB)

CANREB (the CANadian Rare isotope facility with Electron Beam ion source) is dedicated to preparing and delivering beams of highly charged ions (HCI) for experiments. CANREB is part of the larger ARIEL facility [8, 9] and includes the ARQB schematically shown in Fig 2(b) [10]. The ARQB cools and bunches RIB spanning the mass range up to U at $\leq$100 Hz repetition rates containing $< 10^7$ ions/bunch. Bunch energies are then adjusted by a pulsed drift tube to $6 < E_{bunch} < 14$ keV immediately before acceptance into an electron beam ion source (EBIS) for charge breeding [11].

Although the main purpose of the ARQB is to prepare RIB for charge-breeding, RM production in a similar manner to the IRC should be possible in the ion guide. Three advantages of the ARQB over the IRC are that (1) the system was commissioned as of 2021 summer, now allowing RIB from ISAC to be used to produce RM, (2) beam bunching is possible, and (3) options exist for cryogenic cooling of He buffer gas to further reduce beam emittance. However, two major disadvantages are that (a) the ARQB is not designed for gas-reaction chemistry, particularly with reactive chemicals including F_2, Cl_2, Br_2, H_2O, alcohols and other difficult compounds, and (b) cryogenic operation will limit the range of reactants due to freeze-out.

We are exploring possibilities of using the ARQB for RM production with a range of reaction gases. Ion optical simulations are also being developed to determine how best to extract RM beams from CANREB. Options include 'bouncing' out of the EBIS (without electron beam), deflecting off electrostatic lenses before the EBIS, or using the ARQB in reverse-extraction mode after RM formation.

If RM extraction can occur towards the EBIS, mass analysis may be possible by the Nier spectrometer (NIS) [10] before RM injection into ISAC for experiments. However, CANREB beamlines from the NIS can only go to 2 places: the ARIEL yield station, or the ISAC-I linac that cannot accelerate heavy molecules. Therefore, sending RM to the ISAC low energy area will need additional beamline infrastructure to deliver for a low-energy experimental program. The case for reverse-extraction from the ARQB will similarly require beamline infrastructure to deliver to ISAC-I. Options are being considered how to deliver RM to experimenters once the controlled formation of molecules inside both the ARQB and IRC are understood experimentally.

3. Summary

RFQ gas-reaction cells are a robust, simple method for RM production rather than inside ISOL targets where non-equilibrium, high temperature conditions exist in an environment abundant with reactants from all elements and isotopes produced in the target by spallation, fragmentation and fission (actinide targets). Currently, no other laboratory has a dedicated rare RM beam facility. Such a facility at TRIUMF has the potential to (1) significantly expand the portfolio of available beams in the ARIEL operation era for research, (2) strengthen the science opportunities for low-energy precision experiments on radioactive molecules, (3) offer many interdisciplinary opportunities for students, and (4) develop new scientific collaborations with external partners.

4. Acknowledgments

TRIUMF is funded by the National Research Council of Canada (NRC). The CANREB project is funded by grants from the Canadian Foundation for Innovation (CFI), the Natural Sciences and Engineering Research Council of Canada (NSERC), and collaboratively with Saint Mary's University and the University of Manitoba.

5. References

[1] R. F. Garcia Ruiz, R. Berger, J. Billowes, C. L. Binnersley, M. L. Bissell, A. A. Breier, A. J. Brinson, K. Chrysalidis, T. E. Cocolios, B. S. Cooper, K. T. Flanagan, T. F. Giesen, R. P. de Groote, S. Franchoo, F. P. Gustafsson, T. A. Isaev, Á. Koszorus, G. Neyens, H. A. Perrett, C. M. Ricketts, S. Rothe, L. Schweikhard, A. R. Vernon, K. D. A. Wendt, F. Wienholtz, S. G. Wilkins and X. F. Yang (2020). *Spectroscopy of short-lived radioactive molecules.* Nature 581, 396–400.

[2] P.G. Bricault (2016). Nuclear Instruments and Methods in Physics Research B, 376, 3–7.

[3] M. Dombsky, D. Bishop, P. Bricault, D. Dale, A. Hurst, K. Jayamanna, R. Keitel, M. Olivo, P. Schmor, and G. Stanford (2000). *Commissioning and initial operation of a radioactive beam ion source at ISAC.* Rev. Sci. Instrum. 71, 2.

[4] A. Gottberg (2016). Nuclear Instruments and Methods in Physics Research B, 376, 8–15.

[5] C.R.J. Charles, R.J. Cornett, X.-L. Zhao, A.E. Litherland, W.E. Kieser (2015). *On-line I^-/Te^- separation for the AMS analysis of ^{125}I.* NIMB 361, 189-192.

[6] A.E. Litherland, I. Tomski, J.P. Doupe (2003). *Ion reactions for isobar separation in accelerator mass spectrometry.* NIMB 204, 720-724.

[7] D.R. Bandura, V.I. Baranov, A.E. Litherland, S.D. Tanner (2006). *Gas-phase ion-molecule reactions for resolution of atomic isobares: AMS and ICP-MS perspectives.* Int. J. Mass. Spec. 255-256, 312-327.

[8] L. Merminga, F. Ames, R. Baartman, P. Bricault, Y. Bylinski, Y-C. Chao, R. Dawson, D. Kaltchev, S. Koscielniak, R. Laxdal, F. Mammarella, M. Marchetto, G. Minor, A. Mitra, Y.-N. Rao, M. Trinczek, A. Trudel, V. Verzilov, V. Zvyagintsev (2011). *ARIEL: TRIUMF'S Advanced Rare Isotope Laboratory.* Proceedings of IPAC2011, San Sebastian, Spain, 1917-1919.

[9] J. Dilling, R. Krucken, L. Merminga. (2013) ARIEL overview. In: Dilling J., Krucken R., Merminga L. (eds) *ISAC and ARIEL: The TRIUMF Radioactive Beam Facilities and the Scientific Program.* Springer, Dordrecht.

[10] F. Ames, R. Baartman, B. Barquest, C. Barquest, M. Blessenohl, J. R. Crespo López-Urrutia, J. Dilling, S. Dobrodey, L. Graham, R. Kanungo, M. Marchetto, M. R. Pearson, and S. Saminathan (2018). *The CANREB project for charge state breeding at TRIUMF.* AIP Conference Proceedings 2011, 070010-1-3.

[11] M.A. Blessenohl, S. Dobrodey, C. Warnecke, M. K. Rosner, L. Graham, S. Paul, T. M. Baumann, Z. Hockenbery, R. Hubele, T. Pfeifer, F. Ames, J. Dilling and J. R. Crespo López-Urrutia (2018). *An electron beam ion trap and source for re-acceleration of rare-isotope ion beams at TRIUMF.* Reviews of Scientific Instruments 89, 052401-1-16.

[12] A.R. Moberg, S. Warren, C. Munoz Pequeno, J. Cruikshank, T. Giles, D. Hanstorp (2020). *Time-of-Flight study of molecular beams extracted from the ISOLDE RFQ cooler and buncher.* NIMB 463, 522-524.

Operation experience of LION and RHIC-EBIS for RHIC and NSRL

T. Kanesue, E. Beebe, B. Coe, S. Ikeda, S. Kondrashev, A. Lopez-Reyes, M. Okamura, R. Schoepfer, T. Rodowicz

Collider-Accelerator Department, Brookhaven National Laboratory, New York 11973, USA

Email: tkanesue@bnl.gov

Abstract. LION is a laser ablation ion source to provide singly charged heavy ions of various species for RHIC-EBIS. High charge state heavy ion beams from RHIC-EBIS are used for RHIC physics experiments and NASA Space Radiation Laboratory (NSRL) quasi-simultaneously. The demands for heavy ion beams are growing and more ion species are available and more NSRL beam time is used because of unique capability and flexibility of the sources. With the combination of LION and RHIC-EBIS, ion species can be switched on a pulse-by-pulse basis without the effect of previously used species. The present performance and operation experiences of LION and RHIC-EBIS are shown.

1. Introduction

LION and RHIC-EBIS are a part of the RHIC-EBIS pre-injector [1], which has been providing various heavy ion species for collider experiments at Relativistic Heavy Ion Collider (RHIC) and space radiation research at NASA Space Radiation Laboratory (NSRL) simultaneously in sequence since 2010. RHIC is one of two heavy ion collider accelerators, and the only polarized proton collider in the world. NSRL is an accelerator-based research laboratory for space radiation research to study the effect of Galactic Cosmic Ray (GCR) using mixed radiation of heavy ions and protons extracted from the AGS-Booster [2]. The available beam energy is up to 1.5 GeV/u for Fe and 2.5 GeV for proton. For the GCR simulator, ion species for NSRL is required to be switched rapidly while beams for RHIC is supplied.

A layout of the RHIC-EBIS pre-injector to supply 2 MeV/u of heavy ion beams (charge-to-mass ratio 1/2 ~ 1/6) is shown in Fig. 1. An electron beam ion source (RHIC-EBIS) is used as a charge bleeder and all primary ions are provided from the external ion sources to avoid "memory effect" of

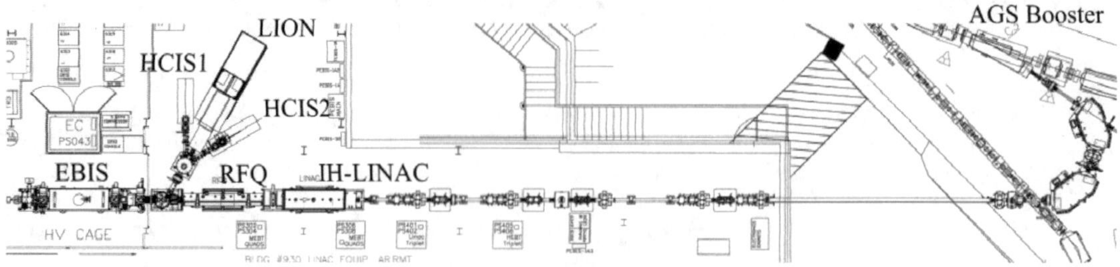

Figure 1. A layout of the RHIC-EBIS pre-injector.

previous species. A laser ion source, named LION, provides 1+ ions from any solid-state targets. Two hollow cathode ion sources (HCIS) produce 1+ ions from gaseous targets.

2. LION

LION is a key device to realize rapid species change. This is a laser ablation ion source and ions are generated by laser irradiation onto a solid target. At LION, low laser power density just above the threshold for plasma generation ($10^8 \sim 10^9$ W/cm^2) is used for 1+ ion production while a laser ion source in general is famous as an intense heavy ion source of highly charged ions using high laser power density above 10^{12} W/cm^2. At LION, the laser energy on a target is 200 mJ to 500 mJ with defocused laser spot of about 3 mm in diameter. Importantly, LION does not require warm-up, and full beam performance can be achieved from the first pulse though other ion sources typically needs 15 min or more to be stabilized. In addition, LION does not use working-gas, and vacuum pressure is kept below 10^{-4} Pa, which is very important to combine with EBIS. An ion beam out of LION produced by one laser shot is more than enough to provide ion yield required by users.

There are two target systems a target chamber. Each target system is driven by a dedicated laser so that beams for RHIC and NSRL are supplied independently. A rotating Au target is used for RHIC. A XY target supplies multiple ion species for NSRL. The XY target has a 254-mm-by-50-mm target holder driven by a two-dimensional linear stage. The number of species on the holder has been increased over the years, and recently 10~12 species are installed at a time. Typical target dimension is 25 mm by 50 mm or 25 mm by 25 mm except for Si (50 mm in diameter). Usually, NSRL run starts earlier than RHIC run from the end of September. If heavy ions are needed for RHIC, we provide beams later typically around January in the following year. Both NSRL and RHIC operations finish almost same time in July. The installed targets will last for the entire run. It takes 3~6 hours including pumping when new targets are needed.

Beam intensity out of LION is controlled by a 3m-long solenoid on expanding laser ablation plasma as keeping the laser poser density for 1+ ion production [3]. Laser is always shooting on the same spot and target position is changed to switch ion species. The switching time between species depends on the XY target motion and currently it takes 1~25 sec at LION. The beam current of lighter ions such as C and Si is hundreds μA to 1 mA with pulse length of 100 μs while that of heavier ions such as Au and Ta is about 100 μA with pulse length of 300 μs. These ions are injected into EBIS with fast injection scheme [4].

A laser ion source can produce plasma from any solid-state materials. Target purity and abundance of isotopes directly affects species and isotopes in laser produced plasma. A pure solid element with an isotope of 100 % natural abundance such as Au is an ideal target. In addition to pure materials, compound targets and alloys can be used as long as solid flat surface is available. With these targets, ions of all species and isotopes in a target are generated. Since a beam transport line between LION and EBIS uses all electrostatic components, all ions are injected into EBIS. Intended ion species is selected in rf linear accelerators, booster, and a beam line to NSRL target room after a stripper foil. We use alumina (Al_2O_3) targets for O beam [5]. Sintered enriched was ZrO_2 (42 x 3 x 2 mm^3) were used to provide ^{96}Zr beam (natural abundance of ^{96}Zr is low at 2.8 %) for isobar heavy ion collider experiment during RHIC Run-18 [6]. The isobar experiment was successfully completed with the enriched ZrO_2 targets [7]. Commercially available thoriated tungsten targets were used to provide Th and W. Recently in Run-21, we tested ZrO_2 targets for O beam at RHIC-EBIS because it has higher ratio of O compared to alumina. However, degradation of O beam intensity was quick within a week. This was different from Zr beam from the same targets, which lasted enough for RHIC operation. Eventually we used an alumina target to provide O beam for RHIC experiment. It is important to test and analyze the contents of plasma and target properties at test bench.

3. HCIS

Two hollow cathode ion sources (HCIS) are used to produce primary ions from gaseous species such as He, Ne, Kr, Ar, and Xe. Note that HCIS can also produce ions from solid materials though recently all ions from solid targets are provided from LION due to easy handling of the ion source. Each HCIS can provide one gaseous species. It takes about 20 min to replace gas when more gaseous species is needed. Typical ion beam current from HCIS is 5~50 μA and long beam pulse of 10~40 ms. This low current beam is accumulated by an EBIS ion trap using slow injection scheme.

4. RHIC-EBIS

The RHIC-EBIS is a state-of-the-art electron beam ion source. This EBIS has a 188-cm-long ion trap and utilized 7~10 A of electron beam with a 5 T superconducting solenoid. RHIC-EBIS can switch species at faster than 5 Hz (200 ms), which is the basic control cycle. Depending on the status of RHIC and NSRL, the number of pulses and ion species for NSRL change frequently. RHIC-EBIS is required to provide stable beams throughout the complicated operation mode.

RHIC is a collider accelerator. Once beams are injected in two rings, the circulating beams in opposite direction are used to take physics data at collision points until the beam quality became unusable to take data. The storage time can vary from more than 10 hours for nominal 100 GeV/n Au beam to less than 30 min for low energy experiments as low as 3.85 GeV/n. RHIC-EBIS provides typically 12 pulses separated at 5 Hz within a supercycle of about 6 seconds during RHIC injection and tuning for the injection typically 30 min before the injection starts. The supercycle is a repetitive cycle to accelerate ions to inject beam into RHIC. EBIS provides beams at earlier part of the supercycle, and the 12 pulses of EBIS beams become 2 bunches in RHIC. When RHIC-EBIS is not used for RHIC injection nor beam tuning, the number of EBIS pulses is reduced as standby mode to reduce load to ion sources and save lifetime of a cathode and a LION target. We keep at least 1 pulse to keep the system up and to realize any device failures. For short store such as 30 min, the number of the standby cycles is increased to 6 pulses to reduce thermal stress due to frequent change of the number of pulses. The short storage time was frequently used during Beam Energy Scan II (BES-II) started from Run-19 and successfully completed in Run-21 to investigate the first-order phase transition and determine the location of a possible critical point in a Quantum Chromodynamics (QCD) phase diagram with variable collision energy [8,9].

Beam operation for NSRL involves frequent change of species to simulate the effect of GCR. Before beams of multiple species are used for NSRL users, a set of parameters of all related devices is "archived" for each combination of species and charge state by EBIS experts and Main Control Room (MCR). To change species, the archived setting is restored, and ion species from external sources is switched by a sequencer without assistant from experts. When GCR simulator mode is used, a dosimetry system at NSRL automatically cut off beam in ~1 ms to control on sample, and initiate species change. The demand of NSRL beam has been increasing due to its unique capability. RHIC-EBIS initially provided beams for NSRL 5 days per week during day time. Recently we provide more time per day and 6 days per week routinely. The GCR simulator mode is routinely used, which increases the number of species change per day. More species have been added for LION to cover LET range without HCIS to make the source operation easier. The operation of the RHIC-EBIS pre-injector is getting more difficult compared to earlier time of the pre-injector.

Figure 2 shows time structure of the EBIS electron beam within a 6 sec supercycle when RHIC is at injection with NSRL running. We have a combiniation of beams for RHIC (full intensity or standby), and different species/charge state for NSRL. A duty cycle of the EBIS electron beam changes depending on the total number of pulses in a superccyle and confinement time of NSRL species. Cathode heater power needs to be high enough to keep electron beam current of 12 consecutive pulses for RHIC. However, increased heater increases the risk of cathode heater break, reduce cathode heater lifetime, and increase outgassing which can cause breakdown. We try to minimize cathode heater power without sacrificing beam intensity for reliable operation. In addition, electron beam current is affeccted by the duty cycle. The electron beam current decreases when the number of pulses for RHIC is increased or the confinement time for NSRL beam is extended because electron beam emission provides some cooling. This could change the electron beam trajectory, which can cause breakdown. Table 1 shows EBIS confinement time for different species to indicate the change of duty cycle. It takes about 10 seconds or less to respond to the duty cycle change. For reliable operation, we introduced software to automatically adjust cathode heater power depending on the total number of pulses in a supercycle and confinement time for NSRL beam to reduce the variation of the electron beam current.

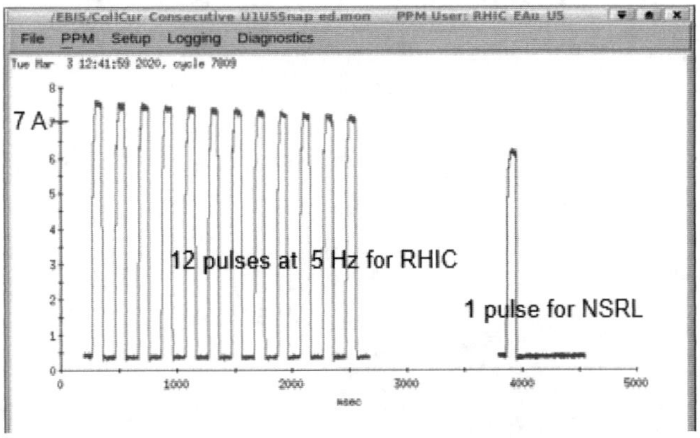

Figure 2. EBIS electron beam when EBIS is pulsing for RHIC injection and NSRL. Here, 5 sec out of 6 sec super cycle is shown.

Table 1. EBIS confinement time for different species

Species	Mass	Charge state	Q/M	Confinement time (ms)	**Species**	Mass	Charge state	Q/M	Confinement Time (ms)
He	4	2	0.5	17	**Fe**	56	24	0.429	822
C	12	5	0.417	51	**Nb**	93	19	0.204	70
O	16	7	0.438	81	**Ag**	107	29	0.271	386
Si	28	11	0.393	118	**Au**	197	32	0.162	70
Ti	48	18	0.375	242	**Au**	197	43	0.218	307
Fe	56	20	0.375	242	**Bi**	209	35	0.167	130

Figure 3 shows an example of dayly beam usage. As providing beams for RHIC, species change between He, O, Si, Fe from EBIS and proton beam from the 200 MeV Linac occurred more than 130 times as a GCR simulator mode. Stable heavy ion beams were delivered throughout the day. The species switching time at EBIS was about 40 sec. This is not limitting factor because it takes about 1 min to adjust Booster and a beam line to NSRL. Because the pre-injector is very reliable, the RHIC-EBIS pre-injector is unattended by experts for routine operation, and NSRL can change ion species at any time without notifying ion source experts.

5. Conclusion and Future plan

The operation of the RHIC-EBIS pre-injector including LION and RHIC-EBIS is getting more complicated with frequent RHIC injection as short as 30 min and more NSRL beam time and species. To increase the reliability, an active cathode heater control was introduced to compensate the effect of duty factor on electron beam current. The RHIC-EBIS pre-injector has been reliably providing various heavy ion beams under such operation mode. The pre-injector is unattended by EBIS experts for routine operation once a set of parameters for each species are "archived" by the experts. MCR can ramp up the number of EBIS pulses for RHIC injection and ramp down to the standby mode safely by a sequencer software. NSRL can change species without any assistance from EBIS experts and MCR.

Next year from 2022 summer to the end of the year, the existing RHIC-EBIS will be replaced with the Extended-EBIS to increase beam intensity and to provide intense polarized $^3He^{2+}$ ions for the future electron-ion collider (EIC) [10,11]. The Extended-EBIS will provide Au beam to Run-25, which is the last year of RHIC operation, and transition to EIC will be started. NSRL operation will continue during the transition. EIC operation is expected to start from FY2031.

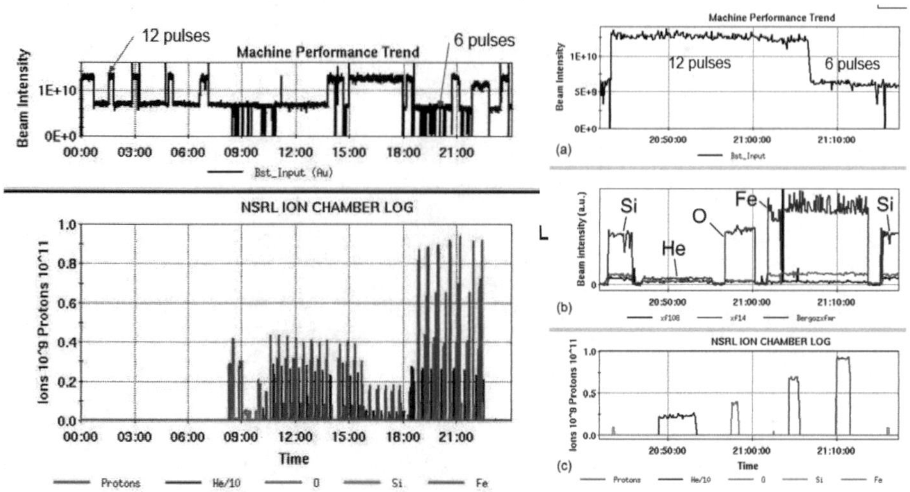

Figure 3. Beams for RHIC and NSRL when EBIS is providing beams for both facilities. Left plot shows beams during a day (Upper plot: total beam in a supercycle for RHIC at Booster input. Lower plot: beam at NSRL). Right plot shows 30 min period on the same day (Top: EBIS beam for RHIC. Middle: EBIS beam for NSRL. Bottom: Beam at NSRL).

Acknowledgement
This work has been supported by Brookhaven Science Associates, LLC under Contract No. DE-SC0012704 with the U.S. Department of Energy and by the National Aeronautics and Space Administration.

References

[1] Alessi J, et al. 2011 *Proc. 2011 Particle Accelerator Conference* (New York, NY, USA, March 2011) p 1966
[2] La Tessa C, Sivertz M, Chiang I, Lowenstein D and Rusek A 2016 *Life Sciences in Space Research* **11** 18-23
[3] Ikeda S, Kumaki M, Kanesue T and Okamura M, 2016 *Rev. Sci. Instrum.* **87** 02A915
[4] Beebe E, Alessi J, Binello S, Kanesue T, McCafferty D, Morris J, Okamura M, Pikin A, Ritter J and Schoepfer R 2015 *AIP Conference Proceedings* **1640** 5
[5] Saquilayan G Q, Ikeda S, Kanesue S, Wada M and Okamura M 2018 *AIP Conference Proceedings* **2011** 020013
[6] Okamura M, Beebe E, Ikeda S, Kanesue T, Raparia D, Muench L, Karino T and Haba H 2020 *Rev. Sci. Instrum.* **91** 013319
[7] Marr G, et al. 2019 *Proc. 10th International Particle Accelerator Conference* (Melbourne, Australia, May 2019) p 28
[8] Odyniec G, et al. 2019 *Proceedings of Corfu Summer Institute 2018 "School and Workshops on Elementary Particle Physics and Gravity"* (Corfu, Greece, August 2018) p 151
[9] Liu C, et al. 2019 *Proc. 10th International Particle Accelerator Conference* (Melbourne, Australia, May 2019) p 540
[10] Zelenski A, Atoian G, Beebe E, Poblaguev A, Raparia D, Ritter J, Maxwell J, Milner R and Musgrave M 2018 *Proc. 9th International Particle Accelerator Conference* (Vancouver, BC, Canada, April 2018) p 644
[11] Ikeda S, Beebe E, Kanesue T, Kondrashev S and Okamura M 2020 *Rev. Sci. Instrum.* **91** 013327

Magnet system design for the new 18 GHz ECR ion source at GSI

A Andreev, F Maimone, J Mäder, R Lang, P T Patchakui, K Tinschert and R Hollinger

GSI Helmholtzzentrum für Schwerionenforschung, Planckstr. 1, 64291, Darmstadt, Germany

E-mail: a.andreev@gsi.de

Abstract. In order to increase intensities and charge states of available ion species, a new room-temperature ECR Ion Source (ECRIS) operating at 18 GHz is currently under development at GSI. The new ECRIS is based on a Heavy Ion Ion Source Injector (HIISI), developed at the Department of Physics, University of Jyväskylä (JYFL), and features three normal conducting coils and a permanent magnet hexapole for plasma confinement. The latter has to be installed inside a refrigerated hexapole chamber, allowing to achieve the required radial confining field and avoiding demagnetization of permanent magnets. Computer simulations are carried out with Opera software package for two Halbach hexapole arrangements and the resulting three-dimensional magnetic fields are compared. The demagnetization of permanent magnets due to the superposition of fields generated by the coils and the hexapole is also simulated for both arrangements.

1. Introduction

The 14.5 GHz CAPRICE ECR Ion Source (ECRIS) installed at the High Charge State Injector (HLI) of GSI provides highly charged ion beams of various gaseous and metallic elements for the accelerator facility. Since its commissioning in 1992, the source has undergone a number of upgrades allowing to improve stability and intensity of the accelerated ion beams. Recently, an optical emission spectrometer and a microwave shielding grid have been applied for routine operation which improved the long time stability and reproducibility of the ^{48}Ca beam [1].

In order to benefit from the technological advances of ECRIS development and from increased intensities and available charge states of heavy ion beams for the FAIR facility, a new ECRIS is currently under development. In accordance with semi-empirical ECR scaling laws, higher intensities require higher plasma densities, which implies an increase of the microwave frequency and of the confining magnetic field.

The new 18 GHz ECRIS is based on the design of Heavy Ion Ion Source Injector (HIISI) developed at the Department of Physics, University of Jyväskylä (JYFL) [2]. Its design incorporates a number of innovative technical solutions allowing to reach the required magnetic field configuration with nonsuperconducting coils and permanent magnets.

2. Magnet system design

The new ECRIS will be based on a room-temperature magnet system. The axial magnetic field is produced by three normal conducting coils. The reverse direction of current in the middle

Table 1. Design parameters of the magnet system of the new ECRIS.

Parameter	Value
Magnetic field at injection B_{inj}	2.8 T
Magnetic field at center B_{min}	0.45 T
Magnetic field at extraction B_{ext}	1.3 T
Nominal currents of coils (Inj. / Center / Ext.)	1000 A / -300 A / 820 A
Inner radius of the hexapole	58 mm
Outer radius of the hexapole	120 mm
Length of the hexapole	420 mm

coil allows to adjust the minimum magnetic field at the center. The radial plasma confinement is provided by a permanent magnet hexapole, which has a Halbach structure with 36 segments. Table 1 summarizes the main parameters of the magnet system.

In order to maximize the radial confining field, NdFeB permanent magnets of grade N45SH have been chosen [3]. This material is characterized by the following nominal values of remanence and coercivity at 20 °C : $B_{\text{rem}} = 1.35\,\text{T}$ and $H_{\text{cJ}} = 1592\,\text{kA/m}$, respectively.

2.1. Hexapole arrangements

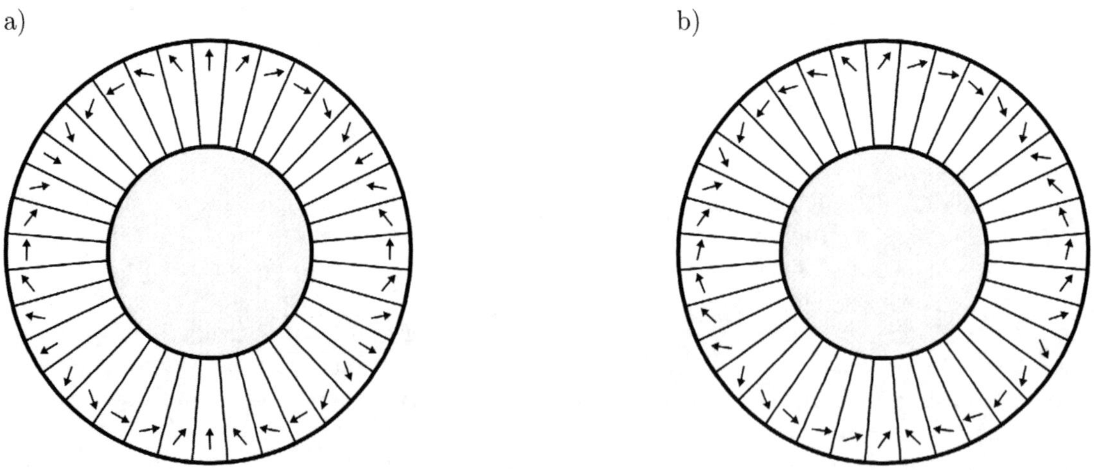

Figure 1. Magnetization directions for typical (a) and offset-Halbach (b) hexapoles. Arrows show magnetization direction of each permanent magnet block.

In this work, two Halbach hexapole arrangements are compared with each other. The first hexapole arrangement which corresponds to a typical Halbach structure is shown in Figure 1 a. The second arrangement studied is a so-called offset-Halbach arrangement (see Figure 1 b) introduced in [4]. In the latter design, the magnetization direction of each of the 36 magnet segments is chosen in a way that the magnetic poles are located between the magnet pieces.

Three-dimensional simulations of the complete magnetic system were performed using the Opera software package [5]. Thus, the resulting magnetic field map represents a superpostion of the solenoid and hexapole fields. All the simulations were done with the magnet grade N45SH and the coil currents given in Table 1. The influence of the offset angle on the configurations of magnetic and demagnetizing fields is discussed in the following.

2.2. Magnetic field structure

The magnetic field along the axis of the plasma chamber is mainly determined by the coil currents. Therefore, as expected, both arrangements provide identical axial magnetic field profile which is shown in Figure 2. The maximum axial magnetic fields are about 2.8 T and 1.3 T at the injection and the extraction, respectively.

Figure 3 shows the radial field profiles obtained for both arrangements. The longitudinal coordinate (z) corresponds to the middle of the hexapole assembly. One can see, that the offset-Halbach structure allows to reach a higher magnetic field at the inner wall of the plasma chamber (see also Figure 4). The maximum radial magnetic fields achieved with typical and offset-Halbach are 1.21 T and 1.31 T, respectively. Figure 6 shows the radial magnetic field at the magnetic pole along the axial direction for both arrangements. One can note, that the field profiles provided by both structures are identical, with overall higher radial magnetic field produced by the offset-Halbach hexapole at injection ($z < 216$ mm) and extraction ($z > 216$ mm) sides. The hexapole design without offset provides a more homogeneous field at the inner surface of the plasma chamber with the maximum $|B| = 1.31$ T (see Figure 5). The absolute value of the magnetic field at the inner surface of the plasma chamber which is produced by the offset-Halbach hexapole is higher ($|B| = 1.39$ T). This design also provides a field profile in which the peak magnetic field periodically occurs in narrow regions. Three-dimensional views of the magnetic field structure for both hexapole arrangements are shown in Figure 7.

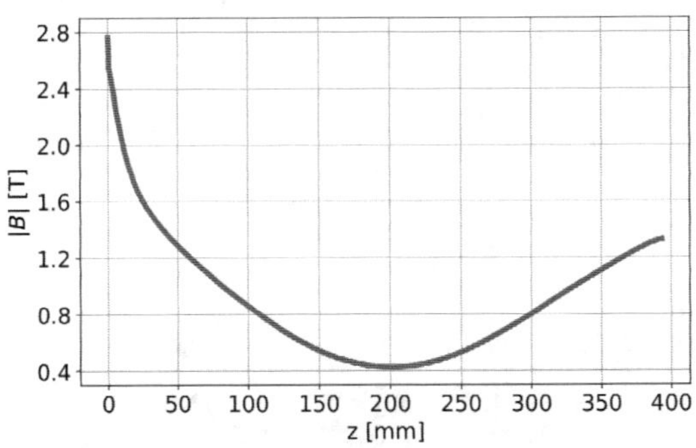

Figure 2. Axial magnetic field. The coordinates $z = 0$ mm and $z = 394$ mm correspond to the injection and extraction ends of the plasma chamber, respectively.

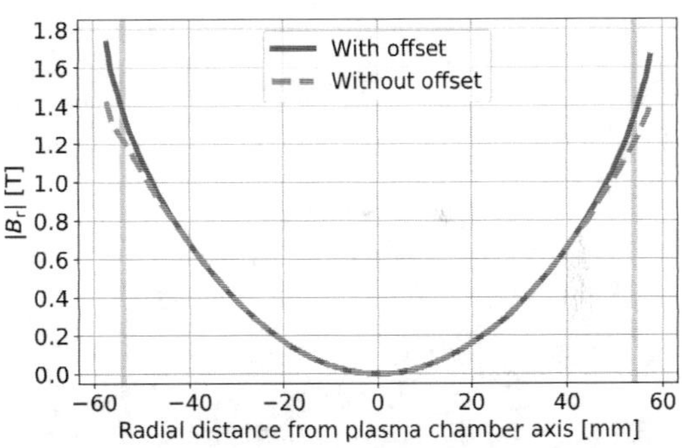

Figure 3. Radial component of magnetic field along radius ($z = 210$ mm). The shaded area marks inner wall of the plasma chamber ($r = 53.5$ mm).

2.3. Demagnetization analysis

A superposition of fields generated by the coils and the hexapole can cause a partial demagnetization of permanent magnets. This phenomenon occurs when the amplitude of the demagnetizing field exceeds the coercivity of a permanent magnet block and its direction is opposite to the magnetization direction of this segment.

In present analysis, a 30 % safety margin is taken for the coercivity of the permanent magnets in order to avoid partial demagnetization of the permanent magnet material. Therefore, the amplitude of the demagnetizing field should not exceed a value of 1.1 kA/m. The demagnetization analysis shows that the offset-Halbach arrangement experiences a larger

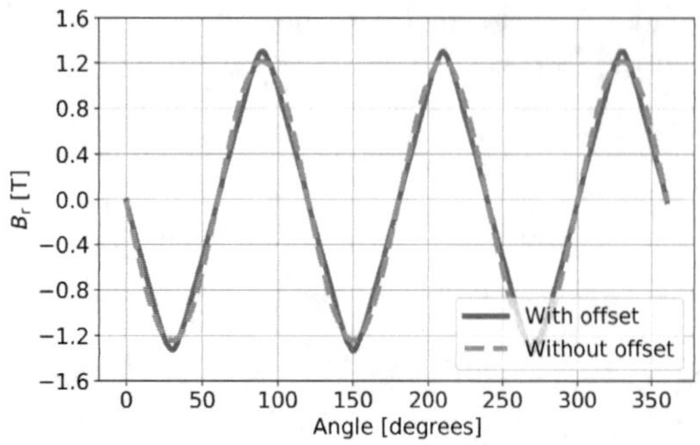

Figure 4. Radial component of magnetic field at the surface of the plasma chamber inner wall ($r = 53.5\,\mathrm{mm}$, $z = 210\,\mathrm{mm}$).

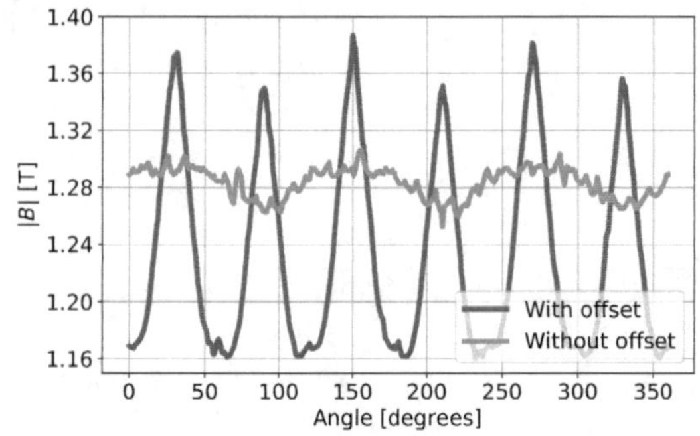

Figure 5. Magnetic field at the surface of the plasma chamber inner wall ($r = 53.5\,\mathrm{mm}$, $z = 210\,\mathrm{mm}$).

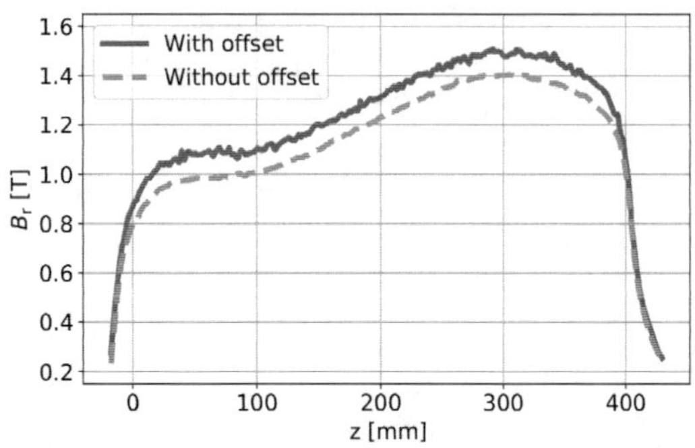

Figure 6. Radial component of the magnetic field at the inner surface of the plasma chamber along the axial direction ($r = 53.5\,\mathrm{mm}$).

a) b)

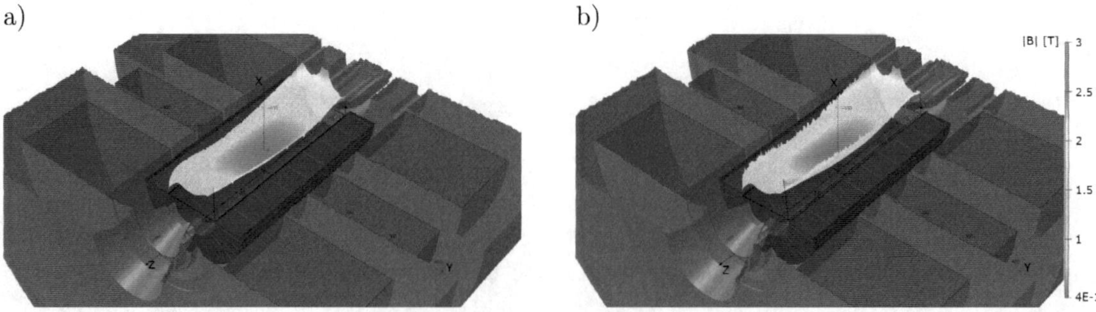

Figure 7. Magnetic field structure along the axial direction for typical (a) and offset-Halbach (b) magnetization directions.

demagnetizing field. From Figure 8 one can note that the maximum amplitude of the demagnetizing field is $1.49\,\mathrm{kA/m}$ and $1.74\,\mathrm{kA/m}$ for typical and for offset-Halbach hexapole, respectively. However, a smaller volume of permanent magnet material is subjected to the demagnetization with the latter design. The volume of permanent magnet material which is exposed to demagnetizing field exceeding $1.1\,\mathrm{kA/m}$ is $306\,\mathrm{cm}^3$ for the offset-Halbach arrangement, compared to $428\,\mathrm{cm}^3$ for the typical one. It should be noted that these values correspond to 2 and 3 percentage of the total permanent magnet volume, respectively.

The hexapole chamber is refrigerated to avoid demagnetization of permanent magnets by increasing coercivity of the magnet material. In addition to the cooling of the permanent

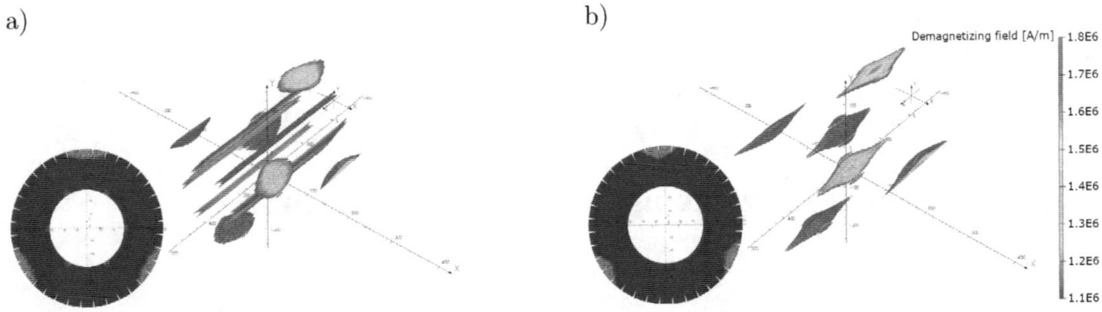

Figure 8. Demagnetizing field which exceeds a value of 1.1 kA/m for typical (a) and offset-Halbach (b) magnetization directions. Note that only the regions where the field is higher than the threshold value are shown. Transverse cross-section views show the largest demagnetizing field from injection side.

magnets, a material grade with higher coercivity can be used at those regions of the hexapole which experience the largest demagnetizing field. For example, a permanent magnet hexapole of the AISHa (Advanced Ions Source for Hadrontherapy) consists of two concentric rows of permanent magnet blocks to compensate for the demagnetizing field at the injection and extraction sides [6]. The analysis indicates that partially demagnetized regions are located along both the inner and outer radii of the typical hexapole assembly. Consequently, more hexapole segments would have to be replaced in this arrangement.

3. Conclusion

In this work, two Halbach hexapole arrangements were investigated with the Opera software package in terms of the produced magnetic and demagnetizing fields. It was found that a so-called offset-Halbach hexapole arrangement provides a 7 % increase of the radial magnetic field compared to a structure without offset, thereby allowing maximize the required radial confining field. Using offset-Halbach stucture allows to reduce the amount of material which is exposed to demagnetization by one percentage point. These features make the offset-Halbach arrangement a more attractive design for the permanent magnet system of the new ECRIS.

References

[1] Maimone F, Mäder J, Lang R, Patchakui P T, Tinschert K and Hollinger R 2019 *Rev. Sci. Instrum.* **90** 123108 URL `https://doi.org/10.1063/1.5127571`

[2] Koivisto H, Tarvainen O, Kalvas T, Ranttila K, Heikkinen P, Xie D, Machicoane G, Thuillier T, Skalyga V and Izotov I 2014 HIISI, new 18 GHz ECRIS for the JYFL accelerator laboratory *Proc. ECRIS'14* (Nizhny Novgorod, Russia) pp 99–103

[3] Neodymium iron boron magnets, Arnold Magnetic Technologies URL `https://www.arnoldmagnetics.com/products/neodymium-iron-boron-magnets/`

[4] Suominen P, Tarvainen O, Koivisto H and Hitz D 2004 *Rev. Sci. Instrum.* **75** 59–63 URL `https://doi.org/10.1063/1.1628846`

[5] Opera electromagnetic and electromechanical simulation software URL `https://www.3ds.com/products-services/simulia/products/opera`

[6] Celona L, Castro G, Castro M, Chines F, Gammino S, Leonardi O, Marletta S and Mascali D 2014 ECR ion source developments at INFN-LNS *Proc. ECRIS'14* (Nizhny Novgorod, Russia) pp 87–93

Electron gun producing beams with controllable current density

Alexander Pikin[1,2,*], Vittorio Bencini[1], Hannes Pahl[1,3], Fredrik Wenander[1]

1. BE Department, CERN, CH-1217 Geneva, Switzerland
2. Centro de Investigaciones Energéticas, Medioambientales y Tecnológicas, Avenida Complutense 40, 28040 Madrid, Spain
3. Heidelberg Graduate School for Physics, Heidelberg University, Im Neuenheimer Feld 226, 69120 Heidelberg, Germany

Corresponding author: alexander.pikin@cern.ch

Abstract. The existing Brillouin-type electron gun at the TwinEBIS test bench is, according to Herrmann theory, capable of producing an electron beam with a current density of 3850 A/cm^2 in the 2 T solenoid. To control the electron beam current density and the magnetic flux inside the beam, the existing electron gun - now using purely electrostatic focusing - can be modified by permitting magnetic flux to reach the cathode. In such a configuration, the stabilizing magnetic flux inside the electron beam can be controlled by changing the current in the magnet coil surrounding the cathode. The radial oscillations of the electron beam, resulting from the increased magnetic field on the cathode, can be significantly reduced by employing a non-adiabatic magnetic field near the electron gun. This method has been recently developed and successfully used at REXEBIS at CERN. We present the computer simulations of such electro-optical system.

1. Introduction

The TwinEBIS device has an electron gun with magnetically shielded cathode and makes use of electrostatic compression in the cathode-anode gap [1]. With a very low magnetic field on the cathode (1.3 Gs), this gun produces practically a Brillouin beam with high compression, and the electron beam current density calculated with the Herrmann formula [2,3] reaches 3850 A/cm^2 in a magnetic field of 2 T for an electron current of 1.0 A. According to refs. [4,5], the stability of the electron-ion system in the ion trap of electron beam ion source (EBIS) depends on the density of electrons and ions, and on the length of the ion trap. Since the capacity of the EBIS is proportional to the length of its ion trap, stabilizing the electron-ion system in the trap by shrinking the ion trap length is not an option. The remaining free parameter to achieve a stable condition is the electron beam current density. For the very low enclosed magnetic flux within the Brillouin electron beam, the radius of the beam depends on the neutralization of the electron beam space charge and with high neutralization in the trap the electron beam current density increases by a factor $A=1/(1-f)$ [6], where f is a beam neutralization. Such a dynamic increase of the electron beam current density is not desirable in EBIS devices because it may trigger instabilities with subsequent loss of electron-ion overlap and significant degradation of the effective current density. A straightforward solution for reducing the electron beam current density and for preventing it from collapsing when it is neutralized is to permit some magnetic flux to reach the cathode. Such an electron gun, switchable between a Brillouin and magnetically immersed flow, by varying the magnetic field on the cathode with a magnetic coil has been proposed by R. Becker [7].

The goal of this study is not to calculate or simulate the stable ion trap conditions, but to find an optical solution for the electron beam formation in a range of magnetic field on the cathode. This solution should allow one to experimentally tune the magnetic field on the cathode with acceptable electron beam ripples, and provide an electron beam current density in a range of 300 to 1000 A/cm^2 in an ion trap with a 2 T magnetic field. Using this option, it would be possible to experimentally find the optimum EBIS conditions where the ion trap is stable and the electron beam has the maximum current density. Such desired flexibility presents a challenge for the design of the electro-optical system because deviating from the optimum Pierce gun parameters comes at a price of increased radial oscillations of the electron beam. Since these oscillations limit the maximum transmittable electron beam current, they should be minimized across the entire operating range of EBIS.

For suppressing the electron beam radial ripples we are planning to use the previously developed method employing a non-adiabatic magnetic field [8]. This method can be applied to a coherent beam by creating a local dip in the magnetic field on the descending phase of the electron cyclotron motion. Such dip can be produced either with a magnet coil or with an iron ring. The latter method is much simpler, but suffers from a limited range of the electron beam parameters.

2. Electron gun design.

The proposed electron gun for TwinEBIS has an electrostatic geometry similar to the previous version described in ref. [1]. Its spherical cathode has 12 mm diameter with a 10 mm radius of curvature. The perveance of the original gun is P=0.98x10 $^{-6}$ A/V$^{1.5}$ and it has not changed in the new design. The already existing magnetic coil around the cathode is planned to be used for controlling the exact magnetic field on the cathode surface. This coil has 16 turns and is made of Kapton-isolated copper wire. To accommodate a larger electron beam emitted by the cathode at the elevated magnetic fields, the diameter of the anode aperture will be increased from 2.4 mm to 5.4 mm. The main difference of the new gun design compared to the previous is a larger aperture in the magnetic shield around the gun: its optimized diameter was increased from 4.4 mm to 20.0 mm to allow magnetic flux to reach the cathode. The design of the proposed TwinEBIS electro-optical system is shown in Figure 1. This electro-optical system includes the electron gun and the iron ring positioned at optimum distance from the gun. The iron ring is positioned on the descending phase of the second cycle of the cyclotron oscillation.

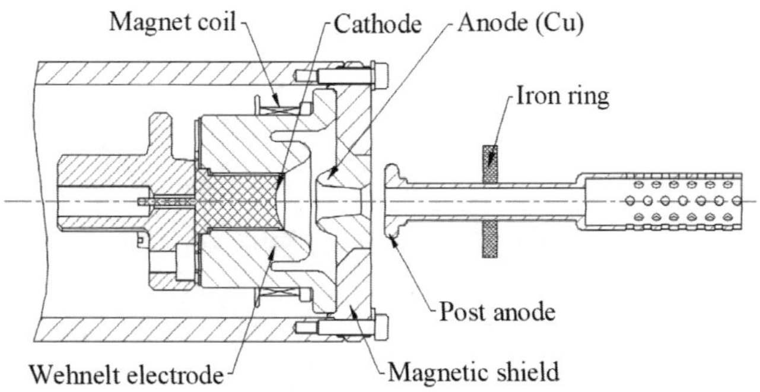

Figure 1. Design of the proposed TwinEBIS electron beam forming system.

3. Simulations

The simulations of the beam transmission with the proposed electro-optical system have been done with the software package TRAK [9] using the Child-Langmuir model for the cathode emission. Figure 2 shows the electron beam current density and the beam parameter K as functions of magnetic flux density on the cathode as computed with the Herrmann formula [2,3]. The parameter K is the ratio of the current density compression factor to the ratio of the magnetic fields on the cathode and inside the ion trap, i,e, $K=(j_{trap}/j_{cath})/(B_{trap}/B_{cath})$. In effect, the parameter K can be treated as a degree of magnetization of the electron beam and for a magnetic field higher than 30 Gs on the cathode, the magnetic flux within the electron beam in the ion trap is approximately constant and approaches the magnetic flux on the cathode. One can see that for the current density range 1000 to 300 A/cm^2, the required magnetic field on the cathode is between 17 and 58 Gs. The control over the magnetic field will be attained by tuning the cathode coil, which superimposes its field onto the residual flux from the main focusing field that penetrates through the widened aperture in the magnetic shield.

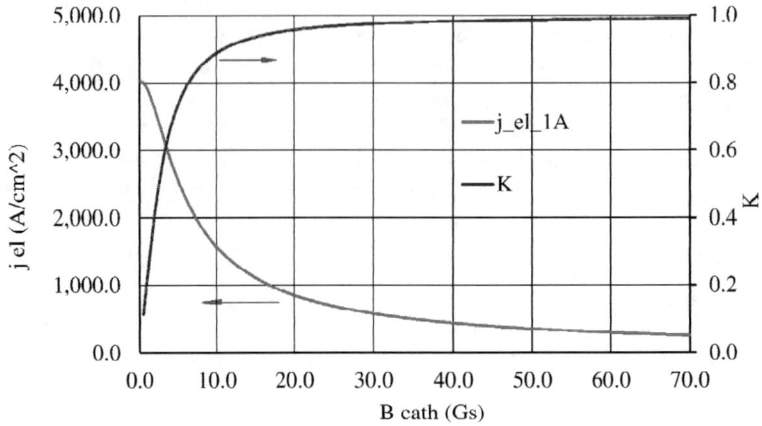

Figure 2. Electron beam current density in the ion trap region predicted with the Herrmann formula and evolvement of the parameter K (explanation in text) as function of the cathode magnetic field.

This magnetic field combination allows us to minimize the electron beam radial ripples at the elevated magnetic field on the cathode and to reduce the power dissipated in the coil, which does not have a water cooling. The dependence of the calculated cathode magnetic field on the coil current is presented in Figure 3. This dependence is very close to linear. If the coil is not energized, the magnetic field on the cathode is 14.9 Gs due to the magnetic flux leaking through the aperture of the gun iron shield.

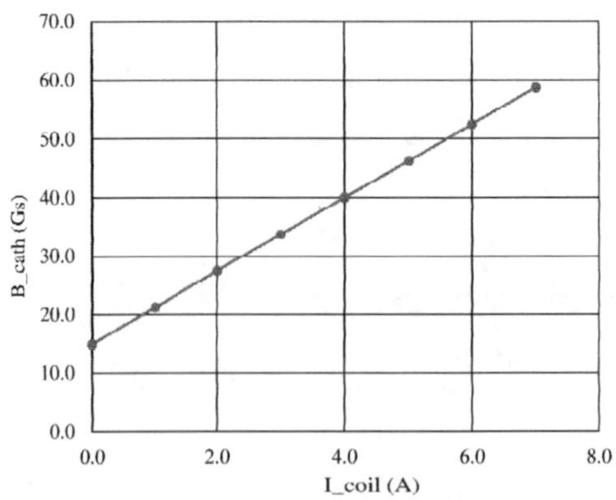

Figure 3. Dependence of the cathode magnetic field on the coil current.

Figure 4 presents a simulation of the electron beam and magnetic field for a case of 7 A current in the magnet coil, producing a magnetic field of 58.7 Gs on the cathode. The electron beam after the iron ring stays laminar with low amplitude of the radial ripples.

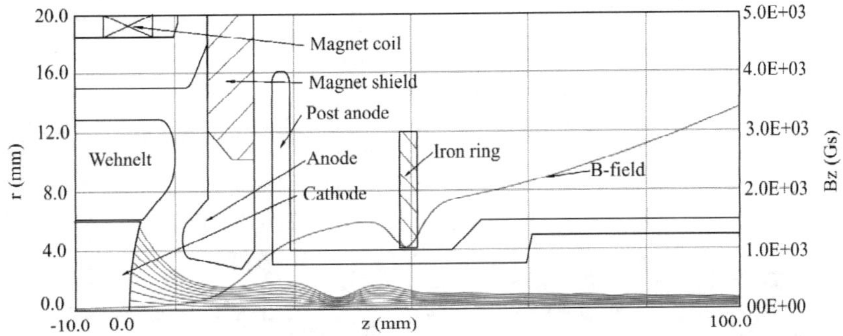

Figure 4. Simulation of the electron beam transmission with 58.7 Gs on the cathode. The electron current is 1.0 A, the cathode at -12 kV, post anode at 0 V and anode at -2 kV.

Figure 5 presents the beam envelopes for three different magnetic fields on the cathode. The beam envelope of the original electron gun is also shown as a dashed line for comparison.

Our simulations showed that the amplitude of the radial ripples depends on the electron energy and to minimize the ripples it is desirable to adjust the energy for each magnetic field on the cathode. The reason for such adjustment is the shift of the radial oscillation phase caused by the variations of the longitudinal electron velocity component at different magnetic fields on the cathode.

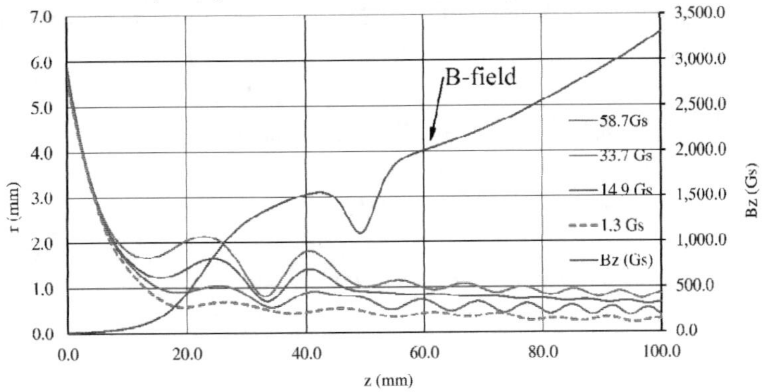

Figure 5. Beam envelope traces of electron beams with different magnetic fields on the cathode. The beam parameters are the same as in Figure 4.

In real devices, it is desirable to keep the voltages on the gun electrodes as stable as possible. To make the needed span of the optimization voltages small, the iron ring is positioned on the back-slope of the second radial oscillation. The iron ring does not eliminate the radial oscillations completely, but noticeably reduces them. The plots in Figure 6 illustrate the effect of this adjustment on the quality of the electron beam generated with a non-adiabatic magnetic field. The relative amplitude of radial beam oscillations is defined as $\frac{\Delta r}{r} = \frac{r_{max}-r_{min}}{r_{max}+r_{min}}$. We estimate the relative error of the simulated ripple amplitude to 5%.

From the left plot, one can see that the radial oscillation amplitude associated with a given magnetic field strength is energy independent for beam optics excluding the iron ring, whereas a pronounced minimum is observed if the magnetic field is modified by the presence of an iron ring. The right plot also shows that with optimized electron energy for each magnetic field on the cathode, the radial ripples of the electron beam remain lower than 10% for the most part of the cathode magnetic field dynamic range.

Journal of Physics: Conference Series **2244** (2022) 012103 doi:10.1088/1742-6596/2244/1/012103

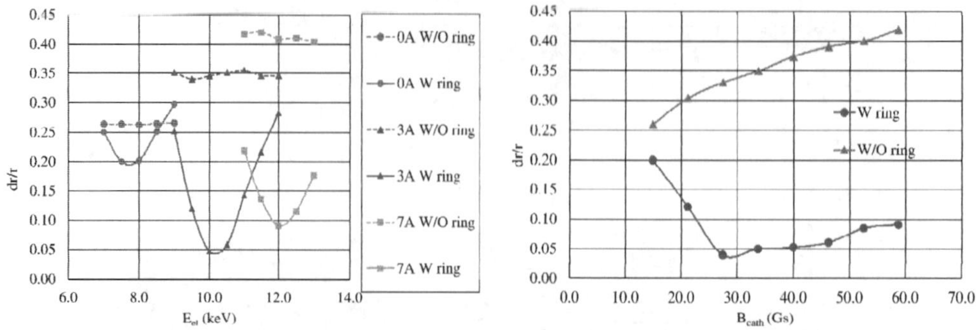

Figure 6. Left: Radial beam oscillation dependence on the electron energy, with and without iron ring present. The beam oscillations are given for three values of magnetic field on the cathode (0 A, 3 A and 7 A). Right: Dependence of radial oscillations on the magnetic field on the cathode. For the case with the iron ring, the electron energy has been optimized at each magnetic field value to achieve minimum oscillations

4. Conclusion

Our simulations demonstrate that the operating range of the magnetic field on the electron gun cathode can be extended to provide an electron beam density in the range of 300 to 1000 A/cm^2. A combination of a nonadiabatic magnetic field on the back slope of the radial beam oscillation and an adjustment of the electron energy allows us to attain an acceptable electron beam quality in a required range of current density in the ion trap.

5. Acknowledgement

Hannes Pahl has received funding through the Wolfgang Gentner Programme of the German Federal Ministry of Education and Research (Grant No. 05E15CHA).

6. References

[1] Mertzig R, Breitenfeldt M, Mathot S, Pitters J, Shornikov A and F Wenander 2017 A high-compression electron gun for C6+ production: concept, simulations and mechanical design, *Nuclear Instruments and Methods in Physics Research* A **859** 102–11

[2] Herrmann G 1958 Optical theory of thermal velocity effects in cylindrical electron beams, *Journal of Applied Physics* **29** number 2 127-36

[3] Amboss K 1969 Studies of a magnetically compressed electron beam *IEEE Transactions on Electron Devices* **ED-16** number 11 897-904

[4] Litwin C, Vella M C and Sessler A 1982 Linear electrostatic instability of the electron beam ion source, *Nuclear Instruments and Methods* **198** 189-92

[5] Mars R 2012 Ion heating and cooling in an EBIT or EBIS https://indico.cern.ch/event/199158/contributions/379787/attachments/295625/413143/T_II_2_Marrs.pdf

[6] Vella M C 1981 Physics of the high current density electron beam ion source (EBIS*), Nuclear Instruments and Methods* **187** 313-21

[7] Becker R and Kester O 2012 The design of an electron gun switchable between immersed and Brillouin flow, *Review of Scientific Instruments* **83** 02A506

[8] Pikin A, Pahl H and Wenander F 2020 Method of controlling the cyclotron motion of electron beam with a nonadiabatic magnetic field *Phys. Rev. Accel. Beams* **23** 103502

[9] https://www.fieldp.com/trak.html

Trapping secondary electrons in ExB drift for an alpha generating penning ion source

N Savard[1,2], A Groutso[3], S Melanson[2], D Potkins[2], M Dehnel[2]

[1] University of British Columbia, Vancouver, BC, V6T 1Z4, CA
[2] Dehnel-Particle Accelerator Components and Engineering Inc, 625 Front St, Nelson, BC, V6T 1Z4, CA
[3] Buckley Systems Ltd, 6 Bowden Road, Mt Wellington, Auckland, 1060, NZ

E-mail: nsavard@phas.ubc.ca

Abstract. A Penning ion source test stand has been developed by D-Pace to characterize alpha ion extraction from a helium discharge with variations in: extraction potential up to 15 kV, gas flow rate up to 30 sccm, and magnetic field confinement up to 1.1 Tesla. A brief summary of the Penning ion source design is described. Attempts at extraction of ions leads to severe sparking events on the high voltage insulator due to ExB drifts of secondary electrons. This paper shows the design of two sets of electrodes which together trap most electrons moving towards this insulator due to ExB drifts. This eliminates the severe high voltage breakdown on the insulator and allows for a more robust operation of the ion source.

1. Introduction

A test stand for a Penning ion source (or PIG source) was built in collaboration with D-Pace Inc and Buckley Systems Ltd [1]. The purpose of this test stand is to characterize the PIG source for α-particle extraction, and to study the plasma physics within the ion source to understand why various operational modes correspond to better or worse ion output. Though there's not much literature on the use of penning ion sources for α production, there are a couple that showed ion extraction of many 100 μA [2, 3]. These experiments may show that such sources can be used as replacements for expensive ECR ion sources currently used in medical accelerators [4].

The design is based on using the confining magnetic field as a control parameter for source operation since PIG sources are usually only tested under constant field strengths. Thus a 270 mm diameter C-magnet was designed for both the confining magnetic field within the source and for 180° mass separation of the extracted He^+ and α particles. A Faraday cup is built to move along the axis perpendicular to the normal vector of the ion source slit surface. This diagnostic is used to measure the ion currents as a function of distance from the ion source extraction which is separated based on its mass-to-charge ratio. The extraction goes up to 15 kV, with the ion source anode at 15 kV surrounded by a grounded vacuum box and puller (or extraction) electrode. The extraction voltage is static, thus producing a DC (or CW) ion beam. Additional operation parameters are the gas flow (internal ion source pressure), and the arc current. As this is a self-heated hot cathode source, the arc voltage is dependent on the other inputs.

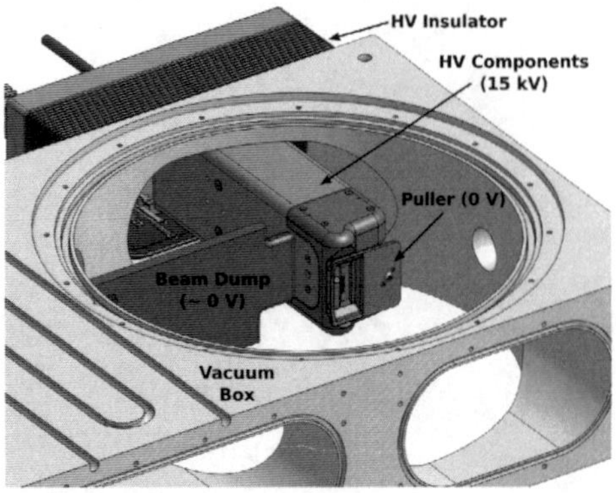

(a) Ion Source CAD design shown in vacuum box

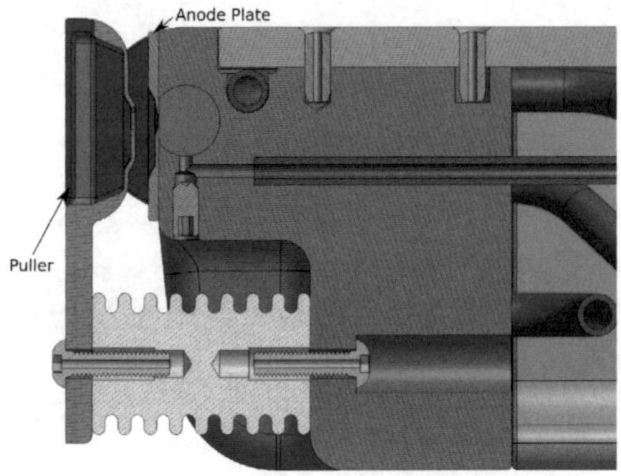

(b) Cross Section of ion source

Figure 1: 3D and cross-sectional view of the Penning ion source design. a) shows the ion source components within the context of the grounded vacuum box. b) shows the asymmetric slants in the anode plate and puller used to offset the magnetic field curvature

2. Test Stand Design

The prototype ion source was designed to have different anodes that could be swapped out to experiment with various anode geometries. This required the stacking of the anode and cathode separated by an electrical insulator in a limited pole gap space of 78 mm. This spacing was selected to prevent over-bowing of the magnetic field lines. The stacked parts and limited spacing required the electrical insulator be exposed to the outer vacuum, in addition to the insulators which separate the ion source from the grounded magnet poles. Unfortunately, this configuration led to many sparking events when attempting to operate at high voltage (HV) even without plasma. This was likely due to triple points in areas where the insulators come into contact with these electrodes, generating high electric fields and electron field emission which generates sparking along these insulators. In addition, attempted ion extraction likely created secondary electrons from ions hitting the puller electrode or ionizing the background gas, which then follow the magnetic field lines and collect on these exposed insulators. Failed attempts to fix these issues through increased spacing and use of various replacement materials led to the design of a new ion source. This ion source was designed to have only vacuum separation between the HV electrodes to decrease the electric field and therefore minimize sparking events. Also, any sparking that does occur does not become a permanent problem due to conductive paths being carved out of insulator surfaces.

Figure 1a shows a CAD model of the new ion source within the vacuum box. The circular flanges on the top and bottom of the vacuum box are where the magnet poles are placed. The HV components come in from the side through a PTFE insulating flange. The anode is made of water-cooled copper, and a beam dump is used to dump the remaining beam that does not land on the movable Faraday cup. Electrostatic simulations were done using the COMSOL™AC/DC module [5]. The limited space and the need for a long plasma column, to increase the capture rate of emitted electrons from the cathode, meant that the lowest fields that could be achieved at 15 kV is about 3.2 MV/m. These were reached by designing the ion source with the largest radius of curvature possible which removes kinks in the electric equipotential lines.

Figure 1b shows the mid-cross section of the ion source. The puller configuration was designed using charged particle tracing including space charge effects. The puller and anode plate have a steeper slant in the direction of the magnetic Lorentz force on the extracted ions to counter this curved trajectory with an electric field. Along with the slight offset of the puller and anode slits centers, this should allow for ions to be extracted from the ion source for magnetic field strengths of 0.4 - 1.0 T.

Figure 2 shows the CAD model of the cathode design, which comes in from the side to then rest above and below the plasma column. This was done due to contraints on vertical space in the pole gap. The insulator surrounds a water-cooled copper block which cools the base of the tantalum cathode. A thin neck on the cathode and a thin MACOR™ piece is used to keep the cathode hot enough for thermionic emission by limiting the conductive heat transfer to the cooling block. A sputter shield was placed inside after experiments showed that the insulator would still get covered with sputtered tantalum regardless of its distance from the cathode surface facing the plasma column. Sputtered tantalum now covers this piece instead of creating potential shorts across the insulator, which was also a significant issue affecting the previous design. The thinner and longer cathode neck also reduced heat conduction to the cooling block compared to the previous design, allowing for stable arc generation down to 5 sccm (previously 10 sccm).

3. Electron Trapping

First experiments showed that the new ion source could sustain up to 15 kV without plasma for magnetic fields up to 1.0 T and inlet gas flows up to the maximum of 40 sccm (background helium pressure of about 5×10^{-4} Torr). This is a significant improvement over the last model which could not hold high voltage. However, when attempting to extract ions, any voltage greater than 6 kV would generate arcs near the back HV insulator, creating severe scorch marks and cracking of the PTFE and causing the HV power supply to be unable to hold extraction potential. Charged particle tracing simulations using the COMSOL™ Charged Particle Tracing module show this is not due to low energy ions bouncing along the bracket wall to the insulator since they'd get pushed vertically to the grounded magnet poles. However, electrons are not completely trapped along the electric field lines, and will undergo an E×B drift defined by [6]:

$$\vec{v}_E = \frac{\vec{E}\times\vec{B}}{\vec{B}^2} \tag{1}$$

where $\vec{v}_E$ is the drift velocity due to this drift. This drift is shown as arrows in the mid-plane of the vacuum box in figure 3, where the grey boundaries and transparent walls are at ground and green is at 15 kV. All arrows go directly to the back insulator on the side, which is exactly where the scorch marks are seen. So it seems that the secondary electrons that are generated, likely through either ion bombardment of the puller or ionization of the background helium, then get transported all the way to the back insulator through the static E×B drift, and collect on it until the charge density becomes large enough for sparking to occur.

Complete elimination of these electrons requires a complete change of the orientation of the ground and high voltage electrodes, though time and cost prevents a design overhaul. However, a novel technique for trapping these electrons was made by adding a set of electrodes to the beam dump (connected to ground through a 510 Ω resistor) and the HV bracket. This is shown in figure 4. The electrodes were designed to minimize the electric fields using electrostatic simulations to prevent further sparking issues. Ultimately, like the ion source design, fields could not be reduced to less than 3.6 kV/mm everywhere due to limited space. The electrodes work as shown in figure 4b, where the E×B drift forces the electrons to either get trapped on the flat part of the wing electrode in the center, or to follow the drift curving between the two electrode until it arrives within the 'wings' of the wing electrode, which has an electric field along the magnetic field thus allowing the electrons to be absorbed. The beam extender electrode is kept vertically larger than the HV ion source components to make sure that the electrons stay near the center rather than escaping around the top of it.

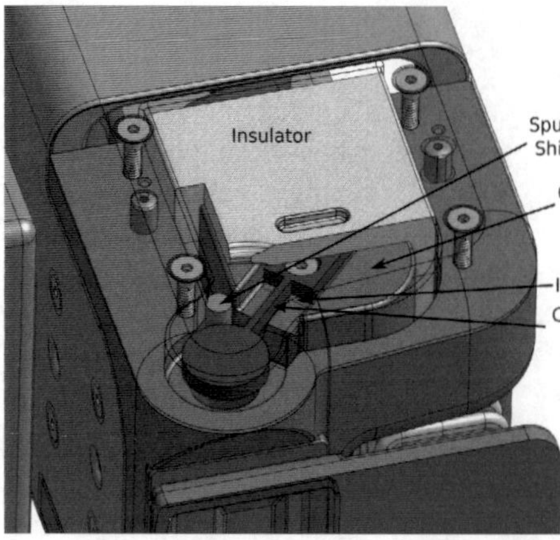

Figure 2: 3D CAD view of cathode design.

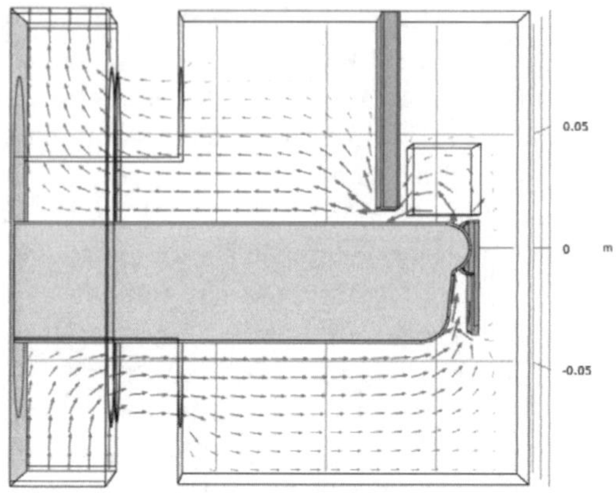

Figure 3: E×B drift direction of electrons

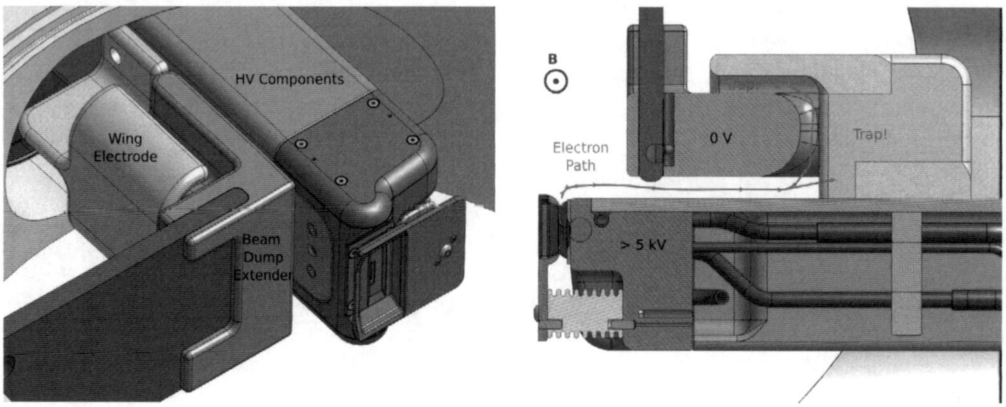

(a) Outer View

(b) Cross section view

Figure 4: Outer and cross sectional view of the added electrodes used as an electron trap. Figure (a) shows the outer view with the two electrodes with respect to the ion source, and (b) shows how the trap mechanism works with a cross sectional view.

In addition, some electrons may come from divergent ions which go through the 2 mm space between the beam dump and magnet poles, which can then emit electrons via secondary emission. A simple aluminum piece was made to shield this spacing to prevent this. After the design and manufacture of the electron trap electrodes and this extra shield, experiments found the ion source can successfully reach up to 15 kV with ion extraction without sparking cutting operation. The total bias current on the HV power supply goes up to 35 mA, though it is unknown how much current comes from secondary electrons or parastic discharges at this time. There were still light scorch marks on the insulator likely due to either electrons produced past the trap or to electrons which escape from the trap. However, ion beams can be extracted without any discharges overloading the extraction power supply, thus allowing for continuous operation during experiments (usually a couple hours duration). Figure 5b shows scorch marks on the wing electrode, which are correctly predicted as electron dumping spots in a COMSOL™ charged particle tracing simulation shown in figure 5a, where electrons are released in a volume near extraction. It should be noted that these electrons can be introduced anywhere on the extraction side of the beam dump and still follow the E×B drift into the trap. The agreement between the visible scorch marks and charged particle tracing simulations are evidence that the reason for the improvement is in fact the trapping of these electrons.

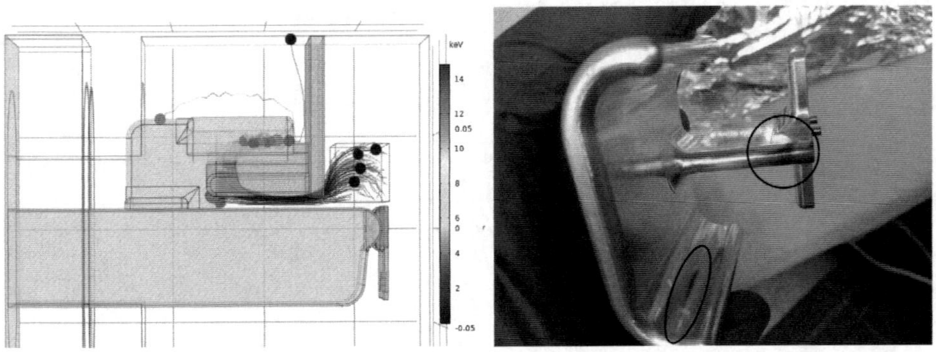

(a) Simulation of Electron Trap (b) Marks on Wing Electrode

Figure 5: Figure (a) shows a charged particle tracing simulations of the electrons born beyond the beam dump over time when 15 kV (green) is applied to the ion source and everything else is grounded (grey and transparent). Color bar shows the final electron energy in keV. 99% of simulated electrons are dumped on the wing electrode. Figure (b) shows a pictures of the wing electrode after use, which shows scorch marks exactly where electrons are predicted to be dumped.

4. Conclusion

A new PIG source was designed with significant improvements over the previous model due to its ability to sustain plasma at lower gas flows and the capacity to hold high voltage up to 15 kV at all operational magnetic fields and gas flows. Ion extraction did cause some other unexpected sparking events due to the E×B drift of secondary electrons to the back insulator. However, the new set of electrodes designed to trap these electrons drifting to the back was successfully able to trap most of these electrons, finally allowing for experiments to be run on the test stand.

Only preliminary tests have been run thus far, which have shown that the ion source is capable of producing He^+ currents up to 35 mA and α-particles up to 300 μA. Future experiments will study the plasma parameters (electron temperature and density) as a function of confining magnetic field, inlet helium gas flow, and arc current. This will allow us to compare which plasma parameters, set by the operational parameters, best optimizes the ion source for maximal α-production. This information can be used to improve on the current design for future prototypes, which will focus on studying the beam dynamics of the α-beam.

Acknowledgements

The author would like to thank D-Pace and Buckley Systems for their technical support, and to the TRIUMF Accelerator Division for use of their staff-license COMSOL™ modules.

References

[1] Savard N, Potkins D, Jackle P, Theroux J, Melanson S, Marcoux G, Dehnel M 2019 Development of a penning ion source test stand for production of alpha particles *10th Int. Particle Accelerator Conf.* (Geneva:JACoW Publishing) p 1932-1935

[2] Kuo T, Laughlin J 1975 Initial Experience in ion production using an ultra-high power density method *7th Int. Conf. on Cyclotrons and their Apps.* (Zurich: Birkhauser Basel) p 381-385

[3] Chen L, Chen N 1985 An intense alpha ion source for INRS cyclotron *IEEE Trans. on Nuc. Sci.* **32** p 1773-1775

[4] Milton B, Dawson R, Dehnel M, Kuo T, Ruegg R, Risler R 1998 Design of an axial injection system for the seattle MC50 cyclotron *15th Int. Conf. on Cyclotrons and their Apps.* p 638-641
[5] COMSOL Multiphysics™v5.5 www.comsol.com COMSOL AB Stockholm Sweden
[6] Chen F 2011 *Introduction to plasma physics and controlled fusion* (New York: Plenum Press)

Developments of control system for ion source using machine learning

Y Morita[#A)], M Fukuda[A)], T Yorita[A)], H Kanda[A)], K Hatanaka[A)], T Saitou[A)], H Tamura[A)], Y Yasuda[A)], T Washio[B)], Y Nakashima[C)], M Iwasaki[A,B,D)], H W Koay[A)], K Takeda[A)], T Hara[A)], T H Chong[A)], H Zhao[A)]

[A)] Research Center for Nuclear Physics, Osaka University
[B)] Department of Reasoning for Intelligence, Osaka University
[C)] Institute for Datability Science, Osaka University
[D)] Osaka City University

Email : morita16@rcnp.osaka-u.ac.jp

Abstract. Various factors influence each other in an ion source. Therefore, when operating an ion source, it is necessary to optimize and adjust various parameters. This time, we performed an experiment to automize adjustment that maximizes the brightness of the beam using machine learning. By automatically adjusting 4 parameters, we succeeded in finding a point with a beam brightness of 4.32×10^{-6} $\mathrm{mA}/(\pi\,\mathrm{mm\,mrad})^2$ in 25 steps. This shows that automatic adjustment using Bayesian optimization is feasible.

1. Introduction

An ECR ion source has various parameters such as RF power and gas pressure, which influence each other. The parameters are varied every time during operation and some of them are highly experieence dependent [1]. Therefore, the adjustment time, beam intensity and beam quality may vary depending on the operator. In accordance to this issue, we developed an automatic control system by applying machine learning technology, which is evolving rapidly in recent years. The realization of automatic control is expected to improve the reproducibility of beam quality and intensity, further enabling the visualization of correlation and multiple parameters.

2. Experimental setup

This chapter describes the experimental setup in which the automatic adjustment experiment was performed.

2.1. 10GHz ECR Ion source 'NANOGAN'

We used a 10 GHz ECR ion source 'NANOGAN' [2] (Fig.1) built by PANTECHIK. The NANOGAN drawer system has multiple electrodes that the voltage can be applied to the intermediate electrodes. By adjusting the voltage of the intermediate electrode, the electric field on the extraction surface is expected to be optimized and the beam emittance will be reduced. As for RF, the reflected wave can be reduced by moving in and out the RF tuner to optimize the injection efficiency.

Journal of Physics: Conference Series **2244** (2022) 012105 doi:10.1088/1742-6596/2244/1/012105

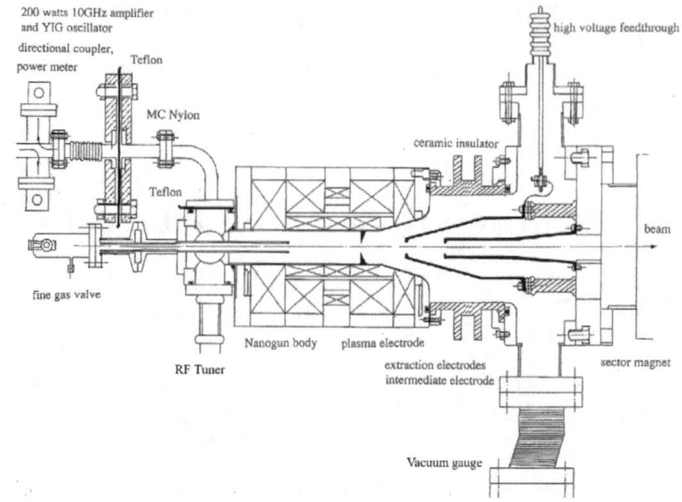

Figure 1. NANOGAN ECR ion source and extraction system. This picture is quoted from T Itahashi, et al., 'Performance of the NANOGUN™ electron cyclotron resonance ion source applied for nuclear astrophysics', Rev. Sci. Instrum. 71, 1075 (2000) [2] and partial modification of connected device and ceramic insulator size.

2.2. Test bench

This time, we built a simple test bench (Fig.2 (a)) and conducted an automatic adjustment experiment. In the test bench, the beam drawn from NANOGAN is bent 90 degrees by a sector electromagnet, focused using two Q elctromagnets, and the beam intensity and emittance are measured at the diagnostic station. At the diagnostic station, the emittance monitor is installed at upstream, but the emittance monitor is retracted from the beam axis when measuring the beam intensity, and is inserted on the axis only when measuring the emittance. Fig.2 (b) shows the results of the orbit calculation of the test bench using the transport matrix calculation code 'MAD-X' [3].

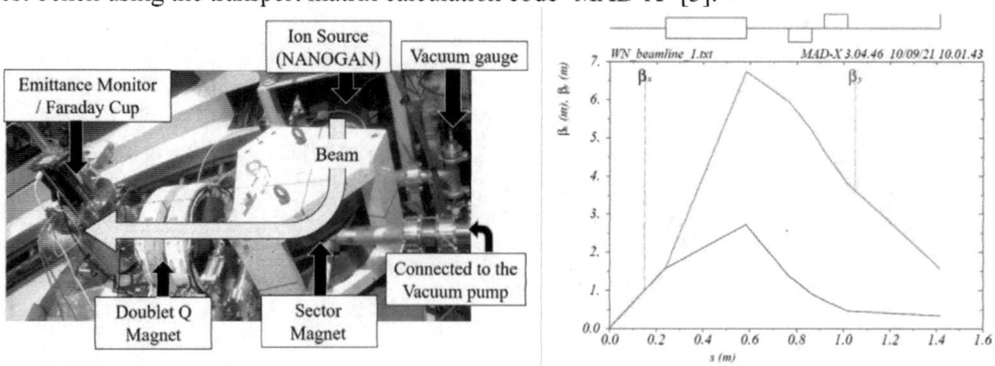

(a) (b)

Figure 2. (a) shows arrangement of magnets and diagnostic stations in the test bench.
(b) shows beam transport calculation by MAD-X.

2.3. Measurement systems

A Pepper-Pot emittance monitor (PPEM) [4] was used for beam emittance measurement. By using PPEM, high-speed measurement in real time is possible, and beam adjustment can be performed smoothly. PPEM measured 90% emittance and used it to calculate beam brightness. The beam intensity was measured using a Faraday cup (Fig. 3 (a)). The Faraday cup has a depth of 35.5 mm, and the emission of secondary electrons can be suppressed by applying voltage to the voltage-applied part.

The relationship between the applied voltage to the voltage applied-part and the measured value is shown in Fig 3 (b). From this result, about -0.3 kV was applied to supresse the emission of secondary electrons during measurement.

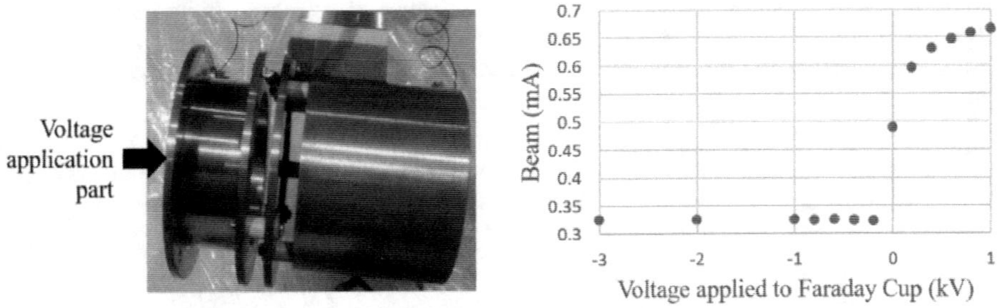

(a) (b)

Figure 3. (a) shows Faraday cup and the voltage applied-part.
(b) shows the relationship between the applied voltage and the measured beam value.

2.4. Parameters

As the parameters of NANOGAN, the power of the RF synthesizer, the position of the RF tuner, how the gas valve opens, the RF frequency, the voltage of the intermediate electrode, and the electromagnet were adjusted. The gas valve and RF tuner were controlled by changing amount of the motor rotation, as in previous research [5]. The power of the RF synthesizer was -14 to -11 dBm, it was adjusted in the range of about 4 to 8 W at the plasma chamber. The RF frequency can be adjusted within the range of 10GHz ± 0.2GHz, which is the limit value of the power amplifier. The voltage applied to the intermediate electrode was adjusted in the range of 0 kV to 15 kV to avoid discharging effect. In addition, each device was controlled using EPICS [6].

3. Adjustment method

3.1. About machine learning

Neural networks (NN) [7], which is one of the machine learning methods, are rapidly expanding due to the improvement of the machine power of computers. But NN cannot be used when reproducibility is low. And ion sources have low reproducibility over a long period. Thus, NN may be unstable during adjustment of the ion source. As a result, we did not use NN but chosen the Bayesian optimization (BO) [8]. The advantage of BO is that we can easily get out of the minimum value and optimize multiple parameters simultaneously. This time, we used 3 data randomly taken before the start of adjustment as prior information.

3.2. Adjustment procedure

As an adjustment procedure, first, the power of the RF synthesizer, the RF frequency, the gas valve, and the voltage parameters of the intermediate electrode are set simultaneously by BO.

Next, the RF tuner and magnet were adjusted in order using independent BO. The RF tuner was adjusted to minimize the power of the reflected waves, and the magnet was adjusted to maximize the beam intensity measured by the Faraday cup. When the magnet was operated without any adjustment, the magnetic field changed due to the change in the temperature of the cooling water. Consequently, the beam intensity could not be measured correctly. To resolve this, we decided to adjust it each step.

After that, the beam intensity I was measured using a Faraday cup. After inserting the PPEM, the emittance of the x-x' (ε_x) and y-y' (ε_y) planes was measured to calculate the beam brightness (eq. 1).

$$Brightness = \frac{I}{\varepsilon_x \varepsilon_y} \quad (1)$$

After the brightness calculation, another parameter was set again and the evaluation was repeated in the same manner. After 25 steps, the parameter was set to the highest brightness and the adjustment was completed. Fig. 4 shows the corresponding flowchart.

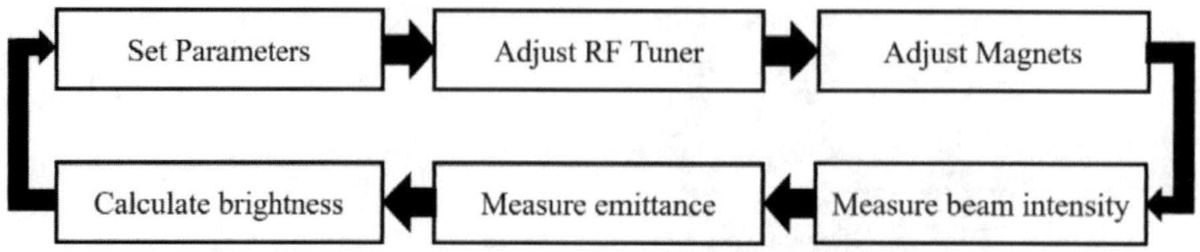

Figure 4. Flow chart of adjustment procedure

4. Results

An experiment was conducted in which a voltage of 30 kV was applied to the plasma electrode to maximize the brightness of $^{4}He^{2+}$. Fig. 5 (a) shows the relationship between the beam brightness at each step and the highest value up to that point. In addition, Fig. 5 (b) shows the magnitude of the parameter change at each step.

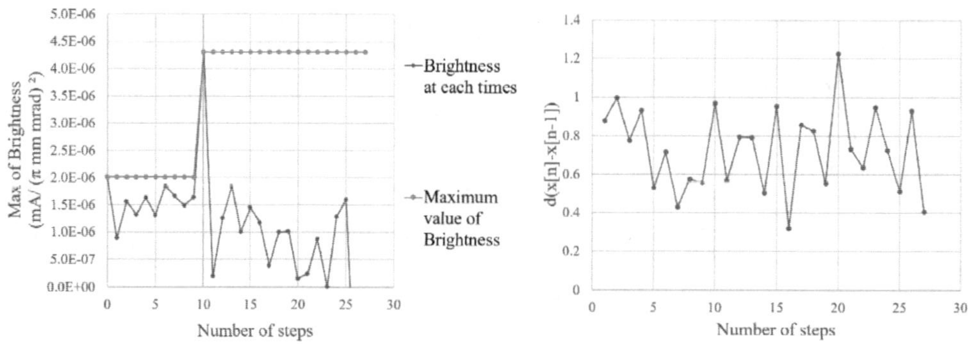

(a) (b)

Figure 5. (a) shows the relationship between the beam brightness and the highest value up to that point. (b) shows the magnitude of the parameter change at each step.

In addition, d is calculated by using eq. 2. x[n] is the value of each parameter at the step n, $RF\ power_{max}$ and $Voltage_{max}$ represent the maximum value of the setting range of each parameter, and $RF\ power_{min}$ and $Voltage_{min}$ represents the minimum value of the setting range of each parameter. If the adjustment is completed, the area around the optimum solution will be searched to find a minimal d. Although the number of adjustments cannot be said to have completed this time, we have already succeeded in finding a local solution.

$$d(x[n] - x[n-1]) = \sqrt{\left(\frac{x_{RF\ power}[n] - x_{RFpower}[n-1]}{RF\ power_{max} - RF\ power_{min}}\right)^2 + \cdots + \left(\frac{x_{Voltage}[n] - x_{Voltage}[n-1]}{Voltage_{max} - Voltage_{min}}\right)^2} \quad (2)$$

Table 1: Parameters at maximum brightness

Power of RF synthesizer	Gas Valve	RF Frequency	Intermediate electrode voltage
-12.9 dBm	5400 pulses	9.87 GHz	4.8 kV

Eventually, the parameters converge to obtain maximum brightness and the process ends after this. Table 1 shows the parameters at maximum brightness in that process. Table 2 shows the beam state using the set of parameters.

Table 2: Beam state at parameters with maximum brightness

Beam intensity	ε_x	ε_y	Brightness
0.138 mA	170 π mm mrad	180 π mm mrad	4.32×10^{-6} mA/(π mm mrad)^2

5. Discussion

The beam intensity was 0.14 mA when the two parameters of the RF synthesizer power and gas valve were manually adjusted. In addition, on another day, the emittance were $\varepsilon_x = 160\,\pi$ mm mrad and $\varepsilon_y = 190\,\pi$ mm mrad when only the intermediate electrode voltage was manually adjusted with a beam intensity of 0.1 mA. Although this is the result of optimizing beam intensity and emittance, individually the result obtained in this experiment is almost the same as the result of manual adjustment.

In previous research [5], the approximate adjustment was completed in about 30 steps, but since the number of parameters increased from 2 to 4 this time, it did not reach the end in 25 steps. On the other hand, finding the local minimum was successful, and there is a good chance that it can be adjusted more accurately by increasing the number of steps. In the coming future, it will be possible to increase the number of steps and conduct experiments that include multiple adjustments under the same conditions to verify the accuracy.

After this experiment, it was confirmed that the bellows of the RF tuner was damaged. We think that this is because trials made possible from automatic adjustment a huge number of compared to manually adjusting. So it is necessary to have a highly durable device and devise to prevent wear when using machine learning.

6. Conclusion

We performed an experiment of automatic adjustment that maximizes the brightness of the beam using machine learning. We automatically adjusted 4 parameters using Bayesian optimization and found a set of parameters that gives a beam brightness of 4.32×10^{-6} mA/(π mm mrad)^2 in 25 steps. This is about the same as the value from manual adjustment. From this result, Bayesian optimization is considered to be a useful method in ion source control.

References

[1] O Tarvainen, et al., 'Plasma diagnostic tools for ECR ion sources—What can we learn from these experiments for the next generation sources', Rev. Sci. Instrum. 90, 113321 (2019)
[2] T Itahashi, et al., 'Performance of the NANOGUN™ electron cyclotron resonance ion source applied for nuclear astrophysics', Rev. Sci. Instrum. 71, 1075 (2000)
[3] https://mad.web.cern.ch/mad/
[4] Y Morita et al., 'Developments of real time emittance monitors', Rev. Sci. Instrum. 91, 043303 (2020)
[5] Y Morita, et al., Proceedings of PASJ in 2021. WEP005.
[6] https://epics.anl.gov/index.php
[7] Lars Kai Hansen, Peter Salamon, 'Neural Network Ensembles ', IEEE TRANSACTIONS ON PATTERN ANALYSIS AND MACHINE INTELLIGENCE, VOL. 12, NO. IO, OCTOBER 1990
[8] M Pelikan, D E Goldberg, E C Paz, 'BOA : The Bayesian Optimization Algorithm', Proceedings of the 1999 Genetic and Evolutionary Computation Conference, (GECCO-99), Morgan Kaufmann, San Francisco, CA (1999)

Journal of Physics: Conference Series **2244** (2022) 012106 doi:10.1088/1742-6596/2244/1/012106

Suppression of damages on cathodes in the negative hydrogen ion source for the stable NBI system

Y Shimabukuro, J Hiratsuka, M Ichikawa, M Murayama, G Saquilayan, M Kisaki, A Kojima, H Tobari and M Kashiwagi

Naka Fusion Institute, National Institutes for Quantum Science and Technology, 801-1 Mukoyama, Naka, Ibaraki 311-0193, Japan

shimabukuro.yuji@qst.go.jp

Abstract. A protection system of ion source components from local abnormal discharge has been developed with ITER-class filament-driven negative ion source for high power and stable long-term operation. Protection from unpredictable discharge is a common issue for filament and RF-driven ion sources. To understand how fast should be cut off the discharge, threshold energy not to cause critical damage on the filament was investigated. The experimental exploration concluded that the energy less than 1.87 J or fast cut-off faster than 343 μs is the threshold for the robust protection of the filaments. A developed system successfully detected and cut off the discharge at 100 μs under noisy environments with the high current power supply. Moreover, the detection system is remotely adjustable. Such versatile protection system can contribute to any abnormal discharges in high current system including various ion sources, gyrotron, and high voltage components.

1. Introduction

Neutral beam injection (NBI) for ITER requires high-power (1 MeV, 40 A) and a long pulse (3600 s) negative ion beams on an annual order [1]. One of the issues for such negative ion sources is damage to the ion source components by unpredictable abnormal discharges. Such damages have been observed in various ion sources and have limited the performance of the ion sources, i.e., damage on filaments in filament-driven large ion source for JT-60SA [2], radio frequency (RF)-antenna in RF-driven ion source for the particle accelerators [3,4], the cathode of the Freeman and Bernas type ion sources for ion implantation [5,6], and high-power gyrotron for RF heating technology [7]. In particular, such abnormal discharges remarkably increase with cesium seeded operation to obtain the large negative ion current required for ITER. Thus, the development of a protection system from the abnormal discharge in the ion sources is a common issue for various ion sources widely used from industry, physics to nuclear fusion reactors.

The protection system from the abnormal discharge on the filament, so-called arcing, has been tested in filament-drive ion sources [2]. Initially, it was experimentally confirmed that impurities such as oxygen and Cs enhances arcing. Secondly, a fast cut-off system of arc current at arcing has been developed [8]. It worked effectively to expand the lifetime of the filament by a factor of three. However, faster cut-off time was limited due to various arcing and surrounding noise.

This paper reports one of the ideal systems to realize no damages on the filament using the actual large plasma volume simulating the ITER-class ion source. As the first step, a target cut-off time was decided by clarifying the threshold energy not to cause damage on the filament by arcing. Then, a high-speed and high-sensitivity protection system that can cut off within the target cut-off time, whenever and wherever arcing occurs, has been developed.

2. Experimental setup

A large volume negative ion source with a multi-cusp magnetic field was used in this experiment as shown in Fig.1 [8]. As the cathode, a total of 48 tungsten filaments composed of eight filament groups with 6 filaments are installed. The arc discharge is ignited between the chamber wall as the anode and these filaments using hydrogen gas at around 0.3 Pa. The typical arc power is around from 150 kW to 250 kW. Typical damages due to an arcing on the filament are shown in Fig.2. Several cathode spots and broken parts were observed. There are a lot of microprotrusions on the periphery of the spot as shown in the enlarged view taken by a Keyence VHX-S660 microscope.

In the NBI, this ion source is installed on the top of the accelerator at the high voltage side at 500-1000 kV. Therefore, arc and filament power supplies are also installed on the same high voltage side. Conventionally, the arc current measured by the current transducer (CT) located in the arc power supply is sent to the control system via optical fiber. Over-current (OC) was judged in the control system, and a signal to cut off the arc current was returned to the arc power supply. Thus, more than 1 ms was taken to cut off the arcing in the original system.

The fast cut-off system of the arcing has been developed by introducing a field-programmable gate array (FPGA) on the high voltage side [8]. In this test, the National Instruments cRIO-9053 FPGA was selected. FPGA can be remotely programmed through a network with optical fiber from the control system, and FPGA can judge the OC and cut off the gate driver signal to turn off the insulated gate bipolar transistor (IGBT) according to this program. A detection program in FPGA judges the arcing in every several μs.

3. Two issues on arcing protection

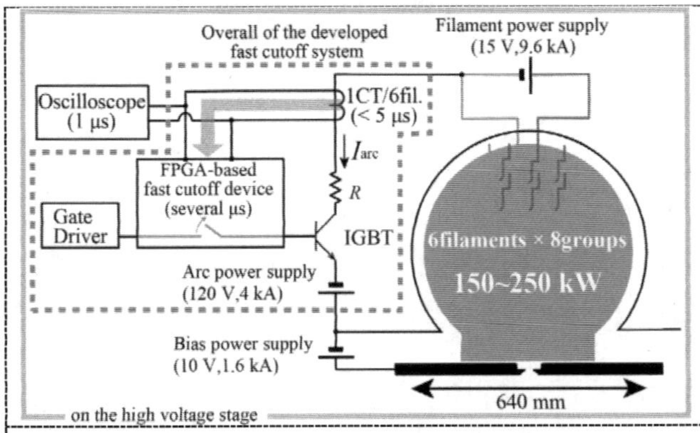

Figure 1. A schematic view of the experimental setup.

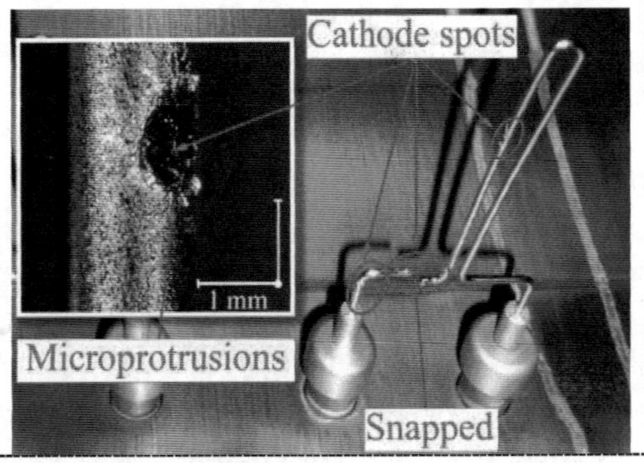

Figure 2. A photograph of the snapped tungsten filament with 1.8 mm in diameter.

Figure 3 shows the cumulative number of arcing with respect to the shot numbers. The arcing occurrence frequency is remarkably increased when the arcing occurs even once. The result implies that the previous cathode spot may be the trigger of the next arcing. Microprotrusions around the spot as shown in Fig.2 may induce the local heating due to its stronger electric field concentration [9,10]. The first issue is that the threshold arcing energy not to create a spot even once should be clarified experimentally to quantify the target energy and the cut-off time.

Figure 4 show the illustration of the typical arc current from the beginning of plasma to the stable plasma. Besides arcing occurred unpredictively, there are constant noise and intermittent noise which disturb the high-speed detection of the arcing. In the conventional system, the OC was detected by the constant OC level. The OC level is too large for the arcing during the beginning of plasma. On the

Journal of Physics: Conference Series **2244** (2022) 012106 doi:10.1088/1742-6596/2244/1/012106

contrary, the marginal OC level detects the noise, and frequently stopped the operation. The second issue is to detect the arcing properly and quickly according to various types of arcing under noisy environments even in high current power supply.

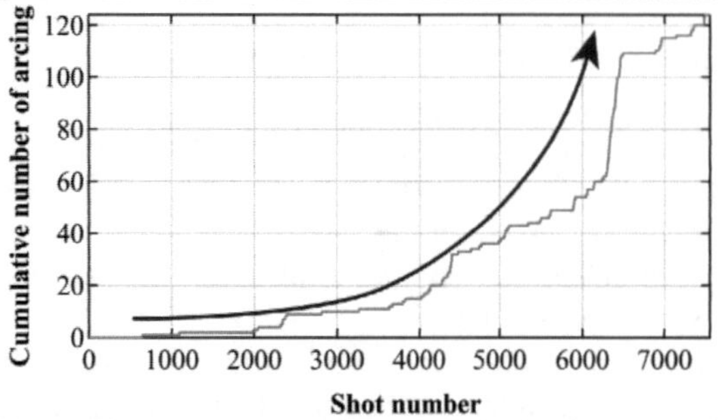

Figure 3. A cumulative number of arcing regarding shot numbers

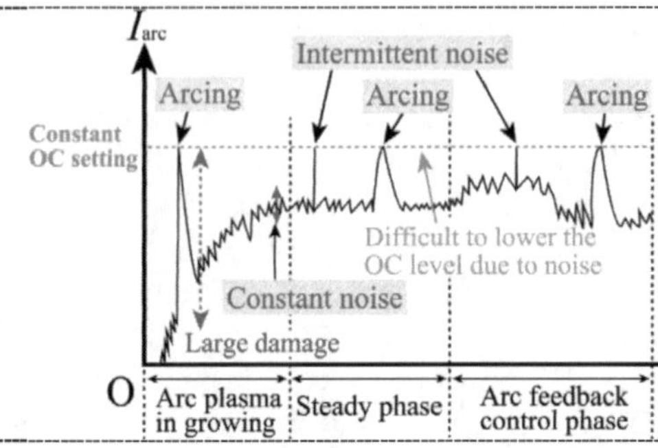

Figure 4. A schematic illustration of typical arc current behaviour for several operating phases.

4. Threshold of input energy not to create cathode spot

The relationship between the arcing energy and cathode spot size has been investigated to solve the first issue, that is, clarify the threshold arcing energy not to create a cathode spot. The diameter of the cathode spot was measured by the microscope and the arcing energy was calculated by the arc voltage and current at the arcing.

Figure 5 shows the typical growing of the arc current with the arc voltage of 120 V. The operation was stopped by the arcing around 1800 A. Figure 6 shows the expansion view which corresponds to arcing of Fig.5 by the oscilloscope, where the vertical axis indicates the net arcing current which is calculated as the constant current subtracted from total current and the horizontal axis corresponds the elapsed time since arcing occurs in the unit of μs.

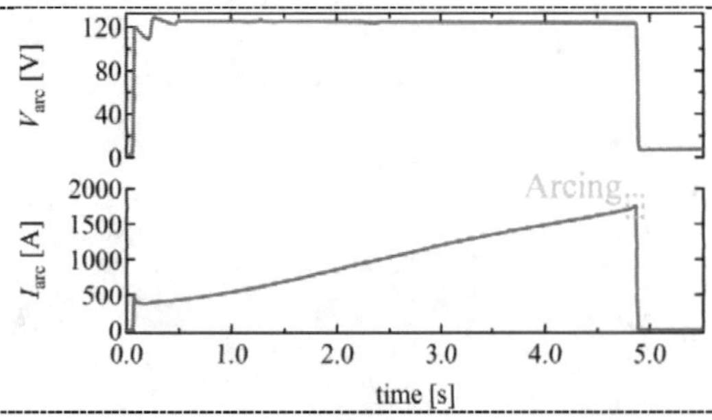

Figure 5. A typical waveform of arc voltage and arc current.

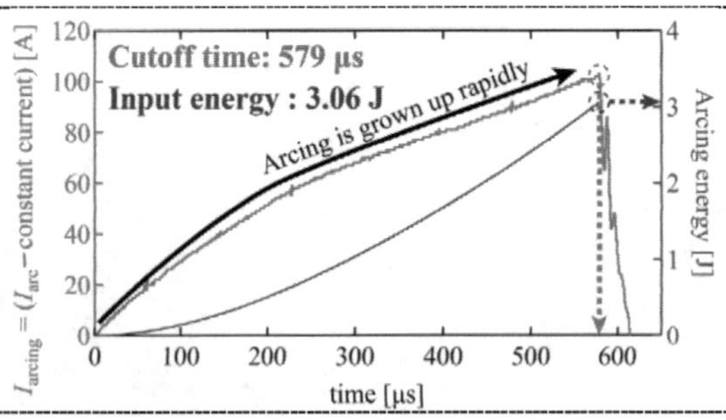

Figure 6. The net arcing current and energy flowing into the filament with respect to the elapsed time since arcing occurs.

The energy flowing into the filament at arcing is calculated by the following equation as the summation of the product of arc voltage, net arcing current, and time in each sampling.

$$\Delta P\ [\mathrm{J}] \sim I_{\text{arcing}}(t)[\mathrm{A}] \times V_{\text{arc}}(t)[\mathrm{V}] \times \Delta t[\mathrm{s}], \Delta t = 1\ \mu\mathrm{s}$$

In Fig.6, the arcing was cut at 579 μs with the energy input of 3.06 J.

Figure 7 shows the investigation result of cathode spot diameter with respect to the energy flowed into the corresponding filament at arcing. Although the arcing history was recorded, no filaments that had a visible cathode spot for the energy input lower than 1.87 J. Meanwhile, some filaments with an energy input of more than 2.23 J had a cathode spot and microprotrusions around there. Thus, the threshold energy not to create a cathode spot on the filament at arcing has been revealed experimentally as 1.87 J.

To clarify the target cut-off time for filament protection, measured arcing energy until the cut-off was plotted for all 29 arcing in one experimental campaign of 2 months in Fig. 8. The cut-off time corresponding to 1.87 J was 343 μs in the fastest arcing.

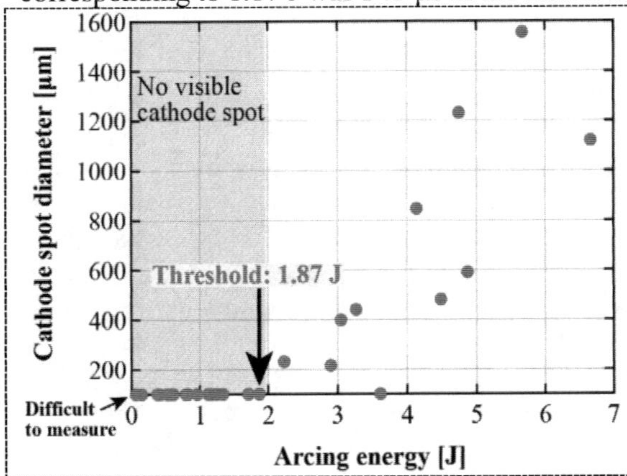

Figure 7. Cathode spot size for arcing energy.

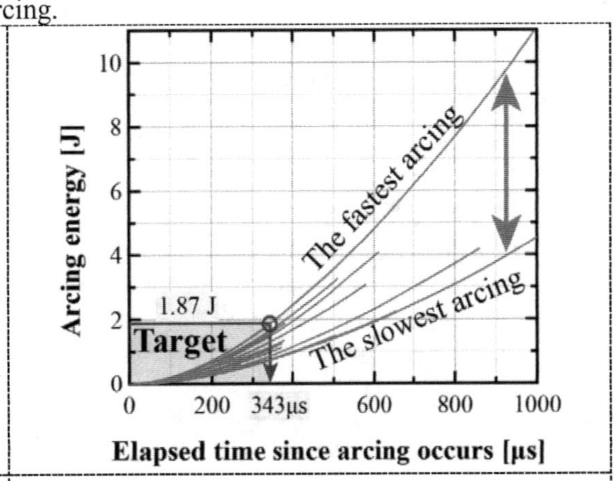

Figure 8. The growing of the arcing since occurrence for various arcing.

5. High sensitivity detection of arcing

The protection system has been improved to solve the second issue, that is, to detect the arcing sensitively and quickly according to various types of arcing under noisy environments. Figure 9 shows the procedure to improve the arcing detection step by step. Firstly, the constant noise was removed by applying a digital low path filter (LPF) in the FPGA. This LPF can be adjusted remotely *in-situ* of the operation. However, the intermittent noise cannot be removed sufficiently.

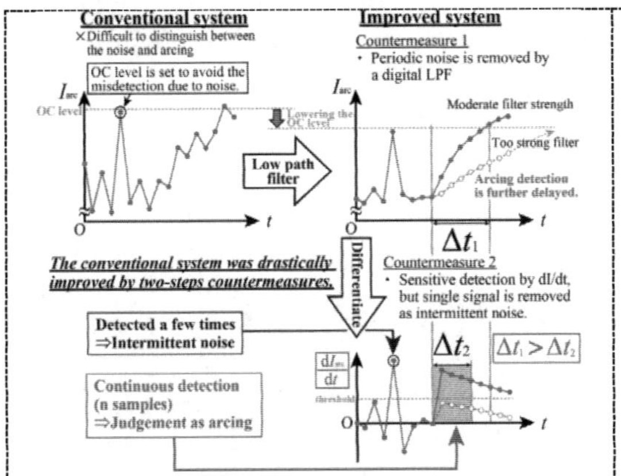

Figure 9. The system improvement flow of the fast cut-off logic.

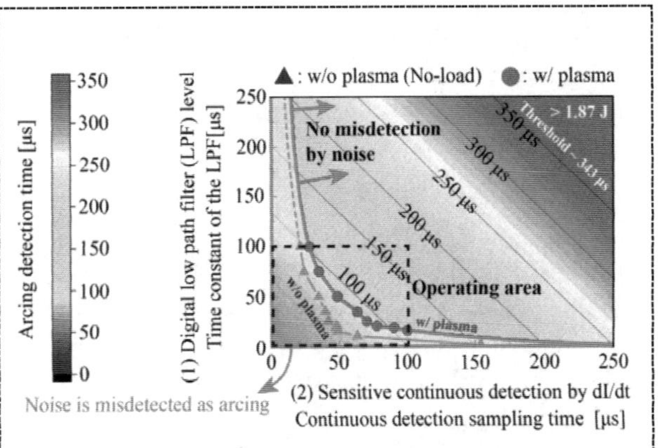

Figure 10. The total arcing detection time obtained from the time constant of digital LPF and the continuous detection sampling time of dI/dt.

In order to catch a small variation of arc current at the arcing sensitively, the time differential of the arc current (dI/dt) was applied instead of the conventional constant OC. This has another advantage to distinguish the noise from the arcing. The dI/dt will be the positive value both at the arcing and the intermittent noise. However, the increase of the dI/dt is intermittent at the intermittent noise and is continued at the arcing. Hence, the conventional current-based logic takes the time Δt_1 from occurrence to detection of the grown arcing current, while the dI/dt-based logic combining with the counting system requires only a few samples of time Δt_2. The new system is obviously faster and more sensitive than the conventional one as $\Delta t_1 > \Delta t_2$.

The protection system must be working as fast as possible. The required cut-off time depends upon the LPF strength and the sampling number for the dI/dt detection. The optimization process of the arcing detection has been established by adjusting those two parameters in the actual ion source operation as shown in Fig.10.

Firstly, the adjustment of the protection system was performed with only the operation of each power supply without plasma. When the power supply is energized, constant noise and intermittent noise were observed. The stable operation level was checked by changing the LPF and the sampling number, respectively. Secondly, the adjustment was performed with plasma as shown by the red circles in Fig.10. The boundary for the stable operation is slightly moved to the slower side under the high-power plasma existence situation. From these results, the actual parameters with plasma can be roughly predicted from the parameters without plasma. Because the high-power plasma usually requires a long conditioning time and risk of arcing, this prediction of the parameter before actual operation can contribute the risk mitigation.

Finally, Figure 10 shows that the arcing is cut off stably around 100 μs, which is the one-third of the target cut-off time of 343 μs, which required not to create a cathode spot even once. The arcing energy at 100 μs is only 0.2 J estimated in Fig.8, which is one-ninth of the threshold arcing energy.

6. Conclusion

The protection system of the ion source component from unpredictable abnormal discharge has been successfully developed with the filament-driven large negative ion source. The threshold energy not to cause the visible damage on the filament at the arcing was experimentally clarified as 1.87 J for the first time. The cut-off time to satisfy this threshold energy was 343 μs. In order to cut-off the arcing within this duration stably, the high-speed and high-sensitive arcing detection system has been established introducing the dI/dt instead of the conventional constant OC. The sensitivity was improved by combining the LPF and continuous detection of the dI/dt to distinguish from the intermittent noise. Finally, the arcing could be cut-off at 100 μs corresponding to only 0.2 J in the energy, which is one-ninth of the threshold energy. The reliable system can be tuned by the parameters remotely according to the noise of each system and be worked *in-situ* in a high-voltage environment. The protection settings are completed before the occurrence of abnormal discharge because the optimized parameters are roughly predictable by pre-adjustment under the no-load condition. The versatile system and procedure have wide applicability not limited to filament and RF antenna of high-power NBI system.

References

[1] Hemsworth R S, Boilson D, Blatchford P, Palma M D, Chitarin G, de Esch H P L, Geil F, Dremel M, Graceffa J and Marcuzzi D 2017 *New J. Phys.* **19** 025005

[2] Ichikawa M, Yoshida M, Kojima A, Hanada M, Kashiwagi M, Watanabe K, Umeda N, Tobari H, Nishikiori R, Hiratsuka J and NB Heating Technology Group 2016 *Plasma Fusion Res.* **11** 2405108

[3] Leung K N, DeVries G J, Kunkel W B, Perkins L T, Pickard D S, Saadatmand K, Wengrow A B and Wiliams M D 1996 *5th European Particle Accelerator Conference* **e-proc** 1513

[4] Welton RF, Dudnikov V G, Han B X, Murray S N, Pennisi T R, Pillar C, Santana M, Stockli M P and Turvey M W 2014 *Rev. Sci. Instrum.* B 02B135

[5] Pelletier J and Anders A 2005 *IEEE Trans. Plasma Sci.* **33** 1944

[6] Okano Y, Miyamoto N, Kasuya T and Wada M 2015 *AIP Conf. Proc.* **1655** 070001

[7] Brand P and Mueller G A 2003 *Fusion Eng. Des.* **66** 573

[8] Ichikawa M, Kojima A, Hiratsuka J, Yoshida M, Umeda N, Saquilayan G M, Watanabe K, Tobari H and Kashiwagi M 2020 *Rev. Sci. Instrum.* **91** 023502

[9] Robson A E and Thonemann P C 1958 *Proc. Phys. Soc.* **73** 508

[10] Barengolts S A, Mesyats G A and Tsventoukh M M 2012 *25th Internatonal Symposium on Discharges and Electrical Insulation in Vacuum (ISDEIV)* 376

Dynamics and reactivity of thermalized ions in a radiofrequency quadrupole gas-reaction cell used for the production of radioactive molecular ions

J-F Alary[1], L M Cousins[1], R Javahery[1], W E Kieser[2], E L Flannigan[2] and C Charles[3]

[1]Isobarex Corp., Vaughan, ON, L6A 2Y9, Canada
[2]Andre E. Lalonde AMS Laboratory, University of Ottawa, Ottawa, ON, K1N 6N5, Canada
[3]TRIUMF, Vancouver, BC, V6T 2A3, Canada

alaryjf@isobarex.ca

Abstract. Two-dimensional radiofrequency quadrupole (RFQ) ion guides are versatile tools that are used over a wide range of ion energies. Properly tuned, they can provide an ideal environment for ion-molecule chemistry to proceed by allowing thermalized ions to react on-line with gaseous reactants. We exploit this capability in the Ion Reaction Cell (IRC), an RFQ-based system designed to form radioactive molecules (RM) from radioactive ion beams (RIB). Functionally, the IRC accepts RIB produced upstream by the isotope separation on-line (ISOL) method, decelerates it to eV energies for production of the RM and then re-accelerates RM to their incident energy. This system promises to form a highly versatile source of complex inorganic or organic RM that can be used in fundamental symmetry studies or as precursors to new radiopharmaceuticals. We discuss the challenges of transferring ions from a RIB into a gas cell at room temperature while constraining ions within the RFQ energy well, and of reforming a beam from synthesized RM. Various factors influencing ion dynamics and reactivity in thermal and non-thermal zones inside the IRC are considered, using results obtained recently with the Isobar Separator for Anions, an analogous RFQ technique used in Accelerator Mass Spectrometry, to illustrate key points.

1. Introduction

A large variety of chemical reactions between ions and neutral molecules can be accomplished in a pressurized radiofrequency quadrupole (RFQ) ion guide [1-4]. However, these reactions are highly dependent on the translational energy of ions when they interact with neutrals. This factor becomes important when ions are injected at relatively high velocity (> 1 keV) into the RFQ ion guide. As typical RIBs are accelerated to energy up to 60 keV at TRIUMF [5, 6], the Langevin model of reactional cross-section [7] predicts that only a vanishingly small reaction rate between neutrals and ions will be achieved. To obtain efficient chemical conversion, near thermal conditions must be achieved by combining electrostatic deceleration of ions, followed by collisions with a reactive or non-reactive gas near room temperature (~300 K). Approaches such as mounting the pressurized RFQ ion guide on a high voltage platform matching the energy of the incoming beam combined with the use of focusing and stopping electrodes, have successfully been applied in Accelerator Mass Spectrometry (AMS) to partially thermalize ions [1-4]. This approach enables the selective suppression of atomic isobars from several radionuclides of interest via highly specific ion-molecule reactions. We will review recent results obtained with a similar RFQ-based system installed on a 3 MV AMS at the A. E. Lalonde AMS laboratory at University of Ottawa [8]. This system is used to selectively react S^- with gaseous NO_2 at a pressure of about 10 μbar in the ion guide, in the determination of the radionuclide ^{36}Cl at low isotopic ratios (10^{-15}) in sulfur containing samples [1]. This reaction, exothermic by 0.2 eV [9], is favoured, while Cl^- is transmitted intact as the electron transfer reaction involving the chlorine anion is endothermic by 1.34 eV. A relative suppression ratio of 10^6 for sulfur can be achieved, implying a near complete electron transfer to NO_2 from S^-. This system offers an excellent platform to explore the dynamic and reactivity of ions in an RFQ-based system that can be used to form radioactive molecules (RM) directly from a positive RIB.

2. Experimental

2.1. RFQ ion guide

A schematic of the RFQ ion guide as well as its decelerating and accelerating electrostatic electrodes are shown in figure 1. Ions exiting the RFQ ion guide are reaccelerated toward the main tandem accelerator of the AMS system. The system uses a segmented RFQ configuration in which a specific offset voltage (± 240 V) can be applied to individual segments. Einzel lenses focus the ion beam in the deceleration and acceleration stages. It uses deceleration and acceleration cone electrodes with a 6 mm and 4 mm i.d. aperture at their center, respectively. The Einzel lenses are referenced to lab ground while the cone electrodes and RFQ ion guide are controlled from the high voltage platform.

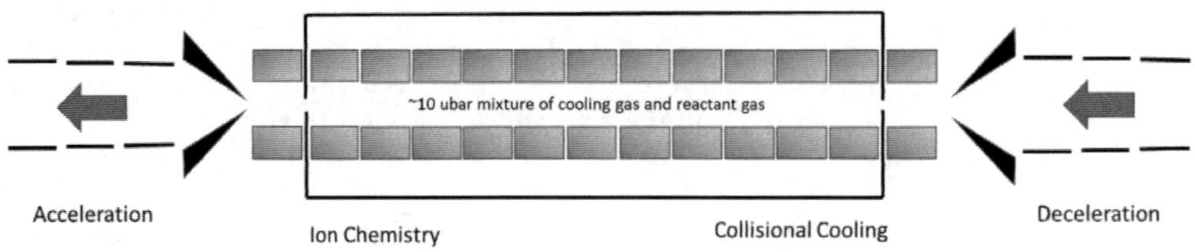

Figure 1. Schematic of the segmented RFQ ion guide with pressurized cell and deceleration and acceleration stages. Ions enter the ion guide from the right.

The pressurized section covers 11 RFQ segments and has a 6 mm i.d. aperture at the entrance and a 4 mm i.d. aperture at the exit. The cell is filled with a mixture of helium and nitrogen dioxide optimized to achieve best transmission for chlorine anions (~30%) and near complete suppression of sulfur anions. The system is differentially pumped to ensure a vacuum of 10^{-6} mbar or less in the Einzel lens sections and about 5 x 10^{-5} mbar in the central section surrounding the pressurized cell. The system is integrated into the injection beam line of the AMS system after the low energy mass selection magnet [1-4, 8]. The dimensionless Mathieu parameters q and a describe the coupling of ions with the quadrupolar RF field generated inside the ion guide [10]. The system can maintain a value of q of 0.4, typical for RFQ ion guides, up to a singly charged ion mass of 370 amu. Since we do not perform mass filtering, no potential is applied between opposite pairs of pole and thus a is set to zero

The main parameters of the RFQ ion guide and Einzel lenses are shown in table 1.

Table 1. Main parameters of the RFQ ion guide and Einzel lenses.

Parameter	Value
RF Frequency	2.5 MHz
RF Amplitude	Up to 7.5 kV_{pp}
RFQ Inscribed Diameter (r_0)	6.9 mm
Segment Length	38 mm
RFQ Path Length	494 mm
Gas Cell Length	418 mm
Beam Energy (in/out)	20 keV
Einzel Lenses i.d.	60 mm

2.2. Chemical processes in a pressurized RFQ ion guide

Chemical processes that have been studied in a Selected-Ion Flow Tube (SIFT), a low energy system similar to a pressurized RFQ ion guide, are listed in Table 2 [11].

Table 2. Chemical reactions reported in a SIFT apparatus.

Reaction	Classification
$A^+ + B \rightarrow B^+ + A$	Charge Transfer
$AC^- + B \rightarrow A^- + C + B$	Ion Dissociation
$A^- + B \rightarrow B^- + A$	Electron Transfer
$AH^+ + B \rightarrow BH^+ + A$	Proton Transfer
$A^+ + B \rightarrow [A.B]^+$	Adduct Formation
$A^+ + BC \rightarrow AB^+ + C$	Cluster Reaction
$AC^+ + BDE \rightarrow$ Products	Allowed Chemistry

For specific isobar suppression in AMS, electron transfer and ion dissociation reactions are used. The reaction of S^- with NO_2, reported here, is an electron transfer reaction with a favourable energy balance, as mentioned above. The on-line synthesis of RM from RIB will be realized via reaction classes that may include cations such as adduct formation, cluster reaction and allowed chemistry.

3. Results and Discussion

3.1. Energy profile in the RFQ ion guide

In figure 2, we present ion energy profiles obtained by first optimizing all voltages applied to the RFQ ion guide for maximum transmission of chlorine anion, then decreasing the voltage on individual segment or aperture until ion transmissions is halted. The results are a series of curves providing the energy profile of ions at various stages of the ion guide, as they are cooled by helium (~12 μbar.)

a)

b)

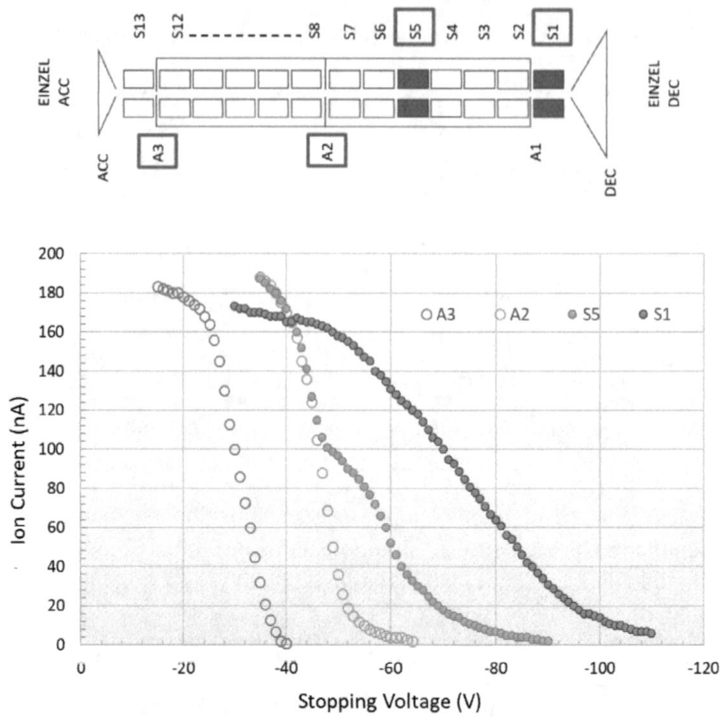

Figure 2. Stopping potential required to halt the transmission of chlorine anions for selected electrodes of the RFQ ion guide: a) schematic of the RFQ ion guide where RFQ segments S1 and S5 and middle and exit apertures A2 (12 mm i.d.) and A3 are highlighted; b) stopping transmission curves obtained by decreasing the voltage applied to the electrodes highlighted in a).

The stopping curves in figure 2 are showing a gradual thermalization of anions caused by successive collisions with helium atoms (about 50 collisions along the length of the RFQ ion guide at the pressure used.) The energy spread is also narrowing. The curve observed at the exit aperture A3 is getting closer to a typical Maxwell-Boltzmann energy distribution. Stopping voltages at 50% and 10% ion transmission on each of these curves, plus on curves generated from aperture A1 and segments S2, S12 and S13 (not shown on figure 2), were used to calculate the energy of collision of $^{37}Cl^-$ ions with helium at center of mass *Ecm*:

$$E_{cm} = E_{lab} \frac{m_n}{(m_n+m_i)} \quad (1)$$

Where E_{lab} is the energy of ions (stopping voltage) at 50% and 10% transmission, and m_n and m_i are the mass of the neutral (He) and colliding ion ($^{37}Cl^-$), respectively. The resulting E_{cm} values are shown on figure 3 in relation to the distance from the tip of the deceleration electrode.

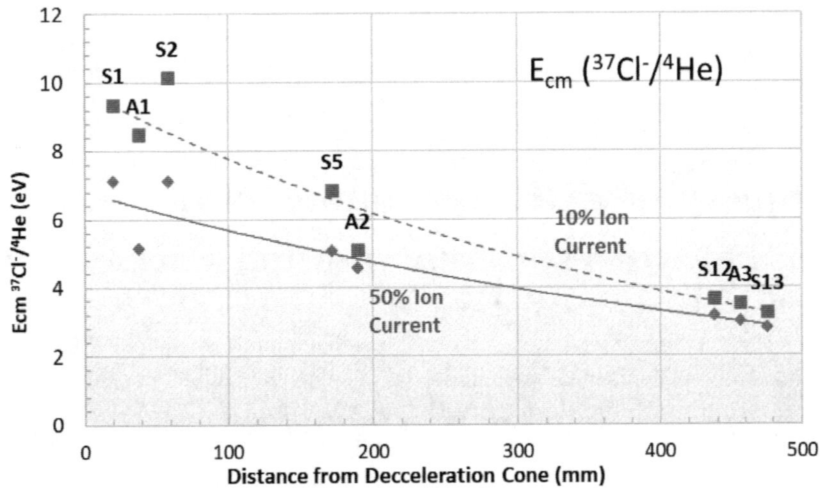

Figure 3. Energy of collision at center of mass (*Ecm*) between ^{4}He and ^{37}Cl$^-$, calculated from the stopping voltages required to achieve 50% and 10% ion transmission at segment S1, S2, S5, S12 and S13, and aperture A1, A2 and A3, in relation to distance from the deceleration electrode.

On figure 3, the E_{cm} for the 10% most energetic collisions decrease from about 10 eV in the first RFQ segment, to 2.6 eV in the last segment. The median E_{cm} (at 50% transmission) decreases from 6.6 eV to 2.4 eV. This graph provides a measure of the ion cooling occurring along the ion guide.

4. Conclusion

A gas filled segmented RFQ ion guide is a highly versatile tool achieving efficient cooling of ions and reaction with a neutral reactant. Future improvements will aim at lowering E_{cm} further along the ion guide. Achieving E_{cm} below 1 eV would enhance both ion transmission and reactivity with neutrals.

References

[1] Eliades J, Litherland A E, Kieser W E, Cousins L, Ye S J and Zhao X-L 2010 *Nuclear Instruments and Methods in Physics Research* B **268** 839-842
[2] Charles C R J, Cornett R J, Zhao X-L, Litherland A E and Kieser W E 2015 *Nuclear Instruments and Methods in Physics Research* B **361** 189–192
[3] Alary J-F, Javahery G, Kieser W, Zhao X-L, Litherland A 2015 *Nuclear Instruments and Methods in Physics Research* B **361** 197–200
[4] MacDonald C, Charles C R J, Cornett R J, Zhao X-L, Kieser W E and Litherland A E 2015 *Rapid Commun. Mass Spectrom.* **29** 115–118
[5] Merminga L *et al* 2011 *Proceedings of IPAC 2011* San Sebastian, Spain 1917-1919
[6] Dilling J, Krucken R, Merminga L 2013 *ARIEL overview* In: Dilling J., Krucken R., Merminga L. (eds) *ISAC and ARIEL: The TRIUMF Radioactive Beam Facilities and the Scientific Program* Springer, Dordrecht
[7] Eichelberger B R, Snow T P and Bierbaum V M 2003 *J Am Soc Mass Spectrom* **14** 501–505
[8] Kieser W E, Zhao X-L, Clark I D, Cornett R J, Litherland A E, Klein M, Mous D J W, Alary J-F 2015 *Nuclear Instruments and Methods in Physics Research* B **361** 110-114
[9] Dunkin D B, Fehsenfeld F C and Ferguson E E, 1972 *Chemical Physics Letters* **15** 257-259
[10] Gerlich D, *Inhomogeneous RF fields: A versatile tool for the study of processes with slow ions*, in *State Selected and State to State Ion Molecule Reaction Dynamics, Part 1: Experiment*, Volume 82 Cheuk- Yiu Ng and Michael Baer (eds) 1992 John Wiley & Sons, Inc.
[11] Bierbaum V M, *Instrumentation: Flow Tubes* In: *Encyclopedia of Mass Spectrometry*, 1, 98-109 P.B. Armentrout (ed), 2003 Elsevier

Development of ECR ion source with high-temperature superconducting REBCO coils

T H Chong[1], M Fukuda[1], T Yorita[1], H Kanda[1], Y Yasuda[1], H W Koay[1], Y Morita[1], K Takeda[1], T Hara[1], M Hisamatsu[1], H Zhao[1], A Ishiyama [2], S Noguchi[3], H Ueda[4], S Fukui[5] Y Matsubara[6], Y Mikami[6] N Takahashi[6], J Yoshida[6], T Hirayama[6], S Nagaya[7] and T Watanabe[7]

[1] Research Center for Nuclear Physics(RCNP), Osaka University, Osaka 567-0047, Japan
[2] Waseda University, Toyko 169-8555, Japan
[3] Hokkaido University, Sapporo Hokkaido 060-0808, Japan
[4] Okayama University, Okayama 700-8530, Japan
[5] Niigata University, Niigata 950-2181, Japan
[6] Technology Research Center, Sumitomo Heavy Industries Ltd., Kanagawa 237-8555, Japan
[7] Chubu Electric Power, Aichi 459-8522, Japan

E-mail: oscar@rcnp.osaka-u.ac.jp

Abstract. A new High-Temperature Superconducting ECR (HTS-ECR) ion source is under development in Research Center for Nuclear Physics (RCNP), Osaka University. This ion source will be used for production of high intensity proton, deuteron and He^{2+} ion beams. The HTS-ECR magnets are composed of three solenoid coils and a set of sextupole coils made of REBCO tapes, a high-temperature superconductor. The HTS-ECR ion source is designed to operate at frequency of 2.45 GHz and 10 GHz. Performance test of the HTS coils had been carried out at 31 K and 77 K. This HTS coil technology will be applied to development of a meter-size HTS coil system of a high intensity compact AVF cyclotron. This paper introduces the basic design of the HTS-ECR ion source. The performance test results showed that REBCO solenoids remain superconducting state with a current up to 400 A. Simulation results of the magnetic field and electromagnetic field distributions in a plasma chamber fulfilled the requirements of electron cyclotron resonance conditions at 2.45 GHz and 10 GHz. Simulation result of mirror ratios and electromagnetic field amplitudes are also presented in this paper.

1. Introduction

A new ECR ion source using REBCO(Rare-Earth Barium Copper Oxide, $GdBa_2Cu_3O_{7-x}$) tape, a second generation high temperature superconductor, is under development in Research Center for Nuclear Physics(RCNP), Osaka University. A set of mirror coils and sextupole coils using a superconducting tape enables operators to optimize the magnetic field distribution for increasing a beam current. While low temperature superconductor coil required operation temperature lower than 4 K, high temperature supercondutor (HTS) coil has critical temperature of around 90 K, and typical operation temperature is at 30 K. This makes the ECR ion source with HTS coils more reliable due to a sufficient temperature margin against a high critical temperature. Furthermore, while critical current in HTS coil is limited by the external magnetic field, REBCO tapes have the advantage of a higher critical current density than BISCCO tapes under a high magnetic field[1, 2]. This makes REBCO tapes a promising material for any type of coils in an ion source and accelerator. However, applications of REBCO tapes may be strictly limited when the HTS coils are subject to severe mechanical stress in operation. In RCNP, we constructed new solenoid coils and sextupole coils using REBCO tapes and carried out perfromance tests under 77 K, in order to test their reproducibility after reaching critical current a several times. We have also done performance tests under 31 K, in order to test the reliability and capability of REBCO coils as superconducting magnets for an ECR ion source. In addition, we carried out magnetic and electric field simulation to examine the capability of HTS-ECR as an ion source.

2. REBCO coils

The HTS-ECR magnet consists of a set of solenoids and sextupole coils, which was designed to produce the magnetic field necessary for electron cyclotron resonance and electron confinement. Figure 1 shows the REBCO coil configuration in HTS-ECR. Starting from the injection end, there will be three solenoid coils, namely M1 coil, PC coil and M2 coil. An inner diameter of the three solenoid coils is 190 mm, and the sextupole coils are placed inside of the solenoid coil at the radial position of 90 mm. The width of the REBCO($GdBa_2Cu_3O_{7-x}$) tapes was 12 mm for the solenoid coils and 6 mm for the sextupole coils. The M1, PC and M2 coils comprises 4 pancakes with 106 turns, 2 pancakes with 103 turns and 2 pancakes with 68 turns, respectively. The sextupole coils of racetrack type have 2 pancakes with 30 turns.

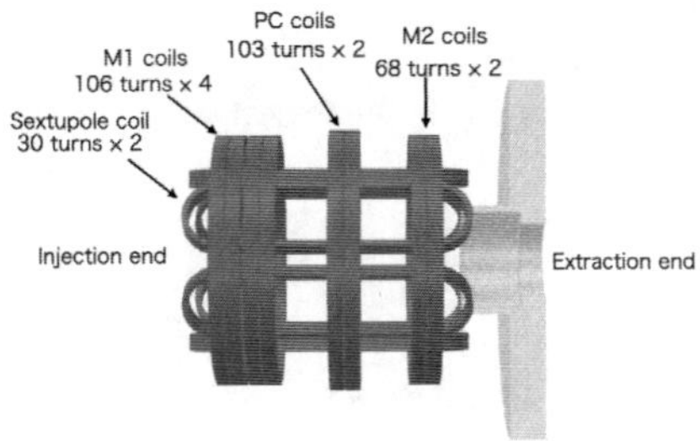

Figure 1. REBCO solenoids and sextupole coils in HTS-ECR.

The REBCO tapes we used are products from SuperOx Japan. The tapes have a multi-layer structure consisting of Hastelloy with oxide coatings, silver coating, and both side copper plating. The full thickness of the REBCO tapes is approximately 0.1 mm. The superconducting layer is 1 ~ 3 μm in thickness. Solenoids and sextupole coils using this REBCO tapes have been constructed in collaboration with Chubu Electric Power Co. Inc.. The individually fabricated solenoid and sextupole coils are shown in Fig. 2. The main parameters of the solenoid and sextupole coils are listed in Table 1.

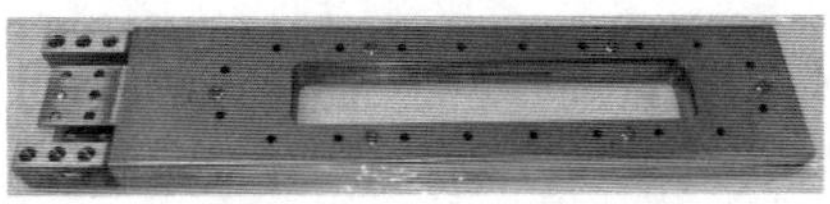

Figure 2. REBCO solenoid (left) and sextupole (right).

Journal of Physics: Conference Series **2244** (2022) 012108 doi:10.1088/1742-6596/2244/1/012108

Table 1. Characteristics of HTS-ECR coils.

Characteristics	Solenoid	Sextupole coil
REBCO tape Width	12 mm	6 mm
Critical current, I_c	>500 A(30 K, self field)	>250 A(30 K, self field)
Inner diameter/Length	~190 mm	180 mm
Minimum radius of curvature	95 mm	25 mm
Winding insulation	Metal insulation	No insulation

3. Performance test results

To investigate stability and reliability of the REBCO coils we carried out performance tests multiple times. We first carried out tests at 77K (liquid nitrogen temperature) for it can be easily done without a cryostat. We increased the current in the coils until the voltage increase rapidly, which intimated a critical current and normal conducting state transition. The test was done multiple times at RCNP and coil factory. Figure 3 shows the test results of I-V characteristics for one of the sextupole coils. A coil current was supplied from a DC power supply (KEYSIGHT 6683A). The voltage between two electrodes for coil current input and output was measured with a digital multimeter (KEYSIGHT 34470A). Four line represents two results from RCNP and two results from the coil factory. Here we found that all the test refults show a similar I-V charactreistics, implying that the critical current was not changed by the multiple excitations and repeated heat cycle between a room temperature and 77 K. Solenoid coils and the other sextupole coils share a similar test result and showed well reproducibility.

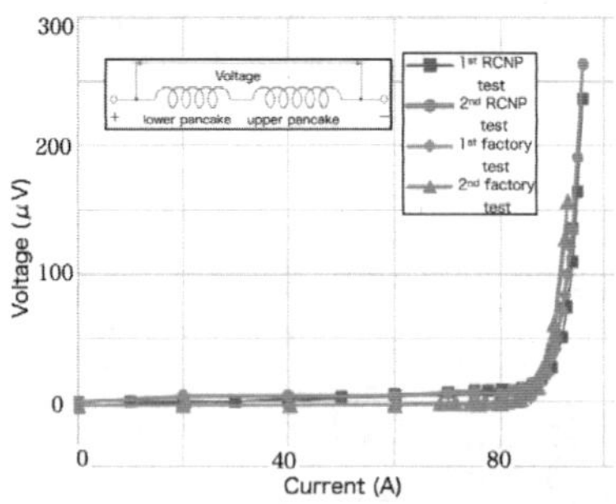

Figure 3. I-V characterstic of one of the sextupole in 77 K.

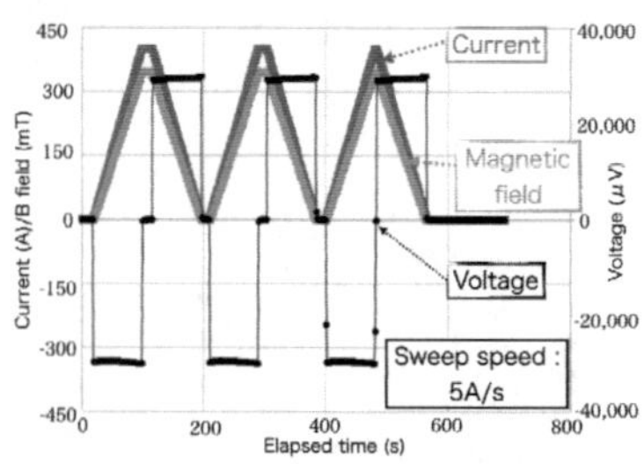

Figure 4. Magnetic field induction of one of the solenoid in 31 K.

In order to investigate the characteristics of the HTS solenoid coil in low temperature less than 40 K, we carried out second performance test at 31 K. The M2 coil was placed inside cryostat of a single-coil test bench and cooled down to 31 K by a GM cryocooler. While varying the current at a sweep speed of 5 A/s, we recorded the magnetic field at the center of the coil with a Hall sensor and a voltage over the coil. Figure 4 shows the test result of M2 solenoid coil at 31 K. The coil current was limited to 400 A because of the rated voltage of a power supply used for this test bench. Still, here we found that M2 coil is capable of inducing axial magnetic field up to 345 mT stably. Futhermore, the test results showed that M2 solenoid has a self inductance of 0.006 H.

A self inductance of a coil can be approximately obtained through Eq. 1 [3].

$$L = K_{\text{nagaoka}} \times \frac{\mu_0 \times \pi \times r^2 \times N^2}{l} \tag{1}$$

Where μ_0 is the permeability of vacuum, r is the raidus of coil, N is the number of turns, l is the axial length of the coil and K_{nagaoka} stands for Nagaoka coefficient[4], a function depanding on r and l. By using Eq. 1, we expected the self inductance of M2 coil to be approximately 0.0068 H. Since the test result gave a close self inductance, we assume that there was no voltage increase other than self induction. The test result intimating that no normal conduction state transition occured through out the iterative performance test below 400 A at 31 K.

The other solenoids and sextupole coils showed similar test results. These results suggest that ECR ion source with REBCO solenoids is capable of operation with high reliability. We intend to carry out the test with longer elapsed time and higher maximum current in the future.

4. Magnetic field configuration of HTS-ECR

To investigate the capability of the REBCO coils of inducing magnetic field required for an ECR ion source, we calculated the magnetic field configuration inside HTS-ECR using OPERA-3D. The magnetic field distribution in the axial direction is shown in Fig. 5. In Fig. 5, two dot lines represent B_{ECR}, the magnetic field needed for electron cyclotron resonances at 2.45 GHz and 10 GHz operation. In 10 GHz mode, we construct a min-B configuration for stable plasma confinement[5] by applying 500 A in each solenoids and 250 A in each sextupole coils. The axial magnetic field is shown in Fig. 5 represented by the blue line. In min-B configuration, by comparing B_{ECR} with the axial field at injection end B_{inj}, extraction end B_{ext}, minimum axial field B_{min} and radial field at chamber wall B_{rad}, the mirror ratio would be as below.

$$\frac{B_{\text{inj}}}{B_{\text{ECR}}} \sim 3, \quad \frac{B_{\text{ext}}}{B_{\text{ECR}}} \sim 1.2, \quad \frac{B_{\text{min}}}{B_{\text{ECR}}} \sim 0.65, \quad \frac{B_{\text{rad}}}{B_{\text{ECR}}} \sim 2 \tag{2}$$

On the other hand, the optimum mirror ratio for heavy ion production has been found to be as below[6].

$$\frac{B_{\text{inj}}}{B_{\text{ECR}}} \sim 4, \quad \frac{B_{\text{ext}}}{B_{\text{ECR}}} \sim 2, \quad \frac{B_{\text{min}}}{B_{\text{ECR}}} \sim 0.8, \quad \frac{B_{\text{rad}}}{B_{\text{ECR}}} > 2 \tag{3}$$

Since one of the main application of the ion source is to produce high intensity He^{2+} beam, it is necessary to find out optimum magnetic field configuration for increasing the He^{2+} beam intensity during the operation in the near future. Study regarding to this topic will be done in the future.

For operation at 2.45 GHz, Fig. 5 shows an almost flat magnetic field distribution, instead of mirror field, represented by the lower flat line. We tend to deliver performance test on the HTS-ECR in the future, to determine whether mirror field or flat axial magnetic field is more effective in production of an intense high quality beam of proton and He^{2+}, considering an extraction system as well.

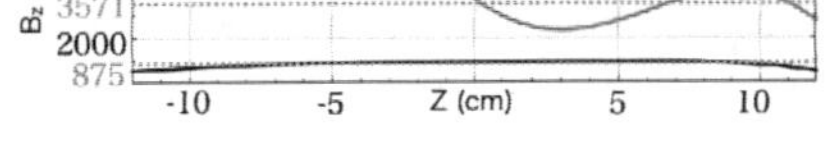

Figure 5. Axial magnetic field configuration induced by REBCO solenoids.

5. Chamber dimension

The plasma chamber dimension of the ion source was determined by using electromagneitc field simulation code ANSYS/HFSS, with the goal of maximizing the electric field inside. The electric field configuration inside the chamber for 2.45 GHz and 10 GHz operation are shown in Fig. 6. For 10 GHz operation, the chamber length would be 170 mm, with a radius of 36.4 mm. With 1 W RF input, the peak electric field magnitude would be lager than 26000 V/m. For the 2.45 GHz operation, we will use a different chamber with length of 170 mm and radius of 38.3 mm, the peak electric field magnitude would be 5400 V/m. Since the dimension of the chamber is limited by the space surrounded by sextupoles and cryostat, we would input the RF power through a tapered waveguide for the 2.45 GHz chamber. However, the tapered part for impedance matching section should be optimized thoroughly to improve the electric field amplitude. The simulation results suggest that the HTS-ECR would have sufficient power for ionization of H^+ and He^{2+} ion[7].

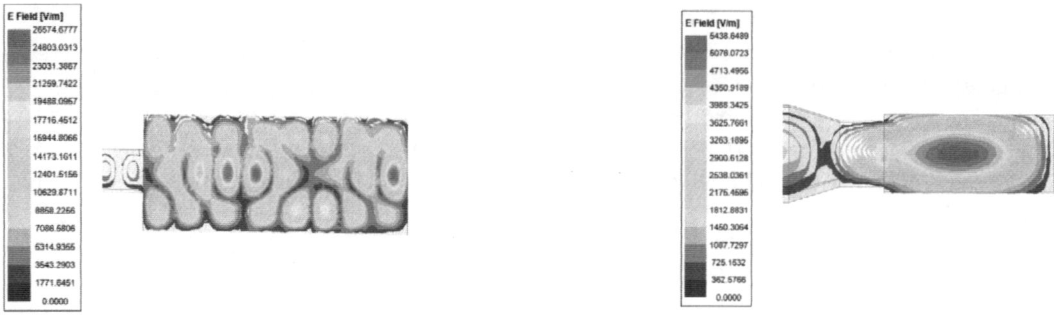

(a) (b)

Figure 6. Section view of electric field in HTS-ECR for (a)10 GHz and (b)2.45 GHz operation, with 1 W RF input.

6. Conclusion

Solenoids and sextupole coils using REBCO tapes were constructed, and performance tests were carried out on them. Electronmagnetic properties of the coils remain the same throughout the test, suggesting that REBCO solenoids and sextupole coils are capable of producing magnetic field with high reliability and reproducibility. With these properties, we expect superconducting magnet with REBCO tapes would be applied on future ECR ion source, accelerator and MRI machine. We intended to construct a ECR ion source with these coils, and carry out performance test on the ion source in next year. Performance tests with longer elapsed time, higher maxium current and lower temperature will also be done in the future. Ion source with these REBCO coils are capable of inducing mirror field with mirror ratio as shown in Eq. 2. Study regarding to optimum magnetic field configuration for proton, deuteron and He^{2+} production will also be done in future.

References

[1] Benkel T, Miyoshi T, Chaud X, Badel A and Tixador A 2017 *Eur. Phys. J. Appl. Phys.* **79** 30601
[2] Senatore C, Alessandrini M, Lucarelli A, Tediosi R, Ugliette D and Iwasa Y 2014 *Superconductor Science and Technology* **27** 103001
[3] Lundin R 1985 *Proceedings of the IEEE* **73** 1428-29
[4] Nagaoka H 1909 *J. College Sci.* **27** 1-33
[5] Antaya T A and Gammino S 1994 *Rev. Sci. Instrum.* **65** 1723-7
[6] Gammino S, Ciavola G, Harkewicz R, Harrison K, Srivastava A and Briand P 1996 *Rev. Sci. Instrum.* **67** 4109-13
[7] Geller R *Electron Cyclotron Resonance Ion Sources and ECR Plasmas* (Bristol and Philadelphia: Institute of Physics Publishing) p 179

Journal of Physics: Conference Series **2244** (2022) 012109 doi:10.1088/1742-6596/2244/1/012109

First injector commissioning results with helium beam at MedAustron Ion Therapy Center

Nadia Gambino[1,3], Matthias Kausel[1], Greta Guidoboni[1], , Liviu Penescu[2], Valeria Rizzoglio[1], Claus Schmitzer[1] and Christoph Kurfürst[1]

[1] EBG MedAustron GmbH, Marie-Curie Straße 5, 2700 Wiener Neustadt, Austria
[2] Abstract Landscapes, 20 place du Millénaire, 34000 Montpellier, France
[3] Corresponding author: nadia.gambino@medaustron.at

Abstract. MedAustron is a synchrotron-based Particle Therapy Accelerator located in lower Austria which delivers clinical proton and carbon beams in the range of 62-252.7 MeV/u and 120-400 MeV/u respectively in two clinical treatment rooms. A proton Gantry was recently commissioned in a third clinical treatment room and a fourth experimental beamline is dedicated to non-clinical research activities. Within the latter, the injector commissioning has recently started for He^{2+} beam generation. The long-term goal is to use helium for cancer treatment due to its favourable physical and biological properties. The helium beam properties from the source up to injection further downstream into the accelerator, i.e. from the Low Energy Beam Transfer line to the Linear Accelerator are discussed in terms of beam emittance, intensity and transmission efficiency. Furthermore, a comparison with simulated data is presented.

1. Introduction

MedAustron is a hadron therapy center for cancer treatment based on the PIMMS study which was founded in 2007 [1]. Medical treatment with proton ions started in December 2016 and with carbon ions in July 2019. Currently, about 35 patients (fractions) per day are treated and it is foreseen to complete 378 patients by the end of 2021. The medical energies range from of 62-252 MeV/u and 120-400 MeV/u for proton and carbon ions, respectively. The accelerator complex comprises an injector delivering a 7 MeV/u beam which is then sent to a Medium Energy Beam Transfer (MEBT) line for injection into a synchrotron. The synchrotron has a 25.4 m diameter and the beam is extracted via 3rd order resonance using a Betatron core. Via the High Energy Beam Transfer (HEBT) line the beam is sent to four available irradiation rooms: IR1 with horizontal beam-line for non-clinical research, IR2 with a horizontal and a vertical beamline, IR3 with a horizontal beam-line and IR4 with a proton Gantry. Currently medical treatment is performed in two of the four treatment rooms, i.e. in IR2 and IR3. The second room delivers both carbon and proton ions in the horizontal and in the vertical plane. The third room currently delivers only protons, but it is foreseen to start the commissioning of carbon ions by 2022. The fourth clinical room (IR4) was recently commissioned with a proton Gantry and it is foreseen to start medical treatment in it by 2022. Recently, the commissioning of helium beams has been started on the third ion source branch. The motivation for the helium commissioning relies on the fact that it has favourable physical and biological properties for certain tumour treatments (higher RBE) [2]. In this work, the first commissioning results with helium ions beams at MedAustron will be presented at the injector level.

2. Helium Injector Commissioning

The MedAustron injector layout is shown in Figure 1. It features three identical ECRIS from Pantechnik. One is reserved for proton beams production (S1) and one for carbon beams production (S2). The third source (S3) is foreseen as the future ion source for helium beam production. The extracted beam from the source is of 8 keV/u for protons and carbon ions and is of 12 keV/u for helium ions beams. The beam is transported through the Low Energy Beam Transfer (LEBT) line to the Linear Accelerator (LINAC) section. The LINAC contains a Radio Frequency Quadrupole (RFQ) module operated at 216.816 MHz which is able to accelerate the beam up to 400 keV/u for proton and carbon beams. The RFQ is followed

Journal of Physics: Conference Series **2244** (2022) 012109 doi:10.1088/1742-6596/2244/1/012109

by a Buncher and an IH-cavity (Interdigital H-mode Drift-Tube Linac), where the energy of the proton and carbon ions reach 7 MeV/u.

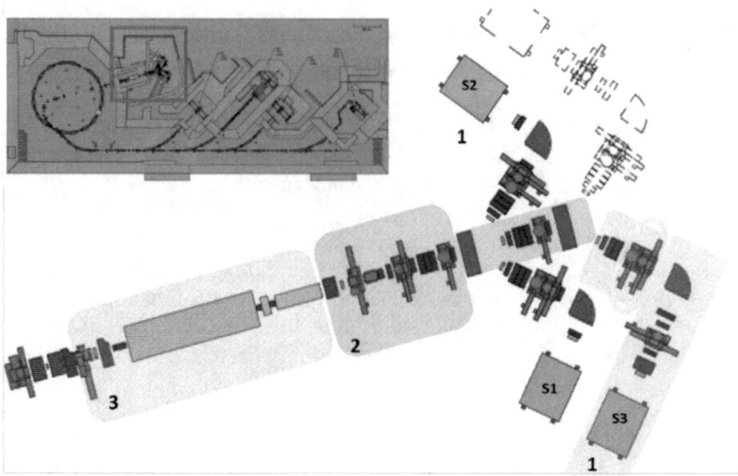

Figure 1: Injector Layout at MedAustron. Sector 1 indicates the third source branch dedicated to helium beam production. The Beam Diagnostic Tank where the source beam has been characterized is highlighted. Sector 2 indicates the LEBT section and sector 3 the LINAC section.

2.1. Source Commissioning

The ion beam in the source is produced via a plasma through the Electron Cyclotron Resonance (ECR) heating mechanism [3]. The Supernanogan of Pantechnik operates at 14.5 GHz heating frequency and is entirely equipped with permanent magnets both for the radial magnetic field and for the longitudinal magnetic field with a B_{ECR} of 0.5 T. An axial mirror ratio B_{max}/B_{min} about two times higher than the ECR resonance magnetic field is obtained. The plasma has limited contact to the chamber walls and the high charge state ions concentrate in the center of the extracted beam with a triangular intensity distribution. The source body is placed at 24 kV, while a puller electrode is placed at negative potential in order to accelerate the beam towards the focus. The focus electrode adapts the beam size to the focal point of the dipole magnet for a good transmission into the beam line and optimized matching into the RFQ. A DC Bias tip, introduced from the backside of the vacuum chamber into the plasma adds electrons to the plasma and favours the extraction of the ions as it reduces the plasma potential. The general commissioning approach of the source relies on starting from conservative initial source parameters to then find the optimal source set-point from single and combined source parameter scans.

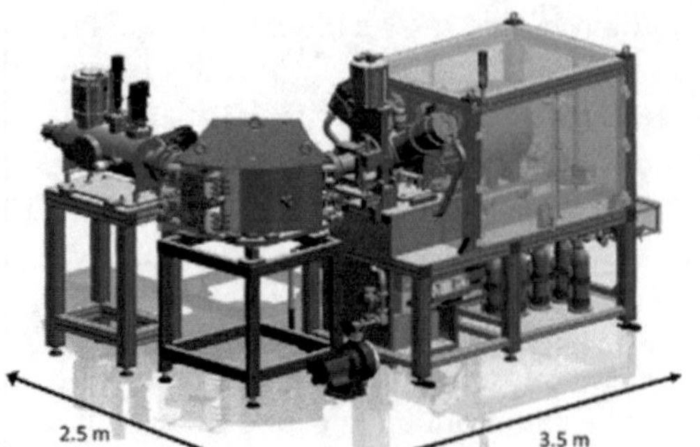

Parameters	He^{2+}
RF Frequency (GHz)	14.45
RF Power (W)	30-32
Gas Type	He_4
Extracted Current (μA)	670
Focus Voltage (V)	750
Puller Voltage (V)	1850
DC Bias Voltage (V)	100

Figure 2: Layout of Supernanogan from Pantechnik with commissioned helium parameters on S3 branch at MedAustron

The source parameters for the He^{2+} ion beam are indicated in Figure 2: Layout of Supernanogan from Pantechnik with commissioned helium parameters on S3 branch at MedAustron. The extracted current measured at the Faraday Cup placed immediately after the dipole spectrometer is shown in Figure 3a. One can see that the required current stability for medical treatments is met, which is on the order of ±2.5% with respect to the average current [4]. The absolute current intensity of 670 µA is not yet optimized for medical treatments as the requirements have not yet been defined. Since the S3 beamline is longer compared to the S1 and S2 beamlines, 2D scans of the focus voltage versus the solenoid current (placed immediately after the extraction system of the source, marked in violet in sector 1 of Figure 1) and of the solenoid current versus the first quadrupole current (marked in red in sector 1 of Figure 1) were necessary to properly focus the beam on the Faraday Cup. In Figure 3b the measured ion spectrum from 500-2000 Gauss is shown. The two helium peaks are measured at the expected magnet fields with a He2+ peak current of 670 µA. Very small contribution of contaminants are measured between the two helium peaks, which are on the order of 1-2 µA and therefore negligible with respect to the main peaks.

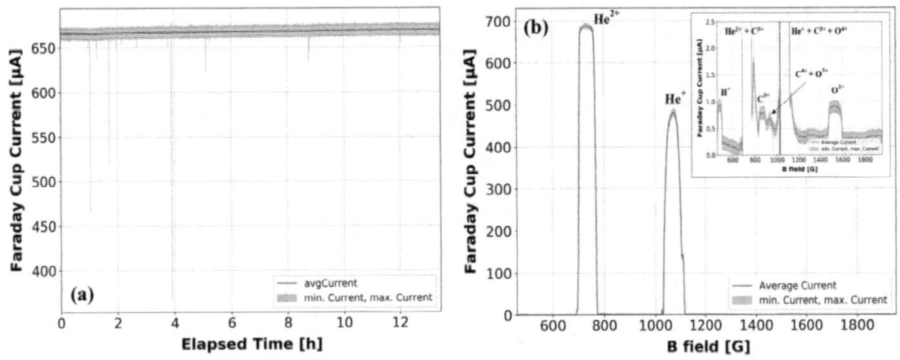

Figure 3: Long-term stability current for He^{2+}beam (a) and measured helium spectrum (b) both acquired on the Faraday Cup located in the second Beam Diagnostic Tank highlighted in Figure 1. A zoom into helium spectrum is also shown where a negligible contribution of contaminants is detected.

2.2. *Source Emittance Measurements*

A full characterization of the He^{2+} ion beam in the phase space has been performed. The emittance was measured on the same Beam Diagnostic Tank as the Faraday Cup intensity and ion spectrum measurements shown in Figure 3. The emittance was measured in both horizontal and vertical plane through a slit-profile-scan method on a Wire Scanner. The measured transverse emittances are shown in Figure 4.

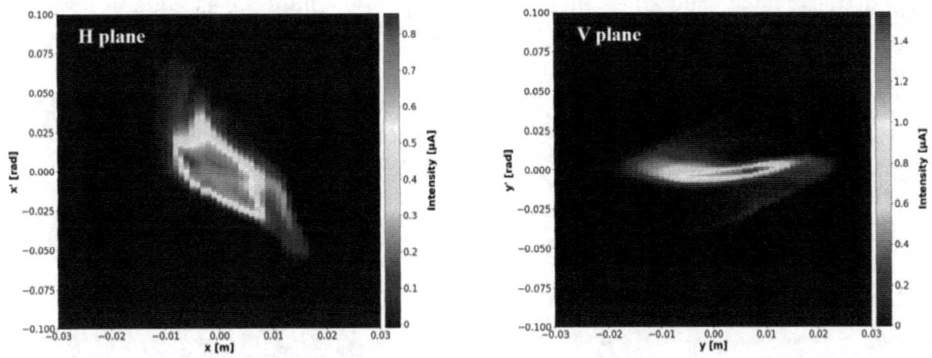

Figure 4: Horizontal and vertical emittance measured at the diagnostic tank highlighted in yellow in Figure 1.

Repetitive beam emittance measurements for several months were performed in order to assess the reproducibility of the beam stability in the phase space. This is crucial in order to have a proper evaluation of the extracted Twiss (Courant-Snyder) parameters as several factors can impact on their evaluation (for example a source shut down, source warm-up time etc.). The measurements from 2021-08-05 were not reproducible exactly for such reasons and therefore not taken into account for the standard deviation calculation. The Twiss parameters are the initial input values for the beam dynamic simulations. The measured emittance is analysed by a script that uses statistical algorithms to estimate the Twiss parameters and the RMS emittance of the beam.

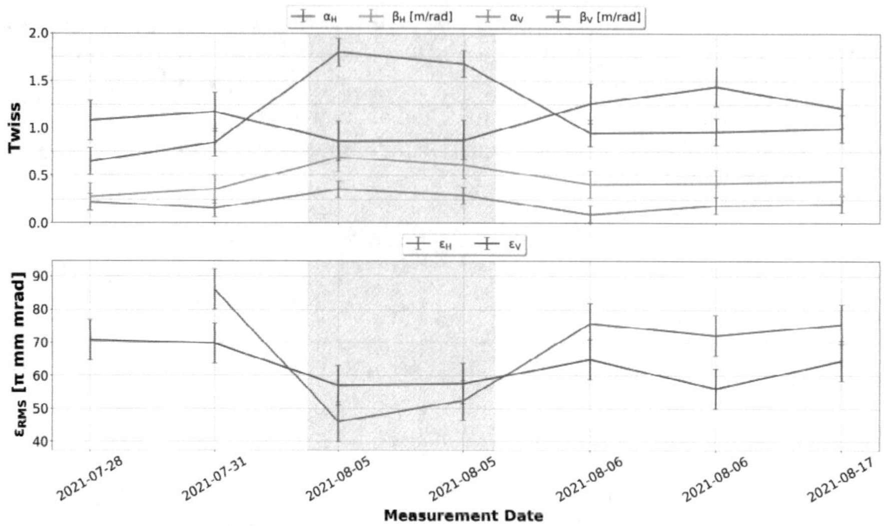

Figure 5: Reproducibility analysis for the Twiss parameters and the RMS emittance.

These emittance measurements were possible during offline commissioning, i.e. also during clinical operation, thanks to the transfer of the whole third source branch in a so-called test system, which is completely independent from the medical accelerator.

2.3 LEBT Commissioning and Comparison with Simulations

The objective of the LEBT commissioning is to transport the beam from the source to the entrance of the RFQ while minimizing the losses and matching the acceptance parameters of the RFQ. Since the MedAustron particle therapy accelerator is already in operation. The beam optics cannot be measured anymore directly before the RFQ entrance via the installation of a test bench as in the case of the initial commissioning phase. As a consequence, beam dynamics simulations are essential to have an optimal commissioning. In this work the analysis of the transverse Twiss parameters and RMS emittance at the source level are taken as input parameters for TRACE3D and MADX simulations [5] [6]. The matching routine of TRACE3D was used to obtain the initial optics of the LEBT up to the RFQ as shown in Figure 6.

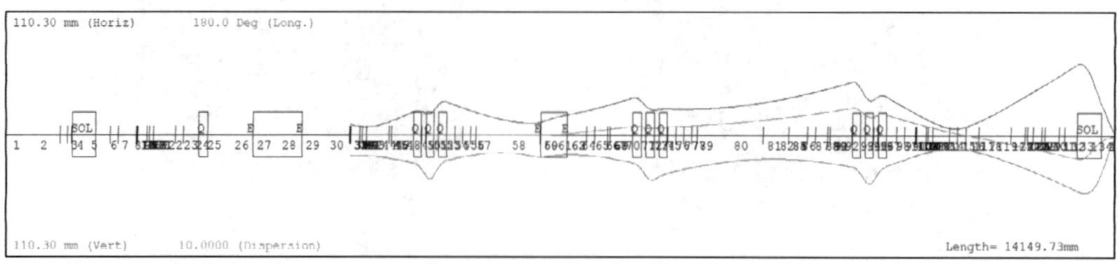

Figure 6: Trace3D simulation for a continuous beam, i.e. simulated with a very small bunching frequency and a very large "envelope" in the longitudinal direction. Linear space charge effects are taken into account.

With the applied simulated strengths on the real machine a 60 - 65% transmission through the LEBT line was obtained, i.e. from the source section to the entrance of the RFQ. Further beam optimizations and steering allowed to reach almost 90% transmission as the measured current on the LEBT current transformer placed before the RFQ was on the order of 590 μA. Additionally, multi-particle tracking has been performed via MADX/PTC allowing a direct comparison with the measured Wire Scanner Profiles along the LEBT line. A preliminary comparison between the simulated and the measured profiles is shown in Figure 7. The predicted profiles are in good agreement with the measured data.

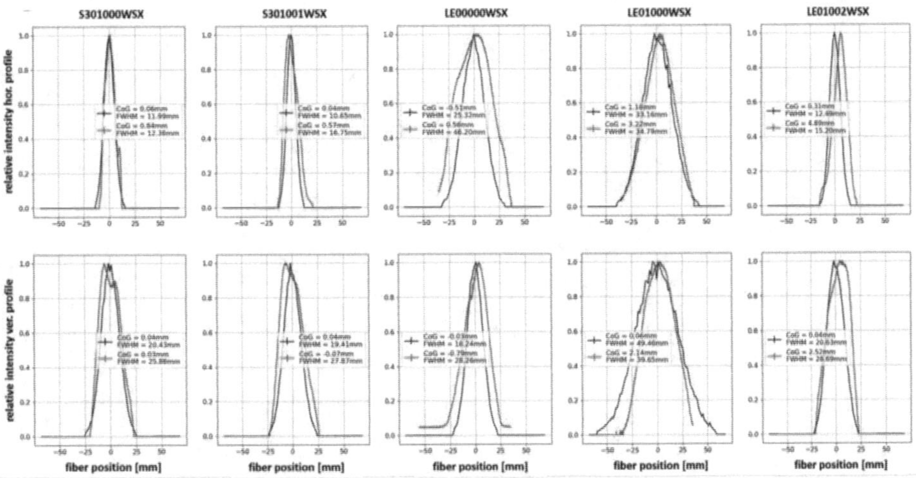

Figure 7: Comparison between measured (black) and simulated (red) wire scanner (WSX) profiles in both H and V planes via MADX and PTC tracking from the source sector (S3) to the LEBT sector (LE)

3. Conclusions and future perspectives

The commissioning of helium ion beams has started at MedAustron. The source commissioning was completed and by the end of 2021 we foresee to finalize the LEBT commissioning. A 90% transmission efficiency from the Source to LEBT, up to the RFQ entrance was reached. Deep emittance analysis studies allows to better understand beam fluctuations at the source level and to make reliable comparisons with simulations and beam transport optimization. Preliminary beam transmission efficiencies through the RFQ are on the order of 40%. LINAC and MEBT Commissioning is planned for 2022. Further improvements of the simulation tools are also planned in order to optimize and support the beam commissioning.

References

[1] P. J. Bryant et. al., Developments in the design of proton and ion accelerators for medical use, Proc. of EPAC'98, paper FRX01A, 1998.

[2] S. Mein et. al., Biophysical modeling and experimental validation of relative biological effectiveness (RBE) for 4He ion beam therapy, Radiat Oncol 14, 123, 2019.

[3] R. Geller, Electron Cyclotron Resonance Ion Sources, IOP Publishing, 1996.

[4] N. Gambino et. al., Impact of ion source stability for a medical accelerator, J. of. Instrumentation, ECRIS2018, 14, 2019.

[5] K. R. Crandall et. al., "TRACE 3-D," Los Alamos Nat. Lab. Report LA-UR-97-886, 200

[6] http://mad.web.cern.ch/mad/

[7] L. Penescu et. al., Commissioning and Operation of the MedAustron Injector, IPAC2014, Dresden, Germany JACoW Publishing.

Acknowledgements

The authors would like to thank the whole MTA department of MedAustron and the Non Clinical Research Team of MedAustron for supporting the helium beam commissioning.

AUTHOR INDEX